Plant Stress Biology

Bhoopander Giri • Mahaveer Prasad Sharma
Editors

Plant Stress Biology

Strategies and Trends

Springer

Editors
Bhoopander Giri
Swami Shraddhanand College
University of Delhi
New Delhi, Delhi, India

Mahaveer Prasad Sharma
ICAR-Indian Institute of Soybean Research
Indore, Madhya Pradesh, India

ISBN 978-981-15-9379-6 ISBN 978-981-15-9380-2 (eBook)
https://doi.org/10.1007/978-981-15-9380-2

This Springer imprint is published by the registered company Springer Nature Singapore Pte Ltd.
The registered company address is: 152 Beach Road, #21-01/04 Gateway East, Singapore 189721, Singapore

Foreword

Adapting Crop Plants to Stress in a Changing Climate

The earth is undergoing rapid warming due to increased greenhouse gasses in the atmosphere. Climate changes resulting in increased temperature and decreased rainfall will lead to increased abiotic and biotic stresses in crop plants. It is clear that it is left to scientists and agriculturists to find ways to mitigate the coming increased stresses on crop plants. We must find solutions to the increased stress in crops that do not result in environmental degradation due to fertilizer runoff into streams, lakes, and oceans or fungicides and pesticides that have adverse impacts on nontarget organisms and processes. One solution to confer stress is the use of microbes that enter into plant tissues as endophytes. Over several years, it has become clear that endophytic microbes modify a plant's tolerance to biotic and abiotic stresses (Hardoim et al. 2015; Khan et al. 2012; White et al. 2019). In some cases, increased stress tolerance may stem from microbial capacity to produce ACC deaminase that reduces the buildup of ethylene in tissues of plants and reduces stress reactions (Bacon and White 2016). In other cases, endophyte-mediated stress reduction is thought to be the result of increased oxidative stress tolerance in plants. The endophyte-mediated antioxidants and other oxidative stress mitigation molecules act as biostimulants which elicit an oxidative response in plant cells and tissues (Irizarry and White 2017; White et al. 2019).

Endophytic microbes are not the only solution to stress in crop plants. In this book, Dr. Bhoopander Giri and Dr. Mahaveer Prasad Sharma have assembled chapters from many authors that further explore how stress is manifested in plants and how it may be mitigated. The chapter topics include: (1) microbe-mediated abiotic stress protection; (2) breeding for stress resistance; (3) nutrient used efficiency and its relationship to stress resistance; (4) use of plant hormones to increase stress tolerance in plants; (5) the role of synthetic biology in future stress adaptation in crop plants; and (6) signaling molecule involvement in crop tolerance of stress. Additional topics are covered that highlight increased stress tolerance as a means to increase the hardiness of crop plants. This book is an important assemblage of chapters that provide an understanding of stress tolerance and potential solutions to the problem of increasing stress in crop plants.

Department of Plant Biology, Rutgers University
New Brunswick, NJ, USA

James White

References

Bacon CW, White JF (2016) Functions, mechanisms and regulation of endophytic and epiphytic microbial communities of plants. Symbiosis 68:87–98. https://doi.org/10.1007/s13199-015-0350-2

Hardoim PR, van Overbeek LS, Berg G, Pirttilä AM, Compant S, Campisano A, Döring M, Sessitsch A (2015) The hidden world within plants: ecological and evolutionary considerations for defining functioning of microbial endophytes. Microbiol Mol Biol Rev 79(3):293–320. https://doi.org/10.1128/MMBR.00050-14

Irizarry I, White JF (2017) Application of bacteria from non-cultivated plants to promote growth, alter root architecture and alleviate salt stress of cotton. J Appl Microbiol 122(4):1110–1120. https://doi.org/10.1111/jam.13414

Khan Z, Guelich G, Redman R, Doty S (2012) Bacteria and yeast endophytes from poplar and willow promotes growth in crop plants and grasses. ISRN Agronomy 2012:1–11. https://doi.org/10.5402/2012/890280

White JF, Kingsley K, Verma R, Obi N, Dvinskikh S, Elmore M, Verma S, Gond S, Kowalski K, Zhang Q (2019) Review: endophytic microbes and their potential applications in crop management. Pest Manag Sci 75(10):2543–2548. https://doi.org/10.1002/ps.5527

Preface

In the recent agriculture scenario, climate change and food security are the two prominent challenges faced by the scientists to cater to the needs of the burgeoning global population. The global climate is predicted to change drastically over the next century and different plant growth parameters will be affected due to these aberrations. The struggle for survival is a natural and continuous process. Plants growing in the natural environment confront with a variety of biotic (pathogen infections) and abiotic stresses (salinity, drought, heat and cold stresses, etc.) that drastically affect plant growth and productivity under field conditions. These challenges are likely to grow as consequences of global climate change.

Gradual increase in the global warming of this planet has grown incidence of extreme weather conditions such as increasing periods of drought, variation in temperature, incidents of flooding and accumulation of soluble salts that all are liable to drastically impact food production, because these extreme weather conditions largely affect plant growth and thereby lowering crop production. Indeed, plants have evolved various direct and indirect mechanisms to respond or adapt to extreme environmental conditions. Therefore, the acquaintance with underlying signalling pathways, physiological, biochemical and molecular mechanisms in plants and the role of beneficial soil microorganisms in plant's stress tolerance are pivotal for sustainable crop production. Indeed, microbes present in phyllosphere, rhizosphere (bacteria and fungi) and inside the plants (endophytes) could play a significant role in the alleviation of biotic and abiotic stresses in plants due to the activation of the ISR system and improve on the PGP traits through enhanced nitrogen fixation, exopolysaccharides (EPS) secretion, phytohormones (IAA, GA, CK), 1-aminocyclopropane-1-carboxylate (ACC) deaminases, trehalose synthase gene, volatile compounds, inducing accumulation of osmolytes (proline), antioxidants, upregulation or downregulation of stress responsive genes (acdS). Thus, the use of such beneficial microbes could play a vital role in mitigating abiotic and biotic stresses in plants.

The present volume "Plant Stress Biology: Strategies and Trends" written by the experts in the field covers the latest research on plants tolerance to abiotic (such as drought, waterlogging, salinity, temperature, cold stresses) and biotic stresses and the potential of plant–microbe interactions to avoid the damage caused by these stresses. With a fortune of information on theoretical, technical and experimental aspects, this extensive volume is a valuable resource for researchers, academicians and students in the broad field of plant stress biology, physiology, microbiology, environmental and agricultural science.

We are highly delighted and thankful to all our contributing authors for their endless support and outstanding cooperation to write altruistically these authoritative and valuable chapters. We extend our sincere thanks to all our colleagues who helped us in the preparation and compilation of this generous volume. We also thank Springer officials, especially William Achauer, Anil Chandy, Sabine Schwarz, Aakanksha Tyagi and Suraj Kumar for their generous support and efforts to accomplishing this wide volume. We specially thank our families for consistent support and encouragement.

New Delhi, Delhi, India — Bhoopander Giri
Indore, Madhya Pradesh, India — Mahaveer Prasad Sharma

Contents

Editors and Contributors

About the Editors

Bhoopander Giri received his Ph.D. in Mycorrhizal Research from the Department of Botany, University of Delhi in 2001. He has worked as a post-doctoral fellow at the Department of Plant Pathology, North Carolina State University, USA and is currently working as an Assistant Professor at the Department of Botany, Swami Shraddhanand College, University of Delhi, Delhi. He is a recipient of CSIR Research Associateship (2003), DST Young Scientist Fellowship (2005), and Raman Post-Doctoral Fellowship. Dr. Giri has served as a General Secretary for International Symbiosis Society (ISS), USA. He is a life-member/member of many prestigious academic societies. Besides, he serves as a reviewer for several international and national journals, including Experimental and Environmental Botany, Planta, Applied Soil Ecology, Journal of Plant Growth Regulation, Agroforestry System, Colloids and Surfaces B: Biointerfaces, Scientia Horticulturae, Acta Physiologiae Plantarum, Spanish Journal of Agricultural Research, Journal of Experimental Botany, Plant Science Today, Institute for Life Long Learning, University of Delhi (e-content) and UGC-CEC pre-view subject expert. Dr. Giri has published more than 40 papers in the national and international peer-reviewed journals and books. He has presented research papers in several national and international conferences (Denmark, Switzerland, Canada, U.K., USA, etc.) and also served as session chair and organizing secretary for a couple of conferences. Dr. Giri has edited a few books, namely, Root Biology, Microorganisms in Saline Environments: Strategies and Functions, Biofertilizers for Sustainable Agriculture and Environment, and Soil Health published by Springer International Publishing AG-A Part of Nature.

Mahaveer Prasad Sharma is currently working in Agricultural Research Service as Principal Scientist (Agricultural Microbiology) at Indian Institute of Soybean Research, Indore (Under ICAR (Indian Council of Agricultural Research)-DARE, Ministry of Agriculture & Farmers Welfare, Govt. of India). Dr. Sharma has started his career in mycorrhizal research at University of Delhi and TERI New Delhi and while serving TERI he obtained Ph.D. in Microbiology from Jiwaji University Gwalior in 2002. He has specialized in soil microbiological research involving the uses of plant growth promoting microbes particularly arbuscular mycorrhizal fungi in improved plant growth, soil carbon sequestration, drought tolerance, and overall productivity of crops. He has been awarded many awards like gold medal award during his master's course, best paper presentation, and travel grant awards for participating abroad in various scientific meetings/conferences. He owned several external research grants (DBT, DST, ICAR) for his research on applied aspects of plant–AMF–microbe interactions. He has also been awarded the prestigious DBT-postdoctoral CREST Award-2013 under which he worked in ARS-USDA, MD USA during 2013–2014 on signature fatty acid biomarkers in soil. He is life member of several professional societies, member of scientific bodies, reviewer of many journals related to agri-life sciences (Frontiers, Elsevier, Springer, Scientific Reports, etc.,) and deputed abroad to Iran, Australia, Portugal, Switzerland, USA at various occasions to participate in meetings and conferences. There are about 80 important research articles published in refereed journals, magazines, and review articles in books of international repute and has many microbial accessions with NCBI database and cultures deposited in International Microbial Repository Authorities to his credit.

Contributors

Sumit Kumar Aggarwal ICAR-Indian Institute of Maize Research, Ludhiana, Punjab, India

Imtiaz Ahmad Department of Entomology, The Pennsylvania State University, University Park, PA, USA

Mamta Arya ICAR-NBPGR Regional Station, Bhowali, Nainital, India

Sonia Balyan National Institute of Plant Genome Research, Delhi, India

Chandni Bansal National Institute of Plant Genome Research, Delhi, India

Mary E. Barbercheck Department of Entomology, The Pennsylvania State University, University Park, PA, USA

Pooja Baweja Department of Botany, Maitreyi College, University of Delhi, New Delhi, India

Karima Bencherif Université de Djelfa, Djelfa, Algeria
Université du Littoral Côte d'Opale, Unité de Chimie Environnementale et Interactions sur le Vivant (UCEIV), Calais Cedex, France

Ekta Bhardwaj Department of Botany, University of Delhi, Delhi, India

J. K. Bisht Crop Production Division, ICAR-Vivekananda Parvatiya Krishi Anusandhan Sansthan, Almora, Uttarakhand, India

Shekhar Chand Bisht PlantbiotiX, Bio Agri Division of Zytex Biotech Pvt. Ltd, Mumbai, India

Subhash Chandra ICAR-Indian Institute of Soybean Research, Indore, Madhya Pradesh, India

Anu Chitikineni Center of Excellence in Genomics and Systems Biology, ICRISAT, Patancheru, India

Dipanti Chourasiya ICAR-Indian Institute of Soybean Research, Indore, Madhya Pradesh, India

Yolande Dalpè Agriculture et agroalimentaire Canada, Centre de recherche et développement d'Ottawa, Ottawa, ON, Canada

Jaishri Rubina Das National Institute of Plant Genome Research, Delhi, India

Sandip Das Department of Botany, University of Delhi, Delhi, India

María del Mar Jiménez-Gasco Department of Plant Pathology and Environmental Microbiology, The Pennsylvania State University, University Park, PA, USA

Ramgopal Devdas ICAR-National Research Centre for Orchids, Pakyong, India

Rahul Gajghate ICAR-Indian Grassland and Fodder Research Institute, Jhansi, Uttar Pradesh, India

Mansi Gogna Laboratory of Plant Physiology, Department of Botany, University of Delhi, Delhi, India

Divyanshu Goswami ICAR-Indian Institute of Soybean Research, Indore, Madhya Pradesh, India

Apoorva Gupta National Institute of Plant Genome Research, Delhi, India

Ravi Gupta Department of Botany, School of Chemical and Life Science, New Delhi, India

Sanjay Gupta ICAR-Indian Institute of Soybean Research, Indore, Madhya Pradesh, India

Anissa Lounès Hadj-Sahraoui Université du Littoral Côte d'Opale, Unité de Chimie Environnementale et Interactions sur le Vivant (UCEIV), Calais Cedex, France

Mohamed Hamidi Université de Djelfa, Djelfa, Algeria
Laboratoire de chimie organique et de substances naturelles, Université de Djelfa, Djelfa, Algeria

Harikrishna Division of Genetics, Indian Agricultural Research Institute, New Delhi, India

Sarita Jha National Institute of Plant Genome Research, Delhi, India

Viraj Gangadhar Kamble ICAR-Indian Institute of Soybean Research, Indore, Madhya Pradesh, India

Neha Katru Department of Botany, Central University of Jammu, Samba, Jammu and Kashmir, India

Rucha Kavishwar ICAR-Indian Institute of Soybean Research, Indore, Madhya Pradesh, India

Nisha Khatri Department of Botany, Dyal Singh College, University of Delhi, New Delhi, India

Sun Tae Kim Department of Plant Bioscience, Pusan National University, Miryang, South Korea

Mrinal K. Kuchlan ICAR-Indian Institute of Soybean Research, Indore, Madhya Pradesh, India

Amit Kumar Department of Botany, Hansraj College, University of Delhi, New Delhi, India

Deepak Kumar Department of Botany, Institute of Science, Banaras Hindu University, Varanasi, Uttar Pradesh, India

Gaurav Kumar Department of Environmental Studies, PGDAV College, University of Delhi, New Delhi, India

Sanjeev Kumar ICAR-Indian Institute of Soybean Research, Indore, Madhya Pradesh, India

Giriraj Kumawat ICAR-Indian Institute of Soybean Research, Indore, Madhya Pradesh, India

Ayushi Mahajan Department of Botany, Central University of Jammu, Samba, Jammu and Kashmir, India

D. Mahanta Crop Production Division, ICAR-Vivekananda Parvatiya Krishi Anusandhan Sansthan, Almora, Uttarakhand, India

Hemant Singh Maheshwari ICAR-Indian Institute of Soybean Research, Indore, Madhya Pradesh, India

Saloni Mathur National Institute of Plant Genome Research, Delhi, India

V. S. Meena Crop Production Division, ICAR-Vivekananda Parvatiya Krishi Anusandhan Sansthan, Almora, Uttarakhand, India

Sahil Mehta Crop Improvement Group, International Centre for Genetic Engineering and Biotechnology, New Delhi, India
Department of Microbiology, Faculty of Life sciences, Institute of Advanced Management and Research, Ghaziabad, Uttar Pradesh, India

Qingfeng Meng Department of Instrument Science & Engineering, School of Electronic Information and Electrical Engineering, Shanghai Jiao Tong University, Shanghai, China

Pankaj Kumar Mishra Crop Production Division, ICAR-Vivekananda Parvatiya Krishi Anusandhan Sansthan, Almora, Uttarakhand, India

Vennampally Nataraj ICAR-Indian Institute of Soybean Research, Indore, Madhya Pradesh, India

Henry Nguyen Division of Plant Science, University of Missouri, Columbia, MO, USA

B. M. Pandey Crop Production Division, ICAR-Vivekananda Parvatiya Krishi Anusandhan Sansthan, Almora, Uttarakhand, India

Ruby Panwar Department of Botany, Hansraj College, University of Delhi, New Delhi, India

M. Parihar Crop Production Division, ICAR-Vivekananda Parvatiya Krishi Anusandhan Sansthan, Almora, Uttarakhand, India

Anupam Patra International Centre for Genetic Engineering and Biotechnology, New Delhi, India

A. Pattanayak Crop Production Division, ICAR-Vivekananda Parvatiya Krishi Anusandhan Sansthan, Almora, Uttarakhand, India
ICAR-Indian Institute of Agricultural Biotechnology, Ranchi, Jharkhand, India

Maddur Puttaswamy Raghavendra Maharani's Science College for Women, Mysuru, Karnataka, India

Meenakshi Raina Department of Botany, Central University of Jammu, Samba, Jammu and Kashmir, India

Vangala Rajesh ICAR-Indian Institute of Soybean Research, Indore, Madhya Pradesh, India

Laxman Singh Rajput ICAR-Indian Institute of Soybean Research, Indore, Madhya Pradesh, India

Sombir Rao National Institute of Plant Genome Research, Delhi, India

Milind B. Ratnaparkhe ICAR-Indian Institute of Soybean Research, Indore, Madhya Pradesh, India

Gyanesh Kumar Satpute ICAR-Indian Institute of Soybean Research, Indore, Madhya Pradesh, India

Akanksha Sharma Department of Botany, Central University of Jammu, Samba, Jammu and Kashmir, India

Mahaveer Prasad Sharma ICAR-Indian Institute of Soybean Research, Indore, Madhya Pradesh, India

Ram Kumar Sharma Division of Genetics, Indian Agricultural Research Institute, New Delhi, India

M. Shivakumar ICAR-Indian Institute of Soybean Research, Indore, Madhya Pradesh, India

Richa Shukla Department of Botany, University of Delhi, Delhi, India

Ajay Kumar Singh ICAR-National Institute of Abiotic Stress Management, Pune, India

Archana Singh Department of Botany, Hansraj College, University of Delhi, New Delhi, India

Baljinder Singh National Institute of Plant Genome Research, New Delhi, India

Indrakant Kumar Singh Molecular Biology Research Lab, Department of Zoology, Deshbandhu College, University of Delhi, New Delhi, India

Ishwar Singh Department of Botany, Hansraj College, University of Delhi, Delhi, India

Maharaj Singh ICAR-Indian Institute of Soybean Research, Indore, Madhya Pradesh, India

Priyanka Singla Department of Botany, Mount Carmel College, Bengaluru, Karnataka, India

Manoj Kumar Srivastava ICAR-Indian Institute of Soybean Research, Indore, Madhya Pradesh, India

Sameena Ejaz Tanwir Citrus Research and Education Centre, University of Florida, Lake Alfred, FL, USA

Lebin Thomas Department of Botany, Hansraj College, University of Delhi, Delhi, India

Fares Trodi Université de Djelfa, Djelfa, Algeria

Rajeev Varshney Center of Excellence in Genomics and Systems Biology, ICRISAT, Patancheru, India

Sanjay Yadav ICAR-Indian Institute of Soybean Research, Indore, Madhya Pradesh, India

Abiotic Stress in Plants: An Overview

1

Pooja Baweja and Gaurav Kumar

Abstract

Various abiotic stress factors affect plants negatively, affecting their growth and development. These stress factors can be natural or anthropogenic and may have short-term or long-term effects on vegetation. Due to stress, many changes occur in plants, which may be reversible or irreversible and even cause acute damage. The abiotic stress factors are considered as one of the major factors for crop loss worldwide. Plants avoid such stress factors either by acclimatization or by avoidance and have with various defence mechanisms. In this chapter, authors have summarized the effects of temperature, drought, flood, light, and heavy metals as abiotic stress factors on plants along with the molecular mechanisms and signal transduction during such stresses.

Keywords

Abiotic stress · Temperature stress · Heavy metals stress · Salinity stress · Light stress · Signal transduction

1.1 Introduction

Plants are generally exposed to extremes of biotic and abiotic environmental factors, which have become the integral part of their life cycle too. For its optimum growth a plant requires carbon, energy, light, mineral nutrients, and water. Whenever any of

P. Baweja
Department of Botany, Maitreyi College, University of Delhi, New Delhi, India

G. Kumar (✉)
Department of Environmental Studies, PGDAV College, University of Delhi, New Delhi, India

B. Giri, M. P. Sharma (eds.), *Plant Stress Biology*,
https://doi.org/10.1007/978-981-15-9380-2_1

Table 1.1 Various stress factors encountered by plants

S. no.	Type of stress			
I	Natural	Abiotic	Light	High irradiance—photoinhibition and photooxidation
				Low irradiance
			Temperature	High/heat
				Low (chilling/frost)
			Drought	
			Flood	
			Mineral deficiency	
		Biotic	Insects	
			Viruses	
			Fungi	
			Bacteria	
II	Anthropogenic	Pollutants	Air	SO_2, CO_2, NOx, O_3
			Water	
			Soil	Heavy metals, chemicals
		Herbicides		
		Pesticides		
		Fungicides		
		Climate change		

these requirements are either in limited or in excessive amount, there is stress. The stress can be defined as any change in environmental condition having undesirable effects on normal growth and development of plants (Levitt 1980). Stress can also be defined as a non-ideal condition for survival. The first publication on "General Adaptation Syndrome" now known as "Biological Stress" was published in Nature in 1936 by Hans Selye (Lichtenthaler 1998). The concept of stress was originally developed by him and this term has also been used for unfavourable environmental constraints in plants. Plants may experience Eu-stress, which is a positive and growth promoting stress or they may experience De-stress which affects plants negatively (Lichtenthaler 1998). The biological species flourish best in optimum conditions and can tolerate the environmental factors as per Law of Limiting factors (Shelford 1952). Once these factors are in extremes they become a condition which promotes stress. The stress can be also categorized as natural or anthropogenic and can also be distinguished into biotic stress and abiotic stress based on which factor is causing stress (Table 1.1).

The biotic stress is caused by factors such as attacks by herbivores, predators, parasites, pathogens, etc., on the other hand, light, temperature, wind, water (flooding or drought), salinity, relative humidity, nutrition, etc., are responsible for abiotic stress. The effect each factor will have on plant depends on its quality and intensity. Although plants are adapted to changing environmental conditions, but as they cannot move or hide being sessile, they are vulnerable to stress. This co-occurrence of different stresses causes extensive loss to agricultural production;

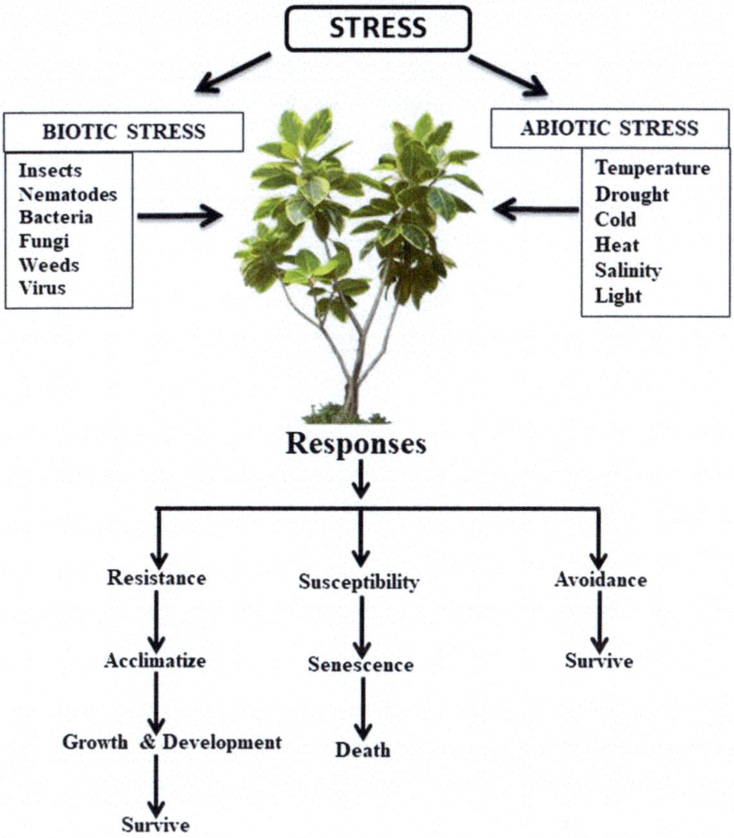

Fig. 1.1 Biotic and abiotic stresses affecting plant growth and development (Modified after Hopkins and Huner 2009)

for example, if there is high temperature stress, then the probability of drought and high salinity is high in that area, due to availability of water at low levels. When such conditions prevail there is a cascade of effects at molecular, biochemical, and physiological levels (Mittler 2006).

Once a plant is in stress condition, it responds in different ways as depicted in Fig. 1.1, either it shows resistance and survives or it is susceptible to the stressful environment and dies and the last is avoidance, i.e. ephemeral plants complete their lifecycle before the inception of stress, thus avoids stress and survive (Hopkins and Huner 2009). All these responses of plants are due to complex serial changes at molecular, cellular, and physiological levels. Plants show such responses either to acclimatize with certain adaptations or withstand it. There are changes at biochemical, molecular, physiological, anatomical, morphological levels to survive in a stressful environment (Koyro et al. 2012). Plants have developed various perception and signalling pathways to survive such harsh environmental conditions. Various

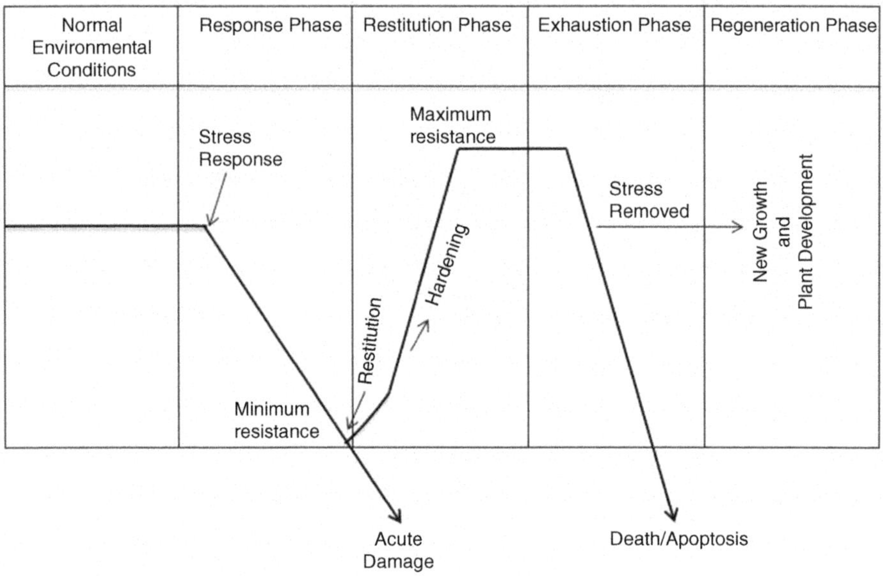

Fig. 1.2 Different response phases shown by plants during stress conditions (Modified after Lichtenthaler 1998)

studies are on-going on *Arabidopsis thaliana* and *Oryza sativa* (Rice) as model organisms to understand the insights of plant stress biology.

Amongst various abiotic stresses, heat and drought affect the plant's growth the most. Various studies have been carried out on these two factors. Popova et al. (2013) studied the RNA dependent DNA methylation pathway and concluded that this pathway is involved in heat stress resistance in *Arabidopsis.* Lee et al. (2013) studied the consequences of ABA production during drought conditions. The responses shown by plants towards stress are dynamic, complex and can be plastic (irreversible) or elastic (reversible) (Cramer 2010; Skirycz and Inze 2010; Cramer et al. 2011). Inhibition of protein synthesis, protein folding, processing are one of the first responses to abiotic stresses at molecular level. Plants also show some responses due to abiotic stress at hormone level and produce ABA or ethylene (Cramer et al. 2011).

The stress responses in plants are differentiated into four stages (Lichtenthaler 1998) as mentioned below (Fig. 1.2):

1. Response—considered as commencement of stress and is also considered as Alarm phase.
2. Restitution—stage of confrontation or resistance and in this stage stress continues.
3. End—this is the stage of exhaustion, where stress intensity is too high and cell death may occur.

4. Regeneration—In this phase if stress has not been so prolonged and extensive, regeneration may occur.

1.2 Types of Abiotic Stress

The abiotic stress includes inanimate environmental components as mentioned above. These abiotic stress factors can occur naturally or may be caused by anthropogenic activities. As plants are sessile and their surrounding environment is ever-changing the occurrence of physiological plasticity is essential for their survival (Gaspar et al. 2002). Plants respond to different types of stresses in different ways as described in different sections:

1.2.1 Temperature as Stress Factor

Plants respond differently to changes in temperature as for every plant species there is an optimum range in which it survives best. In a same temperature range one species may survive and other species may be sensitive as their thermo-sensitive metabolic processes may get hampered due to changes in temperature leading to change in cell processes. Plant species which are growing in hot and dry deserts or high altitudes experience very high temperature during day and low during night. Plants growing in tropical forests or northern forests grow in altogether different temperature ranges so when such plants grow in different climatic conditions they get stressed and are prone to injuries.

1.2.1.1 Low Temperature Stress

Low temperature is considered as one of the major abiotic stress limiting the growth and development. Plants growing in temperate regions are adapted for low temperature conditions, and based on temperature range plants can survive, i.e., cold tolerant or get injury and are classified as chilling-sensitive (10–15 °C); freezing sensitive (0 °C); and freezing resistant plants (able to survive sub-zero temperature) (Pollock and Eagles 1988; Hughes and Dunn 1990). When exposed to low temperature, plants exhibit three types of injuries: desiccation, chilling injury, and freezing injury. Freezing in plants can be induced either from freeing in soil water or freezing of water inside the cells or at intracellular level. Some plants develop freezing tolerance due to formation of antifreeze proteins, which along with some sugars (sucrose) inhibits crystallization at cellular level. Woody or hardy plants are reported to be highly resistant to chilling injuries. The common symptoms of chilling injury are changes in membrane structure, composition, protoplast streaming reduced, increased/decreased respiration, reduced photosynthesis, production of abnormal metabolites, reduced plants growth, curling of leaves, necrosis, die back of stems, abnormal ripening of fruit, decreased N partitioning in young shoots, altered reproductive behaviour (Hashimoto and Komatsu 2007). Generally tropical plants are more susceptible to chilling than the temperate plants; for example, *Zea mays*, *Lycopersicum esculentum*, *Glycine max* show chilling injuries at 10–15 °C, there are some temperate plants which experience chilling injury at 0 °C–5 °C; for example, apple (*Malus* sp.) and *Asparagus* sp. (Hopkins and Huner 2009).

Plants also acclimatize themselves for chilling injuries. The acclimatization in plants is associated with changes in several biochemical and physiological processes such as altered gene expressions dehydrins (*LEA*, late embryogenesis abundant) are induced in various plants, changes in hormone levels (increased ABA), increase in soluble sugars, amino acids, organic acids, increased levels of osmoprotectants, etc. (Palva et al. 2002).

1.2.1.2 High Temperature Stress

Change in global mean temperature and light conditions have significant impact on distribution, abundance, phenology, and physiology of various crop species (Djanaguiraman and Prasad 2014). It has been estimated that climate change decreases the average suitable cultivable area for many plant species such as *Arachis*, *Solanum*, and *Vigna* by 63–100%. It is also predicted that some species might extinct in near future due to reduction in cultivable area (Jarvis et al. 2008). Soil temperature ranges between 45 and 80 °C, where forest soil temperature remains between 40 and 50 °C and in deserts it may reach up to 70–80 °C. Depending on the temperature range, responses exhibited by plants can be categorized as over temperature, intermediate and under temperature responses. The high temperature or heat stress damages the plants and causes the altered phenology, reduced growth and development, scorching of leaves and fruits, sun scabs, abrasions, etc. (Nahar et al. 2015). At cellular and molecular level, high temperature affects stability of various proteins, thus prohibiting or altering many enzymatic activities leading to metabolic disorders, change in membrane and cyto-skeleton structure, altered RNA species. Effect of high temperature has been observed in many crop species such as *Oryza sativa* (rice), *Capsicum annuum* (capsicum), *Triticum aestivum* (wheat), *Hordeum vulgare* (barley), *Zea mays* (maize), *Glycine* max (soybean), *Abelmoschus esculentus* (okra), etc. (Hasanuzzaman et al. 2013). Also, there is formation of Heat Shock Proteins (HSPs), whenever temperature is elevated beyond the threshold limits, which varies from species to species.

1.2.2 Light Stress

Light is an important environmental factor which is essential for CO_2 assimilation through photosynthesis, but as it crosses a plants tolerance level it becomes a stress factor and photo inhibition occurs in plants. Plants are exposed to two types of radiations: UV-A (315–400 nm) and UV-B (280–315 nm). UV-A is photo-oxidative and UV-B is photo-oxidative as well as causes photo lesions in bio-membranes. UV damages disulphide bridges in the proteins, dimerization of thymine groups in DNA, and disrupts xanthophyll cycle. When exposed to excessive light stress, plants may sense high photon flux. Shade loving plants get damaged even by brief exposure of strong light. In plants the excessive light damage photosystem II by destroying photosensitive pigments and thylakoid structures, thus inhibiting photosynthesis. There is disruption in electron transport chain and breaking of protein sub-units (Hopkins and Huner 2009). To avoid damage by strong light, plants show some adaptive features such as orienting leaves at an angle to receive less radiation, rolling up of shoots, dense covering of trichomes on upper surface of leaves, thickened cell walls, and production of anthocyanin pigment to shield mesophyll (Balfagón et al. 2019).

In tomato (*Solanum lycopersicum*) lipidomics analysis identified lipophilic antioxidant molecules, which protects PS II from photodamage (Spicher et al. 2017). The study conducted by Balfagón et al. (2019) showed that in *Arabidopsis*, in high light intensity combined with heat stress, the stomata opens to increase transpiration and cool the leaf instead of reducing stomatal aperture (Devireddy et al. 2018).

1.2.3 Water Stress

Water which is present in 80–90% in non-woody plants is key molecules for all metabolic processes. It is essential and when it becomes limiting and is not available to roots or when transpiration rate is very high it becomes a stress factor. Water stress may arise when there is water deficit or excess of water (flooding). During flooding oxygen becomes limiting and is not supplied to roots as per requirement. Due to water deficit stress there is desiccation; protoplasm dries up, solute concentration increases, leading to loss in integrity of membranes. Selectivity of membranes is lost. Proteins get denatured and displaced, and finally leading to loss of cellular compartmentalization. Due to water deficit the water potential inside the cells decreases leading to stomatal closure. It is also known that hydropassive stomatal closure occurs in plants to avoid transpirational loss and is evident in all tropical, temperate, or dessert plants (Hopkins and Huner 2009). Stomatal closure limits gaseous exchange, reduced transpiration and photosynthesis, uptake of mineral nutrients and disruption of homeostasis and ions. Drought stress also causes various other adverse effects such as decreased water potential leading to hydro active closure of stomata, lesser number of stomata, thickening of leaf cell wall, decreased cell enlargement, slow or inhibited growth and reproduction, cutinization of leaf surface, poorly developed conducting system, increase in root–shoot ratio, production of osmolytes such as proline, formation of reactive oxygen species (ROS), ascorbate, and glutathione accumulates in cell and further aggravates the condition (Lisar et al. 2012). Water deficits also alter the cell wall confirmations, non-enzymatically through the interaction of pectate and calcium (Boyer 2009; Cramer et al. 2011). Although photosynthesis gets disrupted in both C3 and C4 plants but studies show that C4 plants are more sensitive to drought. Water stress also alters the mRNA expression and thus new proteins are synthesized. Different proteins which are synthesized under stress are late embryogenesis abundant (LEA), dehydrins, desiccation stress proteins, protease, etc. (Lisar et al. 2012).

When there is flooding plants undergo hypoxia, a stage when oxygen concentration is less. During flooding or excessive irrigation water logging occurs in soil, thus limiting the oxygen availability to roots. When water logging is for a prolonged period damage occurs depending on the extent of adaptability and soil conditions. The damage also depends on the type, developmental stage, genotype of plants, extent of severity, and duration of water stress (Fukao and Bailey-Serres 2004; Mariani and Ferrante 2017).

1.2.4 Salinity Stress

Salinity stress is another major problem of world, which decreases crop productivity. High concentration of salt accumulates in the soil either due to improper irrigation or poor drainage. Soil salinity is measured by accumulation of soluble salts such as Cl^-, SO_4^{2-}, Na^+, Ca^{2+}, etc., and it harms the plant growth and development through water stress and cytotoxicity (Isayenkov and Maathuis 2019). Soil salinity may be categorized into primary (natural salinity; for example, marshes, salt lake, tidal swamps, etc.) or secondary (due to anthropogenic activities). High saline condition decreases the water potential and causes osmotic stress. Osmotically active compounds or osmolytes (Glycine-Betaine) also known as compatible metabolites accumulate in cells to lower the osmotic potential (Rasool et al. 2013). Salinity is also many times accompanied by the formation of ROS, polyamines (putrescene), and proline. Saline soils also limit the uptake of nutrients by root from soil (Sharma et al. 2016). During high salinity stress, the photosynthetic activities gets affected as thylakoid membranes of chloroplast get damaged, reduction of photosynthetic pigments takes place, further inactivation of electron transport chain, photophosphorylation occurs (Srivastava et al. 2019). Also, the turgor pressure of leaves changes reducing the stomatal conductance and rate of transpiration (Srivastava et al. 2019). ABA accumulates in guard cells, excessive production of hydrogen peroxide occurs which leads to triggering of oxidative stress by influencing the production of reactive oxygen species (Foyer and Noctor 2005). It has been reported that legumes and cereal crops are very sensitive to salinity. Salinity affects nitrogen fixation as it suppresses the formation of root nodules (Ramana et al. 2012). Nutritional imbalance enhances vegetative growth over reproductive growth, thus decreasing crop yield (Munns and Tester 2008). It also causes chlorosis and senescence, DNA, RNA content decreases, mitotic and enzymatic activity slows down (Ali et al. 2004).

The physiological drought condition in plants can be overcome by the application of certain soil microorganisms such as arbuscular mycorrhizal fungi (AMF). Symbiotic association established between AMF and plant roots, mainly in saline environments, could withstand plants under excessive saline conditions as the AMF hyphae run several metres away from the stressed area or from the zone of nutrient depletion, thereby increasing root surface area and facilitating nutrient uptake by the plant. Due to this, physiological processes and various metabolic activities get improved and adverse effects of physiological drought could minimize in plants (Saxena et al. 2017).

1.2.5 Heavy Metal Stress

Extensive activities related to agriculture, industrialization, and urbanization are adding enormous amount of heavy metals; for example, Arsenic (As), Cadmium (Cd), Chromium, Lead (Pb), Mercury (Hg) in environment. Addition of excessive amount of heavy metals damage and alter the soil texture, pH, thus affecting plant growth and development. These toxic elements promote morphological, metabolic abnormalities and thus disrupt cell homeostasis (Amari et al. 2017; Tiwari and Lata 2018).

Some of these metals are vital micronutrients, and are accountable for numerous metabolic processes in a plant, but, if they exceed the threshold limit, these metals can have detrimental effects on metabolic pathways, physiological processes, plant growth and development, and finally senescence (Ghori et al. 2019). As a first line of resistance exhibited, plants reduce the heavy metal uptake, through cellular and root exudates that restrict entry of metals inside the cells (Shahid et al. 2015; Ghori et al. 2019). The second line of resistance involves adopting other mechanisms for detoxification by transports, sequestration of these heavy metal ions, and chelates in the plant's vacuoles (Ghori et al. 2019). One of the most destructive effects caused by heavy metals in plants is disruption of bio-membranes by lipid per-oxidation (Demiral and Türkan 2005; Yadav 2010). Some of the harmful effects induced by heavy metals are described below:

High concentration of Zinc results in stunted growth, chlorosis, senescence, and changed root and shoot ratio (Fontes and Cox 1998; Yadav 2010). Another heavy metal chromium is released from tanning industry. The excess availability of chromium can bring chlorosis in young leaves, inhibit plant growth, nutrient imbalance, root injury, and wilting of tips (Scoccianti et al. 2006; Yadav 2010). Lead in soil resulting from various sources like municipal sewage sludge, mining and smelting activities, paints, gasoline and explosives, etc. It deteriorates plant growth, morphology, and photosynthesis. Many enzymatic activities are inhibited when lead concentration is very high, membrane permeability is lost, disturbs uptake of mineral, and causes water imbalance (Sharma and Dubey 2005; Yadav 2010). Recently, several studies revealed the physiological and molecular mechanisms of arsenic toxicity, its accumulation, detoxification, and tolerance in various concentrations in plants such as carrot, lettuce, rice, and spinach can cause several physiological disorders (Kumar et al. 2015; Tiwari and Lata 2018).

1.3 Molecular Mechanisms and Signal Transduction in Stress

The three systematic approaches or three "*OMICS*" (such as Metabolomics, Transcriptomics, Proteomics) studies have enhanced the knowledge on how a plant response at molecular level and what are the complex regulatory mechanisms. The transcriptomics deals with the RNAs and their expressions. Proteomics studies about how proteins modify and metabolomics is a tool to study about the metabolites. The integration of all these three omics has shown various clear pathways related to concepts of plant responses towards stress (Cramer et al. 2011). Signal perception, transduction, and finally expression of stress response genes are the three steps at molecular level once the stress is present in environment (Fig. 1.3). The cascades are associated with reactive oxygen species, Ca^{2+}, Abscisic acid (ABA), Jasmonates (JA), Salicylic acid (SA), Ethylene (ET) which coordinate various signal pathways. Once the signal is perceived, transduction pathways such as calcium dependent protein kinase (CDPK), mitogen activated protein kinase (MAPK) get activated, further down regulating various transcription factors (Baillo et al. 2019). One example of Ca^{2+} signal pathway can be understood in mechanism of closing of guard cells in drought stress. During drought stress, changes occur in cytosolic concentration of Ca^{2+} which initiates CPK activity resulting into the release of ABA, which further activates the calcium chelator (BAPTA). When guard cells receive BAPTA, signal comes to control the transpiration and guard cell closes (Zou et al. 2015; Atif et al. 2019). Experimental studies on wheat have shown that CPKs show functional diversity and complexity. The isoforms of closely

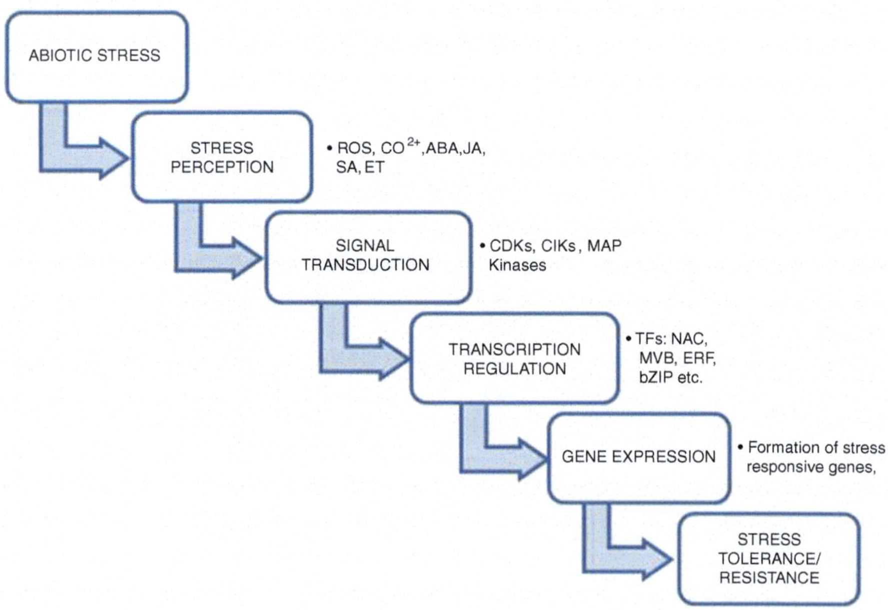

Fig. 1.3 Signalling pathway in response to abiotic stress (Modified after Baillo et al. 2019)

related CPK genes such as TaCPK 7 and TaCPK 12 in wheat show diversity wherein, the former responds to various stresses such as temperature (low), salinity, drought, H_2O_2, whereas the latter responds only through ABA signalling (Geng et al. 2011; Atif et al. 2019).

As transcription factors act as key stress tolerance mediators, they can be modified to increase the stress tolerance in various crops. Many studies have been conducted on five major cereal crops such as barley, maize, sorghum, rice, and wheat. Gene modification through these transcription factors (TFs) can improve stress tolerance in transgenic plants. Although boosting tolerance level through TFs is not so easy and is a complicated procedure, as one TF gene during down-regulation may promote or suppress other genes. Various TFs families may be involved in one or multiple stress responses, thus making TF responses very complex and complicated and might have cross-talk between different signal pathways. There are millions of TF at molecular level, and identifying, modifying each TF is a bit challenging task. Application of CRISPR/Cas 9, a gene editing tool is being used to improve the stress tolerance in plants (Baillo et al. 2019).

Abiotic stress also enhances the production, synthesis, and transcription of heat shock proteins (HSPs) in comparison to other normal proteins. Post transcriptional modifications of proteins such as microRNA and alternating splicing also help to cope with abiotic stress. The HSPs have been categorized based on molecular weight into Large (68,000–104,000 Da), Intermediate (20,000–23,000 Da), and Small (15,000–18,000 DA) (Hughes and Dunn 1990). It has now been established that HSPs help newly synthesized protein's folding and protect them during stress. Therefore HSPs are also known as molecular chaperones (Ul Haq et al. 2019). HSPs are also present in cell and cellular compartments during standard environmental conditions and studies have confirmed their role in normal growth and development apart from being stress responsive (Eck et al. 2007). HSPs not only develop during heat stress but are also found to be present in other abiotic stresses.

HSPs express differentially in different species. Genes encoding HSPs, are present in different cell compartments and thus they are expressed differently and may also be specific depending on the stress intensity (Liu et al. 2006; Ul Haq et al. 2019). Under temperature stress high molecular weight HSPs (HSP 118, −114, −110, −108, −104, −103, −101, −100, and −97) are formed. In *Arabidopsis* and maize HSPs −100 and −101 are expressed and responsive against the high temperature stress and thermo-tolerance (Queitsch et al. 2000; Nieto-Sotelo et al. 2002). In Pea low molecular weight HSPs 18.1 and −17.9 accumulates after 4 h treatment (Dupuis and Dumas 1990). Low temperature also induces the expression of HSPs in rice, maize, and *Arabidopsis* to protect them against cold stress (Bae et al. 2003; Kosova et al. 2011; Hlavackova et al. 2013). HSPs have also found to be present during drought stress. Mostly HSPs 70 are upregulated as evident from studies on rice, *Arabidopsis,* sugar cane, cotton, maize, *Cicer* sp. (Subba et al. 2013; Reddy et al. 2014; Yer et al. 2018). In *Chenopodium rubrum under* high light stress HSP 23 gets upregulated (Korotaeva et al. 2001). In marine ecosystems low light stress induced accumulation of HSP 70, ClpBi, and HSP 60 (Kumar et al. 2017). Similarly

HSPs have also been found to be upregulated by various other abiotic stresses such as heavy metals, flooding, oxidative stress, etc.

1.4 Conclusions

Various abiotic stresses affect plant's phenology, disrupt metabolic activities at physiological, cellular, and molecular levels. Such changes unfavourably affect the growth and development of plants and in field crop species even significantly reduce the crop yield. The various stress factors are transduced through signal transduction pathways. The components involved in such pathways are calcium, ROS, protein kinase, etc. The stress signals further regulate transcription factors and thereby controlling gene regulation. Future studies are required focussing on different crop plants, to explore regulatory mechanisms; for example, MiRNA, alternative splicing, and cross-talk between different pathways. Such biotechnological studies using molecular approach would help to develop stress resistant and tolerant crops.

References

Ali Y, Aslam Z, Ashraf MY, Tahir GR (2004) Effect of salinity on chlorophyll concentration, leaf area, yield and yield components of rice genotypes grown under saline environment. Int J Environ Sci Technol 1(3):221–225

Amari T, Ghnaya T, Abdelly C (2017) Nickel, cadmium and lead phytotoxicity and potential of halophytic plants in heavy metal extraction. S Afr J Bot 111:99–110

Atif RM, Shahid L, Waqas M, Ali B, Rashid MAR, Azeem F, Nawaz MA, Wani SH, Chung G (2019) Insights on calcium-dependent protein kinases (CPKs) signaling for abiotic stress tolerance in plants. Int J Mol Sci 20(21):5298. https://doi.org/10.3390/ijms20215298

Bae MS, Cho EJ, Choi E, Park OK (2003) Analysis of the *Arabidopsis* nuclear proteome and its response to cold stress. Plant J 36:652–663

Baillo EH, Kimotho RN, Zhang Z, Xu P (2019) Transcription factors associated with abiotic and biotic stress tolerance and their potential for crops improvement. Genes (Basel) 10(10):771. https://doi.org/10.3390/genes10100771

Balfagón D, Sengupta S, Gómez-Cadenas A, Fritschi FB, Azad RK, Mittler R, Zandalinas SI (2019) Jasmonic acid is required for plant acclimation to a combination of high light and heat stress. Plant Physiol 181(4):1668–1682

Boyer JS (2009) Evans review: cell wall biosynthesis and the molecular mechanism of plant enlargement. Funct Plant Biol 36(5):383–394

Cramer GR (2010) Abiotic stress & plant responses from the whole vine to the genes. Aust J Grape Wine Res 16:86–93

Cramer GR, Urano K, Delrot S, Pezzotti M, Shinozaki K (2011) Effects of abiotic stress on plants: a systems biology perspective. BMC Plant Biol 11:163

Demiral T, Türkan I (2005) Comparative lipid peroxidation, antioxidant defense systems and proline content in roots of two rice cultivars differing in salt tolerance. Environ Exp Bot 53:247–257

Devireddy AR, Zandalinas SI, Gómez-Cadenas A, Blumwald E, Mittler R (2018) Coordinating the overall stomatal response of plants: rapid leaf-to-leaf communication during light stress. Sci Signal 11(518):1–9. https://doi.org/10.1126/scisignal.aam9514

Djanaguiraman M, Prasad PVV (2014) High temperature stress. In: Jackson M, Ford-Lloyd B, Parry M (eds) Plant genetic resources and climate change. CABI, Wallingford, pp 201–220

Dupuis I, Dumas C (1990) Influence of temperature stress on in vitro fertilization and heat shock protein synthesis in maize (*Zea mays* L.) reproductive tissues. Plant Physiol 94:665–670. https://doi.org/10.1104/pp.94.2.665

Eck ERHB, Ettig SEF, Nake CLK, Artig KAH, Hattarai TRB (2007) Specific and unspecific responses of plants to cold and drought stress. J Biosci 32:501–510

Fontes RLS, Cox FR (1998) Zinc toxicity in soybean grown at high iron concentration in nutrient solution. J Plant Nutr 21:1723–1730

Foyer CH, Noctor G (2005) Oxidant and antioxidant signalling in plants: a re-evaluation of the concept of oxidative stress in a physiological context. Plant Cell Environ 28:1056–1071

Fukao T, Bailey-Serres J (2004) Plant responses to hypoxia-is survival a balancing act? Trends Plant Sci 9(9):459–456

Gaspar T, Franck T, Bisbis B, Kevers C, Jouve L, Hausman JF, Dommes J (2002) Concepts in plant stress physiology. Application to plant tissue cultures. Plant Growth Regul 37:263–285

Geng S, Zhao Y, Tang L, Zhang R, Sun M, Guo H, Kong X, Li A, Mao L (2011) Molecular evolution of two duplicated CDPK genes CPK7 and CPK12 in grass species: a case study in wheat (*Triticum aestivum* L.). Gene 475:94–103

Ghori NH, Ghori T, Hayat MQ, Imadi SR, Gul A, Altay V, Ozturk M (2019) Heavy metal stress and responses in plants. Int J Environ Sci Technol 1:1–22

Hasanuzzaman M, Nahar K, Alam MM, Roychowdhury R, Fujita M (2013) Physiological, biochemical, and molecular mechanisms of heat stress tolerance in plants. Int J Mol Sci 14:9643–9684

Hashimoto M, Komatsu S (2007) Proteomic analysis of rice seedlings during cold stress. Proteomics 7:1293–1302

Hlavackova I, Vitamvas P, Santrueek J, Kosova K, Zelenkova S, Prasil IT, Ovesna J, Hynek R, Kodicek M (2013) Proteins involved in distinct phases of cold hardening process in frost resistant winter barley (*Hordeum vulgare* L.) Cv Luxor. Int J Mol Sci 14:8000–8024

Hopkins WG, Huner NP (2009) Introduction to plant physiology, 4th edn. Wiley, Hoboken, NJ, p 503

Hughes MA, Dunn MA (1990) The effect of temperature on plant growth and development. Biotechnol Genet Eng Rev 8(1):161–188. https://doi.org/10.1080/02648725.1990.10647868

Isayenkov SV, Maathuis FJM (2019) Plant salinity stress: many unanswered questions remain. Front Plant Sci 10:80. https://doi.org/10.3389/fpls.2019.00080

Jarvis A, Lane A, Hijmans RJ (2008) The effect of climate change on crop wild relatives. Agric Ecosyst Environ 126:13–23

Korotaeva NE, Antipina AI, Grabelnykh OI, Varakina NN, Borovskii GB, Voinikov VK (2001) Mitochondrial low-molecular-weight heat-shock proteins and the tolerance of cereal mitochondria to hyperthermia. Russ J Plant Physiol 48:798–803

Kosova K, Vitamvas P, Prasil IT, Renaut J (2011) Plant proteome changes under abiotic stress-contribution of proteomics studies to understanding plant stress response. J Proteome 74:1301–1322

Koyro HW, Ahmad P, Geissler N (2012) Abiotic stress responses in plants: an overview. In: Ahmad P, Prasad MNV (eds) Environmental adaptations and stress tolerance 1 of plants in the era of climate change. Springer, New York. https://doi.org/10.1007/978-1-4614-0815-4_1

Kumar S, Dubey RS, Tripathi RD, Chakrabarty D, Trivedi PK (2015) Omics and biotechnology of arsenic stress and detoxification in plants: current updates and prospective. Environ Int 74:221–230

Kumar M, Padula MP, Davey P, Pernice M, Jiang Z, Sablok G, Contreras-Porcia L, Ralph PJ (2017) Proteome analysis reveals extensive light stress-response reprogramming in the seagrass *Zostera muelleri* (Alismatales, Zosteraceae) metabolism. Front Plant Sci 7:2023

Lee SC, Lim CW, Lan WZ, He K, Luan S (2013) ABA signaling in guard cells entails a dynamic protein–protein interaction relay from the PYL–RCAR family receptors to ion channels. Mol Plant 6:528–538

Levitt J (1980) Responses of plants to environmental stresses. In: Chilling, freezing, and high temperature stresses, vol 1, 2nd edn. Academic Press, New York, p 497
Lichtenthaler HK (1998) The stress concept in plants: an introduction. Ann N Y Acad Sci 851:187–198
Lisar SYS, Motafakkerazad R, Hossain MM, Rahman IMM (2012) Water stress in plants: causes, effects and responses. Intech, Rijeka. https://doi.org/10.5772/39363
Liu D, Zhang X, Cheng Y, Takano T, Liu S (2006) RHSP90 gene expression in response to several environmental stresses in rice (*Oryza sativa* L.). Plant Physiol Biochem 44:380–386
Mariani L, Ferrante A (2017) Agronomic management for enhancing plant tolerance to abiotic stresses-drought, salinity, hypoxia, and lodging. Horticulturae 3:52
Mittler R (2006) Abiotic stress, the field environment and stress combination. Trends Plant Sci 11:15–19
Munns R, Tester M (2008) Mechanisms of salinity tolerance. Annu Rev Plant Biol 59:651–681
Nahar K, Hasanuzzaman M, Ahamed KA, Hakeem KR, Ozturk M, Fujita M (2015) Plant responses and tolerance to high temperature stress: role of exogenous phytoprotectants. In: Hakeem KR (ed) Crop production and global environmental issues. Springer, Cham. https://doi.org/10.1007/978-3-319-23162-4_17
Nieto-Sotelo J, Martinez LM, Ponce G, Cassab GI, Alagon A, Meeley RB, Ribaut JM, Yang R (2002) Maize HSP101 plays important roles in both induced and basal Thermotolerance and primary root growth. Plant Cell 14:1621–1633
Palva ET, Htiharju ST, Tamminen I, Puhakainen T, Laitinen R, Svensson J, Helenius E, Heino P (2002) Biological mechanisms of low temperature stress response: cold acclimation and development of freezing tolerance in plants. JIRCAS Work Ret 23:9–15
Pollock CJ, Eagles CF (1988) Low temperature and the growth of plants. In: Long SP, Woodward FI (eds) Plants and temperature, vol 42. Society for Experimental Biology Symposium, Company of biologists, Cambridge, pp 157–180
Popova OV, Dinh HQ, Aufsatz W, Jonak C (2013) The RdDM pathway is required for basal heat tolerance in *Arabidopsis*. Mol Plant 6:396–410. https://doi.org/10.1093/mp/sst023
Queitsch C, Hong SW, Vierling E, Lindquist S (2000) Heat shock protein 101 plays a crucial role in Thermotolerance in *Arabidopsis*. Plant Cell 12:479–492. https://doi.org/10.1105/tpc.12.4.479
Ramana GV, Padhy SP, Chaitanya KV (2012) Differential responses of our soybean (*Glycine max* L.) cultivars to salinity stress. Legume Res 35:185–193
Rasool S, Hameed S, Azooz MM, Rehman MU, Siddiqi TO, Ahmed P (2013) Salt stress: causes, types and responses of plants. In: Ahmad P, Azooz MM, Prasad MNV (eds) Ecophysiology and responses of plants under salt stress. Springer, New York, pp 1–24. https://doi.org/10.1007/978-1-4614-4747-4_1
Reddy PS, Kishor PBK, Seiler C, Kuhlmann M, Eschen-Lippold L, Lee J, Reddy MK, Sreenivasulu N (2014) Unraveling regulation of the small heat shock proteins by the heat shock factor HvHsfB2c in barley: its implications in drought stress response and seed development. PLoS One 9(3):e89125. https://doi.org/10.1371/journal.pone.0089125
Saxena B, Shukla K, Giri B (2017) Arbuscular mycorrhizal fungi and tolerance of salt stress in plants. In: Arbuscular mycorrhizas and stress tolerance of plants. Springer, Singapore, pp 67–97. https://doi.org/10.1007/978-981-10-4115-0_4
Scoccianti V, Crinelli R, Tirillini B, Mancinelli V, Speranza A (2006) Uptake and toxicity of Cr (Cr^{3+}) in celery seedlings. Chemosphere 64:1695–1703
Shahid M, Khalid S, Abbas G, Shahid N, Nadeem M, Sabir M, Aslam M, Dumat C (2015) Heavy metal stress and crop productivity. In: Hakeem K et al (eds) Crop production and global environmental issues. Springer, Cham
Sharma P, Dubey RS (2005) Lead toxicity in plants. Braz J Plant Physiol 17:35–52
Sharma V, Rena V, Kumar D, Pandey RN, Singh B (2016) Sulfur regulates iron uptake and iron use efficiency in bread and durum wheat. Ind J Plant Physiol 21:189–196
Shelford VE (1952) Paired factors and master factors in environmental relations. Ill Acad Sci Trans 45:155–160

Skirycz A, Inze D (2010) More from less: plant growth under limited water. Curr Opin Biotechnol 21(2):197–203

Spicher L, Almeida J, Gutbrod K, Pipitone R, Dörmann P, Glauser G, Rossi M, Kessler F (2017) Essential role for phytol kinase and tocopherol in tolerance to combined light and temperature stress in tomato. J Exp Bot 68:5845–5856

Srivastava P, Wu Q-S, Giri B (2019) Salinity: an overview. In: Giri B, Varma A (eds) Microorganisms in saline environments: strategies and functions. Soil biology, vol 56, pp 3–18. https://doi.org/10.1007/978-3-030-18975-4_1

Subba P, Barua P, Kumar R, Datta A, Soni KK, Chakraborty S, Chakraborty N (2013) Phosphoproteomic dynamics of chickpea (*Cicer arietinum* L.) reveals shared and distinct components of dehydration response. J Proteome Res 12:5025–5047

Tiwari S, Lata C (2018) Heavy metal stress, signalling, and tolerance due to plant associated microbes: an overview. Front Plant Sci 9:452

Ul Haq S, Khan A, Ali M, Khattak AM, Gai WX, Zhang HX, Wei AM, Gong ZH (2019) Heat shock proteins: dynamic biomolecules to counter plant biotic and abiotic stresses. Int J Mol Sci 20(21):5321. https://doi.org/10.3390/ijms20215321

Yadav SK (2010) Heavy metals toxicity in plants: an overview on the role of glutathione and phytochelatins in heavy metal stress tolerance of plants. S Afr J Bot 76(2):167–179

Yer EN, Baloglu MC, Ayan S (2018) Identification and expression profiling of all Hsp family member genes under salinity stress in different poplar clones. Gene 678:324–336

Zou JJ, Li XD, Ratnasekera D, Wang C, Liu WX, Song LF, Zhang WZ, Wu WH (2015) Arabidopsis calcium-dependent protein Kinase8 and CATALASE3 function in abscisic acid-mediated signaling and H_2O_2 homeostasis in stomatal guard cells under drought stress. Plant Cell 27:1445–1460

Silicon: A Plant Nutritional "Non-Entity" for Mitigating Abiotic Stresses

2

Sahil Mehta, Mansi Gogna, Baljinder Singh, Anupam Patra, Indrakant Kumar Singh, and Archana Singh

Abstract

In the present era, the progression of plant development in the environment is adversely affected by increasing incidences of abiotic and biotic stresses. These stressors singly or in combination negatively pose pressure on plants. As a result, the plants display a wide range of morphological, physiological, biochemical, metabolic, molecular as well as epigenetic responses that help them in averting stress-triggered alterations. In addition to "Omics," plant breeding, functional genomics, transgenic technology, and genome editing approaches, better mineral nutrition coupled with soil-health amendments is still considered as the key management practice. Use of Silicon (Si), the second most predominant and quasi-essential element, has been recommended in the recent past. Si not only promotes growth and development of plants, it also works as "anti-stress agent." Si mitigates this alleviating effect mainly by ROS detoxification, immobilization, and compartmentation of toxic metal ions, modification in water and nutrients

S. Mehta · A. Patra
International Centre for Genetic Engineering and Biotechnology, New Delhi, India

M. Gogna
Laboratory of Plant Physiology, Department of Botany, University of Delhi, New Delhi, India

B. Singh
National Institute of Plant Genome Research, New Delhi, India

I. K. Singh (✉)
Molecular Biology Research Lab, Department of Zoology, Deshbandhu College, University of Delhi, New Delhi, India
e-mail: iksingh@db.du.ac.in

A. Singh (✉)
Department of Botany, Hansraj College, University of Delhi, New Delhi, India
e-mail: archanasingh@hrc.du.ac.in

B. Giri, M. P. Sharma (eds.), *Plant Stress Biology*,
https://doi.org/10.1007/978-981-15-9380-2_2

uptake, alteration in gene expression and phytohormone biosynthesis, maintenance of osmotic potential and gaseous exchange, and formation of Si–cuticle double layer. Moreover, being non-corrosive and non-pollutive, Si-supplementation has proven to be the most economic as well as eco-friendly method. The present chapter is an attempt to primarily address the involvement of Si in minimizing the negative effects of abiotic stresses.

Keywords

Abiotic stress · Silicon · Drought · Salinity · Nanoparticles

2.1 Introduction

The term "stress" signifies a set of conditions that deviate plant growth, development, and other characteristics adversely from its "normal state" (Lal et al. 2018; Mehta et al. 2019a). Plants encounter these stresses during their entire life cycle initiating from germination and seedling stage to the post-harvest stage (Singh et al. 2018; Sharma et al. 2020). All these growth-limiting "stresses" can be categorized into two major categories, namely abiotic and biotic stresses (Singh et al. 2019; Mehta et al. 2019b; Ramegowda et al. 2020). Abiotic stress conditions arise due to fluctuation in plant's physical environment (naturally occurring inanimate factors) like rain, drought, floods, salinity, metal/metalloid toxicity, nutrient paucity, dwindling seasonal patterns, and temperature shifts (Lal et al. 2018; Mohammadi et al. 2020). On the other hand, the latter one is a consequence of living disturbances, such as fungi, bacteria, virus, nematodes, rodents, oomycetes, etc., that negatively affect the plant's well-being (Rahman et al. 2019; Ali et al. 2020).

The most frightening fact is that the frequency and incidences of these stresses have increased in the last 50 years (Xu 2016; Surówka et al. 2020). It is actually because of amalgamated effects of growing human population and anthropogenic activities (Cripps 2016; Mahmoud and Gan 2018; Tamburino et al. 2020). These activities include over-exploitation of resources, deforestation, desertification, pollution, and global warming (Mahmoud and Gan 2018; Mona et al. 2019; Baldos et al. 2019). All these factors either individually or in combination have affected the total food productivity negatively and the whole scenario of food production will turn worse in the near future (Kamanga and Mndala 2019; Rafique et al. 2020). There is a huge gap between food productivity, demand, and supply (Grafton et al. 2015; Müller et al. 2020). Therefore, the most serious challenge at present is to refine the scientific research as well as administrative strategies, so as to feed the every-minute increasing population in future (Conceição et al. 2016; Martin-Shields and Stojetz 2019; Tyagi 2020). Additionally, the focus must be given on enhancing the already declined soil fertility as the soil provides habitat, nutrients, and beneficial microbes, which is required for proper growth and development of plants (Shahid et al. 2019).

In order to resolve the issue on food security in a best possible manner, it is highly important to understand the responses as well as adjustments that occur during averting stress-triggered alterations (Pecinka and Mittelsten Scheid 2012; Goswami et al. 2020). As an effort, various groups of researchers are keenly focusing on understanding the mechanism through various newly developed tools and techniques (Anamika et al. 2019; Shahbazy et al. 2020). These efforts have resulted in generating knowledge regarding the "changes and adjustments" and their associated mechanisms up to an extent (Gilliham et al. 2017; Vakilian 2020). Furthermore, many crossbred and transgenic plants have also been developed in the last 15 years (Hasanuzzamam et al. 2018; Dixit et al. 2020). However, time, efforts, and environment suitability are primary factors that are considered majorly. Hence, there is a need to look for reliable environment-friendly methods for sustainable agriculture (Kawalekar 2013; Ahirwar et al. 2020).

In this context, one of the most reliable practices has been to supply adequate mineral nutrition coupled with maintenance of sound soil-health (Pandey et al. 2015; Fresno et al. 2018; Lu et al. 2020). This method assures both environmental and economic benefits and crop plants can be supplemented with required components directly in the form of fertilizers and its effect can be observed at morphology, physiology, biochemistry, and metabolome levels (Ma 2004; Marschner 2012; Liang et al. 2015; Mu et al. 2020). Till date, few elements have been studied for promoting a range of tolerance mechanisms for alleviating various stresses in several important agricultural and horticultural crops (Kaur et al. 2016; Chauhan et al. 2017; Salgado et al. 2020). One such studied element is Si, a "multi-talented" quasi-essential element that has been established as a stimulant to trigger growth and development in stressed plants at an optimal concentration (Malhotra and Kapoor 2019; Ahanger et al. 2020; Singh et al. 2020). This is because the Si is being directly supplemented by the small and marginal farmers in their fields since 1840s in the form of non-corrosive, non-pollutive, regular fertilizer for economic as well as ecological benefits (von Liebig 1843). Now, due to its positive effects, the status of Si has shifted from "beneficial but non-essential" to "quasi-essential" by the International Plant Nutrition Institute (IPNI) (http://www.ipni.net/). Furthermore, in 2013, the Association of American Plant Food Control Officials (AAPFCO) also officially announced Si as a plant "beneficial substance" (http://www.ipni.net/publication/bettercrops.nsf/0/26A7E8FDB7F2FBBF85257C28007A07BB/$FILE/BC%202013-4%20p14.pdf). Besides, the beneficial effects of Si for imparting stress tolerance is also well documented in the form of Si nanoparticles and Si priming (Abdel Latef and Tran 2016; Rastogi et al. 2019; Parveen et al. 2019; Siddiqui et al. 2020). This chapter focuses on highlighting the significance of Si as a growth regulator and anti-stress agent.

2.2 Silicon: Occurrence and Sources

In accordance with multiple sources, the soils inherit their element composition primarily from parent rocks that were subjected to geochemical as well as pedochemical weathering processes primarily. As per the data evaluated, silicon is the second most abundant element in the earth's crust in terms of quantity after oxygen, i.e. 27.7%. (Mitra 2015; Malhotra and Kapoor 2019). In the earth's crust, Si has been deposited in the form of quartz (SiO_2), sand, and sandstone (Rédei 2008; Malhotra and Kapoor 2019). Within the soil, it comprises even up to 70% of soil mass in the form of monosilicic acid, polysilicic acid as well as complexes with organic and inorganic compounds such as aluminum oxides and hydroxides (Rao and Susmitha 2017). Out of all, the most important form is plant-available form, i.e. silicic acid (H_4SiO_4), a non-charged plant-available molecule that considerably ranges between 10 ppm to over 100 ppm (Epstein 2009; Liang et al. 2015; Zargar et al. 2019). During the crop use, the polysilicic acid, and inorganic and organic complexes act as important sinks/sources that replenish the monosilicic acid (Rao and Susmitha 2017). In soil solution, the concentration of Si is equivalent even up to many macroelements such as potassium (K), calcium (Ca), and phosphorus (P) (Epstein 1994; Malhotra and Kapoor 2019). Silicon solubility in the soil is affected by a variety of factors, which include irrigation water, the particle size of the silicon fertilizer employed, critical soil characteristics (moisture, temperature, and pH), presence of organic complexes concentration of iron (Fe), phosphate (P), and aluminum (Al) ions as well as dissolution reactions occurring (Gérard et al. 2002; Tavakkoli et al. 2011; Rao and Susmitha 2017; Zargar et al. 2019). As per one report, the range of silicon present in the soil around the globe lies within 50–400 g silicon/kg of soil (Matichenkov and Calvert 2002; Haynes 2019). The silicon present within soil improves water absorption capacity, soil physical and chemical properties as well as maintain other nutrients in plant-available form by creating silica bridges (Rao and Susmitha 2017; Malhotra and Kapoor 2019; Zargar et al. 2019). Compared to the soil, silicon comprises 0.0001% and 0.026% in terms of quantity in oceans and humans, respectively (http://www.elementalmatter.info/element-silicon.htm).

On the other hand, the availability of Si in plants is low but ranges from 0.1 to 10% of dry weight (Hodson et al. 2005; Balakhnina and Borkowska 2013; Malhotra and Kapoor 2019). In here, the most important fact to be noted is that concentration of Si ranges distinctly within and among plant species which depend directly on the capability of the roots to uptake Si (Hodson et al. 2005; Ma and Yamaji 2006; Malhotra and Kapoor 2019). Plants have been categorized into three major classes such as accumulators, intermediate, and excluders based on the Si accumulation in their tissues (Mitani and Ma 2005; Marschner 2012; Luyckx et al. 2017). The best examples of accumulators are members of Equisetales, Cyperales, and Poales (especially rice, sugarcane, and maize) with a value of more than 1 for silicon/calcium ratio. On the contrary, tomato and soybean are examples of Si excluders that show the value of <0.5 for silicon/calcium. However, nettle and snapdragon are examples of the intermediate type (Mitani and Ma 2005; Luyckx et al. 2017;

Malhotra and Kapoor 2019). Within angiosperms, the monocots tend to accumulate more silicon in their aerial parts due to the presence of silicon transporters (Henriet et al. 2006; Malhotra and Kapoor 2019). For a long time, it is a noted fact that the silicon levels in the soil is enhanced by fertilization. The agricultural wastes such as silicate slag, bagasse furnace ash, lignite fly ash, and rice straw are considered as rich silicon sources that are being employed mostly. The other sources of silicon employed these days include wollastonite, calcium silicate, potassium silicate, garnet, silica gel, diopside, calcium silicate hydrate, etc. (Kalra et al. 2003; Daniel Maxim et al. 2008; Malhotra and Kapoor 2019; Zargar et al. 2019).

2.3 Silicon: Uptake, Transportation, and Accumulation

The majority of the investigations regarding silicon uptake and transportation has been focused more on monocots; however, the speed for enhancing knowledge regarding dicots has also increased in the last 5 years. As per literature, Si is absorbed by lateral roots actively in the form of neutral, monomeric monosilicic acid, $Si(OH)_4$, whose concentration ranges between 0.1–0.6 mM (Knight and Kinrade 2001; Rao and Susmitha 2017). The ability of monosilicic acid to cross the plasma membrane of lateral root depends highly on the physiological pH and water (Raven 2001). The vehicle for its uptake and distribution is a simple molecule, i.e. water; however, both molecules vary in size (Exley et al. 2020). Therefore, the pace of both water uptake and Si adsorption has been classified into three possible situations, namely (1) active (Si-uptake > water uptake), (2) passive (where Si-uptake = water uptake), and lastly (3) rejective (Si-uptake < water uptake) in higher plants (Cornelis et al. 2011; Zargar et al. 2019). Ostensibly, upon the entry with water via the symplastic route, silicic acid encounters a myriad of different enumerable water channels that control the movement of silicic acid further. For example, in rice, a high silicon accumulating plant, the silicon transportation is highly governed by majorly three low silicon rice genes, i.e. *Os*LSi1, *Os*LSi2, and *Os*LSi6 (Ma et al. 2006, 2007; Yamaji and Ma 2009; Dhakate et al. 2019). Among these genes, LSi1 (influx transport activity) and LSi2 (efflux transport activity) have been shown to be involved in silicon transport from root cells to the apoplast (Ma and Yamaji 2008; Rao and Susmitha 2017) (Fig. 2.1).

*Os*LSi1 gene belonging to the NIP-III (nodulin26-like intrinsic proteins) subfamily of aquaporin is primarily found to be constitutively located in the basal zones of roots. Within this, the *Os*LSi1 gene is found to be localized exactly on the plasma membrane of the distal side of both exodermis and endodermis cells where casparian stripes are located (Yamaji and Ma 2007; Ma and Yamaji 2008; Dhakate et al. 2019). On the other hand, expression patterns and cellular localization studies have revealed that the *Os*LSi2 gene (efflux Si-transporter) is localized on the proximal side of the same cells (Yamaji and Ma 2009; Yamaji and Ma 2011; Dhakate et al. 2019). This rice *Os*Lsi2 gene is found to be responsible for reloading and diffusing Si into the vascular bundles (Yamaji and Ma 2011; Ma and Yamaji 2015). After reaching the apoplast, monosilicic acid in xylem sap needs to be unloaded so as to prevent the Si

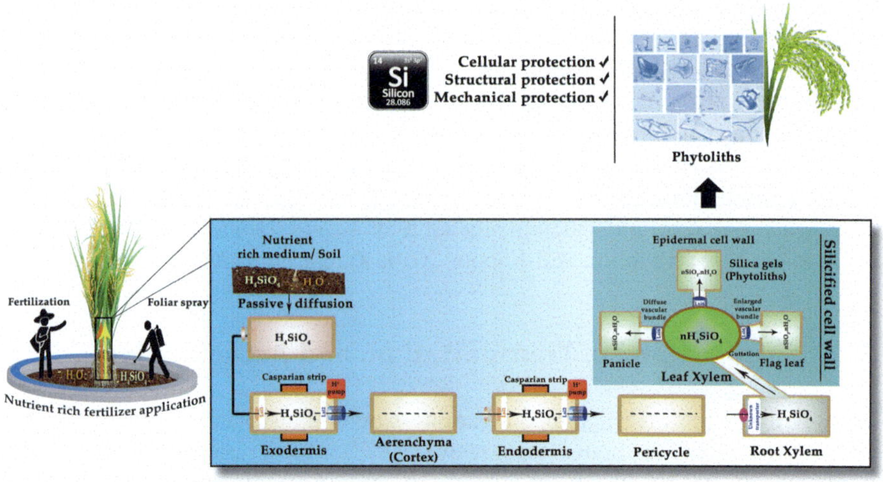

Fig. 2.1 Diagrammatic representation of silicon uptake, transportation, and accumulation

deposition within xylem. In this regard, *Os*LSi6 gene plays role in transferring Si from the large vascular bundles to the panicles (Yamaji and Ma 2009; Feng et al. 2011; Rao and Susmitha 2017). The knock-out and localization studies revealed that the *Os*Lsi6 gets localized on the adaxial side of xylem parenchyma cells in the leaf sheaths as well as leaf blades (Feng et al. 2011; Ma and Yamaji 2015). Therefore, it is important to keep a note that both apoplastic and symplastic route operates for silicic acid (Exley et al. 2020). In addition to the rice, homologs of Si-transporters have also been observed in other plant species. The list includes barley (Mitani et al. 2009a, b; Chiba et al. 2009), maize (Mitani et al. 2009a, b), crookneck pumpkin (Mitani et al. 2011), wheat (Montpetit et al. 2012), soybean (Deshmukh et al. 2013), field horsetail (Vivancos et al. 2016), cucumber (Sun et al. 2017, 2018), tobacco (Zellner et al. 2019), poinsettia (Hu et al. 2020), and tomato (Sun et al. 2020). The possible reason for the identification of Si-transporters lies in the ability to mine the well-annotated plant genome sequences available for more than 100 species.

Upon successful transport, the silicon gets deposited under the cuticle and in intercellular spaces or vascular bundles (Heckman 2013). Beneath the cuticle, the silicon gets deposited as a cuticle–silicon double layer (silicic acid) (Rao and Susmitha 2017; Rao et al. 2017). Furthermore, with the age, the concentration of monosilicic acid increases which results in polymerization to form silica gel (SiO_2. nH_2O) through a non-enzymatic reaction (Mitani and Jian 2005; Zargar et al. 2019). In addition, amorphous silica particles polymerize in plant cells to form phytoliths without incurring any energy as soon as its concentration exceeds a critical mark of 2 mM. These phytoliths are found as silica cells in vascular bundles and silica bodies in bulliform cells, and fusoid cells (Rao and Susmitha 2017; Nawaz et al. 2019).

2.4 Silicon and Abiotic Stresses

As per the literature, the important roles attributed to Si in plants include elevation in growth, crop yield, and quality, photosynthesis, nitrogen (N_2) fixation as well as protection against abiotic and biotic stresses (Balakhnina and Borkowska 2013; Steiner et al. 2018; Malhotra and Kapoor 2019; Zargar et al. 2019; Ahanger et al. 2020; Singh et al. 2020). Interestingly, the abiotic stresses are the first and foremost reason that besets the annual productivity rate. Abiotic stresses include salinity, extreme temperature, UV-B radiation, heavy metal/metalloid toxicity, mechanical injury, nutrient deficit, nutrient toxicity, and drought (Sharma et al. 2020). All these stresses result in oxidative, osmotic as well as an ionic inconvenience in plants that ultimately culminate into reactive oxygen species (ROS) accumulation, altered metabolic and physiological processes (Sharma et al. 2020). This ultimately results in hampered growth and net productivity which affect the farmer fields, farmer pockets, and overall economy (Sharma et al. 2020).

A plethora of lab-scale, greenhouse-level, and field experiments have been conducted that endorses the numerous benefits of silicon on the plants growing under adverse conditions (Soundararajan et al. 2014; Manivannan et al. 2016; Luyckx et al. 2017; Liu et al. 2019; Malhotra and Kapoor 2019; Zargar et al. 2019; Ahanger et al. 2020). It has also been deduced that it is not single but an amalgamation of key mechanisms that alleviate stresses which include: (1) attunement of antioxidant systems and osmolytes for harmful ROS detoxification, (2) immobilization or complex formation/co-precipitation or compartmentation of noxious metal ions and enhanced uptake of nutrients, (3) modifying water as well as nutrients uptake processes, (4) regulating expression of various genes, phytohormone biosynthesis, maintenance of osmotic potential, photosynthetic apparatus, and gaseous exchange, and (5) formation of Si–cuticle double layer (Tripathi et al. 2016; Etesami and Jeong 2018; Etesami and Jeong 2020; Souri et al. 2020) (Fig. 2.2).

2.4.1 Drought

Drought imposes a grave threat to plant population on earth as the morphological as well as physiological functions of a plant get affected (Kusaka et al. 2005; Shao et al. 2008). Drought stress can be injurious to physiological and metabolic events such as turf quality, growth rate, root/shoot ratio, leaf carbon/nitrogen content, photosynthesis, transpiration, and stomatal conductance (Saud et al. 2014). Application of Si ameliorates stressful drought conditions by elevating photosynthesis, total leaf water content, chlorophyll content, and turf rate by 44%, 33%, 42%, and 44%, respectively (Saud et al. 2014). It often results in disrupted water supply via the xylem, which consequently results in lost turgor pressure and reduced stomatal closure (Taiz and Zeiger 2006). It also convulses the photosynthetic framework via its association with UV or visible rays (Garcia-Plazaola and Becerril 2000). The effectiveness of Si in combating drought stress has been noted in many plants; for example, exogenous

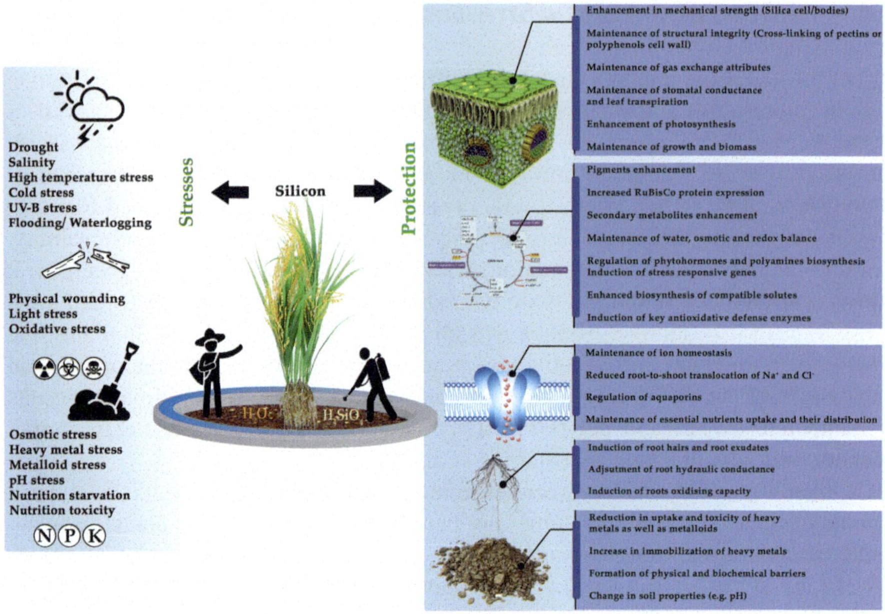

Fig. 2.2 Key mechanisms involved in the ameliorative effect of silicon on plants facing various abiotic stresses

supply of Si regulate, leaf structure, water potential, erectness, and structure of xylem vessels (Gong et al. 2005; Hattori et al. 2005a, b).

An interesting study was carried out to comprehend the differences between drought-tolerant and drought-sensitive tomato lines in response to exogenous application of Si. Elevation in amino acid biosynthesis is noted in drought-tolerant tomato lines due to Si-induced increase in sulfur (S) and ammonia (NH^{4+}) levels. Whereas in drought-sensitive tomato lines, the application of Si results in accumulation of gamma-aminobutyric acid (GABA) proline and, which is key to the maintenance of cellular ionic redox equilibrium (Ali et al. 2018). *Brassica napus* faces extreme oxidative stress. Silicon application provides defense against antioxidant enzymes such as glyoxalase, ascorbate-glutathione, proline, and so on (Hasanuzzamam et al. 2018).

The deposition of exogenously supplemented Si on leaf epidermis helps to generate a higher water potential under scarce conditions (Lux et al. 2003). Similarly, suberin-containing endodermis also accumulates Si in drought-stressed cereal cultivars. In Si-treated plants, the water uptake is higher and faster from the rhizosphere to roots (Hattori et al. 2003, 2005a, b). Si strengthens plant's tolerance against water stress by elevating root silicification, lignification, and suberization (Guerriero et al. 2016). During water scarcity, Si forms a complex with hemicellulose (He et al. 2013; Ma et al. 2015). Si also enhances Casparian strip development leading to an increase in the level of suberization in roots of rice plants (Fleck et al. 2015).

Moreover, Si-supplementation also enhances transpirational bypassing of toxic ions from symplast streamflow (Coskun et al. 2016). This theory allows to propose that the movement of Na^+ and Cl^- is limited in rice plants by suberized exodermis and endodermis of roots, thereby bypassing the step of xylem loading via symplast (Coskun et al. 2016). Much like in salinity stress, foliar supplementation of Si brings about an increase in the expression of different aquaporins (AQPs) located in the membranes of root cells.

2.4.2 Salinity Stress

About 20% of global crop production is affected by salinity stress (Hussain et al. 2018). Under salt stress, rice plants display an interesting deviation in the apoplastic movement of noxious Na^+ and Cl^-, on exposure to Si (Shi et al. 2013). Si-mediated amelioration of salt stress has been studied and deciphered at various biochemical and physiological levels. Starting from its impact on roots, the primary site of ion uptake to serial tissue like leaves wherein salt stress has a drastic effect on several proteins and enzymes related to photosynthesis and stomatal opening (Liu et al. 2019; Gogna and Bhatla 2019, 2020). Silicon protects photosynthetic machinery of plants under stress due to persisting soil salinity by preventing pigment degradation and regulation of several photosystems and chloroplast-related proteins (Muneer et al. 2014; Soundararajan et al. 2017). Exogenous application of Si modulates enzymatic antioxidant machinery constituting enzymes like catalase, superoxide dismutase, and guaiacol/ascorbate peroxidase (Zhu et al. 2004; Manivannan et al. 2015). The primary action undertaken by a stressed plant to overcome salinity is the restriction of Na^+/Cl^- uptake via roots (Liu et al. 2019; Gogna et al. 2020). Exogenously supplied Si not only limits the uptake of toxic ions by plant roots but also regulates several other essential biochemical aspects like photosynthesis, maintenance of redox equilibrium, and effective distribution of nutrients to the plant (Liu et al. 2019). Similar to Ca^{2+}, the application of Si not only restricts excess uptake of Na^+ but also mediates accumulation of K^+, thereby impacting tolerance against salinity. This study has been extensively carried out across sugarcane, aloe, zinnia, and rose (Ashraf et al. 2010; Manivannan et al. 2015; Garg and Bhandari 2016; Soundararajan et al. 2018). The foremost role of Si in alleviating salt stress is its possible interaction with noxious ions responsible for oxidative stress and disrupted ionic homeostasis. K^+ is the most essential element necessary for plant growth, development, and yield. However, uptake of Na^+ under salt stress often results in K^+ deficiency (Liebersbach et al. 2004). Thus, the addition of Si not only negates the competition between Na^+ and K^+ but also alleviates K^+ distribution in salt-stressed wheat and blueberry (Tuna et al. 2008).

Plant root aquaporins are involved in the facilitation of water and mineral nutrition transportation (Liu et al. 2019). NIP family of AQPs has been found to play a significant role in the uptake and transport of Si and other metalloids (Wu and Beitz 2007). Studies pertaining to rice plants have revealed that efflux and influx of Si are carried out via the NIP family of AQPs (Ma et al. 2006). Apart from NIPs,

PIPs (plasma membrane intrinsic proteins) also regulate Si levels in the cell under abiotic stress conditions in roots of Sorghum plants (Liu et al. 2014, 2015). Salinity stress often disrupts the hydraulic conductance of roots. Recent studies on sorghum reveal that hydraulic conductance can be restored by exogenous application of Si (Liu et al. 2014, 2015). Morphologically, Si application revived suppressed lateral root growth and enhances the mechanical strength of the primary root (Liu et al. 2019).

Some other major consequences of salinity stress include reduction in photosynthetic efficiency, diminished stomatal conductance and level of transpiration, and lastly, damaged photosynthetic apparatus (photosystem I and II) (Hetherington and Woodward 2003; Gupta and Huang 2014; Yang et al. 2015). Recent studies in wide varieties of plants such as sorghum, maize, tomato, tobacco, and pumpkin have shown that application of Si not only improves stomatal conductance in plants but also elevates the capacity of leaves to fix CO_2 (Parveen and Ashraf 2010; Nabati et al. 2013; Hajiboland and Cheraghvareh 2014; Hu et al. 2014; Li et al. 2015). Stomatal malfunction under salt stress conditions leads to a loss in levels of reduced CO_2 and disrupts the process of gaseous exchange. Thus, exogenously applied Si repairs stomatal conductance (Hetherington and Woodward 2003; Abbas et al. 2015; Parveen and Ashraf 2010). Similarly, in salt-stressed leaves of *Capsicum annuum*, stomata remained open when supplemented with exogenous Si (Manivannan et al. 2016). Recent reports also show that salinity is deleterious to photosystem I and II of tomato plants. PS I and II can be revived by foliar application of Si (Mateos-Naranjo et al. 2015). Similarly, Si application also improves pigment quality and efficiency in PS II of C4 grass, *Spartina densiflora* under salt stress (Gorbe and Calatayd 2012; Oukarroum et al. 2015; Mateos-Naranjo et al. 2015). Si exhibits the same mitigating effects as 24-Epibrassinolide on *Brassica juncea* under salt stress (Siddiqui et al. 2018). Si (Na_2SiO_3) has also been noted to carry out biofortification and reduce water loss in salt-tolerant and -sensitive cultivars of rice. The impact of Si application has been found to vary with varying sensitivities of salt (Das et al. 2018).

2.4.3 Heavy Metal Stress

Silicon has the ability to restrict and ameliorate heavy metal toxicity by several mechanisms. It increases the rate of chelation in cells via stimulation of plant root exudates which play a role in limiting uptake of heavy metals (Adrees et al. 2015). It can quench the free heavy metallic ions from its apoplastic region which results in reduced translocation (Adrees et al. 2015). Biosilicification is another silicon-mediated tolerance mechanism wherein silicic acid undergoes polymerization in the apoplast and a barrier of amorphous Si is formed which prevents penetration of toxic heavy metals such as aluminum (Al), manganese (Mn), cadmium (Cd), zinc (Zn), arsenic (As), and sodium (Na) into symplast or water transportation stream (Ma et al. 2015; Exley 2015; Guerriero et al. 2016). Another mechanism of counteracting heavy metals is via lignification. Lignified cell walls are good metal binders and therefore prevent metal movement from roots to plant aerial tissue

(Ma and Yamaji 2006; Ye et al. 2012). Silicon holds the unique ability to form complexes with metal ions in the cell wall and eventually, forming a precipitate of metal ions as co-factors (Pontigo et al. 2017). Silicon can react to form silicates and oxides with heavy metals (Exley 2015) keeping the toxicants out of any plant metabolic process (Exley 2015). Application of silicon to the soil is beneficial since it balances the disrupted soil and immobilizes heavy metals like Cd making them unavailable to plants (Wu et al. 2013).

2.4.3.1 Cd Toxicity

Presence of Cd in soil inhibits root growth of rice plants. The toxicity can be identified by the appearance of black spots in the cortex and pericycle of roots (Kim et al. 2014). In wheat and maize crops, it affects seed germination, nutrient content, and lowers shoot and root length. (Ma et al. 2015). In barley, photosynthetic apparatus, pigments, and lipids are affected by Cd-induced toxicity (Hodson et al. 2005). Heavy metal stress studies have revealed that silicon has the ability to decrease cadmium uptake and further limits its translocation to plant aerial tissue like shoots. Cd and Mn are often precipitated on the epidermis of the shoot or leaf blade by forming amorphous silica (Ma et al. 2015). Cd is often compartmentalized in root cell walls by Si, leading to its lowered accumulation in shoots of rice (Bhat et al. 2019). In maize plants, Cd forms colloidal silicon embedded in the cell walls to prevent its uptake or transport to the aerial parts (Bhat et al. 2019). Similarly, the application of Si to *Poa annua* seedlings imparts tolerance to cadmium toxicity (Zama et al. 2018).

2.4.3.2 As Toxicity

Arsenic toxicity is majorly seen in rice and spinach plants. In rice plants, it is overcome by competition with other heavy metal ions at the point of entry/site of uptake in roots. Dry biomass of leaves is regulated positively upon the application of Si to spinach plants. A subsequent increase in levels of glutaredoxin (GRX) is also noted (Dubey et al. 2018). Si-biochars are components composed by coupling bamboo with Si. The element has been used to reduce bioaccumulation of arsenic in spinach leaves by ~38% (Li et al. 2017) (Table 2.1).

2.4.3.3 Al Toxicity

Recent studies have revealed that silicon has the ability to regulate malic and formic acid formation in plants. The formation of these cellular byproducts is helpful in regulating uptake of aluminum (Pontigo et al. 2017). Phenolic compounds of maize have also been investigated in relation to their ability to reduce Al-uptake (Adrees et al. 2015). Si often complexes with Al to form Si–Al or aluminum silicate localized in the plant cell wall, primarily in epidermis and hypodermis. Complex formation makes toxic Al unavailable to the plants (Horst et al. 2010; Liu et al. 2013). Another mechanism to combat Al toxicity is to form hydroaluminosilicates in root apoplast, thereby, reducing mobility of noxious Al (Rogalla and Romheld 2002).

Table 2.1 Si-mediated mitigation of major abiotic stress responses in different plants

Stress	Plant species	Effect of stress	Effect of Silicon supplementation	References
Salinity	*Sorghum bicolor*	Inhibits uptake of noxious ions, photosynthesis as well as stomata opening	Modulation of catalase, peroxidase, SOD; restricts ion uptake via roots	Soundararajan et al. (2017), Liu et al. (2019)
Drought	*Brassica napus, Solanum lycopersicum*	Oxidative stress, Decreased photosynthesis	Accumulation of proline, GABA and ascorbate-glutathione cycle members	Ali et al. (2018), Hasanuzzamam et al. (2018)
Heavy metals	*Zea mays, Oryza sativa, Spinacia oleracea, Bambusa vulgaris*	Reduced toxic ion uptake and accumulation	Lignification, suberization	Li et al. (2017), Dubey et al. (2018), Bhat et al. (2019)
Cold stress	*Hordeum vulgare*	Loss of membrane integrity	Accumation of soluble carbohydrates and other osmolytes	Joudmand and Hajibolan (2019)

2.4.4 Thermal Stress

Thermal stress is often caused by extreme fluctuations in temperature such as heat, chilling, and freezing stress. High temperature or heat stress causes the burning of aerial plant tissues (mostly, leaves), scorched twigs, senescence, and discoloration of leaves (Fahad et al. 2017). A rise in temperature can cause a loss in germination vigor of seeds and therefore poor growth and yield. They may also lead to reduced flower and seed-set in sorghum and several cereal crops (Fahad et al. 2017). Similar observations have also been made in maize and sugarcane. Heat stress leads to reduction in oil, protein, and starch contents of oilseed crops. Several physiological and biochemical processes are damaged on exposure to heat stress leading to water scarcity, reduction in leaf tissue, reduced root conductance, and increased transpiration (Huang et al. 2012). It also impacts nutrient metabolism in plants. Nitrate reductase activity is drastically reduced under temperature stress (Klimenko et al. 2006). As observed in drought and salt stress, heat stress also visibly affects photosynthesis and associated apparatus. Low CO_2 availability, stomatal closure, reduced moisture, and changes in photosynthetic pigments are noted under heat stress (Fahad et al. 2017). Heat stress also impairs photosystem II along with the regeneration capacity of RuBP (Wise et al. 2004).

Foliar application of Si is an effective method for protecting rice and grapevine plants growing in chilling and freezing growth conditions, respectively (Habibi 2015; Azeem et al. 2016). Temperature and salinity stress modulate catalase activity in wheat and *Salvia,* which is effectively countered by the application of exogenous silicon (Liang et al. 2003). Electrolyte leakage is an important indicator of thermal stress in plants. Si-supplementation causes reduced electrolyte leakage due to high temperature (Ma et al. 2015). Thus, Si may have a role in generating thermal stability in cell membranes although further investigations need to be carried out to decipher the mechanisms and pathways involved.

In congruence with heat stress, chilling temperature conditions limit the growth and development of plants drastically, by diminishing root proliferation and early plant growth (Moradtalab et al. 2018). Cold stress and tolerance studies have been extensively studied using maize as a model plant. Chilling stress results in chlorosis, necrosis of leaf tissue, and inhibits root and shoot extension growth (Imran et al. 2013). These morphological deviations are often accompanied by physiological stress responses such as elevated production of ROS (Pei et al. 2010). Mechanisms involved in mitigating chilling stress still remain unexplored. Recent studies reveal that cold stress amelioration is often carried out by Zn, Mn, and Si, commonly known as the "cold stress protectants" (Bradáčová et al. 2016). Si helps to translocate micronutrients of seed reserved to seedlings under cold stress (Moradtalab et al. 2018). Silicon has the ability to prevent leaching due to cold stress by maintaining Zn/Mn reserves in the seed, which also act as cold stress suppressants. The most significant role of Si in mitigating cold stress is the restoration of root growth in maize plants. A similar role of Si under cold stress has been noted in soybean plants as well (Pascual et al. 2016).

Recent studies have shed some light on the existence of the ICE–CBF–COR pathway, which plays a key role in imparting cold stress acclimatization (Ritonga and Chen 2020). Activation of C-binding repeats (CBF) via cold induction by the inducer CBF-expression (ICE) results in the activation of cold responsive genes. Activation and regulation of the ICE–CBF–COR pathway result in the expression of several downstream genes, necessary for imparting tolerance against cold and chilling environmental conditions (Ritonga and Chen 2020). Low-temperature stress can often be categorized into chilling stress ($<20\ ^\circ C$) and freezing stress ($<0\ ^\circ C$) (Mickelbart et al. 2015; Guo et al. 2017; Liu and Zhou 2018; Shi et al. 2018). Crops such as tomato, soybean, cotton, corn, rice, and potato are intolerant to cold environmental conditions (both chilling and freezing stress). However, plants like oats, barley, rye, and wheat show good adaptability to cold stress (Zhang et al. 2011, 2017). Under low-temperature stress conditions, ICE mediated regulation of Gibberellic acid (GA) level is essential for cold tolerance. Plants facing cold stress have to maintain the stability of cell membranes and structural integrity for survival (Chen et al. 2018). Exposure to freezing conditions can often lead to the formation of ice nucleators and crystals in plant cell apoplast which ultimately results in dehydration, electrolyte leakage, and membrane disintegration (Puhakainen 2004). Under extreme situations, these ice crystals can puncture the cell leading to plants death (Demidchik et al. 2014; Sun et al. 2019). The most commonly observed mechanism for studying cold temperature tolerance is the accumulation of cryoprotective polypeptides, sugars, and osmolytes (Khan et al. 2015).

Furthermore, application of silicon to leaves of barley under cold (chilling as well as freezing stress) leads to an increase in levels of antioxidant enzymes, soluble carbohydrates, and osmolytes, especially in leaf apoplasm (Joudmand and Hajibolan 2019) (Table 2.1). Being the first site to be exposed to the cues of low temperature, overwintering plants often accumulate anti-freeze amino acids and compatible sugars in their apoplasm (Liang et al. 2015).

2.4.5 Nutrition Stress

Since a long time, there is a common observation that tremendous use of nitrogen fertilizers is being practiced by farmers for the multiple benefits, however, this also results in adverse effects such as enhanced lodging coupled with susceptibility to both pests and diseases (Ma 2004; Thomidis et al. 2016; Khan et al. 2018; Hosseini et al. 2015). There are many reports in which silicon has been used to minimize this offside issues as the silicate crystals provide mechanical strength as well as hinder insect feeding and inhibit penetration of fungal germ tube on the plant surface (Elsherbiny and Taher 2018; Singh et al. 2020). Besides, there are many reports which endorse the enhanced uptake and assimilation of an important nutrient, i.e. nitrogen in the presence of silicon (Pati et al. 2016; Malav Jugal and Ramani 2017; Patil et al. 2018; Haddad et al. 2018; Laîné Haddad et al. 2019; Gou et al. 2020; Raj et al. 2020). In addition, the use of silicon has also resulted in improved nodulation, better N_2 fixation, increased N use efficiency, and stimulated amino acid remobilization (Detmann et al. 2012; Steiner et al. 2018; Kurdali et al. 2019; Mohanty et al. 2020). Furthermore, excessive concentration of nitrogen is also toxic as it negatively affects plant and the quality of their products (Neilsen et al. 2008; Hilbert et al. 2015; Abrol et al. 2017). Si also mitigates excessive nitrogen stress (Singh et al. 2006; Liang et al. 2015; Campos et al. 2016; Barreto et al. 2016, 2017; Viciedo et al. 2019, 2020).

Next to nitrogen, phosphorus is another essential mineral element required for plant vigor in higher amount but the contrasting point is low availability of plant-available phosphorus in soil (Achary et al. 2017; Chu et al. 2020). In this regard, the application of soluble silicon fertilizers has resulted in an increased amount of bioavailable phosphorus as well as water-soluble phosphorus concentration (Owino-Gerroh and Gascho 2005; Singh et al. 2006; Liang et al. 2015; Tripathi et al. 2016; Agostinho et al. 2017; Zia et al. 2017; Kostic et al. 2017; Rezakhani et al. 2019; Zhang et al. 2019; Liao et al. 2020). The mechanism operating for how the silicon influence phosphorus uptake in plants was solved before the onset of the twenty-first century (Etesami and Jeong 2018, 2020).

In addition, potassium is also one of the major macronutrients that play important role in plant's growth, development, and metabolism as well as even range up to 2–10% of the dry mass (Cruz et al. 2019; Etesami and Jeong 2020). Silicon application has been also reported to alleviate the K-deficiency stress in stressed plants by modifying K-availability in both plants as well as soil (Mali and Aery 2008; Miao et al. 2010; Chen et al. 2016; Cuong et al. 2017). Sulfur-deficient barley crops were alleviated by supplementation of Si which helps regulate the action of ABA metabolism-related genes (Maillard et al. 2018). Additionally, application of silicon has been also found to support the plant in nutrient deficiency stress conditions as well as toxicity; for example, calcium (Mali and Aery 2008; Etesami and Jeong 2018; Dong et al. 2018; do Nascimento et al. 2020), magnesium (Hosseini et al. 2019; do Nascimento et al. 2020), boron (Savić and Marjanović-Jeromela 2013; Liu et al. 2017; Metwally et al. 2018; Pereira de Souza Junior et al. 2019; Oliveira et al. 2020), iron (You-Qiang et al. 2012; Pavlovic et al. 2013; Bityutskii et al. 2014; Patil et al. 2018; do Nascimento et al. 2020; dos Santos et al. 2020), manganese (Dragišić Maksimović et al. 2007; Patil et al. 2018; de Oliveira et al. 2019; do Nascimento et al. 2020), zinc (Gu et al. 2012; Bityutskii et al. 2014; Hernandez-Apaolaza 2014; Pascual et al. 2016; do Nascimento et al. 2020; Raj et al. 2020), copper (Frantz et al. 2011; Patil et al. 2018; Raj et al. 2020; El-Beltagi et al. 2020), and sulfur (Maillard et al. 2018; Réthoré et al. 2020). The interaction of silicon with the above-mentioned nutrient elements has not been explored extensively and further investigation in this direction is highly recommended.

2.4.6 UV-B Radiation Stress

Apart from the above-mentioned stresses, UV-B stress is considered as harmful stress for both plants and animals including humans (Jordan 2002; Yin and Ulm 2017; Chakraborty et al. 2017). The reason lies in its ability to influence biochemistry, physiology, and genetic changes in plants (Jordan 2002; Tripathi et al. 2017; Etesami and Jeong 2018; Azarafshan et al. 2020). The exogenous application of silicon to the plants has also resulted in alleviating the effects of UV-B stress on many plants (Fang et al. 2011; Yao et al. 2011; Schaller et al. 2013; Tripathi et al. 2017). All these studies together in combination have revealed that the exogenous application of silicon results in the formation of a cuticle–Si double layer, which acts as a glass layer and reduces the further transmission of UV radiation from the epidermis (Gatto et al. 1998; Currie and Perry 2007; Etesami and Jeong 2018). In addition, the silicon application induces resistance in plants by modifying ROS consumption (Shen et al. 2010), levels of UV absorbing compounds (Liang et al. 2015), and antioxidative enzyme activities (Fang et al. 2019).

2.4.7 Wounding Stress

One of the interesting stresses among the variety of non-biological stresses is wounding stress that results in physical injury in the plants (Malhotra and Kapoor 2019). This arises actually from strong winds or water and due to the attack by herbivores (insects, birds, and nematodes) (Malhotra and Kapoor 2019; Singh et al. 2020; Souri et al. 2020). Primarily, these physical injuries increase the vulnerability to pathogenic attack by creating openings in plant organs as well as initiate oxidative stress at the secondary level that ultimately leads to death in serious cases via cell apoptosis (Malhotra and Kapoor 2019). In order to cope-up with wounding stress, the silicon treatment results in modulation in the levels of antioxidant enzymes (such as phenylalanine ammonia-lyase, peroxidase, polyphenol oxidase, and catalase), malondialdehyde, hormones, and changes in gene expression pattern (Kim et al. 2011, 2014, 2016; Hajiboland et al. 2017; Hall et al. 2019).

2.4.8 High pH Stress

Over the last century, due to the continuous anthropogenic activities and farming, there has been a paradigm shift in the pH of soil which either results in acidic or alkaline conditions (Wang et al. 2017; Shen et al. 2019). This pH changes in the pH of soil affect not only the plant roots but their functionality too. As a result of it, many modifications are induced frequently in the root and shoot physiology due to change in the pH of the xylem sap. Even in this regard, exogenous application of silicon has also resulted in alleviating the effects of pH stress on many plants (Abdel Latef and Tran 2016; Wang et al. 2017; Liu et al. 2018; Khan et al. 2019; Ju et al. 2020) (Table 2.2).

2.5 Necessity of Silicon in Agriculture

With increasing incidences and severity of climate change, the challenges to the crop production will also increase in a long way to the future (Mahmoud and Gan 2018; Tamburino et al. 2020). This will be truly reflected in the form of more and more cases of diseases, pest attack, salinity, and drought conditions (Bashyal 2018; Rathee and Dalal 2018; Balamurugan et al. 2019; Shiru et al. 2020). Moreover, the intensified cum successive cropping has resulted in the elimination of basic cations from the soil (Jaiyeoba 2003; Xiao et al. 2013; Hao et al. 2019; Macedo et al. 2020). In addition, the continuous fertilization by the farmers necessitates the liming programs for maintenance of yields (Tubana et al. 2016; Hao et al. 2019; Xu et al. 2020). In this case, the high liming potential of silicon sources like silicate slags serves as a good agronomic option to correct soil pH (Tubana et al. 2016; Keeping et al. 2017; Etesami and Jeong 2018; Haynes 2019). In addition, many silicon and fertilizers act as a low-cost, good source of some important nutrients in plants such as Ca^{2+} and Mg^{2+} as well as fertilization enhance uptake of N, P, and K (Tubana et al. 2016; Zhao et al. 2020). Furthermore, various silicon sources that are being employed are actually low-cost byproducts from industries (Chanchal Malhotra et al. 2016). Even in few cases, many farmers have used straws of wheat and rice straw along with other small grain crops as a potential Si source (Marxen et al. 2016; Seleiman et al. 2019). Additionally, there are many reports on a foliar spray of Si containing solution (Haynes et al. 2013; Rodrigues et al. 2015; Oliveira et al. 2019). In the recent past, many liquids as well as granulated forms of Si fertilizers have been also generated due to the advances in the field of nanotechnology (Tubana et al. 2016; Rastogi et al. 2019; Siddiqui et al. 2020). These nano-size scale high-bioavailability fertilizers easily penetrate and create a thick-silicate layer on the surfaces of leaves (Chanchal Malhotra et al. 2016; Siddiqui et al. 2020). All these

Table 2.2 Highlights of silicon-mediated mitigation of minor abiotic stress conditions

Stress	Experimental study	Plant species	Silicon application	Effect of silicon application	References
Wounding	Effect of silicon on mitigation of wounding stress	*O. sativa*	0.5 and 2 mM Na_2SiO_3	Improved leaf chlorophyll content and reduced oxidative stress	Kim et al. (2016)
	Effect of silicon on local and systemic response to mechanical stress in tobacco	*Nicotiana rustica*	1 mM Na_2SiO_3	Decreased production of peroxidases, polyphenol oxidase and augmented lignin accumulation in local response	Hajiboland et al. (2017)
UV-B	Effect of silicon on alleviating UV-B radiation damage	*Glycine max*	1.7 mM K_2SiO_3	Reduced accumulation of cyclobutane pyrimidine dimers (CPDs), increased production of UV-B absorbing compounds, increased photolyase gene expression	Chen et al. (2016)
	Role of silicon nanoparticles on alleviating UV-B stress	*Triticum aestivum*	10 µM SiNPs	Reduced oxidative stress, enhanced antioxidant production, increased nitric oxide production, overall SiNP appear to be better than Si	Tripathi et al. (2017)
Nutrient	Role of silicon in decreasing iron deficiency responses in soybean plants	*G. max*	1 mM Na_2SiO_3	Improved biomass and iron concentration, improved photosynthesis	Muneer and Jeong (2015)
	Effect of silicon on potassium deficiency in sorghum	*S. bicolor*	1 mM H_2SiO_3	Improved potassium concentration in xylem sap and improved growth and water status	Chen et al. (2016)
	Role of silicon in phosphorus uptake under low P conditions	*T. aestivum*	Na_2SiO_3 (400 mg Si kg^{-1} dry soil)	Improved P uptake and increased expression of Pi transporter genes	Kostic et al. (2017)
	Role of silicon in nitrogen uptake under low N conditions	*B. napus*	1.7 mM Na_2SiO_3	Improved uptake of N and increase expression of nitrate transporter gene	Haddad et al. (2018)
	Effect of silicon on plants growing under phosphorus deficiency conditions	*S. lycopersicum*	1.5 mM K_2SiO_3	Increased photosynthesis, biomass, reduced ROS and malondialdehyde levels and enhanced uptake of most essential elements	Zhang et al. (2019)
	Effect of silicon on rice plants growing under sulphur deficiency	*O. sativa*		Lower accumulation of stress phytohormones, enhanced growth and	Réthoré et al. (2020)

(continued)

Table 2.2 (continued)

Stress	Experimental study	Plant species	Silicon application	Effect of silicon application	References
			1 mM Monosilicic acid $[Si(OH)_4]$	balanced source to sink metabolic homeostasis	
pH	Effect of silicon on tolerance to alkaline stress	*Z. mays*	1.5 mM Na_2SiO_3	Improved growth of stressed plants, enhanced relative water content and photosynthetic pigments	Abdel Latef and Tran (2016)
	Effect of silicon on mitigating pH stress	*Festuca arundinacea*	0, 2, 8 mM Na_2SiO_3	Under acid stress, low Si concentration is effective in alleviating pH stress and under alkali stress it was opposite	Wang et al. (2017)
	Effect of silicon priming on high alkaline stress tolerance	*Medicago sativa*	0.075–3.75 mM Na_2SiO_3	Increased biomass as well as chlorophyll content and reduced oxidative stress	Liu et al. (2018)
	Effect of silicon and salicylic acid on high pH tolerance in tomato	*S. lycopersicum*	1 mM Na_2SiO_3	Plants displayed high chlorophyll content, reduced accumulation of ROS and high relative water content	Khan et al. (2019)
	Effect of silicon on rice plants under acid rain stress	*O. sativa*	0, 1, 2 and 4 mM Na_2SiO_3	Improved plant growth, chloroplast structure and photosynthesis	Ju et al. (2020)

factors establish the importance of a common agronomic practice worldwide, i.e. silicon fertilization. Therefore, owing to the above-described points along with the benefits of silicon in alleviating both non-biological stresses, it is a point to admit that silicon is a plant nutritional "non-entity" for mitigating plethora of abiotic stresses.

2.6 Future Prospects

According to the current status of Si-research, it has been deciphered that there are multiple unaddressed questions related to many avenues which can be explored in a long way to the future, including

1. Detailed deduction of the complete "omics network" that operates during both non-biological and biological stresses in presence of silicon,
2. Deducing the effect of silicon amendments on plants exposed to combinatorial stresses,
3. Deducing the transport mechanisms that work during foliar uptake of Si in leaves,
4. Determining the effect of silicon fertilizers on the plant "whole microbiome" as well as plant–microbe interactions,
5. Effect of Si on non-accumulator plant species for enhancing stress resistance,
6. Understanding the detailed effects of Si on root anatomy around the whole plant kingdom,
7. Evaluating the economic feasibility of various Si sources,
8. Complete chemical analysis of the products made from slag,
9. Developing a prediction model that correlates the Si-mediated recovery with carbon accumulated and amino acid metabolism during stresses,
10. Evaluating the complete potential of SiNPs for alleviating abiotic stresses in farmer fields on large scale, and
11. Quantification of the content of monosilicic acid and polysilicic acid as well as grain size to develop an optimized system that works well for every crop that is being cultivated by humans.

References

Abbas T, Balal RM, Shahid MA, Pervez MA, Ayyub CM, Aqueel MA (2015) Silicon-induced alleviation of NaCl toxicity in okra (Abelmoschus esculentus) is associated with enhanced photosynthesis, osmoprotectants and antioxidant metabolism. Acta Physiol Plant 37:6

Abdel Latef AA, Tran LS (2016) Impacts of priming with silicon on the growth and tolerance of maize plants to alkaline stress. Front Plant Sci 7:243

Abrol Y, Adhya TK, Aneja V, Raghuram N, Pathak H, Kulshrestha U, Sharma C, Singh B (2017) The Indian nitrogen assessment: sources of reactive nitrogen, environmental and climate effects, management options, and policies, 1st edn. Elsevier, Amsterdam

Achary VMM, Ram B, Manna M, Datta D, Bhatt A, Reddy MK, Agrawal K (2017) Phosphite: a novel P fertilizer for weed management and pathogen control. Plant Biotechnol 15:1493–1508
Adrees M, Ali S, Rizwan M, Zia-ur-Rehman M, Ibrahim M, Abbas F, Farid M, Qayyum MF, Irshad MK (2015) Mechanisms of silicon-mediated alleviation of heavy metal toxicity in plants: a review. Ecotoxicol Environ Safety 119:186–197
Agostinho FB, Tubana BS, Martins MS, Datnoff LE (2017) Effect of different silicon sources on yield and silicon uptake of rice grown under varying phosphorus rates. Plan Theory 6:35
Ahanger MA, Bhat JA, Siddiqui MH, Rinklebe J, Ahmad P (2020) Silicon and secondary metabolites integration in plants: a significant association in stress tolerance. J Exp Bot: eraa291. https://doi.org/10.1093/jxb/eraa291
Ahirwar NK, Singh R, Chaurasia S, Chandra R, Prajapati S, Ramana S (2020) Effective role of beneficial microbes in achieving the sustainable agriculture and eco-friendly environment development goals: a review. Front Microbiol 5:111–123
Ali N, Schwarzenberg A, Yvin JC, Hosseini SA (2018) Regulatory role of silicon in mediating differential stress tolerance responses in two contrasting tomato genotypes under osmotic stress. Front Plant Sci 9:1475
Ali M, Javaid A, Naqvi SH, Batcho A, Kayani WK, Lal A, Sajid IA, Nwogwugwu JO (2020) Biotic stress triggered small RNA and RNAi defense response in plants. Mol Biol Rep 47:5511–5522. https://doi.org/10.1007/s11033-020-05583-4
Anamika MS, Singh B, Patra A, Islam MA (2019) Databases: a weapon from the arsenal of bioinformatics for plant abiotic stress research. In: Wani SH (ed) Recent approaches in Omics for plant resilience to climate change. Springer, Cham, pp 135–169
Ashraf M, Rahmatullah Afzal M, Ahmed R, Mujeeb F, Sarwar A, Ali L (2010) Alleviation of detrimental effects of NaCl by silicon nutrition in salt-sensitive and salt-tolerant genotypes of sugarcane (*Saccharum officinarum* L.). Plant Soil 326:381–391
Azarafshan M, Peyvandi M, Abbaspour H, Noormohammadi Z, Majd A (2020) The effects of UV-B radiation on genetic and biochemical changes of *Pelargonium graveolens* L′ her. Physiol Mol Biol Plants 26:605–616
Azeem S, Li Z, Zheng H, Lin W, Arafat Y, Zhang Z, Lin X, Lin W (2016) Quantitative proteomics study on LSi1 in regulation of rice (*Oryza sativa* L.) cold resistance. Plant Growth Regul 78:307–323
Balakhnina T, Borkowska A (2013) Effects of silicon on plant resistance to environmental stresses. Int Agrophys 27:225–232
Balamurugan A, Sakthivel K, Kumar A, Muthamilan M (2019) First report of *Curvularia hominis* inciting fruit rot of ridge gourd (Luffa acutangula) in Tamil Nadu. India J Plant Pathol 102:529–529
Baldos UL, Hertel TW, Moore FC (2019) Understanding the spatial distribution of welfare impacts of global warming on agriculture and its drivers. Am J Agric Econ 101:1455–1472
Barreto RF, Prado RM, Leal AJF, Troleis MJB, Junior GS, Monteiro CC, Santos LCN, Carvalho RF (2016) Mitigation of ammonium toxicity by silicon in tomato depends on the ammonium concentration. Acta Agric Scand B Soil Plant Sci 66:483–488
Barreto RF, Júnior AAS, Maggio MA, de Mello Prado R (2017) Silicon alleviates ammonium toxicity in cauliflower and in broccoli. Sci Hortic 225:743–750
Bashyal BM (2018) Etiology of an emerging disease: Bakanae of rice. Indian Phytopathol 71:485–494
Bhat JA, Shivaraj SM, Singh P, Navadagi DB, Tripathi DK, Dash PK, Solanke AU, Sonah H, Deshmukh R (2019) Role of silicon in mitigation of heavy metal stresses in crop plants. Plan Theory 8:71
Bityutskii N, Pavlovic J, Yakkonen K, Maksimović V, Nikolic M (2014) Contrasting effect of silicon on iron, zinc and manganese status and accumulation of metal-mobilizing compounds in micronutrient-deficient cucumber. Plant Physiol Biochem 74:205–211

Bradáčová K, Weber NF, Morad-Talab N, Asim M, Imran M, Weinmann M, Neumann G (2016) Micronutrients (Zn/Mn), seaweed extracts, and plant growth-promoting bacteria as cold-stress protectants in maize. Chem Biol Technol Agric 3:19

Campos CNS, de Mello Prado R, Caione G, de Lima Neto AJ, Mingotte FALC (2016) Silicon and excess ammonium and nitrate in cucumber plants. Afr J Agric Res 11:276–283

Chakraborty S, Mali K, Chatterjee S, Banerjee S, Roy K, Dutta N, Bhaumik N, Mazumdar S (2017) Dermatological effect of UV rays owing to ozone layer depletion. In: 2017 4th international conference on opto-electronics and applied optics (optronix). IEEE, pp 1–6

Chanchal Malhotra C, Kapoor R, Ganjewala D (2016) Alleviation of abiotic and biotic stresses in plants by silicon supplementation. Scientia 13:59–73

Chauhan R, Awasthi S, Tripathi P, Mishra S, Dwivedi S, Niranjan A, Mallick S, Tripathi P, Pande V, Tripathi RD (2017) Selenite modulates the level of phenolics and nutrient element to alleviate the toxicity of arsenite in rice (*Oryza sativa* L.). Ecotoxicol Environ Saf 138:47–55

Chen D, Cao B, Wang S, Liu P, Deng X, Yin L, Zhang S (2016) Silicon moderated the K deficiency by improving the plant-water status in sorghum. Sci Rep 6:1–14

Chen L, Zhao Y, Xu S, Zhang Z, Xu Y, Zhang J, Chong K (2018) OsMAD57 together with OsTB1 coordinates transcription of its target OsWRKY94 and D14 to switch its organogenesis to defense for cold adaptation in rice. New Phytol 218:219–231

Chiba Y, Mitani N, Yamaji N, Ma JF (2009) HvLsi1 is a silicon influx transporter in barley. Plant J 57:810–818

Chu Q, Zhang L, Zhou J, Yuan L, Chen F, Zhang F, Feng G, Rengel Z (2020) Soil plant-available phosphorus levels and maize genotypes determine the phosphorus acquisition efficiency and contribution of mycorrhizal pathway. Plant Soil 449:357–371

Conceição P, Levine S, Lipton M, Warren-Rodríguez A (2016) Toward a food secure future: ensuring food security for sustainable human development in Sub-Saharan Africa. Food Policy 60:1–9

Cornelis JT, Delvaux B, Georg R, Lucas Y, Ranger J, Opfergelt S (2011) Tracing the origin of dissolved silicon transferred from various soil-plant systems towards rivers: a review. Biogeosciences 8(1):89–112. https://doi.org/10.5194/bg-8-89-2011

Coskun D, Britto DT, Huynh WQ, Kronzucker HJ (2016) The role of silicon in higher plants under salinity and drought stress. Front Plant Sci 7:1072

Cripps E (2016) Population and environment. In: Gardiner SM, Thompson A (eds) The Oxford handbook of environmental ethics. Oxford University Press, New York, pp 380–390

Cruz FJR, de Mello Prado R, Felisberto G, Santos ÁS, Barreto RF (2019) Potassium nutrition in fruits and vegetables and food safety through hydroponic system. In: Asaduzzaman M, Asao T (eds) Improvement of quality in fruits and vegetables through hydroponic nutrient management. IntechOpen, London, pp 23–44

Cuong TX, Ullah H, Datta A, Hanh TC (2017) Effects of silicon-based fertilizer on growth, yield and nutrient uptake of rice in tropical zone of Vietnam. Rice Sci 24:283–290

Currie HA, Perry CC (2007) Silica in plants: biological, biochemical and chemical studies. Ann Bot 100:1383–1389

Daniel Maxim L, Niebo R, LaRosa S, Johnston B, Allison K, McConnell EE (2008) Product stewardship in wollastonite production. Inhal Toxicol 20:1199–1214

Das P, Manna I, Biswas AK, Bandyopadhyay M (2018) Exogenous silicon alters ascorbate-glutathione cycle in two salt-stressed indica rice cultivars (MTU 1010 and Nonabokra). Environ Sci Pollut Res 25:226625–226642

de Oliveira RLL, de Mello Prado R, Felisberto G, Checchio MV, Gratão L (2019) Silicon mitigates manganese deficiency stress by regulating the physiology and activity of antioxidant enzymes in sorghum plants. J Soil Sci Plant Nutr 19:524–534

Demidchik V, Straltsova D, Medvedev SS, Pozhvanov GA, Sokolik A, Yurin V (2014) Stress-induced electrolyte leakage: the role of K+- permeable channels and involvement in programmed cell death and metabolic adjustment. J Exp Bot 65:1259–1270

Deshmukh RK, Vivancos J, Guérin V, Sonah H, Labbé C, Belzile F, Bélanger RR (2013) Identification and functional characterization of silicon transporters in soybean using comparative genomics of major intrinsic proteins in Arabidopsis and rice. Plant Mol Biol 83:303–315
Detmann KC, Araújo WL, Martins SC, Sanglard LM, Reis JV, Detmann E, Rodrigues FÁ, Nunes-Nesi A, Fernie AR, DaMatta FM (2012) Silicon nutrition increases grain yield, which, in turn, exerts a feed-forward stimulation of photosynthetic rates via enhanced mesophyll conductance and alters primary metabolism in rice. New Phytol 196:752–762
Dhakate P, Sharma P, Mehta S, Akter J, Bhatt V, Chandanshive S, Chakravarty D, Rahman M, Islam MA (2019) Beneficial role of metalloids in plants: molecular understanding and applicability. In: Wani SH (ed) Recent approaches in Omics for plant resilience to climate change. Springer, Cham, pp 357–387
Dixit S, Singh UM, Singh AK, Alam S, Venkateshwarlu C, Nachimuthu VV, Yadav S, Abbai R, Selvaraj R, Devi MN, Ramayya J (2020) Marker assisted forward breeding to combine multiple biotic-abiotic stress resistance/tolerance in rice. Rice 13:29
do Nascimento CWA, de Souza Nunes GH, Preston HAF, da Silva FBV, Preston W, Loureiro FLC (2020) Influence of silicon fertilization on nutrient accumulation, yield and fruit quality of melon grown in Northeastern Brazil. SILICON 12:937–943
Dong Z, Li Y, Xiao X, Chen Y, Shen X (2018) Silicon effect on growth, nutrient uptake, and yield of peanut (*Arachis hypogaea* L.) under aluminum stress. J Plant Nutr 41:2001–2008
dos Santos MS, Sanglard LM, Barbosa ML, Namorato FA, de Melo DC, Franco WC, Pérez-Molina J, Martins SC, DaMatta FM (2020) Silicon nutrition mitigates the negative impacts of iron toxicity on rice photosynthesis and grain yield. Ecotoxicol Environ Saf 189:110008
Dragišić Maksimović J, Bogdanović J, Maksimović V, Nikolic M (2007) Silicon modulates the metabolism and utilization of phenolic compounds in cucumber (Cucumis sativus L.) grown at excess manganese. J Plant Nutr Soil Sci 170:739–744
Dubey AK, Kumar N, Ranjan R, Gautam A, Pande V, Sanyal I, Mallick S (2018) Application of glycine reduces arsenic accumulation and toxicity in Oryza sativa L. by reducing the expression of silicon transporter genes. Ecotoxicol Environ Saf 148:410–417
El-Beltagi HS, Sofy MR, Aldaej MI, Mohamed HI (2020) Silicon alleviates copper toxicity in flax plants by up-regulating antioxidant defense and secondary metabolites and decreasing oxidative damage. Sustainability 12:4732
Elsherbiny EA, Taher MA (2018) Silicon induces resistance to postharvest rot of carrot caused by Sclerotinia sclerotiorum and the possible of defense mechanisms. Postharvest Biol Tech 140:11–17
Epstein E (1994) The anomaly of silicon in plant biology. PNAS 91:11–17
Epstein E (2009) Silicon: its manifold roles in plants. Ann Appl Biol 155:155–160
Etesami H, Jeong BR (2018) Silicon (Si): review and future prospects on the action mechanisms in alleviating biotic and abiotic stresses in plants. Ecotoxicol Environ Saf 147:881–896
Etesami H, Jeong BR (2020) Importance of silicon in fruit nutrition: agronomic and physiological implications. In: Srivastava AK, Hu C (eds) Fruit crops: diagnosis and management of nutrient constraints. Elsevier, Amsterdam, pp 255–277
Exley C (2015) A possible mechanism of biological silicification in plants. Front Plant Sci 6:853
Exley C, Guerriero G, Lopez X (2020) How is silicic acid transported in plants? SILICON. https://doi.org/10.1007/s12633-019-00360-w
Fahad S, Bajwa AA, Nazir U, Anjum SA, Farooq A, Zohaib A, Sadia S, Wajid N, Adkins S, Shah S, Ihsan MZ, Alharby H, Wu C, Wang D, Jianliang H (2017) Crop production under drought and heat stress: plant responses and management options. Front Plant Sci 8:1147
Fang C, Wang Q, Yu Y, Huang L, Wu X, Lin W (2011) Silicon and its uptaking gene Lsi1 in regulation of rice UV-B tolerance. Acta Agron Sin 37:1005–1011
Fang C, Li L, Zhang Wang D, Yang L, Reza BM, Lin W (2019) Lsi1 modulates the antioxidant capacity of rice and protects against ultraviolet-B radiation. Plant Sci 278:96–106
Feng J, Yamaji N, Mitani-Ueno N (2011) Transport of silicon from roots to panicles in plants. PJA Ser B 87:377–385

Fleck AT, Schulze S, Hinrichs M, Specht A, Wassmann F, Schreiber L (2015) Silicon promotes exodermal Casparian band formation in Si-accumulating and Si-excluding species by forming phenol complexes. PLoS One 10:e0138555

Frantz JM, Khandekar S, Leisner S (2011) Silicon differentially influences copper toxicity response in silicon-accumulator and non-accumulator species. J Am Soc Hortic Sci 136:329–338

Fresno T, Moreno-Jiménez E, Zornoza M, Peñalosa JM (2018) Aided phytostabilisation of As-and Cu-contaminated soils using white lupin and combined iron and organic amendments. J Environ Manag 205:142–150

Garcia-Plazaola JI, Becerril JM (2000) Effects of drought on photoprotective mechanisms on photoprotective mechanisms in European beech (*Fagus sylvatica* L.) seedlings from different provenances. Trees 14:485–490

Garg N, Bhandari PJ (2016) Interactive effects of silicon and arbuscular mycorrhiza in modulating ascorbate-glutathione cycle and antioxidant scavenging capacity in differentially salt-tolerant *Cicer arietinum* L. genotypes subjected to long-term salinity. Protoplasma 253:1325–1345

Gatto A, Escoubas L, Roche K, Commandré M (1998) Simulation of the degradation of optical glass substrates caused by UV irradiation while coating. Opt Commun 148:347–354

Gérard F, François M, Ranger J (2002) Processes controlling silica concentration in leaching and capillary soil solutions of an acidic brown forest soil (Rhône, France). Geoderma 107:197–226

Gilliham M, Able JA, Roy SJ (2017) Translating knowledge about abiotic stress tolerance to breeding programmes. Plant J 90:898–917

Gogna M, Bhatla SC (2019) Biochemical mechanisms regulating salt tolerance in sunflower. Plant Signal Behav 14:1670597

Gogna M, Bhatla SC (2020) Salt-tolerant and -sensitive seedlings exhibit noteworthy differences in lipolytic events in response to salt stress. Plant Signal Behav 15:1737451

Gogna M, Choudhary A, Mishra G, Kapoor R, Bhatla SC (2020) Changes in lipid composition in response to salt stress and its possible interaction with intracellular Na+-K+ ratio in sunflower (*Helianthus annuus* L.). Environ Exp Bot 178:104147. https://doi.org/10.1016/j.envexpbot.2020.104147

Gong H, Zhu X, Chen K, Wang S, Zhang C (2005) Silicon alleviates oxidative damage of wheat plants in pods under drought. Plant Sci 169:313–321

Gorbe E, Calatayud A (2012) Applications of chlorophyll fluorescence imaging technique in horticultural research: a review. Sci Hortic 138:24–35

Goswami K, Mittal D, Gautam B, Sopory SK, Sanan-Mishra N (2020) Mapping the salt stress-induced changes in the root miRNome in pokkali rice. Biomol Ther 10:498

Gou T, Yang L, Hu W, Chen X, Zhu Y, Guo J, Gong H (2020) Silicon improves the growth of cucumber under excess nitrate stress by enhancing nitrogen assimilation and chlorophyll synthesis. Plant Physiol Biochem 152:53–61. https://doi.org/10.1016/j.plaphy.2020.04.031

Grafton RQ, Williams J, Jiang Q (2015) Food and water gaps to 2050: preliminary results from the global food and water system (GFWS) platform. J Food Secur 7:209–220

Gu HH, Zhan SS, Wang SZ, Tang YT, Chaney RL, Fang XH, Cai XD, Qiu RL (2012) Silicon-mediated amelioration of zinc toxicity in rice (*Oryza sativa* L.) seedlings. Plant Soil 350:193–204

Guerriero G, Hausmann JF, Legay S (2016) Silicon and the plant extracellular matrix. Front Plant Sci 7:463

Guo X, Zhang L, Zhu J, Liu H, Wang A (2017) Cloning and characterization of SiDHN, a novel dehydrin gene from *Saussurea involucrata* Kar. Et. Kir. that enhances cold and drought tolerance in tobacco. Plant Sci 256:160–169

Gupta B, Huang B (2014) Mechanism of salinity tolerance in plants: physiological, biochemical, and molecular characterization. Int J Genom 2014:701596

Habibi G (2015) Effects of soil- and foliar-applied silicon on the resistance of grapevine plants to freezing stress. Acta Biol Szeged 59:109–117

Haddad C, Arkoun M, Jamois F, Schwarzenberg A, Yvin JC, Etienne P, Laîné P (2018) Silicon promotes growth of *Brassica napus* L. and delays leaf senescence induced by nitrogen starvation. Front Plant Sci 9:516

Hajiboland R, Cheraghvareh L (2014) Influence of Si supplementation on growth and some physiological and biochemical parameters in salt-stressed tobacco (*Nicotiana rustica* L.) plants. J Plant Nutr 25:205–207

Hajiboland R, Bahrami-Rad S, Poschenrieder C (2017) Silicon modifies both a local response and a systemic response to mechanical stress in tobacco leaves. Biol Plant 61:187–191

Hall CR, Waterman JM, Vandegeer RK, Hartley SE, Johnson SN (2019) The role of silicon in antiherbivore phytohormonal signalling. Front Plant Sci 10:1132

Hao T, Zhu Q, Zeng M, Shen J, Shi X, Liu X, Zhang F, de Vries W (2019) Quantification of the contribution of nitrogen fertilization and crop harvesting to soil acidification in a wheat-maize double cropping system. Plant Soil 434:167–184

Hasanuzzamam M, Nahar K, Anee TI, Khan MIR, Fujita M (2018) Silicon-mediated regulation of antioxidant defense and glyoxalase systems confers drought stress tolerance in *Brassica napus* L. S Afr J Bot 115:50–57

Hattori T, Inanaga S, Tanimoto E, Lux A, Luxova M, Sugimoto Y (2003) Silicon-induced changes in viscoelastic properties of Sorghum root cell walls. Plant Cell Physiol 44:743–749

Hattori T, Inanaga S, Araki H, Morita S, Luxová M, Lux A (2005a) Application of silicon enhanced drought tolerance in Sorghum bicolor. Physiol Plant 123:459–466

Hattori T, Inanaga S, Areki H (2005b) Application of silicon enhanced drought tolerance in *Sorghum bicolor*. Physiol Plant 123:459–466

Haynes RJ (2019) What effect does liming have on silicon availability in agricultural soils? Geoderma 337:375–383

Haynes RJ, Belyaeva ON, Kingston G (2013) Evaluation of industrial wastes as sources of fertilizer silicon using chemical extractions and plant uptake. J Soil Sci Plant Nutr 176:238–248

He C, Wang L, Liu X, Li X, Ma J (2013) Evidence for 'silicon' within the cell walls of suspension-cultured rice cells. New Phytol 200:700–709

Heckman J (2013) Silicon: a beneficial substance. Better Crops 97:14–16

Henriet C, Draye X, Oppitz I, Swennen R, Delvaux B (2006) Effects, distribution and uptake of silicon in banana (Musa sp.) under controlled conditions. Plant Soil 287:359–374

Hernandez-Apaolaza L (2014) Can silicon partially alleviate micronutrient deficiency in plants? A review. Plant 240:447–458

Hetherington AM, Woodward FI (2003) The role of stomata in-sensing and driving environmental change. Nature 424:901

Hilbert G, Soyer J, Molot C, Giraudon J, Milin M, Gaudillere J (2015) Effects of nitrogen supply on must quality and anthocyanin accumulation in berries of cv. Merlot. Vitis 42:69

Hodson M, White L, Mead A, Broadley M (2005) Phylogenetic variation in the silicon composition of plants. Ann Bot 96:1027–1046

Horst WJ, Wang Y, Eticha D (2010) The role of the root apoplast in aluminium-induced inhibition of root elongation and in aluminium resistance of plants: a review. Ann Bot 106:185–197

Hosseini A, Hosseini M, Goldani M, Karimi J, Madadi H (2015) Effect of nitrogen fertilizer on biological parameters of the Aphis craccivora (Hemiptera: Aphididae) and associated productivity losses in common globe Amaranth. J Agric Sci Technol 17:1517–1528

Hosseini SA, Naseri Rad S, Ali N, Yvin JC (2019) The ameliorative effect of silicon on maize plants grown in Mg-deficient conditions. Int J Mol Sci 20:969s

Hu L, Xiang L, Zhang L, Zhou X, Zou Z, Hu X (2014) The photoprotective role of spermidine in tomato seedlings under salinity-alkalinity stress. PLoS One 9:e110855

Hu J, Li Y, Jeong BR (2020) Putative silicon transporters and effect of temperature stresses and silicon supplementation on their expressions and tissue silicon content in poinsettia. Plant 9:569

Huang B, Rachmilevitch S, Xu J (2012) Root carbon and protein metabolism associated with heat tolerance. J Exp Bot 63:3455–3465

Hussain M, Ahmad S, Hussain S, Lal R, Ul-Allah S, Nawaz A (2018) Rice in saline soils. Physiology, biochemistry, genetics and management. In: Sparks DL (ed) Advances in agronomy, vol 148. Academic Press, Cambridge, pp 231–287

Imran M, Mahmood A, Romheld V, Neumann G (2013) Nutrient seed priming improves seedling development of maize exposed to low root zone temperature during early growth. Eur J Agron 49:141–148

Jaiyeoba IA (2003) Changes in soil properties due to continuous cultivation in Nigerian semiarid Savannah. Soil Till Res 70:91–98

Jordan BR (2002) Molecular response of plant cells to UV-B stress. Func Plant Biol 29:909–916

Joudmand A, Hajibolan R (2019) Silicon mitigates cold stress in barley plants via modifying the activity of apoplasmic enzymes and concentration of metabolites. Acta Physiol Plant 41:29

Ju S, Wang L, Chen J (2020) Effects of silicon on the growth, photosynthesis and chloroplast ultrastructure of *Oryza sativa* L. seedlings under acid rain stress. SILICON 12:655–664

Kalra N, Jain MC, Joshi HC, Chaudhary R, Kumar S, Pathak H, Sharma SK, Kumar V, Kumar R, Harit RC, Khan SA (2003) Soil properties and crop productivity as influenced by flyash incorporation in soil. Environ Monit Assess 87:93–109

Kamanga RM, Mndala L (2019) Crop abiotic stresses and nutrition of harvested food crops: a review of impacts, interventions and their effectiveness. Afr J Agric Res 14:118–135

Kaur S, Kaur N, Siddique KH, Nayyar H (2016) Beneficial elements for agricultural crops and their functional relevance in defence against stresses. Arch Agron Soil Sci 62:905–920

Kawalekar JS (2013) Role of biofertilizers and biopesticides for sustainable agriculture. J Innov Biol 2:73–78

Keeping MG, Miles N, Rutherford RS (2017) Liming an acid soil treated with diverse silicon sources: effects on silicon uptake by sugarcane (*Saccharum* hybrids). J Plant Nutr 40:1417–1436

Khan H, Shah SH, Uddin N, Azhar N, Asim M, Syed S, Ullah F, Tawab F, Inayat J (2015) Biochemical and physiological changes of different plant species in response to heat and cold stress. ARPN J Agric Biol Sci 10:213–216

Khan S, Anwar S, Kuai J, Noman A, Shahid M, Din M, Ali A, Zhou G (2018) Alteration in yield and oil quality traits of winter rapeseed by lodging at different planting density and nitrogen rates. Sci Rep 8:1–12

Khan A, Kamran M, Imran M, Al-Harrasi A, Al-Rawahi A, Al-Amri I, Lee IJ, Khan AL (2019) Silicon and salicylic acid confer high-pH stress tolerance in tomato seedlings. Sci Rep 9:1–16

Kim YH, Khan AL, Hamayun M, Kang SM, Beom YJ, Lee IJ (2011) Influence of short-term silicon application on endogenous physiohormonal levels of *Oryza sativa* L. under wounding stress. Biol Trace Elem Res 144:1175–1185

Kim YH, Khan AL, Waqas M, Jeong HJ, Kim DH, Shin JS, Kim JG, Yeon MH, Lee IJ (2014) Regulation of jasmonic acid biosynthesis by silicon application during physical injury to *Oryza sativa* L. J Plant Res 127:525–532

Kim YH, Khan AL, Waqas M, Shahzad R, Lee IJ (2016) Silicon-mediated mitigation of wounding stress acts by up-regulating the rice antioxidant system. Cereal Res Commun 44:111–121

Klimenko S, Peshkova A, Dorofeev N (2006) Nitrate reductase activity during heat stress in winter wheat. J Stress Physiol Biochem 2:50–55

Knight CT, Kinrade SD (2001) A primer on the aqueous chemistry of silicon. In: Datnoff LE, Snyder GH, Korndorfer GH (eds) Studies in plant science, vol 8. Elsevier, Amsterdam, pp 57–84

Kostic L, Nikolic N, Bosnic D, Samardzic J, Nikolic M (2017) Silicon increases phosphorus (P) uptake by wheat under low P acid soil conditions. Plant Soil 419:447–455

Kurdali F, Al-Chammaa M, Al-Ain F (2019) Growth and N 2 fixation in saline and/or water stressed *Sesbania aculeata* plants in response to silicon application. SILICON 11(2):781–788

Kusaka M, Ohta M, Fujimura T (2005) Contribution of inorganic components to osmotic adjustment and leaf folding for drought tolerance in pearl millet. Physiol Plant 125:474–489

Laîné Haddad C, Arkoun M, Yvin JC, Etienne P (2019) Silicon promotes agronomic performance in *Brassica napus* cultivated under field conditions with two nitrogen fertilizer inputs. Plan Theory 8:137

Lal SK, Kumar S, Sheri V, Mehta S, Varakumar Ram B, Borphukan B, James D, Fartyal D, Reddy MK (2018) Seed priming: an emerging technology to impart abiotic stress tolerance in crop plants. In: Rakshit A, Singh HB (eds) Advances in seed priming. Springer, Singapore, pp 41–50

Li H, Zhu Y, Hu Y, Han W, Gong H (2015) Beneficial effects of silicon in alleviating salinity stress of tomato seedlings grown under sand culture. Acta Physiol Plant 37:71

Li P, Zhao CZ, Zhand YQ, Wang XM, Wang JF, Wang F, Bi YR (2017) Silicon enhances the tolerance of *Poa annua* to cadmium by inhibiting its absorption and oxidative stress. Biol Plant 61:741–750

Liang Y, Chen Q, Liu Q, Zhang W, Ding R (2003) Exogenous silicon (Si) increases antioxidant enzyme activity and reduces lipid peroxidation in roots of salt-stressed barley (*Hordeum vulgare* L.). J Plant Physiol 160:1157–1164

Liang Y, Nikolic M, Bélanger R, Gong H, Song A (2015) Silicon in agriculture. Springer, Dordrecht

Liao M, Fang Z, Liang YQ, Huang XH, Yang X, Chen SS, Xie XM, Xu CX, Guo JW (2020) Effects of supplying silicon nutrient on utilization rate of nitrogen and phosphorus nutrients by rice and its soil ecological mechanism in a hybrid rice double-cropping system. J Zhejiang Univ Sci B 21:474–484

Liebersbach H, Steingrobe B, Claassen N (2004) Roots regulate ion transport in the rhizosphere to counteract reduced mobility in dry soil. Plant Soil 260:79–88

Liu Y, Zhou J (2018) MAPping kinase regulation of ICE1 in freezing tolerance. Trends Plant Sci 23:91–93

Liu J, Ma J, He C, Li X, Zhang W, Xu F, Lin Y, Wang L (2013) Inhibition of cadmium ion uptake in rice (*Oryza sativa*) cells by a wall-bound form of silicon. New Phytol 200:691–699

Liu YL, Deng X, Wang S, Tanaka K, Zhang S (2014) Aquaporin-mediated increase in root hydraulic conductance is involved in silicon-induced improved root water uptake under osmotic stress in *Sorghum bicolor* L. J Exp Bot 65:4747–4756

Liu YL, Wang S, Zhang M, Deng X, Zhang S, Tanaka E (2015) Enhanced root hydraulic conductance by aquaporin regulation accounts for silicon alleviated salt-induced osmotic stress in *Sorghum bicolor* L. Environ Exp Bot 111:42–51

Liu C, Lu W, Ma Q, Ma C (2017) Effect of silicon on the alleviation of boron toxicity in wheat growth, boron accumulation, photosynthesis activities, and oxidative responses. J Plant Nutr 40:2458–2467

Liu D, Liu M, Liu XL, Cheng XG, Liang ZW (2018) Silicon priming created an enhanced tolerance in alfalfa (*Medicago sativa* L.) seedlings in response to high alkaline stress. Front Plant Sci 9:716

Liu B, Soundararaja P, Manivannan A (2019) Mechanisms of silicon-mediated amelioration of salt stress in plants. Plants 8:307

Lu H, Wu Y, Liang Song Q, Zhang H, Wu J, Wu W, Liu X, Dong C (2020) Alkaline amendments improve the health of soils degraded by metal contamination and acidification: crop performance and soil bacterial community responses. Chemosphere 257:127309

Lux A, Luxova M, Abe J, Morita S, Inanaga S (2003) Silification of bamboo (*Phyllostachys heterocycla* Mitf.) root and leaf. Plant Soil 255:85–91

Luyckx M, Hausman JF, Lutts S, Guerriero G (2017) Silicon and plants: current knowledge and technological perspectives. Front Plant Sci 8:411

Ma JF (2004) Role of silicon in enhancing the resistance of plants to biotic and abiotic stresses. Soil Sci Plant Nutr 50:11–18

Ma JF, Yamaji N (2006) Silicon uptake and accumulation in higher plants. Trends Plant Sci 11:392–339

Ma JF, Yamaji N (2008) Functions and transport of silicon in plants. Cell Mol Life Sci 65:3049–3057

Ma JF, Yamaji N (2015) A cooperative system of silicon transport in plants. Trends Plant Sci 20:435–442

Ma JF, Tamai K, Yamaji N, Mitani N, Konishi S, Katsuhara M, Ishiguro M, Murata Y, Yano M (2006) A silicon transporter in rice. Nature 440:688–691

Ma JF, Yamaji N, Mitani N, Tamai K, Konishi S, Fujiwara T, Katsuhara M, Yano M (2007) An efflux transporter of silicon in rice. Nature 448:209–212

Ma J, Cai H, He C, Zhang W, Wang L (2015) A hemicellulose-bound form of silicon inhibits cadmium ion uptake in rice (*Oryza sativa*) cells. New Phytol 206:1063–1074

Macedo FG, Santos EF, Lavres J (2020) Agricultural crop influences availability of nickel in the rhizosphere, a study on base cation saturations, Ni dosages and crop succession. Rhizosphere 13:100182

Mahmoud SH, Gan TY (2018) Impact of anthropogenic climate change and human activities on environment and ecosystem services in arid regions. Sci Total Environ 633:1329–1344

Maillard A, Ali N, Schwarzenberg A, Jamois F, Yvin JC, Hosseini SA (2018) Silicon transcriptionally regulates sulfur and ABA metabolism and delays leaf senescence in barley under combined sulfur deficiency and osmotic stress. Environ Exp Bot 155:394–341

Malav Jugal K, Ramani V (2017) Effect of silicon on nitrogen use efficiency, yield and nitrogen and silicon contents in rice under loamy sand soil. Res J Chem Environ 21:4

Malhotra C, Kapoor RT (2019) Silicon: a sustainable tool in abiotic stress tolerance in plants. In: Hasanuzzaman M, Hakeem KR, Nahar K, Alharby HF (eds) Plant abiotic stress tolerance: agronomic in molecular and biotechnological approaches. Springer, Cham, pp 333–356

Mali M, Aery NC (2008) Influence of silicon on growth, relative water contents and uptake of silicon, calcium and potassium in wheat grown in nutrient solution. J Plant Nutr 31:1867–1876

Manivannan A, Soundararajan Arrum LS, Ko CH, Muneer S, Jeong BR (2015) Silicon-mediated enhancement of physiological and biochemical characteristics of *Zinnia elegans* 'dreamland yellow' green under salinity stress. Hortic Environ Biotechnol 56:721–731

Manivannan A, Soundararajan P, Muneer S, Ko CH, Jeong BR (2016) Silicon mitigates salinity stress by regulating the physiology, antioxidant enzyme activities, and protein expression in *Capsicum annuum* 'Bugwang'. Biomed Res Int 2016:3076357

Marschner H (2012) Marschner's mineral nutrition of higher plants, 3rd edn. Academic press, London

Martin-Shields C, Stojetz W (2019) Food security and conflict: empirical challenges and future opportunities for research and policy making on food security and conflict. World Dev 119:150–164

Marxen A, Klotzbücher T, Jahn R, Kaiser K, Nguyen VS, Schmidt A, Schädler M, Vetterlein D (2016) Interaction between silicon cycling and straw decomposition in a silicon deficient rice production system. Plant Soil 398:153–163

Mateos-Naranjo E, Galle A, Florez-Sarasa I, Perdomo JA, Galmes J, Ribas-Carbo M, Flexas J (2015) Assessment of the role of silicon in the Cu-tolerance of the C4 grass *Spartina densiflora*. J Plant Physiol 178:74–83

Matichenkov VV, Calvert DV (2002) Silicon as a beneficial element for sugarcane. J Am Soc Sugarcane Technol 22:21–30

Mehta S, James D, Reddy MK (2019a) Omics technologies for abiotic stress tolerance in plants: current status and prospects. In: Wani SH (ed) Recent approaches in omics for plant resilience to climate change. Springer, Cham, pp 1–34

Mehta S, Singh B, Dhakate P, Rahman M, Islam MA (2019b) Rice, marker-assisted breeding, and disease resistance. In: Wani SH (ed) Disease resistance in crop plants. Springer, Cham, pp 83–111

Metwally AM, Radi AA, El-Shazoly RM, Hamada AM (2018) The role of calcium, silicon and salicylic acid treatment in protection of canola plants against boron toxicity stress. J Plant Res 131:1015–1028

Miao BH, Han XG, Zhang WH (2010) The ameliorative effect of silicon on soybean seedlings grown in potassium-deficient medium. Ann Bot 105:967–973

Mickelbart MV, Hasegawa PM, Bailey-Serres J (2015) Genetic mechanisms of abiotic stress tolerance that translate to crop yield stability. Nat Rev Genet 16:337–251

Mitani N, Jian FM (2005) Uptake system of silicon and accumulation in higher plants. Trends Plant Sci 11:392–397

Mitani N, Ma JF (2005) Uptake system of silicon in different plant species. J Exp Bot 56:1255–1261

Mitani N, Chiba Y, Yamaji N, Ma JF (2009a) Identification and characterization of maize and barley Lsi2-like silicon efflux transporters reveals a distinct silicon uptake system from that in rice. Plant Cell 21:2133–2142

Mitani N, Yamaji N, Ma JF (2009b) Identification of maize silicon influx transporters. Plant Cell Physiol 50:5–12

Mitani N, Yamaji N, Ago Y, Iwasaki K, Ma JF (2011) Isolation and functional characterization of an influx silicon transporter in two pumpkin cultivars contrasting in silicon accumulation. Plant J 66:231–240

Mitra GN (2015) Regulation of nutrient uptake by plants. Springer, New Delhi

Mohammadi H, Hazrati S, Ghorbanpour M (2020) Tolerance mechanisms of medicinal plants to abiotic stresses. In: Tripathi DK, Singh VP, Chauhan DK, Sharma S, Prasad SM, Dubey NK, Ramawat N (eds) Plant life under changing environment. Academic Press, London, pp 663–679

Mohanty S, Nayak AK, Swain CK, Dhal B, Kumar A, Tripathi R, Shahid M, Lal B, Gautam Dash GK, Swain P (2020) Silicon enhances yield and nitrogen use efficiency of tropical low land rice. Agron J 112:758–777

Mona MH, El-Naggar HA, El-Gayar EE, Masood MF, Mohamed ESN (2019) Effect of human activities on biodiversity in nabq protected area, south sinai, Egypt. Egypt J Aquat Res 45:33–43

Montpetit J, Vivancos J, Mitani-Ueno N, Yamaji N, Rémus-Borel W, Belzile F, Ma JF, Bélanger RR (2012) Cloning, functional characterization and heterologous expression of TaLsi1, a wheat silicon transporter gene. Plant Mol Biol 79:35–46

Moradtalab N, Weinmann M, Walker F, Hoglinger B, Ludewig U, Neumann G (2018) Silicon improves chilling tolerance during early growth of maize by effect on micronutrients and homeostasis and hormonal balances. Front Plant Sci 9:420

Mu J, Hu Z, Huang L, Xie Z, Holm E (2020) Preparation of a silicon-iron amendment from acid-extracted copper tailings for remediating multi-metal-contaminated soils. Environ Pollut 257:113565

Müller B, Hoffmann F, Heckelei T, Müller C, Hertel TW, Polhill JG, van Wijk M, Achterbosch T, Alexander Brown C, Kreuer D (2020) Modelling food security: bridging the gap between the micro and the macro scale. Glob Environ Chang 63:102085

Muneer S, Jeong BR (2015) Silicon decreases Fe deficiency responses by improving photosynthesis and maintaining composition of thylakoid multiprotein complex proteins in soybean plants (*Glycine max* L.). J Plant Growth Regul 34:485–498

Muneer S, Park YG, Manivannan A, Soundararajan P, Jeong BR (2014) Physiological and proteomic analysis in chloroplasts of *Solanum lycopersicum* L. under silicon efficiency and salinity stress. Int J Mol Sci 15:21803–21824

Nabati J, Kafi M, Masoumi A, Mehrjerdi Z (2013) Effect of salinity and silicon application on photosynthetic characteristics of sorghum (*Sorghum bicolor* L.). Int J Agric Sci 3:483–492

Nawaz MA, Zakharenko AM, Zemchenko IV, Haider MS, Ali MA, Imtiaz M, Chung G, Tsatsakis A, Sun S, Golokhvast KS (2019) Phytolith formation in plants: from soil to cell. Plan Theory 8:249

Neilsen GH, Neilsen D, Peryea FJ, Fallahi E, Fallahi B (2008) Effects of mineral nutrition on fruit quality and nutritional disorders in apples. VI Int Symp Miner Nutr Fruit Crops 868:49–60

Oliveira TB, Aucique-Pérez CE, Rodrigues FDÁ (2019) Foliar application of silicon decreases wheat blast symptoms without impairing photosynthesis. Bragantia 78:423–443

Oliveira KR, Junior JS, Bennett SJ, Checchio MV, de Cássia Alves R, Felisberto G, de Mello Prado R, Gratão L (2020) Exogenous silicon and salicylic acid applications improve tolerance to boron toxicity in field pea cultivars by intensifying antioxidant defense systems. Ecotoxicol Environ Saf 201:110778

Oukarroum A, Bussotti F, Goltsev V, Kalkaji HM (2015) Correlation between reactive oxygen species production and photochemistry of photosystems I and II in *Lemna gibba* L. plants under salt stress. Environ Exp Bot 109:80–88

Owino-Gerroh C, Gascho GJ (2005) Effect of silicon on low pH soil phosphorus sorption and on uptake and growth of maize. Commun Soil Sci Plant Anal 35:2369–2378

Pandey V, Patel A, Patra DD (2015) Amelioration of mineral nutrition, productivity, antioxidant activity and aroma profile in marigold (*Tagetes minuta* L.) with organic and chemical fertilization. Ind Crop Prod 76:378–385

Parveen N, Ashraf M (2010) Role of silicon in mitigating the adverse effects of salt stress on growth and photosynthetic attributes of two maize (*Zea mays* L.) cultivars grown hydroponically. Pak J Bot 42:1675–1684

Parveen A, Liu W, Hussain S, Asghar J, Perveen S, Xiong Y (2019) Silicon priming regulates morpho-physiological growth and oxidative metabolism in Maize under drought stress. Plan Theory 8:431

Pascual MB, Echevarria V, Gonzalo MJ, Hernández-Apaolaza L (2016) Silicon addition to soybean (*Glycine max* L.) plants alleviate zinc deficiency. Plant Physiol Biochem 108:132–138

Pati S, Pal B, Badole S, Hazra GC, Mandal B (2016) Effect of silicon fertilization on growth, yield, and nutrient uptake of rice. Commun Soil Sci Plant Anal 47:284–290

Patil AA, Durgude AG, Pharande AL (2018) Effect of silicon application along with chemical fertilizers on nutrient uptake and nutrient availability for rice plants. IJCS 6:260–266

Pavlovic J, Samardzic J, Maksimović V, Timotijevic G, Stevic N, Laursen KH, Hansen TH, Husted S, Schjoerring JK, Liang Y, Nikolic M (2013) Silicon alleviates iron deficiency in cucumber by promoting mobilization of iron in the root apoplast. New Phytol 198:1096–1107

Pecinka A, Mittelsten Scheid O (2012) Stress-induced chromatin changes: a critical view on their heritability. Plant Cell Physiol 53:801–808

Pei ZF, Ming DF, Liu D, Wan GL, Geng XX, Gong HJ, Zhou WJ (2010) Silicon improves the tolerance to water-deficit stress induced by polyethylene glycol in wheat (*Triticum aestivum* L.) seedlings. J Plant Growth Regul 29:106–115

Pereira de Souza Junior J, de Mello Prado R, Machado dos Santos Sarah M, Felisberto G (2019) Silicon mitigates boron deficiency and toxicity in cotton cultivated in nutrient solution. J Plant Nutr Soil Sci 182:805–814

Pontigo S, Godoy K, Jimenez H, Gutierrez-Moraga A, Mora MDLL, Cartes P (2017) Silicon-mediated alleviation of aluminium toxicity by modulation of Al/Si uptake and antioxidant performance in ryegrass plants. Front Plant Sci 8:642

Puhakainen T (2004) Short-day potential of low temperature-induced gene expression of a C-repeat binding factor-controlled gene during cold acclimation in silver birch. Plant Physiol 136:4299–4307

Rafique S, Abdin MZ, Alam W (2020) Response of combined abiotic stresses on maize (Zea mays L.) inbred lines and interaction among various stresses. Maydica 64:8

Rahman M, Sultana S, Nath D, Kalita S, Chakravarty D, Mehta S, Wani SH, Islam MA (2019) Molecular breeding approaches for disease resistance in sugarcane. In: Wani SH (ed) Disease resistance in crop plants. Springer, Cham, pp 131–155

Raj GB, Rajamani K, Kumara BH (2020) Influence of silicon fertilization on nitrogen fractions and nutrient status of Rice grown soils in Telangana state. Curr J Appl Sci Technol 39:26–34

Ramegowda V, Da Costa MVJ, Harihar S, Karaba NN, Sreeman SM (2020) Abiotic and biotic stress interactions in plants: a cross-tolerance perspective. In: Hossain MA, Liu F, Burritt DJ, Fujita M, Huang B (eds) Priming-mediated stress and cross-stress tolerance in crop plants. Academic Press, Cambridge, pp 267–302

Rao GB, Susmitha P (2017) Silicon uptake, transportation and accumulation in Rice. J Pharmacogn Phytochem 6:290–293

Rao GB, Pi Y, Syriac EK (2017) Silicon nutrition in rice: a review. J Pharmacogn Phytochem 6:390–392

Rastogi A, Tripathi DK, Yadav S, Chauhan DK, Živčák M, Ghorbanpour M, El-Sheery NI, Brestic M (2019) Application of silicon nanoparticles in agriculture. 3 Biotech 9:90

Rathee M, Dalal P (2018) Emerging insect pests in Indian agriculture. Indian J Entomol 80:267–281

Raven JA (2001) Silicon transport at the cell and tissue level. In: Datnoff LE, Snyder GH, Korndörfer GH (eds) Studies in plant science, vol 8. Elsevier, Amsterdam, pp 41–55

Rédei G (2008) Encyclopedia of genetics, genomics, proteomics, and informatics. Springer, Dordrecht

Réthoré E, Ali N, Yvin JC, Hosseini SA (2020) Silicon regulates source to sink metabolic homeostasis and promotes growth of rice plants under sulfur deficiency. Int J Mol Sci 21:3677

Rezakhani L, Motesharezadeh B, Tehrani MM, Etesami H, Hosseini HM (2019) Phosphate–solubilizing bacteria and silicon synergistically augment phosphorus (P) uptake by wheat (*Triticum aestivum* L.) plant fertilized with soluble or insoluble P source. Ecotoxicol Environ Saf 173:504–513

Ritonga FN, Chen S (2020) Physiological and molecular mechanism involved in cold stress tolerance in plants. Plan Theory 5:560

Rodrigues FA, Dallagnol LJ, Duarte HSS, Datnoff LE (2015) Silicon control of foliar diseases in monocots and dicots. In: Rodrigues FA, Datnoff LE (eds) Silicon and plant diseases. Springer, Cham, pp 67–108

Rogalla H, Romheld V (2002) Role of leaf apoplast in silicon-mediated manganese tolerance of *Cucumis sativus* L. Plant Cell Environ 25:549–555

Salgado OGG, Teodoro JC, Alvarenga J, de Oliveira C, de Carvalho TS, Domiciano D, Marchiori ER, Guilherme LRG (2020) Cerium alleviates drought-induced stress in Phaseolus vulgaris. J Rare Earth 38:324–331

Saud S, Li X, Chen Y, Zhang L, Fahad S, Hussain S, Sadiq A, Chen Y (2014) Silicon application increases drought tolerance of Kentucky bluegrass by improving plant water relations and morphophysiological functions. Sci World J 368694:1–10

Savić J, Marjanović-Jeromela A (2013) Effect of silicon on sunflower growth and nutrient accumulation under low boron supply. Helia 36:61–68

Schaller J, Brackhage C, Bäucker E, Dudel EG (2013) UV-screening of grasses by plant silica layer? J Biosci 38:413–416

Seleiman MF, Refay Y, Al-Suhaibani N, Al-Ashkar I, El-Hendawy S, Hafez EM (2019) Integrative effects of rice-straw biochar and silicon on oil and seed quality, yield and physiological traits of *Helianthus annuus* L. grown under water deficit stress. Agronomy 9:637

Shahbazy M, Moradi P, Ertaylan G, Zahraei A, Kompany-Zareh M (2020) FTICR mass spectrometry-based multivariate analysis to explore distinctive metabolites and metabolic pathways: a comprehensive bioanalytical strategy toward time-course metabolic profiling of *Thymus vulgaris* plants responding to drought stress. Plant Sci 290:110257

Shahid M, Ahmed T, Noman M, Manzoor N, Hussain S, Mahmood F, Muhammad S (2019) Microbe-mediated reclamation of contaminated soils: current status and future perspectives. In: Singh PD, Gupta VK, Ratna P (eds) Microbial interventions in agriculture and environment. Springer, Singapore, pp 261–279

Shao HB, Chu LY, Shao MA, Jaleel CA, Hongmei M (2008) Higher plant antioxidants and redox signaling under environmental stresses. Comptes Rendus Biol 331:433–441

Sharma P, Sharma MMM, Patra A, Vashisth M, Mehta S, Singh B, Tiwari M, Pandey V (2020) The role of key transcription factors for cold tolerance in plants. In: Wani SH (ed) Transcription factors for abiotic stress tolerance in plants. Elsevier Academic Press, Cambridge, MA

Shen X, Zhou Y, Duan L, Li Z, Eneji AE, Li J (2010) Silicon effects on photosynthesis and antioxidant parameters of soybean seedlings under drought and ultraviolet-B radiation. J Plant Physiol 167:1248–1252

Shen C, Shi Y, Fan K, He JS, Adams JM, Ge Y, Chu H (2019) Soil pH dominates elevational diversity pattern for bacteria in high elevation alkaline soils on the Tibetan plateau. FEMS Microbiol Ecol 95:fiz003

Shi Y, Wang Y, Flowers TJ, Gong H (2013) Silicon decreases sodium chloride transport in rice in saline conditions. J Plant Physiol 170:847–853

Shi Y, Ding Y, Yang S (2018) Molecular recognition of CBF signaling in cold acclimation. Trends Plant Sci 23:623–637

Shiru MS, Shahid S, Dewan A, Chung ES, Alias N, Ahmed K, Hassan QK (2020) Projection of meteorological droughts in Nigeria during growing seasons under climate change scenarios. Sci Rep 10:1–18

Siddiqui H, Yusuf M, Faraz A, Farizan M, Sami F, Hayat S (2018) 24-Epibrassinolide supplemented with silicon enhances the photosynthetic efficiency of *Brassica juncea* under salt stress. S Afr J Bot 118:120–128

Siddiqui H, Ahmed KBM, Sami F, Hayat S (2020) Silicon nanoparticles and plants: current knowledge and future perspectives. In: Hayat S, Pichtel J, Faizan M, Fariduddin Q (eds) Sustainable agriculture reviews, vol 41. Springer, Cham, pp 129–142

Singh K, Singh R, Singh J, Singh Y, Singh KK (2006) Effect of level and time of silicon application on growth, yield and its uptake by rice (*Oryza sativa*). Indian J Agric Sci 76:410–413

Singh B, Mehta S, Tiwari M, Bhatia S (2018) Legume breeding for fungal resistance: a lesson to learn. In: Islam MA, Dhakate P (eds) Molecular approaches for plant improvement. Kalpaz Publications, New Delhi, pp 159–180

Singh B, Mehta S, Aggarwal SK, Tiwari M, Bhuyan SI, Bhatia S, Islam MA (2019) Barley, disease resistance, and molecular breeding approaches. In: Wani SH (ed) Disease resistance in crop plants. Springer, Cham, pp 261–299

Singh A, Kumar A, Hartley S, Singh IK (2020) Silicon: its ameliorative effect on plant defense against herbivory. J Exp Bot:eraa300. https://doi.org/10.1093/jxb/eraa300

Soundararajan P, Sivanesan L, Jana S, Jeong BR (2014) Influence of silicon supplementation on the growth and tolerance to high temperature in *Salvia splendens*. Hortic Environ Biotech 55:271–279

Soundararajan P, Manivannan A, Cho YS, Jeong BR (2017) Exogenous application of silicon improved the recovery of hyperhydric shoots in *Dianthus caryophyllus* L. by stabilizing the physiology and protein expression. Front Plant Sci 8:738

Soundararajan P, Manivannan A, Ko CH, Jeong BR (2018) Silicon enhanced redox homeostasis and protein expression to mitigate the salinity stress in Rosa hybrid 'rock fire'. Plant Growth Regul 37:16–34

Souri Z, Khanna K, Karimi N, Ahmad P (2020) Silicon and plants: current knowledge and future prospects. J Plant Growth Regul. https://doi.org/10.1007/s00344-020-10172-7

Steiner F, Zuffo AM, Bush A, Santos DMDS (2018) Silicate fertilization potentiates the nodule formation and symbiotic nitrogen fixation in soybean. Pesqui Agropecu Trop 48:212–221

Sun H, Guo J, Duan Y, Zhang T, Huo H, Gong H (2017) Isolation and functional characterization of CsLsi1, a silicon transporter gene in *Cucumis sativus*. Physiol Plant 159:201–214

Sun H, Duan Y, Qi X, Zhang L, Huo H, Gong H (2018) Isolation and functional characterization of CsLsi2, a cucumber silicon efflux transporter gene. Ann Bot 122:641–648

Sun X, Zhu Z, Zhang L, Fang L, Zhang J, Wang Q, Li S, Liang Z, Xin H (2019) Overexpression of ethylene response factors VaERF080 and VAERF087 from *Vitis amurensis* enhances cold tolerance in Arabidopsis. Sci Hortic 243:320–326

Sun H, Duan Y, Mitani-Ueno N, Che J, Jia J, Liu J, Guo J, Ma JF, Gong H (2020) Tomato roots have a functional silicon influx transporter but not a functional silicon efflux transporter. Plant Cell Environ 43:732–744

Surówka E, Rapacz M, Janowiak F (2020) Climate change influences the interactive effects of simultaneous impact of abiotic and biotic stresses on plants. In: Hasanuzzaman M (ed) Plant ecophysiology and adaptation under climate change: mechanisms and perspectives I. Springer, Singapore, pp 1–50

Taiz L, Zeiger E (2006) Plant physiology, 4th edn. Sinauer, Sunderland, MA

Tamburino L, Bravo G, Clough Y, Nicholas KA (2020) From population to production: 50 years of scientific literature on how to feed the world. Glob Food Secur 24:100346

Tavakkoli E, Lyons G, English P, Guppy CN (2011) Silicon nutrition of rice is affected by soil pH, weathering and silicon fertilisation. J Plant Nutr Soil Sci 174:437–446

Thomidis T, Zioziou E, Koundouras S, Karagiannidis C, Navrozidis I, Nikolaou N (2016) Effects of nitrogen and irrigation on the quality of grapes and the susceptibility to *Botrytis* bunch rot. Sci Hortic 212:60–68

Tripathi DK, Singh V, Ahmad P, Chauhan DK, Prasad SM (2016) Silicon in plants: advances and future prospects. CRC Press, Boca Raton

Tripathi DK, Singh S, Singh V, Prasad SM, Dubey NK, Chauhan DK (2017) Silicon nanoparticles more effectively alleviated UV-B stress than silicon in wheat (*Triticum aestivum*) seedlings. Plant Physiol Biochem 110:70–81

Tubana BS, Babu T, Datnoff LE (2016) A review of silicon in soils and plants and its role in US agriculture: history and future perspectives. Soil Sci 181:393–411

Tuna AL, Kaya C, Higgs D, Murillo-Amador B, Aydemir S, Girgin AR (2008) Silicon improves salinity tolerance in wheat plants. Environ Exp Bot 62:10–16

Tyagi NK (2020) Managing water–energy–food security nexus under changing climate: implementation challenges and opportunities in India. Trans Indian Natl Acad Eng. https://doi.org/10.1007/s41403-020-00117-7

Vakilian KA (2020) Machine learning improves our knowledge about miRNA functions towards plant abiotic stresses. Sci Rep 10:1–10

Viciedo DO, de Mello Prado R, Toledo RL, dos Santos LCN, Hurtado AC, Nedd LLT, Gonzalez LC (2019) Silicon supplementation alleviates ammonium toxicity in sugar beet (*Beta vulgaris* L.). J Soil Sci Plant Nutr 19:413–419

Viciedo DO, de Mello Prado R, Toledo RL, Aguilar DS, Dos Santos LCN, Hurtado AC, Calzada KP, Aguilar CB (2020) Physiological role of silicon in radish seedlings under ammonium toxicity. J Sci Food Agric. https://doi.org/10.1002/jsfa.10587

Vivancos J, Deshmukh R, Grégoire C, Rémus-Borel W, Belzile F, Bélanger RR (2016) Identification and characterization of silicon efflux transporters in horsetail (*Equisetum arvense*). J Plant Physiol 200:82–89

von Liebig J (1843) Chemistry in its applications to agriculture and physiology. Taylor and Walton, London

Wang Y, Shao S, Ju S (2017) The effects of silicon on *Festuca arundinacea* under PH stress. In: Xin N, El-Hami K, Kun Z (eds) 7th international conference on management, education, information and control (MEICI 2017). Atlantis Press, Amsterdam

Wise RR, Olson AJ, Schrader SM, Sharkey TD (2004) Electron transport is the functional limitation of photosynthesis in field-grown Pima cotton plants at high temperature. Plant Cell Environ 27:717–724

Wu B, Beitz E (2007) Aquaporins with selectivity for unconventional permeants. Cell Mol Life Sci 64:2413–2421

Wu JW, Shi Y, Zhu YX, Wang YC, Gong HJ (2013) Mechanisms of enhanced heavy metal tolerance in plants by silicon: a review. Pedosphere 23:815–823

Xiao K, Xu J, Tang C, Zhang J, Brookes C (2013) Differences in carbon and nitrogen mineralization in soils of differing initial pH induced by electrokinesis and receiving crop residue amendments. Soil Biol Biochem 67:70–84

Xu Y (2016) Envirotyping for deciphering environmental impacts on crop plants. Theor Appl Genet 129:653–673

Xu D, Carswell A, Zhu Q, Zhang F, de Vries W (2020) Modelling long-term impacts of fertilization and liming on soil acidification at Rothamsted experimental station. Sci Total Environ 713:136249

Yamaji N, Ma JF (2007) Spatial distribution and temporal variation of the rice silicon transporter Lsi1. Plant Physiol 143:1306–1313

Yamaji N, Ma JF (2009) A transporter at the node responsible for intervascular transfer of silicon in rice. Plant Cell 21:2878–2883

Yamaji N, Ma JF (2011) Further characterization of a rice silicon efflux transporter, Lsi2. Soil Sci Plant Nutr 57:259–264

Yang Y, Yu L, Wang L, Guo S (2015) Bottle gourd rockstock-grafting promotes photosynthesis by regulating the stomata and non-stomata performances in leaves of watermelon seedlings under NaCl stress. J Plant Physiol 186:50–58

Yao X, Chu J, Cai K, Liu L, Shi J, Geng W (2011) Silicon improves the tolerance of wheat seedlings to ultraviolet-B stress. Biol Trace Elem Res 143:507–517

Ye J, Yan C, Liu J, Lu H, Liu T, Song Z (2012) Effects of silicon on the distribution of cadmium compartmentation in root tips of *Kandelia obovata* Yong. Environ Pollut 162:369–373

Yin R, Ulm R (2017) How plants cope with UV-B: from perception to response. Curr Opin Plant Biol 37:42–48

You-Qiang FU, Hong SHEN, Dao-Ming WU, Kun-Zheng CAI (2012) Silicon-mediated amelioration of Fe2+ toxicity in rice (Oryza sativa L.) roots. Pedosphere 22:795–802

Zama EF, Reid BJ, Sun GX, Yuan HY, Li XM, Zhu YG (2018) Silicon (Si) biochar for the mitigation of arsenic (As) bioaccumulation in spinach (*Spinacia oleracea*) and improvement in the plant growth. J Clean Prod 189:386–395

Zargar SM, Mahajan R, Bhat JA, Nazir M, Deshmukh R (2019) Role of silicon in plant stress tolerance: opportunities to achieve a sustainable cropping system. 3 Biotech 9:73

Zellner W, Lutz L, Khandekar S, Leisner S (2019) Identification of NtNIP2, 1: an Lsi1 silicon transporter in *Nicotiana tabacum*. J Plant Nutr 42:1028–1035

Zhang F, Jiang Y, Bai L, Zhang L, Chen LJ, Li HG, Yin Y, Yan WW, Yi Y, Gua ZF (2011) The ICE-CBF-COR pathway in cold acclimation and AFPs in plants. Middle East J Sci Res 8:493–498

Zhang X, Da Silva JAT, Niu M, Li M, He C, Zhao J, Zeng S, Duan J, Ma G (2017) Physiological and transcriptomic analyses reveal a response mechanism to cold stress in Santalum album L. leaves. Sci Rep 7:42165

Zhang Y, Liang Y, Zhao X, Jin X, Hou L, Shi Y, Ahammed GJ (2019) Silicon compensates phosphorus deficit-induced growth inhibition by improving photosynthetic capacity, antioxidant potential, and nutrient homeostasis in tomato. Agronomy 9:733

Zhao Q, Li X, Wu Q, Liu Y, Lyu Y (2020) Evolution of mineral phases and microstructure of high efficiency Si–Ca–K–Mg fertilizer prepared by water-insoluble K-feldspar. J Sol-Gel Sci Technol 94:3–10

Zhu Z, Wei G, Li J, Qian Q, Yu J (2004) Silicon alleviates salt stress and increases antioxidant enzymes activity in leaves of salt-stressed cucumber (*Cucumis sativus* L.). Int J Mol Sci 167:527–533

Zia Z, Bakhat HF, Saqib ZA, Shah GM, Fahad S, Ashraf MR, Hammad HM, Naseem W, Shahid M (2017) Effect of water management and silicon on germination, growth, phosphorus and arsenic uptake in rice. Ecotoxicol Environ Saf 144:11–18

Plant Morphological, Physiological Traits Associated with Adaptation Against Heat Stress in Wheat and Maize

3

Rahul Gajghate, Dipanti Chourasiya, Harikrishna, and Ram Kumar Sharma

Abstract

Increase in global temperature adversely affects the growth, development, and productivity of crops worldwide. Among major cereals, wheat and maize are important crops grown across the world which are adversely affected by heat stress. High temperature leads to poor germination, reduced water and nutrient uptake and increase in evapotranspiration. The major impact of heat stress severely seen in the reproductive stage leads to sterility, incomplete filling of grains, and loss of grain yield. The reduction in photosynthesis by deactivating photosynthetic enzymes and the release of reactive oxygen species in the cell organelle are the constraints due to high temperature. The small effects can be suppressed by the plant itself by releasing phytohormones, which helps to maintain homeostasis and increase the recovery rate from damage. To develop heat-tolerant high yielding varieties responses to high temperature must be systematically understood. Several screening methods and selection indices under field trial help the breeder to identify the tolerant lines. For obtaining efficient heat-tolerant genotypes, the transgenic, gene editing approaches consolidated with marker aided selection programs for heat stress-related genes with the application of molecular breeding techniques like marker-assisted recurrent selection and genomic selection will help in enhancing genetic gains in these crops.

R. Gajghate (✉)
ICAR-Indian Grassland and Fodder Research Institute, Jhansi, Uttar Pradesh, India

D. Chourasiya
ICAR-Indian Institute of Soybean Research, Indore, Madhya Pradesh, India

Harikrishna · R. K. Sharma
Division of Genetics, Indian Agricultural Research Institute, New Delhi, India

B. Giri, M. P. Sharma (eds.), *Plant Stress Biology*,
https://doi.org/10.1007/978-981-15-9380-2_3

Keywords

Heat stress · Temperature stress · Crop productivity · Heat tolerance

3.1 Introduction

By the year 2050, the world population grows exponentially and will exceed nine billion resulting in more food demand, despite the reduction of soil fertility and less availability of productive land, water, and other variable factors incidental to climate change (Mickelbart et al. 2015; Govindaraj et al. 2018). In recent years climate change and its variability are emerging as major challenges for the world and its efficacy on crop yield is highly diverse (Deryng et al. 2014). One of the major changes seen in the decade is an increase in the intensity of heat stress (HS), which is one of the inimical stresses that are rising constantly. The rise in temperature makes alternation in crops growing periods and the distribution and thus cause serious threat to crop production and productivity worldwide (Smith 1996; Hall 2001; Stone 2001; Porter 2005; Lesk et al. 2016). The global air temperature is reported to be increasing at the rate of 0.18 °C per decade (IPCC 2014), which will lead to an increase the temperature as 1.5–4.5 °C (IPCC 2012) higher than the current level by 2100 (Hansen et al. 2012).

The rise in temperature even by a single degree beyond a certain threshold level for a period sufficient to induce irreversible damage to plant growth and development is referred to as HS (Hall 2001; Wahid et al. 2007; Hasanuzzaman et al. 2013). The direct effect of HS can be seen at the protein level where its aggregation and denaturation increases the cell membrane fluidity, while indirectly it affects through enzymes inactivation in chloroplasts and mitochondria cell organelle, either inhibiting synthesis or degradation of protein and loss of cell membrane integrity (Smertenko et al. 1997; Howarth 2005). All these alterations result in catastrophic collapse of cellular organization leads to cell injury and death within a few minutes (Schoffl et al. 1999). The occurrence of high temperature (HT) is common during anthesis and grain filling stages in many cereals crops and in maize especially during the flowering and grain filling period (Giaveno and Ferrero 2003; Barnabas et al. 2008) and wheat crops (Rahman et al. 2009) which ultimately affect the production and productivity (Wahid et al. 2007; Hansen et al. 2012).

Wheat and maize are the most widely grown cereal crops grown across the world. Wheat is grown about 30% of the world's cereal area in temperate environments, and also as winter season crop in many tropical cropping system areas with over 220 million ha cultivated worldwide, therefore it is known as the "King of cereals" Ramdas et al. (2019). It shares about 20% of the total dietary calories and proteins around the globe (Lobell and Gourdji 2012; Shiferaw et al. 2013). Whereas, Maize is known as the "Queen of cereals" and is the second most widely cultivated crop after wheat, which is grown over a range of agroclimatic zones around the world (FICCI 2014). Maize is not only an important food source for the human diet but also a basic element of animal feed and raw material for the manufacture of many industrial products (Klopfenstein et al. 2013).

Wheat (*Triticum aestivum* L. em Thell) and maize (*Zea mays* L.) crops are very sensitive to HT (Gupta et al. 2013; Tao et al. 2016). Zhao et al. (2017) observed with no addition of fertilization, adaptive methods, and genetic improvement each degree rise in above threshold temperature leads to decrease in average global yields by 6% and 7.4% in wheat and maize, respectively. HS induces flower abortion, fertilization failure, and shrink seed size in maize (Dupuis and Dumas 1990; Begcy et al. 2019). End-of-season or "terminal" HS resulted in lower yields due to the direct effect on grain number and dry weight (Macas et al. 2000; Guedira et al. 2002; Wollenweber et al. 2003) in most severe case lead to complete kernel abortion and sterility in maize (Shah et al. 2011).

The HS effects on the root zone of wheat, temperatures above 30 °C inhibit plant growth by reducing chlorophyll content, nutrient uptake (Huang et al. 2012), obstruct plant water relations, decreases stomatal conductance (Behboudian et al. 1994), and altering chemical signaling (Wang et al. 2014). In maize root temperature strongly affect the root tip growth (Nagel et al. 2009). Moreover, in various studies showed the effect of HS on different growth stages of wheat and maize and their response on various stages (Porter and Gawith 1999; Akman 2009). This chapter focal point is to get an idea about the impact of HT on traits associated with heat tolerance and to formulate management strategies for yield improvement in wheat and maize crops under HS to develop HT tolerant varieties.

3.2 Plant Responses to Heat Stress

HS causes myriad, and often adverse, alterations in plant parts and processes leading to morphophysiological changes, prohibiting the growth and yield in the plant (McClung and Davis 2010). The requirement of optimum temperature for plants varied with growth stages, crop season, and environments. Therefore, the response to HS differs significantly with the rise in temperature (Table 3.1), encountering

Table 3.1 Optimum and maximum temperature for different growth stages in wheat and maize

Crop growth stages	Optimum temperature (°C)	Maximum temperature (°C)	References
Wheat			
Growth	20–30	–	Kobza and Edwards (1987)
Anthesis	23.0 ± 1.15	32.0 ± 1.74	Farooq et al. (2011)
Grain filling	21.3 ± 1.27	34.3 ± 2.66	Farooq et al. (2011)
Maize			
Growth	28–31 °C	–	Wahid et al. (2008)
Anthesis	30.5 ± 2.5	37.3 ± 1.3	Sanchez et al. (2014)
Grain filling	26.4 ± 2.1	36.0 ± 1.4	Sanchez et al. (2014)

growth stages, and plant type (Ruelland and Zachowski 2010). Some similar effects of HS on morphology and physiology changes in wheat and maize are given below.

3.2.1 Morphological Responses

HS affects the whole plant from cellular to organ level. There is a general propensity towards the reduction in cell size, curtailed water loss, and stomatal closure at the cellular level. The impact of HS can be visualized from root to shoot and the reproductive parts resulting in poor growth and productivity at the whole plant level.

3.2.1.1 Growth

The primary effect of HS impedes seedling establishment and poor germination in the plant (Fahad et al. 2017). In *T. aestivum* HT ranging from 28 °C to 30 °C may change the growth period by reducing seed germination and physiological maturity (Yamamoto et al. 2008; Akter and Rafiqul Islam 2017) Temperature beyond (45 °C) in the early development stage (first 6 days of growth) in wheat attributed cell death and embryo damage leads to germination failure (Essemine et al. 2010). Because, HT retrograde mitochondria, alter the protein expression profiles, decreases ATP (adenosine triphosphate) accumulation and oxygen uptake in imbibing embryos, illation reduced germination percentage, plant emergence, abnormal seedlings, poor seedling vigor, reduced radicle and plumule growth (Piramila et al. 2012), which turn into degrading seed quality related to loss in mass, vigor, and germination in wheat (Essemine et al. 2010; Balla et al. 2012; Hampton et al. 2013) and maize (Iloh et al. 2014). In maize, there are dissident reports about post-emergence seedling growth under HS. Momcilovic and Ristic (2007) study showed that the maize coleoptile tolerant to HS at all stages of seedling development while Weaich et al. (1996) and Akman (2009) reported that exposure to 40 °C verily reduced coleoptile growth and growth completely stopped at ≥45 °C. Moreover, seedling grew under HT (45–50 °C) to promote leaf senescence and abscission (Kosova et al. 2011), hindrance to gain height results in shorter and more bush-like morphology in maize (Iloh et al. 2014; Laghri et al. 2012).

Day and night temperature (30/25 °C) may have stiff effects on the development of leaf viz., leaf area (Shah and Paulsen 2003), and leaf growth in wheat (Savicka and Skute 2012). Wheat plant exposed to HS (>34 °C) accentuated loss of chloroplast integrity, inhibits chlorophyll biosynthesis and denial of photosystem-II-mediated electron transport (Haque et al. 2014) which is amenable to increase flag leaf senescence in wheat (Ciuca and Petcu 2009; Asseng et al. 2013) and, also attributed to a reduction in photosynthetic pigments therefore decline in photosynthetic activity in wheat (Balla et al. 2019). HT significantly declines relative growth rate, shoot dry mass and internode length leads to stunted growth in maize (Weaich et al. 1996; Ashraf and Hafeez 2004; Wahid et al. 2007). Ferris et al. (1998) studied on spring wheat grew for 70 days at ambient temperature, rise in temperature for 12 days from 16 °C to 25 °C, resulted in a decrease in root mass without affecting shoot mass but plants at 25 °C found a well-developed root system in wheat (Huang et al. 1991). It implies that HS indirectly affects the root by translocation of shoot carbon to root or change in root water relations driven by shoot water demand (Jordan and Nobel 1984). In severe HS condition reduction in mass, number, and growth of root leads to limit the supply of nutrients and water to the plants (Wahid et al. 2007; Huang et al. 2012). Nagel et al. (2009) in maize found that root zone temperature below 25 °C manifest 46% decrease in the primary root length as compare to controlled condition. However, above threshold temperature showed a decline in root growth since further growth in root tip required a balance between cell division and cell elongation in the meristem (Beemster and Baskin 1998; Beemster et al. 2003; Fiorani and Beemster 2006), the later decline in root growth was due to insufficient division of new cells in apical meristem to recoup the enhanced cell elongation rate. In wheat growth pattern of crown roots were found to be inhibited by lower temperature (10 °C), while the growth of seminal axes was restricted by HT (30 °C) (Huang et al. 1991). Nagel et al. (2009), studied the branching pattern of lateral roots in maize. They reported an increase in temperature widens the angle between a main and lateral root thus more volume of the substrate can be accessed by a root system in HS. Elevated root zone temperature above air temperature by up to 6 °C to 8 °C caused shorter stature, reduced leaf area as well as increased carbon allocation to reproductive structures as compared to plants grown in a controlled environment (Monje et al. 2007).

3.2.1.2 Reproductive Development

The reproductive stage was found to be the most sensitive and crucial stage affected in wheat (Farooq et al. 2011; Nawaz et al. 2013; Dwivedi et al. 2017) and maize by HS (Cairns et al. 2012). One degree rise in temperature during the reproductive stage caused severe yield loss in both wheat and maize (Lobell et al. 2011; Bennett et al. 2012; Yu et al. 2014). HT impede pollen tube germination, pollen viability, ovule viability, an anomaly in position of stigmatic and style, number of pollens per silk during fertilization, poor growth of endosperm, and development of the barren embryo (Giorno et al. 2013). HS hinder the cell division process leads to abruption in micro and mega sporogenesis process in wheat and maize (Saini et al. 1983, 1984; Zaidi et al. 2016) responsible for the production of a sterile plant due to absent in flower or fruit at the reproductive stage (Yun-Ying et al. 2008). However, the sensitivity for HT is more in male than female gametophytic tissue (Devasirvatham et al. 2012, 2013).

HT (>30 °C) at the tetrad stage of pollen development in maize (Begcy et al. 2019) and wheat decreases the viability of pollen grains (Begcy et al. 2018), pollen tube growth (Hedhly et al. 2009; Yang et al. 2013). Begcy et al. (2019) in maize examined the closer look of pollen and found that heat-stressed pollen contained fewer starch granules which can be considered as one of the reasons for pollen development and did not develop a functional pollen tube. Cossani and Reynolds (2012) exposed wheat crop at HT (30 °C) for 3 days led to abnormalities both structurally and functionally in anther (80% of florets). The increased temperature at mid-anthesis period was also reported to be sensitive to HS in spring wheat (Ferris et al. 1998) and maize (Sanchez et al. 2014). Minimum temperature (Tmin), maximum temperature (Tmax), and Optimum temperature (Topt) at the anthesis stage of maize and wheat are presented in Table 3.1. HT (30 °C) impaired abnormal ovary growth with reduced nucellus and short sized or no embryo sac in wheat during meiosis (Saini et al. 1983). Overall, the effect of HS results in reduced grain number per spike in wheat (Prasad and Djanaguiraman 2014; Dwivedi et al. 2017), kernel number, and kernel weight in maize (Sanchez et al. 2014).

3.2.1.3 Grain and Yield

The duration of grain development experiencing 6–8 °C increased in temperatures effects the grain number, grain weight, and grain growth rate in wheat (Viswanathan and Khanna-Chopra 2001; Dias and Lidon 2009; Yin et al. 2009; Farooq et al. 2011; Johkan et al. 2011; Lukac et al. 2011; Ottman et al. 2012; Balla et al. 2019). In maize reduction of fertilized structure and ear, growth rate lead to a reduction in kernel number and ultimately affect crop yield (Cicchino et al. 2010; Khodarahmpour 2011). Furthermore, in wheat flour and bread quality including changes in protein content (increases the ratio between gliadin and glutenin) of the flour was observed (Wardlaw et al. 2002) which produces a weak dough (Li et al. 2013), But the minimum effect of HS observed in protein concentration of grain (Lizana and Calderini 2013). However, HS in the early stage of grain filling stage has a high concentration of protein in wheat (Castro et al. 2007). HT at day and night (31/20 °C) changed the structure of the aleurone layer and cell endosperm resulting in shrinking of grain (Dias et al. 2008). In maize impact of HT (33–40 °C) seen at the flowering stage results in higher yield reduction than at the grain filling period (Edreira and Otegui 2012). The changes in the number and size of the grain depending upon the growth stage encountering HS. HT $\geq$20 °C during anthesis and spike initiation promote spike development with the cost of reduction in the number of grains per spike (Semenov 2009). HS after spike initiation reduces grain size resulting in shriveled grains (Macas et al. 2000). In another study, Schittenhelm et al. (2020) in wheat showed that thousand grain weight, one of the main grain yield traits, strongly affected by terminal HS. In their studies, a 49% decrease in 1000 grain weight was observed as it was reduced from 47 to 24 g. Thus, the cumulative effects of all morphophysiological changes under HT stress may result in lowering grain number per spike, grain weight per spike, harvest index, and yield (Guedira et al. 2002; Shah and Paulsen 2003; Dias and Lidon 2009; Taghizadeh and Shrifi 2010; Refay 2011; Talukder et al. 2014).

3.2.2 Physiological Response

Heat stress tolerance is a complex phenomenon that aims to reduce the damaged plant system. The damage may be due to cellular structure and modification to several physiological processes viz. photosynthesis, respiration, cell membrane fluidity, osmolytes accumulation, etc. (Wahid et al. 2007; Waqas et al. 2017). Variation in physiological traits for thermotolerance in wheat was analyzed by

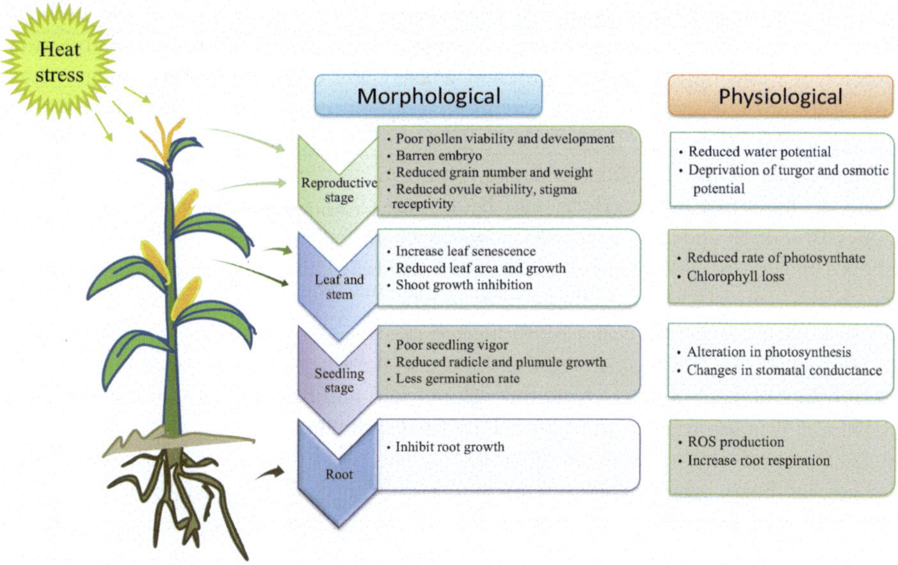

Fig. 3.1 Morphological and physiological changes in the plant under heat stress

Singh (2009) and Balla et al. (2019). Some of the key effects of HS on the physiological process (Fig. 3.1) are described below.

3.2.2.1 Respiration

Respiration response to HT differs with the phenological stage and crop species (Xu et al. 1995). Mitochondria are the central organelle for the respiration process, during HS its activity either increases or gets retarded (Paulsen 1994; Stone 2001). With the rise in temperatures from 0 to 35 or 40 °C the rate of respiration increases exponentially up to the threshold level, further increase in temperature beyond 50 °C decreases respiration due to impairment in the respiratory apparatus (Prasad et al. 2008b; Almeselmani et al. 2009).

Almeselmani et al. (2012) in wheat compared the rate of respiration in flag leaf under HT stress (35/25 °C day/night) and control condition (23/18 °C, day/night) and found significantly higher in heat susceptible varieties under HT. As temperature rises, cost in terms of ATP for respiration increases and reaches up to the point where the rate of photosynthesis cannot make amends in respiratory losses, resulting in turn down the availability of assimilates for crop growth and development justifying the decline in grain weight (Levitt 1980; Peng et al. 2004). In wheat, night temperature against average daily temperature increases the respiratory rate, cell metabolism, and the growth rate of leaves (Chen et al. 2014), suggesting leaf night respiration correlated with an increase in leaf area (Kanno et al. 2009; Fan et al. 2015). In the rhizosphere, HT increases the rate of respiratory carbon loss ultimately reduces the production of ATP, and enhanced the generation of reactive oxygen species (ROS) especially at night which results in cell damage (Huang et al. 2012).

3.2.2.2 Photosynthesis and Photosystems

Photosynthesis, one of the most sentient physiological processes, which is adversely affected at supra-optimal temperatures in wheat (Dias et al. 2010; Feng et al. 2014). The optimum temperature range for photosynthetic processes in most of the crop species is 30 °C to 35 °C, however, as the temperature reaches to >40 °C, it influences on the photosynthetic capacity of plants especially of C3 plants (wheat) than C4 plants like maize (Berry and Bjorkman 1980; Schuster and Monson 1990; Crafts-Brandner and Salvucci 2002; Naidu et al. 2003; Shah and Paulsen 2003; Cheikh and Jones 2006; Yang et al. 2006; Massad et al. 2007; Tao et al. 2016). There is a sharp decline in photosynthetic rates in wheat and maize crops when exposed to HS in both the vegetative and reproductive phases. Al-Khatib and Paulsen (1984) and Grover et al. (1986) observed that failure in the supply of photosynthates during grain filling stage conducive to reduce grain yield and biomass (Edreira and Otegui 2012) and it is more prone at night HT (>14 °C) in wheat (Prasad et al. 2008a) and maize (Crafts-Brandner and Salvucci 2002). The rise in photosynthesis rate during the post-anthesis stage resulted in increases in grain filling (Martinez et al. 2014). Fan et al. (2015) in wheat, illustrated that warmer temperature in the night at post-anthesis stage in winter increased the flag leaf photosynthetic carbon assimilation ability resulting in the accumulation of photosynthetic products, confirmed by the relatively high content of Rubisco (ribulose1,5-bisphosphate carboxylase) and soluble protein at the post-anthesis stage, which was conducive to increase grain yield. HS impairs the process of photosynthesis comprises various components, including the photosystems (Camejo et al. 2005), photosynthetic pigments (Camejo et al. 2006), and CO_2 reduction pathways (Wise et al. 2004).

Damage and disorder caused by HS in the stroma and thylakoid lamellae of chloroplast inhibit the activities of membrane-associated electron carriers and enzymes, and cessation of photophosphorylation (Wise et al. 2004; Marchand et al. 2005; Ristic et al. 2008; Wang et al. 2009, 2010). HT leads to swelling, increased leakiness, and disruption of all photochemical reactions especially from light-harvesting complex II from the photosystem (PS) II core complex, and disintegration of PS II-mediated electron transfer of thylakoid (Ristic et al. 2008; Marutani et al. 2012) and thus, impairment of thylakoids caused chlorophyll loss (Ristic et al. 2007, 2008). At HT dissociation of Rubisco activase enzyme (Prasad et al. 2004; Ahmad et al. 2010) resulted in a reduction of the photosynthetic capacity in both light and dark conditions but it is irreversible under the dark condition in wheat (Mathur et al. 2011; Raines 2011). Thus, suppression of carbon assimilation resulted in a reduction of ROS generation (Camejo et al. 2006; Guo et al. 2009) in response, reduction in protein synthesis and prevent recovery of impaired PS II (Murata et al. 2007; Allakhverdiev et al. 2008). At very high level, it may result in severe cell injury and even cell death (Apel and Hirt 2004). Amirjani (2012) reported that in the case of the wheat crop, the amount of chlorophyll (Chl) a, Chl b, and carotenoids did not significantly change at control condition (30/25 °C), but reduced at HT (35/30 °C), while the Chl a/b and Chlorophyll to carotenoids ratios remained unaltered under HS. In maize, HS promotes the degradation of chlorophyll affects the contents of Chl a, Chl b, and total Chl (Hussain et al. 2019), and resulted in reducing the acceptance of light quanta, avoiding excessive free radicals, and causing damage to plants (Havaux and Tardy 1999). To identify the tolerant line, the increased affinity of Rubisco for CO_2 versus oxygen, and a better catalytic rate in photosynthesis are likely to be key targets beneficial at warmer temperatures (Parry et al. 2011).

3.2.2.3 Water Relations

In field conditions deficit of soil water and HT occur simultaneously hindrance the growth and physiological process of the plant in tropical and sub-tropical environments (Wahid et al. 2007; De Boeck et al. 2015; Zandalinas et al. 2016). The stable tissue water status was found to be severely impaired under HS with limited soil water (Nicolas et al. 1984; Machado and Paulsen 2001) leads to a reduction of osmotic potential in leaf (Huve et al. 2005). Increased water use efficiency (net CO_2 assimilation rate/transpiration) causes stomatal closure with a reduction in the net photosynthetic rate (Ruggiero et al. 2017). In wheat, under stress condition increase in the leaf temperature resulted in a decrease in the relative water content, water potential in leaves, and reduction in photosynthetic productivity affecting the rate of transpiration and plant growth (Feng et al. 2014). Hussain et al. (2019) observed that under HS stomatal conductance and transpiration rate in maize leaves increased with high water losses. Water and nutrient availability aboveground parts of the plant depended upon the turgor pressure within cells and other components such as root number, mass, and growth of the roots (Wahid et al. 2007; Huang et al. 2012). HT increases evaporation in younger plants especially under the shorter period of heat stress to cool the tissue, depending upon the availability of water in soil (Balla et al. 2019). Sharma et al. 2015 reported that stomatal conductance decreases with increase in temperature and transpiration was increase while Balla et al. 2019 reported that out of total genotypes studied only a few showed the same reaction type and clearly understood that no strong alliance exists among stomatal conductance, evaporation and heat stress tolerance. The reproductive stage is highly susceptible to water stress and is more difficult with a gradual increase in temperature. To maintain an upper limit of water status during flowering the optimum temperature should not increase above 31 °C (Atkinson and Urwin 2012). Almeselmani et al. (2009) in wheat, noticed that HT (35/25 °C) inflicts after the tillering stage resulted in decline in water potential and is more severe in susceptible genotypes to HS. The HT in plant increases the transpiration and reduction of the osmotic potential in leaf resulted in the release of several antioxidants linked to dehydration tolerance (Ahmad et al. 2010). HT seems to cause a rise in hydraulic conductivity to facilitate the aquaporin activity in the plant cell membrane (Martinez-Ballesta et al. 2009) and reduced water viscosity (Cochard et al. 2007). In heat-tolerant durum and bread wheat genotypes, maintenance of high stomatal conductance is a prerequisite to promote transpiration for heat dissipation (Dias et al. 2008).

3.2.2.4 Phytohormones

Plants tend to cope with unfavorable environmental conditions, though the degree of tolerance dependent upon the type of stress to different crop species. Phytohormones are the natural molecule in plants conciliate growth, development, nutrient allocation (Fahad et al. 2015), and acts as a natural defense molecule in plants by maintaining antioxidants level under stress condition. The involvement of abscisic acid (ABA) in biochemical pathways is found to cope up with the adverse effects of the HT (Maestri et al. 2002). In wheat, ABA plays an important role in the regulation of heat tolerance (Lata and Prasad 2011). Walker-Simmons and Sesing (1990) found that ABA accumulation gradually increases in the grain at HT (25 °C compared to 15 °C), however, the increasing level of ABA did not rise because of HS, its role has been seen after the recovery from stress (Larkindale and Huang 2005). ABA act as a major chemical for root-to-shoot stress signal (Schachtman and Goodger 2008), including stomatal closure and inhibition in leaf expansion. Exogenous application of ABA increases filling rate and sink capacity in grain by regulating endogenous hormone molecule to accelerate endosperm cell division and accumulation of photosynthate (Yang et al. 2014), as well as induces Heat shock protein (Hsp101) transcripts in wheat (Campbell et al. 2001). Jones and Setter (2000) reported a link between ethylene production and heat susceptibility in wheat and identify several heat tolerance genes in ethylene biosynthesis/signaling. Ethylene production in response to HT differs in different crop species (Arshad and Frankenberger 2002) which helps to regulate seed germination, flowering, and fruiting in addition to provide tolerance to plant. Lower the content of ethylene in spike correlates with higher grain yield (Valluru et al. 2017). Fenglu et al. (1997) in maize, studied the production of ethylene maximum at the top ear and minimum at the middle ear, indicates the important role of ethylene in assimilate partitioning to grain filling. Salicylic acid (SA) is a major component of the signaling pathway that react to hypersensitive response and systemic acquired resistance (Kawano et al. 1998). It also stabilizes the heat shock transcription factors and helps to bind the heat shock element with the promoter of heat shock-related genes thus regulate the signaling pathway in HS condition and promote the growth of the plant. Persistent of HS for long term withhold by SA along with Ca^{2+} homeostasis and antioxidant systems (Wang and Li 2006). Cytokinin also helps to mitigate heat tress by changing grain cytokinin content under HS conditions. In dwarf wheat variety at HT decrease in cytokinin content correlates with reduced grain filling and grain weight (Banowetz et al. 1999).

3.2.2.5 Oxidative Stress and Antioxidant Defense

Oxidative damage is usually a subsequent stage of most of the abiotic stresses in plants. Exposure of plants to heat initially causes oxidative damage by the formation of ROS (Wahid et al. 2007; Liu and Huang 2000). Continual HS in crops may cause accumulation of ROS involved in proteolysis of protein with depolarization of cell membrane and trigger of programmed cell death (Qi et al. 2010; Mittler et al. 2011) which causes premature leaf senescence and root growth inhibition (Miller et al. 2009). Under HS condition, majority of ROS are produced in the PS I and PS II of chloroplasts, mitochondria, plasma membrane, apoplasts, peroxisomes, and endoplasmic reticulum while a little amount is also produced by microbodies (Soliman et al. 2011). There are two ways for the production of ROS either enzymatic or nonenzymatic (Apel and Hirt 2004). The increased level of ROS in the cell acts as toxic element which induces oxidation of protein and damages nucleotides. This causes lipid peroxidation of cell membrane and pigments destruction which makes a serious threat to the cell functioning (Apel and Hirt 2004; Xu et al. 2006; Hasanuzzaman et al. 2012). HS induced membrane peroxidation and aggravated membrane injury in wheat (Savicka and Skute 2010) and maize (Kumar et al. 2012).

Oxidative stress is an aversive effect of HS in plant cells due to the production of singlet oxygen (O^2), superoxide radical (O^{2-}), hydrogen peroxide (H_2O_2), and hydroxyl radical (OH^-) notably enhance in membrane peroxidation and reduction in membrane thermostability in many crops including wheat (Savicka and Skute 2010), leads to cell injury and even cell death (Marutani et al. 2012; Suzuki et al. 2012). Among the ROS, O^{2-} is formed by photo-oxidation reactions (flavoprotein, redox-cycling) through Mehler reaction in chloroplasts, during mitochondrial electron transport chain reactions and glyoxisomal photorespiration, by nicotinamide adenine dinucleotide phosphate (NADPH) oxidase in plasma membranes, xanthine oxidase, and membrane polypeptides. The scavenging of O^{2-} by superoxide dismutase (SOD) results in the production of H_2O_2, which is removed by ascorbate peroxidase (APX) or catalase (CAT). However, both O^{2-} and H_2O_2 are not much toxic as the OH^-, which is formed by the combination of O^{2-} and H_2O_2 in the presence of trace amounts of Fe^{2+} and Fe^{3+} by the Haber–Weiss reaction. OH^- react with almost all constituents of cells and can impaired chlorophyll, lipids, nucleic acid, protein, and other macromolecules cause premature leaf senescence, root growth inhibition, affects plant metabolism, reduce growth and yield (Sairam and Tyagi 2004; Miller et al. 2009; Qi et al. 2010; Mittler et al. 2011).

HS tolerant plants mediate antioxidant defense mechanism which works either enzymatic or nonenzymatic way. The enzymatic defense system is considered to be the most effective (Farooq et al. 2008). Major enzymes involved as protectants are SOD, APX, CAT, and peroxidase. SOD is an elementary antioxidant enzyme, which converts O^{2-} into H_2O_2 and O_2, APX neutralizes H_2O_2 by using ascorbate as a substrate, CAT breaks down H_2O_2, glutathione reductase which reduces glutathione disulfide to the sulfhydryl from glutathione, and peroxidase has ameliorating effects of HS in wheat and maize (Suzuki et al. 2011; Caverzan et al. 2016). The nonenzymatic defense system involves reduction in glutathione, tocopherols, ascorbic acid, and carotenoids. Tiwari and Yadav (2019) describe the role of the ascorbate-glutathione cycle in maize crops in terminal (reproductive) HS. Therefore, the enrichment of antioxidants in cell is a better approach by the crop to conflict the effects of ROS (Sharma and Dubey 2005). Reassurance against oxidative stress is a major key to determine the endurance of a crop under HS. Understanding the expression, accumulation, and developmental pathway of antioxidants under environmental stress condition helps to improvise and make a significant step towards the development of heat-tolerant lines.

3.3 Screening Methodologies for Heat Tolerance

Identification and efficient screening procedure of traits associated with HS in germplasm are preliminary actions taken before starting breeding for heat tolerance. Challenges are faced throughout the field during screening for heat tolerance due to the interaction of traits with other environmental factors. Even though there are few relevant traits that help to select a tolerant line in field conditions (Fig. 3.2) are described below:

Canopy Temperature Depression (CTD) It is the difference between the temperature of the ambient environment around the crop and its canopy. Calculated as CTD = Ta − Tc, where, Ta is ambient temperature, Tc is canopy temperature of crop plants. Canopy temperature measured by handheld infrared thermometer, best time to measure CTD is afternoon (13:00 and 14:30 h) at 40° of viewing angle parallel to the surface of the earth and aims infrared thermometer directly into the canopy to escape the confounding effect of soil temperature. Lowering the canopy temperature is a kind of heat escape mechanism to maintain the homeostasis in the plant by evapotranspiration cooling and other physiological mechanisms viz, metabolism, partitioning, and vascular transport. CTD shows a high genetic correlation with yield in wheat (Reynolds et al. 1994; Narayanan et al. 2018).

Early Heading The early heading is a heat escape mechanism as it escapes the risks of heading at HT stress. In wheat HT at heading reduces the grain yield (Balla et al. 2019). It significantly increases the duration of grain filling by completing the life cycle earlier in the season before the onset of HS, however, this trait is restricted for

Fig. 3.2 Different screening measures to identify heat tolerance lines

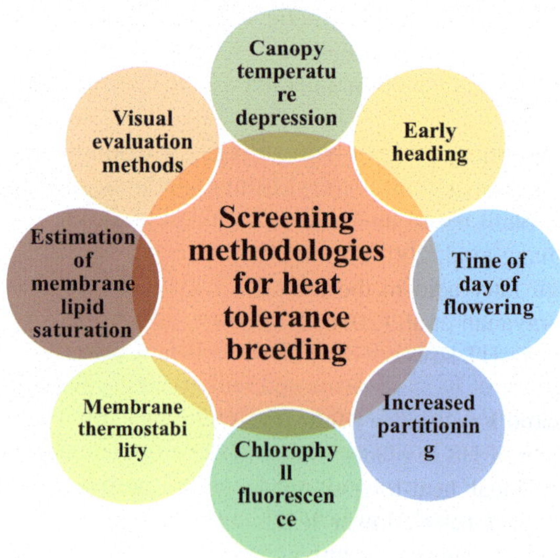

the area where the temperature is cool or frost limits the early heading trait (Tewolde et al. 2006; Narayanan et al. 2018).

Time of Day of Flowering It is the time of the day at which anthesis begins in a crop (Sheehy et al. 2007). After sunrise, the air temperature increase along with duration exceeds the threshold temperature. Increase the chances of impede in anther dehiscence, pollen tube growth, thus anthesis in the early hours of the morning before sunrise helps to escape the HS and reduces spikelet sterility in plants.

Increased Partitioning It is a tolerance mechanism, in which the duration of grain filling during HS severely affects the quality and quantity of grain as HS affects the utilization of mobilized stem reserved contributes to grain development. Increases partitioning of reserves from different plant parts viz., leaves, stem, etc., is a potential approach to improve grain filling and enhance yield in wheat (Narayanan et al. 2018).

Chlorophyll Fluorescence Measurement It is a high-throughput screening method that is used to identify the damage in photosynthetic tissue, especially in thylakoid which is closely associated with chlorophyll loss. Tolerant genotypes retain the chlorophyll content in HT genotypes. It is a simple quick rapid method measured by the portable fluorometer to find the ration of a variable to minimum chlorophyll fluorescence (Fv/Fm).

Membrane Thermostability (MTS) Denaturation of protein and increments of unsaturated fatty acids in HS leads the electrolyte leakage, confirms the reduction in cell membrane thermostability (Blum 2017). It is a highly heritable trait, measures the reduced cell MTS and reflects stress-induced changes made it as an indirect measure of HS tolerance in diverse plant species (Liu and Huang 2000; Xu et al. 2006). Calculated as $\text{MTS} = (1 - T_1/T_2) \times 100$, where T_1 is conductivity meter reading afterword heat treatment, and T_2 is conductivity reading afterword autoclaving (Ibrahim and Quick 2001).

Estimation of Membrane Lipid Saturation Greater amount of saturated fatty acids in membrane lipids maintain the membrane fluidity at HT stress. Thus, during HS plant try to increase the content of saturated and monounsaturated fatty acids (Zhang et al. 2005). In maize Karim et al. (1997, 1999) reported that leaves with a high fatty acid in the membrane are less prone to damage through HS. In wheat Narayanan et al. (2016, 2018) identified changes in lipid metabolism with changes during HT stress.

Visual Evaluation Methods/Morphological Methods Ease method to lookout the effect of HS in plants when they are subjected to the same environmental condition. The ideal heat-tolerant maize plant should have a higher amount of leaf wax, compact tassel, lower leaf, and cob angle. Other visual assessments like pollen viability, stigma receptivity (Dupuis and Dumas 1990; Ottaviano et al. 1992) leaf

Table 3.2 Selection indices in wheat and maize

Selection indices	Formula	References
Stress susceptibility index	$\mathrm{SSI} = \frac{1-(\mathrm{ys}/\mathrm{yp})}{1-(\overline{\mathrm{ys}}/\overline{\mathrm{yp}})}$	Fischer and Maurer (1978)
Mean productivity	$\mathrm{MP} = \frac{\mathrm{yp}+\mathrm{ys}}{2}$	Rosielle and Hamblin (1981), Hossain et al. (1990)
Tolerance	$\mathrm{TOL} = \mathrm{yp} - \mathrm{ys}$	Rosielle and Hamblin (1981), Hossain et al. (1990)
Stress tolerance index	$\mathrm{STI} = \frac{\mathrm{yp}\times\mathrm{ys}}{\mathrm{yp}^2}$	Fernandez (1992)
Geometric mean productivity	$\mathrm{GMP} = \sqrt{\mathrm{yp}\times\mathrm{ys}}$	Fernandez (1992), Ramirez and Kelly (1998)
Yield index	$\mathrm{YI} = \frac{\mathrm{ys}}{\overline{\mathrm{ys}}}$	Lin et al. (1986), Gavuzzi et al. (1997)
Yield stability index	$\mathrm{YSI} = \frac{\mathrm{ys}}{\mathrm{yp}}$	Bouslama and Schapaugh (1984)

ys yield of genotypes under HS condition, *yp* yield of genotypes under timely sowing condition, $\overline{\mathrm{ys}}$ and $\overline{\mathrm{yp}}$ the mean yields of all genotypes under HS and timely sowing conditions, $1 - (\overline{\mathrm{ys}}/\overline{\mathrm{yp}})$ the stress intensity

firing, tassel sterility and silk receptivity (Zaidi et al. 2016) also considered best-suited heat tolerance related traits at terminal HS. In wheat, early ground cover, waxy leaves (Richards 1996), stay green (Kumar et al. 2010) assist in the selection of tolerance genotype (Govindaraj et al. 2018). In conventional breeding, screening methods as above validates with the selection indices are very cost-effective and easy-to-assay technique to find genotypes tolerance to HS. In this regard, researchers suggested several indices are based either on stress tolerance or susceptibility of genotype (Kamrani et al. 2018) (Table 3.2). Hossain et al. (1990) defined stress tolerance (TOL) as the differences in yield between the non-stress (Y_p) and stress (Y_s) environments and mean productivity (MP) as the average yield of Y_p and Y_s.

Fischer and Maurer (1978) proposed a stress susceptibility index (SSI) in genotype as a ratio of genotypic performance under stress and non-stress conditions. Fernandez (1992) introduced a stress tolerance index (STI) to identify the performance of genotype in both stresses condition. Geometric mean productivity (GMP) is used to find the relative performance of genotypes in the stressed and non-stressed environment (Ramirez and Kelly 1998). Rosielle and Hamblin (1981); Hossain et al. (1990) reported a positive correlation between MP and Ys, however, these indices do not permit discrete genotype performance in stress conditions from the performance in both stressed and non-stressed condition. Kamrani et al. (2018) in wheat and Khodarahmpour (2011) in maize studied stress indices and concluded as a reliable indicator of yield stability and a proxy for heat tolerance (Paliwal et al. 2012).

Heat Susceptibility Index (HSI) HSI is calculated using the formula, $\mathrm{HSI} = (1 - Y/Y_p)/D$, where Y denotes the average yield per plant or plot at HT, Y_p is the average yield per plant or plot at optimum temperature, D is the stress intensity, which is calculated as $1 - X/X_p$, in which X is the mean of Y in all genotypes, and X_p is the mean of Y_p in all genotypes. Majorly use in maize and wheat to identify the tolerant genotype. Minimum HSI means more tolerant the genotype to

HS. Generally, genotypes with HSI ≤ 0.5 are highly heat tolerant, HSI value lies in between 0.5 and ≤ 1.0 then genotype is moderately tolerant, if HSI > 1.00 then the genotype is susceptible.

3.4 Approaches to Mitigate Heat Stress

Growth and development stages of the plants were adversely affected by HS depending upon the extent, stress duration, and the growth stage (Ruelland and Zachowski 2010). Development of heat-tolerant line through screening in germplasm line and identification of genes/quantitative trait loci (QTL) involved in heat tolerance mechanism seems to be the most appropriate strategies against HS (Asseng et al. 2011; Chapman et al. 2012; Feng et al. 2014; Akter and Rafiqul Islam 2017) (Fig. 3.3).

3.4.1 Screening and Breeding for Heat Tolerance

In breeding program the progress of stress tolerance actually depends upon the availability of the good genetic stock, screening methods and selection criteria rely

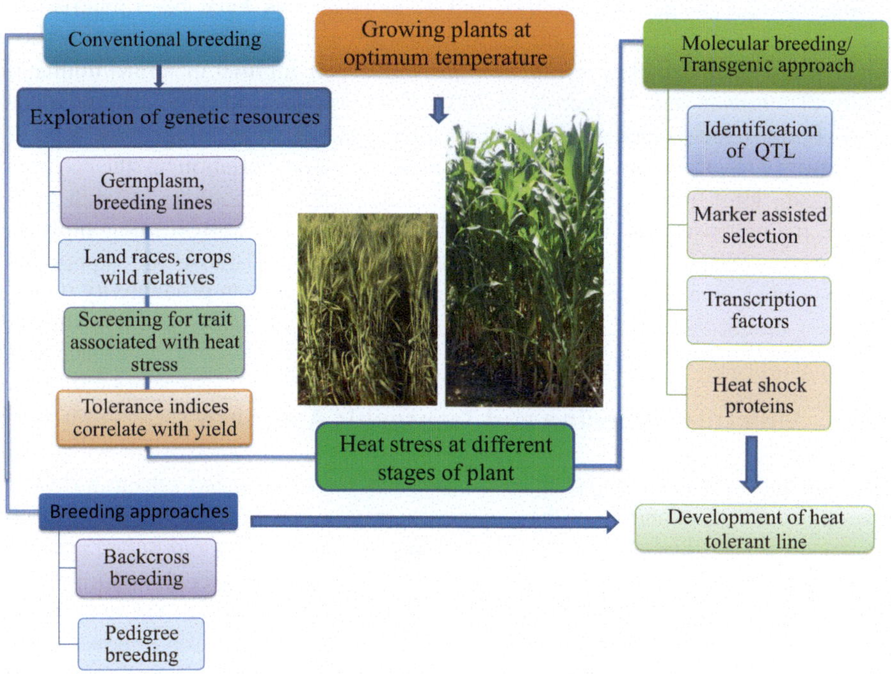

Fig. 3.3 A schematic diagram showed identification of crop tolerance to heat stress and the use of conventional and molecular breeding strategies under heat stress environment

on traits correlate with higher grain yield under HS condition (Reynolds and Langridge 2016). Direct selection can be done by the abovementioned screening methods and selection indices (Table 3.2). However, separate evaluations on each stage throughout the ontogeny of the plant and uncontrolled environmental factors adversely impact the precision and repeatability of the trials (Chen et al. 1982). Besides, for selection for indirect or associated traits few facilities like high-resolution thermal imaging system to measure leaf temperature (Jones and Sirault 2014), image analysis phenotypic platform "HTpheno" (Hartmann et al. 2011) and "Rootscope" used to measured heat shock responses in plants (Kast et al. 2013). Soon, the new-generation phenomics platforms allow to screen huge germplasm for HT and are very cost-effective.

For the identification of a tolerant genotype, genetic diversity is the necessities in the breeding program (Chapman et al. 2012; Balla et al. 2019). Landraces and wild relatives are the best sources for genetic diversity in the breeding program. To maintain the genetic gain, it was suggested to screen and make crosses with exotic material (Giaveno and Ferrero 2003) considered as the best method in maize to increase heat tolerance. In wheat, the recurrent selection method has been useful to develop heat tolerance lines (Gororo et al. 2002; Machado et al. 2010). Also, the stable transfer of chromosomes from wild relative *Leymus racemosus* helps to develop heat tolerance in the hot and arid area (Mohammed et al. 2014). Few genus viz., *Aegilops tauschii, Aegilops geniculata, A. speltoides, A. searsii, A. longissimi, Triticum dicoccoides*, and *T. monococcum* have been reported as potential sources of HS tolerant germplasms in wheat (Zaharieva et al. 2001; Pradhan et al. 2012). Few synthetic wheat also had high-temperature tolerance ability (Trethowan and Mujeeb-Kazi 2008; Kurahashi et al. 2009) and which can act as bridges for introgression of many alien genes into cultivated wheat pools (Siddiqui 1976; Rajaram et al. 1977). Limited genetic variation in the existing germplasm, long time to screen genotypes, and complexity associated with traits are the major reason for few successes in conventional plant breeding program (Parmar et al. 2017). To fasten the program molecular and transgenic approaches can be used as an effective method to improve HS tolerance.

3.4.2 Molecular Breeding and Transgenic Approach

The application of QTL mapping and subsequent marker-assisted selection (MAS) gives a better understanding of the genetic association among tolerances to environmental stress (Heffner et al. 2009). Many major or minor QTLs and linked markers for HT are available in wheat (Pinto et al. 2010; Vijayalakshmi et al. 2010; Paliwal et al. 2012; Mason et al. 2011; Beecher et al. 2012) and maize (Ottaviano et al. 1991; Frova and Sari-Gorla 1994; Sanguineti et al. 1999) and can be used to develop HS tolerant lines. All gathered information helps to facilitate targeted breeding for heat tolerance in wheat and maize crops (Fig. 3.3). Summary of such efforts and progress is presented and discussed in the following paragraphs.

QTLs for HT related traits have been discovered for example cellular membrane stability, pollen germination, and pollen tube growth in maize (Ottaviano et al. 1991; Frova and Sari-Gorla 1994). Which identified for different traits involved in heat stress tolerance like grain filling duration (Yang et al. 2002; Mason et al. 2010; Barakat et al. 2012; Paliwal et al. 2012), grain number (Pinto et al. 2010), leaf senescence (Mason et al. 2010; Vijayalakshmi et al. 2010), reproductive stage (Farooq et al. 2011), tillering, stay green character (Kumar et al. 2010; Li et al. 2010) and canopy temperature (Paliwal et al. 2012) have been identified in wheat. Mason et al. (2010) also identified many QTLs associated with yield and yield attribute traits. Some QTLs have a small effect on the phenotype, to overcome this marker-assisted recurrent selection (MARS) or genomic selection (GS) approach is used to pyramiding several QTLs using large populations (Tester and Langridge 2010). The identification of QTLs and their significant use in MAS, MARS, and marker-assisted backcross breeding is helping to improve heat tolerance in different crops (Zheng et al. 2012). Incorporation of genes of interested traits into the desired plants to improve tolerance against HS utilized through genetic engineering and transgenic approaches can be followed (Chapman et al. 2012).

Transgenic approaches involve the incorporation of superior genes to elite cultivars and thus helps to avoid the issue of linkage drag, to develop plant tolerance to HS (Zheng et al. 2012). In maize, gene phosphoenolpyruvate carboxylase (*ZmPEPC*) overexpressed improved heat tolerance in wheat by enhancing activities of the photochemical enzyme, delayed chlorophyll degradation, upregulated expression of genes related to photosynthesis, adjusted contents of proline (Qi et al. 2017). In the wheat transfer of maize elongation factor (*EFTu*) *EFTu1*gene overexpressed and improved tolerance to HT stress. Fu et al. (2008) reported in wheat, the increase in the protein synthesis *EFTu* in the chloroplast. The constitutive expression of *EFTu* in transgenic wheat protects leaf protein from denaturation, lowering thylakoid membranes disruption, and enhanced photosynthetic capability. According to Ristic et al. (2008), transgenic wheat accruing more EFTu has exceptional tolerance to HS than those exposed to less *EFTu* transgene of certain heat shock protein also improved tolerance against HS. In maize transgenic overexpression of certain heat shock factors (HSF) and HSF-fusion proteins transfer from *A. thaliana* results in an expression of Hsp101 results increase in thermotolerance (Iba 2002; Queitsch et al. 2000).

In addition to conventional, molecular breeding, transgenic approach, and agronomic management practices like soil moisture management, application of fertilizer, sowing methods, maintaining proper time and use of exogenous protectants can also attenuate the effect of HS (Ortiz et al. 2008; Singh et al. 2011). Applications of organic mulches conserve soil moisture and increase nitrogen use efficiency in wheat (Chakraborty et al. 2008; Singh et al. 2011). The $CaCl_2$ application increase antioxidants like SOD, guaiacol peroxidase, and CAT (Dias et al. 2009). In maize and wheat, early sowing prevents HS at the flowering stage (Perego et al. 2014; Kajla et al. 2015). Use of *Bacillus amyloliquefaciens* (UCMB5113) and *Azospirillum brasilense* (NO40) strains for seed treatment reduces ROS generation provides tolerance in the wheat seedling (Abd El-Daim et al. 2014). However, the use of these practices is time bounded and tedious.

3.5 Conclusion and Future Perspective

An increase in the concentration of greenhouse gases leads to rising temperature globally. HT is the limiting factor for wheat and maize production, because it affects plant growth at all developmental stages, especially at the reproductive and grain filling stage. Although physiological process viz., respiration, photosynthesis, transpiration is well known but studies of osmolytes, assimilate portioning from source to sink are not very well understood. Knowledge of the sensitivity of phenology of crop to heat stress required to begin suitable strategies for inducing HS tolerance. HT is a quantitative trait controlled by many genes and difficult to quantify, the screening traits MTS, CTD, chlorophyll fluorescence along with stress indices like SSI, TOL, GMP appears to be useful indicators of heat tolerance and can be used in breeding programs, however tolerance to HT is difficult to access in the field conditions due to changes in timing and heatwave so, easy and fast screening method should be in place for applying in breeding programs. The major challenge in breeding program is linkage drag. To overcome this, molecular breeding help in the identification of QTL related to heat stress, and its transfer through marker-assisted breeding program appeared to be the most appropriate approach to develop heat-tolerant lines of different crops. Transgenic and molecular breeding approaches are also useful to transfer trait-specific genes and to develop heat tolerance genotypes in wheat and maize.

References

Abd El-Daim IA, Bejai S, Meijer J (2014) Improved heat stress tolerance of wheat seedlings by bacterial seed treatment. Plant Soil 379:337–350. https://doi.org/10.1007/s11104-014-2063-3

Ahmad P, Jaleel CA, Salem MA, Nabi G, Sharma S (2010) Roles of enzymatic and nonenzymatic antioxidants in plants during abiotic stress. Crit Rev Biotechnol 30:161–175. https://doi.org/10.3109/07388550903524243

Akman Z (2009) Comparison of high temperature tolerance in maize, rice and sorghum seeds by plant growth regulators. J Anim Vet Adv 8:358–361

Akter N, Rafiqul Islam M (2017) Heat stress effects and management in wheat. A review. Agron Sustain Dev 37:37. https://doi.org/10.1007/s13593-017-0443-9

Al-Khatib K, Paulsen GM (1984) Mode of high temperature injury to wheat during grain development. Physiol Plant 61:363–368

Allakhverdiev SI, Kreslavski VD, Klimov VV, Los DA, Carpentier R, Mohanty P (2008) Heat stress: an overview of molecular responses in photosynthesis. Photosynth Res 98:541–550. https://doi.org/10.1007/s11120-008-9331-0

Almeselmani M, Deshmukh PS, Sairam RK (2009) High temperature stress tolerance in wheat genotypes: role of antioxidant defence enzymes. Acta Agron Hung 57:1–14

Almeselmani M, Deshmukh PS, Chinnusamy V (2012) Effect of prolonged high temperature stress on respiration, photosynthesis and gene expression in wheat (*Triticum aestivum* L.) varieties differing in their thermotolerance. Plant Stress 6:25–32

Amirjani M (2012) Estimation of wheat responses to "high" heat stress. Am Eurasian J Sustain Agric 6:222–233

Apel K, Hirt H (2004) Reactive oxygen species: metabolism, oxidative stress, and signal transduction. Ann Rev Plant Biol 55:373–399. https://doi.org/10.1146/annurev.arplant.55.031903.141701

Arshad M, Frankenberger WT (2002) Ethylene in agriculture: synthetic and natural sources and applications. In: Ethylene. Springer, Boston, MA. https://doi.org/10.1007/978-1-4615-0675-1_8

Ashraf M, Hafeez M (2004) Thermo tolerance of pearl millet and maize at early growth stages: growth and nutrient relations. Biol Plant 48:81–86

Asseng S, Foster I, Turner NC (2011) The impact of temperature variability on wheat yields. Glob Chang Biol 17:997–1012. https://doi.org/10.1111/j.1365-2486.2010.02262.x

Asseng S, Royce R, Cammarano D (2013) Temperature routines in wheat, workshop modeling wheat response to high temperature, vol 8. CIMMYT, Mexico, p 128

Atkinson NJ, Urwin PE (2012) The interaction of plant biotic and abiotic stresses: from genes to the field. J Exp Bot 63:3523–3543. https://doi.org/10.1093/jxb/ers100

Balla K, Karsai I, Bencze S, Veisz O (2012) Germination ability and seedling vigour in the progeny of heat-stressed wheat plants. J Acta Agron Hung 60:299–308

Balla K, Karsai I, Bonis P, Kiss T, Berki Z et al (2019) Heat stress responses in a large set of winter wheat cultivars (*Triticum aestivum* L.) depend on the timing and duration of stress. PLoS One 14 (9):e0222639. https://doi.org/10.1371/journal.pone.0222639

Banowetz GM, Ammar K, Chen DD (1999) Temperature effects on cytokinin accumulation and kernel mass in a dwarf wheat. Ann Bot 83:303–307

Barakat MN, Al-Doss AA, Elshafei AA, Moustafa KA (2012) Bulked segregant analysis to detect quantitative trait loci (QTL) related to heat tolerance at grain filling rate in wheat using simple sequence repeat (SSR) markers. Afr J Biotechnol 11:12436–12442

Barnabas B, Jager K, Feher A (2008) The effect of drought and heat stress on reproductive processes in cereals. Plant Cell Environ 31:11–38

Beecher FW, Mason E, Mondal S, Awika J, Hays D, Ibrahim A (2012) Identification of quantitative trait loci (QTLs) associated with maintenance of wheat (*Triticum aestivum* Desf) quality characteristics under heat stress conditions. Euphytica 188:361–368

Beemster GTS, Baskin TI (1998) Analysis of cell division and elongation underlying the developmental acceleration of root growth in *Arabidopsis thaliana*. Plant Physiol 116:1515–1526. https://doi.org/10.1104/pp.116.4.1515

Beemster GTS, Fiorani F, Inze D (2003) Cell cycle: the key to plant growth control? Trends Plant Sci 8:154–158. https://doi.org/10.1016/S1360-1385(03)00046-3

Begcy K, Weigert A, Egesa AO, Dresselhaus T (2018) Compared to Australian cultivars, European summer wheat (*Triticum aestivum*) overreacts when moderate heat stress is applied at the pollen development stage. Agronomy 8:99

Begcy K, Nosenko T, Zhou LZ, Fragner L, Weckwerth W, Dresselhaus T (2019) Male sterility in maize after transient heat stress during the tetrad stage of pollen development. Plant Physiol 181:683–700

Behboudian MH, Lawes GS, Griffiths KM (1994) The influence of water deficit on water relations, photosynthesis and fruit growth in Asian pear (*Pyrus serotinia* Rehd.). Sci Hortic 60:89–99

Bennett D, Izanloo A, Reynolds M, Kuchel H, Langridge P, Schnurbusch T (2012) Genetic dissection of grain yield and physical grain quality in bread wheat (*Triticum aestivum* L.) under water limited environments. Theor Appl Genet 125:255–271. https://doi.org/10.1007/s00122-012-1831-9

Berry J, Bjorkman O (1980) Photosynthetic response and adaptation to temperature in higher plants. Ann Rev Plant Physiol 31:491–543

Blum A (2017) Osmotic adjustment is a prime drought stress adaptive engine in support of plant production: osmotic adjustment and plant production. Plant Cell Environ 40:4–10. https://doi.org/10.1111/pce.12800

Bouslama M, Schapaugh WT (1984) Stress tolerance in soybean. Part 1: evaluation of three screening techniques for heat and drought tolerance. Crop Sci 24:933–937

Cairns JE, Sonder K, Zaidi PH, Verhulst N, Mahuku G, Babu R et al (2012) Maize production in a changing climate: impacts, adaptation, and mitigation strategies. Adv Agron 114:1–65. https://doi.org/10.1016/B978-0-12-394275-3.00006-7

Camejo D, Rodriguez P, Morales MA, Dell'amico JM, Torrecillas A, Alarcon JJ (2005) High temperature effects on photosynthetic activity of two tomato cultivars with different heat susceptibility. J Plant Physiol 162:281–289. https://doi.org/10.1016/j.jplph.2004.07.014

Camejo D, Jimenez A, Alarcon JJ, Torres W, Gomez JM, Sevilla F (2006) Changes in photosynthetic parameters and antioxidant activities following heat-shock treatment in tomato plants. Funct Plant Biol 33:177–187. https://doi.org/10.1071/FP05067

Campbell JL, Klueva NY, Zheng HG, Nieto-Sotelo J, Ho TD, Nguyen HT (2001) Cloning of new members of heat shock gene family in wheat (*Triticum aestivum* (L) Moench) inducible by heat, dehydration, and ABA. Biochim Biophys Acta 1517(2):270–277

Castro M, Peterson CJ, Rizza MD, Dellavalle PD, Vazquez D, Ibanez V, Ross A (2007) Influence of heat stress on wheat grain characteristics and protein molecular weight distribution. In: Buck HT, Nisi JE, Salomon N (eds) Wheat production in stressed environment. Springer, Dordrecht, pp 365–371

Caverzan A, Casassola A, Brammer SA (2016) Antioxidant responses of wheat plants under stress. Genet Mol Biol 39:1–6. https://doi.org/10.1590/1678-4685-GMB-2015-010

Chakraborty D, Nagarajan S, Aggarwal P, Gupta VK et al (2008) Effect of mulching on soil and plant water status, and the growth and yield of wheat (*Triticum aestivum* L.) in a semi-arid environment. Agric Water Manag 95:1323–1334. https://doi.org/10.1016/j.agwat.2008.06.001

Chapman SC, Chakraborty S, Dreccer MF, Howden SC (2012) Plant adaptation to climate change-opportunities and priorities in breeding. Crop Pasture Sci 63:251–268

Cheikh N, Jones RJ (2006) Heat stress effects on sink activity of developing maize kernels grown in vitro. Physiol Plant 95:59–66

Chen HH, Shen ZY, Li PH (1982) Adaptability of crop plants to high temperature stress. Crop Sci 22:719–725

Chen J, Tian Y, Zhang X, Zheng C, Song Z, Deng A, Zhang W (2014) Night time warming will increase winter wheat yield through improving plant development and grain growth in North China. J Plant Growth Regul 33:397–407

Cicchino M, Rattalino Edreira JI, Uribelarrea M, Otegui ME (2010) Heat stress in field grown maize: response of physiological determinants of grain yield. Crop Sci 50:1438–1448. https://doi.org/10.2135/cropsci2009.10.0574

Ciuca M, Petcu E (2009) SSR markers associated with membrane stability in wheat (*Triticum aestivum* L.). Romanian Agric Res 26:21–24

Cochard H, Venisse JS, Barigah TS, Brunel N, Herbette S, Guilliot A, Tyree MT, Sakr S (2007) Putative role of aquaporins in variable hydraulic conductance of leaves in response to light. Plant Physiol 143:122–133. https://doi.org/10.1104/pp.106.090092

Cossani CM, Reynolds MP (2012) Physiological traits for improving heat tolerance in wheat. Plant Physiol 160:1710–1718

Crafts-Brandner SJ, Salvucci ME (2002) Sensitivity of photosynthesis in a C4 plant maize to heat stress. Plant Physiol 129:1773–1780. https://doi.org/10.1104/pp.002170

De Boeck HJ, Bassin S, Verlinden M, Zeiter M, Hiltbrunner E (2015) Simulated heat waves affected alpine grassland only in combination with drought. New Phytol 209:531–541. https://doi.org/10.1111/nph.13601

Deryng D, Conway D, Ramankutty N, Price J, Warren R (2014) Global crop yield response to extreme heat stress under multiple climate change futures. Environ Res Lett 9:1–13. https://doi.org/10.1088/1748-9326/9/3/034011

Devasirvatham V, Tan DKY, Gaur PM, Raju TN, Trethowan RM (2012) High temperature tolerance in chickpea and its implications for plant improvement. Crop Pasture Sci 63:419–428. https://doi.org/10.1071/CP11218

Devasirvatham V, Gaur PM, Mallikarjuna N, Raju TN, Trethowan RM, Tan DK (2013) Reproductive biology of chickpea response to heat stress in the field is associated with the performance in controlled environments. Field Crops Res 142:9–19. https://doi.org/10.1016/j.fcr.2012.11.011

Dias AS, Lidon FC (2009) Evaluation of grain filling rate and duration in bread and durum wheat, under heat stress after anthesis. J Agron Crop Sci 195(2):137–147. https://doi.org/10.1111/j.1439-037X.2008.00347.x

Dias AS, Bagulho AS, Lidon FC (2008) Ultrastructure and biochemical traits of bread and durum wheat grains under heat stress. Braz J Plant Physiol 20:323–333. https://doi.org/10.1590/S1677-04202008000400008

Dias AS, Lidon FC, Ramalho JC (2009) Heat stress in *Triticum*: kinetics of Ca and Mg accumulation. Braz J Plant Physiol 21:123–134. https://doi.org/10.1590/S1677-04202009000200005

Dias AS, Semedo J, Ramalho JC, Lidon FC (2010) Bread and durum wheat under heat stress: a comparative study on the photosynthetic performance. J Agron Crop Sci 197:50–56

Dupuis I, Dumas C (1990) Influence of temperature stress on *in vitro* fertilization and heat shock protein synthesis in maize (*Zea mays* L.) reproductive tissues. Plant Physiol 94:665–670. https://doi.org/10.1104/pp.94.2.665

Dwivedi R, Prasad S, Jaiswal B, Kumar A, Tiwari A, Patel S et al (2017) Evaluation of wheat genotypes (*Triticum aestivum* L.) at grain filling stage for heat tolerance. Int J Pure App Biosci 5:971–975

Edreira JIR, Otegui ME (2012) Heat stress in temperate and tropical maize hybrids: differences in crop growth, biomass partitioning and reserves use. Field Crops Res 130:87–98

Essemine J, Ammar S, Bouzid S (2010) Impact of heat stress on germination and growth in higher plants: physiological, biochemical and molecular repercussions and mechanisms of defence. J Biol Sci 10:565–572. https://doi.org/10.3923/jbs.2010.565.572

Fahad S, Hussain S, Bano A et al (2015) Potential role of phytohormones and plant growth-promoting rhizobacteria in abiotic stresses: consequences for changing environment. Environ Sci Pollut Res 22:4907–4921. https://doi.org/10.1007/s11356-014-3754-2

Fahad S, Bajwa AA, Nazir U, Anjum SA, Farooq A, Zohaib A et al (2017) Crop production under drought and heat stress: plant responses and management options. Front Plant Sci 8:1147. https://doi.org/10.3389/fpls.2017.01147

Fan Y, Tian M, Jing Q, Tian Z, Han H, Jiang D, Cao W, Dai T (2015) Winter night warming improves pre-anthesis crop growth and post-anthesis photosynthesis involved in grain yield of winter wheat (*Triticum aestivum* L.). Field Crops Res 178:100–108. https://doi.org/10.1016/j.fcr.2015.04.001

Farooq M, Aziz T, Basra SMA, Cheema MA, Rehman H (2008) Chilling tolerance in hybrid maize induced by seed priming with salicylic acid. J Agron Crop Sci 194:161–168. https://doi.org/10.1111/j.1439-037X.2008.00300.x

Farooq M, Bramley H, Palta JA, Siddique KHM (2011) Heat stress in wheat during reproductive and grain-filling phases. Crit Rev Plant Sci 30:491–507

Feng B, Liu P, Li G, Dong ST, Wang FH, Kong LA, Zhang JW (2014) Effect of heat stress on the photosynthetic characteristics in flag leaves at the grain-filling stage of different heat-resistant winter wheat varieties. J Agron Crop Sci 200:143–155. https://doi.org/10.1111/jac.12045

Fenglu Z, Ming Z, Zhimin W, Shuan WZJ, Jinglum G (1997) Maize kernel development and release of ethylene by grains. J China Agric Univ 2:85–89

Fernandez GCJ (1992) Effective selection criteria for assessing plant stress tolerance. In: Kuo CG (ed) Proceedings of the international symposium on adaptation of vegetables and other food crops in temperature and water stress. AVRDC Publication, Shanhua, pp 257–270

Ferris R, Ellis RH, Wheeler TR, Hadley P (1998) Effect of high temperature stress at anthesis on grain yield and biomass of field grown crops of wheat. Plant Cell Environ 34:67–78

FICCI (2014) Federation of Indian Chambers of Commerce and Industry, India maize summit 2014. http://ficci.in/spdocument/20386/India-Maize-2014_v2.pdf

Fiorani F, Beemster GTS (2006) Quantitative analyses of cell division in plants. Plant Mol Biol 60:963–979

Fischer RA, Maurer R (1978) Drought resistance in spring wheat cultivars. I. Grain yield response. Aust J Agric Res 29:897–907

Frova C, Sari-Gorla M (1994) Quantitative trait loci (QTLs) for pollen thermo tolerance detected in maize. Mol Gen Genet 245:424–430

Fu J, Momcilovic I, Clemente TE, Nersesian N, Trick HN, Ristic Z (2008) Heterologous expression of a plastid EF-Tu reduces protein thermal aggregation and enhances CO2 fixation in wheat (*Triticum aestivum*) following heat stress. Plant Mol Biol 68:277–288

Gavuzzi P, Rizza F, Palumbo M, Campaline RG, Ricciardi GL, Borghi B (1997) Evaluation of field and laboratory predict ors of drought and heat tolerance in winter cereals. Plant Sci 77:523–531

Giaveno C, Ferrero J (2003) Introduction of tropical maize genotypes to increase silage production in the central area of Santa Fe, Argentina. Crop Breed Appl Biotechnol 3:89–94

Giorno F, Wolters-Arts M, Mariani C, Rieu I (2013) Ensuring reproduction at high temperatures: the heat stress response during anther and pollen development. Plants 2:489–506. https://doi.org/10.3390/plants2030489

Gororo NN, Eagles HA, Eastwood RF, Nicolas ME, Flood RG (2002) Use of *Triticum tauschii* to improve yield of wheat in low-yielding environments. Euphytica 123:241–254

Govindaraj M, Pattanashetti SK, Patne N, Kanatti AA (2018) Breeding cultivars for heat stress tolerance in staple food crops. In: Ciftci YO (ed) Next generation plant breeding. Intech Open, London, pp 45–74. https://doi.org/10.5772/intechopen.76480

Grover A, Sabat SC, Mohanty P (1986) Effect of temperature on photosynthetic activities of senescing detached wheat leaves. Plant Cell Physiol 27:117–126

Guedira M, McCluskey PJ, MacRitchie F, Paulsen GM (2002) Composition and quality of wheat grown under different shoot and root temperatures during maturation. Cereal Chem 79:397–403. https://doi.org/10.1094/CCHEM.2002.79.3.397

Guo FL, Yu J, Tao J, Su J, Zhang CH (2009) Effects of high temperature stress on photosynthesis and chlorophyll fluorescence of *Euphorbia pulcherrima*. J Yangzhou Univ Agric Life Sci 30:71–74

Gupta NK, Agarwal S, Agarwal VP, Nathawat NS, Gupta S, Singh G (2013) Effect of short-term heat stress on growth, physiology and antioxidative defence system in wheat seedlings. Acta Physiol Plant 35:1837–1842. https://doi.org/10.1007/s11738-013-1221-1

Hall AE (2001) Crop responses to environment. CRC Press LLC, Boca Raton, FL. https://doi.org/10.1201/9781420041088

Hampton JG, Boelt B, Rolston MP, Chastain TG (2013) Effects of elevated CO_2 and temperature on seed quality. J Agric Sci 151:154–162. https://doi.org/10.1017/S0021859612000263

Hansen J, Sato M, Ruedy R (2012) Perception of climate change. Proc Natl Acad Sci 109:14726–14727. https://doi.org/10.1073/pnas.1205276109

Haque MS, Kjaer KH, Rosenqvist E, Sharma DK, Ottosen CO (2014) Heat stress and recovery of photosystem II efficiency in wheat (*Triticum aestivum* L.) cultivars acclimated to different growth temperatures. Environ Exp Bot 99:1–8. https://doi.org/10.1016/j.envexpbot.2013.10.017

Hartmann A, Czauderna T, Hoffmann R, Stein N, Schreiber F (2011) HTPheno: an image analysis pipeline for high-throughput plant phenotyping. BMC Bioinform 12:148

Hasanuzzaman M, Hossain MA, da Silva JAT, Fujita M (2012) Plant responses and tolerance to abiotic oxidative stress: antioxidant defenses is a key factor. In: Bandi V, Shanker AK, Shanker C, Mandapaka M (eds) Crop stress and its management: perspectives and strategies. Springer, Berlin, pp 261–316

Hasanuzzaman M, Nahar K, Alam MM, Roychowdhury R, Fujita M (2013) Physiological, biochemical, and molecular mechanisms of heat stress tolerance in plants. Int J Mol Sci 14:9643–9684. https://doi.org/10.3390/ijms14059643

Havaux M, Tardy F (1999) Loss of chlorophyll with limited reduction of photosynthesis as an adaptive response of Syrian barley landraces to high-light and heat stress. Funct Plant Biol 26:569–578

Hedhly A, Hormaza JI, Herrero M (2009) Global warming and sexual plant reproduction. Trends Plant Sci 14:30–36. https://doi.org/10.1016/j.tplants.2008.11.001

Heffner EL, Sorrells ME, Jannink JL (2009) Genomic selection for crop improvement. Crop Sci 49:1–12. https://doi.org/10.2135/crop-sci2008.08.0512

Hossain AB, Sears AG, Cox TS, Paulsen GM (1990) Desiccation tolerance and its relationship to assimilate partitioning in winter wheat. Crop Sci 30:622–627

Howarth CJ (2005) Genetic improvements of tolerance to high temperature. In: Ashraf M, Harris HJC (eds) Abiotic stresses—plant resistance through breeding and molecular approaches. The Haworth Press, New York, pp 277–300

Huang BR, Taylor HM, McMichael BL (1991) Growth and development of seminal and crown roots of wheat seedlings as affected by temperature. Environ Exp Bot 31:471–477

Huang B, Rachmilevitch S, Xu J (2012) Root carbon and protein metabolism associated with heat tolerance. J Exp Bot 63:3455–3465. https://doi.org/10.1093/jxb/ers003

Hussain HA, Men S, Hussain S, Chen Y, Ali S, Zhang S et al (2019) Interactive effects of drought and heat stresses on morphophysiological attributes, yield, nutrient uptake and oxidative status in maize hybrids. Sci Rep 9:3890. https://doi.org/10.1038/s41598-019-40362-7

Huve K, Bichele I, Tobias M, Niinemets U (2005) Heat sensitivity of photosynthetic electron transport varies during the day due to changes in sugars and osmotic potential. Plant Cell Environ 29:212–218

Iba K (2002) Acclimative response to temperature stress in higher plants: approaches of gene engineering for temperature tolerance. Annu Rev Plant Biol 53:225–245. https://doi.org/10.1146/annurev.arplant.53.100201.160729

Ibrahim AMH, Quick JS (2001) Genetic control of high temperature tolerance in wheat as measured by membrane thermal stability. Crop Sci 41(5):1405–1407

Iloh AC, Omatta G, Ogbadu GH, Onyenekwe PC (2014) Effects of elevated temperature on seed germination and seedling growth on three cereal crops in Nigeria. Sci Res Essays 9:806–813

IPCC (2014) Summary for policymakers. In: Stocker TF, Qin D, Plattner GK, Tignor M, Allen SK, Boschung J, Nauels A, Xia Y, Bex V, Midgley PM (eds) Climate change 2013: the physical science basis. Contribution of working group I to the fifth assessment report of the intergovernmental panel on climate change. Cambridge University Press, Cambridge

IPCC (Intergovernmental Panel on Climate Change) (2012) Managing the risks of extreme events and disasters to advance climate change adaptation. In Field CB, Barros V, Stocker TF, Qin D, Dokken DJ, Ebi KL, Mastrandrea MD, Mach KJ, Plattner GK, Allen SK, Tignor M, Midgley PM (eds.), A special report of working groups I and II of the intergovernmental panel on climate change. Cambridge University Press, Cambridge, pp. 582

Johkan M, Oda M, Maruo T, Shinohara Y (2011) Crop production and global warming. In: Casalegno S (ed) Global warming impacts-case studies on the economy, human health, and on urban and natural environments. InTech, Rijeka, pp 139–152

Jones RJ, Setter TL (2000) Hormonal regulation of early kernel development. In: Westgate M, Boote K (eds) Physiology and modeling kernel set in maize. Crop Science Society of America, Madison, WI, pp 25–42

Jones HG, Sirault XRR (2014) Scaling of thermal images at different spatial resolution: the mixed pixel problem. Agron 4:380–396

Jordan PW, Nobel PS (1984) Thermal and water relations of roots of desert succulents. Ann Bot 54:705–717

Kajla M, Yadav VK, Chhokar RS, Sharma RK (2015) Management practices to mitigate the impact of high temperature on wheat. J Wheat Res 7:1–12

Kamrani M, Hoseini Y, Ebadollahi A (2018) Evaluation for heat stress tolerance in durum wheat genotypes using stress tolerance indices. Arch Agron Soil Sci 64(1):38–45

Kanno K, Mae T, Makino A (2009) High night temperature stimulates photosynthesis, biomass production and growth during the vegetative stage of rice plants. Soil Sci Plant Nutr 55:124–131

Karim MA, Fracheboud Y, Stamp P (1997) Heat tolerance of maize with reference of some physiological characteristics. Ann Bangladesh Agric 7:27–33

Karim MA, Fracheboud Y, Stamp P (1999) Photosynthetic activity of developing leaves of *Zea mays* is less affected by heat stress than that of developed leaves. Physiol Plant 105:685–693. https://doi.org/10.1034/j.1399-3054.1999.105413.x

Kast EJ, Nguyen MDT, Lawrence RE, Rabeler C, Kaplinsky NJ (2013) The rootscope: a simple high-throughput screening system for quantitating gene expression dynamics in plant roots. BMC Plant Biol 13:158

Kawano T, Sahashi N, Takahashi K, Uozumi N, Muto S (1998) Salicylic acid induces extracellular superoxide generation followed by an increase in cytosolic calcium ion in tobacco suspension culture: the earliest events in salicylic acid signal transduction. Plant Cell Physiol 39:721–730

Khodarahmpour Z (2011) Genetic analysis of tolerance to heat stress in maize (*Zea mays* L.). Afr J Agric Res 6:2767–2773

Klopfenstein TJ, Erickson GE, Berger LL (2013) Maize is a critically important source of food, feed, energy and forage in the USA. Field Crops Res 153:5–11. https://doi.org/10.1016/j.fcr.2012.11.006

Kobza J, Edwards GE (1987) Influences of leaf temperature on photosynthetic carbon metabolism in wheat. Plant Physiol 83:69–74

Kosova K, Vítamvas P, Prasil IT, Renaut J (2011) Plant proteome changes under abiotic stress-contribution of proteomics studies to understanding plant stress response. J Proteome 74:1301–1322

Kumar U, Joshi AK, Kumari M, Paliwal R, Kumar S, Roder MS (2010) Identification of QTLs for stay green trait in wheat (*Triticum aestivum* L.) in the 'Chirya 3' x 'Sonalika' population. Euphytica 174:437–445. https://doi.org/10.1007/s10681-010-0155-6

Kumar S, Gupta D, Nayyar H (2012) Comparative response of maize and rice genotypes to heat stress: status of oxidative stress and antioxidants. Acta Physiol Plant 34:75–86

Kurahashi Y, Terashima A, Takumi S (2009) Variation in dehydration tolerance, ABA sensitivity and related gene expression patterns in D-genome progenitor and synthetic hexaploid wheat lines. Int J Mol Sci 10(6):2733–2751

Laghri KA, Mahboob AS, Arain MA (2012) Effect of high temperature stress on grain yield and yield components of wheat (*Triticum aestivum* L.). Sci Technol Dev 31:83–90

Larkindale J, Huang B (2005) Effects of abscisic acid, salicylic acid, ethylene and hydrogen peroxide in thermotolerance and recovery for creeping bentgrass. Plant Growth Regul 47:17–28

Lata C, Prasad M (2011) Role of DREBs in regulation of abiotic stress responses in plants. J Exp Bot 62:4731–4748

Lesk C, Rowhani P, Ramankutty N (2016) Influence of extreme weather disasters on global crop production. Nature 529:84–87. https://doi.org/10.1038/nature16467

Levitt J (1980) Response of plants to environmental stresses. Chilling, freezing and high temperature stresses, vol I. Academic Press, New York, p 497

Li Z, Peng T, Xie Q, Han S, Tian J (2010) Mapping of QTL for tiller number at different stages of growth in wheat using double haploid and immortalized F2 populations. J Genet 89:409–415. https://doi.org/10.1007/s12041-010-0059-1

Li YF, Wu Y, Hernandez-Espinosa N, Pena RJ (2013) Heat and drought stress on durum wheat: responses of genotypes, yield, and quality parameters. J Cereal Sci 57:398–404. https://doi.org/10.1016/j.jcs.2013.01.005

Lin CS, Binns M, Lefkovitch LP (1986) Stability analysis where do we stand? Crop Sci 26:894–900

Liu X, Huang B (2000) Carbohydrate accumulation in relation to heat stress tolerance in two creeping bentgrass cultivars. J Am Soc Hortic Sci 125:442–447

Lizana XC, Calderini DF (2013) Yield and grain quality of wheat in response to increased temperatures at key periods for grain number and grain weight determination: considerations for the climatic change scenarios of Chile. J Agric Sci 151:209–221

Lobell DB, Gourdji SM (2012) The influence of climate change on global crop productivity. Plant Physiol 160(4):1686–1697

Lobell DB, Banziger M, Magorokosho C, Vivek B (2011) Nonlinear heat effects on African maize as evidenced by historical yield trials. Nat Clim Chang 1:42–45

Lukac M, Gooding MJ, Griffiths S, Jones HE (2011) Asynchronous flowering and within plant flowering diversity in wheat and the implications for crop resilience to heat. Ann Bot 109:843–850. https://doi.org/10.1093/aob/mcr308

Macas B, Gomes MC, Dias AS, Coutinho J (2000) The tolerance of durum wheat to high temperatures during grain filling. In: Royo C, Nachit MM, Di Fonzo N, Araus JL (eds) Options Mediterraneennes. Durum wheat improvement in the Mediterranean region: new challenges. CIHEAM, Zaragoza, pp 257–261

Machado S, Paulsen GM (2001) Combined effects of drought and high temperature on water relations of wheat and sorghum. Plant Soil 233:179–187. https://doi.org/10.1023/A:1010346601643

Machado JC, de Souza MA, de Oliveira DM, Cargnin A, Pimentel AJB, de Assis JC (2010) Recurrent selection as breeding strategy for heat tolerance in wheat. Crop Breed Appl Biotechnol 10:9–15

Maestri E, Klueva N, Perrotta C, Gulli M, Nguyen HT, Marmiroli N (2002) Molecular genetics of heat tolerance and heat shock proteins in cereals. Plant Mol Biol 48:667–681. https://doi.org/10.1023/A:1014826730024

Marchand FL, Mertens S, Kockelbergh F, Beyens L, Nijs I (2005) Performance of high arctic tundra plants improved during but deteriorated after exposure to a simulated extreme temperature event. Glob Change Biol 11:2078–2089

Martinez CA, Bianconi M, Silva L, Approbato A, Lemos M, Santos L, Curtarelli L et al (2014) Moderate warming increases PSII performance, antioxidant scavenging systems and biomass production in *Stylosanthes capitata* Vogel. Environ Exp Bot 102:58–67

Martinez-Ballesta MC, Lopez-Perez L, Muries B, Munoz-Azcarate O, Carvajal M (2009) Climate change and plant water balance: the role of aquaporins a review. In: Lichtfouse E (ed) Climate change, intercropping, pest control and beneficial microorganisms. Springer, Dordrecht, pp 71–89. https://doi.org/10.1007/978-90-481-2716-05

Marutani Y, Yamauchi YKY, Mizutani M, Sugimoto Y (2012) Damage to photosystem II due to heat stress without light-driven electron flow: involvement of enhanced introduction of reducing power into thylakoid membranes. Planta 236:753–761

Mason RE, Mondal S, Beecher FW et al (2010) QTL associated with heat susceptibility index in wheat (*Triticum aestivum* L.) under short-term reproductive stage heat stress. Euphytica 174:423–436. https://doi.org/10.1007/s10681-010-0151-x

Mason RE, Mondal S, Beecher F, Hays D (2011) Genetic loci linking improved heat tolerance in wheat (*Triticum aestivum* L) to lower leaf and spike temperatures under controlled conditions. Euphytica 180:181–194

Massad RS, Tuzet A, Bethenod O (2007) The effect of temperature on C4-type leaf photosynthesis parameters. Plant Cell Environ 30:1191–1204

Mathur S, Jajoo A, Mehta P, Bharti S (2011) Analysis of elevated temperature-induced inhibition of photosystem II using chlorophyll a fluorescence induction kinetics in wheat leaves (*Triticum aestivum*). Plant Biol 13:1–6. https://doi.org/10.1111/j.1438-8677.2009.00319.x

McClung CR, Davis SJ (2010) Ambient thermometers in plants: from physiological outputs towards mechanisms of thermal sensing. Curr Biol 20:1086–1092

Mickelbart MV, Hasegawa PM, Bailey-Serres J (2015) Genetic mechanisms of abiotic stress tolerance that translate to crop yield stability. Nat Rev Genet 16:237–251

Miller G, Schlauch K, Tam R, Cortes D et al (2009) The plant NADPH oxidase RbohD mediates rapid, systemic signaling in response to diverse stimuli. Sci Signal 2(84):ra45

Mittler R, Vanderauwera S, Suzuki N, Miller G, Tognetti VB, Vandepoele K, Gollery M, Shulaev V, Van Breusegem F (2011) ROS signaling: the new wave? Trends Plant Sci 16:300–309

Mohammed YSA, Tahir ISA, Kamal NM, Eltayeb AE, Ali AM, Kamal NM (2014) Impact of wheat-*Leymus racemosus* added chromosomes on wheat adaptation and tolerance to heat stress. Breed Sci 3:450–460

Momcilovic I, Ristic Z (2007) Expression of chloroplast protein synthesis elongation factor, EF-Tu, in two lines of maize with contrasting tolerance to heat stress during early stages of plant development. J Plant Physiol 164:90–99. https://doi.org/10.1016/j.jplph.2006.01.010

Monje O, Anderson S, Stutte GW (2007) The effects of elevated root zone temperature on the development and carbon partitioning of spring wheat. J Am Soc Hortic Sci 132(2):178–184

Murata N, Takahashi S, Nishiyama Y, Allakhverdiev SI (2007) Photoinhibition of photosystem II under environmental stress. Biochim Biophys Acta Bioenerg 1767(6):414–421. https://doi.org/10.1016/j.bbabio.2006.11.019

Nagel KA, Kastenholz B, Jahnke S, Dusschoten DV et al (2009) Temperature responses of roots: impact on growth, root system architecture and implications for phenotyping. Funct Plant Biol 36:947–959

Naidu SL, Moose SP, AL-Shoaibi AK, Raines CA, Long SP (2003) Cold tolerance of C4 photosynthesis in *Miscanthus giganteus*: adaptation in amounts and sequence of C4 photosynthetic enzymes. Plant Physiol 132:1688–1697. https://doi.org/10.1104/pp.103.021790

Narayanan S, Prasad PV, Welti R (2016) Wheat leaf lipids during heat stress: II. Lipids experiencing coordinated metabolism are detected by analysis of lipid co-occurrence. Plant Cell Environ 39:608–617

Narayanan S, Prasad PVV, Welti R (2018) Alterations in wheat pollen lipidome during high day and night temperature stress. Plant Cell Environ 41:1749–1761

Nawaz A, Farooq M, Cheema SA, Wahid A (2013) Differential response of wheat cultivars to terminal heat stress. Int J Agric Biol 15:1354–1358

Nicolas ME, Gleadow RM, Dalling MJ (1984) Effects of drought and high temperature on grain growth in wheat. Aust J Plant Physiol 11:553–566

Ortiz R, Sayre KD, Govaerts B, Gupta R, Subbarao GV, Ban T, Hodson D, Dixon JM, Ortiz-Monasterio JI, Reynolds M (2008) Climate change: can wheat beat the heat? Agric Ecosyst Environ 126:46–58. https://doi.org/10.1016/j.agee.2008.01.019

Ottaviano E, Sari-Gorla M, Pe E, Frova C (1991) Molecular markers (RFLPs and HSPs) for the genetic dissection of thermo tolerance in maize. Theor Appl Genet 81:713–719

Ottaviano E, Mulcahy DL, Sari Gorla M, Bergamini Mulcahy G (1992) Angiosperm pollen and ovules. Springer-Verlag, New York, pp 355–358

Ottman MJ, Kimball BA, White JW, Wall GW (2012) Wheat growth response to increased temperature from varied planting dates and supplemental infrared heating. Agron J 104:7–16

Paliwal R, Roder MS, Kumar U, Srivastava JP, Joshi AK (2012) QTL mapping of terminal heat tolerance in hexaploid wheat (*T. aestivum* L.). Theor Appl Genet 125:561–575. https://doi.org/10.1007/s00122-012-1853-3

Parmar N, Singh KH, Sharma D, Singh L, Kumar P, Nanjundan J et al (2017) Genetic engineering strategies for biotic and abiotic stress tolerance and quality enhancement in horticultural crops: a comprehensive review. 3 Biotech 7:239. https://doi.org/10.1007/s13205-017-0870-y

Parry MAJ, Reynolds M, Salvucci ME et al (2011) Raising yield potential of wheat. II. Increasing photosynthetic capacity and efficiency. J Exp Bot 62(2):453–467. https://doi.org/10.1093/jxb/erq304

Paulsen GM (1994) High temperature responses of crop plants. In: Boote KJ, Bennett JM, Sinclair TR, Paulsen GM (eds) Physiology and determination of crop yield. ASA, CSSA, and SSSA, Madison, WI, pp 365–389

Peng S, Huang J, Sheehy JE, Laza RC et al (2004) Rice yields decline with higher night temperature from global warming. Proc Natl Acad Sci U S A 101:9971–9975

Perego A, Sanna M, Giussani A, Chiodini ME, Fumagalli M, Pilu SR et al (2014) Designing a high yielding maize ideotype for a changing climate in Lombardy plain (northern Italy). Sci Total Environ 499:497–509. https://doi.org/10.1016/j.scitotenv.2014.05.092

Pinto RS, Reynolds MP, Mathews KL, McIntyre CL, Olivares-Villegas JJ, Chapman SC (2010) Heat and drought adaptive QTL in a wheat population designed to minimize confounding agronomic effects. Theor Appl Genet 121:1001–1021. https://doi.org/10.1007/s00122-010-1351-4

Piramila BHM, Prabha AL, Nandagopalan V, Stanley AL (2012) Effect of heat treatment on germination, seedling growth and some biochemical parameters of dry seeds of black gram. Int J Pharm Phytopharmacol Res 1:194–202

Porter JR (2005) Rising temperatures are likely to reduce crop yields. Nature 436:174. https://doi.org/10.1038/436174b

Porter JR, Gawith M (1999) Temperatures and the growth and development of wheat: a review. Eur J Agron 10:23–36

Pradhan GP, Prasad PVV, Fritz AK, Kirkham MB, Gill BS (2012) High temperature tolerance in *Aegilops* species and its potential transfer to wheat. Crop Sci 52:292–304

Prasad PVV, Djanaguiraman M (2014) Response of floret fertility and individual grain weight of wheat to high temperature stress: sensitive stages and thresholds for temperature and duration. Funct Plant Biol 41:1261–1269. https://doi.org/10.1071/FP14061

Prasad PVV, Boote KJ, Vu JCV, Allen LH (2004) The carbohydrate metabolism enzymes sucrose-P synthase and ADG-pyrophosphorylase in phaseolus bean leaves are up-regulated at elevated growth carbon dioxide and temperature. Plant Sci 166:1565–1573. https://doi.org/10.1016/j.plantsci.2004.02.009

Prasad PVV, Pisipati SR, Ristic Z et al (2008a) Impact of nighttime temperature on physiology and growth of spring wheat. Crop Sci 48(6):2372–2380

Prasad PVV, Staggenborg SA, Ristic Z (2008b) Impacts of drought and/or heat stress on physiological, developmental, growth, and yield processes of crop plants. In: Ahuja LH, Saseendran SA (eds) Response of crops to limited water: understanding and modeling water stress effects on plant growth processes. ASA, CSSA, Madison, WI, pp 301–355

Qi Y, Wang H, Zou Y, Liu C, Liu Y, Wang Y, Zhang W (2010) Over-expression of mitochondrial heat shock protein 70 suppresses programmed cell death in rice. FEBS Lett 585:231–239

Qi X, Xu W, Zhang J, Rui G, Zhao M, Lin H, Wang H, Dong H, Yan L (2017) Physiological characteristics and metabolomics of transgenic wheat containing the maize C4 phosphoenolpyruvate carboxylase (PEPC) gene under high temperature stress. Protoplasma 254:1017–1030

Queitsch C, Hong SW, Vierling E, Lindquist S (2000) Hsp101 plays a crucial role in thermotolerance in *Arabidopsis*. Plant Cell 12:479–492

Rahman MA, Chikushi J, Yoshida S, Karim AJMS (2009) Growth and yield components of wheat genotypes exposed to high temperature stress under control environment. Bangladesh J Agric Res 34:361–372

Raines CA (2011) Increasing photosynthetic carbon assimilation in C3 plants to improve crop yield: current and future strategies. Plant Physiol 155:36–42. https://doi.org/10.1104/pp.110.168559

Rajaram S, Mujeeb-Kazi A, Singh RP (1977) Wheat (*Triticum asetivum*). In: Bahl PN (ed) Genetics, cytogenetics and breeding of crop plants. Sci Publ, Enfield, NH, pp 77–144

Ramadas S, Kiran Kumar TM, Singh GP (2019) Wheat production in India: trends and prospects recent advances in grain crops research. In: Shah F (ed) Recent advances in grain crops research publisher. Publisher Intech Open. https://doi.org/10.5772/intechopen.86341

Ramirez P, Kelly JD (1998) Traits related to drought resistance in common bean. Euphytica 99:127–136

Refay YA (2011) Yield and yield component parameters of bread wheat genotypes as affected by sowing dates: middle-east. J Sci Res 7:484–489

Reynolds M, Langridge P (2016) Physiological breeding. Curr Opin Plant Biol 31:162–171. https://doi.org/10.1016/j.pbi.2016.04.005

Reynolds MP, Balota M, Delgado MIB, Amani I, Fischer RA (1994) Physiological and morphological traits associated with spring wheat yield under hot, irrigated conditions. Aust J Plant Physiol 21:717–730

Richards RA (1996) Defining selection criteria to improve yield under drought. Plant Growth Regul 20:157–166. https://doi.org/10.1007/BF00024012

Ristic Z, Bukovnik U, Prasad PVV (2007) Correlation between heat stability of thylakoid membranes and loss of chlorophyll in winter wheat under heat stress. Crop Sci 47:2067–2073

Ristic Z, Bukovnik U, Momcilovic I, Fu J, Prasad PVV (2008) Heat induced accumulation of chloroplast protein synthesis elongation factor, EF-Tu, in winter wheat. J Plant Physiol 165:192–202. https://doi.org/10.1016/j.jplph.2007.03.003

Rosielle AA, Hamblin J (1981) Theoretical aspects of selection for yield in stress and non-stress environments. Crop Sci 21:943–946

Ruelland E, Zachowski A (2010) How plants sense temperature. Environ Exp Bot 69:225–232. https://doi.org/10.1016/j.envexpbot.2010.05.011

Ruggiero A, Punzo P, Landi S, Costa A, Van Ooosten M, Grillo S (2017) Improving plant water use efficiency through molecular genetics. Horticulture 3:31. https://doi.org/10.3390/horticulturae3020031

Saini HS, Sedgley M, Aspinall D (1983) Effect of high temperature stress during floral development on pollen tube growth and ovary anatomy in wheat (*Triticum aestivum* L.). Aust J Plant Physiol 10(2):137–144

Saini HS, Sedgley M, Aspinall D (1984) Developmental anatomy in wheat of male sterility induced by high temperature stress, water deficit or abscisic acid. Aust J Plant Physiol 11(4):243–253

Sairam RK, Tyagi A (2004) Physiology and molecular biology of salinity stress tolerance in plants. Curr Sci 86:407–421

Sanchez B, Rasmussen A, Porter JR (2014) Temperatures and the growth and development of maize and rice: a review. Glob Change Biol 20:408–417. https://doi.org/10.1111/gcb.12389

Sanguineti MC, Tuberosa R, Landi P, Salvi S, Maccaferri M, Casarini E, Conti S (1999) QTL analysis of drought-related traits and grain yield in relation to genetic variation for leaf abscisic acid concentration in field-grown maize. J Exp Bot 50(337):1289–1297. https://doi.org/10.1093/jxb/50.337.1289

Savicka M, Skute N (2010) Effects of high temperature on malondialdehyde content, superoxide production and growth changes in wheat seedlings (*Triticum aestivum* L.). Ekologija 56:26–33. https://doi.org/10.2478/v10055-010-0004-x

Savicka M, Skute N (2012) Some morphological, physiological and biochemical characteristics of wheat seedling *Triticum aestivum* L. organs after high temperature treatment. Ekologija 58:9–21

Schachtman D, Goodger J (2008) Chemical root to shoot signaling under drought. Trends Plant Sci 13:281–287. https://doi.org/10.1016/j.tplants.2008.04.003

Schittenhelm S, Langkamp-Wedde T, Kraft M, Kottmann L, Matschiner K (2020) Effect of two-week heat stress during grain filling on stem reserves, senescence, and grain yield of European winter wheat cultivars. J Agron Crop Sci 00:1–12. https://doi.org/10.1111/jac.12410

Schoffl F, Prandl R, Reindl A (1999) Molecular responses to heat stress. In: Shinozaki K, Yamaguchi-Shinozaki K (eds) Molecular responses to cold, drought, heat and salt stress in higher plants. RG Landes, Austin, TX, pp 81–98

Schuster WS, Monson RK (1990) An examination of the advantages of C3-C4 intermediate photosynthesis in warm environments. Plant Cell Environ 13:903–912. https://doi.org/10.1111/pce.1990.13.issue-9

Semenov MA (2009) Impacts of climate change on wheat in England and Wales. J Royal Soc Interface 6:343–350

Shah NH, Paulsen GM (2003) Interaction of drought and high temperature on photosynthesis and grain-filling of wheat. Plant Soil 257:219–226. https://doi.org/10.1023/A:1026237816578

Shah F, Huang J, Cui K, Nie L, Shah T, Chen C, Wang K (2011) Impact of high-temperature stress on rice plant and its traits related to tolerance. J Agric Sci 149:545–556

Sharma P, Dubey RS (2005) Drought induces oxidative stress and enhances the activities of antioxidant enzymes in growing rice seedlings. Plant Growth Regul 46:209–221

Sharma DK, Andersen SB, Ottosen CO, Rosenqvist E (2015) Wheat cultivars selected for high Fv/Fm under heat stress maintain high photosynthesis, total chlorophyll, stomatal conductance, transpiration and dry matter. Physiol Plant 153:284–298. https://doi.org/10.1111/ppl.12245

Sheehy JE, Mabilangan AE, Dionora KJA et al (2007) Time of day of flowering in wild species of the genus *Oryza*. Int Rice Res Notes 32:12–13

Shiferaw B, Smale M, Braun HJ, Duveiller E, Reynolds M, Muricho G (2013) Crops that feed the world 10. Past successes and future challenges to the role played by wheat in global food security. Food Secur 5:291–317
Siddiqui KA (1976) Synthetic amphyploids in breeding-genetic and evolutionary studies in wheat. In: Muhammed A et al (eds) Genetic diversity in plants. Plenum Press, New York, pp 97–102
Singh S (2009) Variation in physiological traits for thermotolerance in wheat. Indian J Plant Physiol 14:407–412
Singh A, Singh D, Kang JS, Aggarwal N (2011) Management practices to mitigate the impact of high temperature on wheat: a review. IIOABJ 2:11–22
Smertenko A, Draber P, Viklicky V, Opatrny Z (1997) Heat stress affects the organization of microtubules and cell division in Nicotiana tabacum cells. Plant Cell Environ 20:1534–1542. https://doi.org/10.1046/j.1365-3040.1997.d01-44.x
Smith KL (1996) Ohio agronomy guide for corn production. Ohio State University, Columbus, GA
Soliman WS, Fujimori M, Tase K, Sugiyama S (2011) Oxidative stress and physiological damage under prolonged heat stress in C3 grass *Lolium perenne*. Grassl Sci 57:101–106. https://doi.org/10.1111/grs.2011.57.issue-2
Stone P (2001) The effects of heat stress on cereal yield and quality. In: Basra AS (ed) Crop responses and adaptations to temperature stress. Food Products Press, Binghamton, NY, pp 243–291
Suzuki N, Miller G, Morales J, Shulaev V, Torres MA (2011) Respiratory burst oxidases: the engines of ROS signaling. Curr Opin Plant Biol 14:691–699
Suzuki N, Koussevitzky S, Mittler R, Miller G (2012) ROS and redox signalling in the response of plants to abiotic stress. Plant Cell Environ 35:259–270
Taghizadeh R, Shrifi RS (2010) Effects of post anthesis heat stress on yield and yield attributes in wheat (*Triticum aestivum*) genotypes. J Food Agric Environ 8:775–777
Talukder SK, Babar MA, Vijaylakshmi K, Poland J, Prasad PVV, Bowden R, Fritz A (2014) Mapping QTL for the traits associated with heat tolerance in wheat (*Triticum aestivum* L.). BMC Genet 15:97. https://doi.org/10.1186/s12863-014-0097-4
Tao ZQ, Chen YQ, LI C et al (2016) The causes and impacts for heat stress in spring maize during grain filling in the North China plain - a review. J Integr Agric 15(12):2677–2687
Tester M, Langridge P (2010) Breeding technologies to increase crop production in a changing world. Science 327:818–822
Tewolde H, Fernandez CJ, Erickson CA (2006) Wheat cultivars adapted to post-heading high temperature stress. J Agron Crop Sci 192(2):111–120
Tiwari YK, Yadav SK (2019) High temperature stress tolerance in maize (*Zea mays* L.): physiological and molecular mechanisms. J Plant Biol 62:93–102. https://doi.org/10.1007/s12374-018-0350-x
Trethowan RM, Mujeeb-Kazi A (2008) Novel germplasm resources for improving environmental stress tolerance of hexaploid wheat. Crop Sci 48(4):1255–1265
Valluru R, Reynolds MP, Davies WJ, Sukumaran S (2017) Phenotypic and genome-wide association analysis of spike ethylene in diverse wheat genotypes under heat stress. New Phytol 214:271–283
Vijayalakshmi K, Fritz AK, Paulsen GM, Bai G, Pandravada S, Gill BS (2010) Modeling and mapping QTL for senescence-related traits in winter wheat under high temperature. Mol Breed 26:163–175. https://doi.org/10.1007/s11032-009-9366-8
Viswanathan C, Khanna-Chopra R (2001) Effect of heat stress on grain growth, starch synthesis and protein synthesis in grains of wheat (*Triticum aestivum* L.) varieties differing in grain weight stability. J Agron Crop Sci 186:1–7
Wahid A, Gelani S, Ashraf M, Foolad M (2007) Heat tolerance in plants: an overview. Environ Exp Bot 61:199–223
Wahid A, Sehar S, Perveen M, Gelani S, Basra SMA, Farooq M (2008) Seed pretreatment with hydrogen peroxide improves heat tolerance in maize at germination and seedling growth stages. Seed Sci Technol 36:633–645. https://doi.org/10.15258/sst.2008.36.3.13

Walker-Simmons M, Sesing J (1990) Temperature effects on embryonic abscisic acid levels during development of wheat grain dormancy. J Plant Growth Regul 9:51–56
Wang LJ, Li SL (2006) Salicylic acid-induced heat or cold tolerance in relation to Ca2+ homeostasis and antioxidant systems in young grape plants. Plant Sci 170:685–694
Wang JZ, Cui LJ, Wang Y, Li JL (2009) Growth, lipid peroxidation and photosynthesis in two tall fescue cultivars differing in heat tolerance. Biol Plant 53:247–242
Wang LJ, Fan L, Loescher W, Duan W, Liu GJ et al (2010) Salicylic acid alleviates decreases in photosynthesis under heat stress and accelerates recovery in grapevine leaves. BMC Plant Biol 10:34. https://doi.org/10.1186/1471-2229-10-34
Wang X, Cai J, Liu F, Dai T, Cao W, Wollenweber B, Jiang D (2014) Multiple heat priming enhances thermo-tolerance to a later high temperature stress via improving subcellular antioxidant activities in wheat seedlings. Plant Physiol Biochem 74:185–192
Waqas MA, Khan I, Akhter MJ, Noor MA, Ashraf U (2017) Exogenous application of plant growth regulators (PGRs) induces chilling tolerance in short-duration hybrid maize. Environ Sci Pollut Res 24:11459–11471
Wardlaw IF, Blumenthal C, Larroque O, Wrigley CW (2002) Contrasting effects of chronic heat stress and heat shock on kernel weight and flour quality in wheat. Funct Plant Biol 29:25–34. https://doi.org/10.1071/PP00147
Weaich K, Briston KL, Cass A (1996) Modeling preemergent maize shoot growth. II. High temperature stress conditions. Agron J 88:398–403
Wise RR, Olson AJ, Schrader SM, Sharkey TD (2004) Electron transport is the functional limitation of photosynthesis in field-grown pima cotton plants at high temperature. Plant Cell Environ 27:717–724. https://doi.org/10.1111/j.1365-3040.2004.01171.x
Wollenweber B, Porter JR, Schellberg J (2003) Lack of interaction between extreme high temperature events at vegetative and reproductive growth stages in wheat. J Agron Crop Sci 189:142–150
Xu Q, Paulsen AQ, Guikema JA et al (1995) Functional and ultrastructural injury to photosynthesis in wheat by high temperature during maturation. Environ Exp Bot 35(1):43–54
Xu S, Li J, Zhang X, Wei H, Cui L (2006) Effects of heat acclimation pretreatment on changes of membrane lipid peroxidation, antioxidant metabolites, and ultrastructure of chloroplasts in two cool-season turfgrass species under heat stress. Environ Exp Bot 56:274–285. https://doi.org/10.1016/j.envexpbot.2005.03.002
Yamamoto Y, Aminaka R, Yoshioka M, Khatoon M, Komayama K, Takenaka D, Yamashita A, Nijo N, Inagawa K, Morita N, Sasaki T, Yamamoto Y (2008) Quality control of photosystem II: impact of light and heat stresses. Photosynth Res 98:589–608
Yang J, Sears RG, Gill BS et al (2002) Genotypic differences in utilization of assimilate sources during maturation of wheat under chronic heat and heat shock stresses. Euphytica 125 (2):179–188
Yang X, Chen X, Ge Q, Li B, Tong Y, Zhang A, Li Z, Kuang T, Lu C (2006) Tolerance of photosynthesis to photoinhibition, high temperature and drought stress in flag leaves of wheat: a comparison between a hybridization line and its parents grown under field conditions. Plant Sci 171:389–397
Yang X, Tang X, Chen B, Tian Z, Zhong H (2013) Impacts of heat stress on wheat yield due to climatic warming in China. Prog Geogr 32:1771–1779
Yang DQ, Wang ZL, Ni YL, Yin YP et al (2014) Effect of high temperature stress and spraying exogenous ABA post-anthesis on grain filling and grain yield in different types of stay-green wheat. Sci Agric Sin 47:2109–2125
Yin XY, Guo W, Spiertz JH (2009) A quantitative approach to characterize sink-source relationships during grain filling in contrasting wheat genotypes. Field Crops Res 114:119–126. https://doi.org/10.1016/j.fcr.2009.07.013
Yu Q, Li L, Luo Q, Eamus D, Xu S, Chen C, Wang E, Liu J, Nielsen DC (2014) Year patterns of climate impact on wheat yields. Int J Climatol 34:518–528. https://doi.org/10.1002/joc.3704

Yun-Ying C, Hua D, Li-Nian Y, Zhi-Qing W, Shao-Chuan Z, Jian-Chang Y (2008) Effect of heat stress during meiosis on grain yield of rice cultivars differing in heat tolerance and its physiological mechanism. Acta Agron Sin 34(12):2134–2214

Zaharieva M, Gaulin E, Havaux M et al (2001) Drought and heat responses in the wild wheat relative *Aegilops geniculata* Roth: potential interest for wheat improvement. Crop Sci 41 (4):1321–1329

Zaidi PH, Zaman-Allah M, Trachsel S, Seetharam K, Cairns JE, Vinayan MT (2016) Phenotyping for abiotic stress tolerance in maize – heat stress. A field manual. CIMMYT, Hyderabad

Zandalinas SI, Rivero RM, Martínez V, Gomez-Cadenas A, Arbona V (2016) Tolerance of citrus plants to the combination of high temperatures and drought is associated to the increase in transpiration modulated by a reduction in abscisic acid levels. BMC Plant Biol 16:105. https://doi.org/10.1186/s12870-016-0791-7

Zhang M, Barg R, Yin M, Gueta-Dahan Y et al (2005) Modulated fatty acid desaturation via overexpression of two distinct omega-3 desaturases differentially alters tolerance to various abiotic stresses in transgenic tobacco cells and plants. Plant J 44:361–371

Zhao C, Liu B, Piao S, Wang X, Lobell DB, Huang Y et al (2017) Temperature increase reduces global yields of major crops in four independent estimates. Proc Natl Acad Sci U S A 114:9326–9331. https://doi.org/10.1073/pnas.1701762114

Zheng B, Chenu K, Dreccer MF, Chapman SC (2012) Breeding for the future: what are the potential impacts of future frost and heat events on sowing and flowering time requirements for Australian bread wheat (*Triticum aestivum*) varieties? Glob Chang Biol 18:2899–2914. https://doi.org/10.1111/j.1365-2486.2012.02724.x

Breeding and Molecular Approaches for Evolving Drought-Tolerant Soybeans

4

Gyanesh Kumar Satpute, Milind B. Ratnaparkhe, Subhash Chandra, Viraj Gangadhar Kamble, Rucha Kavishwar, Ajay Kumar Singh, Sanjay Gupta, Ramgopal Devdas, Mamta Arya, Maharaj Singh, Mahaveer Prasad Sharma, Giriraj Kumawat, M. Shivakumar, Vennampally Nataraj, Mrinal K. Kuchlan, Vangala Rajesh, Manoj Kumar Srivastava, Annapurna Chitikineni, Rajeev K. Varshney, and Henry T. Nguyen

Abstract

Soybean [*Glycine max* (L.) Merr] is an agronomically important oilseed crop in the world and an important source of protein and oil for both humans and animals. In addition, soybean is also becoming a major crop for bio-diesel production. Therefore, demand for soybean is increasing continuously worldwide. Soybean enriches the soil by fixing atmospheric nitrogen through symbiotic interaction with Rhizobia. With increasing challenges posed by climate change, it is predicted that incidents of drought will be more frequent and severe and it will further reduce crop yields. Abiotic stresses such as drought cause severe losses to soybean productivity worldwide by adversely affecting the plant growth, development, and yield. Introgression of genes controlling drought adaptive traits,

G. K. Satpute (✉) · M. B. Ratnaparkhe · S. Chandra · V. G. Kamble · R. Kavishwar · S. Gupta · M. Singh · M. P. Sharma · G. Kumawat · M. Shivakumar · V. Nataraj · M. K. Kuchlan · V. Rajesh · M. K. Srivastava
ICAR-Indian Institute of Soybean Research, Indore, Madhya Pradesh, India

A. K. Singh
ICAR-National Institute of Abiotic Stress Management, Baramati, India

R. Devdas
ICAR-National Research Centre for Orchids, Pakyong, India

M. Arya
ICAR-NBPGR Regional Station, Bhowali, Nainital, India

A. Chitikineni · R. K. Varshney
Center of Excellence in Genomics and Systems Biology, ICRISAT, Patancheru, India

H. T. Nguyen
Division of Plant Science, University of Missouri, Columbia, MO, USA

B. Giri, M. P. Sharma (eds.), *Plant Stress Biology*,
https://doi.org/10.1007/978-981-15-9380-2_4

yields related traits and root system architecture traits by breeding and molecular approaches will be very useful for enhancing drought stress tolerance in soybean, leading to cultivar development. Elucidation of function of genes and their integration in soybean genotypes by molecular breeding and genomic approaches and utilizing robust phenotyping tools to evaluate drought adaptive traits will be crucial for understanding response of soybean plants to drought stress. Recent advances in genomics lead identification, functional characterization, and introgression of genes associated with adaptation of soybean plants to drought stress. In order to perform genetic and genomic analysis, molecular markers have been employed on RIL or F_2 populations. In addition, the genome typified with single nucleotide polymorphisms (SNPs) and its utilization in molecular breeding applications like QTL mapping, positional cloning, association mapping studies, genomic selection and genome editing is gaining impetus. Thus, the rapid development of soybean genomics and transcriptomics has provided tremendous opportunity for the genetic improvement of soybean for drought tolerance with yield stability.

Keywords

Drought tolerance · Abiotic stress · Quantitative trait loci · Breeding · Genetic engineering · Signal transduction · Transcriptomic approaches

4.1 Introduction

Soybean is one of the most important crops in the world and provides oil and protein for both humans and livestock. Global food security is a major challenging task for world agricultural community since the world population is growing exponentially and crop cultivable land is decreasing due to adverse climatic conditions (Foley et al. 2011). The USA, Brazil, Argentina, China, and India account for about 93% of global soybean production. It is one of the most economical sources of good quality protein (40%), edible oil (20%), essential amino acids, dietary minerals, vitamins, and nutraceuticals like isoflavones, tocopherols, etc. of immense health benefits. Such diverse uses of soybean make it a wildly desired crop, and demand for soybean is rapidly increasing (Ray et al. 2013; Deshmukh et al. 2014). However, soybean yield is threatened by various abiotic stresses mainly drought (Phang et al. 2008; Manavalan et al. 2009). Adverse environmental factor mainly drought leads to morphological, physiological, biochemical, and molecular changes that adversely affect plant growth and productivity (Wang et al. 2001). Thus, understanding responses of soybean plants to drought stress and enhancing abiotic stress resilience to maintain genetic yield potential are extremely demanding areas in agricultural research. To improve drought stress tolerance in soybean, a wide range of approaches, including gene discovery, QTL mapping, genome-wide association studies (GWAS), and biotechnologcal approaches such as genomic selection and gene editing can be used to develop soybean varieties with improved drought tolerance (Deshmukh et al. 2014; Phang et al. 2008; Manavalan et al. 2009).

Climate models have indicated that drought episodes will become more frequent because of the long-term effects of global warming (Salinger et al. 2005; Cook et al. 2007) and may significantly affect soybean yield in many countries (Long et al. 2005; Easterling et al. 2007). While increase in CO2 under climate change might be expected to increase soybean productivity, vicious effects of frequent droughts and associated diseases and herbivores infestation may counteract such benefits. Drought is considered one of the most devastating among abiotic stress factors (Manavalan et al. 2009; Tran and Mochida 2010) reducing about 40% soybean yield annually (Specht et al. 1999) and depending upon the intensity of its occurrence at critical growth stage, the losses could be as high as 80% (Oya et al. 2004; Dias et al. 2012). Efficient resource acquisition and remobilization are challenges for soybean productivity under water-limited scenario. The hydrostatic pressure created by transpiration from the shoot is transmitted to the xylem vessels of the shoot and the roots, which drives water in the root cylinder toward the xylem vessels (Steudle 1995; Tyree 1997) and are governed by hydraulic mechanisms such as leaf conductance (Sinclair et al. 2010), leaf canopy size (Ratnakumar and Vadez 2011; Vadez et al. 2011), control of leaf expansion (Simonneau et al. 2009), and transpiration response in soybean to high vapor pressure deficit (VPD) (Sinclair et al. 2008; Ocheltree et al. 2014). In soybean, vegetative growth is sensitive to water deficits. Besides, usual inhibitory effects on leaf expansion, transpiration and photosynthesis, water deficit also inhibit nitrogen fixation in soybean. Water stress, occurring during the beginning of pod setting and full seed-filling, has a greater negative irreversible impact on yield through reducing seed size considerably (Doss and Thurlow 1974), as compared to other stages, and thus invite designing of breeding strategy targeting the final expression of yield under drought, i.e. seed size by integrating constitutive plant traits and stress-responsive processes (Blum 2011). A drought resistance index in terms of yield can be developed by comparing yield between stress and non-stress conditions, which is gaining popularity as a useful criterion in selection for drought resistance (Fukai et al. 1999). Drought resistance is measured by phenotyping the specific and relevant attributes of dehydration tolerance and dehydration avoidance, where a concrete breeding program integrates robust, reliable, relatively fast, and economical phenotyping facilities. Dehydration tolerance, phenotyped only on the basis of similar plant water status in all the genotypes, includes the most prominent feature of whole plant dehydration tolerance assessing the capacity for stem reserve utilization for seed filling by chemical desiccation method. Protocols for dehydration avoidance include measuring plant water status, in terms of visual symptoms of leaf senescence, relative water content, and constitutive traits without exposure to drought stress such as root system architecture traits. (Blum 2011).

An important component of reproductive success of the crop under drought stress is the capacity for seed filling from stem reserve, when transient photosynthesis is inhibited by stress. This is a dehydration tolerance mechanism since the transport of reserves from stem to seed takes place in dehydrated plants, in the case of severe drought in the field. It can be phenotyped in large populations by the chemical desiccation method. The method was developed by Blum et al. (1983a, b) as a fast and relatively simple field assay for revealing the capacity for seed filling from stem reserves. The method is based on the application of a chemical desiccant to plant canopies after flowering as a means for inhibiting plant photosynthesis and thus revealing the capacity for seed filling by stem reserves. The treatment does not simulate drought stress. It simulates the effect of stress by inhibiting current assimilation. With this method a chemical desiccant (potassium iodide; 0.2% w/v) is sprayed to complete wetting over the whole canopy, at seed filling stage (R5 plus 8–10 days) in soybean (Bhatia et al. 2014), which mainly destroys chlorophyll and simulates natural senescence. Chemical desiccation can be incorporated into breeding programs in two ways: (1) it can be used to assess responses of individual advanced lines or families, always compared with non-treated controls under non-stress conditions and (2) the method can be used for early generation advancement through mass selection where F_2 bulks are chemically desiccated and selections are made for seed size divergently by mechanical sieving. After two cycles of early generation selection, vigorous lines were selected and tested for their response to chemical desiccation stress. Mass selection for large grains under chemical desiccation significantly improved grain weight and grain yield of the population under desiccation stress, as compared to control where selection for grain size was performed without chemical desiccation (Blum et al. 1991; Haley and Quick 1993; Annual Report 2019).

Physiological processes such as delayed leaf senescence, water status, and canopy temperature are very crucial for drought stress tolerance capability in plants. Delayed leaf senescence in a flowering plant induces extreme drought tolerance and evidenced by suppression of drought-induced leaf senescence in transgenic tobacco plants expressing *isopentenyltransferase* (IPT), an enzyme that catalyzes the rate-limiting step in cytokinin synthesis, resulted in outstanding drought tolerance as shown by vigorous growth after a long drought period, among other responses of high water contents, retained photosynthetic activity albeit at a reduced level, and displayed minimal yield loss when watered with only 30% of the amount of water used under control conditions during the drought (Rivero et al. 2007). A "slow-wilting" line has been recognized in soybean (Fletcher et al. 2007). The visual scoring of delayed leaf senescence of a given genotype must be based on an integrated impression of the symptom in the whole plant or even the whole plot. Scoring is performed on 1–5 arbitrary rating scale with 1 being sensitive with dried leaves chaffy matured pods and 5 being delayed leaf senescence with well-filled matured pods and green leaves. Very small variations in leaf senescence score, even if they are statistically significant, are of no real consequence in breeding for dehydration avoidance and large and prominent differences are sought (Blum 2011).

Leaf relative water content (RWC) is a simple, standard, and effective estimate and a reliable indicator of water status in plants (Sinclair and Ludlow 1985) with respect to dehydration avoidance. Usually the top-most fully expanded sun-lit leaf must be sampled to determine leaf RWC as per Blum (2011). Boyer et al. (2008) cautioned against excessive rehydration of samples which can result in excessive absorption of water by the leaf sample, beyond its normal full turgor capacity. This would bias estimated RWC downward. In order to control the severity and timing in the field, drought stress is affected by stopping water supply, be it by terminating irrigation or by activating the rainout shelter. Stress will then develop gradually and it is crucial to be able to translate the number of days without watering into the desired level of plant stress. When grown on deep soil of good water holding capacity soybean may take around 2–3 weeks to reach midday RWC of about 60–70%. In phenotyping and selection work, it is not absolutely necessary to measure all relevant atmospheric and soil variables, in order to estimate daily crop water-use, and thus to predict the timing and rate of the planned imposed drought, where one can access the plant and estimate directly or indirectly its water status with only a minimal reference to environmental variables. Since drought phenotyping is usually repeated in the same location during the course of breeding, experience gained can be an important lead for gauging stress treatments (Blum 2011).

Canopy temperature depression, an indirect measure of plant-water status of the crop, is the difference between air temperature and plant canopy temperature (Tuberosa 2012). Genotypes use more available soil moisture to cool the canopy by transpiration under drought stress (Reynolds et al. 2009). Further, plant leaves emit long-wave infrared radiation according to their temperature. Low water status in stressed plants leading to reduced transpiration raises canopy temperature. The infrared thermometer is designed to sense long-wave infrared radiation emitted from its target, converting it to an average temperature display which can be related to transpiration and to the genetic potential of roots in exploring soil moisture (Pinto and Reynolds 2015) and drought susceptibility index in stressed environments (Blum 1989).

4.2 Exploring Roots of Drought Tolerance

The food security in the 21st century will rely increasingly on the release of cultivars with improved resistance to drought conditions and with high yield stability (Swaminathan 2005; Borlaug 2007; Pennisi 2008) and demands attention of plant scientists belowground (Bishopp and Lynch 2015). The development of crops with root systems that can capture water and nutrients efficiently would contribute in improving the economic development in poor nations and the sustainability of agriculture in rich nations by reducing reliance on intensive fertilization and irrigation (Bishopp and Lynch 2015). Carbon through CO2 from air is stored in the roots and leaves. Varieties having root systems extended around one meter limit their access to available water in deeper layers, particularly during drought stress at pod filling stage. Doubling root biomass to a nominal two meters would lock away more carbon in soil down to 2 meters in the croplands which could reduce the annual rise in global CO2 levels in the atmosphere helping fight global warming (Kell 2011) and develop drought resilient varieties. Root system architecture traits are measured in soybean, morphologically, to capture narrow root angle to the soil surface, which promotes lateral root development in the upper root regions where light penetration is the greatest and at the same time number of forks and number of crossings could be significantly important traits for soil penetration, enhancing rooting depth away from the soil surface and root length density (RLD), with profuse fine roots accommodating large surface area and root volume (Satpute et al. unpublished), the traits deemed essential for water extraction during soil moisture stress. However, deep or profuse rooting would have no effect in shallow soil, in soil where there is no water at depth, or under conditions of mild water stress (Vadez 2014). A modeling study in soybean has shown that increasing the rate of rooting depth would lead to faster soil depletion and yield penalties, especially in the driest quartile of the years, and there would be no benefit, but even a penalty from faster and deeper rooting (Sinclair et al. 2010). A relationship between water extraction and RLD could be resolved to some extent, using root development model that is capable of reconstituting root system architecture in a 3-D context (de Dorlodot et al. 2007; Draye et al. 2010; Pages et al. 2010; Lobet et al. 2011), which gives power to interpret water extraction data for harnessing the genetics of the components of this architecture such as root angles, different types of roots, branching patterns, etc. (Draye et al. 2010; Lobet et al. 2011; Lynch and Brown 2012). The conditions to the success of this breeding strategy are that water would be available at depth (deep soil and water available at depth); deep-water extraction would have an increased benefit if it took place during the grain filling period and that might imply searching for genetic material capable of sustaining root growth during reproductive development; and cropping conditions of moderate VPD in crops where this potential extra water uptake from deep rooting would represent a large proportion of the total transpirational water needs (Vadez 2014).

An alternative way to approach the role of root for water stress adaptation, using a lysimetric system, is assessing water extraction by roots as a way to harness the functionality of root systems. Roots need to be looked at with a view to the whole plant (Comas et al. 2013) and resource availability in time and space (Lynch 2013). Programs need to focus on traits regulating the rate at which plants use the available water before and during stress, involving roots in the sensing mechanism of water stress. The capacity to extract the available water at depth is probably critical and may come from deep roots having a high hydraulic conductivity (Vadez 2014). Certain root anatomical traits, including xylem vessel size and abundance, root cortical aerenchyma, the number of root cells or the number of root cell files, contributing to drought adaptation as the building blocks of its hydraulic properties, eventually affect critical plant water-budget traits.

The current model of water uptake through the root cylinder to the xylem, the composite transport model, is such that water is taken up via two major pathways. In an apoplastic pathway, a large part of that water travels across the intercellular space between cells (apoplast) in the root cortex, toward the endodermis and the xylem vessels. The exodermis could represent a variable apoplastic barrier that plants could use to modulate their water transport characteristics (Hose et al. 2001). The resistance to water flow usually increases under water deficit (Steudle 2000). Most of that resistance is located in the root cylinder (radial resistance), whereas xylem vessels normally offer much less resistance (axial resistance) (Steudle 2000). Another pathway is symplastic water transfer. During the night in the absence of transpiration, water can be taken up by roots through an osmotic gradient (Steudle 2000) and has high resistance because water goes through cells, traveling in the membrane continuum (endoplasmic reticulum and plasmodesmata) using membrane transporters (aquaporins—AQPs) (Steudle and Peterson 1998; Steudle 2000) highlighting a possible role of AQPs to alter the hydraulic properties of the roots (Tyerman et al. 2002; Maurel et al. 2009). AQPs are integral membrane proteins that increase membrane permeability to water and other small molecules (Kaldenhoff and Fischer 2006). Root water uptake in soybean can be enhanced or reduced by the overexpression or loss of one or more PIP genes, respectively (Javot 2003; Zhou et al. 2014b). In addition to water transport in roots, a variety of AQPs are expressed in the coats of developing seeds (Schuurmans et al. 2003). Nutrient and water transport across plasma membranes (PMs) in seed coats is highly coordinated by regulatory mechanisms and integrates the activities of many nutrient transporters and facilitators. Thus, it is expected that plasma membrane intrinsic proteins (PIPs) that are specifically expressed in native PMs of seed coats are important for seed filling (Zhou et al. 2007). A soybean GmPIP2 subfamily member, GmPIP2;9, was found predominately expressed in roots and developing seeds (Lu et al. 2018). The soybean genome contains a total of 22 PIP genes (Sakurai et al. 2005). A recent study showed that altered plant transpiration led to rapid changes in root expression of soybean PIP1;6 (GmPIP1;6) that correlated with changes in root hydraulic conductance (Vandeleur et al. 2014). Notwithstanding evidence for importance of root traits in drought tolerance (Garay and Wilhelm 1982; Chen et al. 2007b; Manavalan et al. 2010; Sinclair et al. 2010; Fenta et al. 2014; Fried et al. 2018), little work has been done in breeding for drought-tolerant soybean varieties.

4.3 Breeding Approaches for Drought Tolerance in Soybean

Drought-tolerant traits, introduced through breeding approaches, resulted in soybean transpiration rates that plateau at VPD levels above 1.4–2.1 kPa (Mourtzinis et al 2019). Developing high productivity genotypes under water-limited scenario by introgressing traits explaining plant water relations and hydraulic processes into a single genetic background either through breeding and/or genomic approaches is a way forward in realizing genetic combinations supported by plant genetic resource activities identifying candidate drought-tolerant parental lines and genomic resources (Satpute et al. 2020). Advance phenotyping-based breeding approaches are pre-requisite and being adopted systematically by developing early generation biparental, backcross, or multi-parent intercross populations (Shivakumar et al. 2018) using identified candidate drought-tolerant exotic and/or indigenous parental lines and wider-adaptability high yielding variety(-ies). The populations are advanced through F_2 generation by mass selection where bulks are chemically desiccated with potassium iodide 0.2% (Blum et al. 1991; Bhatia et al. 2014; Satpute et al. 2019b) followed by selections made divergently for seeds size by mechanical sieving. Mass selection for large seeds under chemical desiccation significantly improved seed weight and grain yield under chemical desiccation stress, as compared to controls where selection was performed without chemical desiccation (Blum 2011). After two cycles of selection, intensive selection of candidate genotypes for multiple drought tolerance related traits is practiced in advance generations using three-tier selection scheme (Satpute et al. 2018) followed by multi-traits indexing (Satpute et al. 2020) for analyzing traits function.

4.3.1 Three-Tier Selection Scheme

Selection for yield per se under water-limited environment confounding the complexity of breeding implores identifying less complex traits related to the drought tolerance and separating them into major components which may ease further understanding of the genetic basis. A large-scale three-tier sequence of selection scheme, which utilizes a set of drought related traits, has previously been employed by a research group (Sinclair 2011) in recognizing drought-tolerant soybean lines. From a screen of about 3500 well-watered soybean plant introduction (PI) lines, about 250 lines (slightly <10%) with low petiole ureide concentrations were selected. High leaf ureide concentrations had been shown to be associated with sensitivity to nitrogen fixation to drought. The lines in the second-tier screen were grown for a field test of nitrogen accumulation under dry conditions on a sandy soil, which had both a low nitrogen content and low water holding capacity. The lines were lightly irrigated approximately every other day to maintain them for about 3 weeks at or slightly above the soil water content resulting in slight leaf wilting. Selected 10% of these field-tested lines (24 lines) were screened in the third tier for intensive measurement of nitrogen fixation response to soil drying in the greenhouse test. Ultimately, 11 PI lines were identified that had substantial superiority in nitrogen fixation tolerance to soil drying. The group identified candidate parental soybean lines for use in breeding efforts to sustain nitrogen fixation activity during soil water deficit.

On the similar line, a three-tier selection scheme has been reported in order to evaluate large number of germplasm accessions (Fig. 4.1) and breeding populations (Satpute et al. 2019a, 2020). Unlike previous approach, with choice of traits, viz. low petiole ureide concentrations under well-watered condition, field test of nitrogen accumulation and greenhouse test of intensive nitrogen fixation response under dry conditions, 1000 soybean germplasm accessions, procured from Medium Term Storage (MTS) facility at ICAR-Indian Institute of Soybean Research, Indore, India were evaluated for delayed leaf senescence trait, stem reserve mobilization, drought tolerance related above-ground traits, viz. drought tolerance index, canopy temperature depression, relative water content, which are reliable and speedy in measurement in large breeding populations, and root system architecture traits (Fig. 4.1).

In the first tier of three-tier selection scheme, about 86 accessions (slightly less than 10%) with 4-5 scores were selected for delayed leaf senescence, in a field trial during summer season. Accessions were sown in early, medium and late flowering groups and irrigations were withheld at pod initiation stage for imposing 21 days

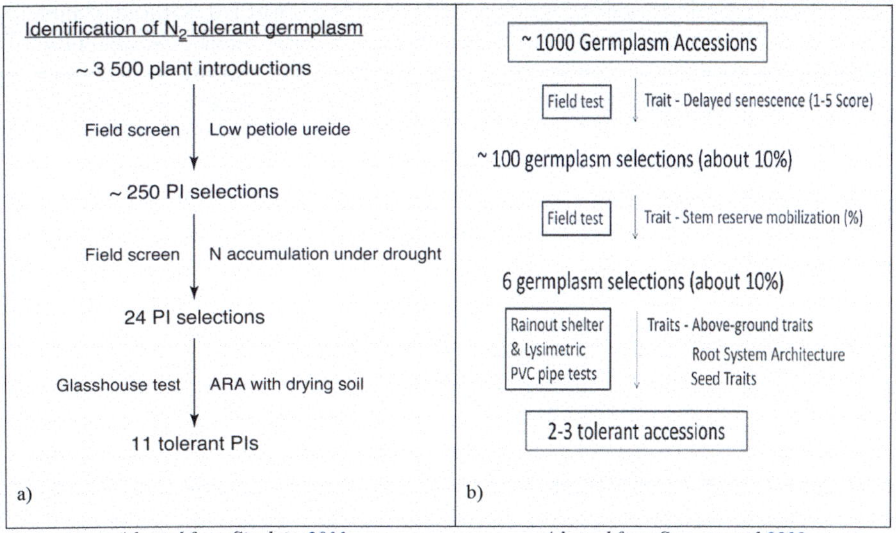

Fig. 4.1 Three-tier selection scheme is applicable for different sets of drought tolerance related traits and was developed to identify candidate soybean lines with (**a**) high tolerance of nitrogen fixation (Adopted from Sinclair 2011) or (**b**) high productivity under drought conditions (Adopted from Satpute et al. 2020). The broadest and the least accurate screen was based on (**a**) concentration of ureides in the petiole or (**b**) delayed leaf senescence trait; and from a large number of germplasm accessions ~10% accessions were selected in the first tier. In the second-tier screen about 10% of these field-tested accessions were selected for (**a**) N accumulation under drought or (**b**) stem reserve mobilization trait. The third tier of screening was performed for intensive measurement of (**a**) nitrogen fixation response to soil drying in the greenhouse or (**b**) above-ground plant traits and seed traits in rainout-shelters and root system architecture traits in lysimetric PVC pipes for identifying drought-tolerant germplasm. Ultimately, novel accessions were identified that had substantial superiority in (**a**) nitrogen fixation tolerance or in (**b**) productivity traits to soil drying

stress and inducing plant level stress of relative water content (RWC) $\leq$70% arriving at pod filling stage (R5 plus 8–10 days) (Blum 2011; Bhatia et al. 2014). In the semi-arid tropics, setting up the managed stress environment out of the normal season i.e. during the dry off-season is a common approach to avoid unpredictable seasonal rainfall (Mahalakshmi and Blum 2006) with due care to all possible off-season effects for effective drought phenotyping. When planted off-season, the humidity, temperature and photoperiod may not allow reasonably normal crop growth and development, crop cycle, phenology. Besides, certain diseases and pests like YMV and white flies can be highly prevalent and birds are attracted to grains outside its normal season which 2require protection measures. In such a case data on plant water status and plant responses to stress like delayed leaf senescence may be targeted while actual selection of seed is performed from the preferred lines grown also during the normal season (Blum 2011). In the second tier, accessions were evaluated for stem reserve mobilization trait (SRM) during rainy season in an field test conducted in sets of: (1) sprayed chemical desiccant, potassium iodide

(KI) (0.2%), at 8–10 days after R_5 plus 8–10 days stage and (2) unsprayed control (Bhatia et al. 2014) and ~10% of the accessions (six accessions) were selected for the trait. These promising six accessions were evaluated in the third tier for a number of above-ground plant traits at soil drying plant stress of ≤70% RWC at pod filling stage in the rainout shelter test and below-ground traits by morphologically measuring root system architecture traits, adopting standard lysimetric procedure under well-watered condition using PVC pipes (Vadez 2014) and *Win*Rhizo Arabidopsis (Regents, Canada) root image analysis. The approach of lysimetric system for root studies is a set of long and large PVC tubes, in which plants are grown individually and have plant spacing and soil volume available for soil exploration close to what is practiced under field conditions (Vadez et al. 2008, 2013). Root growth is known to stop its downward movement around anthesis (Robertson et al. 1993), although maintenance of growth can be found (e.g., Hafner et al. 1993), a trait worthy of screening, provided water is available at depth. The importance of deep-water extraction would be more if its timing coincided with the time of most critical water demand, i.e. reproduction and grain filling. Water extraction after anthesis with restricted vegetative growth completely contributes to grain development leading to high water use efficiency (kg grain mm^{-1}).

In breeding populations, the three-tier selection scheme is applicable in advance generations where approximately half of the seeds from selected F_4 generation single plants are used for screening lines for delayed senescence trait in F_5 generation during summer field trial in the first tier. The remaining half of the seeds from those F_4 single plants, of which lines were selected for delayed senescence trait, serves as a population for selecting lines for SRM trait during rainy season in the second tier. Thus, the scheme offers selection for delayed senescence trait followed by SRM trait in consecutive summer and rainy seasons, respectively, in the same year and selection can be practiced effectively for both the traits at 10% intensity in each selection cycle. Selected lines are evaluated for above-ground plant traits in rainout shelter-induced water stress condition vis-à-vis root architecture traits in the PVC pipes. Multiplexing drought tolerance related traits, using principal component analysis for correlation matrix in SAS (Version 3.0), provides a powerful multi-trait index which helps in identifying a set of drought-tolerant accessions or elite breeding lines (Satpute et al. 2020) for developing climate-smart drought resilient soybean varieties, understanding the role of surrogate traits and discovering unique drought tolerance related genes/QTLs.

4.3.2 Genetic and Genomic Resources

In crops such as soybean, drought resistance is translated to several related traits enhancing yield stability rather than that increasing survivability under drought (Blum 2009; Passioura 2010; Sinclair 2011; Valliyodan et al. 2016). These related traits are correlated with yield under drought and have no yield penalty under non-stress conditions. The success of soybean improvement through molecular approaches under drought stress depends on the discovery of genetic variations for

Table 4.1 Genetic resource for drought tolerance

Tolerant genetic resource/genotype	Basis of drought tolerance	Source	References
PI 416937	High Relative Water Content (RWC), higher lateral root spreading, increased Water Use Efficiency (WUE), low leaf hydraulic conductance, slow wilting	USDA, ARS	Sloane et al. (1990), Patterson and Hudak (1996), Sinclair et al. (2007), King et al. (2009)
Young	High WUE	USDA, ARS	Mian et al. (1996)
Jackson	High biomass and total N_2	–	Purcell et al. (1997)
PI 407155	Low electrolyte leakage, high biomass accumulation, high root moisture content	–	Chen et al. (2006)
R01-416F and R01-581F	Higher N_2 fixation	USA	Chen et al. (2007a, b)
93705-36 and PI 471938	Slow wilting	USDA	King et al. (2009)
PI 468917	Lower transpiration efficiency, greater root length	–	Seversike et al. (2014)
C12 and W05	Long root and high biomass, high leaf expansion	CUHK, China	Hossain et al. (2014)
PI 567690 and PI 567731	Leaf wilting	China	Pathan et al. (2014)
EC 538828, JS 9752 and EC 602288	Remobilization of stem reserve at terminal drought, drought resistance index based on yield	India	Bhatia et al. (2014), Bhatia and Jumrani (2016)
NTCPR94-5157, N09-13890, NC-Raleigh, SC07-1518RR	High WUE and greater root penetration	USA	Fried et al. (2019)
PK 1180 and SL 46	Seedling survivability	India	Sreenivasa et al. (2020)

drought related traits present in the germplasm and efficient utilization of available genomic resources. Identification of genetic diversities for traits related to drought tolerance, viz. root system architecture (RSA), water use efficiency, canopy wilting, and sustained N-fixation under drought have helped in discovery of genetic resources (Table 4.1) in soybean which are routinely being used as donor/check genotypes for deciphering soybean genetic response for drought stress through marker and genomics assisted strategies.

In the past decades, accelerated development in genetics, genomics, and soybean genome sequence information has resulted in the identification of SNPs, copy number variation, and structural variation in soybean germplasm (Kim et al. 2010; Schmutz et al. 2010; Ratnaparkhe et al. 2014) (Table 4.2). The advancement in next-generation sequencing approaches (NGS) and cheap sequencing cost have

Table 4.2 Details of whole-genome sequencing efforts in soybean (Chaudhary et al. 2019)

Genotype(s)/no.	Sequencing depth	Method	No. of SNPs	References
G. max var. Williams 82 (1)	–	De novo sequencing and assembly	–	Schmutz et al. (2010)
G. soja var. IT182932 (1)	~52.07×	Resequencing De novo sequencing and assembly	~2.5 Million	Kim et al. (2010)
17 *G. soja* and 14 *G. max* (31)	×5 depth	Resequencing	6,318,109	Lam et al. (2010)
8 *G. soja*, 17 *G. max* (8 landraces, 9 cultivars) (25)	–	SOAP	5,102,244	Li et al. (2013a, b)
10 *G. Max*, 6 *G. soja* (16)	>14×	Resequencing	3,871,469	Chung et al. (2014)
G. soja (7)	~111.9×	De novo sequencing and assembly	3.62–4.72 M SNP per line	Li et al. (2014)
10 Semi-wild, 1 *G. soja* (11)	9 Semi-wild at ~3× while 1 Semi-wild at ~41×, and 1 Wild at ~55×	Resequencing De novo sequencing and assembly	7,704,637	Qiu et al. (2014)
G. soja W05 (1)	~1×	De novo sequencing and Assembly	1,798,504	Qi et al. (2014)
62 *G. soja*, 240 *G. max* (130) landraces, 110 improved cultivars) (302)	>11×	Resequencing	9,790,744	Zhou et al. (2015a, b, c)
G. max cv. Enrei (1)	22.2×	Reference-based assembly	1659,041	Shimomura et al. (2015)
Wild, Landraces, Elite Lines (106)	17×	Resequencing	10,417,285	Valliyodan et al. (2016)

revolutionized soybean research in various forms of molecular tools, viz. de novo sequencing, whole-genome resequencing (WGR), genotyping by sequencing (GBS), and transcriptomic analysis (Liu et al. 2020b). These advances have made a significant impact on molecular breeding strategies through marker development such as SSRs (Hwang et al. 2009), SNPs (Kim et al. 2010; Lam et al. 2010; Chung et al. 2014; Zhou et al. 2015a; Valliyodan et al. 2016; Ratnaparkhe et al. 2020), insertion/deletion (INDEL) markers (Song et al. 2015), specific-locus amplified fragment (SLAF) markers (Zhang et al. 2016). Furthermore, the technical progress and availability of millions of SNPs have facilitated the development of high-density array-based genotyping chips such as Illumina Infinium array (SoySNP50 K iSelect Bead Chip) for _50,000 SNPs (Song et al. 2013), Soy SNP 6 K Infinium Bead Chip

(Akond et al. 2013), and the Axiom Soya SNP array for approximately 180,000 SNPs (Lee et al. 2015), which are being used for the genotyping of soybean lines (Chaudhary et al. 2019). GBS is becoming one of the popular sequencing-based genotyping approaches which has significantly reduced labor and time and improved precision in the identification of key genes as compared to the conventional PCR-based genotyping methods and being utilized in several crop species and soybean (Poland and Rife 2012; Sonah et al. 2013). Additionally, GBS also allows the detection of new variants in the population of interest, which can be utilized in future breeding programs (Chaudhary et al. 2019).

4.4 Quantitative Trait Loci for Drought Tolerance Related Traits

Different type of molecular markers has been used to map genomic location of major genes and quantitative trait loci (QTLs) for many traits in soybean. More than thousand QTLs representing more than 100 agronomically important traits have been mapped in soybean (Grant et al. 2010). Current information on all mapped QTLs in soybean is available on the USDA-ARS soybean genetic database *SoyBase* (http://soybase.org). Although a number of QTLs were mapped in the soybean but introgression and pyramiding of genes or QTLs affecting the same trait is a great challenge to breeding programs. Due to increasing necessity to develop drought-tolerant soybean with enhanced yield, breeding strategies with molecular tools have progressed at a massive rate in the past decade. Since molecular markers identified genetic variants for different drought related traits more precisely, markers are important in developing genetic linkage maps, genetic resource evaluation, and selection of desired alleles and mapping of genes/QTLs. Since microsatellites (SSRs) are less abundant in the genome, SNP markers became more popular and facilitated QTL analysis for nearly every agronomic trait in soybean (https://soybase.org, http://soykb.org). Gene/QTLs mapping in soybean has become more standard with the availability of whole-genome sequence (WGS) (Schmutz et al. 2010). This ground-breaking change in genome sequencing made available of the development of thousands of SSRs and millions of SNP markers. QTL analysis plays a significant role in identifying genomic regions which control over phenotypic variation and it requires a large segregating population (biparental mapping population) such as an F_2 population or recombinant inbred lines (RILs). In general, QTL mapping uses a large number of RILs, which are established for at least several generations of Selfing (typically up to F_6 or F_7) (Takuno et al. 2012). However, RILs are helpful for the QTL detection, but it estimates the influence of single QTL depending on population size. Moreover, the results are highly population specific for multigenic traits like drought tolerance traits (Deshmukh et al. 2014). On the other hand, plants that are homozygous for the unfavorable allele are eliminated in an F_2 population and frequencies of favorable alleles increase during inbred development (Bernardo 2010). There are a number of important QTL studies (Table 4.3) for traits related to drought tolerance reported in the past three decades. Although QTL mapping has

Table 4.3 List of QTLs related to drought tolerance traits

Parents involved	Population/size	Traits related to drought tolerance	Linked marker/s	Type of marker	LG/Ch No.	QTLs	PVE (%)	References
Young/PI416937	F_4 (120)	Water use efficiency (WUE), leaf ash	cr497-1	RFLP	J	3	13.2	Mian et al. (1996)
S100/Tokyo	F_2 (116)	WUE	A489H	RFLP	L	2	14	Mian et al. (1998)
Minsoy/Noir 1	RIL (236)	Transpiration efficiency (TE), carbon isotope discrimination (CID), yield	Satt205, Satt489	SSR	C2	3	7–12.8	Specht et al. (2001)
Kefeng 1/Nannong 1138-2	RIL	Dry root weight, total root length and root volume, canopy wilting	–	–	N6-C2 N8- D1b + WN11-E, N18-K	13	–	Liu et al. (2005)
Jackson/KS4895	RIL (81)	Leaf wilting	Sat_044	SSR	K	1	17	Bhatnagar et al. (2005)
Hutcheson/PI471938	F_4 (140)	Yield and wilting	Satt226, Sat_375, Sat_074	SSR	D2, F1, F2	6	–	Monteros et al. (2006)
KS 4895/Jackson	RIL (92)	Canopy wilting	Satt177, Satt362	SSR	A2, B2, D2, F	4	47	Charlson et al. (2009)
Kefeng1/Nannong1138-2	RIL (184)	Seed yield per plant, drought susceptibility index	–	–	C2, H	10	–	Du et al. (2009a)
Kefeng1/Nannong1138-2	RIL (184)	Leaf water, seed yield	–	–	C2, D1b, H, A2	17	–	Du et al. (2009b)

PI 416937/ Benning	RIL (128)	Root traits	Satt383, Satt339, Sct_191, Satt429 Sat_299	SSR	1, 3, 4, 8, 20	5	51	Abdel-Haleem et al. (2011)
Hongfeng 11/Clark	BC	Relative water content (RWC), water holding capacity, root traits, canopy wilting	Sat_136, Satt167, Satt398, Satt694, GMSL514	SSR	A1,A2, B1, B2, D1a,C2, E, F, K, G, I, L, M, N, O	40	–	Li et al. (2011)
Hongfeng 11/Clark	BC (46)	Germination rate and seedling stage drought	Satt449, Satt499, Satt440, Sat_180	SSR	A1, K I, H	31	–	Qiu et al. (2011)
PI 416937/ Benning	RIL (150)	Canopy wilting	Satt 302	SSR	2, 4, 5, 12, 14, 17,19	7	75	Abdel-Haleem et al. (2012)
Hongfeng 11/Harosoy	BC2F3 (95)	–	Satt253, Satt513, Satt693, Satt240, Satt323, Satt255	SSR	–	18	–	Zhang et al. (2012)
SNWS 0048/ Jinda73	BIL	Physiological trait	–	–	D2, G, M, N	9	–	Yang et al. (2014)
Jingdou 23/ZDD 2315	RIL (447)	Root and shoot traits of seedlings	Satt333, Satt327, Satt519, Satt597	SSR	–	24	7.05–38.91	Liang et al. (2014)
373 genotypes	–	WUE	39 MTAs	SNP	–	21	–	Dhanapal et al. (2015)
93,705 KS4895/ Jackson, 08705 KS4895/ Jackson	–	Canopy-wilting trait	–	–	2, 5, 11, 14,17, 19	13	–	Hwang et al. (2015)

(continued)

Table 4.3 (continued)

Parents involved	Population/ size	Traits related to drought tolerance	Linked marker/s	Type of marker	LG/Ch No.	QTLs	PVE (%)	References
KS4895/PI 424140, A5959/PI 416937, Benning/PI 416937								
Dunbar/PI 326582A,	BIL (251)	Root and shoot traits	Satt315, Satt253, Satt142	SSR	8, 12	4	7.7–20.8	Manavalan et al. (2015)
CPI 26671/G 2120; Valder/G 2120	RIL	Epidermal conductance, RWC, plant survival after stress	soPt-856,602	DArT	6	106 34	–	Vu et al. (2015)
M8206/ TongShan, ZhengYang/ M8206	–	Drought tolerance	–	SNP	Chromosomes from 01 to 20	111	–	Khan et al. (2019)
PK 1180/ UPSL 298	F_2/BSA	Seedling survivability	Satt277	SSR	6	One gene	–	Sreenivasa et al. (2020)
PI 416997/PI 567201D	RIL (196)	CID	–	SNP	6,7,10, 11, 15, 17, 20	16	2.5–29.9	Bazzer et al. (2020)

advanced swiftly in the past few years, a large number of mapped QTLs cannot be utilized in the breeding program because of false-positive QTLs and low accuracy. However, the accuracy can be enhanced by adapting different QTL mapping methods and effective statistical analysis such as single marker analysis (SMA), simple interval mapping (SIM), composite interval mapping (CIM), multiple interval mapping (MIM), and Bayesian interval mapping (BIM). Also, a number of QTL mapping software have been developed such as Mapmaker/QTL, QTL Cartographer, MapQTL, PLABQTL, PGRI, MapManager, QTLMAPPER, QGene, QTLSTA, Ici Mapping, and QTL network. Further utilization of QTL information for marker-assisted breeding has become challenging due to complex inheritance of unstable QTLs (Deshmukh et al. 2014). Statistical tools such as "Meta-QTL analysis" have been advanced that compile QTL data from different reports together on the same map for identification of precise QTL region (Deshmukh et al. 2012; Sosnowski et al. 2012). Hwang et al. (2015) identified various QTLs related to canopy wilting, during "Meta-QTL" study on five different populations (RILs), among identified QTLs, one QTL on chromosome 8 in the 93,705 KS4895 × Jackson population co-segregated with a QTL for wilting published previously in a Kefeng1 × Nannong 1138-2 population. The advances in sequencing technologies, statistical approaches, and software resulted in exponential intensification in soybean studies to understand plants response to extreme climatic conditions importantly drought stress.

Identification of genes underlying root system architecture and canopy characteristics is critical to develop soybean that is suited to water-limited environments. Prince et al. (2015a) identified four significant QTLs associated with different root architectural traits on Gm06 and Gm 07 in an interspecific RILs population of *G. max* (V71–370) × *G. soja* (PI407162). In an another study, Manavalan et al. (2015) identified a major QTL on Gm08 that governed root traits (tap root length and lateral root number) and shoot length and identified six transcription factors (MYBHD, TPR, C2H2 Zn, bZIP, GRAS, and Ring finger) and two key cell wall expansion-related genes which encode xyloglucan endotransglycosylases as candidate genes in the confidence interval of the QTL. These are key candidate genes for validation and to develop a better root ideotype in soybean.

4.5 Genome-Wide Association Studies for Drought Tolerance Related Traits

QTL mapping using biparental populations has limitations because of restricted allelic diversity and genomic resolution. The allelic diversity can be increased to some extent by using populations derived from multi-parental crosses (Deshmukh et al. 2014). Recently, Multi-parent Advanced Generation Inter-Cross populations (MAGIC) has been used to identify QTL for blast and bacterial blight resistance, salinity and submergence tolerance, and grain quality traits in rice (Bandillo et al. 2013). Such multi-parental populations have mapping resolution limitations since it depends on meiotic events (crossing-over) (Kover et al. 2009). In contrast, the genome-wide association study (GWAS) approach provides opportunities to explore the tremendous allelic diversity existing in natural soybean accessions. Mapping resolution of GWAS is also higher since millions of crossing events have been accumulated in the germplasm during evolution. The recent advances in high-throughput genotyping have played important role in the genome-wide association studies in soybean. The large data sets generated from NGS and high-density genotyping require sound computational algorithms for detection of minor QTLs as well as rare alleles with major effect phenotype. GWAS is routinely being used in many plant species, but only a few studies have been reported in soybean in regard to drought tolerance. GWAS for quantitative traits like drought tolerance is predictable to be affected by a confounding population. Different models have been developed for population stratification and spurious allelic associations like MLM and CMLM which takes into account the population structure and kinship (Deshmukh et al. 2014). Development in statistical tools, genotyping approaches, and studies involving larger set will definitely improve GWAS power. Recently, a large number of QTLs associated with shoot ureide were mapped in both biparental populations and genome-wide association studies (GWAS) in diverse lines (Hwang et al. 2013; Ray et al. 2015), which indicated the complexity of N-fixation under drought and suggested that genomic selection should be better suited to improve such complex trait. Dhanapal et al. (2015) analyzed a population of 373 genotypes with 12,347 single nucleotide polymorphisms (SNPs) in four environments for carbon isotope ratio (δ13C), an important physiological trait acting as surrogate for water use efficiency (WUE) and found association of 39 SNPs, which are likely tagged to 21 different loci with this drought-tolerant trait (Table 4.4).

Likewise Kaler et al. (2017) also reported 54 environment-specific SNPs associated with $\delta^{13}C$ and 47 SNPs associated with $\delta^{18}O$, which are tagged with 46 putative loci and 21 putative loci for $\delta^{13}C$ and $\delta^{18}O$, respectively. There are many loci reported for several drought related traits, viz. chlorophyll fluorescence (Hao et al. 2012; Herritt et al. 2018), canopy temperature (Kaler et al. 2018), delayed canopy wilting (Steketee et al. 2020; Ye et al. 2020), and drought susceptibility index (Chen et al. 2020) (Table 4.4). Latest updates on GWAS in soybean for drought tolerance were reported in germplasm association panel containing 259 soybean released Chinese cultivars for drought related traits based on germinating soybean seeds. The enquiry was based on a total of 4616 SNPs, and 15 SNP trait associations were identified by GWAS, among which three SNPs were suggestively linked with two of the drought-tolerance indices (Liu et al. 2020a).

4.6 Transcriptomic Approaches

Transcriptome analysis provides gene function information under various conditions, which differs in dissimilar environments, cell types, developmental stages, and cell states. There has been a tremendous progress in the application of transcriptome analysis for the abiotic stress tolerance. During abiotic stresses, a wide

Table 4.4 Details of genome-wide association studies (GWAS) performed for traits related to drought tolerance in soybean

Trait	GWAS loci	Markers	Genotypes	Method	References
Chlorophyll and chlorophyll fluorescence parameters	51	1536 SNP	168	MLM	Hao et al. (2012)
Ureide concentration	53	33,957SNP	374	PROC GLIMMIX	Ray et al. (2015)
Carbon isotope ratio (δ13C)	21	12,347 SNP	373	GLM and MLM	Dhanapal et al. (2015)
Chlorophyll contents	27	31,253 SNP	332	MLM	Dhanapal et al. (2016)
Carbon isotope ratio (δ13C) Oxygen isotope ratio (δ13C)	46 21	31,260 SNP	373	Farm-CPU	Kaler et al. (2017)
Canopy temperature	34	31,260 SNP	345	Farm-CPU	Kaler et al. (2018)
Chlorophyll fluorescence	53	32,453 SNP	189	CMLM	Herritt et al. (2018)
Delayed canopy wilting	44	34,379 SNP	162	MLM	Steketee et al. (2020)
Germination under drought	15	4616 SNP	259	MLM	Liu et al. (2020)
Drought susceptibility index and yield traits	302	105,970 SNP	136	MLM	Chen et al. (2020)

range of defense mechanisms are activated that increase tolerance against adverse situations in order to avoid damage caused by abiotic stresses such as drought. The first step toward stress response is stress signal recognition and subsequent molecular, biochemical, and physiological reactions activated through signal transduction (Le et al. 2012). Earlier, strategies using expressed sequence tags (ESTs) and techniques, i.e. suppression subtractive hybridization (SSH), have been extensively used for transcriptome profiling of soybean under abiotic stress conditions (Clement et al. 2008). These techniques are competent but do not give analysis of entire genes in the soybean genome. Several high-throughput techniques have been developed for transcriptome investigation due to the advancement in sequencing technology and the availability of the whole soybean genome sequence (Schmutz et al. 2010; Cheng et al. 2013). These platforms have been extensively used for transcriptome profiling to understand drought stress tolerance mechanisms in soybean (Table 4.5).

Microarray is a high-throughput technology where thousands of probes representing different genes are hybridized with RNA samples. The Affymetrix

Table 4.5 List of differentially expressed genes (DEGs) related to drought tolerance in soybean

Genotype name	Number of DEGs/ candidate genes	Putative functions	Platform used	References
Glycine max L. Merr	9148 genes	Genes related to membrane transport, defense signaling, metabolism associated with roots	Affymetrix chips containing 37,500 probe sets	Haerizadeh et al. (2011)
Williams 82	3276 for V_6 3270 for R_2	Expression of many *GmNAC* and hormone-related genes	61 K Affymetrix Soybean Array GeneChip	Le et al. (2012)
DT 2008 and W82	822 and 632 genes	Genes related to osmoprotectant biosynthesis cell wall-related proteins phosphatase 2C proteins TFs (NAC, AP2)	66 K Affymetrix, Soybean Array GeneChip	Ha et al. (2015)
Pana and PI 567690	1914 and 670	Hydrolase activity carbohydrate/lipid metabolism apoplast, and chlorophyll a/b binding proteins xyloglucan endo-transglycosylases	–	Prince et al. (2015b)
Williams 82	6609	Genes related to cell wall modification, lipid metabolism, carbohydrate metabolism, hormonal pathways, and TFs	Illumina HiSeq 2000	Song et al. (2016)
Glycine max L. Merr	49, 148, and 1576 genes, respectively	RNASeq analysis was performed on seed coat which plays a crucial role in controlling carbon and nitrogen transfer to developing seed set	RNASeq	Leisner et al. (2017)
Williams 82	*GmWRKY12*	Increase in proline (Pro) content and decreased malondialdehyde content	RNA-Seq, qRT-PCR	Shi et al. (2018)
Williams	*NAC4*, *NAC29*, *NAC25*, *NAC72*	2771 DEGs 1798 genes were upregulated and 973 were downregulated and related to ABA biogenesis, secondary metabolite synthesis, etc.	RNA-Seq, HiSeq4000	Xu et al. (2018)

Gene Chip representing 61K probe sets is frequently being used for transcriptome analysis of soybean RNA samples under drought stress (Haerizadeh et al. 2011; Le et al. 2012). The normalized expression data generated using the Affymetrix Gene Chip can be utilized to compare soybean experiments performed across the world. RNA-seq, another cost-effective and high-throughput technique, analyzes transcriptome by sequencing. The RNA-seq approach has several advantages over the microarray technology where available genomic information is used. For instance, RNA-seq is being used for transcription start site mapping, strand-specific measurements, gene fusion detection, small RNA characterization, and detection of alternative splicing events (Ozsolak and Milos 2010). Transcriptome profiling revealed through massively parallel RNA sequencing has offered new insights into gene networks that respond to drought stress (Table 4.5), such as NAC, etc. (Xu et al. 2018). These efforts can be used to generate an expression atlas for soybean genes related to drought tolerance which may serve as a useful genomic resource.

4.7 Molecular Events During Drought Stress in Soybean

Understanding the molecular mechanism of stress tolerance and developing stress-tolerant cultivars is important to achieve optimal yield from soybean crop. Modulations in gene expression are the earliest responses in plants, and a number of stress-responsive genes have been noted to have important functions in drought and salt resistance. Drought tolerance property of soybean involves complex network of genes and metabolites. Calcium channels, calcium binding proteins, receptor like protein kinases (RLKs), G-protein coupled receptors, histidine kinases are proposed to act as potential osmosensors in plants. Expression of several transcription factors (TFs), receptor like kinases (RLKs), calcium signaling components were upregulated in roots under drought stress (Tripathi et al. 2016a). It was shown that the genes involved in hormone, carbohydrate, and cell wall metabolism were differentially regulated in soybean roots under water stress. In another study, the level of expression of two auxin-responsive factors (ARFs), GmARF3 and GmARF50, was increased in roots and shoots under dehydration stress (Ha et al. 2015). Proteomic analysis of root suggested the involvement of osmoprotectants, kinases, and transcription factors in drought response (Mohammadi et al. 2012).

4.7.1 Signal Transduction Under Drought Stress

Abscisic acid (ABA) biosynthesis and accumulation in response to drought is reported in several plants (Sachdeva et al. 2020). ABA regulates the stomatal closure and other metabolic pathways during abiotic stress. Ca^{2+} and ROS are reported to participate in ABA mediated signal transduction pathway. Increased cytosolic Ca^{2+} level induces several Ca^{2+} binding proteins, viz. calmodulins (CaMs), calmodulin like (CML), calcium-dependent protein kinases (CDPKs), and calcineurin B like proteins (CBLs). Ca^{2+} application affects the nodulation process in soybean. Under drought stress, ROS generation is enhanced leading to accumulation of H_2O_2, which activates ROS scavenging mechanism. Mitogen activated protein (MAP) kinase cascade is involved in signaling pathway of many TFs under both biotic and abiotic stresses (Fujita et al. 2006). Degradation of proteins mediated by ubiquitination is another pathway involved in abiotic stress tolerance (Lyzenga and Stone 2012). Metabolic engineering using TFs may regulate several genes of the downstream pathway leading to improved tolerance to abiotic stresses. To reduce the undesired phenotypes of constitutively expressed TFs, tissue and developmental stage specific promoters may be used (Kasuga et al. 2004). In soybean, the expression level of various calcium-dependent protein kinases (GmCDPKs) was studied in response to wounding, stimulated herbivory, aphid feeding, treatment of jasmonic acid (JA), ethylene (ET), and salicylic acid (SA) and in response to drought and abscisic acid (ABA). Also, many GmCDPKs were induced after drought or ABA treatment (Hettenhausen et al. 2016).

4.7.2 Role of Transcription Factors in Drought Tolerance

Transcription factors (Tfs) are DNA-binding proteins that interact with other transcriptional regulators, including chromatin remodeling/modifying proteins, to initiate or inhibit the transcription by RNA polymerase. TFs have the capacity to act as a tool to improve the multigenic traits like drought tolerance (Rabara et al. 2014). Multigenic control makes the development of drought-tolerant varieties a difficult task using conventional breeding methods. TFs act as master regulators of many physiological processes and play an important role in the regulation of gene expression under abiotic stress. Alteration in expression levels of TFs regulates the expression of many downstream genes resulting in several different phenotypic effects. A particular TF can control the expression of several target genes of a specific pathway, thereby have the potential to modulate multigenic traits, such as drought tolerance. Forty genes belonging to six major families of TFs were upregulated during various abiotic stresses (Seki et al. 2002).

WRKY proteins are involved in the signal transduction of plant hormones, like abscisic acid (ABA), jasmonic acid (JA), and gibberellin (GA). Expression levels of four WRKY family of transcription factor (GmWRKY2, GmWRKY15, GmWRKY50, and GmWRKY55) were induced in drought treatment. In this regard, several reports strongly establish the potential of WRKY TFs as effective tool to engineer abiotic stress tolerance, such as drought in plants (Rushton et al. 2010; Chen et al. 2012; Rabara et al. 2013; Tripathi et al. 2013).

The potential of basic leucine zipper (bZIP) family of TFs as tools to improve crop responses to drought was shown by heterologous expression of soybean GmbZIP1 in Arabidopsis (Gao et al. 2011). Soybean bZIP (GmbZIP1) overexpression was shown to increase drought tolerance in *Arabidopsis* (Gao et al. 2011). Fifteen bZIP genes were induced by drought and salt stress (Yang et al. 2020). Among these, the expression of GmbZIP2 was significantly induced under stress conditions. It was shown to improve drought and salt tolerance upon overexpression by enhancing the expression of stress-responsive genes, such as GmMYB48, GmWD40, GmDHN15, GmGST1, and GmLEA. Another bZIP transcription factor, GmFDL19 has also been reported to enhance drought tolerance in soybean (Li et al. 2017b). Chlorophyll content and activities of several antioxidant enzymes were more in over-expressors while malonaldehyde content was lower than wild-type plants.

Four NAC family TFs (GmNAC4, GmNAC25, GmNAC29, and GmNAC72) were reported to be increased significantly in drought stress. In another study, 28 dehydration-responsive GmNAC genes were analyzed and it was revealed that eight of these genes were found to be induced in drought-tolerant soybean varieties under drought conditions (Hussain et al. 2017). Also, four of these (GmNAC4, GmNAC5, GmNAC20, and GmNAC21) were more dehydration resistant than others. The NAC family TFs are also proposed to be master regulators of various metabolic pathways in plants and have potential to manipulate the drought tolerance in transgenic plants.

ERF TFs are plant-specific TFs regulating a number of developmental and stress-related processes (Dietz et al. 2010). The AP2/ERF family consists of several subfamilies: the AP2, ERF, dehydration responsive element binding protein (DREB), and RAV (Mizoi et al. 2012). Several studies have reported that DREB TFs have potential to engineer drought tolerance. GmDREB2 improved salt and drought tolerance in *Arabidopsis* (Chen et al. 2007a). Two different types of transgenic soybean plants over-expressing AtDREB1D gene under constitutive and ABA-inducible promoters were raised. The transgenic plants showed increased drought tolerance by maintaining higher membrane stability (Guttikonda et al. 2014).

Transgenic soybean harboring GmDREB6 transcription factor was raised and the expression of P5CS gene and proline content was studied (Nguyen et al. 2019). Under normal condition, proline content was slightly higher in the transgenic plants. However, under salt stress the proline content increased to a large extent in transgenic plants. GmDREB6 has been proposed to bind GT-1 region in the promoter of P5CS gene and activate its expression (Zhang et al. 2013). Expression level of soybean DREB TF was studied in drought-sensitive and drought-tolerant cultivars. Also, the expression level of some known DREB regulated target genes were also investigated under water stress conditions (Stolf-Moreira et al. 2010). The drought-tolerant genotype had increased expression of aquaporin (Gmpip1), defensin (Gmdefensin), and galactinol synthase (Gmgols) under drought stress conditions.

In soybean, an R1 MYB transcription factor, GmMYB176, is reported to regulate isoflavone synthesis by affecting the expression of GmCHS8. GmMYB118 was significantly regulated by salt and drought treatment, and over-expression of GmMYB118 improved tolerance to drought and salt in both *Arabidopsis* and soybean. GmMYB expression was induced by drought, salt, ABA, and H_2O_2. The transgenic lines over-expressing GmMYB84 exhibited enhanced drought tolerance than WT plants. The over-expressors have longer primary root length, greater proline and ROS content, higher antioxidant enzyme activity, lower dehydration rate, and reduced MDA content (Wang et al. 2017). The activities of antioxidant enzymes were induced by ROS in over-expressor lines. The GmMYB84 was shown to bind the *cis* elements in the promoter of GmRBOHB1 and GmRBOHB2 which results in increased ROS levels leading to increased root growth under drought stress conditions (Wang et al. 2017). In addition, several members of bHLH, SRS, VOZ, NFYA family of transcription factors are shown to be involved in regulation of expression of various abiotic stress-related genes in soybean.

4.8 Genomics Assisted Breeding

Marker-assisted selection (MAS) is the indirect selection method where the linked molecular marker is used to transfer important agronomical traits from one genotype to another genotype. Marker-assisted backcrossing is an important approach in soybean for transferring trait of interest (Lee et al. 2006). The high-throughput genotyping technologies eased the process of marker identification and QTL mapping for different traits in soybean. The molecular breeding approaches such as marker-assisted backcrossing (MABC) and marker-assisted recurrent selection (MARS) aided in the introgression of the trait of interest in soybean. The soybean cyst nematode-resistant line, LDX01-1-65(PI636464) was developed using MABC (Chaudhary et al. 2019). Gene pyramiding involves combining favorable alleles controlling the same attribute from more than two parental lines (Melchinger 1990). Marker-assisted gene pyramiding was successfully carried out to develop durable resistance to several pathogens causing diseases in soybean (Walker et al. 2010). Although drought tolerance is accompanied by many traits which governed by mainly polygenes/QTLs, introgression of minor QTLs from donor to recipient cultivar is not an easy task. In QTL mapping in five biparental populations, a total of ten genomic regions or QTLs (Table 4.3) were mapped to be associated with canopy wilting under drought, with varied phenotypic contributions, and the majority of these QTLs (9/10) have donor alleles for slow wilting phenotypes from PI 416937, Jackson, or both (Charlson et al. 2009; Abdel-Haleem et al. 2012; Hwang et al. 2015). Molecular markers associated with these QTLs (identified in Meta-QTL studies) can be used to perform MAS to introgress the slow canopy wilting phenotypes from the donor in elite backgrounds. However, transferring these QTLs is a challenging job for breeders due to complex, quantitative nature and sensitivity to environmental factors of canopy-wilting trait under drought. Most of the mapped minor QTLs were found to be unstable across independent environments and populations. For instance, even major QTLs on chromosome 12 ($R2 = 0.27$) were identified in all five environments from Benning × PI 416937 (Abdel-Haleem et al. 2012) but were not detected in any populations or site-years, including the Benning × PI 416937 cross reported by Hwang et al. (2015). For this reason, QTL confirmation in more advanced generations should be performed to validate each individual QTL. This also indicates that stacking all confirmed QTLs in the same elite background by MAS is necessary to build the drought tolerance shown in the donor (Valliyodan et al. 2016).

Marker-assisted breeding for simple Mendelian traits is effective, but it can be challenging for complex traits such as drought stresses that are generally polygenic. Even major QTLs linked to drought tolerance traits can explain only a small fraction of phenotypic variation and may show unexpected trait expression in new genetic backgrounds because of epistatic interactions or GE interaction. These limitations can be effectively overcome by the use of strategy called "Genomic-selection" (GS). GS is a relatively simple and more powerful approach since it uses all marker information simultaneously to develop a prediction model avoiding biased marker effects (Heffner et al. 2009). In soybean, some efforts have been made to evaluate

GS using different models. A GS study conducted in soybean has used a panel of 288 accessions and 79 SCAR markers to predict 100 seed weight (Shu et al. 2012). In this report, high correlation ($r2 = 0.9$) has been observed among the genomic estimated breeding value (GEBV) and the phenotypic value. Predicting the precision of GS will need more investigations involving high-throughput genotyping of larger populations evaluated with multi-environment. These multi-environmental trials not only include the effect of G × E but also increase the number of breeding cycles per year. The challenge for GS is to get accurate GEBV with respect to the G × E effect. Improved factorial regression models have been proposed for GS that consider stress covariates derived from daily weather data, which revealed increased accuracy by 11.1% for predicting GEBV in unobserved environments where weather data is available (Heslot et al. 2014). This study suggests possible utilization of phenotypic data and historical data of weather conditions accumulated over decades in different soybean breeding programs. Similar information can be used for drought tolerance improvement in soybean (Deshmukh et al. 2014). Most of the GS studies have used RIL populations to train the prediction model. Therefore, GS and QTL mapping can be performed simultaneously. A set of diverse cultivars can be used for GS and GWAS altogether, so GWAS, GS, and QTL mapping can be combined together for marker-assisted breeding for drought tolerance related traits (Deshmukh et al. 2014). QTL or GWAS loci possess hundreds of genes which make the identification of candidate genes difficult (Sonah et al. 2012). This is similar in transcriptome profiling where thousands of genes have been found to be differentially expressed even with genetically similar isogenic lines (Table 4.5). Therefore, combining QTL mapping or GWAS with transcriptome profiling can complement each other (Deshmukh et al. 2014). Recently, several sequences based data sets have been generated by resequencing efforts (Lam et al. 2010; Li et al. 2013b, 2014; Chung et al. 2014; Qiu et al. 2014; Zhou et al. 2015c; Valliyodan et al. 2016). The availability of well-annotated soybean genome sequence and resequencing based data sets also facilitates development of large number of SNP and Indel markers which are being utilized in QTL mapping and molecular breeding for drought tolerance in soybean.

4.9 Genetic Engineering Approaches for Developing Drought Tolerance in Soybean

The complexities of mechanisms controlling drought adaptive traits and the limited availability of germplasm for tolerance to drought stress have restricted genetic advances in soybean for increase in yield and improvement of other traits associated with drought stress tolerance. Understanding the mechanisms by which plants perceive and transduce the stress signals to initiate adaptive responses and their engineering using molecular biology and genomic approaches is essential for improving drought stress tolerance in soybean crops. Attempts have been made to enhance drought stress tolerance through biotechnological approaches and drought-tolerant varieties of soybean have been produced. Zhang et al. (2019) reported that over-expression of C-5 sterol desaturase gene from *Flammulina velutipes* (*FvC5SD*) improves drought tolerance in soybean. In this study, FvC5SD gene was introduced into the soybean variety Shennong9 through the *Agrobacterium*-mediated transformation in order to enhance drought stress tolerance. Under drought conditions, the transgenic soybean plants accumulated lower levels of reactive oxygen species and exhibited higher activities and expression levels of enzymes than wild-type soybean. The basic leucine zipper (bZIP) family of transcription factors plays an important role in the growth and developmental process as well as responds to various abiotic stresses, such as drought. Li et al. (2017b) demonstrated that *GmFDL19* also enhances tolerance to drought and salt stress in soybean at the seedling stage. Wei et al. (2019) generated transgenic soybean plants and further investigated roles and biological mechanisms of *GmWRKY54* in response to drought stress. Wei demonstrated that expression of *GmWRKY54*, driven by either a constitutive promoter (pCm) or a drought-induced promoter (RD29a), confers drought tolerance. *GmWRKY54* is a transcriptional activator and affects a large number of stress-related genes as revealed by RNA sequencing.

Efforts have been made to study response of soybean plants to drought stress using advanced genetic engineering and genomic approaches like high-throughput sequencing technologies, chip-based analysis, RNA seq, etc., which enabled researchers to utilize enormous nucleotide database to find genes involved in various metabolic pathways dealing with abiotic stress tolerance. But still, there are many drought-responsive genes that have been identified but their function is still unknown. Therefore, genetic engineering through reverse genetics approaches could be useful for identification and functional elucidation of drought stress tolerance related genes (Azevedo et al. 2011). Identification, a functional characterization and introgression of stress-related genes through advanced genetic engineering techniques are important to provide long-term tolerance against drought stress (Jan et al. 2016, 2017).

The first commercial cultivation of genetically engineered soybean was started in 1996 which has spread in the area of 95.9 million hectares (mHa) till 2018. This area covers about 78% of the global soybean cultivated area, i.e., 123.5 mHa. The USA is the world's top producer of soybean whereas Brazil is the top exporter of soybean in the world. To bring suitable changes into the molecular structure of soybean, 38 transgenic events were approved in around 31 countries, mostly in the North and South American continent. These events were attributed to traits like herbicide resistance, insect resistance, drought tolerance, and pyramiding of two or more gene traits. All these events account for 50% of the world's biotech crop area. Among them, there are two specific transgenic events, commercially available for farmers to grow drought-tolerant soybeans, namely HB4 (popular as trade name Verdeca HB4 Soybean) and HB4 × GTS 40-3-2. In these events, gene Hahb, isolated from *Helianthus annuus,* has been genetically engineered to produce transcription factor Hahb-4 which binds to a dehydration transcription regulating region of the plant responsible for better performance under drought condition. In addtion, new advances in functional genomics studies in soybean using VIGS, RNAi and genome editing approaches has played important role in soybean improvement and gene function studies.

4.9.1 Virus-Induced Gene Silencing: A Potential Biotechnological Tool for Rapid Elucidation of Genes Function

Although comparative and functional genomic strategies have provided initial clues about function of abiotic stress-responsive genes in soybean and many other crop species (Gorantla et al. 2007; Tran and Mochida 2010; Soares-Cavalcanti et al. 2012), comprehensive functional characterization tools are necessary for understanding the precise role of these genes in combating drought stress. One such tool is virus induced gene silencing (VIGS) which has emerged as a potential gene knock-down technique in several crop species because it does not require transformation (Baulcombe 1999; Burch-Smith et al. 2004; Senthil-Kumar and Mysore 2011). Virus-induced gene silencing (VIGS) is a reverse genetic tool for functional elucidation of genes involving gene transcript suppression. In VIGS system, recombinant virus carrying a partial sequence of a host gene is used to infect the plant. When the virus spreads systemically, the endogenous gene transcripts, which are homologous to the insert in the VIGS vector, are degraded by post-transcriptional gene silencing (PTGS) (Baulcombe 1999).

With increased identification of differentially expressed genes employing high-throughput transcript profiling under various abiotic stresses including drought stress, functional elucidation of stress-responsive genes is crucial to understand their role in stress tolerance. In recent years, VIGS has been successfully used as a versatile tool for gene function analysis in various model plants and also in crop plants like soybean. Viral vector-based silencing of gene of interest and studying the gene knock-down in plants under stress can be one of the potential options for assessing functional significance of stress-responsive genes in soybean. Analysis of stress downregulated as well as stress upregulated genes is crucial for understanding molecular responses of crop plants to abiotic stresses. A large number of genes whose expression altered during various abiotic stresses have been identified through expression profiling, expressed sequence tags (ESTs), and cDNA library generated from various plant species (Seki et al. 2002; Govind et al. 2009; Marques et al. 2009; Bohnert et al. 2006; Becker and Lange 2010; Chen et al. 2015; Ramegowda et al. 2017; Abd El-Daim et al. 2018). However, identifying the functional significance of individual differentially expressed genes during drought stress is a challenging task. It is utmost important to elucidate the function of these stress-responsive genes to understand the mechanism of stress tolerance and also for characterizing candidate genes contributing tolerance of susceptible species by genetic engineering. An inventory of genes showing altered expression under several abiotic stresses has been established for many crop species employing expressed sequence tag (EST) analysis (Gorantla et al. 2007; Wani et al. 2010; Blair et al. 2011). In contrast to the enormous progress made in generating sequence information, functional analysis of stress-responsive genes is lagging behind.

VIGS technology has been extensively used to investigate function of genes responsive to various kinds of abiotic stresses (Senthil-Kumar et al. 2007; Cho et al. 2008; Govind et al. 2009; Kuzuoglu-Ozturk et al. 2012; Manmathan et al. 2013; Bao et al. 2015; Wang et al. 2016; Li et al. 2017b; Park et al. 2017; Ramegowda et al. 2017; Ullah et al. 2018). Recent development in VIGS vectors has extended the application of VIGS for functional characterization of abiotic stress-responsive genes and also enhancing abiotic stress tolerance in several crops including soybean. Rao et al. (2014) studied functional relevance of *GmCam4* (*Calmodulin*) gene by silencing and over-expression using Bean Pod Mottle Virus (BPMV)-based vector. Silencing of *GmCam4* resulted susceptible response to salt stress while over-expression resulted salinity tolerance in soybean plants at 200 mM NaCl level (Rao et al. 2014). In the recent past, VIGS has been successfully used to unravel the abiotic stress tolerance mechanisms in crop plants (Senthil-Kumar and Udayakumar 2006; Senthil-Kumar et al. 2008; Manmathan et al. 2013). Zhou et al. (2020) found that over-expressing gma-miR398c in soybean decreased *GmCSD1a/b*, *GmCSD2a/b/c*, and *GmCCS* expression, which weakened the ability to scavenge $O_2^{\cdot -}$ and by their means negatively regulates drought tolerance in soybean.

4.9.2 RNAi Approach: A Powerful Tool for Gene Function Studies and Enhancing Drought Tolerance in Soybean

RNA interference (RNAi) is a versatile tool frequently used for gene function studies in soybean. RNAi phenomenon involves small interfering RNA (siRNA) or short hairpin or microRNA (miRNA) to suppress the expression of sequence-specific gene at post-transcriptional or translational level. This technology has been extensively used to study functional relevance of genes, enhancing crop yield, improvement of nutritional quality, and increasing crop productivity through suppression of expression of genes responsive to abiotic stress, involved in biomass and grain yield.

In future, there will be huge demand for genetically improved crops with ability to maintain yield stability under adverse environmental conditions. Drought stress tolerance and adaptation of crop plants to drought stress have been improved through RNAi approach for manipulating expression of transcription factor genes, genes associated with signaling and biosynthetic pathways, and accumulation of antioxidants (Gupta et al. 2014; Wang et al. 2015; Pradhan et al. 2015; Meena et al. 2017; Li et al. 2017a). Several genes associated with drought stress-related physiologies and pathways have been functionally characterized to understand stress tolerance mechanisms and for improving abiotic stress tolerance in crop plants (Zhou et al. 2015b; Guo et al. 2016; Ji et al. 2016; Ma et al. 2017; Li et al. 2017a; Cai et al. 2018). It is utmost important to elucidate the role of transcription factors or genes by genetic manipulation for higher yield and also yield stability under various abiotic stress conditions. Several researchers tried to identify and characterize various genes responsive to drought and salinity stress by using genomics, transcriptomics, proteomics, and metabolomics approaches (Wang et al. 2015; Tripathi et al. 2016b; Ji et al. 2016; Qin et al. 2016; Li et al. 2017a). Therefore, it is essential to know the role of specific small RNA followed by genetic manipulation for improvement of drought stress tolerance in soybean crop. The RNAi technology has been successfully used for improvement of soybean crop in terms of enhancing abiotic stress tolerance (Wang et al. 2015; Srivastava et al. 2017; Li et al. 2017a; Mao et al. 2018). RNAi has been effectively utilized for incorporating desired traits for abiotic stress tolerance in various plant species (Jagtap et al. 2011; Pradhan et al. 2015; Meena et al. 2017; Li et al. 2017a). Wang et al. (2015) studied the interaction of *GmWRKY27* with *GmMYB174* and reported that these two cooperatively inhibit transcription of *GmNAC29* by binding to the core sequences in its promoter. The downregulation of expression of *GmNAC29* leads to reduced intracellular ROS levels. The *GmWRKY27* may also increase proline content by indirectly suppressing the transcription of PDH which ultimately led to improvement in stress tolerance in soybean (Wang et al. 2015). Li constructed soybean *GmRACK1* silenced (RNA interference, RNAi) and over-expressing plants. The GmRACK1-RNAi lines showed significantly improved drought stress tolerance while the over-expressing seedlings were hypersensitive to drought stress when compared to wild-type in terms of plant survival rates after 10 days of drought. GmRACK1-RNAi plants were found to be more sensitive to ABA when seeds germinated and root grew.

4.9.3 Genome Editing Based Techniques

The availability of soybean wild species and genetic variations in soybean germplasm is crucial for soybean improvement programs targeting drought toleance. However, the lack of enough natural germplasm, genetic diversity, and mutant collections limits both basic and applied research, particularly in soybean. The genome editing tools provide opportunity to overcome these limitations via creation of such variations in the genome of crop plants. Such approaches can reduce breeding or gene transformation time greatly for production of new varieties/transgenic plants with desired traits, such as abiotic stress tolerance. The CRISPR technology is being seen as an advancement of plant breeding technologies. Non-transgenic approaches are also available for delivery of such nucleases to produce mutant plants (Marton et al. 2010). As a result, crop varieties produced using these technologies may qualify as non-GM and would have enormous impact on plant biotechnology and breeding. There are four genome editing tools, meganucleases, zinc-finger nucleases (ZFN), Transcription Activator-like Effector Nucleases (TALEN) and the Clustered Regularly Interspaced Short Palindromic Repeat (CRISPR)/CRISPR-ASsociated nuclease protein (Cas) system, which have provided targeted gene modification in plants (Cermak et al. 2015; Gao et al. 2010; Li et al. 2012, 2013a, b; Shukla et al. 2009). Among these, the CRISPR-Cas9 system is the easiest to implement and is highly efficient. The system consists of a Cas9 endonuclease derived from *Streptococcus pyogenes* and a chimeric single guide RNA that directs Cas9 to a target DNA sequence in the genome. The CRISPR-Cas9 genome editing is accomplished by introducing a DNA double-strand break in the target locus by nuclease enzyme named Cas9, followed by DNA repair through either the endogenous imprecise Nonhomologous End-Joining (NHEJ) or the high-fidelity Homology-Directed Repair (HDR) pathways. NHEJ can induce small insertions or deletions at the repair junction while HDR stimulates precise sequence alterations, including programmed sequence correction as well as DNA fragment insertion, when a DNA repair template is exogenously supplied. The system has been successfully tested in staple crops, such as maize, wheat, rice, and soybean (Cai et al. 2015; Du et al. 2016; Sun et al. 2016; Svitashev et al. 2015; Wang et al. 2014; Zhou et al. 2014a, 2015c).

The recent availability of genome editing tools provides ample opportunity to introduce targeted modifications in the genome efficiently to study the functional aspects of various components of the genome in diverse plants and offers potential avenues for production of drought-tolerant soybeans. Genome editing tools provide a method for introducing targeted mutation, insertion/deletion (indel), and precise sequence modification using customized nucleases in a wide variety of organisms. CRISPR-Cas9 mediated genome engineering can enable manipulation of nearly any sequence in the genome. Abiotic stress is a complex trait, which is governed by multiple genes. There is a substantial interaction between components of several signaling, regulatory and metabolic pathways, which lead to abiotic stress response/adaptation (Nakashima et al. 2009; Garg et al. 2014; Mickelbart et al. 2015). Further, plants have undergone whole genome duplication events and a large fraction of genes are represented by multi-gene families with functional redundancy. Many times knock-out of a single gene may not produce desired phenotype, thus making it difficult to reveal its function. Due to ease of design and high efficiency of sgRNAs, multiple genes can be targeted simultaneously using CRISPR-Cas9 system, which can overcome the problem posed by functional redundancy of genes. Multiplex genome editing has been successfully implemented in model and crop plants (Li et al. 2013a; Mao et al. 2013; Zhou et al. 2014a). Such approaches can allow deciphering the role of multiple and functionally redundant genes involved in the same biological process such as drought stress response. Another approach could be the pyramiding/stacking of multiple genes involved in a stress response pathway or regulatory network via HDR-mediated gene targeting. The genes involved in drought stress-related gene regulatory network, signal transduction, and metabolite production may be targeted via CRISPR-Cas9 technologies for production of drought tolerant soybeans.

4.9.4 Rhizobial Inoculation to Enhance The Drought Stress Tolerance in Soybean

Symbiotic association between legume plants and N2 fixing microbes like rhizobium has always been one of the fascinating areas for the researchers across the world since decades. Soybean, a legume plant makes the symbiotic association with soil bacterium, Rhizobium species by which plant can harness the benefit of biologically fixed nitrogen which helps the host plant for achieving growth and development. In return, rhizobia species get food and shelter inside the root nodules of a legume plant. In this way, both the participating species get mutually benefitted from each other. In the recent studies, it is also found that such symbiotic association not only contributes to the growth and vigor development of the host plant but it can also enhance the drought stress tolerance capacity. The bacterial endophytes producing high trehalose inside the nodules or assimilating trehalose in bacteria could be potential strategy to enhanced survival and stability (Sharma et al., 2020). Bradyrhizobia imparts growth and drought alleviation in soybean through various direct and indirect mechanisms (Bharti et al., 2018). Soybean plants inoculated with Bradyrhizobium improved plant fitness and physiological parameters besides improving the plant rhizospheric health has been reported (Sharma et al., 2012). This opens the door for various possible opportunities where host specific rhizobium strains can be identified and can be used to study the host–microbe interactions at the physiological and molecular level to understand the mechanism of enhanced capacity of plant against the abiotic stresses. A similar kind of study has been conducted at Helsinki Institute of Sustainability Science (HELSUS), University of Helsinki, Helsinki, Finland, and Southern Agricultural Research Institute, Hawassa, Ethiopia where Aserse et al. (2019) came up with the identification of some elite rhizobial strains (HAMBI3562 and HAMBI3570 for common bean and HAMBI3513 for soybean) that significantly increased shoot dry length and nitrogen content in common bean and soybean (Table 4.6).

In another study conducted differently by Bado et al. (2013) and Zerihun and Haile (2017) they mentioned the other benefits like improvement of N and C content of the soil which in turn benefit the sustainability of the soil fertility and enhance subsequent cereal production cultivated in rotation with legumes.

Table 4.6 List of rhizobium strains specific for common bean and soybean to enhance drought tolerance

Sr. no.	Host plant	HAMBI code for *rhizobium* strain	*Rhizobium* strain
1	Common bean	HAMBI3562	*Rhizobium phaseoli HBR10*
2	Common bean	HAMBI3570	*R. phaseoli HBR53*
3	Common bean	HAMBI3556	*R. Etli HBR5*
4	Soybean	HAMBI3524	*Bradyrhizobium japonicum TAL379*
5	Soybean	HAMBI3520	*B. elkanni SBR2B*
6	Soybean	HAMBI3513	*B. elkanni SBR8B*

4.9.5 Application of Nanotechnology to Increase Drought Tolerance in Soybean

Nanotechnology is one of the promising approaches that has been extensively studied as biotechnological tools where metal-based nanoparticles are being applied in the crop system to study the effectiveness and targeted delivery of the molecular product stimulating the regulation pathways (Tripathi et al. 2018). Nanoparticles (nano-scale particles = NSPs) are atomic or molecular aggregates with at least one dimension between 1 and 100 nm (Roco 2003) and are used in low quantity as the replacement of plant mineral nutrients. As compared to conventional fertilization the amendment of nanoparticles improved the plant response to drought stress (Saxena et al. 2016). The application of these micronutrient-based nanoparticles, such as copper, iron, cobalt, manganese, magnesium, nickel, and zinc, helps to increase the crop yield, even under environmental stress conditions (Ashraf et al. 2012).

ZnO nanoparticles (NP) application increased seed germination of soybean under water stress (Sedghi et al. 2013). The use of extremely low concentrations of ZnO NP, lower than 500 ppm, can guarantee the enhancement of the Zn content in the seed without toxicity to plant cells (Hossain et al. 2016). The physiological traits, viz. drought tolerance index, relative water content, and biomass reduction rate, were significantly improved, especially in Fe NP-treated plants. Fe and Cu NP-treated plants maintained relative water content (RWC) at 71%, which was significantly higher than the RWC of control plants (64%) (Linh et al. 2020). Root architecture plays an important role in resistance to drought. Iron oxide NPs at the concentration of 50 to 2000 mgL^1 increased root growth by 6–40% in soybean (Alidoust and Isoda 2013). The efficacy of iron oxide NPs was higher due to higher solubility, higher surface area, higher penetration through seed coat, and subsequently emerging roots and better availability to root radicals (Denher et al. 2010). Quantitative PCR analysis of drought-responsive genes showed a gene-, tissue-, and nanoparticle-dependent upregulation of gene expression. In addition to this the expression of three drought-responsive genes promoted in leaves OD nanoparticle-treated plants. The Fe NP triggered the expression of all tested genes in roots. The expression of the selected drought tolerance marker genes, GmRD20A, GmDREB2, GmERD1, GmFDL19, GmNAC11, GmWRKY27, GmMYB118, and GmMYB174, was found to be upregulated in roots or shoots (or both) of NP-treated plants under drought. This suggests that nanoparticle application can increase drought tolerance of soybean by promoting the expression of genes associated with drought (Linh et al. 2020).

4.10 Conclusions and Future Perspectives

In just the past few years we have witnessed tremendous progress in soybean genomics and an explosive expansion of new resources. We have seen the development of high-density soybean genetic maps, construction of physical and transcript maps, EST sequencing and analysis, development of high-density cDNA and oligo arrays, and large scale re-sequencing of soybean genomes and comparison of homologous segments. These resources and the resultant studies have shed much light on the structure, organization, and evolution of the soybean genome. Tremendous progress has been made in the mapping and molecular breeding for various abiotic stresses in soybean. However, genes for many quantitative traits of economic interest are yet to be identified. A lot of genotypic and phenotypic information needs to be generated for identification and characterization of gene associated with abiotic stresses. Once the underlying gene sequence is fully characterized, haplotype analysis of the structural variants identified in the underlying genes for quantitative trait of interest could discover novel and useful alleles. The knowledge of gene function generated through QTL mapping, gene identification and characterization and resulted development of functional markers through allele mining may be translated in to a useful product using genomics assisted breeding approaches for drought tolerance. The latest genome editing tools provide opportunity to overcome certain limitations via creation of variations in the soybean genome. Such approaches can reduce soybean breeding time greatly for production of new varieties/ transgenic plants with desired traits such as abiotic stress tolerance. Due to these advances, we will be able to further explore genomic approaches to the elucidation of key genes or functional components that control complex drought related agronomical and physiological traits in soybean. Breeding approaches with potentially effective plant genetic resources (PGR), high hybridization efficiencies, and precise phenotyping facilities help realize climate-smart drought-tolerant varieties with adaptability to the Target Environment, sustaining long-term profitability of farmers in soybean cultivation, facing worse drought-trodden situation globally.

Acknowledgements We are thankful to the Director, ICAR-Indian Institute of Soybean Research, Indore, for useful suggestions and inputs. We acknowledge ICAR-National Agricultural Science Fund (NASF) for financial support.

References

Abd El-Daim IA, Bejai S, Fridborg I, Meijer J (2018) Identifying potential molecular factors involved in Bacillus amyloliquefaciens 5113 mediated abiotic stress tolerance in wheat. Plant Biol. https://doi.org/10.1111/plb.12680

Abdel-Haleem H, Lee GJ, Boerma RH (2011) Identification of QTL for increased fibrous roots in soybean. Theor Appl Genet 122(5):935–946

Abdel-Haleem H, Carter TE Jr, Purcell LC, King CA, Ries LL, Chen P, Schapaugh W Jr, Sinclair TR, Boerma HR (2012) Mapping of quantitative trait loci for canopy-wilting trait in soybean (*Glycine max* L. Merr). Theor Appl Genet 125(5):837–846. https://doi.org/10.1007/s00122-012-1876-9

Akond M, Schoener L, Kantartzi S, Meksem K, Song Q, Wang D et al (2013) A SNP-based genetic linkage map of soybean using the soy SNP 6K Illumina Infinium bead Chip genotyping array. J Plant Genom Sci 1:80–89. https://doi.org/10.5147/jpgs.2013.0090

Alidoust D, Isoda A (2013) Effect of γFe_2O_3 NPs on photosynthetic characteristics of soybean (Glycine max (L.) Merr): foliar spray versus soil amendments. Acta Physiol Plant 35:3365–3375

Annual Report (2019) ICAR – Indian Institute of Soybean Research, Indore, India

Aserse AA, Markos D, Getachew G, Yli-Halla M, Lindström K (2019) Rhizobial inoculation improves drought tolerance, biomass and grain yields of common bean (Phaseolus vulgaris L.) and soybean (Glycine max L.) at Halaba and Boricha in southern Ethiopia. Arch Agron Soil Sci 66(4):488–501. https://doi.org/10.1080/03650340.2019.1624724

Ashraf MY, Mahmood K, Ashraf M, Akhter J, Hussain F (2012) Optimal supply of micronutrients improves drought tolerance in legumes. In: Crop production for agricultural improvement. Springer, Dordrecht, pp 637–657. https://doi.org/10.1007/978-94-007-4116-4_25

Azevedo H, Silva-Correia J, Oliveira J, Laranjeira S, Barbeta C, Amorim-Silva V, Botella MA, Lino-Neto T, Tavares RM (2011) A strategy for the identification of new abiotic stress determinants in Arabidopsis using web-based data mining and reverse genetics. OMICS 15 (12):935–947. https://doi.org/10.1089/omi.2011.0083

Bado BV, Cescas MP, Bationo A, Sedogo MP, Cescas MP, Traore T (2013) Influence of legumes on nitrogen (N) fertilizer recommendations for succeeding sorghum in the Guinea savannah of West Africa. Afr J Agric Res 8:6416–6421

Bandillo N, Raghavan C, Muyco PA, Sevilla MAL, Lobina IT, Dilla-Ermita CJ, Tung C-W, McCouch S, Thomson M, Mauleon R, Singh RK, Gregorio G, Redoña E, Leung H (2013) Multi-parent advanced generation inter-cross (MAGIC) populations in rice: progress and potential for genetics research and breeding. Rice 6:1–15. https://doi.org/10.1186/1939-8433-6-11

Bao H, Chen X, Lv S, Jiang P, Feng J, Fan P, Nie L, Li Y (2015) Virus-induced gene silencing reveals control of reactive oxygen species accumulation and salt tolerance in tomato by γ-aminobutyric acid metabolic pathway. Plant Cell Environ 38:600–613

Baulcombe DC (1999) Fast forward genetics based on virus-induced gene silencing. Curr Opin Plant Biol 2:109–113

Bazzer SK, Kaler AS, Ray JD, Smith JR, Fritschi FB, Purcell LC (2020) Identification of quantitative trait loci for carbon isotope ratio (δ 13 C) in a recombinant inbred population of soybean. Theor Appl Genet 15:1–5

Becker A, Lange M (2010) VIGS genomics goes functional. Trends Plant Sci 15:1–4

Bernardo R (2010) Genomewide selection with minimal crossing in self-pollinated crops. Crop Sci 50:624–627

Bharti A, Richa Agnihotri, Hemant S Maheshwari, Anil Prakash and Sharma MP (2018) Bradyrhizobia-mediated drought tolerance in soybean and mechanisms involved. Springer Nature Singapore Pte Ltd. D. K. Choudhary et al (eds), In silico approach for sustainable agriculture, https://doi.org/10.1007/978-981-13-0347-0_7. pp 121-139

Bhatia VS, Jumrani K (2016) A maximin-minimax approach for classifying soybean genotypes for drought tolerance based on yield potential and loss. Plant Breed 136:691–700

Bhatia VS, Jumrani K, Pandey GP (2014) Evaluation of the usefulness of senescing agent potassium iodide as a screening tool for tolerance to terminal drought in soybean. Plant Knowl J 3(1):23–30

Bhatnagar S, King CA, Purcell L, Ray JD (2005) Identification and mapping of quantitative trait loci associated with crop responses to water-deficit stress in soybean [Glycine max (L.) Merr.]. The ASACSSA-SSSA International annual meeting poster abstract, November 6–10, 2005, Salt lake city, UT, USA

Bishopp A, Lynch JP (2015) The hidden half of crop yields. Nat Plant 1:15117. https://doi.org/10.1038/nplants.2015.117

Blair MW, Fernandez AC, Ishitani M, Moreta D, Seki M, Ayling S, Shinozaki K (2011) Construction and EST sequencing of full-length, drought stress cDNA libraries for common beans (*Phaseolus vulgaris* L.). BMC Plant Biol 11:171

Blum A (1989) Osmotic adjustment and growth of barley genotypes under drought stress. Crop Sci 29(1):230–233

Blum A (2009) Effective use of water (EUW) and not water-use efficiency (WUE) is the target of crop yield improvement under drought stress. Field Crops Res 112:119–123

Blum A (2011) Plant breeding for water-limited environments. Springer. ISBN 978-1-4419-7490-7. https://doi.org/10.1007/978-1-4419-7491-4

Blum A, Poyarkova H, Golan MJ (1983a) Chemical desiccation of wheat plants as a simulator of post-anthesis stress. I. Effects on translocation and kernel growth. Field Crops Res 6:51–58

Blum A, Mayer J, Golan G (1983b) Chemical desiccation of wheat plants as a simulator of post-anthesis stress. II. Relations to drought stress. Field Crops Res 6:149–155

Blum A, Shpiler L, Golan G, Mayer J, Sinmena B (1991) Mass selection of wheat for grain filling without transient photosynthesis. Euphytica 54:111–116

Bohnert HJ, Gong Q, Li P, Ma S (2006) Unraveling abiotic stress tolerance mechanisms – getting genomics going. Curr Opin Plant Biol 9:180–188

Borlaug NE (2007) Sixty-two years of fighting hunger: personal recollections. Euphytica 157:287–297

Boyer JS, James RA, Munns R, Codon TAG, Passioura JB (2008) Osmotic adjustment leads to anomalously low estimates of relative water content in wheat and barley. Funct Plant Biol 35:1172–1182

Burch-Smith TM, Anderson JC, Martin GB, Dinesh-Kumar SP (2004) Applications and advantages of virus-induced gene silencing for gene function studies in plants. Plant J 39:734–746

Cai Y, Chen L, Liu X, Sun S, Wu C, Jiang B, Han T, Hou W (2015) CRISPR/Cas9-mediated genome editing in soybean hairy roots. PLoS One 10:e0136064. https://doi.org/10.1371/journal.pone.0136064

Cai B, Li Q, Liu F, Bi H, Ai X (2018) Decreasing fructose 1,6-bisphosphate aldolase activity reduces plant growth and tolerance to chilling stress in tomato seedlings. Physiol Plant. https://doi.org/10.1111/ppl.12682

Cermak T, Baltes NJ, Cegan R, Zhang Y, Voytas DF (2015) High-frequency, precise modification of the tomato genome. Genome Biol 16:232. https://doi.org/10.1186/s13059-015-0796-9

Charlson DV, Bhatnagar S, King CA, Ray JD, Sneller CH, Carter TE Jr, Purcell L (2009) Polygenic inheritance of canopy wilting in soybean [*Glycine max* (L.) Merr.]. Theor Appl Genet 119:587–594

Chaudhary J, Shivaraj SM, Khatri P, Ye H, Zhou L, Klepadlo M, Dhakate P, Kumawat G, Patil G, Sonah H, Ratnaparkhe M (2019) Approaches, applicability, and challenges for development of climate-smart soybean. In: Genomic designing of climate-smart oilseed crops. Springer, Cham, pp 1–74

Chen Y, Chen P, de los Reyes BG (2006) Differential responses of the cultivated and wild species of soybean to dehydration stress. Crop Sci 46:2041–2046. https://doi.org/10.2135/cropsci2005.12.0466

Chen M, Wang QY, Chen XG et al (2007a) GmDREB2, a soybean DRE-binding transcription factor, conferred drought and high salt tolerance in transgenic plants. Biochem Biophys Res Commun 353:299–305

Chen P, Sneller CH, Purcell LC, Sinclair TR, King CA, Ishibashi T (2007b) Registration of soybean germplasm lines R01-416F and R01-581F for improved yield and nitrogen fixation under drought stress. J Plant Regist 1:166–167

Chen L, Song Y, Li S, Zhang L, Zhou C, Yu D (2012) The role of WRKY transcription factors in abiotic stresses. Biochim Biophys Acta 1819:120–128

Chen RG, Jing H, Guo WL, Wang SB, Ma F, Pan BG, Gong ZH (2015) Silencing of dehydrin CaDHN1 diminishes tolerance to multiple abiotic stresses in Capsicum annuum L. Plant Cell Rep 34:2189–2200

Chen L, Fang Y, Li X, Zeng K, Chen H, Zhang H, Yang H, Cao D, Hao Q, Yuan S, Zhang C (2020) Identification of soybean drought-tolerant genotypes and loci correlated with agronomic traits contributes new candidate genes for breeding. Plant Mol Biol 102(1–2):109–122

Cheng YQ, Liu JF, Yang X, Ma R, Liu C, Liu Q (2013) RNA-seq analysis reveals ethylene-mediated reproductive organ development and abscission in soybean (*Glycine max* L. Merr.). Plant Mol Biol Rep 31:607–619. https://doi.org/10.1007/s11105-012-0533-4

Cho SM, Kang BR, Han SH, Anderson AJ, Park JY, Lee YH, Cho BH, Yang KY, Ryu CM, Kim YC (2008) 2R, 3R-butanediol, a bacterial volatile produced by *Pseudomonas chlororaphis* O6, is involved in induction of systemic tolerance to drought in *Arabidopsis thaliana*. Mol Plant-Microbe Interact 21:1067–1075

Chung W-H, Jeong N, Kim J, Lee WK, Lee Y-G, Lee S-H, Yoon W, Kim J-H, Choi I-Y, Choi H-K (2014) Population structure and domestication revealed by high-depth resequencing of Korean cultivated and wild soybean genomes. DNA Res 21:153–167

Clement M, Lambert A, Herouart D, Boncompagni E (2008) Identification of new up-regulated genes under drought stress in soybean nodules. Gene 426:15–22. https://doi.org/10.1016/j.gene.2008.08.016

Comas LH, Becker SR, Cruz VMV, Byrne PF, Dierig DA (2013) Root traits contributing to plant productivity under drought. Front Plant Sci 4:1–16

Cook ER, Seager R, Cane MA, Stahle DW (2007) North American drought: reconstructions, causes, and consequences. Earth Sci Rev 81:93–134

de Dorlodot S, Forster B, Pages L, Price A, Tuberosa R, Draye X (2007) Root system architecture: opportunities and constraints for genetic improvement of crops. Trends Plant Sci 12:474–481

Denher CA, Barton L, Maurice PA, Dubais JL (2010) Size dependent bio availability of hematite (α Fe_2O_3) nanoparticles to a common aerobic bacterium. Environ Sci Technol 45:977–983

Deshmukh RK, Sonah H, Kondawar V, Tomar RSS, Deshmukh NK (2012) Identification of meta quantitative trait loci for agronomical traits in rice (*Oryza sativa*). Ind J Genet Plant Breed 72:264–270

Deshmukh R, Sonah H, Patil G, Chen W, Prince S, Mutava R, Vuong T, Valliyodan B, Nguyen HT (2014) Integrating omic approaches for abiotic stress tolerance in soybean. Front Plant Sci 5:244

Dhanapal AP, Ray JD, Singh SK, Hoyos-Villegas V, Smith JR, Purcell LC, Andy King C, Cregan PB, Song Q, Fritschi FB (2015) Genome-wide association study (GWAS) of carbon isotope ratio (d13C) in diverse soybean [*Glycine max* (L.) Merr.] genotypes. Theor Appl Genet 128:73

Dhanapal AP, Ray JD, Singh SK, Hoyos-Villegas V, Smith JR, Purcell LC, Fritschi FB (2016) Genome-wide association mapping of soybean chlorophyll traits based on canopy spectral reflectance and leaf extracts. BMC Plant Biol 16(1):174

Dias FG, Borges ACN, Viana AAB, Mesquita RO, Romano E, Grossi de Sa MF, Nepomuceno AL, Loureiro ME, Ferreira MA (2012) Expression analysis in response to drought stress in soybean: shedding light on the regulation of metabolic pathway genes. Genet Mol Biol 35:222–232

Dietz KJ, Vogel MO, Viehhauser A (2010) AP2/EREBP transcription factors are part of gene regulatory networks and integrate metabolic, hormonal and environmental signals in stress acclimation and retrograde signalling. Protoplasma 245:3–14

Doss BD, Thurlow DL (1974) Irrigation, row width and plant population in relation to growth characteristics of two soybean varieties. Agron J 65:620–623

Draye X, Kim Y, Lobet G, Javaux M (2010) Model-assisted integration of physio-logical and environmental constraints affecting the dynamic and spatial patterns of root water uptake from soils. J Exp Bot 61:2145–2155

Du W, Wang M, Fu S, Yu D (2009a) Mapping QTLs for seed yield and drought susceptibility index in soybean (Glycine max) across different environments. J Genet Genom 36:721–731

Du W, Yu D, Fu S (2009b) Detection of quantitative trait loci for yield and drought tolerance traits in soybean using a recombinant inbred line population. J Integr Plant Biol 51:868–878

Du H, Zeng X, Zhao M, Cui X, Wang Q, Yang H, Cheng H, Yu D (2016) Efficient targeted mutagenesis in soybean by TALENs and CRISPR/Cas9. J Biotechnol 217:90–97. https://doi.org/10.1016/j.jbiotec.2015.11.005

Easterling WE, Aggarwal PK, Batima P, Brander KM, Erda L, Howden SM, Kirilenko A, Morton J, Soussana JF, Schmidhuber J, Tubiello FN (2007) Food, fibre and forest products. In: Parry ML, Canziani OF, Palutikof JP, van der Linden PJ, Hanson CE (eds) Climate change 2007: impacts, adaptation and vulnerability. Contribution of working group II to the fourth assessment report of the intergovernmental panel on climate change. Cambridge University Press, Cambridge, pp 273–313

Fenta BA, Beebe SE, Kunert KJ, Burridge JD, Barlow KM, Lynch JP, Foyer CH (2014) Field phenotyping of soybean roots for drought stress tolerance. Agronomy 4:418–435

Fletcher AL, Sinclair TR, Allen LH Jr (2007) Transpiration responses to vapor pressure deficit in well-watered 'slow-wilting' and commercial soybean. Environ Exp Bot 61:145–151

Foley JA, Ramankutty N, Brauman KA, Cassidy ES, Gerber JS, Johnston M, Mueller ND, Connell C, Ray DK, West PC, Balzer C, Bennett EM, Carpenter SR, Hill J, Monfreda C, Polasky S, Rockström J, Sheehan J, Siebert S, Tilman D, Zaks DPM (2011) Solutions for a cultivated planet. Nature 478:337–342. https://doi.org/10.1038/nature10452

Fried HG, Narayanan S, Fallen B (2018) Characterization of a soybean (Glycine max L. Merr.) germplasm collection for root traits. PLoS One 13(7):e0200463

Fried HG, Narayanan S, Fallen B (2019) Evaluation of soybean [Glycine max (L.) Merr.] genotypes for yield, water use efficiency, and root traits. PLoS One 14(2):e0212700. https://doi.org/10.1371/journal.pone.0212700

Fujita M, Fujita Y, Noutoshi Y, Takahashi F, Narusaka Y, Yamaguchi-Shinozaki K, Shinozaki K (2006) Crosstalk between abiotic and biotic responses: a current view from the point of convergence in the stress signalling networks. Curr Opin Plant Biol 9:436–442

Fukai S, Pantuwan G, Jongdee B, Cooper M (1999) Screening for drought resistance in rainfed lowland rice. Field Crop Res 64:61–74

Gao H, Smith J, Yang M, Jones S, Djukanovic V, Nicholson MG, West A, Bidney D, Falco SC, Jantz D, Lyznik LA (2010) Heritable targeted mutagenesis in maize using a designed endonuclease. Plant J 61:176–187. https://doi.org/10.1111/j.1365-313X.2009.04041.x

Gao SQ, Chen M, Xu ZS et al (2011) The soybean GmbZIP1 transcription factor enhances multiple abiotic stress tolerance in transgenic plants. Plant Mol Biol 75:537–553

Garay AF, Wilhelm WW (1982) Root system characteristics of two soybean isolines undergoing water stress conditions. Agron J 75:973–977

Garg R, Verma M, Agrawal S, Shankar R, Majee M, Jain M (2014) Deep transcriptome sequencing of wild halophyterice, Porteresiacoarctata, provides novel insights into the salinity and submergence tolerance factors. DNA Res 21:69–84. https://doi.org/10.1093/dnares/dst042

Gorantla M, Babu PR, Lachagari VBR, Reddy AMM, Wusirika R, Bennetzen JL, Reddy AR (2007) Identification of stress-responsive genes in an indica rice (Oryza sativa L.) using ESTs generated from drought stressed seedlings. J Exp Bot 58:253–265

Govind G, Thammegowda HV, Kalaiarasi PJ, Iyer DR, Muthappa SK, Nese S, Makarla UK (2009) Identification and functional validation of a unique set of drought induced genes deferentially expressed in response to gradual water stress in peanut. Mol Gen Genomics 281:591–605. https://doi.org/10.1007/s00438-009-0432-z

Grant D, Nelson RT, Cannon SB, Shoemaker RC (2010) SoyBase, the USDA-ARS soybean genetics and genomics database. Nucleic Acids Res 38(suppl_1):D843–D846

Guo C, Yao L, You C, Wang S, Cui J, Ge X, Ma H (2016) MID1 plays an important role in response to drought stress during reproductive development. Plant J 88:280–293

Gupta K, Sengupta A, Saha J, Gupta B (2014) The attributes of RNA interference in relation to plant abiotic stress tolerance. Gene Technol 3:110. https://doi.org/10.4172/2329-6682.1000110

Guttikonda SK et al (2014) Overexpression of AtDREB1D transcription factor improves drought tolerance in soybean. Mol Biol Rep 41:7995–8008

Ha CV, Watanabe Y, Tran UT, Le DT, Tanaka M, Nguyen KH, Seki M, Nguyen DV, Tran LSP (2015) Comparative analysis of root transcriptomes from two contrasting drought-responsive Williams 82 and DT 2008 soybean cultivars under normal and dehydration conditions. Front Plant Sci 6:551

Haerizadeh F, Singh MB, Bhalla PL (2011) Transcriptome profiling of soybean root tips. Funct Plant Biol 38(6):451–461

Hafner H, George E, Bationo A, Marschner H (1993) Effect of crop residues on root growth and phosphorus acquisition of pearl millet in an acid sandy soil in Niger. Plant Soil 150:117–127

Haley SD, Quick JS (1993) Early-generation selection for chemical desiccation tolerance in winter wheat. Crop Sci 33:1217–1223

Hao D, Chao M, Yin Z, Yu D (2012) Genome-wide association analysis detecting significant single nucleotide polymorphisms for chlorophyll and chlorophyll fluorescence parameters in soybean (Glycine max) landraces. Euphytica 186:919–931

Heffner EL, Sorrells ME, Jannink JL (2009) Genomic selection for crop improvement. Crop Sci 49:1–12. https://doi.org/10.2135/cropsci2008.08.0512

Herritt M, Dhanapal AP, Purcell LC, Fritschi FB (2018) Identification of genomic loci associated with 21chlorophyll fluorescence phenotypes by genome-wide association analysis in soybean. BMC Plant Biol 18(1):312

Heslot N, Akdemir D, Sorrells ME, Jannink JL (2014) Integrating environmental covariates and crop modeling in to the genomic selection frame work to predict genotype by environment interactions. Theor Appl Genet 127:463–480. https://doi.org/10.1007/s00122-013-2231-5

Hettenhausen C, Sun G, He Y, Zhuang H, Sun T, Qi J, Wu J (2016) Genome-wide identification of calcium-dependent protein kinases in soybean and analyses of their transcriptional responses to insect herbivory and drought stress. Sci Rep 6:18973

Hose E, Clarkson DT, Steudle E, Schreiber L, Hartung W (2001) The exodermis: a variable apoplastic barrier. J Exp Bot 52:2245–2264

Hossain MM, Liu X, Qi X, Lam HM, Zhang J (2014) Differences between soybean genotypes in physiological response to sequential soil drying and rewetting. Crop J 2(6):366–380. https://doi.org/10.1016/j.cj.2014.08.001

Hossain Z, Mustafa G, Sakata K, Komatsu S (2016) Insights into the proteomic response of soybean towards Al2O3, ZnO, and Ag nanoparticles stress. J Hazard Mater 304:291–305

Hussain RM, Ali M, Feng X, Li X (2017) The essence of NAC gene family to the cultivation of drought-resistant soybean (Glycine max L Merr.) cultivars. BMC Plant Biol 17:55

Hwang T-Y, Sayama T, Takahashi M, Takada Y, Nakamoto Y, Funatsuki H, Hisano H, Sasamoto S, Sato S, Tabata S (2009) High-density integrated linkage map based on SSR markers in soybean. DNA Res 16:213–225

Hwang S, King C, Davies M, Ray JD, Cregan P, Purcell LC (2013) QTL analysis of shoot ureide and nitrogen concentrations in soybean [Glycine max (L.) Merr.]. Crop Sci 53:2421–2433

Hwang S, King CA, Ray JD, Cregan PB, Chen P, Carter TE, Li Z, Abdel-Haleem H, Matson KW, Schapaugh W (2015) Confirmation of delayed canopy wilting QTLs from multiple soybean mapping populations. Theor Appl Genet 128:2047–2065

Jagtap UB, Gurav RG, Bapat VA (2011) Role of RNA interference in plant improvement. Naturwissenschaften 98:473–492

Jan SA, Shinwari ZK, Shah SH et al (2016) In-planta transformation recent advances. Rom Biotechnol Lett 21(1):11085–11091

Jan SA, Bibi N, Shinwari ZK et al (2017) Impact of salt, drought, heat and frost stresses on morpho-biochemical and physiological properties of Brassica species: an updated review. J Rural Dev Agric 2(1):1–10

Javot H (2003) Role of a single aquaporin isoform in root water uptake. Plant Cell 15:509–522

Ji X, Nie X, Liu Y, Zheng L, Zhao H, Zhang B, Huo L, Wang Y (2016) A bHLH gene from Tamarix hispida improves abiotic stress tolerance by enhancing osmotic potential and decreasing reactive oxygen species accumulation. Tree Physiol 36:193–207

Kaldenhoff R, Fischer M (2006) Aquaporins in plants. Acta Physiol (Oxf) 187:169–176

Kaler AS, Dhanapal AP, Ray JD, King CA, Fritschi FB, Purcell LC (2017) Genome-wide association mapping of carbon isotope and oxygen isotope ratios in diverse soybean genotypes. Crop Sci 6:3085–3100

Kaler AS, Ray JD, Schapaugh WT, Asebedo AR, King CA, Gbur EE, Purcell LC (2018) Association mapping identifies loci for canopy temperature under drought in diverse soybean genotypes. Euphytica 214(8):135

Kasuga M, Miura S, Shinozaki K, Yamaguchi-Shinozaki K (2004) A combination of the Arabidopsis DREB1A gene and stress-inducible rd29A promoter improved drought-and low-temperature stress tolerance in tobacco by gene transfer. Plant Cell Physiol 45(3):346–350

Kell DB (2011) Breeding crop plants with deep roots: their role in sustainable carbon, nutrient and water sequestration. Ann Bot 108(3):407–418. https://doi.org/10.1093/aob/mcr175

Khan MA, Tong F, Wang W, He J, Zhao T, Gai J (2019) Correction to: analysis of QTL–allele system conferring drought tolerance at seedling stage in a nested association mapping population of soybean [Glycine max (L.) Merr.] using a novel GWAS procedure. Planta 249:1653. https://doi.org/10.1007/s00425-019-03143-0

Kim MY, Lee S, Van K, Kim T-H, Jeong S-C, Choi I-Y, Kim D-S, Lee Y-S, Park D, Ma J (2010) Whole-genome sequencing and intensive analysis of the undomesticated soybean (Glycine soja Sieb. and Zucc.) genome. Proc Natl Acad Sci 107:22032–22037

King CA, Purcell LC, Brye KR (2009) Differential wilting among soybean genotypes in response to water deficit. Crop Sci 49(1):290–298. https://doi.org/10.2135/cropsci2008.04.0219

Kover PX, Valdar W, Trakalo J, Scarcelli N, Ehrenreich IM, Purugganan MD, Durrant C, Mott R (2009) A multi parent advanced generation inter-cross to fine-map quantitative traits in Arabidopsis thaliana. PLoS Genet 5:e1000551. https://doi.org/10.1371/journal.pgen.1000551

Kuzuoglu-Ozturk D, Cebeci Yalcinkaya O, Akpinar BA, Mitou G, Korkmaz G, Gozuacik D, Budak H (2012) Autophagy-related gene, TdAtg8, in wild emmer wheat plays a role in drought and osmotic stress response. Planta 236:1081–1092

Lam HM, Xu X, Liu X, Chen W, Yang G, Wong FL, Li MW, He W, Qin N, Wang B (2010) Resequencing of 31 wild and cultivated soybean genomes identifies patterns of genetic diversity and selection. Nat Genet 42:1053

Le DT, Nishiyama R, Watanabe Y, Tanaka M, Seki M, Ham LH, Yamaguchi-Shinozaki K, Shinozaki K, Tran LSP (2012) Differential gene expression in soybean leaf tissues at late developmental stages under drought stress revealed by genome-wide transcriptome analysis. PLoS One 7:e49522. https://doi.org/10.1371/journal.pone.0049522

Lee G, Wu X, Shannon J, Sleper D, Nguyen H (2006) Genome mapping and Mol Breed in plants: soybean. In: Kole C (ed) Genome mapping and mol breed in plants, vol 2 (oilseeds). Springer, New York, pp 1–45

Lee YG, Jeong N, Kim JH, Lee K, Kim KH, Pirani A, Ha BK, Kang ST, Park BS, Moon JK (2015) Development, validation and genetic analysis of a large soybean SNP genotyping array. Plant J 81:625–636

Leisner CP, Yendrek CR, Ainsworth EA (2017) Physiological and transcriptomic responses in the seed coat of field-grown soybean (Glycine max L. Merr.) to abiotic stress. BMC Plant Biol 17:242

Li CD, Jiang HW, Liu CY, Guo T, Xin WZ, Hong WX et al (2011) QTL identification of drought tolerance to soybean in selection population. Acta Agron Sin 37:603–611

Li T, Liu B, Spalding MH, Weeks DP, Yang B (2012) High-efficiency TALEN-based gene editing produces disease-resistant rice. Nat Biotechnol 30:390–392

Li JF, Norville JE, Aach J, McCormack M, Zhang D, Bush J, Church GM, Sheen J (2013a) Multiplex and homologous recombination-mediated genome editing in Arabidopsis and Nicotiana benthamiana using guide RNA and Cas9. Nat Biotechnol 31:688–691. https://doi.org/10.1038/nbt.2654

Li YH, Zhao SC, Ma JX, Li D, Yan L, Li J, Qi XT, Guo XS, Zhang L, He WM (2013b) Molecular footprints of domestication and improvement in soybean revealed by whole genome re-sequencing. BMC Genomics 14:579

Li YH, Zhou G, Ma J, Jiang W, Jin LG, Zhang Z, Guo Y, Zhang J, Sui Y, Zheng L (2014) De novo assembly of soybean wild relatives for pan-genome analysis of diversity and agronomic traits. Nat Biotechnol 32:1045

Li S, Castillo-González C, Yu B, Zhang X (2017a) The functions of plant small RNAs in development and in stress responses. Plant J 90:654–670

Li Y, Chen Q, Nan H, Li X, Lu S, Zhao X, Liu B, Guo C, Kong F, Cao D (2017b) Overexpression of GmFDL19 enhances tolerance to drought and salt stresses in soybean. PLoS One 12: e0179554. https://doi.org/10.1371/journal.pone.0179554

Liang H, Yu Y, Yang H, Xu L, Dong W, Du H, Cui W, Zhang H (2014) Inheritance and QTL mapping of related root traits in soybean at the seedling stage. Theor App Genet 10:2127–2137
Linh TM, Mai NC, Hoe PT, Lien LQ, Ban NK, Hien LTT, Chau NH, Van NT (2020) Metal-based nanoparticles enhance drought tolerance in soybean. J Nanomater 2020:1–13. https://doi.org/10.1155/2020/4056563
Liu Y, Gai JY, Lü HN, Wang YJ, Chen SY (2005) Identification of drought tolerant germplasm and inheritance and QTL mapping of related root traits in soybean (Glycine max (L.) Merr.). Yi Chuan Xue Bao 32(8):855–863
Liu Z, Li H, Gou Z, Zhang Y, Wang X, Ren H, Wen Z, Kang BK, Li Y, Yu L, Gao H (2020a) Genome-wide association study of soybean seed germination under drought stress. Mol Genet Genomics 1:1–3
Liu Y, Du H, Li P et al (2020b) Pan-genome of wild and cultivated soybeans. Cell 182:162–176.e13
Lobet G, Pages L, Draye X (2011) A novel image-analysis toolbox enabling quantitative analysis of root system architecture. Plant Physiol 157:29–39
Long SP, Ainsworth EA, Leakey ADB, Morgan PB (2005) Global food insecurity. Treatment of major food crops with elevated CO2 or ozone under large-scale fully open-air conditions suggests recent models may have overestimated future yields. Philos Trans R Soc Lond B Sci 360:2011–2020
Lu L, Dong C, Liu R, Zhou B, Wang C, Shou H (2018) Roles of soybean plasma membrane intrinsic protein GmPIP2;9 in drought tolerance and seed development. Front Plant Sci 9:530
Lynch JP (2013) Steep, cheap and deep: an ideotype to optimize water and N acquisition by maize root systems. Ann Bot 112:347–357
Lynch JP, Brown KM (2012) New roots for agriculture: exploiting the root phenome. Philos Trans R Soc B Biol Sci 367:1598–1604
Lyzenga WJ, Stone SL (2012) Abiotic stress tolerance mediated by protein ubiquitination. J Exp Bot 63:599–616
Ma H, Chen J, Zhang Z, Ma L, Yang Z, Zhang Q, Li X, Xiao J, Wang S (2017) MAPK kinase 10.2 promotes disease resistance and drought tolerance by activating different MAPKs in rice. Plant J 92:557–570
Mahalakshmi V, Blum A (2006) Phenotyping in the field: global capacity accessible to the GCP – Inventory of phenotyping resources and capacity for the GCP. Final report, Generation Challenge Program, El Batan, Mexico
Manavalan LP, Guttikonda SK, Tran LSP, Nguyen HT (2009) Physiological and molecular approaches to improve drought resistance in soybean. Plant Cell Physiol 50:1260–1276
Manavalan LP, Guttikonda SK, Nguyen VT, Shannon JG, Nguyen HT (2010) Evaluation of diverse soybean germplasm for root growth and architecture. Plant Soil 330:503–514
Manavalan LP, Prince SJ, Musket TA, Chaky J, Deshmukh R, Vuong TD, Song LI, Cregan PB, Nelson JC, Shannon G, Specht JE, Nguyen HT (2015) Identification of novel QTL governing root architectural traits in an interspecific soybean population. PLoS One 10(3):e0120490. https://doi.org/10.1371/journal.pone.0120490
Manmathan H, Shaner D, Snelling J, Tisserat N, Lapitan N (2013) Virus-induced gene silencing of Arabidopsis thaliana gene homologues in wheat identifies genes conferring improved drought tolerance. J Exp Bot 64:1381–1392
Mao Y, Zhang H, Xu N, Zhang B, Gou F, Zhu JK (2013) Application of the CRISPR-Cas system for efficient genome engineering in plants. Mol Plant 6:2008–2011. https://doi.org/10.1093/mp/sst121
Mao C, Ding J, Zhang B, Xi D, Ming F (2018) OsNAC2 positively affects salt-induced cell death and binds to the OsAP37 and OsCOX11 promoters. Plant J 94(3):454–468. https://doi.org/10.1111/tpj.13867
Marques MC, Alonso-Cantabrana H, Forment J, Arribas R, Alamar S, Conejero V, Perez-Amador M (2009) A new set of ESTs and cDNA clones from full-length and normalized libraries for gene discovery and functional characterization in citrus. BMC Genomics 10:428

Marton I, Zuker A, Shklarman E, Zeevi V, Tovkach A, Roffe S, Ovadis M, Tzfira T, Vainstein A (2010) Non transgenic genome modification in plant cells. Plant Physiol 154:1079–1087. https://doi.org/10.1104/pp.110.164806

Maurel C, Santoni V, Luu DT, Wudick MM, Verdoucq L (2009) The cellular dynamics of plant aquaporin expression and functions. Curr Opin Plant Biol 12:690–698

Meena AK, Verma LK, Kumhar BL (2017) RNAi, Its mechanism and potential use in crop improvement: a review. Int J Pure App Biosci 5:294–311

Melchinger A (1990) Use of molecular markers in breeding for oligogenic disease resistance. Plant Breed 104:1–19

Mian M, Bailey M, Ashley D, Wells R, Carter T, Parrott W, Boerma H (1996) Molecular markers associated with water use efficiency and leaf ash in soybean. Crop Sci 36:1252–1257

Mian M, Ashley D, Boerma H (1998) An additional QTL for water use efficiency in soybean. Crop Sci 38:390–393

Mickelbart MV, Hasegawa PM, Bailey-Serres J (2015) Genetic mechanisms of abiotic stress tolerance that translate to crop yield stability. Nat Rev Genet 16:237–251. https://doi.org/10.1038/nrg3901

Mizoi J, Shinozaki K, Yamaguchi-Shinozaki K (2012) AP2/ERF family transcription factor in plant abiotic stress responses. Biochim Biophys Acta 1819:86–96

Mohammadi PP, Moieni A, Hiraga S, Komatsu S (2012) Organ specific proteomic analysis of drought stressed soybean seedlings. J Proteome 75:1906–1923

Monteros MJ, Lee G, Missaoui AM, Carter TE, Boerma HR (2006) Identification and confirmation of QTL conditioning drought tolerance in Nepalese soybean. In: The 11th biennial conference on the molecular and cellular biology of the soybean, abstract PI471938. August 5–8, Lincoln, NE

Mourtzinis S, Specht JE, Conley SP (2019) Defining optimal soybean sowing dates across the US. Sci Rep 9:2800

Nakashima K, Ito Y, Yamaguchi-Shinozaki K (2009) Transcriptional regulatory networks in response to abiotic stresses in Arabidopsis and grasses. Plant Physiol 149:88–95. https://doi.org/10.1104/pp.108.129791

Nguyen QH, VuL TK, Nguyen LTN, Pham NTT, Nguyen YTH, Le SV, Chu MH (2019) Overexpression of the *GmDREB6* gene enhances proline accumulation and salt tolerance in genetically modified soybean plants. Sci Rep 9:1966

Ocheltree TW, Nippert JB, Prasad PVV (2014) Stomatal responses to changes in vapor pressure deficit reflect tissue-specific differences in hydraulic conductance. Plant Cell Environ 37:132–139

Oya T, Nepomuceno AL, Farias JRB, Tobita S, Ito O (2004) Drought tolerance characteristics of Brazilian soybean cultivars – evaluation and characterization of drought tolerance of various Brazilian soybean cultivars in the field. Plant Prod Sci 7:129–137

Ozsolak F, Milos PM (2010) RNA sequencing: advances, challenges and opportunities. Nat Rev Genet 12:87–98. https://doi.org/10.1038/nrg2934

Pages L, Serra V, Draye X, Doussan C, Pierret A (2010) Estimating root elongation rates from morphological measurements of the root tip. Plant Soil 328:35–44

Park C, Lim CW, Baek W, Kim JH, Lim S, Kim SH, Kim KN, Lee SC (2017) The pepper WPP domain protein, CaWDP1, acts as a novel negative regulator of drought stress via ABA signaling. Plant Cell Physiol 58:779–788

Passioura J (2010) Scaling up: the essence of effective agricultural research. Funct Plant Biol 37:585–591

Pathan SM, Lee JD, Sleper DA, Fritschi FB, Sharp RE, Carter TE Jr, Nelson RL, King CA, Schapaugh WT, Ellersieck MR, Nguyen HT, Shannon JG (2014) Two soybean plant introductions display slow leaf wilting and reduced yield loss under drought. J Agron Crop Sci 200(3):231–236. https://doi.org/10.1111/jac.12053

Patterson PR, Hudak CM (1996) Drought-avoidant soybean germplasm maintains nitrogen fixation capacity under water stress. Plant Soil 186(1):39–43. https://doi.org/10.1007/BF00035053

Pennisi E (2008) The blue revolution, drop by drop, gene by gene. Science 320:171–173

Phang TH, Shao GH, Lam HM (2008) Salt tolerance in soybean. J Integr Plant Biol 50:1196–1212

Pinto RS, Reynolds MP (2015) Common genetic basis for canopy temperature depression under heat and drought stress associated with optimized root distribution in bread wheat. Theor Appl Genet 128(4):575–585

Poland JA, Rife TW (2012) Genotyping-by-sequencing for plant breeding and genetics. Plant Genome 5:92–102

Pradhan A, Naik N, Sahoo KK (2015) RNAi mediated drought and salinity stress tolerance in plants. Am J Plant Sci 6:1990–2008

Prince SJ, Song L, Qiu D, dos Santos JVM, Chai C, Joshi T, Patil G, Valliyodan B, Vuong TD, Murphy M (2015a) Genetic variants in root architecture-related genes in a Glycine soja accession, a potential resource to improve cultivated soybean. BMC Genomics 16:132

Prince SJ, Joshi T, Mutavaa RN, Syed N, dos Santos JVM, Patil G, Songa L, Wang JJ, Lina L, Chena W, Shannona JG, Valliyodana B, Xub D, Nguyen HT (2015b) Comparative analysis of the drought responsive transcriptome in soybean lines contrasting for canopy wilting. Plant Sci 240:65–78. https://doi.org/10.1016/j.plantsci.2015.08.017

Purcell LC, DeSilva M, King CA, Kim WH (1997) Biomass accumulation and allocation in soybean associated with genotypic differences in tolerance of nitrogen fixation to water deficits. Plant Soil 196(1):101–113. https://doi.org/10.1023/A:1004289609466

Qi X, Li MW, Xie M, Liu X, Ni M, Shao G, Song C, Kay-Yuen Yim A, Tao Y, Wong FL, Isobe S, Wong CF, Wong KS, Xu C, Li C, Wang Y, Guan R, Sun F, Fan G, Xiao Z, Zhou F, Phang TH, Liu X, Tong SW, Chan TF, Yiu SM, Tabata S, Wang J, Xu X, Lam HM (2014) Identification of a novel salt tolerance gene in wild soybean by whole-genome sequencing. Nat Commun 5:4340

Qin H, Wang Y, Wang J, Liu H, Zhao H, Deng Z, Zhang Z, Huang R, Zhang Z (2016) Knocking down the expression of GMPase gene OsVTC1-1 decreases salt tolerance of rice at seedling and reproductive stages. PLoS One 11(12):e0168650. https://doi.org/10.1371/journal.pone.0168650

Qiu PC, Zhang WB, Li CD, Jiang HW, Liu CY, Fan DM, Zeng QL, Hu HG, Chen QS (2011) Genetic overlap of drought-tolerance loci between germination stage and seedling stage analyzed using introgression lines in soybean. Acta Agron Sin 37(3):477–483. https://doi.org/10.3724/SP.J.1006.2011.00477

Qiu J, Wang Y, Wu S, Wang Y-Y, Ye C-Y, Bai X, Li Z, Yan C, Wang W, Wang Z (2014) Genome re-sequencing of semi-wild soybean reveals a complex soja population structure and deep introgression. PLoS One 9:e108479

Rabara RC, Tripathi P, Lin J, Rushton PJ (2013) Dehydration-induced WRKY genes from tobacco and soybean respond to jasmonic acid treatments in BY-2 cell culture. Biochem Biophys Res Commun 431:409–414

Rabara RC, Tripathi P, Rushton PJ (2014) The potential of transcription factor-based genetic engineering in improving crop tolerance to drought. OMICS J Integr Biol 18(10):601–614

Ramegowda V, Gill US, Sivalingam PN, Gupta A, Gupta C, Govind G, Nataraja KN, Pereira A, Udayakumar M, Mysore KS, Senthil-Kumar M (2017) GBF3 transcription factor imparts drought tolerance in Arabidopsis thaliana. Sci Rep 7:9148. https://doi.org/10.1038/s41598-017-09542-1

Rao SS, El-Habbak MH, Havens WM, Singh AK, Zheng D, Vaughn L, Haudenshield JS, Hartman GL, Korban SS, Ghabrial SA (2014) Overexpression of GmCaM4 in soybean enhances resistance to pathogens and tolerance to salt stress. Mol Plant Pathol 15:145–160

Ratnakumar P, Vadez V (2011) Groundnut (Arachis hypogaea) genotypes tolerant to intermittent drought maintains a high harvest index and have small leaf canopy under stress. Funct Plant Biol 38:1016–1102

Ratnaparkhe MB, Ramesh SV, Giriraj Kumawat, Husain SM, Sanjay Gupta (2014) Soybean Genomics. In Legumes in the Omic Era, Gupta S eds, Springer Heidelberg pp 41-72

Ratnaparkhe MB, Marmat N, Kumawat G, Shivakumar M, Viraj Kamble, Nataraj V, Deshmukh MP, Ramesh SV, Singh AK, Sonha H, Deshmukh R, Prasad M , Chand S, S Gupta (2020)

Whole genome re-sequencing of soybean accession EC241780 providing genomic landscape of candidate genes involved in rust resistance. Current Genomics 21(7): 504-511. DOI: 10.2174/1389202921999200601142258

Ray DK, Mueller ND, West PC, Foley JA (2013) Yield trends are insufficient to double global crop production by 2050. PLoS One 8:e66428

Ray JD, Dhanapal AP, Singh SK, Hoyos-Villegas V, Smith JR, Purcell LC, King CA, Boykin D, Cregan PB, Song Q, Fritschi FB (2015) Genome-wide association study of ureide concentration in diverse maturity group IV soybean [Glycine max (L.) Merr.] accessions. G3: Genes, Genomes, Genet 5(11):2391–2403

Reynolds M, Manes Y, Izanloo A, Langridge P (2009) Phenotyping approaches for physiological breeding and gene discovery in wheat. Ann Appl Biol 155(3):309–320

Rivero RM, Kojima M, Gepstein A, Sakakibara H, Mittler R, Gepstein S, Blumwald E (2007) Delayed leaf senescence induces extreme drought tolerance in a flowering plant. PNAS 104 (49):19631–19636

Robertson MJ, Fukai S, Ludlow MM, Hammer GL (1993) Water extraction by grain sorghum in a sub-humid environment. 2. Extraction in relation to root growth. Field Crops Res 33:99–112

Roco MC (2003) Nanotechnology: convergence with modern biology and medicine. Curr Opin Biotechnol 14:337–346. https://doi.org/10.1016/S0958-1669(03)00068-5

Rushton PJ, Somssich IE, Ringler P, Shen QJ (2010) WRKY transcription factors. Trends Plant Sci 15:247–258

Sachdeva S, Bharadwaj C, Singh RK, Jain PK, Patil BS, Roorkiwal M, Rajeev Varshney (2020) Characterization of ASR gene and its role in drought tolerance in chickpea (Cicer arietinum L.). PLoS ONE 15(7): e0234550. https://doi.org/10.1371/journal.pone.0234550

Sakurai J, Ishikawa F, Yamaguchi T, Uemura M, Maeshima M (2005) Identification of 33 rice aquaporin genes and analysis of their expression and function. Plant Cell Physiol 46:1568–1577

Salinger MJ, Sivakumar MVK, Motha R (2005) Reducing vulnerability of agriculture and forestry to climate variability and change: workshop summary and recommendations. Clim Chang 70:341–362

Satpute GK, Arya M, Gupta S, Bhatia VS, Devdas R, Ratnaparkhe MB, Kumawat G, Maharaj S (2018) Root system architecture for improvement of moisture deficit stress tolerance traits in soybean (Glycine max L Meril.). In: Extended summaries national conference on enhancing productivity of oilseeds in changing climate scenario held during 7–9 April, 2018 at ICAR-DGR, Junagadh, India, pp 8–9

Satpute GK, Gupta S, Bhatia VS, Devdas R, Kumawat G, Tiwari VK, Purwar JP, Agrawal A (2019a) Multi-tiered selection scheme for breeding drought tolerance in soybean. In: Abstr. golden jubilee international salinity conference (gjisc): resilient agriculture in saline environments under changing climate: challenges & opportunities, held during February 7–9, 2019 at ICAR – CSSRI, Karnal, Haryana, pp 127

Satpute GK, Gupta S, Bhatia VS, Maharaj S, Chandra S, Nagar S, Kamble V, Ratnaparkhe MB (2019b) Improving seed filling under drought stress by stem reserve mobilization in soybean, glycine max. In: Proc. 5th Intl. conf. plant genetics & genomics: germplasm to genome engineering, held during 17–18, October 2019 at New Delhi, India

Satpute GK, Arya M, Gupta S, Bhatia VS, Ramgopal D, Ratnaparkhe MB, Chandra S, Singh M, Nagar S, Kamble VG, Pandey S, Kumawat G, Shivakumar M, Nataraj V, Rajesh V (2020) Identifying drought tolerant germplasm through multiplexing polygenic traits in soybean (Glycine max L. Merrill). J Oilseeds Res 37(SI):56–57

Saxena R, Tomar RS, Kumar M (2016) Exploring nanobiotechnology to mitigate abiotic stress in crop plants. J Pharm Sci Res 8(9):974–980

Schmutz J, Cannon SB, Schlueter J, Ma J, Mitros T, Nelson W, Hyten DL, Song Q, Thelen JJ, Cheng J (2010) Genome sequence of the palaeopolyploid soybean. Nature 463:178

Schuurmans JA, van Dongen JT, Rutjens BP, Boonman A, Pieterse CM, Borstlap AC (2003) Members of the aquaporin family in the developing pea seed coat include representatives of the PIP, TIP, and NIP subfamilies. Plant Mol Biol 53:633–645

Sedghi M, Mitra H, Sahar T (2013) Effect of nano zinc oxide on the germination of soybean seeds under drought stress. Ann West Univ Timi Ser Biol 16(2):73–78

Seki M, Narusaka M, Ishida J, Nanjo T, Fujita M, Oono Y, Kamiya A, Nakajima M, Enju A, Sakurai T, Satou M, Akiyama K, Taji T, Yamaguchi-Shinozaki K, Carninci P, Kawai J, Hayashizaki Y, Shinozaki K (2002) Monitoring the expression profiles of 7000 Arabidopsis genes under drought, cold and high-salinity stresses using a full-length cDNA microarray. Plant J 31:279–292

Senthil-Kumar M, Mysore KS (2011) New dimensions for VIGS in plant functional genomics. Trends Plant Sci 16:656–665

Senthil-Kumar M, Udayakumar M (2006) High-throughput virus induced gene silencing approach to assess the functional relevance of a moisture stress-induced cDNA homologous to Lea4. J Exp Bot 57:2291–2302

Senthil-Kumar M, Govind G, Kang L, Mysore KS, Udayakumar M (2007) Functional characterization of Nicotiana benthamiana homologs of Peanut water deficit-induced genes by virus induced gene silencing. Planta 225:523–539

Senthil-Kumar M, Ramegowda HV, Hema R, Mysore KS, Udayakumar M (2008) Virus-induced gene silencing and its application in characterizing genes involved in water-deficit-stress tolerance. J Plant Physiol 165:1404–1421

Seversike TM, Sermons SM, Sinclair TR, Carter TE Jr, Rufty TW (2014) Physiological properties of a drought-resistant wild soybean genotype: transpiration control with soil drying and expression of root morphology. Plant Soil 374(1–2):359–370. https://doi.org/10.1007/s11104-013-1757-2

Sharma MP, Jaisinghani K, Sharma SK, Bhatia, VS (2012) Effect of native soybean rhizobia and AM fungi in the improvement of nodulation, growth, soil enzymes and physiological status of soybean under microcosm conditions. Agri Res 1(4):346–351

Sharma MP, Grover M, Chourasiya D, Bharti A, Agnihotri R, Maheshwari HS, Pareek, A, Buyer JS, Sharma SK, Schütz L, Mathimaran N, Singla-Pareek S, Grossman J, Bagyaraj DJ (2020) Deciphering the role of trehalose in tripartite symbiosis among Rhizobia, arbuscular mycorrhizal fungi, and legumes for enhancing abiotic stress tolerance in crop plants. Front Microbiol 11:509919. doi: 10.3389/fmicb.2020.509919

Shi WY, Du YT, Ma J, Min DH, Jin LG, Chen J, Chen M, Zhou Y-B, Ma Y-Z, Xu Z-S, Zhang X-H (2018) The WRKY transcription factor GmWRKY12 confers drought and salt tolerance in soybean. Int J Mol Sci 19:4087

Shimomura M, Kanamori H, Komatsu S, Namiki N, Mukai Y, Kurita K, Kamatsuki K, Ikawa H, Yano R, Ishimota M, Kaga A, Katayose Y (2015) The Glycine max cv. Enrei genome for improvement of Japanese soybean cultivars. Int J Genom. https://doi.org/10.1155/2015/358127

Shivakumar M, Kumawat G, Gireesh C, Ramesh SV, Husain SM (2018) Soybean MAGIC population: a novel resource for genetics and plant breeding. Curr Sci 114(4):906–908

Shu Y, Yu D, Wang D, Bai X, Zhu Y, Guo C (2012) Genomic selection of seed weight based on low-density SCAR markers in soybean. Genet Mol Res 12:2178–2188. https://doi.org/10.4238/2013.July.3.2

Shukla VK, Doyon Y, Miller JC, DeKelver RC, Moehle EA, Worden SE, Mitchell JC et al (2009) Precise genome modification in the crop species Zea mays using zinc-finger nucleases. Nature 459:437–441

Simonneau T, Ehlert C, Maurel C, Tardieu F (2009) Integrated control of leaf growth by cell turgor in response to combinations of evaporative demands and aquaporin-mediated reductions in root hydraulic conductivity. Comp Biochem Physiol A Mol Integr Physiol 153A:S228

Sinclair TR (2011) Challenges in breeding for yield increase for drought. Trends Plant Sci 16 (6):289–293

Sinclair TR, Ludlow MM (1985) Who taught plants thermodynamics? The unfulfilled potential of plant water potential. Funct Plant Biol 12(3):213–217

Sinclair TR, Purcell LC, King CA, Sneller CH, Chen P, Vadez V (2007) Drought tolerance and yield increase of soybean resulting from improved symbiotic N2 fixation. Field Crop Res 101 (1):68–71. https://doi.org/10.1016/j.fcr.2006.09.010
Sinclair TR, Zwieniecki MA, Holbrook NM (2008) Low leaf hydraulic conductance associated with drought tolerance in soybean. Physiol Plant 132:446–451
Sinclair TR, Messina CD, Beatty A, Samples M (2010) Assessment across the United States of the benefits of altered soybean drought traits. Agron J 102:475–482
Sloane RJ, Patterson RP, Carter TE Jr (1990) Field drought tolerance of a soybean plant introduction. Crop Sci 30(1):118–123. https://doi.org/10.2135/cropsci1990.0011183X003000010027x
Soares-Cavalcanti NM, Belarmino LC, Kido EA, Wanderley-Nogueira AC, Bezerra-Neto JP, Cavalcanti-Lira R et al (2012) In silico identification of known osmotic stress responsive genes from Arabidopsis in soybean and Medicago. Genet Mol Biol 35:315–321
Sonah H, Deshmukh RK, Chand S, Srinivasprasad M, Rao GJ, Upreti HC, Singh AK, Singh NK, Sharma TR (2012) Molecular mapping of quantitative trait loci for flag leaf length and other agronomic traits in rice (Oryza sativa). Cereal Res Commun 40:362–372. https://doi.org/10.1556/CRC.40.2012.3.5
Sonah H, Bastien M, Iquira E, Tardivel A, Légaré G, Boyle B, Normandeau É, Laroche J, Larose S, Jean M (2013) An improved genotyping by sequencing (GBS) approach offering increased versatility and efficiency of SNP discovery and genotyping. PLoS One 8:e54603
Song Q, Hyten DL, Jia G, Quigley CV, Fickus EW, Nelson RL, Cregan PB (2013) Development and evaluation of SoySNP50K, a high-density genotyping array for soybean. PLoS One 8: e54985
Song X, Wei H, Cheng W, Yang S, Zhao Y, Li X, Luo D, Zhang H, Feng X (2015) Development of INDEL markers for genetic mapping based on whole-genome re-sequencing in soybean. G3: Genes, Genomes, Genetics 5(12):2793–2799
Song L, Prince S, Valliyodan B, Joshi T, Maldonado dos Santos JV, Wang J, Lin L, Wan J, Wang Y, Xu D, Nguyen HT (2016) Genome-wide transcriptome analysis of soybean primary root under varying water deficit conditions. BMC Genomics 17:57. https://doi.org/10.1186/s12864-016-2378-y
Sosnowski O, Charcosset A, Joets J (2012) Bio Mercator V3: an upgrade of genetic map compilation and quantitative trait loci meta-analysis algorithms. Bioinformatics 28:2082–2083. https://doi.org/10.1093/bioinformatics/bts313
Specht JE, Hume DJ, Kumudini SV (1999) Soybean yield potential-a genetic and physiological perspective. Crop Sci 39(6):1560–1570. https://doi.org/10.2135/cropsci1999.3961560x
Specht JE, Chase K, Macrander M, Graef GL, Chung J, Markwell JP, German M, Orf JH, Lark KG (2001) Soybean response to water: a QTL analysis of drought tolerance. Crop Sci 41 (2):493–509. https://doi.org/10.2135/cropsci2001.412493x
Sreenivasa V, Lal SK, Babu PK, Swamy HKM, Yadav RR, Talukdar A, Rathod DR (2020) Inheritance and mapping of drought tolerance in soybean at seedling stage using bulked segregant analysis. Plant Genet Resour:1–8. https://doi.org/10.1017/S1479262120000052
Srivastava AK, Zhang C, Caine RS, Gray J, Sadanandom A (2017) Rice SUMO protease overly tolerant to salt 1 targets the transcription factor, OsbZIP23 to promote drought tolerance in rice. Plant J 92:1031–1043
Steketee CJ, Schapaugh WT, Carter TE, Li Z (2020) Genome-wide association analyses reveal genomic regions controlling canopy wilting in soybean. G3 Genes Genom Genet 10 (4):1413–1425
Steudle E (1995) Water transport across roots. Plant Physiol 108:6
Steudle E (2000) Water uptake by roots: effects of water deficit. J Exp Bot 51:1531–1542
Steudle E, Peterson CA (1998) How does water get through roots? J Exp Bot 49:775–788
Stolf-Moreira R, Medri ME, Marcelino FC, de Olivira MLN, Farias JRB, Abdelnoor RV, Nepomuceno AL (2010) Cloning and quantitative expression of drought induced genes in soybean. Genet Mol Res 9:858–867

Sun Y, Zhang X, Wu C, He Y, Ma Y, Hou H, Guo X, Du W, Zhao Y, Xia L (2016) Engineering herbicide-resistant rice plants through CRISPR/Cas9-mediated homologous recombination of acetolactate synthase. Mol Plant 9:628–631

Svitashev S, Young JK, Schwartz C, Gao H, Falco SC, Cigan AM (2015) Targeted mutagenesis, precise gene editing and site-specific gene insertion in maize using Cas9 and guide RNA. Plant Physiol 2:931–945

Swaminathan MS (2005) Towards an ever-green revolution. In: Tuberosa R, Phillips RL, Gale M (eds) Proceedings of the international congress: in the wake of the double Helix: from the green revolution to the gene revolution. Avenue Media, Bologna, pp 25–36

Takuno S, Terauchi R, Innan H (2012) The power of QTL mapping with RILs. PLoS One 7:e46545

Tran LS, Mochida K (2010) Identification and prediction of abiotic stress responsive transcription factors involved in abiotic stress signaling in soybean. Plant Signal Behav 5:255–257

Tripathi P, Rabara RC, Lin J, Rushton PJ (2013) GmWRKY53, a water and salt inducible soybean gene for rapid dissection of regulatory elements in bY-2 cell culture. Plant Signal Behav 8:e2427

Tripathi AK, Pareek A, Singla-Pareek SL (2016a) A NAP-family histone chaperone functions in abiotic stress response and adaptation. Plant Physiol 171:2854–2868

Tripathi P, Rabara RC, Reese RN, Miller MA, Rohila JS, Subramanian S et al (2016b) A toolbox of genes, proteins, metabolites and promoters for improving drought tolerance in soybean includes the metabolite coumestrol and stomatal development genes. BMC Genomics 17:102

Tripathi DK, Singh S, Gaur S, Singh S, Yadav V, Liu S, Singh VP, Sharma S, Srivastava P, Prasad SM, Dubey NK, Chauhan DK, Shivendra S (2018) Acquisition and homeostasis of iron in higher plants and their probable role in abiotic stress tolerance. Front Environ Sci 5:86. https://doi.org/10.3389/fenvs.2017.00086

Tuberosa R (2012) Phenotyping for drought tolerance of crops in the genomics era. Front Physiol 3:347

Tyerman SD, Niemietz CM, Bramley H (2002) Plant aquaporins: multifunctional water and solute channels with expanding roles. Plant Cell Environ 25:173–194

Tyree MT (1997) The cohesion–tension theory of sap ascent: current controversies. J Exp Bot 48:1753–1765

Ullah A, Sun H, Hakim YX, Zhang X (2018) A novel cotton WRKY gene, GhWRKY6-like, improves salt tolerance by activating the ABA signaling pathway and scavenging of reactive oxygen species. Physiol Plant 162(4):439–454. https://doi.org/10.1111/ppl.12651

Vadez V (2014) Root hydraulics: the forgotten side of roots in drought adaptation. Field Crop Res 165:15–24

Vadez V, Rao S, Kholova J, Krishnamurthy L, Kashiwagi J, Ratnakumar P, Sharma KK, Bhatnagar-Mathur P, Basu PS (2008) Roots research for legume tolerance to drought: quo vadis? J Food Legume 21(2):77–85

Vadez V, Deshpande SP, Kholova J, Hammer GL, Borrell AK, Talwar HS, Hash CT (2011) Stay-green quantitative trait loci's effects on water extraction, transpiration efficiency and seed yield depend on recipient parent background. Funct Plant Biol 38:553–566

Vadez V, Kholova J, Yadav RS, Hash CT (2013) Small temporal differences in water uptake among varieties of pearl millet (Pennisetum glaucum (L.) R. Br.) are critical for grain yield under terminal drought. Plant Soil 371:447–462

Valliyodan B, Qiu D, Patil G, Zeng P, Huang J, Dai L, Chen C, Li Y, Joshi T, Song L (2016) Landscape of genomic diversity and trait discovery in soybean. Sci Rep 6:23598

Vandeleur RK, Sullivan W, Athman A, Jordans C, Gilliham M, Kaiser BN, Tyerman SD (2014) Rapid shoot-to-root signalling regulates root hydraulic conductance via aquaporins. Plant Cell Environ 37:520–538

Vu HTT, Kilian A, James AT, Bielig LM, Lawn RJ (2015) Use of DArT molecular markers for QTL analysis of drought-stress responses in soybean. II. Marker identification and QTL analyses. Crop Pasture Sci 66:817–830. https://doi.org/10.1071/CP14304

Walker D, Monteros M, Yates J (2010) Mol breed. In: Kole C (ed) Genetics, genomics and breeding of soybean. CRC Press, New York, pp 123–159

Wang WX, Vinocur B, Shoseyov O, Altman A (2001) Biotechnology of plant osmotic stress tolerance: physiological and molecular considerations. Acta Hortic 560:285–292

Wang Y, Cheng X, Shan Q, Zhang Y, Liu J, Gao C, Qiu JL (2014) Simultaneous editing of three homoeoalleles in hexaploid bread wheat confers heritable resistance to powdery mildew. Nat Biotechnol 32:947–951. https://doi.org/10.1038/nbt.2969

Wang F, Chen HW, Li QT, Wei W, Li W, Zhang WK, Ma B, Bi YD, Lai YC, Liu XL, Man WQ, Zhang JS, Chen SY (2015) GmWRKY27 interacts with GmMYB174 to reduce expression of GmNAC29 for stress tolerance in soybean plants. Plant J 83:224–236

Wang C, Lu W, He X, Wang F, Zhou Y, Guo X, Guo X (2016) The cotton mitogen-activated protein kinase kinase 3 functions in drought tolerance by regulating stomatal responses and root growth. Plant Cell Physiol 57:1629–1642

Wang N, Zhang W, Qin M, Li S, Qiao M, Liu Z, Xiang F (2017) Drought tolerance conferred in soybean (*Glycine max.* L) by *GmMYB84*, a novel R2R3-MYB transcription factor. Plant Cell Physiol 58(10):1764–1776

Wani SH, Singh NB, Saini HK, Devi LP, Monalisa P (2010) Expressed sequenced tags (ESTs)- a functional genomic approach for gene discovery. Int J Curr Res 5:74–79

Wei W, Liang DW, Bian XH, Shen M, Xiao JH, Zhang WK, Ma B, Lin Q, Lv J, Chen X, Chen SY, Zhang JS (2019) GmWRKY54 improves drought tolerance through activating genes in abscisic acid and Ca2+ signaling pathways in transgenic soybean. Plant J 100:384–398. https://doi.org/10.1111/tpj.14449

Xu C, Xia C, Xia Z, Zhou X, Huang J, Huang Z, Liu Y, Jiang Y, Casteel S, Zhang C (2018) Physiological and transcriptomic responses of reproductive stage soybean to drought stress. Plant Cell Rep 37(12):1611–1624. https://doi.org/10.1007/s00299-018-2332-3

Yang W, Wang M, Yue A, Wu J, Li S, Li G, Du W (2014) QTLs and epistasis for drought-tolerant physiological index in soybean (Glycine max L.) across different environments. Caryologia 67 (1):72–78. https://doi.org/10.1080/00087114.2014.892278

Yang Y, Yu T-F, Ma J, Chen J, Zhou Y-B, Chen M, Ma Y-Z, Wei W-L, Xu Z-S (2020) The soybean bZIP transcription factor gene GmbZIP2 confers drought and salt resistances in transgenic plants. Int J Mol Sci 21:670

Ye H, Li S, Schapaugh WT, Ali ML, Sinclair TR, Riar MK, Mutava RN, Li Y, Vuong T, Valliyodan B, Neto AP, Klepadlo M, Song Q, Shannon JG, Chen P, Nguyen HT (2020) The importance of slow canopy wilting in drought tolerance in soybean. Journal of Experimental Botany 71(2): 642–652. https://doi.org/10.1093/jxb/erz150

Zerihun A, Haile S (2017) The effect of organic and inorganic fertilizers on the yield of two contrasting soybean varieties and residual nutrient effects on a subsequent finger millet crop. Agronomy 7:42

Zhang WB, Qiu PC, Jiang HW, Liu CY, Li CD, Hu GH, Chen QS (2012) Dissection of genetic overlap of drought and low-temperature tolerance QTLs at the germination stage using back-cross introgression lines in soybean. Mol Biol Rep 39(5):6087–6094

Zhang X-X, Tang Y-J, Ma Q-B, Yang C-Y, Mu Y-H, Suo H-C, Luo L-H, Nian H (2013) OsDREB2A, a rice transcription factor significantly affects salt tolerance in transgenic soybean. PLoS One 8:e83011

Zhang Z, Shang H, Shi Y, Huang L, Li J, Ge Q, Gong J, Liu A, Chen T, Wang D (2016) Construction of a high-density genetic map by specific locus amplified fragment sequencing (SLAF-seq) and its application to quantitative trait loci (QTL) analysis for boll weight in upland cotton (Gossypium hirsutum.). BMC Plant Biol 16:79

Zhang L, Li T, Wang Y, Zhang Y, Dong YS (2019) FvC5SD overexpression enhances drought tolerance in soybean by reactive oxygen species scavenging and modulating stress-responsive gene expression. Plant Cell Rep 38:1039–1051

Zhou Y, Qu H, Dibley KE, Offler CE, Patrick JW (2007) A suite of sucrose transporters expressed in coats of developing legume seeds includes novel pH-independent facilitators. Plant J 49:750–764

Zhou H, Liu B, Weeks DP, Spalding MH, Yang B (2014a) Large chromosomal deletions and heritable small genetic changes induced by CRISPR/Cas9 in rice. Nucl Acids Res 42:10903–10914. https://doi.org/10.1093/nar/gku806

Zhou L, Wang C, Liu R, Han Q, Vandeleur RK, Du J et al (2014b) Constitutive overexpression of soybean plasma membrane intrinsic protein GmPIP1;6 confers salt tolerance. BMC Plant Biol 14:181

Zhou J, Peng Z, Long J, Sosso D, Liu B, Eom J-S, Huang S, Liu S, Cruz CV, Frommer WB, White FF, Yang B (2015a) Gene targeting by the TAL effector PthXo2 reveals cryptic resistance gene for bacterial blight of rice. Plant J 82:632–643

Zhou X, Li L, Xiang J, Gao G, Xu F, Liu A, Zhang X, Peng Y, Chen X, Wan X (2015b) OsGL1-3 is involved in Cuticular wax biosynthesis and tolerance to water deficit in Rice. PLoS One 10(1): e116676. https://doi.org/10.1371/journal.pone.0116676

Zhou Z, Jiang Y, Wang Z, Gou Z, Lyu J, Li W, Yu Y, Shu L, Zhao Y, Ma Y (2015c) Resequencing 302 wild and cultivated accessions identifies genes related to domestication and improvement in soybean. Nat Biotechnol 33:408

Zhou Y, Liu W, Li X, Sun D, Xu K, Feng C, Foka ICK, Ketehouli T, Gao H, Wang N, Dong Y, Wang F, Li H (2020) Integration of sRNA, degradome, transcriptome analysis and functional investigation reveals gma-miR398c negatively regulates drought tolerance via GmCSDs and GmCCS in transgenic Arabidopsis and soybean. BMC Plant Biol 20:190. https://doi.org/10.1186/s12870-020-02370-y

Plant Roots and Mineral Nutrition: An Overview of Molecular Basis of Uptake and Regulation, and Strategies to Improve Nutrient Use Efficiency (NUE)

5

Ekta Bhardwaj, Richa Shukla, and Sandip Das

Abstract

Health, productivity, and reproductive success of plants as primary producer are dependent on soil mineral nutrients, among many other biotic and abiotic factors. Mineral nutrients have been categorized into major and minor nutrients. Both deficiency and excess of nutrient availability cause deficiency stress and nutrient toxicity, respectively; "sufficiency range" being the amount needed for optimal growth and productivity. Plants deploy a variety of developmental and adaptive mechanisms, root architecture being one, to maximize its chances of acquiring mineral nutrition in the sufficiency range. Past approaches to maximize crop yield and productivity relied on abundant application of mineral nutrition as chemical fertilizers and formed the basis of green revolution. The degradation of environment became an unintended collateral damage. The present strategy, therefore, is to unravel the molecular basis of nutrient uptake by plant roots, mobilization, assimilation within the plant, and the source–sink relationship in order to improve nutrient uptake efficiency, and nutrient utilization efficiency (NUE). Given the vastness of the subject involving developmental and adaptive traits, complicated and interconnected nature of the various factors that regulate mineral nutrition, the present endeavor is limited to providing an overview of key transporters that are involved in uptake of the sixteen soil-derived mineral nutrients. The roles of transcriptional regulators such as microRNAs are just beginning to be unraveled. Together with transporters, small RNAs and transcription factors hold the key to future crop breeding and improvement programs through improved root system architecture and nutrient uptake, mobilization, and assimilation.

E. Bhardwaj · R. Shukla · S. Das (✉)
Department of Botany, University of Delhi, Delhi, India
e-mail: sdas@botany.du.ac.in

B. Giri, M. P. Sharma (eds.), *Plant Stress Biology*,
https://doi.org/10.1007/978-981-15-9380-2_5

Keywords

Mineral nutrients · Nutrient stress · Root architecture · Molecular regulators · Nutrient use efficiency

Abbreviations

KUE	Potassium use efficiency
NRT	Nitrate transporters
NUE	Nutrient use efficiency
NitUE	Nitrate use efficiency
PHT	Phosphate transporter
PUE	Phosphate use efficiency
RSA	Root system architecture
SUE	Sulfate use efficiency
SULTR	Sulfate transporters

5.1 Introduction

Plants are aptly called the miners of earth crust as they are one of the first link through which minerals are captured from soil and introduced into biosphere. Most other life forms are directly or indirectly dependent on plants for all their requirement of mineral nutrients. Plants uptake mineral nutrients through root systems through intricate and finely balanced molecular and physiological mechanisms; once inside the root, minerals are transported to other plant parts for assimilation and storage. Several investigations have established direct positive correlation between the presence of nitrogen, phosphorus, potassium, and sulfur (NPKS) along with other mineral nutrients, with quality as well as quantity of plant growth, adaptability, and productivity. In the past decades, the goal of increasing plant productivity has led agriculturists and farmers towards excessive use of mineral fertilizers. However, their excessive and indiscriminate use of chemical fertilizers has caused extensive damage to soil and water through run-offs (Guignard et al. 2017). In order to minimize further environmental damage, the present current emphasis is towards identification of molecular genetic components involved in the process of mineral nutrient uptake and assimilation, and through manipulation of plant root architecture to achieve optimal growth and productivity. Study of mineral nutrition is, therefore, now central to agriculture, environment conservation, and health.

Elements which have clear physiological roles and absence of which adversely hampers the completion of life cycle of plants are termed as essential elements. A total of nineteen essential elements have been described; out of which three are obtained from water or air (carbon, hydrogen, and oxygen) while rest of them are

Table 5.1 Essential elements from soil classified into macro- and micro-elements, forms taken up by plants, and function

Macro-elements	Form available for plants	Function
Nitrogen (N)	NO^{3-} and NH^{4} +	nucleic acids (RNA, DNA), amino acids, hormones
Phosphorus (P)	PO^{3-}	DNA, RNA, proteins, lipids, sugars, ATP, ADP, and NADPH
Potasium (K)	K^+	cofactor for many enzymes, maintain cell electronutrility and turgour
Sulfur (S)	SO_4^{2-}	amino acids (cysteine, methionine), coenzymes, vitamins, several secondary metabolites
Calcium (Ca)	Ca^{2+}	Secondary messanger, part of cell wall, cofactor for several enzymes
Magnesium (Mg)	Mg^{2+}	Part of chlorophyll molecule and important enzymes
Silicon (Si)	Silicic acid	Contributes to the mechanical strength of cell wall
Micro-nutrients		
Iron (Fe)	Fe^{2+} and Fe^{3+}	enzyme cofactors , components of electron transport chains (FeS and haem)
Chlorine (Cl)	Cl^-	Stomatal Regulation, Photosynthetic O_2 evolution, Osmoregulation
Boron (Bo)	H_3B_3	plant cell wall,Cell division ,movement of sugar,pollination and seed set,nitrogen and carbohydrate metabolism
Molybdenum (Mo)	MoO_4^{2-}	Cofactor in various enzymes ,nitrogenase and nitrate reductase enzyme
Manganese (Mn)	Mn^{2+}	carbohydrate metabolism,electron transport in photosynthesis,cofactor of various enzymes
Copper (Cu)	Cu^+, Cu^{2+}	plastocyanin,cofactor of a large number of oxidases
Zinc (Zn)	Zn^{2+}, Zinc complexes	maintaining the activity of DNA polymerase enzymes, act as cofactors in metalloenzymes and required at active sites of various enzymes
Nickel (Ni)	Ni^+	Constituent of Urease and hydrogenase
Sodium (Na)	Na^+	plays role in regenertaion of phosphoenolpyruvate in C4 and CAM plants

obtained from the soil. Soil essential elements have been classified on the basis of (Evans and Sorger 1966, Mengel and Kirkby 1987; Table 5.1; Fig 5.1) the following:

1. *Mobility in plants*: mobile elements (N, P, K, Cl, Mg, Mo) and immobile elements (Ca, Cu, Fe, Mn, Ni, S, Zn)
2. *Requirements*: macro-nutrients (N, P, K, S, Ca, Mg, Si) and micro-nutrients (Cl, Fe, B, Mn, Na, Zn, Cu, Ni, Mo)
3. *Functions*:
   (a) group1: part of carbon compounds (N, S)
   (b) group 2: in energy storage (P, Si, B)

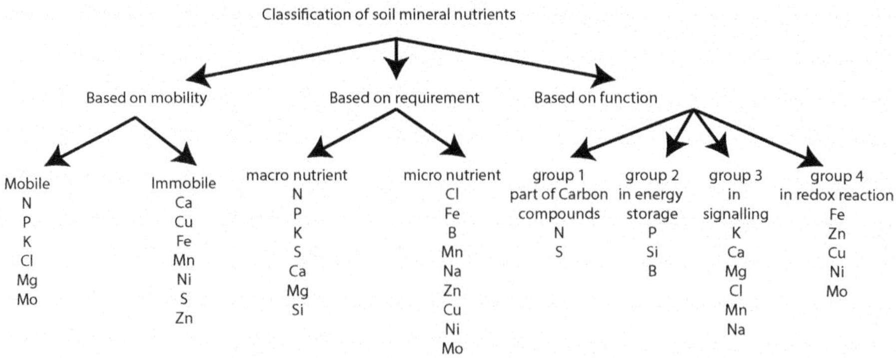

Fig 5.1 Classification of soil mineral nutrients (based on Evans and Sorger 1966 and Mengel and Kirkby 1987)

(c) group 3: ions for signaling (K, Ca, Mg, Cl, Mn, Na), and
(d) group 4: involved in redox reactions (Fe, Zn, Cu, Ni, Mo)

It would be prudent to ask "*how much mineral concentration is sufficient for plant growth?*" It has been observed that there is a range of nutrient amount required for normal plant growth and maximum yield, and is termed as *sufficiency range* (Brady and Weil 1999; Sahrawat 2006). The value of sufficiency range varies across plant species and is dependent on multiple biotic and abiotic factors which are both internal and external in nature. For example, *Oryza sativa* has sufficiency range of 2.80–3.60% (N), 0.10–0.18 (P), and 1.20–2.40 (K), while for *Zea mays* it is 3.00–3.50 (N), 0.25–0.45 (P), and 2.00–2.50 (K) (Sahrawat 2006). Deviations from the sufficiency range are responsible for sub-optimal growth, development, and yield. The genetic constituent, physiology, and architecture (especially root system architecture) are critical plant factors that determine the efficiency of nutrient uptake, assimilation, transport, thus determining the sufficiency range, and contributing to growth and yield. Table 5.2 provides a comprehensive list of optimum macro- and micronutrient requirements of various crops (FAO fertilizer and plant nutrition bulletin 16; 2006).

Before we proceed further, it must be kept in mind that mineral nutrient availability below sufficiency range causes plant nutrient deficiency, whereas availability above sufficiency range is responsible for nutrient toxicity. Either way, both deficiency and toxicity interfere with normal growth and hamper crop quality and yield. Nutrient deficiency is more common as compared to nutrient toxicity that occurs due to excessive application of fertilizers.

Monitoring of soil health including analysis of levels of mineral nutrients can provide clues to mineral deficiency as well toxicity. However, such analyses are often inaccurate as not all mineral nutrients present in soil are available, and taken up by plants. Therefore, phenotypic, physiological, and molecular diagnosis features of plants are considered better way of assessing nutrient deficiency. The most obvious

Table 5.2 Nutrient requirement[a], and absorbed by representative crop plants (given tonnes per hectare; top row) and recommended macro- and micro-nutrient doses/application[a] (lower row) for major crops that can be considered as sufficiency range (FAO fertilizer and plant nutrition bulletin 16; 2006). Each nutrient application is based on requirement per growing season

Crops/plants (top row is nutrient requirement and bottom row is nutrient to be supplied)	Macro nutrients (in kg/Ha unless mentioned otherwise)						Micronutrients (in kg/Ha, unless mentioned otherwise)							
	N	P	K	S	Ca	Mg	B	Fe	Mn	Cu	Zn	Mo	Cl	Si
Triticum aestivum (6.7 tonnes)	200	55	250	22	27	19	n.a	1.8	0.5	0.15	0.5	n.a	n.a	n. a
	120–150	90	160	n.a	n.a	25	n.a	n.a	0.5%; 2–3 times	n.a	62.5kg every 2–3 years	n.a	n.a	n. a
Oryza sativa (1 tonne paddy)	20	11	30	3	7	3	0.015	0.15	0.675	0.018	0.040	0.002	n.a	52
	50–160	20–80	20–60	n.a	n.a	n.a	n.a	1%; 2–3 sprays	n.a	n.a	10–12	n.a	n.a	n. a
Zea mays (9.5 tonnes of grain)	191	89	235	21	57	73	240	2.130	0.340	0.110	0.380	0.009	0.081	n. a
	50–300	30–100	30–100	30–100	n.a	n.a	n.a	n.a	n.a	n.a	25kg $ZnSO_4$ mixed with 25kg soil	n.a	n.a	n. a

(continued)

Table 5.2 (continued)

Crops/plants (top row is nutrient requirement and bottom row is nutrient to be supplied)	Macro nutrients (in kg/Ha unless mentioned otherwise)						Micronutrients (in kg/Ha, unless mentioned otherwise)							
Cicer arietinum (1.5 tonnes)	91	14	60	9	39	18	n.a	1.302	0.105	0.017	0.057	n.a	n.a	n. a
	15–20	40–50	n.a	20–30	n.a	n.a	n.a	Foliar sprays of 2% $FeSO_4$	n.a	n.a	25	n.a	n.a	n. a
Cajanus cajan (L). Millsp. (1.2 tonnes)	85	18	75	9	32	25	n.a	1.440	0.128	0.031	0.038	n.a	n.a	n. a
	15–20	40–50	n.a	20–30	n.a	n.a	n.a	n.a	n.a	n.a	5	n.a	n.a	n. a
Arachis hypogaea L. (1 tonne)	58.1	19.6	30.1	7.9	20.5	13.3	n.a	2.284	0.093	0.036	0.109	n.a	n.a	n. a
	20–30kg	40–70	20–50	n.a	300–500 kg gypsum		5 kg borax	0.5–1	n.a	n.a	n.a	0.5–1	n.a	n. a
Glycine max (L.) (1 tonne of grain)	146	25	53	5	28	22	0.055	0.476	0.123	0.041	0.104	0.013	n.a	n. a

	20–100	50–70	60–100	n.a	n.a	n.a	n.a	n.a	15	n.a	5–10	n.a	n.a	n.a
Brassica napus (per tonne of seed)	32.8	16.4	41.8	17.3	42	8.7	n.a	1.123	0.095	0.017	0.100	n.a	n.a	n.a
	25–240	80–100	150–200	20–80	n.a	30	0.5	n.a	1.5	n.a	10	0.010–0.015	n.a	n.a
Helianthus annuus (3.5 tonnes of seed)	131	87	385	n.a	210	70	0.396	0. 732	0.412	0.059	0.348	n.a	n.a	n.a
	50–80	60–80	50–150	n.a	n.a	n.a	1–2	n.a	n.a	n.a	n.a	n.a	n.a	n.a
Solanum tuberosum L (36 tonnes tubers)	117	32	224	14	37	63	n.a	0.8	n.a	n.a	n.a	n.a	n.a	n.a
	80–300	60–300	60–300	n.a	n.a	n.a	n.a	n.a	n.a	n.a	n.a	n.a	n.a	n.a
Ipomoea batatas Lam. (10 tonnes of tubers)	51.6	17.2	71.0		6.3	6.1	n.a	n.a	n.a	n.a	n.a	n.a	n.a	n.a
	50	50–90	80–120	n.a	n.a	n.a	9–26	n.a	n.a	n.a	n.a	n.a	n.a	n.a
Manihot esculenta Crantz (37 tonnes of fresh tuber)	198	70	220	19	143	47	0.200	0.900	1.090	0.080	0.660	n.a	n.a	n.a
	40–120	40–80	Equivalent to N (1:1)	n.a	n.a	n.a	10	n.a	n.a	n.a	12.5	1.0	n.a	n.a

(continued)

Table 5.2 (continued)

Crops/plants (top row is nutrient requirement and bottom row is nutrient to be supplied)	Macro nutrients (in kg/Ha unless mentioned otherwise)						Micronutrients (in kg/Ha, unless mentioned otherwise)							
Saccharum officinarum L (1 tonne of cane)	0.8	.30	1.32	0.25	0.42	0.50	0.0020	0.031	0.011	0.002	0.0045	0.00001	n.a	n. a
	100–300	60–200	80–200	n.a	n.a	n.a	n.a	1.6% Spray	n.a	n.a	25	n.a	n.a	n. a
Beta vulgaris L. (10 tonnes)	40–50	15–20	45–70	5	n.a	12–15	n.a	n.a	n.a	n.a	n.a	n.a	n.a	n. a
	130	50–200	100–400	n.a	n.a	100	1–2	n.a	6–12	n.a	n.a	n.a	n.a	n. a
Gossypium spp. (2.5 tonnes of seed cotton)	156	36	151	10	168	40	0.320	2.960	0.250	0.120	0.116	n.a	n.a	n. a
	50–300	30–100	30–100	10	n.a	n.a	0.1–0.15% on the leaves	n.a	n.a	n.a	25	n.a	n.a	n. a
Corchorus olitorius L./ *Corchorus capsularis* L. (1 tonne of dry–fibre)	35.2/42	20.3/18.5	63.2/88.5	n.a	55.6/60	13.3/24.5	n.a	0.368/0.784	0.119/0.251	0.018/0.0195	0.139 / 0.214	n.a	n.a	n. a
	30–45	20	10–20	n.a	n.a	40	n.a	n.a	n.a	n.a	n.a	n.a	n.a	n. a

[a]range in requirement and doses reflect variation encountered for crop varieties including high yielding varieties (HYVs) and various environmental conditions

symptoms of nutrient deficiencies and toxicity are visible on above ground parts, namely the leaves and overall plant architecture. Chlorosis, necrosis, stunting, and purple reddish coloring are some of the key diagnostic symptoms (McCauley et al. 2009). However, even before the symptoms are visible on above ground parts, altered growth and developmental patterns in roots are manifested. This is because roots are the primary organs that are involved in initial sensing, assimilation, and uptake. The present chapter, therefore, focuses on few aspects of mineral nutrients, including phenotypic indicators, and molecular regulators of mineral nutrient uptake and assimilation with special reference on roots. In order to retain focus and because of paucity of space, the present chapter does not cover many important aspects on mineral nutrition, most notably role of phytohormones in uptake, transport, and assimilation, and in shaping root system architecture (RSA); and role of root and soil microbiome except in a few passing references and mentions.

5.2 Effect of Nutrient Stress on Root System Architecture

Plant roots have been classified based on several criteria. A developmental origin based nomenclature groups them as primary, secondary, and tertiary; whereas a classification based on morphology and architecture groups them as taproot, fibrous root, lateral root, and shoot-borne (adventitious) roots (Zobel and Waisel 2010). Taproot and fibrous roots are often encountered in dicots and monocots, respectively. The 3-D spatial and temporal arrangement of plant root axes present underground is referred to as root system architecture (RSA; Lynch 1995). This architecture is influenced by complex interaction among myriad host and non-host factors such as environment including geotropism, light, water, soil temperature, moisture, mineral nutrients; biotic factors such as microbes, soil fungi, roots, and root exudates of other plants; endogenous genetic and molecular regulators, and stage of plant growth and development (Morris et al. 2017; Van Gelderen et al. 2018). Variations in RSA and morphology are commonly encountered both at between- and within-species level exhibiting developmental and phenotypic plasticity; developmental and phenotypic plasticity are thus considered an integral part of RSA. (Malamy 2005; Phung et al. 2016; Sanchez et al. 2018; Van Gelderen et al. 2018). As with all other phenotypes, RSA is influenced by multiple overlapping factors and it is extremely difficult to dissect out the effect of a specific factor (Malamy 2005; Van Gelderen et al. 2018). Among all factors, RSA is most dramatically influenced by mineral nutrition and is a definite indicator of stress (Malamy 2005; Gruber et al. 2013).

5.2.1 Altered RSA Under Mineral Nutrient Stress in Various Plants

As mentioned previously, nutrient deficiency is caused when the available mineral nutrients fall below the sufficiency range which is highly variable across plants. For instance, for the three major nutrients—N, P, and K, sufficiency ranges are 3.00%,

0.25%, and 2.0%, respectively, for *Zea mays* prior to tasselling; 2.8% (N), 0.1% (P), and, 1.2% (K), for *Oryza sativa* at tillering stage; 1.75% (N), 0.2% (P), and 1.5% (K) for *Hordeum*; 3.2% (N), 0.15% (P), and 2.0% (K) for *Sorghum bicolor* (Sahrawat 2006). The sufficiency range also is developmental stage dependent. This is exemplified by the level of *N* being 1.75% before heading, and 2.0% at emergence of head from boot in *Triticum aestivum* (Sahrawat 2006).

The alteration in root system architecture including change in root length, surface area, number and density of lateral roots, root mass under nutrient enriched or nutrient deficient soil has been termed as compensatory response (Russell 1977; Fageria and Moreira 2011). While discussing alterations in root and root system architecture in responses to nutrients deficiencies and toxicity, the native RSA of the plant should be kept in mind. It must also be made clear that the uptake and assimilation of mineral nutrients are subject to complex interactions, and inter-dependence as has been demonstrated in several studies (Kellermeier et al. 2014; Raddatz et al. 2020; Iqbal et al. 2020). Some of the key alterations in root characteristics under various mineral nutrient stresses are discussed below, and listed in Table 5.3.

Nitrogen: The availability of Nitrogen is one of the most limiting mineral nutrients for crop yield. In *Zea mays*, response of root to *N*-stress was evaluated in three genotypes with differing Nitrogen Utilization Efficiency (NitUE). Two of the three genotypes, B73XLH51 and Arens885855 have high NitUE, whereas Mo17 has low NitUE. Root dry weight in all genotypes increased under *N*-deficient (no *N*) conditions as compared to when *N* was available. As the *N*-availability was increased, root dry weight decreased for B73XLH51 and Arens885855, the genotypes with high NitUE, but not significantly for Mo17. Under field conditions, root length and dry weight reduced as the availability of *N* was increased; Mo17 showed the most dramatic decrease among the three genotypes (Eghball and Maranville 1993). In a separate study, root length and root surface area, but not root diameter of maize, were found to be reduced both under absence of *N* and high levels of *N* (255 kg/ha) as compared to when *N* was applied at 128 kg/ha (Costa et al. 2002).

In rice, in an early study, effect of *N* on root architecture was studied on four different cultivars—Gin-nen, Tamanishiki, Habataki and Norin 25. In all the cutlivars, similar response of reduced root length and root dry weight was observed with when *N*-level was increased from 5 kg/1000 m^2 to 30 kg/1000 m^2. This was, however, compensated by increase in number of nodal roots (Tanaka et al. 1995). Similar response of reduced root length and root dry weight upon increase in *N*-application has also been observed in other rice cultivars (Fageira and Moreira 2011). Ogawa et al. (2014), however, observed variation when cultivars of *O. sativa* and several other species of rice, namely *O. barthii*, *O. glaberrima*, and *O. rufipogon* derived from different ecosystems were subjected to different concentrations of N (0, 50, and 500 μMd). Majority of the cultivars showed a continuous declining trend in root length from 0, 50, and 500 μM, whereas a few cultivars showed increased root length from 0 to 50 μM and then decrease in length from 50 to 500 μM concentration (Ogawa et al. 2014). However, all cultivars

Table 5.3 Representative examples of root phenotype and architecture under various nutrient stresses (PR-Primary root, LR-lateral roots)

	Plant	Effect on root	Reference
Nitrogen deficiency			
1	Rice	Deeper roots, longer root length and density	Ogawa et al. (2014)
2	*Arabidopsis*	Longer LRs and shorter PR	Gruber et al. (2013)
3	Maize	Low lateral root (LR) branching density, longer LRs,	Postma et al. (2014)
4	Cereal, legume , spinach , tomato	Increase root hair length, increase , decrease and no change in root hair density, less branching	Baligar et al. (1998)
Phosphate deficiency			
1	*Arabidopsis*	Reduction of PR growth, increase in the growth of secondary roots, increase in root hair length and density	Williamson et al. (2001) and Lopez-Bucio et al. (2003)
2	*Phaseolus vulgaris*	Root shallowness, adventitious root formation, and increased dispersion of lateral branching from the basal roots, higher root hair length and density	Strock et al. (2018)
3	*Lupinus albus*	Formation of cluster so flateral roots known as proteoid roots	Péret et al. (2014)
4	Rice	More LRs	Péret et al. (2014)
5	Maize	High LR branching density, shorter LRs	Postma et al. (2014)
6	Spinach, Rape, tomato	Longer roots and root hairs, increased 2nd order laterals, increase or no change in root hair density, decreased shoot/root, induced proteoid roots	Föhse and Jungk (1983)
Potassium deficiency			
1.	Maize, cowpea, common bean, rice, barley	Reduced root mass and length, reduced volume of root axes and 1st order laterals, complete prevention of 2nd order lateral development	Baligar et al. (1998)
2	*Arabidopsis*	Reduced length of PR and LR	Gruber et al. (2013)
Sulphate deficiency			
1	*Arabidopsis*	Branched root system, elongated PR,LRs closer to root tips	Gruber et al. (2013) and Lopez-Bucio et al. (2003)
2	Rice, maize	Significant increase in root length	Pariasca-Tanaka et al. (2019) and Maniou et al. (2014)
Boron deficiency			
1.	Rice, common bean, soybean, wheat and squash	Inhibition of root elongation, stubby and bushy appearance	Baligar et al. (1998)
2.	*Arabidopsis*	Decrease in root length and increase in LR density	Gruber et al. (2013)

(continued)

Table 5.3 (continued)

	Plant	Effect on root	Reference
Zinc deficiency			
1.	*Arabidopsis*	Highly branched root system	Gruber et al. (2013)
Calcium deficiency			
1.	Cereals, legumes, peas and watermelon	Shorter and denser root system, translucent extensions of main root tip	Baligar et al. (1998)
2	*Arabidopsis*	Shallow and branched root system	Gruber et al. (2013)
3	*Glycine max*	Reduced PR length, no secondary roots	Spehar and Galway (1997)
Mg deficiency			
1.	Rice, common bean, cowpea	Reduced dry mass, increased shoot/root	Baligar et al. (1998)
2	*Arabidopsis*	Decrease in PR and LR length, decrease in LR density	Gruber et al. (2013)
Mn deficiency			
1.	Tomato	Reduction in length of main axis, complete absence of lateral roots	Baligar et al. (1998)
2.	*Arabidopsis*	Decrease in PR and LR density	Gruber et al. (2013)
Fe deficiency			
1.	Dicots and monocots, lupines	Inhibiion of root elongation, increased diameter of apical root zones, abundant root hair formation, formation of proteoid roots	Baligar et al. (1998)
2.	*Arabidopsis*	Moderate deficiency (5 and 10 mM) -increase in PR, severe deficiency-decrease in PR and LR length	Gruber et al. (2013)

showed significant increase in total root number when N supply was augmented from 0 to 50, and from 50 to 500 μM (Ogawa et al. 2014). The variation observed was ascribed to plasticity of the trait because of being influenced by other edaphic factors. In another investigation, seminal root length increased by as much as 25%, whereas lateral root density decreased under low Nitrogen condition in comparison to control conditions after 16 days of N-deficiency stress (Sun et al. 2014, 2016).

Another plant where the effects of N-deficiency and toxicity have been well studied is *Arabidopsis thaliana.* In a study, N was supplied either in a homogenous (uniform across media) or heterogeneous (in patches of varying N-concentration) manner. Under homogenous N-supply, primary root length decreases as N-availability is increased, whereas lateral root density was found to remain unchanged under varying N-supply (Linkohr et al. 2002). When N-supply was administered in a heterogeneous manner, absolute primary root length was unchanged from homogeneous N-supply except under extremely low N-condition where the root length was reduced. However, under heterogeneous N-supply, lateral root density increased as compared to homogeneous N-supply. The length of lateral root is reduced in N-deficient patch and increases in N-rich patches (Linkohr et al. 2002). When N is available to plants uniformly at high levels it causes suppression of primary root growth, whereas lateral root remained unaffected. In contrast, heterogeneous *N*-supply as patches reduces lateral root initiation under low *N*-patch, and increases under high *N*-patch (Linkohr et al. 2002). The plasticity of root system architecture was revealed when *N* was supplied at 110, 275, 550, and 11,400 μM to recapitulate gradation of deficiencies from moderate to high, and effect on root growth was observed. The total root length, average primary lateral root length, and average secondary lateral root length increased when *N*-concentration changed from 11,400 to 550 μM, but shows decreasing trend when *N* supply was reduced to 275 and 110 μM. While the density of primary lateral root remained unchanged, density of secondary lateral root increased when N-concentration was reduced from 11,400, to 275, and to 110 μM. Longer lateral roots (LR) were observed under all concentrations of nitrogen deficiency (Gruber et al. 2013). In other studies, severe N deficiency has been shown to cause shorter primary root and lesser number of lateral roots (Araya et al 2015).

Phosphorus: Generally, roots of plants growing on low P tend to explore upper layer of soil which have most of the Pi rich matters (Lopez-Arredondo et al. 2014). To achieve that plants try to attain shallower and broader root system. In *Phaseolus vulgaris*, a change in the angle of lateral root growth under low P conditions was observed so that they grow outward (shallow roots) rather than downward (Bonser et al. 1996). In *A. thaliana*, reduction in primary roots (PR) growth and increase in number of lateral roots (LR) are the visible changes in RSA (Williamson et al. 2001; Lopez-Bucio et al. 2002).

Changes in density and length of root hairs are another feature of RSA which alters with nutrient stresses, especially P and K stresses (Jungk 2001). In *A. thaliana*, root hairs are longer and denser under low P conditions (Williamson et al. 2001). Zhu et al. (2010) proved that root hair length helps in P acquisition. The results indicate that genotypes which have long root hairs (0.8 mM) perform better under low *P* conditions than genotypes with short root hairs. The former one has better plant growth, greater biomass, lower metabolic cost–benefit ratios, and higher *P* acquisition (Zhu et al. 2010). A cottony root tomato strain was found to perform better in P uptake efficiency which also had more dense root hairs and shorter roots (Hochmuth et al. 1985). In an earlier study on *Brassica napus, Spinacia oleracea*, and tomato, low P concentration was found to induce the root hair growth (Föhse and Jungk 1983).

Potassium: Root hairs have the potential to enhance root surface area as it contributes significant amount of root surface, and changes in root hairs is one of the major change in RSA which can be observed under K and P stresses (Jungk 2001). Plants growing in K deficient media try to achieve more root volume by enhancing root hair length for enhanced uptake. In *Arabidopsis thaliana*, 28 h deficiency of K leads to root hair elongation via ethylene or ROS mediated pathway (Jung et al. 2009; Shin et al. 2005). It was observed that root hair length increases several folds in response to *K*-deficiency in rye, ryegrass, oilseed rape, lucerne, barley, pea, and red clover (Høgh-Jensen and Pedersen 2003) without significant difference in number, density of root hairs and overall root length. Decrease in shoot and root biomass along with length of primary root and primary lateral root, and increase in density of secondary lateral roots in *A. thaliana* was observed under *K* deficiency (Gruber et al. 2013).

Sulfur: *A. thaliana* plants grown in nutrient media with low sulfate concentration tend to grow lateral roots closer to the root tip (Lopez-Bucio et al. 2003). Gruber et al. (2013) observed complete withdrawal of S from nutrient media reduced the root biomass by 28%, increased the PR root length by up to 12%, led to reduced density of 1° LR; whereas a concentration of 50 μM leads to significant increase in the average length of 1° LRs (Gruber et al. 2013). Complete withdrawal of sulfate from the media reduced shoot and root biomass by 40% and 28%, respectively. There was a significant decrease in length of PR with decreasing concentration of S in the media (Gruber et al. 2013). In *Oryza sativa*, low S conditions lead to increase in root biomass, increase in root elongation and root:shoot ratio compared to plants in high S conditions (Pariasca-Tanaka et al. 2019). Along with the elongation of root length, formation of aerenchyma in adventitious roots by lysis of cortical cells was observed in *Zea mays* plants growing under S deficiency (Maniou et al. 2014).

5.3 Molecular Regulators in Nutrient Uptake and Transport

5.3.1 Nitrogen

Nitrogen is available for uptake by plants in two major ionic forms in the soil, as nitrate (NO_3^-) and ammonium (NH^{4+}). NO_3^- is the major form available under aerobic conditions, and readily used up by most of the plants (Andrew et al. 2013), whereas NH^{4+} is more efficiently taken up by plants such as rice which grow under anaerobic conditions (Cai et al. 2008). Most plants take up nitrate, which is first converted into nitrite, and then converted into ammonium with the help of nitrate reductase and nitrite reductase, respectively. These two ionic forms use different set of transporters for uptake and transport.

The uptake and transport of nitrate involve members of four families of transporter proteins. These are NPF family (NITRATE TRANSPORTER1 (NRT1)/PEPTIDE TRANSPORTER (PTR), NITRATE TRANSPORTER1 (NRT2) family, CHLORIDE CHANNEL (CLC) family, and SLOWLY ACTIVATING ION CHANNEL (SLAC) family (Wang et al. 2018). Of these four families, NRT1 and NRT2 are considered key transporters of NO_3^-. The genome of *Arabidopsis thaliana* has 53 homologs of NRT1 and seven of NRT2, whereas that of *O. sativa* has 93 homologs of NRT1/NPF and four of NRT2 (Wang et al. 2018). In *Medicago truncatula*, one member of NPF, NPF7.6, has been found to be co-opted for nodulation and localizes to nodule transfer cell (NTC) and acts as a high affinity nitrate uptake (Wang et al. 2020a). A member of the NPF family, NPF4.5, has been found to be involved in a mycorrhizal NO3-uptake in several members of Gramineae including *O. sativa* (OsNOPF4.5), *Z. mays* (ZmNPF4.5), and *S. bicolor* (SbNPF4.5;

Wang et al. 2020b). Seven members of CLC family and five of SLAC family are present in *Arabidopsis thaliana* (Wang et al. 2012).

Members of the NRT1 family are known as Low Affinity Transporters (LATs), whereas those of NRT2 are High Affinity Transporters (HATs). Members of the NRT1 and NRT2 gene family have well-partitioned function in uptake, transport, and assimilation. Some of these have as specific role in uptake or in transport, whereas some have dual roles in uptake and transport. For example, the uptake of NO_3^- from soil is mediated by two LATs—NRT1.1, NRT1.2, and three HATs—NRT2.1, NRT2.2, and NRT2.4, along with other transporters such as NAXT1. Each of these transporters is present in the several cell layers of root and facilitates the uptake from soil to outermost epidermis, via cortex, endodermis, pericycle/parenchyma, and finally into the xylem. This process involves NRT1.1, NRT1.2, NRT2.1, and NRT2.4 in epidermis, NRT1.1 and NRT2.1 in cortical cells, only NRT1.1 in endodermis, and finally NRT1.5 in pericycle/parenchymatous cells that are immediately adjacent to xylem. The transport of NO_{3-} via xylem from root-to-shoot is performed by LATs—NRT1.5, NRT1.8, and NRT1.9. Phloem loading of nitrate occurs with the help of NRT1.9 and NRT1.7. Two transporters—NAXT1 (Nitrate excretion transporter 1) and NRT1.8 are involved in efflux. Other than these, NRT1.4, NRT1.7, and NRT1.6 are involved in transport of NO_3^- from shoot to leaves, remobilization from old to new leaves, and transport to seed for storage, respectively (Wang and Tsay 2011; Wang et al. 2012; Fan et al. 2009). Interestingly, the transport of NO_{3-} mediated by NRT1.9 in xylem and phloem is negatively correlated (Wang and Tsay 2011). The uptake of nitrate is tightly regulated and induced by several factors including N-starvation, light, pH, sugar, auxin (Loqué et al. 2003; Huang et al. 1996; Liu et al. 1999; Wang et al. 2012). Similar partitioning of function of NRT transporters and induction by several factors have also been observed in *Oryza sativa* (Cai et al. 2008; Fan et al. 2016; Hu et al. 2015; Wang et al. 2018). Other transporters that are involved in nitrate transport include amino acid transporters (AAT) such as AlaAT (Shrawat et al. 2008) and AspAT (Schultz and Coruzzi 1995).

Transport of ammonium (NH^{4+}) is driven by AMT/MEP/Rh (Ammonium Transporter/Methylammonium Permease/mammalian Rhesus proteins) family of transmembrane proteins which are found in organisms ranging from bacteria, cyanobacteria to animals and plants. In plants, two sub-families of AMT, viz. AMT1 and AMT2, both having high affinity for ammonium are found (Pantoja 2012). The first ammonium transporter, AMT1.1, was isolated from *Arabidopsis thaliana* and subsequent analysis identified a total of six homologues of AMT from *A. thaliana* (Ninnemann et al. 1994; Gazzarrini et al. 1999; Sohlenkamp et al. 2002). Of the six AMT homologs in *A. thaliana*, AMT1.1 to AMT1.5 form the AMT family and AMT2 is similar to MEP family; members of Rh protein family are limited to mammal in distribution (Loqué and von Wirén 2004). Till date, ammonium transporters have been characterized from several other plant species including *Populus* (Couturier et al. 2007), *Lotus japonicus* (D'Apuzzo et al. 2004), *Triticum* (Liu et al. 2015), *Lycopersicon esculentum* (Ludewig et al. 2002), Tomato (Von Wirén et al. 2000), to name a few examples.

While nitrate reductase (NIA) reduces nitrate to nitrite, nitrite reductase (NIR), glutamine synthetase (GST), and glutamate synthase (GOGAT) help in further assimilation of N. Glutamate dehydrogenase (GDH), aspartate aminotransferase (AspAT), and asparagine synthetase (AS) are enzymes involved in assimilation of *N* and amino acid synthesis (Cheng et al. 1991; Li et al. 2017c; Bernard and Habash 2009; Wong et al. 2004).

NIN-like protein 7 (NLP7), a transcription factor belonging to RWP-RK family, is regulated through availability of nitrate, via a nuclear retention mechanism to further control downstream genes involved in N signaling, assimilation, and metabolism (Marchive et al. 2013). NLP7 has been identified as the major critical regulator and can induce and repress downstream targets. It works upstream of NRT1;1 in the presence of ammonia, and independent of NRT1.1 in its absence (Zhao et al. 2018). Another gene NLA which encodes RING-type ubiquitin E3 ligase is known to play role in adaptive response to N limitation via anthocyanin accumulation (Peng et al. 2007). Expression of several other transcription factors such as ZmDof1 in wheat, OsRDD1 in rice, HY5 and bZIP in *A. thaliana* has been deciphered to play role in Nitrogen uptake (Zhao et al. 2018). Members of protein kinases such as Ca^{2+} CIPK, CPK also play role in inducing primary nitrogen responsive genes (Riveras et al. 2015).

5.3.2 Phosphorus

Phosphate in soil is present in multitude of forms—organic, inorganic, and mineral P; mostly immobile, majority of P is, therefore, unavailable for uptake by plant roots. The inorganic form of Phosphorus, Pi, is the most accessible form for plants and generally ranges up to 10 μM in soil. The forms of Pi available for uptake are $H_2PO_4^-$ and HPO_4^{2-} which are dissociation products of H_3PO_4 (Bieleski 1973; Schachtman et al. 1998). H_3PO_4 disassociates, in a pH dependent manner, first into the monovalent form $H_2PO_4^-$, which then further disassociates into the divalent form HPO_4^{2-}. Studies have shown that most phosphate uptake occurs between pH5.0 and 6.0 where $H_2PO_4^-$ is the predominant monovalent ionic form present. In addition to molecular components of the host plant, arbuscular mycorrhizal associations are known to play critical roles in mobilization of phosphate from outside the rhizosphere, thus facilitating acquisition and uptake, an aspect that is currently beyond the purview of this chapter (Bucher 2007).

The uptake of Phosphate is performed by transporters that have been grouped in four distinct families—PHT1, PHT 2, PHT 3, and PHT4. Members of PHT1 are high affinity transporters and belong to phosphate:H^+ symporter (PHS) family and have conserved function (Muchhal et al. 1996; Smith et al. 1997; Pao et al. 1998; Hasan et al. 2016; Gho and Jung 2019). PHT1 was initially identified as high affinity transporter but later some members such as PHT1;2 in *O. sativa* (OsPT2) and PHT1;6 in *H. vulgare* (HORvu-Pht1;6) were found to be low affinity phosphate transporters. In both these plants, the low affinity transporters from *PHT1* family are suggested to be responsible for translocation and remobilization of stored Pi, rather than direct uptake by roots which is performed by high affinity transporters such as OsPht1;6 (OsPT6) and HORvu-Pht1;1 (Rae et al. 2003; Ai et al. 2009). Members of the PHT1 family are the primary ion channel transporters present on the outer cortical cells and epidermal cells of root, and responsible for uptake of Pi. In addition to this primary role, members of PHT1 family are also believed to be involved in translocation and distribution owing to their presence in various other plant organs such as young and mature leaves, stems, anther, silk, young seed, cobs, stages of pollen grains (Davies et al. 2002; Nagy et al. 2006; Ai et al. 2009).

PHT2 family are low-affinity proton symporters (H+/Pi cotransporter) that have been shown to be up-regulated under *P*-starvation stress, help in translocating phosphate to green shoot tissues and chloroplast (Daram et al. 1999; Versaw and Harrison 2002; Guo et al. 2013). PHT2 has been suggested to be a key regulator of signaling under both *P*-deficient and *P*-sufficient condition, and may control other phosphate transporters responsible for Pi acquisition and translocation. The third family of transporters, PHT3, is involved in uptake of *P* in mitochondria (Takabatake et al. 1999; Rausch and Bucher 2002); and the fourth family—PHT4 is known to play role in intra-cellular Pi translocation, viz. within cytosol, chloroplast, plastids, and the Golgi apparatus (Guo et al. 2008; Cubero et al. 2009; Liu et al. 2011). PHT4;6 in *A. thaliana* has been shown to be localized to Golgi body, is regulated by circadian rhythm, and plays roles in both biotic and abiotic stress responses (Cubero et al. 2009; Wang et al. 2011).

Apart from transporters, several other molecular regulators play important role in *P*-sensing, uptake, transport, and acquisition. Products of PDR2 (Phosphate Deficiency Response 2), LPR1 (Low Phosphate Root 1), and LPR2 are involved in local sensing of available phosphate levels proximal to *Arabidopsis thaliana* roots (Lopez-Arredondo et al. 2014). The local sensing of phosphate availability and uptake is modulated by regulating the root meristem activity through interaction of PDR2, LPR1, SCR, SHR (Ticconi et al. 2009). PHR1 (phosphate starvation responsive 1), which is a R2-R3 MYB transcription factor, also plays a critical role in sensing P availability (Rubio et al. 2001). It binds to P1BS DNA motif (GNATATNC; Rubio et al. 2001) which is conserved in monocots as well as dicots (Zhou et al. 2008). This motif is found in several putative phosphate starvation responsive genes such as PHT1 (Bustos et al. 2010) and *INDUCED BY PHOSPHATE STARVATION* gene (IPS, Rubio et al. 2001). These constitute the primary sensing mechanism, present upstream of the Pi uptake by high affinity PHT1 transporter. Several other studies have investigated post-transcriptional regulation of PHT1 such as through intracellular trafficking by PHF1 (González et al. 2005; Bayle et al. 2011), phosphorylation at C'-terminal end (Bayle et al. 2011), and endocytosis followed by subsequent degradation of transporter protein by PHO84 (Lagerstedt et al. 2002; Persson et al. 2003). The two genes *PHO1* and *PHO2* are responsible for Pi homeostasis (Hamburger et al. 2002; Aung et al. 2006; Bari et al. 2006). Several secondary messengers such as Ca^{2+}, IPs (Stevenson-Paulik et al. 2005), ROS (Tyburski et al. 2009) are known to be induced by phosphate starvation and may play role in sensing and signaling of phosphate levels.

5.3.3 Potassium

The uptake and transport of potassium take place via both active and passive mechanisms. Unlike N, P, and S which have dedicated transporters, potassium has been shown to be transported from soil to the roots via several families of both transporters and ion channels. In *Arabidopsis thaliana* where the first plant potassium ion channels were identified, a total of six different families of ion channels and transporters are now known. These are Shaker family, TPK family, Kir-like family, KUP/HAK/KT family, HKT family, and CPA family (Anderson et al. 1992; Sentenac et al. 1992; Chen et al. 2008a). Subsequently, five families of ion channels and transporters for potassium have been proposed—Shaker family, TPK/KCO-2P family, TPK/KCO-1P family, KUP/HAK/KT family, and HKT family. In addition to these five families, animal cyclic nucleotide-gated channels (CNGCs), glutamate receptors, LCT1, cation proton antiporters (CPA), and cation chloride co-transporter family (CCC) are other probable candidates that participate in potassium uptake, transport, and assimilation (Very and Sentenac 2003). The members of the Shaker family have been further sub-divided into three subfamilies as inward rectifying (IR), weakly inward rectifying (WR), and outward rectifying (OR). AKT1/KC1 is an example of IR-type of the Shaker family, which is upregulated under low K-conditions through signaling via CIPK, CBK, PP2CA; AKT2 is an example of the WR-type, whereas SKOR is an example of the OR-type of K-transporter (Wang and Wu 2013)). Apart from K+ absorption at the root, these are also likely to play role in loading of K^+ in xylem and phloem, guard cell and stomatal movement, pollen tube development and stress regulation (Sentenac et al. 1992; Ache et al. 2000; Pilot et al. 2001; Mouline et al. 2002; Gaymard et al. 1998).

The second family, viz. KT/KUP/HAK includes low affinity (*KUP2*, *KUP3*, *KUP4*), high affinity (*HAK5*, *OsHAK1*), and dual affinity transporters (*KUP1*). In plants, members of this family are further sub-divided into four groups based on affinity and active/passive transport ability. Group I is primarily high affinity transporters, may be involved in active transport, and may not be able to discriminate between other cations such as K^+, Rb^+, and Cs^+ but are less likely to transport Na^+ and $NH4^+$. Members of group II are low affinity and able to transport ions such as K^+, Rb^+ Na^+, and Cs^+. Members of KT/KUP/HAK family have been shown to be present in nearly all plant parts including roots, shoots, leaf, flowers (Fu and Luan 1998; Gierth et al. 2005; Grabov 2007; Elumalai et al. 2002), and critical for seedling establishment (Pyo et al. 2010).

The TPK/KCO family includes KCO-2P and KCO-1P subfamilies that are divided based on number of transmembrane domains (TMS) and pore (P)-domains. KCO-2P contains four TMS and two P-domains, whereas KCO-1P contains two TMS and one P-domain (Very and Sentenac 2003). Members of the KCO family are membrane localized, involved in K+ transport in nearly all plant parts, and exhibit Ca^{2+} dependency (Czempinski et al. 2002; Voelker et al. 2006; Rehman et al. 2017). In a recent study, TPK/KCO family members have also been shown to be involved in nodulation in soybean (Rehman et al. 2017).

The family of HKT transporters are involved in both influx and efflux of K^+ as well as Na^+ or H^+ (Schachtman and Schroeder 1994; Rodríguez-Navarro 2000; Waters et al. 2013; Almeida et al. 2013). The affinity and ability to permit influx and efflux of these ions has been observed to be concentration dependent in some species such as *Triticum aestivum* but not in others as in *Arabidopsis thaliana* (Gassman et al. 1996; Rubio et al. 1995; Uozomi et al. 2000; Mäser et al. 2002a). In multiple species where HKT gene families have been characterized, such as *A. thaliana*, *O. sativa*, *T. aestivum*, *H. vulgare* their expression has been observed in roots, in root and stem in *A. thaliana* and, in root, leaf, and stem in *Eucalyptus* (Rus et al. 2001; Uozomi et al. 2000; Fairbairn et al. 2000; Sharif Shohan et al. 2019; Wang et al. 1998; Horie et al. 2001; Tada 2019). Members of this family are divided into two classes, among which class 2 is found only in monocots (Platten et al. 2006). Members of HKT family have been shown to be associated with salt tolerance in both *A. thaliana* and *O. sativa* (Sharif Shohan et al. 2019; Tada 2019).

Members of cyclic nucleotide gated channels (CNGCs) (CNGC1, CNGC2, CNGC4, and CNGC10) are known to work as K+ rectifying inward channel. Similarly, members of CATION/Hþ EXCHANGERs CHX family (CHX13, CHX17, CHX20, CHX21, CHX23) (Cellier et al. 2004; Padmanaban et al. 2007) and GORK family in the root hairs (Ivashikina et al. 2001) are also known to help in K^+ transport. Long distance transport is proposed to be carried out by K EFFLUX ANTIPORTERs (KEAs) which play role in K^+ efflux into xylem sap, whereas KAT2 and AKT2/3 regulate phloem loading and K homeostasis in the phloem (Aranda-Sicilia et al. 2012; Deeken et al. 2002; Philippar et al. 2004). Intracellular transport is carried out by TPK1 (K efflux from the vacuole), NHX acts in early development stage of plant (Liu et al. 2010). HAK10 is also known to mediate K release from the vacuole to the cytosol in rice (Banuelos et al. 2002).

5.3.4 Sulfur

Sulfur is taken up as sulfate ion (SO_4^{2-}) and reduced to sulfide prior to assimilation into metabolites by plants. There are four types of sulfate transporters found from bacteria to plants and are exemplified by proton/sulfate co-transporter (SUL, SULTR family), sodium/sulfate co-transporter, sulfate/anion exchanger family, and ABC-type transporter complex (Takahashi et al. 2012). Structural feature of sulfate transporter includes presence of twelve transmembrane domain, and, sulfate transporter and antistigma factor antagonists (STAS) domain which is involved in sulfate uptake and flux, (Lass and Ullrich-Eberius 1984; Rouached et al. 2005; Baraniecka and Kopriva 2014). Sulfate is transported from soil to the roots with the help of proton/sulfate co-transporters which were initially divided into five groups—Sultr1, Sultr2, Sultr3, Sultr4, and Sultr5, based on their expression pattern and kinetics of transport mechanism. Expression pattern of these transporter proteins depends on soil/intracellular sulfate amount, cysteine and glutathione concentration. High affinity transporters are involved in soil-to-root uptake and intracellular transport, whereas low affinity transporters participate in vascular transport within the plants.

The genome of *A. thaliana* was found to encode a total of fourteen sulfate transporter genes belonging to five groups. Two genes (SULTR5;1 and SULTR5;2) have MOT domain instead of STAS domain, which is involved in Molybdenum transport, are not affected by levels of sulfate, and thus are not considered sulfate transporters. These have been re-annotated as MOT1 and MOT2 and now are considered as molybdenum transporters (Gasber et al. 2011; Baxter et al. 2008). There are thus now only four groups of sulfate transporters. Group 1 have two homologs in *A. thaliana* (SULTR1;1, SULTR1;2) both of them are high affinity transporters and are involved in sulfate uptake from soil. Group 2 transporters (SULTR2;1, SULTR2;2) are involved in long distance transport and in sulfate flux through the plant (Takahashi et al. 2000). Members of group 3 include chloroplast sulfate transporters (SULTR3;2, SULTR3;3, SULTR3;4; but not SULTR3;5) and low affinity long distance transporter (SULTR3;1). Group 4 transporters are involved in efflux of sulfate from the vacuole in pericycle and xylem parenchyma cells. Few group 1 transporters are also involved in long distance transport within plant (Yoshimoto et al. 2003; Howarth et al. 2003; Takahashi et al. 2011). Members of group 1, 2, and 4 are regulated during sulfate deprived conditions (Yoshimoto et al. 2003; Takahashi et al. 2000. Other than *Arabidopsis,* sulfate transporters have been identified and characterized from a variety of plants such as *G. max* (Ding et al. 2016), *Brassica* (Buchner et al. 2004b), *M. truncatula* (Casieri et al. 2012), *O. sativa* (Kumar et al. 2019; Godwin et al. 2003), *T. aestivum* (Buchner et al. 2010), *Z. mays* (Huang et al. 2018), *Populus* (Dürr et al. 2010), to name a few.

The first step in assimilation and metabolism of sulfate is adenylation to adenosine 5' phosphosulfate (APS) by ATP sulfurylase (ATPS), which functions as a homotetramer and is localized in cytosol as well as chloroplast. Four copies of gene encoding ATP sulfurylase are present in *A. thaliana.* APS can act via two pathways: (1) convert SO_4^{2-} into sulfite by the action of APS reductase (APR) and (2) generate 3'-phosphoadenosine 50-phosphosulfate (PAPS) with the action of APS kinase (APK). The enzyme APR is encoded by three paralogous genes in *A. thaliana* and acts a dimer; APK is encoded by four genes in *A. thaliana*. Cysteine is the first stable product in sulfate assimilation. It is formed by the action of O-acetylserine (thiol) lyase (OAS-TL) for which O-acetylserine (OAS) and sulfide are substrates (Wirtz and Hell 2006). Serine acetyltransferase (SAT) is involved in OAS synthesis through serine and acetyl coenzyme A. The first pathway results in production of cysteine that acts as precursor for synthesis of methionine. Methionine in turn is a precursor for several amino acids and glutathione, essential for several plant metabolites.

There are several other key regulators of sulfate uptake and assimilation. SULFUR LIMITATION 1 (SLIM1), a member of the ethylene insensitive 3-like (EIL) transcription factor family, is considered a key transcriptional regulator of sulfate uptake and metabolism, and acts via miR395 (Maruyama-Nakashita et al. 2006; Kawashima et al. 2011; Mathewman et al. 2012). SLIM1 has been shown to also act under Cadmium stress and promote sulfate uptake (Yamaguchi et al. 2020). *SLIM1* also negatively regulates members of the MYB pathway such as *MYB34* and may influence glucosinolate biosynthesis, an assimilation product of sulfate in Brassicaceae (Maruyama-Nakashita et al. 2006; Takahashi et al. 2011). Sulfate uptake, transfer, and partitioning have also been shown to be positively regulated via arbuscular mycorrhizal associations of *Glomus intraradices* in roots of *Daucus carota* (Allen and Shachar-Hill 2009).

Apart from the transporters associated with N, P, K, and S discussed in the previous sections, there are others that are involved in uptake, transport, and assimilation of other mineral nutrients and a representative list of key transporters and molecular regulators are provided in Tables 5.4 and 5.5.

5.4 MicroRNAs in Nutrient Uptake and Stress

Several reports of small RNA sequencing, and expression analysis of candidate genes has led to the identification of miRNAs involved in nutrient uptake and stress in various plant species. Prominent miRNAs that have been shown to have altered transcript levels under N-stress include miR156, miR169, miR172, miR319, miR395, miR396, miR398, miR399, and others (Pant et al. 2009; Zhao et al. 2011; Liang et al. 2012; Ren et al. 2015; Zuluaga et al. 2017; Zuluaga and Sonnante 2019). In an earlier study, MIR169 family was shown to target NFγA family of TFs which in turn are known to involve in drought resistance (Wang et al. 2003). MIR169 was also found to be significantly downregulated under N deprived conditions in *A. thaliana* (Zhao et al. 2011). Similar results were also observed in *Z. mays* (Yang et al. 2019). Recently TaMIR444a was found to be upregulated under N deficiency, and was found to regulate expression of several other nitrogen responsive genes in wheat (Gao et al. 2016).

Several other miRNAs including miR896 and miR1222 that are upregulated under P stress and, miR1211 that is downregulated in *Lupinus* under *P* stress; microRNAs miR1122, miR1125, miR1135, miR1136 that are upregulated in *P*-deficiency stress in wheat represent unique miRNAs that are specific to a particular nutrient deficiency stress in a single species (Paul et al. 2015). MIR399 is known to be a key regulator and the levels of miR399 are significantly increased in several monocots as well as dicots in response to phosphate starvation. The *MIR399* gene family is a well characterized family which is known to be conserved across monocots and dicots. Mature product of MIR399 has been shown to target and regulate PHO2/UBC24 (a member of ubiquitin conjugase enzyme family) through post-transcriptional gene regulation (PTGS), which is known to play crucial role in degradation of several phosphate transporters (Bari et al. 2006). The MIR395 gene family is known to be involved in sulfate starvation response acting via sulfate transporters (AtSULTR2;1) and ATP sulfurylase (AtAPS4, AtAPS1) (Jones-Rhoades and Bartel 2004; Kawashima et al. 2009; Liang et al. 2010). In a recent study, several miRNAs including novel ones were identified that may be involved in Fe-homeostasis (Paul et al. 2016). Transcript levels of already known microRNAs including those of miR166, miR399, and miR408, and novel microRNAs—miR11, miR26, miR30, and miR31 were found to be significantly altered in transgenic rice over-expressing Soy-*FER1* (encoding ferritin protein). The four novel miRNAs were

Table 5.4 Molecular regulators involved in mineral (N, P, S, K) nutrition uptake, transport, assimilation, and signaling

Molecular regulators and phytohormones	Plant	Role	References
Nitrogen (N)			
NRT/NPF	*A. thaliana*, *O. sativa*, *M. truncatula*	Uptake and transport	Li et al. (2017b), Morère-Le Paven (2011), and Wang et al. (2020a, b)
AMT	*A. thaliana*, *O. sativa*	Uptake and transport	Li et al. (2017b)
NIA, NIR	*A. thaliana*, *O. sativa*, *Zea mays*	Assimilation	Cheng et al. (1991), Hamat et al. (1989), and Shaner and Boyer (1976)
GDH	*A. thaliana*, *O. sativa*		Ameziane et al. (2000)
AS	*A. thaliana*		Li et al. (2017b)
AspAT	*A. thaliana*		Li et al. (2017b)
AlaAT	*A. thaliana*, *O. sativa*		Li et al. (2017b)
GS	*M. truncatula*, *A. thaliana*, *Saccharum* spp.		Bernard and Habash (2009)
GOGAT	*A. thaliana*, *O. sativa*		Bernard and Habash (2009)
NLA	*A. thaliana*	Regulation of adaptive response	Peng et al. (2007)
NLP7, OsRDD1, LBD37/38/39 HY5, *ZmDof1*	*A. thaliana*, *O. sativa*	Regulation of transport and assimilation	Zhao et al. (2018)
Ca^{2+} CIPK, CPK	*A. thaliana*	Signalling	Riveras et al. (2015)
Phosphorus (P)			
PHT	*A. thaliana*, *S. lycopersicum*, *M. truncatula*, *S. tuberosum*, and *C. roseus*, *H. vulgar*, *T. aestivum*, *N. tabacum*, *N. lucifera*, *T. aestivum*	Uptake and transport	Pao et al. (1998), Hasan et al. (2016), Rae et al. (2003), Ai et al. (2009), and Nussaume et al. (2011)
PDR2, LPR1, LPR2, SCR, SHR	*A. thaliana*	Sensing	Darmer et al. (2014) and Ticconi et al. (2009)
PHO	*A. thaliana*	Pi homeostasis	Bari et al. (2006)
IPS	*A. thaliana*	Sensing and regulation	Rubio et al. (2001)
Ca^{2+}, IPs, ROS	*A. thaliana*	Sensing and signaling	Chiou and Lin (2011)
PHR1/PHL1	*A. thaliana*, *O. sativa*	Regulation of transport and assimilation	Rubio et al. (2001) and Zhou (2008)
PHF	*A. thaliana*		Bayle et al. (2011)
PTF1	*O. sativa*		Yi et al. (2005)
bHLH32	*A. thaliana*, *O. sativa*,		Chen et al. (2007)
ZAT6, *WRKY75*	*A. thaliana*		Devaiah et al. (2007)

(continued)

Table 5.4 (continued)

Molecular regulators and phytohormones	Plant	Role	References
Potassium (K)			
KT/KUP/HAK	*A. thaliana*, *O. sativa*	Transport	Grabov (2007)
AKT2/3	*A. thaliana*		Deekan et al. (2002)
KAT1	*O. sativa*, *A. thaliana*		Philippar (2004)
CHX	*A. thaliana*		Cellier et al. (2004) and Padmanaban et al. (2007)
GORK	*A. thaliana*		Ivashikina et al. (2001)
KEA	*A. thaliana*		Aranda-Sicilia et al. (2012)
LCTs	Wheat, *O. sativa*		Very and Sentenac (2003)
CNGCs	*A. thaliana*		Very and Sentenac (2003)
TPK/KCO	*A. thaliana*, *G. max*		Voelker et al. (2006) and Rehman et al. (2017)
HKT1	*A. thaliana*, *O. sativa*, *T. aestivum*		Schachtman and Schroeder (1994)
CBK-CIPK-AKT	*A. thaliana*	Sensing and signaling	Wang and Wu (2013)
Sulfur (S)			
SULTR	*A. thaliana*, Brassica, *O. sativa*, *S. tuberosum*, *S. bicolor*, *Z. mays*, *Poplar*, *M. truncatula*, *T. aestivum*	Transport	Takahashi et al. (1997), Vidmar et al. (2000), Buchner et al. (2004a, b), Kumar et al. (2015), Akbudak et al. (2018), Huang et al. (2018), Vatansever et al. (2016), Dürr et al. (2010), and Gao et al. (2014)
ATPS	*A. thaliana*	Assimilation	Koprivova (2013)
APK	*A. thaliana*		Mugford et al. (2009)
APR	*A. thaliana*		Kopriva (2001)
OAS-TL	*A. thaliana*		Wirtz and Hell (2006)
SAT	*A. thaliana*		Haas (2008)
SLIM1	*A. thaliana*	Regulation of transport and assimilation	Maruyama-Nakashita et al. (2006)
Phytohormones and mineral nutrients			
Mineral	Hormone/s	Plant	References
N	• Auxin • Cytokinin • ABA • Ethylene • JA	*A.thaliana*, *N. tabacum*	Krouk et al. (2010), Krouk (2016), Ristova et al. (2016), and Ruffel et al. (2016), Ruffel (2018)

(continued)

Table 5.4 (continued)

Molecular regulators and phytohormones	Plant	Role	References
P	• Cytokinin • Strigolactone	*A.thaliana*	Franco-Zorrilla et al. (2005) and Mayzlish-Gati et al. (2012)
K	• ABA • GA • Cytokinin • Auxin	*T. aestivum* *A.thaliana*	Erdel and Dhakal (1988) and Nam et al. (2012)
S	• JA • Ethylene • ABA • Nitric oxide • Auxin • Cytokinin • SA	*A.thaliana*; Several plants	Koprivova and Kopriva (2016)

predicted to target the transcript of Natural Resistance-Associated Macrophage Protein 4 (NRAMP4), a metal transporter (Paul et al. 2016).

Several miRNAs are likely to play a general role in general nutrient homeostasis whereas several miRNAs may have role in maintaining homeostasis of specific mineral nutrients. For example, levels of miR156 are upregulated under deficiency of P, N, S, and Mn in *Lupinus albus* (P), *A. thaliana* (N), *B. napus* (S), and *Phaseolus vulgaris* (Mn). Similarly, levels of miR408 are downregulated in wheat and *Arabidopsis* under P and Fe deficiency stress, respectively, and upregulated in maize and *Arabidopsis* under N and Cu deficiency stress, respectively (Paul et al. 2015). MIR827 is known to play role in *N* and *P* stress, it targets the NLA gene which is known to play adaptive role in *N* stress (Kant et al. 2011). Compared to *N* and *P*, MIRNA in sulfate and potassium stress has not been explored by any genome wide search. MIR444 has been known to slightly alter by nitrate as well as potassium deprived condition in Rice (Yan et al. 2014). Table 5.6 provides a comprehensive representative list of miRNAs that have been shown to be responsive to several mineral nutrients in a variety of plants.

Since several conserved MIRNA families occurring in various plant species are known to also display similar responses to nutrient stress across various plants, they can be exploited to generate as common strategy to enhance nutrient use efficiency (NUE). In addition to that several miRNAs—miR160, miR167, miR171, miR393 which are known to alter root system architecture are also found to be responsive to N and P stresses (Sinha et al, 2015). These miRNAs are, therefore, considered as key candidates, and can prove helpful in modifying root architecture of plant for better adaptation to nutrient stresses.

Table 5.5 Transporters and phytohormones involved in uptake, transport, and assimilation of mineral nutrients other than NPSK in plants

Mineral	Form available/ uptake	Transporter and phytohormones	Comments	References
B	Boric acid	1. Boric acid channel: NIP5;1/ NIP3;1 2. Boric acid/ Borate exporter: BOR1-7	1. NOD26-like intrinsic proteins (NIP) is involved in uptake and import of B 2. BOR1 is involved in loading B into xylem 3. Boron is also taken up though passive diffusion through lipid bilayer, and, through selective and non-selective ion channels	Takano et al. (2002, 2006, 2008), Mosa et al. (2016), and Reid (2014)
Ca	Ca^{2+}	1. Hyperpolarization activated Ca2+ channel (HACC) 2. Depolarization-activated Ca2+ channel (DACC) 3. Ca^{2+} permeable outward rectifying K+ channel (KORC) 4. Voltage-insensitive cation channels (VICC) 5. Ca^{2+}-ATPases 6. H^+/Ca^+ antiporters	1. HACC encoded by *annexin* genes 2. DACC encoded by *TPC1* 3. KORC encoded by *SKOR* and *GORK* genes 4. VICC encoded by *CNGC* (cyclic nucleotide-gated) and *GLR* (glutamate receptor homologs) genes 5. Ca^{2+}-ATPases encoded by *ACA8*	White and Broadley (2003), Thor (2019), and Demidchik et al. (2018)
Cl	Cl^-	Voltage-dependent chloride channel CLC/Chloride transporter channel Cation/Chloride Cotransporters (CCCs)	1. CLC-0 was the first Cl^- transporter from an electric ray fish, *Torpedo californica* , and in plants from *N. tabaccum* and *A. thaliana* 2. Eukaryotic CLC proteins are characterized by presence of two cystathionine b-synthetase (CBS) domains 3. Severn	White and Miller (1979), Lurin et al. (1996), Hechenberger et al. (1996), Zifarelli and Pusch (2010), Herdean et al. (2016), and Li et al. (2017a)

(continued)

Table 5.5 (continued)

Mineral	Form available/ uptake	Transporter and phytohormones	Comments	References
			homologues, CLCa-CLCg, are present in *A. thaliana* genome; rice has five homologues 4. A.thaliana CLCa, and CLCe are selective towards NO_3^- ion, whereas CLCc has dual affinity for Cl^- and NO_3^- 5. CCC may retrieve Cl from the root xylem	
Cu	Cu^+, Cu^{2+}	1. Copper transporter protein (COPT) family (high affinity transporter) 2. P_{1B}-type ATPase transporters 3. ZIP family 4. ATX family 5. CCS family 6. COX family 7. Yellow stripe-like 1 (YS1) family 8. H+/Cu2+ antiporters	1. Being a micronutrient, Cu is required in small amount and excess of Cu causes phytotoxicity 2. Ferric reductase (FRO) converts Cu^{2+} to Cu^+ in soil which is then taken up by COPT 3. COPT1 is responsible for uptake of Cu^+ whereas ZIP2 and YSL are speculated to take up Cu^{2+} 4. P_{1B}-type ATPase transporters are involved in transport of a range of ions such as Cu^+, Cu^{2+}, Zn^{2+}, Cd^{2+} and Pb^{2+} 5. ATX, CCS and COX are chaperone proteins 6. Members of YSL family are part of OPT (oligopeptide transporter) superfamily and function as Cu^{2+}-Nicotinamine (NA; phytosiderophore)	Kampfenkel et al. (1995), Yruela (2009), Printz et al. (2016), Migocka and Malas (2018), and Puig (2014)

(continued)

Table 5.5 (continued)

Mineral	Form available/ uptake	Transporter and phytohormones	Comments	References
			complex 7. H+/Cu2+ antiporters are involved in both influx and efflux of Cu2+ and other divalent cations	
Fe	Fe^{3+}, Fe^{2+} (Fe^{2+} is more soluble and thus readily taken up)	1. IRT (member of a ZIP; ZRT-IRT like ZIP) 2. Uptake via chelation of Fe^{3+} to mugineic acid (MA) family of phyto-siderophores (PS-MA) 3. Yellow-stripe 1 (YS1) of the OPT family are Fe-PS/MA transporters 4. Multidrug and toxin efflux (MATE) family is involved in long distance transport of Fe through xylem 5. Fe transport protein (ITP) 6. Nicotinamine, a phytosiderophore (NA) complexes with Fe2+ (Fe2+-NA complex) and may be involved in long distance transport via phloem 7. VIT1 (Vacuolar Iron Transporter 1), 8. Nramp (natural resistance associated macrophage proteins)	1. Ferric reductase (FRO) converts $Fe3^{+}$ to $Fe2^{+}$ in soil 2. IRT1 is the major transporter of Fe (along with Zn, Mn, Co, Cd) from soil into the root 3. 2'-deoxymugineic acid (DMA), 3-pi-hydroxymugineic acid (epi-HMA), and 3-epihydroxy 2'-deoxymugineic acid (epi-HDMA) are the major PS-MA involved in uptake of Fe^{3+} through the YS1 encoded transporter 4. YS1 also transportsFe2+-PS/NA complex 5. MATE transporter is encoded by FRD (ferric reductase defective) gene 6. ITP is involved in transport through phloem 7. VIT1 (Vacuolar Iron Transporter 1), an Fe2+ functions in vacuolar Fe storage 8. Nramp play a role in Fe remobilization into the cytosol during Fe-deficiency	Tsai and Schmidt (2017), Green and Rogers (2004), Durrett et al. (2007), Krüger et al. (2002), Kim et al. (2006), Kim and Guerinot (2007), and Kobayashi and Nishizawa (2012)

(continued)

Table 5.5 (continued)

Mineral	Form available/ uptake	Transporter and phytohormones	Comments	References
		9. HA proton effux transporter		
Mg	Mg^{2+}	1. MgTR (Mg Transporters such as AtMHX), members of corA family 2. *AtMGT1*-10 family 3. Class II HKT transporter can transport both Mg^{2+} and Ca^{2+}	1. AtMHX is an Mg2 +/H+ transporter 2. AtMGT1 and MtMGT10 are high affinity transporters; AtMGT3, AtMGT7 and AtMGT9 are low affinity; AtMGT5 is dual affinity 3. AtMGT1 and AtMGT6 are expressed in root hair, root elongation zones 4. AtMGT1 is induced by Al^{3+} whereas AtMGT6 is induced by low Mg^{2+}	Shaul et al. (1999), Li et al. (2001), and Guo et al. (2016)
Mn	Mn^{2+}	1. NRAMP (natural resistance-associated macrophage protein) family 2. Yellow stripe-Like (YSL) family 3. CAX (cation exchanger) 4. ZIP (Zinc regulated transporter/iron regulated transporter {ZRT/ IRT1}-related protein 5. CDF/MTP (cation diffusion facilitator/metal tolerance protein) 6. CCX (Calcium cation exchanger) 7. P-type ATPase 8. VIT / CCC1-like family (Vacuolar Iron transporter/ Ca2+-sensitive cross complementer1)	1. Members of NRAMP are also reported to transport Fe, Cd and Zn 2. Members of NRAMP family, YSL-family, ZIP-family, and Calcium-permeable channels are involved in TRANSPORT of Mn into the cytosol 3. CAX family, CCX-family, CDF/MTP-family, P-Type ATPase, VIT/ CCC1-like family are involved in export from cytosol 4. ECA3, involve in transport of Mn and Ca into the Golgi	Socha and Guerinot et al. (2014), Mills et al. (2008), and Wu et al. (2016)

(continued)

Table 5.5 (continued)

Mineral	Form available/ uptake	Transporter and phytohormones	Comments	References
		9. ECA3,P2A-Type ATPase (*Arabidopsis*)		
Mo	MoO_4^{-2}	1. MOT1/ SULTR5;2, and MOT2/SULTR5;1 (*A. thaliana*) 2. MOT2 from *Chlamydomonas*	1. MOT1 and MOT2 from Arabidopsis belong to the sulphate transporter (SULTR) family but lack STAS domain necessary for sulphate transport 2. MOT2 from *Chlamydomonas* is not homologous to AtMOT2, and belongs to major facilitator superfamily	Tomatsu et al. (2007), Tejada-Jiménez et al. (2007, 2011), and Bittner (2014)
Na	Na^+	1. KCO family 2. CNGC family 3. HKT family (high affinity) 4. KUP/HAK/KT family 5. LCT1 family 6. Na+/H+ exchangers (NHXs) 7. K+-ATPase (P-type ATPase) 8. NSCC (non-selective cation channels) 9. SOS1 (Na+/H+ antiporter)	1. KCO family members are dual affinity and involved in influx of Na+ and efflux of K+ 2. Members of CNGC, HKT, KUP/HAK/KT and LCT1 are dual affinity; play role in influx of Na+, K+, Ca2+ 3. Na+/H+ exchangers (NHXs) play role in vacuolar compartmentalization of Na 4. Na+ efflux is performed by K+-ATPase and SOS1	Mäser et al. (2002b), Zhang et al. (2010), and Keisham et al. (2018)
Ni	Ni^{2+}; Ni-$(H2O)_6{}^{2+}$; $NiOH^+$; $Ni^2{}^+(H_2O)_5SO_4{}^{2-}$	1. ZIP family 2. NRAMP family 3. YSL family	1. Both Passive diffusion and active transport occurs	He et al. (2012) and Yusuf et al. (2011)
Si	Silicic acid	1. Low Silicon 1 (Ls1) 2. Low silicon 2 (Ls2)	1. Ls1 and Ls2 are influx and efflux transporters, respectively	Ma and Yamaji (2015) and Ma et al. (2006)
Zn	Zn^{2+}; Zinc-complexes	1. ZIP (ZRT-, IRT-like proteins) family	1. Zinc complexes such as Zinc phosphates, Zinc-	Claus et al. (2013), Gupta et al. (2016),

(continued)

Table 5.5 (continued)

Mineral	Form available/ uptake	Transporter and phytohormones	Comments	References
		2. HMA (Heavy Metal ATPases) family 3. Intake as Zinc-phytosiderophore (nicotianamine/ NA, deoxymuigenic acid/MA, avenic acid) complex 4. H^+-ATPase system 5. MTP (Metal Tolerance Protein) family 6. YSL family 7. OPT family 8. PCR (Plant Cadmium Resistance) family 9. VIT family	hydroxides are also available in soil for uptake	Huang et al. (2020), and Tan et al. (2020)
Phytohormones and nutrient				
	Mineral nutrients	Hormone	Plant	
	Zn, B	Auxin Cytokinin ABA	*S. tuberosum*	Puzina (2004)
	Cu	Auxin GA	*Pisum sativum*	Massoud et al. (2018)
	Fe	Ethylene GA CK	*L. esculentum; A. thaliana; O. sativa*	Lucena et al. (2006), Guo et al. (2015a), and Rubio et al. (2009)
	Mg	ABA Ethylene Auxin	*Arabidopsis*	Guo et al. (2015b)
	Mn	GA	*O. sativa*	Guo et al. (2015b)
	Na, Cl	Cytokinin	*S. bicolor*	Amzallag et al. (1992)

Table 5.6 Nutrient deficiency responsive MIRNA families. Names marked in bold are most studied/functionally validated and conserved MIR families along with their validated targets (in bracket)

Stress	MIRNA family	Plant	References
P-deficiency	**MIR399 (PHO2), MIR827 (NLA)**	Arabidopsis, *Oryza sativa*	Du et al. (2016), Bari et al. (2006), Xu et al. (2013), and Kant et al. (2011)
	166, 169,396, 397,399, 159, 408, 1510, 482, 2109, 1512, 3508, 4416, 3522, 4376	Soybean	Xu et al. (2013)
	156, 159, 166, 167, 164, 169, 395. 399, 171	Maize	Du et al. (2016)
	399, 827, 163, 865, 391, 160	Arabidopsis	Lundmark et al. (2010)
	159, 167, 399, 408, 1122, 1125, 1135, 1136, 1136, 408	Triticum	Zhao et al. (2013) and Zualaga et al. (2017)
	miR399, miR827, miR2111	Nicotiana	Huen et al. (2018)
N-deficiency	**169 (NFYA), MIR827 (NLA)**	Arabidopsis, Maize	Zhao et al. (2011), Yang et al. (2019), and Kant et al. (2011)
	164, 172, 398, 397, 827, 408, 169, 399, 528, 167, 395, 319, 160, 168	Maize	Xu et al. (2011)
	169, 171, 395, 397, 398, 399, 408, 827, 857, 160, 780, 826, 842, and 846	Arabidopsis	Liang et al. (2012)
	159, 159, 160, 164, 167, 399, 408, 1117 and 1120	Triticum	Sinha et al. (2015)
S-deficiency	**395 (AtSULTR2;1, AtAPS4, AtAPS1)**	Arabidopsis	Jones-Rhoades and Bartel (2004) and Kawashima et al. (2009)
K-deficiency	444 (MADS box TFs)	*Oryza sativa*	Yan et al. (2014)
Cu-deficiency	miR398, miR397, miR407, miR857	*A .thaliana*	Yamasaki et al. (2007), Abdel-Ghany and Pilon (2008), and Shahbaz and Pilon (2019)
Fe-deficiency	miR159, miR169, miR172, miR173, miR394	A .thaliana	Waters et al. (2012)
	miR167, miR397, miR398, miR408	Phaseolus vulgaris	Valdes-Lopez et al. (2010)
B-deficiency	miR393, miR160, miR3946, miR159, miR782, miR3946, miR7539, miR164, miR6260, miR5929, miR6214, miR3946 and miR3446, miR5037	*Citrus sinensis*	Lu et al. (2015)
Zn-deficiency	miR171g-5p, miR397b-5p, miR398a-5p, WmiR528-5p, nov-miR1, nov-miR4, nov-miR6, nov-miR23, nov-miR26, nov-mi38	*Oryza sativa*	

(continued)

Table 5.6 (continued)

Stress	MIRNA family	Plant	References
Ca-deficiency	miR159 and miR167, miR3509, miR3511, andmiR351, ahy_novel_miRn112, ahy_novel_miRn23, ahy_novel_miRn62, ahy_novel_miRn132, ahy-miR3515, ahy-miR398, ahy-miR3512,and ahy_novel_miRn9	*Arachis hypogaea L.*	Chen et al. (2019)
Mg-deficiency	miR164 , miR395, miR1077, miR1160, and miR8019 , miR7812, miR8019, miR6218, miR1533, miR6426, miR5256, miR5742, miR5561, miR5158, and miR5818,	*Citrus sinensis*	Ma et al. (2016)
Mn--deficiency	miR395, miR399, miR5026, miR5650, miR781a, miR781b, miR866-3p, miR826, miR5595a, miR5995b, miR863-5p, miR2936, and miR861-3p	*Arabiopsis thaliana*	Gong et al. (2019)

5.5 Nutrient Use Efficiency (NUE)

Ratio of intended output and provided nutrient input is called Nutrient Use Efficiency (NUE). Nutrient Use Efficiency can be studied at three different levels but input and output will vary depending on the level at which NUE is being studied (Nardoto et al. 2006):

1. Individual plant level: It is the most basic level at which input is quantified in terms of nutrients supplied, and output in terms of plant yield (both quantitative and qualitative).
2. Population/crop level: At this level, fertilizers, pesticides as well as labor input are included, and output is measured in terms of total harvest.
3. Ecosystem level: It is the most complex level as it includes impact of the plant/crop growth on other species, environment as well as public health. At this level input is calculated for resources used (both renewable and non-renewable) and negative impacts on environment, while the intended outcome includes agricultural goods.

Maximum NUE can be achieved by either increasing the output or decreasing the input. Understanding NUE at individual level is expected to play most important part in NUE of higher levels population and ecosystem and therefore improving NUE of individual plants can lead to higher agronomic yield under field conditions.

NUE is dependent on genetics and physiology of crops, environmental factors, and the mineral nutrient itself. Broadly, NUE depends on the following three factors:

1. Plant/crop type: Innate ability of plant of uptake and assimilate nutrients is certainly the most important factor influencing the NUE. Internal factors such as growth form (herb, shrub, tree), physiology (C3 and C4; intracellular and intercellular pH), time of harvest (annual, biennial perennial) affect the NUE (Wang et al. 2018). A combination of such internal factors translates into differing NUE of mineral nutrients among cultivars of various crop species such as rice, maize, rapeseed, potato (Zhang et al. 2009; Li et al., 2017c; Stahl et al. 2019; Getahun et al. 2019).
2. Environment: It is the second important factor that directly affects the NUE, and includes of both below ground (soil) and above ground factors. Biotic factors of the soil include microbes, mycorrhiza, pests, and weeds, while abiotic factors include water availability and soil pH (Loqué et al. 2003; Bucher 2007; Hasan et al. 2016; Wang et al. 2018; Penn and Camberato 2019). Environment of above ground includes temperature, light and pollution and humidity. All of these are capable of positively or negatively affecting the NUE directly by altering the nutrient availability or plant growth. For instance, nitrate uptake is dependent on water availability (Buljovcic and Engels 2001); and drought stress hampers P-uptake and P-use efficiency of plants (Garg et al. 2004). Circadian clock has been shown to regulate phosphate transporter PHT4;1, and influence innate immunity against *Pseudomonas syringae* in *A. thaliana* (Wang et al. 2011). Pollutants such as SO_2, H_2S are known to alter the concentration of the nutrients (Aghajanzadeh et al. 2016).
3. Nutrient: Physiochemical properties of individual mineral nutrient and regulatory mechanisms guiding their uptake and assimilation processes in plant are the third most critical factor influencing NUE. Complex interactions exist among the regulatory pathways involved in uptake, transport, and assimilation of mineral nutrients as is evident from the large number of shared transporters (Tables 5.4 and 5.5). Several mineral nutrients also use multiple transporters for uptake and assimilation (Tables 5.4 and 5.5). Some of the nutrients such as Na and K share common uptake transporter (HAK5) and uptake mechanism; and application of NaCl plays negative role in uptake of *K* (Nieves-Cordones et al. 2010). This has also been reported both in *N* and *S*, where *S*-deficiency can be overcome by optimizing N supply in *B. napus* (Abdallah et al. 2010); as well as *P* and *S* where the sulfate transporters are upregulated under phosphate stress (Rouached 2011). Another example is the antiporter family which is shared by H+/Ca2+, H+/Cu2+, and H+/Na+ for uptake and transport. Similarly, ZIP family of transporter is shared by Cu2+, Fe2+, Fe3+, Mn2+, Ni+, and Zn2+. Members of YSL family are shared by Cu2+, Mn2+, Ni+, and Zn2+. Phosphate-starvation has also been shown to reduce concentration of several other mineral nutrients such as Boron,

Calcium, Copper, Magnesium, Manganese, Sodium, and Zinc in *Brassica napus*, and increases endogenous Fe levels (Wang et al. 2020).

5.6 Biotechnological Strategies to Ameliorate Stress and Enhance NUE

Since the beginning of agricultural practices, breeding has been the primary tool for selection of varieties with improved yield and content quality such as fatty acids, carbohydrate, and proteins. Towards this goal, optimum supply of mineral nutrients along with several other promising strategies has been a primary approach to maintain high output/input ratio and improve NUE. However, it is almost difficult to make ideal environment for maximum nutrient use and output using traditional practices. Green revolution was driven by attempting to enhance "output" by increasing input such as fertilizers and irrigation along with breeding (Borlaug 1972). It is because the primary aim of green revolution was to meet food demand for the growing population. However soon it was realized that the practice of applying high level of fertilizers is unsustainable, cannot be practiced for long as this can harm biodiversity. One of the major factors for this was described as the concept of diminishing returns, which states that if fertilizers are used at the same speed, they may not give as much yield as they were giving in the past (Tilman et al. 2002). Also, artificially provided nutrients can cause groundwater and soil pollution/ toxicity which can further lead to eutrophication.

With the advancement in biotechnology, pathways for nutrient uptake and assimilation have been deciphered. This knowledge combined with advancements in genetic breeding and biotechnological tools has the potential to give more sustainable and long-term method for enhancing NUE which aim to reduce "input" component of NUE, without compromising "output."

5.6.1 Genome-Wide Studies

Whole genome sequencing and transcriptome analysis can provide knowledge about most fundamental regulators that can be exploited for enhancing NUE. Identification and functional characterization of candidate genes and regulators which have the ability to increase nutrient uptake and assimilation, or candidates to modify root architecture for better absorption are the primary targets. Also, such studies can be useful in identifying QTLs and biomarkers which are helpful in marker assisted breeding and selection (MAB and MAS). For example, Gua1 has been proposed as a biomarker for nitrate use efficiency (NitUE) in *A. thaliana* (Meyer et al. 2019). Several studies on transcriptome analysis have also been carried out to identify candidate genes for NitUE (Li et al. 2009, Oono et al. 2013). Recently, Gho et al. (2018) identified several promoters for improving phosphate use efficiency (PUE) in rice. Earlier in a separate study, five Pi transporters, two auxin response factors, three SYG1/Pho81/XPR1 (SPX) domain-containing proteins, two MYB-like transcription factors (TFs), an MYB TF, and a phosphatase were identified that affect P uptake Yamamoto et al. (2012). Several QTLs have also been identified in *A. thaliana* which possibly harbor candidate genes for enhancing PU (El-Soda et al. 2019). Multiple QTLs like KUP2, ATK2, KAT2, and TPK3 for potassium use efficiency (KUE) in *B. oleracea* have been identified (White et al. 2010). QTLs have been identified for NitUE in *Arabidopsis* (Loudet et al. 2003), barley (Kindu et al. 2014), and rice (Wei et al. 2012); and QTLs for phosphate-stress in *B. napus* (Wang et al. 2020).

5.6.2 Nutrient Uptake Genes

One of the most direct ways of enhancing NUE is through manipulation of nutrient uptake genes. In *A. thaliana*, overexpression of potassium high affinity transporter, *HAK5,* enhances the tolerance to K-deficiency (Hong et al. 2013). Similarly sulfate transporter SULTR1;2, SULTR2;1, SULTR4;1, SULTR4;2 which are induced upon sulfate deficiency can be candidate genes for altering SUE (Yoshimoto et al. 2003; Howarth et al. 2003; Zuber et al. 2010). *OsNRT1.1B* under the constitutive regulation of 35SCaMV promoter (Hu et al. 2015), *OsNRT2.1* under control of *OsNAR2.1* promoter (Chen et al. 2016), and *OsNPF8.20* overexpressing lines have been shown to increase NitUE where significant increase in yield was achieved (Fang et al. 2013; Wang et al. 2018). Overexpression of *OsNRT2.3b* not only increased nitrate uptake, but NH^{4+}, *P, and* Fe uptake were also enhanced (Fan et al. 2016). In *Pisum sativum,* overexpression of *AMINOACIDPERMEASE1 (AAP1)* causes improved *N* uptake and utilization efficiency (Perchlik and Tegeder 2017).

5.6.3 Nutrient Assimilation Genes

N-assimilation genes such as GS1 have been used for increased NUE in tobacco, maize, rice, and *Arabidopsis* (Eckes et al. 1989; Migge et al. 2000; Man et al. 2011). Deciphering the functions and mechanisms of symbiotic bacteria and other root microbiota for nitrogen fixation will also be helpful in achieving the purpose. AlaAT overexpressing lines of *Arabidopsis* and *O. sativa* show enhanced biomass and seed yield under low nitrogen conditions (Good et al. 2007; Shrawat et al. 2008). GS/GOGAT genes have also proved to be a viable candidate for enhancing NiUE (Lu et al. 2011). Recent studies have also identified role of nitrate transporters NPF4.5 from *O. sativa*, *Z. mays*, and *S. bicolor* in root nodulation (Wang et al. 2020b), and NPF7.6 in root nodule symbiosis in *Medicago truncatula* (Wang et al. 2020a), and may hold potential for increased NitUE.

5.6.4 Regulatory Genes Including Transcription Factors and miRNAs

Overexpression of *PHR1* which is one of major TF led to upregulation of P-uptake (Nilsson et al. 2007). Expression of several other TFs *such as ZmDof1 (TF)* in wheat (Peña et al. 2017), OsRDD1—a Dof1 TF in rice (Iwamoto and Tagiri 2016), HY5—a bZip TF in *A. thaliana* (Chen et al. 2016) has been demonstrated to play role in Nitrogen uptake (Li et al. 2017b). The transcription factor TaNAC2-5A was shown to regulate 106 genes, leading to increased NiUE in wheat (He et al. 2015). As has been elaborated previously (Table 5.6), several miRNAs have been identified as key regulators of nutrient uptake such as miR399 and miR395 for P and S, respectively. As a demonstration, over-expression of MIR528 in rice increases tolerance to N-starvation in *Agrostis stolonifera* (Yuan et al. 2015); and overexpression of miR169o in rice led to increased NitUE (Yu et al. 2018). Several miRNAs that exhibit altered expression pattern under nutrient-starvation are also involved in shaping other key traits such as root system architecture (miR167; miR160), flowering time (miR156; miR172; miR167), plant architecture along with nutrient uptake and homeostasis (miR156) (de Lima et al. 2012; Liu et al. 2018; Song et al. 2019). It should, therefore, be borne in mind that key regulators such as miRNAs are involved in developmental processes and any manipulation of NUE through employing miRNAs genes is also likely to alter developmental processes (Fischer et al. 2013).

5.6.5 Other Genes

Over-production of RUBISCO in transgenic rice led to improved NitUE under sufficient *N* fertilization (Yoon et al. 2020). Members of *Bric-a-Brac/Tramtrack/Broad* (*BTB*)/*BT* gene family encoding BT2 proteins have been shown to be negative regulators of NitUE, and loss-of-function *bt2/bt2* mutants exhibit significantly higher levels of nitrate uptake in *Arabidopsis* (Araus et al. 2016). *Arabidopsis* and rice transgenic lines overexpressing H^+-PPase (*AVP1D*) are shown to have better *P* and *K* uptake, biomass, and seed yield than control plants under limited *P* conditions (Yang et al. 2007). Two isozymes of purple APase (PAP), AtPAP12 and AtPAP26, that are involved in phosphate scavenging have been proposed as useful candidates that have the potential for improving PUE in *A. thaliana* (Robinson et al. 2012). Increased activity of a phosphate-starvation inducible gene, Fructose-1,6-bisphosphatase (FBPase), in root nodules of *Phaseolus vulgaris* was found to be tightly linked and correlated to both increased rhizobial symbiosis and enhanced PUE (Lazali et al. 2016). In another approach, polyploids have been proposed to have higher KUE in Arabidopsis (Chao et al. 2013).

5.7 Modified Root System Architecture (RSA)

Changes in root morphology are primary adaptive features of plants for their survival under various stresses that are linked to soil health including availability of mineral nutrients and nutrient related stresses. Several genes and promoters have been identified which are not directly involved in nutrient uptake or assimilation pathway, but can modulate or alter root architecture and morphology which help plants in coping up and adapting to nutrient stresses. Several transcription factors such as CPC (Tominaga-Wada and Nukumizu 2012; Kirik et al. 2004), MYB77 (Shin et al. 2007), NAC1 (Xie et al. 2000), KNAT6 (Dean et al. 2004), and ANR1 (Montiel et al. 2004; Zhang and Forde 1998) that are involved in altering root architecture by initiating lateral roots under potassium stress have been identified. These TFs under the control of root specific promoters can thus be helpful in enhancing KUE. Similarly, several genes have been identified which are sensitive to low Pi and play role in primary root elongation (SIZ1; Miura et al. 2005, PRD; Camacho-Cristóbal et al. 2008) and induce lateral roots and root hairs (PNP; Marchive 2009, FBX; Chen et al. 2008b). Nitrogen stress is known to induce primary root growth, with complex interactions with phytohormones such as auxin, cytokinin, ethylene, abscisic acid (ABA), brassinosteroids (BRs), strigolactones (SLs), as well as nitric oxide (NO). For example, auxin transporters such as *ZmPIN1a* can lead to higher auxin concentration in roots (Li et al. 2018) and thus promote PR growth (similar to adaptive strategy of plants mild *N* deficiency). Similarly, mutation in auxin biosynthesis gene TRYPTOPHAN AMINOTRANSFERASE RELATED 2 (TAR2) can lead to impaired LR growth.

5.8 Conclusion, Future Prospects, and Scope

Excessive use of chemical fertilizers to supplement and provide mineral nutrition, and irrigation in order to boost crop yield for ever-increasing demand has caused serious damage to soil health, and altered ecosystem dynamics (Guignard et al. 2017). The founding principles of first green-revolution thus have serious environmental limitations and repercussions. Advancements in understanding molecular basis of mineral nutrient uptake by plants, transport and assimilation into the biological system, coupled with insights into plant development such as that of root system architecture, and understanding physiological basis of nutrient uptake have opened up newer vistas. Identification of transporter families, transcription factors, and regulatory microRNAs involved in uptake and plant development, coupled with genomic and breeding strategies such as reverse genetics or transgenic technologies, genome editing, and marker-assisted breeding and selection has the potential to significantly increase nutrient uptake, transport and assimilation efficiencies in not only model plants but in crop species.

Acknowledgements The award of JRF/SRF from UGC and DBT to EB, and JRF from CSIR to RS is gratefully acknowledged. Research in our laboratory is funded by Department of

Biotechnology, Govt. of India; Department of Science and Technology, Govt. of India; University Grants Commission, Govt. of India through extra-mural research grant, and Delhi University through R&D support and Infrastructure grants.

References

Abdallah M, Dubousset L, Meuriot F, Etienne P, Avice JC, Ourry A (2010) Effect of mineral sulphur availability on nitrogen and sulphur uptake and remobilization during the vegetative growth of *Brassica napus* L. J Exp Bot 61:2635–2646

Abdel-Ghany SE, Pilon M (2008) MicroRNA-mediated systemic down-regulation of copper protein expression in response to low copper availability in *Arabidopsis*. J Biol Chem 283:15932–15945

Ache P, Becker D, Ivashikina N, Dietrich P, Roelfsema MR, Hedrich R (2000) GORK, a delayed outward rectifier expressed in guard cells of *Arabidopsis thaliana*, is a KC-selective, KC-sensing ion channel. FEBS Lett 486:93–98

Aghajanzadeh T, Hawkesford MJ, De Kok LJ (2016) Atmospheric H_2S and SO_2 as sulfur sources for *Brassica juncea* and *Brassica rapa*: regulation of sulfur uptake and assimilation. Environ Exp Bot 124:1–10

Ai P, Sun S, Zhao J, Fan X, Xin W, Guo Q, Xu G (2009) Two rice phosphate transporters, OsPht1; 2 and OsPht1; 6, have different functions and kinetic properties in uptake and translocation. Plant J 57:798–809

Akbudak MA, Filiz E, Kontbay K (2018) Genome-wide identification and cadmium induced expression profiling of sulfate transporter (SULTR) genes in sorghum (*Sorghum bicolor* L). Biometals 31:91–105

Allen JW, Shachar-Hill Y (2009) Sulfur transfer through an arbuscular mycorrhiza. Plant Physiol 149:549–560

Almeida P, Katschnig D, De Boer AH (2013) HKT transporters—state of the art. Int J Mol Sci 14:20359–20385

Ameziane R, Bernhard K, Lightfoot D (2000) Expression of the bacterial gdhA gene encoding a NADPH glutamate dehydrogenase in tobacco affects plant growth and development. Plant Soil 221:47–57

Amzallag GN, Lerner HR, Poljakoff-Mayber A (1992) Interaction between mineral nutrients, cytokinin and gibberellic acid during growth of sorghum at high NaCl salinity. J Exp Bot 43:81–87

Anderson JA, Huprikar SS, Kochian LV, Lucas WJ, Gaber RF (1992) Functional expression of a probable Arabidopsis thaliana potassium channel in *Saccharomyces cerevisiae*. Proc Natl Acad Sci 89:3736–3740

Andrews M, Raven A, Lea PJ (2013) Do plants need nitrate? The mechanisms by which nitrogen form affects plants. Ann Appl Biol 163:174–199

Aranda-Sicilia MN, Cagnac O, Chanroj S, Sze H, Rodriguez-Rosales MP, Venema K (2012) *Arabidopsis* KEA2, a homolog of bacterial KefC, encodes a Kþ/Hþ antiporter with a chloroplast transit peptide. Biochim Biophys Acta 1818:2362–2371

Araus V, Vidal EA, Puelma T, Alamos S, Mieulet D, Guiderdoni E, Gutiérrez RA (2016) Members of BTB gene family regulate negatively nitrate uptake and nitrogen use efficiency in *Arabidopsis thaliana* and *Oryza sativa*. Plant Physiol 171(2):1523–1532

Aung K, Lin SI, Wu CC, Huang YT, Su CL, Chiou TJ (2006) pho2, a phosphate overaccumulator, is caused by a nonsense mutation in a microRNA399 target gene. Plant Physiol 141:1000–1011

Baligar VC, Fageria NK, Elrashidi MA (1998) Toxicity and nutrient constraints on root growth. HortScience 33:960–965

Banuelos MA, Garciadeblas B, Cubero B, Rodrıguez-Navarro A (2002) Inventory and functional characterization of the HAK potassium transporters of rice. Plant Physiol 130:784–795

Baraniecka P, Kopriva S (2014) Macronutrient use efficiency – sulfur in *Arabidopsis thaliana*. In: Hawkesford MJ et al (eds) Nutrient use efficiency in plants, Plant ecophysiology, vol 10. Springer, Cham, pp 51–92

Bari R, Pant BD, Stitt M, Scheible WR (2006) PHO2, microRNA399, and PHR1 define a phosphate-signaling pathway in plants. Plant Physiol 141:988–999

Baxter I, Muthukumar B, Park HC, Buchner P, Lahner B, Danku J, Zhao K, Lee J, Hawkesford MJ, Guerinot ML, Salt DE (2008) Variation in molybdenum content across broadly distributed populations of *Arabidopsis thaliana* is controlled by a mitochondrial molybdenum transporter (MOT1). PLoS Genet 4:e1000004

Bayle V, Arrighi JF, Creff A, Nespoulous C, Vialaret J, Rossignol M, Nussaume L (2011) *Arabidopsis thaliana* high-affinity phosphate transporters exhibit multiple levels of posttranslational regulation. Plant Cell 23:1523–1535

Bernard SM, Habash DZ (2009) The importance of cytosolic glutamine synthetase in nitrogen assimilation and recycling. New Phytol 182:608–620

Bieleski RL (1973) Phosphate pools, phosphate transport, and phosphate availability. Annu Rev Plant Physiol 24:225–252

Bittner F (2014) Molybdenum metabolism in plants and crosstalk to iron. Front Plant Sci 5:28

Bonser AM, Lynch J, Snapp S (1996) Effect of phosphorus deficiency on growth angle of basal roots in *Phaseolus vulgaris*. New Phytol 132:281–288

Borlaug NE (1972) The green revolution, peace, and humanity. Speech delivered upon receipt of the 1970 Nobel Peace Prize CIMMYT reprint and translation series No 3 Centro Internacional de Mejoramiento de Maiz y Trigo, El Batan, Mexico

Brady NC, Weil RR (1999) The nature and properties of soils, 12th edn. Prentice-Hall, Upper Saddle River, NJ

Bucher M (2007) Functional biology of plant phosphate uptake at root and mycorrhiza interfaces. New Phytol 173:11–26

Buchner P, Prosser IM, Hawkesford MJ (2004a) Phylogeny and expression of paralogous and orthologous sulphate transporter genes in diploid and hexaploid wheats. Genome 47:526–534

Buchner P, Stuiver CEE, Westerman S, Wirtz M, Hell R, Hawkesford MJ, De Kok LJ (2004b) Regulation of sulfate uptake and expression of sulfate transporter genes in *Brassica oleracea* as affected by atmospheric H_2S and pedospheric sulfate nutrition. Plant Physiol 136:3396–3408

Buchner P, Parmar S, Kriegel A, Carpentier M, Hawkesford MJ (2010) The sulfate transporter family in wheat: tissue-specific gene expression in relation to nutrition. Mol Plant 3:374–389

Buljovcic Z, Engels C (2001) Nitrate uptake ability by maize roots during and after drought stress. Plant Soil 229:125–135

Bustos R, Castrillo G, Linhares F, Puga MI, Rubio V, Pérez-Pérez J, Paz-Ares J (2010) A central regulatory system largely controls transcriptional activation and repression responses to phosphate starvation in *Arabidopsis*. PLoS Genet 6(9):e1001102

Cai C, Wang JY, Zhu YG, Shen QR, Li B, Tong YP, Li ZS (2008) Gene structure and expression of the high-affinity nitrate transport system in rice roots. J Integr Plant Biol 50:443–451

Camacho-Cristóbal JJ, Rexach J, Conéjéro G, Al-Ghazi Y, Nacry P, Doumas P (2008) PRD, an *Arabidopsis* AINTEGUMENTA-like gene, is involved in root architectural changes in response to phosphate starvation. Planta 228:511–522

Casieri L, Gallardo K, Wipf D (2012) Transcriptional response of *Medicago truncatula* sulphate transporters to arbuscular mycorrhizal symbiosis with and without sulphur stress. Planta 235:1431–1447

Cellier F, Conéjéro G, Ricaud L, Luu DT, Lepetit M, Gosti F, Casse F (2004) Characterization of AtCHX17, a member of the cation/H+ exchangers, CHX family, from *Arabidopsis thaliana* suggests a role in K+ homeostasis. Plant J 39(6):834–846

Chao DY, Dilkes B, Luo H, Douglas A, Yakubova E, Lahner B, Salt DE (2013) Polyploids exhibit higher potassium uptake and salinity tolerance in *Arabidopsis*. Science 341:658–659

Chen ZH, Nimmo GA, Jenkins GI, Nimmo HG (2007) BHLH32 modulates several biochemical and morphological processes that respond to Pi starvation in Arabidopsis. Biochem J 405:191–198

Chen YF, Wang Y, Wu WH (2008a) Membrane transporters for nitrogen, phosphate and potassium uptake in plants. J Integr Plant Biol 50:835–848

Chen ZH, Jenkins GI, Nimmo HG (2008b) Identification of an F-box protein that negatively regulates Pi starvation responses. Plant Cell Physiol 49:1902–1906

Chen XB, Yao Q, Gao X, Jiang C, Harberd NP, Fu X (2016) Shoot-to-root mobile transcription factor HY5 coordinates plant carbon and nitrogen acquisition. Curr Biol 26:640–646

Chen H, Yang Q, Chen K, Zhao S, Zhang C, Pan R, Chu W (2019) Integrated microRNA and transcriptome profiling reveals a miRNA-mediated regulatory network of embryo abortion under calcium deficiency in peanut (*Arachis hypogaea* L). BMC Genomics 20:1–17

Cheng CL, Acedo GN, Dewdney J, Goodman HM, Conkling MA (1991) Differential expression of the two *Arabidopsis* nitrate reductase genes. Plant Physiol 96:275–279

Chiou TJ, Lin SI (2011) Signaling network in sensing phosphate availability in plants. Annu Rev Plant Biol 62:185–206

Claus J, Bohmann A, Chavarría-Krauser A (2013) Zinc uptake and radial transport in roots of *Arabidopsis thaliana*: a modelling approach to understand accumulation. Ann Bot 112:369–380

Costa C, Dwyer LM, Zhou X, Dutilleul P, Hamel C, Reid LM, Smith DL (2002) Root morphology of contrasting maize genotypes. Agron J 94:96–101

Couturier J, Montanini B, Martin F, Brun A, Blaudez D, Chalot M (2007) The expanded family of ammonium transporters in the perennial poplar plant. New Phytol 174:137–150

Cubero B, Nakagawa Y, Jiang XY, Miura KJ, Li F, Raghothama KG, Pardo JM (2009) The phosphate transporter PHT4; 6 is a determinant of salt tolerance that is localized to the Golgi apparatus of Arabidopsis. Mol Plant 2:535–552

Czempinski K, Frachisse JM, Maurel C, Barbier-Brygoo H, Mueller-Roeber B (2002) Vacuolar membrane localization of the *Arabidopsis* 'two-pore' K+channel KCO1. Plant J 29:809–820

D'Apuzzo E, Rogato A, Simon-Rosin U, El Alaoui H, Barbulova A, Betti UMK (2004) Characterization of three functional high-affinity ammonium transporters in *Lotus japonicus* with differential transcriptional regulation and spatial expression. Plant Physiol 134:1763–1774

Daram P, Brunner S, Rausch C, Steiner C, Amrhein N, Bucher M (1999) Pht2; 1 encodes a low-affinity phosphate transporter from *Arabidopsis*. Plant Cell 11:2153–2166

Davies TGE, Ying J, Xu Q, Li ZS, Li J, Gordon-Weeks R (2002) Expression analysis of putative high-affinity phosphate transporters in Chinese winter wheat. Plant Cell Environ 25:1325–1339

Dean G, Casson S, Lindsey K (2004) KNAT6 gene of *Arabidopsis* is expressed in roots and is required for correct lateral root formation. Plant Mol Biol 54:71–84

Deeken R, Geiger D, Fromm J, Koroleva O, Ache P, Langenfeld-Heyser R, Sauer N, May ST, Hedrich R (2002) Loss of the AKT2/3 potassium channel affects sugar loading into the phloem of *Arabidopsis*. Planta 216:334–344

Demidchik V, Shabala S, Isayenkov S, Cuin TA, Pottosin I (2018) Calcium transport across plant membranes: mechanisms and functions. New Phytol 220:49–69

Devaiah BN, Nagarajan VK, Raghothama KG (2007) Phosphate homeostasis and root development in Arabidopsis are synchronized by the zinc finger transcription factor ZAT6. Plant Physiol 145 (1):147–159

Ding Y, Zhou X, Zuo L, Wang H, Yu D (2016) Identification and functional characterization of the sulfate transporter gene GmSULTR1; 2b in soybean. BMC Genomics 17:373

Du Q, Wang K, Xu C, Zou C, Xie C, Xu Y, Li WX (2016) Strand-specific RNA-Seq transcriptome analysis of genotypes with and without low-phosphorus tolerance provides novel insights into phosphorus-use efficiency in maize. BMC Plant Biol 16:1–12

Dürr J, Bücking H, Mult S, Wildhagen H, Palme K, Rennenberg H, Dittengou F, Herschbach C (2010) Seasonal and cell type specific expression of sulfate transporters in the phloem of *Populus* reveals tree specific characteristics for SO42 storage and mobilization. Plant Mol Biol 72:499–517

Durrett TP, Gassmann W, Rogers EE (2007) The FRD3-mediated efflux of citrate into the root vasculature is necessary for efficient iron translocation. Plant Physiol 144:197–205

Eckes P, Schmitt P, Daub W, Wengenmayer F (1989) Overproduction of alfalfa glutamine synthetase in transgenic tobacco plants. Mol Gen Genet MGG 217:263–268

Eghball B, Maranville JW (1993) Root development and nitrogen influx of corn genotypes grown under combined drought and nitrogen stresses. Agron J 85:147–152

El-Soda M, Moreira CN, Goredema-Matongera N, Jamar D, Koornneef M, Aarts MG (2019) QTL and candidate genes associated with leaf anion concentrations in response to phosphate supply in *Arabidopsis thaliana*. BMC Plant Biol 19:410

Elumalai RP, Nagpal P, Reed JW (2002) A mutation in the Arabidopsis KT2/KUP2 potassium transporter gene affects shoot cell expansion. Plant Cell 14:119–131

Erdel L, Dhakal MR (1988) Effects of K+ status and phytohormones on K+ transport in wheat. Plant Soil 111:171–175

Evans HJ, Sorger GJ (1966) Role of mineral elements with emphasis on the univalent cations. Annu Rev Plant Physiol 17:47–76

Fageria NK, Moreira A (2011) The role of mineral nutrition on root growth of crop plants. Adv Agron 110:251–331

Fairbairn DJ, Liu W, Schachtman DP, Gomez-Gallego S, Day SR, Teasdale RD (2000) Characterisation of two distinct HKT1-like potassium transporters from *Eucalyptus camaldulensis*. Plant Mol Biol 43:515–525

Fan SC, Lin CS, Hsu PK, Lin SH, Tsay YF (2009) The *Arabidopsis* nitrate transporter NRT1 7, expressed in phloem, is responsible for source-to-sink remobilization of nitrate. Plant Cell 21:2750–2761

Fan X, Tang Z, Tan Y, Zhang Y, Luo B, Yang M, Xu G (2016) Overexpression of a pH-sensitive nitrate transporter in rice increases crop yields. Proc Natl Acad Sci 113:7118–7123

Fang Z, Xia K, Yang X, Grotemeyer MS, Meier S, Rentsch D, Zhang M (2013) Altered expression of the PTR/NRT 1 homologue OsPTR 9 affects nitrogen utilization efficiency, growth and grain yield in rice. Plant Biotechnol J 11:446–458

Fischer JJ, Beatty PH, Good AG, Muench DG (2013) Manipulation of microRNA expression to improve nitrogen use efficiency. Plant Sci 210:70–81

Föhse D, Jungk A (1983) Influence of phosphate and nitrate supply on root hair formation of rape, spinach and tomato plants. Plant Soil 74:359–368

Franco-Zorrilla JM, Martín AC, Leyva A, Paz-Ares J (2005) Interaction between phosphate-starvation, sugar, and cytokinin signaling in Arabidopsis and the roles of cytokinin receptors CRE1/AHK4 and AHK3. Plant Physiol 138:847–857

Fu HH, Luan S (1998) AtKUP1: a dual-affinity K+ transporter from *Arabidopsis*. Plant Cell 10:63–73

Gao Y, Tian Q, Zhang WH (2014) Systemic regulation of sulfur homeostasis in *Medicago truncatula*. Planta 239:79–96

Gao S, Guo C, Zhang Y, Zhang F, Du X, Gu J, Xiao K (2016) Wheat microRNA member TaMIR444a is nitrogen deprivation-responsive and involves plant adaptation to the nitrogen-starvation stress. Plant Mol Biol Reporter 34:931–946

Garg BK, Burman U, Kathju S (2004) The influence of phosphorus nutrition on the physiological response of moth bean genotypes to drought. J Plant Nutr Soil Sci 167:503–508

Gasber A, Klaumann S, Trentmann O, Trampczynska A, Clemens S, Schneider S, Sauer N, Feifer I, Bittner F, Mendel RR, Neuhaus HE (2011) Identification of an *Arabidopsis* solute carrier critical for intracellular transport and inter-organ allocation of molybdate. Plant Biol 13:710–718

Gassman W, Rubio F, Schroeder JI (1996) Alkali cation selectivity of the wheat root high-affinity potassium transporter. HKT1. Plant J 10:869–882

Gaymard F, Pilot G, Lacombe B, Bouchez D, Bruneau D et al (1998) Identification and disruption of a plant shaker-like outward channel involved in KC release into the xylem sap. Cell 94:647–655

Gazzarrini S, Lejay L, Gojon A, Ninnemann O, Frommer WB, von Wirén N (1999) Three functional transporters for constitutive, diurnally regulated, and starvation-induced uptake of ammonium into *Arabidopsis* roots. Plant Cell 11:937–947

Getahun BB, Kassie MM, Visser RG, Van der Linden CG (2019) Genetic diversity of potato cultivars for nitrogen use efficiency under contrasting nitrogen regimes. Potato Res 63:267–290

Gho YS, Jung KH (2019) Comparative expression analyses of rice and *Arabidopsis* phosphate transporter families revealed their conserved roles for the phosphate starvation response. Plant Breed Biotechnol 7:42–49

Gho YS, An G, Park HM, Jung KH (2018) A systemic view of phosphate starvation-responsive genes in rice roots to enhance phosphate use efficiency in rice. Plant Biotechnol Rep 12:249–264

Gierth M, Mäser P, Schroeder JI (2005) The potassium transporter AtHAK5 functions in K+ deprivation-induced high-affinity K+ uptake and AKT1 K+ channel contribution to K+ uptake kinetics in *Arabidopsis* roots. Plant Physiol 137:1105–1114

Godwin RM, Rae AL, Carroll BJ, Smith FW (2003) Cloning and characterization of two genes encoding sulfate transporters from rice (*Oryza sativa* L). Plant Soil 257:113–123

Gong J, Li D, Li H, Zhou H, Xu J (2019) Identification of manganese-responsive microRNAs in *Arabidopsis* by small RNA sequencing. Czech J Genet Plant Breed 55:76–82

González E, Solano R, Rubio V, Leyva A, Paz-Ares J (2005) Phosphate transporter traffic facilitator1 is a plant-specific SEC12-related protein that enables the endoplasmic reticulum exit of a high-affinity phosphate transporter in Arabidopsis. Plant Cell 17:3500–3512

Good AG, Johnson SJ, De Pauw M, Carroll RT, Savidov N, Vidmar J, Stroeher V (2007) Engineering nitrogen use efficiency with alanine aminotransferase. Botany 85:252–262

Grabov A (2007) Plant KT/KUP/HAK potassium transporters: single family–multiple functions. Ann Bot 99:1035–1041

Green LS, Rogers EE (2004) FRD3 controls iron localization in *Arabidopsis*. Plant Physiol 136:2523–2531

Gruber BD, Giehl RF, Friedel S, von Wirén N (2013) Plasticity of the *Arabidopsis* root system under nutrient deficiencies. Plant Physiol 163:161–179

Guignard MS, Leitch AR, Acquisti C, Eizaguirre C, Elser JJ, Hessen DO, Soltis DE (2017) Impacts of nitrogen and phosphorus: from genomes to natural ecosystems and agriculture. Front Ecol Evol 5:70

Guo B, Jin Y, Wussler C, Blancaflor EB, Motes CM, Versaw WK (2008) Functional analysis of the Arabidopsis PHT4 family of intracellular phosphate transporters. New Phytol 177:889–898

Guo C, Zhao X, Liu X, Zhang L, Gu J, Li X, Xiao K (2013) Function of wheat phosphate transporter gene TaPHT2; 1 in Pi translocation and plant growth regulation under replete and limited Pi supply conditions. Planta 237:1163–1178

Guo W, Chen S, Hussain N, Cong Y, Liang Z, Chen K (2015a) Magnesium stress signaling in plant: just a beginning. Plant Signal Behav 10(3):e992287

Guo Y, Zhu C, Gan L, Ng D, Xia K (2015b) Effects of exogenous gibberellic acid3 on iron and manganese plaque amounts and iron and manganese uptake in rice. PLoS One 10(2):e0118177

Guo W, Nazim H, Liang Z, Yang D (2016) Magnesium deficiency in plants: an urgent problem. Crop J 4:83–91

Gupta N, Ram H, Kumar B (2016) Mechanism of zinc absorption in plants: uptake, transport, translocation and accumulation. Rev Environ Sci Bio/Technol 15:89–109

Haas FH, Heeg C, Queiroz R, Bauer A, Wirtz M, Hell R (2008) Mitochondrial serine acetyltransferase functions as a pacemaker of cysteine synthesis in plant cells. Plant Physiol 148:1055–1067

Hamat HB, Kleinhofs A, Warner RL (1989) Nitrate reductase induction and molecular characterization in rice (*Oryza sativa* L). Mol Gen Genet 218:93–98

Hamburger D, Rezzonico E, Petétot JMC, Somerville C, Poirier Y (2002) Identification and characterization of the *Arabidopsis* PHO1 gene involved in phosphate loading to the xylem. Plant Cell 14:889–902

Hasan MM, Hasan MM, da Silva JA, Li X (2016) Regulation of phosphorus uptake and utilization: transitioning from current knowledge to practical strategies. Cell Mol Biol Lett 21:7

He S, He Z, Yang X, Baligar VC (2012) Mechanisms of nickel uptake and hyperaccumulation by plants and implications for soil remediation. Adv Agron 117:117–189

He X, Qu B, Li W, Zhao X, Teng W, Ma W, Tong Y (2015) The nitrate-inducible NAC transcription factor TaNAC2-5A controls nitrate response and increases wheat yield. Plant Physiol 169:1991–2005

Hechenberger M, Schwappach B, Fischer WN, Frommer WB, Jentsch TJ, Steinmeyer K (1996) A family of putative chloride channels from Arabidopsis and functional complementation of a yeast strain with a CLC gene disruption. J Biol Chem 271:33632–33638

Herdean A, Teardo E, Nilsson AK, Pfeil BE, Johansson ON, Ünnep R, Garab G (2016) A voltage-dependent chloride channel fine-tunes photosynthesis in plants. Nat Commun 7:1–11

Hochmuth GJ, Gabelman WH, Gerloff GC (1985) A gene affecting tomato root morphology. Hortscience 20(6):1099–1101

Høgh-Jensen H, Pedersen MB (2003) Morphological plasticity by crop plants and their potassium use efficiency. J Plant Nutr 26:969–984

Hong JP, Takeshi Y, Kondou Y, Schachtman DP, Matsui M, Shin R (2013) Identification and characterization of transcription factors regulating Arabidopsis HAK5. Plant Cell Physiol 54:1478–1490

Horie T, Yoshida K, Nakayama H, Yamada K, Oiki S, Shinmyo A (2001) Two types of HKT transporters with different properties of NaC and KC transport in *Oryza sativa*. Plant J 27:129–138

Howarth JR, Fourcroy P, Davidian JC, Smith FW, Hawkesford MJ (2003) Cloning of two contrasting high-affinity sulfate transporters from tomato induced by low sulfate and infection by the vascular pathogen *Verticillium dahlia*. Planta 218:58–64

Hu B, Wang W, Ou S, Tang J, Li H, Che R, Liang C (2015) Variation in NRT1 1B contributes to nitrate-use divergence between rice subspecies. Nat Genet 47:834

Huang NC, Chiang CS, Crawford NM, Tsay YF (1996) CHL1 encodes a component of the low-affinity nitrate uptake system in *Arabidopsis* and shows cell type-specific expression in roots. Plant Cell 8:2183–2191

Huang Q, Wang M, Xia Z (2018) The SULTR gene family in maize (*Zea mays* L): gene cloning and expression analyses under sulfate starvation and abiotic stress. J Plant Physiol 220:24–33

Huang S, Sasaki A, Yamaji N, Okada H, Mitani-Ueno N, Ma JF (2020) The ZIP transporter family member OsZIP9 contributes to root zinc uptake in rice under zinc-limited conditions. Plant Physiol 183(3):1224–1234

Huen A, Bally J, Smith P (2018) Identification and characterisation of microRNAs and their target genes in phosphate-starved *Nicotiana benthamiana* by small RNA deep sequencing and 5'RACE analysis. BMC Genomics 19:940

Iqbal A, Dong Q, Wang X, Gui H, Zhang H, Zhang X, Song M (2020) Transcriptome analysis reveals differences in key genes and pathways regulating carbon and nitrogen metabolism in cotton genotypes under N starvation and resupply. Int J Mol Sci 21:1500

Ivashikina N, Becker D, Ache P, Meyerhoff O, Felle HH, Hedrich R (2001) K+ channel profile and electrical properties of *Arabidopsis* root hairs. FEBS Lett 508:463–469

Iwamoto M, Tagiri A (2016) MicroRNA-targeted transcription factor gene RDD1 promotes nutrient ion uptake and accumulation in rice. Plant J 85:466–477

Jones-Rhoades MW, Bartel DP (2004) Computational identification of plant microRNAs and their targets, including a stress-induced miRNA. Mol Cell 14:787–799

Jung JY, Shin R, Schachtman DP (2009) Ethylene mediates response and tolerance to potassium deprivation in *Arabidopsis*. Plant Cell 21:607–621

Jungk A (2001) Root hairs and the acquisition of plant nutrients from soil. J Plant Nutr Soil Sci 164:121–129

Kampfenkel K, Kushnir S, Babiychuk E, Inzé D, Van Montagu M (1995) Molecular characterization of a putative *Arabidopsis thaliana* copper transporter and its yeast homologue. J Biol Chem 270:28479–28486

Kant S, Peng M, Rothstein SJ (2011) Genetic regulation by NLA and microRNA827 for maintaining nitrate-dependent phosphate homeostasis in *Arabidopsis*. PLoS Genet 7:e1002021

Kawashima CG, Yoshimoto N, Maruyama-Nakashita A, Tsuchiya YN, Saito K, Takahashi H, Dalmay T (2009) Sulphur starvation induces the expression of microRNA-395 and one of its target genes but in different cell types. Plant J 57:313–321

Kawashima CG, Matthewman CA, Huang S, Lee BR, Yoshimoto N, Koprivova A, Takahashi H (2011) Interplay of SLIM1 and miR395 in the regulation of sulfate assimilation in *Arabidopsis*. Plant J 66:863–876

Keisham M, Mukherjee S, Bhatla SC (2018) Mechanisms of sodium transport in plants—progresses and challenges. Int J Mol Sci 19:647

Kellermeier F, Armengaud P, Seditas TJ, Danku J, Salt DE, Amtmann A (2014) Analysis of the root system architecture of *Arabidopsis* provides a quantitative readout of crosstalk between nutritional signals. Plant Cell 26:1480–1496

Kim SA, Guerinot ML (2007) Mining iron: iron uptake and transport in plants. FEBS Lett 581:2273–2280

Kim SA, Punshon T, Lanzirotti A, Li L, Alonso JM, Ecker JR, Guerinot ML (2006) Localization of iron in *Arabidopsis* seed requires the vacuolar membrane transporter VIT1. Science 314:1295–1298

Kindu GA, Tang J, Yin X, Struik PC (2014) Quantitative trait locus analysis of nitrogen use efficiency in barley (*Hordeum vulgare* L). Euphytica 199:207–221

Kirik V, Simon M, Wester K, Schiefelbein J, Hulskamp M (2004) ENHANCER of TRYand CPC 2 (ETC2) reveals redundancy in the region-specific control of trichome development of Arabidopsis. Plant Mol Biol 55:389–398

Kobayashi T, Nishizawa NK (2012) Iron uptake, translocation, and regulation in higher plants. Annu Rev Plant Biol 63:131–152

Kopriva S, Büchert T, Fritz G, Suter M, Weber M, Benda R, Schaller J, Feller U, Schürmann P, Schünemann V, Trautwein AX, Kroneck PMH, Brunold C (2001) Plant adenosine 5-′-phosphosulfate reductase is a novel iron sulfur protein. J Biol Chem 276:42881–42886

Koprivova A, Kopriva S (2016) Hormonal control of sulfate uptake and assimilation. Plant Mol Biol 91(6):617–627

Koprivova A, Giovannetti M, Baraniecka P, Lee BR, Grondin C, Loudet O et al (2013) Natural variation in the ATPS1 Isoform of ATP sulfurylase contributes to the control of sulfate levels in *Arabidopsis*. Plant Physiol 163:1133–1141

Krouk G (2016) Hormones and nitrate: a two-way connection. Plant Mol Biol 91:599–606

Krouk G, Lacombe B, Bielach A, Perrine-Walker F, Malinska K, Mounier E, Hoyerova K, Tillard P, Leon S, Ljung K et al (2010) Nitrate regulated auxin transport by NRT11 defines a mechanism for nutrient sensing in plants. Dev Cell 18:927–937

Krüger C, Berkowitz O, Stephan UW, Hell R (2002) A metal-binding member of the late embryogenesis abundant protein family transports iron in the phloem of *Ricinus communis* L. J Biol Chem 277:25062–25069

Kumar S, Asif MH, Chakrabarty D, Tripathi RD, Dubey RS, Trivedi PK (2015) Comprehensive analysis of regulatory elements of the promoters of rice sulfate transporter gene family and functional characterization of OsSul1;1 promoter under different metal stress. Plant Signal Behav 10:e990843

Kumar S, Khare R, Trivedi PK (2019) Arsenic-responsive high-affinity rice sulphate transporter, OsSultr1; 1, provides abiotic stress tolerance under limiting sulphur condition. J Hazard Mater 373:753–762

Lagerstedt JO, Zvyagilskaya R, Pratt JR, Pattison-Granberg J, Kruckeberg AL, Berden JA, Persson BL (2002) Mutagenic and functional analysis of the C-terminus of *Saccharomyces cerevisiae* Pho84 phosphate transporter. FEBS Lett 526:31–37

Lass B, Ullrich-Eberius CI (1984) Evidence for proton/sulfate co-transport and its kinetics in *Lemna gibba* G1. Planta 161:53–60

Lazali M, Bargaz A, Brahimi S, Amenc L, Abadie J, Drevon JJ (2016) Expression of a phosphate-starvation inducible fructose-1, 6-bisphosphatase gene in common bean nodules correlates with phosphorus use efficiency. J Plant Physiol 205:48–56

Li L, Tutone AF, Drummond RS, Gardner RC, Luan S (2001) A novel family of magnesium transport genes in *Arabidopsis*. Plant Cell 13:2761–2775

Li L, Qiu X, Li X, Wang S, Lian X (2009) The expression profile of genes in rice roots under low phosphorus stress. Sci China Ser C: Life Sci 52:1055–1064

Li B, Tester M, Gilliham M (2017a) Chloride on the move. Trends Plant Sci 22:236–248

Li H, Hu B, Chu C (2017b) Nitrogen use efficiency in crops: lessons from *Arabidopsis* and rice. J Exp Bot 68(10):2477–2488

Li X, Xu Z, Guo C, Ren T, Cong R, Lu J (2017c) Grain yield and nitrogen use efficiency of various modern rice cultivars grown at different nitrogen levels. J Plant Nutr 40:1125–1132

Li Z, Zhang X, Zhao Y, Li Y, Zhang G, Peng Z, Zhang J (2018) Enhancing auxin accumulation in maize root tips improves root growth and dwarfs plant height. Plant Biotechnol J 16:86–99

Liang G, Yang F, Yu D (2010) MicroRNA395 mediates regulation of sulfate accumulation and allocation in *Arabidopsis thaliana*. Plant J 62:1046–1057

Liang G, He H, Yu D (2012) Identification of nitrogen starvation-responsive microRNAs in *Arabidopsis thaliana*. PloS One 7:e48951

Lima JCD, Loss-Morais G, Margis R (2012) MicroRNAs play critical roles during plant development and in response to abiotic stresses. Genet Mol Biol 35:1069–1077

Linkohr BI, Williamson LC, Fitter AH, Leyser HO (2002) Nitrate and phosphate availability and distribution have different effects on root system architecture of Arabidopsis. Plant J 29:751–760

Liu KH, Huang CY, Tsay YF (1999) CHL1 is a dual-affinity nitrate transporter of *Arabidopsis* involved in multiple phases of nitrate uptake. Plant Cell 11:865–874

Liu H, Tang R, Zhang YUE, Wang C, Lv Q, Gao X, Zhang H (2010) AtNHX3 is a vacuolar K+/H+ antiporter required for low-potassium tolerance in *Arabidopsis* thaliana. Plant Cell Environ 33:1989–1999

Liu F, Chang XJ, Ye Y, Xie WB, Wu P, Lian XM (2011) Comprehensive sequence and whole-life-cycle expression profile analysis of the phosphate transporter gene family in rice. Mol Plant 4:1105–1122

Liu J, Fu J, Tian H, Gao Y (2015) In-season expression of nitrate and ammonium transporter genes in roots of winter wheat (*Triticum aestivum* L) genotypes with different nitrogen-uptake efficiencies. Crop Pasture Sci 66:671–678

Liu H, Yu H, Tang G, Huang T (2018) Small but powerful: function of microRNAs in plant development. Plant Cell Rep 37:515–528

Lopez-Arredondo DL, Leyva-González MA, González-Morales SI, López-Bucio J, Herrera-Estrella L (2014) Phosphate nutrition: improving low-phosphate tolerance in crops. Annu Rev Plant Biol 65:95–123

Lopez-Bucio J, Hernandez-Abreu E, Sanchez-Calderon L, Nieto-Jacobo MF, Simpson J, Herrera-Estrella L (2002) Phosphate availability alters architecture and causes changes in hormone sensitivity in the *Arabidopsis* root system. Plant Physiol 129:244–256

Lopez-Bucio J, Cruz-Ramirez A, Herrera-Estrella L (2003) The role of nutrient availability in regulating root architecture. Curr Opin Plant Biol 6:280–287

Loqué D, von Wirén N (2004) Regulatory levels for the transport of ammonium in plant roots. J Exp Bot 55:1293–1305

Loqué D, Tillard P, Gojon A, Lepetit M (2003) Gene expression of the NO3–transporter NRT1 1 and the nitrate reductase NIA1 is repressed in *Arabidopsis* roots by NO2–, the product of NO3–reduction. Plant Physiol 132:958–967

Loudet O, Chaillou S, Merigout P, Talbotec J, Daniel-Vedele F (2003) Quantitative trait loci analysis of nitrogen use efficiency in *Arabidopsis*. Plant Physiol 131:345–358

Lu Y, Luo F, Yang M, Li X, Lian X (2011) Suppression of glutamate synthase genes significantly affects carbon and nitrogen metabolism in rice (Oryza sativa L.). Sci China Life Sci 54:651–663

Lu YB, Qi YP, Yang LT, Guo P, Li Y, Chen LS (2015) Boron-deficiency-responsive microRNAs and their targets in Citrus sinensis. leaves. BMC Plant Biol 15:271

Lucena C, Waters BM, Romera FJ, García MJ, Morales M, Alcántara E, Pérez-Vicente R (2006) Ethylene could influence ferric reductase, iron transporter, and H+-ATPase gene expression by affecting FER (or FER-like) gene activity. J Exp Bot 57(15):4145–4154

Ludewig U, von Wirén N, Frommer WB (2002) Uniport of NH by the root hair plasma membrane ammonium transporter LeAMT1;. 1. J Biol Chem 277:13548–13555

Lundmark M, Kørner CJ, Nielsen TH (2010) Global analysis of microRNA in *Arabidopsis* in response to phosphate starvation as studied by locked nucleic acid-based microarrays. Physiol Plant 140:57–68

Lurin C, Geelen D, Barbier-Brygoo H, Guern J, Maurel C (1996) Cloning and functional expression of a plant voltage-dependent chloride channel. Plant Cell 8:701–711

Lynch J (1995) Root architecture and plant productivity. Plant Physiol 109:7

Ma JF, Yamaji N (2015) A cooperative system of silicon transport in plants. Trends Plant Sci 20:435–442

Ma JF, Tamai K, Yamaji N, Mitani N, Konishi S, Katsuhara M, Yano M (2006) A silicon transporter in rice. Nature 440:688–691

Ma CL, Qi YP, Liang WW, Yang LT, Lu YB, Guo P, Chen LS (2016) MicroRNA regulatory mechanisms on *Citrus sinensis* leaves to magnesium-deficiency. Front Plant Sci 7:201

Malamy JE (2005) Intrinsic and environmental response pathways that regulate root system architecture. Plant Cell Environ 28:67–77

Man H, Pollmann S, Weiler EW, Kirby EG (2011) Increased glutamine in leaves of poplar transgenic with pine GS1a caused greater anthranilate synthetase α-subunit (ASA1) transcript and protein abundances: an auxin-related mechanism for enhanced growth in GS transgenics? J Exp Bot 62:4423–4431

Maniou F, Chorianopoulou SN, Bouranis DL (2014) New insights into trophic aerenchyma formation strategy in maize (*Zea mays* L) organs during sulfate deprivation. Front Plant Sci 5:581

Marchive C, Yehudai-Resheff S, Germain A, Fei Z, Jiang X, Judkins J, Stern DB (2009) Abnormal physiological and molecular mutant phenotypes link chloroplast polynucleotide phosphorylase to the phosphorus deprivation response in Arabidopsis. Plant Physiol 151:905–924

Marchive C, Roudier F, Castaings L, Bréhaut V, Blondet E, Colot V, Krapp A (2013) Nuclear retention of the transcription factor NLP7 orchestrates the early response to nitrate in plants. Nat Commun 4:1–9

Maruyama-Nakashita A, Nakamura Y, Tohge T, Saito K, Takahashi H (2006) *Arabidopsis* SLIM1 is a central transcriptional regulator of plant sulfur response and metabolism. Plant Cell 18:3235–3251

Mäser P, Gierth M, Schroeder JI (2002a) Molecular mechanisms of potassium and sodium uptake in plants. In: Progress in plant nutrition: plenary lectures of the XIV international plant nutrition colloquium. Springer, Cham, pp 43–54

Mäser P, Hosoo Y, Goshima S, Horie T, Eckelman B, Yamada K, Schroeder JI (2002b) Glycine residues in potassium channel-like selectivity filters determine potassium selectivity in four-loop-per-subunit HKT transporters from plants. Proc Natl Acad Sci 99:6428–6433

Massoud MB, Sakouhi K, Zhu E, Ferjani E, Sheehan D, Chaoui A (2018) Protective role of exogenous phytohormones on redox status in pea seedlings under copper stress. J Plant Physiol *221*:51–61

Matthewman CA, Kawashima CG, Húska D, Csorba T, Dalmay T, Kopriva S (2012) miR395 is a general component of the sulfate assimilation regulatory network in Arabidopsis. FEBS Lett 586:3242–3248

Mayzlish-Gati E, De-Cuyper C, Goormachtig S et al (2012) Strigolactones are involved in root response to low phosphate conditions in *Arabidopsis*. Plant Physiol 160(3):1329–1341

McCauley A, Jones C, Jacobsen J (2009) Plant nutrient functions and deficiency and toxicity symptoms. Nutr Manag Module 9:1–16

Mengel K, Kirkby EA (1987) Principles of plant nutrition. International Potash Institute, Switzerland, p 687

Meyer RC, Gryczka C, Neitsch C, Müller M, Bräutigam A, Schlereth A, Altmann T (2019) Genetic diversity for nitrogen use efficiency in *Arabidopsis thaliana*. Planta 250:41–57

Migge A, Carrayol E, Hirel B, Becker TW (2000) Leaf-specific overexpression of plastidic glutamine synthetase stimulates the growth of transgenic tobacco seedlings. Planta 210:252–260

Migocka M, Malas K (2018) Plant responses to copper: molecular and regulatory mechanisms of copper uptake, distribution and accumulation in plants. In: Plant micronutrient use efficiency. Elsevier, Amsterdam, pp 71–86

Mills RF, Doherty ML, López-Marqués RL, Weimar T, Dupree P, Palmgren MG, Williams LE (2008) ECA3, a Golgi-localized P2A-type ATPase, plays a crucial role in manganese nutrition in *Arabidopsis*. Plant Physiol 146:116–128S

Miura K, Rus A, Sharkhuu A, Yokoi S, Karthikeyan AS, Raghothama KG et al (2005) The Arabidopsis SUMO E3 ligase SIZ1 controls phosphate deficiency responses. Proc Natl Acad Sci 102:7760–7765

Montiel G, Gantet P, Jay-Allemand C, Breton C (2004) Transcription factor networks pathways to the knowledge of root development. Plant Physiol 136:3478–3485

Morère-Le Paven MC, Viau L, Hamon A, Vandecasteele C, Pellizzaro A, Bourdin C, Legros C (2011) Characterization of a dual-affinity nitrate transporter MtNRT1 3 in the model legume Medicago truncatula. J Exp Bot 62:5595–5605

Morris EC, Griffiths M, Golebiowska A, Mairhofer S, Burr-Hersey J, Goh T, Vissenberg K (2017) Shaping 3D root system architecture. Curr Biol 27:R919–R930

Mosa KA, Kumar K, Chhikara S, Musante C, White JC, Dhankher OP (2016) Enhanced boron tolerance in plants mediated by bidirectional transport through plasma membrane intrinsic proteins. Sci Rep 6:21640

Mouline K, Very A-A, Gaymard F, Boucherez J, Pilot G et al (2002) Pollen tube development and competitive ability are impaired by disruption of a Shaker KC channel in Arabidopsis. Genes Dev 16:339–350

Muchhal US, Pardo JM, Raghothama KG (1996) Phosphate transporters from the higher plant *Arabidopsis thaliana*. Proc Natl Acad Sci 93:10519–10523

Mugford SG, Yoshimoto N, Reichelt M, Wirtz M, Hill L, Mugford ST et al (2009) Disruption of adenosine-50-phosphosulfate kinase in *Arabidopsis* reduces levels of sulfated secondary metabolites. Plant Cell 21:910–927

Nagy R, Vasconcelos MJV, Zhao S, McElver J, Bruce W, Amrhein N, Bucher M (2006) Differential regulation of five Pht1 phosphate transporters from maize (*Zea mays* L). Plant Biol 8:186–197

Nam Y-J, Tran L-SP, Kojima M, Sakakibara H, Nishiyama R, Shin R (2012) Regulatory roles of cytokinins and cytokinin signaling in response to potassium deficiency in *Arabidopsis*. PLoS One 7(10):e47797

Nardoto GB, Da Cunha BMM, Pinto AS, Klink CA (2006) Nutrient use efficiency at ecosystem and species level in savanna areas of Central Brazil and impacts of fire. J Tropical Ecol 22:191–201

Nieves-Cordones M, Alemán F, Martínez V, Rubio F (2010) The *Arabidopsis thaliana* HAK5 K+ transporter is required for plant growth and K+ acquisition from low K+ solutions under saline conditions. Mol Plant 3:326–333

Nilsson L, Müller R, Nielsen TH (2007) Increased expression of the MYB-related transcription factor, PHR1, leads to enhanced phosphate uptake in *Arabidopsis thaliana*. Plant Cell Environ 30:1499–1512

Ninnemann O, Jauniaux JC, Frommer WB (1994) Identification of a high affinity NH4+ transporter from plants. EMBO J 13:3464–3471

Nussaume L, Kanno S, Javot H, Marin E, Nakanishi TM, Thibaud MC (2011) Phosphate import in plants: focus on the PHT1 transporters. Front Plant Sci 2:83

Ogawa S, Valencia MO, Ishitani M, Selvaraj MG (2014) Root system architecture variation in response to different NH 4+ concentrations and its association with nitrogen-deficient tolerance traits in rice. Acta Physiol Plant 36:2361–2372

Oono Y, Kawahara Y, Yazawa T, Kanamori H, Kuramata M, Yamagata H, Antonio B (2013) Diversity in the complexity of phosphate starvation transcriptomes among rice cultivars based on RNA-Seq profiles. Plant Mol Biol 83:523–537

Padmanaban S, Chanroj S, Kwak JM, Li X, Ward JM, Sze H (2007) Participation of endomembrane cation/H+ exchanger AtCHX20 in osmoregulation of guard cells. Plant Physiol 144:82–93

Pant BD, Musialak-Lange M, Nuc P, May P, Buhtz A, Kehr J, Walther D, Scheible WR (2009) Identification of nutrient-responsive *Arabidopsis* and rapeseed microRNAs by comprehensive real-time polymerase chain reaction profiling and small RNA sequencing. Plant Physiol 150:1541–1555

Pantoja O (2012) High affinity ammonium transporters: molecular mechanism of action. Front Plant Sci 3:34

Pao SS, Paulsen IT, Saier MH (1998) Major facilitator superfamily. Microbiol Mol Biol Rev 62:1–34

Pariasca-Tanaka J, Baertschi C, Wissuwa M (2019) Identification of loci through genome-wide association studies to improve tolerance to sulfur deficiency in rice. Front Plant Sci 10:1668

Paul S, Datta SK, Datta K (2015) miRNA regulation of nutrient homeostasis in plants. Front Plant Sci 6:232

Paul S, Gayen D, Datta SK, Datta K (2016) Analysis of high iron rice lines reveals new miRNAs that target iron transporters in roots. J Exp Bot 67:5811–5824

Penn CJ, Camberato JJ (2019) A critical review on soil chemical processes that control how soil pH affects phosphorus availability to plants. Agriculture 9:120

Peña PA, Quach T, Sato S, Ge Z, Nersesian N, Changa T, Clemente TE (2017) Expression of the maize Dof1 transcription factor in wheat and sorghum. Front Plant Sci 8:434

Peng M, Hannam C, Gu H, Bi YM, Rothstein SJ (2007) A mutation in NLA, which encodes a RING-type ubiquitin ligase, disrupts the adaptability of Arabidopsis to nitrogen limitation. Plant J 50:320–337

Perchlik M, Tegeder M (2017) Improving plant nitrogen use efficiency through alteration of amino acid transport processes. Plant Physiol 175:235–247

Péret B, Desnos T, Jost R, Kanno S, Berkowitz O, Nussaume L (2014) Root architecture responses: in search of phosphate. Plant Physiol 166:1713–1723

Persson BL, Lagerstedt JO, Pratt JR, Pattison-Granberg J, Lundh K, Shokrollahzadeh S, Lundh F (2003) Regulation of phosphate acquisition in *Saccharomyces cerevisiae*. Current Genet 43:225–244

Philippar K, Ivashikina N, Ache P, Christian M, Luthen H, Palme K, Hedrich R (2004) Auxin activates KAT1 and KAT2, two Kþ-channel genes expressed in seedlings of *Arabidopsis thaliana*. Plant J 37:815–827

Phung NTP, Mai CD, Hoang GT, Truong HTM, Lavarenne J, Gonin M, Courtois B (2016) Genome-wide association mapping for root traits in a panel of rice accessions from Vietnam. BMC Plant Biol 16:64

Pilot G, Lacombe B, Gaymard F, Cherel I, Boucherez J et al (2001) Guard cell inward KC channel activity in *Arabidopsis* involves expression of the twin channel subunits KAT1 and KAT2. J Biol Chem 276:3215–3221

Platten JD, Cotsaftis O, Berthomieu P, Bohnert H, Davenport RJ, Fairbairn DJ, Mäser P (2006) Nomenclature for HKT transporters, key determinants of plant salinity tolerance. Trends Plant Sci 11:372–374

Postma JA, Dathe A, Lynch J (2014) The optimal lateral root branching density for maize depends on nitrogen and phosphorus availability. Plant Physiol 166:590–602

Printz B, Lutts S, Hausman JF, Sergeant K (2016) Copper trafficking in plants and its implication on cell wall dynamics. Front Plant Sci 7:601

Puig S (2014) Function and regulation of the plant COPT family of high-affinity copper transport proteins. Adv Bot 2014:476917

Puzina TI (2004) Effect of zinc sulfate and boric acid on the hormonal status of potato plants in relation to tuberization. Russ J Plant Physiol 51(2):209–215

Pyo YJ, Gierth M, Schroeder JI, Cho MH (2010) High-affinity K+ transport in arabidopsis: AtHAK5 and AKT1 are vital for seedling establishment and post germination growth under low-potassium conditions. Plant Physiol 153:863–875

Raddatz N, Morales De Los Ríos L, Lindahl AM, Quintero FJ, Pardo JM (2020) Coordinated transport of nitrate, potassium and sodium. Front Plant Sci 11:247

Rae AL, Cybinski DH, Jarmey JM, Smith FW (2003) Characterization of two phosphate transporters from barley; evidence for diverse function and kinetic properties among members of the Pht1 family. Plant Mol Biol 53:27–36

Rausch C, Bucher M (2002) Molecular mechanisms of phosphate transport in plants. Planta 216:23–37

Rehman HM, Nawaz MA, Shah ZH, Daur I, Khatoon S, Yang SH, Chung G (2017) In-depth genomic and transcriptomic analysis of five K+ transporter gene families in soybean confirm their differential expression for nodulation. Front Plant Sci 8:804

Reid R (2014) Understanding the boron transport network in plants. Plant Soil 385:1–13

Ren Y, Sun F, Hou J, Chen L, Zhang Y, Kang X, Wang Y (2015) Differential profiling analysis of miRNAs reveals a regulatory role in low N stress response of *Populus*. Funct Integr Genomics 15:93–105

Ristova D, Carré C, Pervent M, Medici A, Kim GJ, Scalia D, Ruffel S, Birnbaum KD, Lacombe B, Busch W, Coruzzi GM, Krouk G (2016) Combinatorial interaction network of transcriptomic and phenotypic responses to nitrogen and hormones in the Arabidopsis thaliana root. Sci Signal 9:rs13

Riveras E, Alvarez JM, Vidal EA, Oses C, Vega A, Gutiérrez RA (2015) The calcium ion is a second messenger in the nitrate signaling pathway of *Arabidopsis*. Plant Physiol 169:1397–1404

Robinson WD, Park J, Tran HT, Del Vecchio HA, Ying S, Zins JL, Plaxton WC (2012) The secreted purple acid phosphatase isozymes AtPAP12 and AtPAP26 play a pivotal role in extracellular phosphate-scavenging by *Arabidopsis thaliana*. J Exp Bot 63:6531–6542

Rodríguez-Navarro A (2000) Potassium transport in fungi and plants. Biochim Biophys Acta 1469:1–30

Rouached H (2011) Multilevel coordination of phosphate and sulfate homeostasis in plants. Plant Signal Behav 6:952–955

Rouached H, Berthomieu P, El Kassis E, Cathala N, Catherinot V, Labesse G, Fourcroy P (2005) Structural and functional analysis of the C-terminal STAS (sulfate transporter and anti-sigma antagonist) domain of the *Arabidopsis thaliana* sulfate transporter SULTR1 2. J Biol Chem 280:15976–15983

Rubio F, Gassmann W, Schroeder JI (1995) Sodium-driven potassium uptake by the plant potassium transporter HKT1 and mutations conferring salt tolerance. Science 270:1660–1663

Rubio V, Linhares F, Solano R, Martín AC, Iglesias J, Leyva A, Paz-Ares J (2001) A conserved MYB transcription factor involved in phosphate starvation signaling both in vascular plants and in unicellular algae. Genes Dev 15:2122–2133

Rubio V, Bustos R, Irigoyen ML, Cardona-López X, Rojas-Triana M, Paz-Ares J (2009) Plant hormones and nutrient signaling. Plant Mol Biol *69*(4):361

Ruffel S (2018) Nutrient-related long-distance signals: Common players and possible crosstalk. Plant Cell Physiol 59:1723–1732

Ruffel S, Poitout A, Krouk G, Coruzzi GM, Lacombe B (2016) Long-distance nitrate signaling displays cytokinin dependent and independent branches. J Integr Plant Biol 58:226–229

Rus A, Yokoi S, Sharkhuu A, Reddy M, Lee BH et al (2001) AtHKT1 is a salt tolerance determinant that controls NaC entry into plant roots. Proc Natl Acad Sci USA 98:14150–14155
Russell RS (1977) Plant root systems: their function and interaction with the soil. McGraw-Hill Book, London
Sahrawat KL (2006) Plant nutrients: sufficiency and requirements. In: Encyclopedia of soil science. Taylor & Francis, Abingdon, pp 1306–1310
Sanchez DL, Liu S, Ibrahim R, Blanco M, Lübberstedt T (2018) Genome-wide association studies of doubled haploid exotic introgression lines for root system architecture traits in maize (Zea mays L). Plant Sci 268:30–38
Schachtman DP, Schroeder JI (1994) Structure and transport mechanism of a high-affinity potassium uptake transporter from higher plants. Nature 370:655–658
Schachtman DP, Reid RJ, Ayling SM (1998) Phosphorus uptake by plants: from soil to cell. Plant Physiol 116:447–453
Schultz CJ, Coruzzi GM (1995) The aspartate aminotransferase gene family of Arabidopsis encodes isoenzymes localized to three distinct subcellular compartments. Plant J 7:61–75
Sentenac H, Bonneaud N, Minet M, Lacroute F, Salmon JM, Gaymard F, Grignon C (1992) Cloning and expression in yeast of a plant potassium ion transport system. Science 256:663–665
Shahbaz M, Pilon M (2019) Conserved Cu-microRNAs in Arabidopsis thaliana function in copper economy under deficiency. Plants 8(6):141
Shaner DL, Boyer JS (1976) Nitrate reductase activity in maize (Zea mays L) leaves: I Regulation by nitrate flux. Plant Physiol 58:499–504
Sharif Shohan MU, Sinha S, Nabila FH, Dastidar SG, Seraj ZI (2019) HKT1; 5 transporter gene expression and association of amino acid substitutions with salt tolerance across rice genotypes. Front Plant Sci 10:1420
Shaul O, Hilgemann DW, de-Almeida-Engler J, Van Montagu M, Inzé D, Galili G (1999) Cloning and characterization of a novel Mg2+/H+ exchanger. EMBO J 18:3973–3980
Shin R, Berg RH, Schachtman DP (2005) Reactive oxygen species and root hairs in Arabidopsis root response to nitrogen, phosphorus and potassium deficiency. Plant Cell Physiol 46:1350–1357
Shin R, Burch AY, Huppert KA, Tiwari SB, Murphy AS, Guilfoyle TJ, Schachtman DP (2007) The Arabidopsis transcription factor MYB77 modulates auxin signal transduction. Plant Cell 19:2440–2453
Shrawat AK, Carroll RT, DePauw M, Taylor GJ, Good AG (2008) Genetic engineering of improved nitrogen use efficiency in rice by the tissue-specific expression of alanine aminotransferase. Plant Biotechnol J 6:722–732
Sinha SK, Rani M, Bansal N, Venkatesh K, Mandal PK (2015) Nitrate starvation induced changes in root system architecture, carbon: nitrogen metabolism, and miRNA expression in nitrogen-responsive wheat genotypes. Appl Biochem Biotechnol 177:1299–1312
Smith FW, Ealing PM, Dong B, Delhaize E (1997) The cloning of two Arabidopsis genes belonging to a phosphate transporter family. Plant J 11:83–92
Socha AL, Guerinot ML (2014) Mn-euvering manganese: the role of transporter gene family members in manganese uptake and mobilization in plants. Front Plant Sci 5:106
Sohlenkamp C, Wood CC, Roeb GW, Udvardi MK (2002) Characterization of Arabidopsis AtAMT2, a high-affinity ammonium transporter of the plasma membrane. Plant Physiol 130 (4):1788–1796
Song X, Li Y, Cao X, Qi Y (2019) MicroRNAs and their regulatory roles in plant–environment interactions. Annu Rev Plant Biol 70:489–525
Spehar CR, Galwey NW (1997) Screening soya beans [Glycine max (L) Merill] for calcium efficiency by root growth in low-Ca nutrient solution. Euphytica 94(1):113–117
Stahl A, Vollrath P, Samans B, Frisch M, Wittkop B, Snowdon RJ (2019) Effect of breeding on nitrogen use efficiency-associated traits in oilseed rape. J Exp Bot 70(6):1969–1986
Stevenson-Paulik J, Bastidas RJ, Chiou ST, Frye RA, York JD (2005) Generation of phytate-free seeds in *Arabidopsis* through disruption of inositol polyphosphate kinases. Proc Natl Acad Sci 102:12612–12617

Strock CF, DelaRiva LM, Lynch JP (2018) Reductioninrootsecondary growth as a strategy for phosphorus acquisition. Plant Physiol 176:691–703

Sun H, Tao J, Liu S, Huang S, Chen S, Xie X et al (2014) Strigolactones are involved in phosphate- and nitrate-deficiency-induced root development and auxin transport in rice. J Exp Bot 65:6735–6746

Sun H, Bi Y, Tao J, Huang S, Hou M, Xue R et al (2016) Strigolactones are required for nitric oxide to induce root elongation in response to nitrogen and phosphate deficiencies in rice. Plant Cell Environ 39:1473–1484

Tada Y (2019) The HKT transporter gene from Arabidopsis, AtHKT1; 1, is dominantly expressed in shoot vascular tissue and root tips and is mild salt stress-responsive. Plants 8(7):204

Takabatake R, Hata S, Taniguchi M, Kouchi H, Sugiyama T, Izui K (1999) Isolation and characterization of cDNAs encoding mitochondrial phosphate transporters in soybean, maize, rice, and Arabidopsis. Plant Mol Biol 40(3):479–486

Takahashi H, Yamazaki M, Sasakura N, Watanabe A, Leustek T, De Almeida EJ, Engler G, Van Montagu M, Saito K (1997) Regulation of sulfur assimilation in higher plants: a sulfate transporter induced in sulfate-starved roots plays a central role in *Arabidopsis thaliana*. Proc Natl Acad Sci USA 94:11102–11107

Takahashi H, Watanabe-Takahashi A, Smith FW, Blake-Kalff M, Hawkesford MJ, Saito K (2000) The roles of three functional sulphate transporters involved in uptake and translocation of sulphate in Arabidopsis thaliana. Plant J 23:171–182

Takahashi H, Kopriva S, Giordano M, Saito K, Hell R (2011) Sulfur assimilation in photosynthetic organisms: molecular functions and regulations of transporters and assimilatory enzymes. Annu Rev Plant Biol 62:157–184

Takahashi H, Buchner P, Yoshimoto N, Hawkesford MJ, Shiu SH (2012) Evolutionary relationships and functional diversity of plant sulfate transporters. Front Plant Sci 2:119

Takano J, Noguchi K, Yasumori M, Kobayashi M, Gajdos Z, Miwa K, Fujiwara T (2002) *Arabidopsis* boron transporter for xylem loading. Nature 420(6913):337–340

Takano J, Wada M, Ludewig U, Schaaf G, Von Wirén N, Fujiwara T (2006) The *Arabidopsis* major intrinsic protein NIP5; 1 is essential for efficient boron uptake and plant development under boron limitation. Plant Cell 18(6):1498–1509

Takano J, Miwa K, Fujiwara T (2008) Boron transport mechanisms: collaboration of channels and transporters. Trends Plant Sci 13(8):451–457

Tan L, Qu M, Zhu Y et al (2020) Zinc transporter5 and zinc transporter9 function synergistically in zinc/cadmium uptake. Plant Physiol 183(3):1235–1249

Tanaka S, Yamauchi A, Kono Y (1995) Root system morphology of four rice cultivars: response of different component roots to nitrogen application. Jpn J Crop Sci 64:148–155

Tejada-Jiménez M, Llamas Á, Sanz-Luque E, Galván A, Fernández E (2007) A high-affinity molybdate transporter in eukaryotes. Proc Natl Acad Sci 104(50):20126–20130

Tejada-Jiménez M, Galván A, Fernández E (2011) Algae and humans share a molybdate transporter. Proc Natl Acad Sci 108(16):6420–6425

Thor K (2019) Calcium—nutrient and messenger. Front Plant Sci 10:440

Ticconi CA, Lucero RD, Sakhonwasee S, Adamson AW, Creff A, Nussaume L, Abel S (2009) ER-resident proteins PDR2 and LPR1 mediate the developmental response of root meristems to phosphate availability. Proc Natl Acad Sci 106(33):14174–14179

Tilman D, Cassman KG, Matson PA, Naylor R, Polasky S (2002) Agricultural sustainability and intensive production practices. Nature 418(6898):671–677

Tomatsu H, Takano J, Takahashi H, Watanabe-Takahashi A, Shibagaki N, Fujiwara T (2007) An Arabidopsis thaliana high-affinity molybdate transporter required for efficient uptake of molybdate from soil. Proc Natl Acad Sci 104:18807–18812

Tominaga-Wada R, Nukumizu Y (2012) Expression analysis of an R3-Type MYB transcription factor CPC-LIKE MYB4 (TRICHOMELESS2) and CPL4-related transcripts in *Arabidopsis*. Int J Mol Sci 13(3):3478–3491

Tsai HH, Schmidt W (2017) One way or another? Iron uptake in plants. New Phytol 214 (2):500–505

Tyburski J, Dunajska K, Tretyn A (2009) Reactive oxygen species localization in roots of Arabidopsis thaliana seedlings grown under phosphate deficiency. Plant Growth Regul 59:27–36
Uozomi N, Kim EJ, Rubio F, Yamaguchi T, Muto S et al (2000) The *Arabidopsis* HKT1 gene homolog mediates inward NaC currents in Xenopus laevis oocytes and NaCl uptake in Saccharomyces cerevisiae. Plant Physiol 122:1249–1259
Valdes-Lopez O, Yang SS, Aparicio-Fabre R, Graham PH, Reyes JL, Vance CP, Hernandez G (2010) MicroRNA expression profile in common bean (*Phaseolus vulgaris*) under nutrient deficiency stresses and manganese toxicity. New Phytol 187:805–818
van Gelderen K, Kang C, Pierik R (2018) Light signaling, root development, and plasticity. Plant Physiol 176(2):1049–1060
Vatansever R, Koc I, Ozyigit II, Sen U, Uras ME, Anjum NA, Filiz E (2016) Genome-wide identification and expression analysis of sulfate transporter (SULTR) genes in potato (Solanum tuberosum L). Planta 244(6):1167–1183
Versaw WK, Harrison MJ (2002) A chloroplast phosphate transporter, PHT2; 1, influences allocation of phosphate within the plant and phosphate-starvation responses. Plant Cell 14 (8):1751–1766
Very AA, Sentenac H (2003) Molecular mechanisms and regulation of K+ transport in higher plants. Annu Rev Plant Biol 54(1):575–603
Vidmar JJ, Tagmount A, Cathala N, Touraine B, Davidian JC (2000) Cloning and characterization of a root specific high-affinity sulfate transporter from *Arabidopsis thaliana*. FEBS Lett 475:65–69
Voelker C, Schmidt D, Mueller-Roeber B, Czempinski K (2006) Members of the *Arabidopsis* AtTPK/KCO family form homomeric vacuolar channels in planta. Plant J 48:296–306
Von Wirén N, Lauter FR, Ninnemann O, Gillissen B, Walch-Liu P, Engels C, Frommer WB (2000) Differential regulation of three functional ammonium transporter genes by nitrogen in root hairs and by light in leaves of tomato. Plant J 21(2):167–175
Wang YY, Tsay YF (2011) Arabidopsis nitrate transporter NRT1 9 is important in phloem nitrate transport. Plant Cell 23(5):1945–1957
Wang Y, Wu WH (2013) Potassium transport and signaling in higher plants. Annu Rev Plant Biol 64:451–476
Wang TB, Gassmann W, Rubio F, Schroeder JI, Glass ADM (1998) Rapid up-regulation of HKT1, a high-affinity potassium transporter gene, in roots of barley and wheat following withdrawal of potassium. Plant Physiol 118:651–659
Wang R, Okamoto M, Xing X, Crawford NM (2003) Microarray analysis of the nitrate response in *Arabidopsis* roots and shoots reveals over 1,000 rapidly responding genes and new linkages to glucose, trehalose-6-phosphate, iron, and sulfate metabolism. Plant Physiol 132:556–567
Wang GY, Shi JL, Ng G, Battle SL, Zhang C, Lu H (2011) Circadian clock-regulated phosphate transporter PHT4; 1 plays an important role in *Arabidopsis* defense. Mol Plant 4(3):516–526
Wang YY, Hsu PK, Tsay YF (2012) Uptake, allocation and signaling of nitrate. Trends Plant Sci 17:458–467
Wang YY, Cheng YH, Chen KE, Tsay YF (2018) Nitrate transport, signaling, and use efficiency. Annu Rev Plant Biol 69:85–122
Wang W, Ding G, White PJ, Wang M, Zou J, Xu F, Shi L (2020) Genetic dissection of the shoot and root ionomes of *Brassica napus* grown with contrasting phosphate supplies. Ann Bot 126:119–140
Wang Q, Huang Y, Ren Z, Zhang X, Ren J, Su J, Zhang C, Tian J, Yu Y, Gao GF, Li L, Kong Z (2020a) Transfer cells mediate nitrate uptake to control root nodule symbiosis. Nat Plants 6:800–808
Wang S, Chen A, Xie K et al (2020b) Functional analysis of the OsNPF45 nitrate transporter reveals a conserved mycorrhizal pathway of nitrogen acquisition in plants. Proc Natl Acad Sci U S A 117(28):16649–16659

Waters BM, McInturf SA, Stein RJ (2012) Rosette iron deficiency transcript and microRNA profiling reveals links between copper and iron homeostasis in *Arabidopsis thaliana*. J Exp Bot 63:5903–5918

Waters S, Gilliham M, Hrmova M (2013) Plant high-affinity potassium (HKT) transporters involved in salinity tolerance: structural insights to probe differences in ion selectivity. Int J Mol Sci 14(4):7660–7680

Wei D, Cui K, Ye G, Pan J, Xiang J, Huang J, Nie L (2012) QTL mapping for nitrogen-use efficiency and nitrogen-deficiency tolerance traits in rice. Plant Soil 359(1-2):281–295

White PJ, Broadley MR (2003) Calcium in plants. Ann Bot 92(4):487–511

White MM, Miller C (1979) A voltage-gated anion channel from the electric organ of Torpedo californica. J Biol Chem 254(20):10161–10166

White PJ, Hammond JP, King GJ, Bowen HC, Hayden RM, Meacham MC, Broadley MR (2010) Genetic analysis of potassium use efficiency in *Brassica oleracea*. Ann Bot 105(7):1199–1210

Williamson LC, Ribrioux SPCP, Fitter AH, Leyser HMO (2001) Phosphate availability regulates root system architecture in Arabidopsis. Plant Physiol 126:875–890

Wirtz M, Hell R (2006) Functional analysis of the cysteine synthase protein complex from plants: structural, biochemical and regulatory properties. J Plant Physiol 163:273–286

Wong HK, Chan HK, Coruzzi GM, Lam HM (2004) Correlation of ASN2 gene expression with ammonium metabolism in Arabidopsis. Plant Physiol 134:332–338

Wu D, Yamaji N, Yamane M, Kashino-Fujii M, Sato K, Feng Ma J (2016) The HvNramp5 transporter mediates uptake of cadmium and manganese but not iron. Plant Physiol 172 (3):1899–1910

Xie Q, Frugis G, Colgan D, Chua NH (2000) Arabidopsis NAC1 transduces auxin signal downstream of TIR1 to promote lateral root development. Genes Dev 14:3024–3036

Xu Z, Zhong S, Li X, Li W, Rothstein SJ, Zhang S, Xie C (2011) Genome-wide identification of microRNAs in response to low nitrate availability in maize leaves and roots. PloS One 6(11): e28009

Xu F, Liu Q, Chen L, Kuang J, Walk T, Wang J, Liao H (2013) Genome-wide identification of soybean microRNAs and their targets reveals their organ-specificity and responses to phosphate starvation. BMC Genomics 14(1):66

Yamaguchi C, Khamsalath S, Takimoto Y, Suyama A, Mori Y, Ohkama-Ohtsu N, Maruyama-Nakashita A (2020) SLIM1 transcription factor promotes sulfate uptake and distribution to shoot, along with phytochelatin accumulation, under Cadmium stress in *Arabidopsis thaliana*. Plants (Basel) 9:163

Yamamoto E, Yonemaru JI, Yamamoto T, Yano M (2012) OGRO: the overview of functionally characterized genes in rice online database. Rice 5(1):1–10

Yamasaki H, Abdel-Ghany SE, Cohu CM, Kobayashi Y, Shikanai T, Pilon M (2007) Regulation of copper homeostasis by micro-RNA in *Arabidopsis*. J Biol Chem 282:16369–16378

Yan Y, Wang H, Hamera S, Chen X, Fang R (2014) miR444a has multiple functions in the rice nitrate-signaling pathway. Plant J 78(1):44–55

Yang H, Knapp J, Koirala P, Rajagopal D, Peer WA, Silbart LK, Murphyand A, Gaxiola RA (2007) Enhanced phosphorus nutrition in monocots and dicots over-expressing a phosphorus-responsive type I H+ -pyrophosphatase. Plant Biotechnol J 5(6):735–745

Yang Z, Wang Z, Yang C, Yang Z, Li H, Wu Y (2019) Physiological responses and small RNAs changes in maize under nitrogen deficiency and resupply. Genes Genomics 41(10):1183–1194

Yi K, Wu Z, Zhou J, Du L, Guo L, Wu Y, Wu P (2005) OsPTF1: A novel transcription factor involved in tolerance to phosphate starvation in rice. Plant Physiol 138:2087–2096

Yoon DK, Ishiyama K, Suganami M, Tazoe Y, Watanabe M, Imaruoka S, Mae T (2020) Transgenic rice overproducing Rubisco exhibits increased yields with improved nitrogen-use efficiency in an experimental paddy field. Nat Food 1(2):134–139

Yoshimoto N, Inoue E, Saito K, Yamaya T, Takahashi H (2003) Phloem-localizing sulfate transporter, Sultr1; 3, mediates re-distribution of sulfur from source to sink organs in Arabidopsis. Plant Physiol 131:1511–1517

Yruela I (2009) Copper in plants: acquisition, transport and interactions. Funct Plant Biol 36 (5):409–430

Yu C, Chen Y, Cao Y, Chen H, Wang J, Bi YM, He C (2018) Overexpression of miR169o, an overlapping microRNA in response to both nitrogen limitation and bacterial infection, promotes nitrogen use efficiency and susceptibility to bacterial blight in rice. Plant Cell Physiol 59 (6):1234–1247

Yuan S, Li Z, Li D, Yuan N, Hu Q, Luo H (2015) Constitutive expression of rice microRNA528 alters plant development and enhances tolerance to salinity stress and nitrogen starvation in creeping bentgrass. Plant Physiol 169(1):576–593

Yusuf M, Fariduddin Q, Hayat S, Ahmad A (2011) Nickel: an overview of uptake, essentiality and toxicity in plants. Bull Environ Contam Toxicol 86:1–17

Zhang H, Forde BG (1998) An Arabidopsis MADS box gene that controls nutrient-induced changes in root architecture. Science 279(5349):407–409

Zhang YL, Jian-Bo FAN, Dong-Sheng WANG, Qi-Rong SHEN (2009) Genotypic differences in grain yield and physiological nitrogen use efficiency among rice cultivars. Pedosphere 19 (6):681–691

Zhang JL, Flowers TJ, Wang SM (2010) Mechanisms of sodium uptake by roots of higher plants. Plant Soil 326:45

Zhao M, Ding H, Zhu JK, Zhang F, Li WX (2011) Involvement of miR169 in the nitrogen-starvation responses in *Arabidopsis*. New Phytol 190(4):906–915

Zhao Y, Xu Z, Mo Q, Zou C, Li W, Xu Y, Xie C (2013) Combined small RNA and degradome sequencing reveals novel miRNAs and their targets in response to low nitrate availability in maize. Ann Bot 112(3):633–642

Zhao L, Zhang W, Yang Y, Li Z, Li N, Qi S, Wang Y (2018) The *Arabidopsis* NLP7 gene regulates nitrate signaling via NRT1 1–dependent pathway in the presence of ammonium. Sci Rep 8 (1):1–13

Zhou J, Jiao F, Wu Z, Li Y, Wang X, He X, Wu P (2008) OsPHR2 is involved in phosphate-starvation signaling and excessive phosphate accumulation in shoots of plants. Plant Physiol 146:1673–1686

Zhu J, Zhang C, Lynch JP (2010) The utility of phenotypic plasticity of root hair length for phosphorus acquisition. Funct Plant Biol 37(4):313–322

Zifarelli G, Pusch M (2010) CLC transport proteins in plants. FEBS Lett 584(10):2122–2127

Zobel RW, Waisel Y (2010) A plant root system architectural taxonomy: a framework for root nomenclature. Plant Biosyst 144(2):507–512

Zuber H, Davidian JC, Wirtz M, Hell R, Belghazi M, Thompson R, Gallardo K (2010) Sultr4; 1 mutant seeds of *Arabidopsis* have an enhanced sulphate content and modified proteome suggesting metabolic adaptations to altered sulphate compartmentalization. BMC Plant Biol 10(1):1–13

Zuluaga DL, Sonnante G (2019) The use of nitrogen and its regulation in cereals: structural genes, transcription factors, and the role of miRNAs. Plants 8(8):294

Zuluaga DL, De Paola D, Janni M, Curci PL, Sonnante G (2017) Durum wheat miRNAs in response to nitrogen starvation at the grain filling stage. PLoS One 12(8):e0183253

Plant Growth Promoting Rhizobacteria: Mechanisms and Alleviation of Cold Stress in Plants

6

Pankaj Kumar Mishra, Shekhar Chand Bisht, B. M. Pandey, V. S. Meena, M. Parihar, D. Mahanta, J. K. Bisht, and A. Pattanayak

Abstract

Microorganisms have a variety of evolutionary adaptations and physiological acclimation mechanisms that allow them to survive and remain active to face environmental stress. Among the mechanisms identified in microbes, tolerance to low temperatures is of paramount significance, considering the area of the earth's surface that is exposed to varying degrees of low temperatures. Based on low temperature preference, microbes have been classified as psychrophiles or cold loving, and psychrotrophs or cold tolerant. Psychrophiles are exposed to extremes of cold for major part of the year, while psychrotrophs are exposed to transient cold conditions in nature. The ability of psychrotolerant bacteria to survive and proliferate at low temperatures implies that they have devised a number of mechanisms that help them to tide over the transient cold. These adaptations include lipid modification to maintain membrane fluidity, induction of specific proteins (Csp and Cap), synthesis and utilization of cryoprotectants, cold adapted enzymes, synthesis of ROS detoxifying enzymes, ice binding proteins, and RNA

P. K. Mishra (✉) · B. M. Pandey · V. S. Meena · M. Parihar · D. Mahanta · J. K. Bisht
ICAR-Vivekananda Parvatiya Krishi Anusandhan Sansthan, Almora, Uttarakhand, India
e-mail: Pankaj.Mishra@icar.gov.in

S. C. Bisht
ICAR-Vivekananda Parvatiya Krishi Anusandhan Sansthan, Almora, Uttarakhand, India

PlantbiotiX, Bio Agri Division of Zytex Biotech Pvt. Ltd, Mumbai, India

A. Pattanayak
ICAR-Vivekananda Parvatiya Krishi Anusandhan Sansthan, Almora, Uttarakhand, India

ICAR-Indian Institute of Agricultural Biotechnology, Ranchi, Jharkhand, India

B. Giri, M. P. Sharma (eds.), *Plant Stress Biology*,
https://doi.org/10.1007/978-981-15-9380-2_6

degradosomes. This review highlights the current knowledge on cold tolerance/adaptation mechanisms operating in psychrotolerant bacteria and their utility in alleviation of cold stress and modern biotechnology.

Keywords

Cold tolerance · Psychrotrophic · Sensing · Cold-active enzymes · Cryoprotectants · Cold shock protein · Cold acclimation protein · Biotechnological applications

Abbreviations

AAT	Aspartate aminotransferase
AFPs	Antifreeze proteins
Caps	Cold acclimation proteins
CRP	Cold resistance protein
CSD	Cold shock domain
Csps	Cold shock proteins
CSR	Cold shock response
DNA	Deoxyribonucleic acid
EDTA	Ethylenediaminetetraacetic acid
EPS	Extracellular polysaccharides
GST	Glutathione S-transferase
hik	Histidine kinase
INAP	Ice nucleation active protein
RIR	Ice recrystallization
kDa	Kilo Dalton
LP	Lipoglycoprotein
MDa	Mega Dalton
PCR	Polymerase chain reaction
PHB	Poly-β-hydroxybutyrate
RI	Recrystallization inhibition
RNA	Ribose nucleic acid
ROS	Reactive oxygen species
TF	Trigger factor
TH	Thermal hysteresis

6.1 Introduction

Microorganisms are ubiquitous in nature and colonize different ecological niches on the earth surface. Due to their ubiquity, they have to encounter several extreme environments. It is this ubiquity and ability to survive the extreme environments that sets microbes distinct from all life forms. The discovery of microorganisms inhabiting extreme environments has made more plausible the search for life in stress environments. Most microorganisms must accommodate a variety of changing conditions and stresses in their environment in order to survive and multiply (Selvakumar et al. 2009). This is due to the fact that most life processes are temperature dependent and life almost comes to a standstill under sub-optimal temperatures. Cold temperatures affect the cell interiors and a myriad of cellular processes, rendering microbial cells inactive or often irreversibly damaged. Since more than 80% of the earth's biosphere is exposed to temperatures below 5°C, throughout the year (Hebraud and Potier 1999), microorganisms capable of coping with low temperature stress have naturally evolved in several environments. Considering their ubiquity and dominance, cold adapted microorganisms are widely regarded as the most successful colonizers of our planet (Russell 1990).

To survive at cold temperature microorganisms require to exhibit at least a few of a range of adaptive behaviors and physiological adjustments. Studies on psychrophiles have revealed that these such adaptations include lipid modification to maintain membrane fluidity, accumulation of polyols, genome adaptations, and production of cold shock proteins (Csps) and cold-active enzymes, oxidative enzymes, including enzymes important for protein synthesis (Berger et al. 1996; Ray et al. 1994a, b, c, d, e; Chattopadhyay et al. 2011). To survive under freezing conditions, some bacteria have developed a variety of strategies, such as the maintenance of membrane fluidity and constant metabolic activities (Kawahara et al. 2001). Also, it has been suggested that trehalose, glycerol, and sorbitol are major cryoprotectants for prokaryotic cells to response to freezing temperature, thereby maintaining some enzyme activities in vivo (Kawahara et al. 2001). However, it is known that cryotolerance in some bacterial species is associated with fatty acid changes in membrane lipids (DeAngelis and Gobbetti 2004), as well as certain culture conditions (Zhao and Zhang 2005). Although some polar bacteria possess antifreeze proteins (AFPs) when they are living at the junction of ice crystals (Deming 2002), the accumulation of molecules that inhibit ice recrystallization (IR) could be part of an adaptive response in Arctic and Antarctic microbes (Muryoi et al. 2004; Xu et al. 1998). In contrast, some microorganisms have ice-nucleating protein complexes that promote the growth of ice growth around the cell surface to avoid lethal intracellular freezing. On the other hand, some bacteria possess antifreeze protein or anti-ice nucleating proteins that are believed to contribute to freeze tolerance of organism (Panicker et al. 2002).

In this context, life in the cold environments has a lot of significance, and adaptations of microorganisms to these extreme cold temperature's conditions are of major interest. Since the cold adaptation requires a complex range of structural and functional adaptations, and these adaptations render cold-adapted organisms very useful for a number of biotechnological applications. The present review describes adaptive strategies and the resulting biotechnological perspectives of bacteria inhabiting low-temperature environments.

6.2 Definition and Ecological Diversity of Cold Tolerant Microorganisms

Cold tolerant microorganisms exhibit distinctly different properties than representatives of mesophiles and thermophiles. According to Morita (1975) microorganisms with low temperature optima are generally referred to as psychrophiles (cold loving). He defined cold-adapted bacteria based on their cardinal growth temperature, viz. lower limit, optimum and upper limit (Russell 2006). A true psychrophile can be defined as an organism whose optimal growth temperature is below 20°C and which is capable to grow at 0°C or below. On the other hand, microorganisms which can grow at 0°C or at subzero temperature but have optimum growth at 20–30°C are referred to as psychrotolerant (also named as cold tolerant or psychrotrophic). However, any such classification is artificial and individual cold-adapted microorganisms may not fit into these definitions. Psychrotolerants may grow as fast as psychrophiles at low temperatures. The main difference between the two groups is that psychrotolerants have a much broader growth temperature range (30–40°C) comparable to psychrophiles (~20°C) (Russell 2006).

The new terms "eurypsychrophile" and "stenopsychrophile" have been proposed to substitute psychrophile and psychrotolerant, respectively (Cavicchioli 2006). "Steno-" and "eury-" are ecological terms derived from Shelford's law of tolerance that respectively describes a narrow and wide tolerance to an environmental determinant. The term stenopsychrophile ("true psychrophile") describes microorganisms with a restricted growth temperature range that cannot tolerate higher temperature for growth. Eurypsychrophile ("psychrotolerant"/cold tolerant or "psychrotrophic") describes microorganisms that like permanently cold environment, but that can tolerate a wide range of temperature extending into the mesophilic range and are, therefore, literally "mesotolerant," not "psychrotolerant." In general, the term psychrophile describes microorganisms that grow in cold environment. It is noteworthy that the term "trophic" pertains to a nutritional state and is not a useful term for clarifying the temperature that can tolerate by any organisms (Cavicchioli 2006). Psychrotolerant microorganisms are much more widely distributed than true psychrophiles. They persist in permanently cold habitats such as polar regions, high altitude areas or regions characterized with low winter temperatures. Diverse microorganisms have remained viable within glacial ice cores for over 120,000 years (Miteva et al. 2004).

The lowest temperature limit for life seems to be around −20°C, which is the value reported for bacteria living in permafrost soil and in sea ice. During the past two decades, microbial life has been extensively investigated in frozen natural habitats (snow, glacial and sea ice, permafrost area, ice clouds, etc.) due to the increasing interest in life on distance frozen planets (astrobiology) (Price 2004). Bacteria, archaea, and eukaryotes like yeast occur in cold environments. While bacteria dominate and are present in greater diversity than archaea in polar environments, archaea are widespread in cold and deep ocean waters (Karner et al. 2001; Deming 2002). Diverse morphological types of microbes encountered in cold environments include spore-formers, non-spore formers, and filamentous bacteria.

Together they cover a wide range of metabolic types ranging from aerobes to anaerobes and include both heterotrophs and autotrophs. The most commonly reported microorganisms are Gram-negative α, β, and γ-proteobacteria (*Pseudomonas* spp. and *Vibrio* spp.), and the cytophaga-*Flavobacterium*-Bacteroides phylum. Coryneforms, *Arthrobacter* sp. and *Micrococcus* sp. are the most frequently found Gram-positive bacteria (D'Amico et al. 2006). Also, diverse cold tolerant microorganisms are widely encountered in refrigerated environments and have become a major cause of concern in the food processing and storage industry.

Cold-adapted microorganisms are known to contribute to the processes of nutrient turnover, biomass production, and litter decomposition in cold ecosystems. There are evidences of a wide range of metabolic activities in cold habitats, e.g. nitrogen fixation, photosynthesis, methanogenesis, and degradation of natural or xenobiotic organic compounds such as proteins, carbohydrates, lignin, and hydrocarbons (Cummings and Black 1999; Margesin et al. 2002; Trotsenko and Khmelenina 2005). Metabolism remains active at subzero temperature, as microbial DNA and protein precursor synthesis are noticed in glacial ice at −15°C (Christner 2002) and in snow at −12 to −17°C (Carpenter et al. 2000). Metabolic activity is found in permafrost bacteria at temperatures up to −20°C (Rivkina et al. 2000). Bacteria are capable of performing basic life functions at temperatures far below 0°C. For example, the Arctic bacterium *Colwellia psychrerythraea* is motile at temperatures of −10°C and its swimming speeds are comparable at −5 and −10°C (Junge et al. 2003). Despite all challenges, life thrives in these environments with a remarkable microbial biodiversity.

6.3 Effect of Temperature on Growth and Metabolic Activity

Decrease in temperature causes an exponential reduction of the reaction rate and the magnitude of which depends on the value of the activation energy. Consequently, most biological systems display a reaction rate 2–3 times lower when the temperature is decreased by 10°C (Q_{10} value). Temperatures outside the linear range of the Arrhenius plot (log of growth rate vs the reciprocal of the absolute temperature) are stress inducing temperatures. For psychrophiles, Arrhenius plots remain linear down to 0°C, for psychrotolerants and mesophiles they deviate from linearity at 5–10 and at 20°C, respectively (Gounot and Russell 1999). Optimal growth temperature is often correlated to the maximum growth rate. The temperature, at which the growth rate is maximum, reflects only kinetic effects and occurs above the linear part of the Arrhenius curve, which means that the physiological conditions are not ideal (Gounot and Russell 1999; Glansdorff and Xu 2002) and growth rate may not be as relevant as growth yield.

Low temperature can influence the response of microorganisms either directly or indirectly. Direct effects include decreased growth rate and enzyme activities, alteration of cell composition, differential nutritional requirements, etc. Indirect effects are usually observed on the solubility of solute molecules, diffusion of nutrients, membrane osmosis, and cell density (Herbert 1986). As temperature falls, the lag phase that precedes growth extends leading to a decrease in the growth rate and the final cell number. The cold temperature largely alters the fluidity of lipid bilayer which in turn affects the solute transport system across it. The lipid bilayer which is the basic structure of the microbial membranes must have proper fluidity to maintain the cell permeability and movement of essential solutes. The functional state of this bilayer is a liquid-crystalline phase, but a decline in temperature induces a gel phase transition and a drastic loss of the membrane properties. The effect of the rapid cold shock on the membrane was found to correlate with high rates of cell inactivation (90 and 70%) in *E. coli* and *Bacillus subtilis*, respectively. Thus, membrane alternation seems to be the principal cause of cold shock injury in *E. coli* and *Bacillus subtilis* (Hoang et al. 2007). Cold stress also induces a shift in the carbon source utilization and enhances the susceptibility of bacteria to antibiotics (Ponder et al. 2005). In some bacteria, production of pigments and other enzymatic activities are enhanced at low temperatures, e.g. lipase and proteinase production by *Pseudomonas* and certain other genera occur preferentially at low temperatures (Witter et al. 1966; Olson and Nottingham 1980). The prior temperature history of the cell has been found to be an important factor for the survival and growth of organisms because of its effects on the extent of lag phase before the onset of growth (Dufrenne et al. 1997).

6.4 Determinants of Cold Tolerance in Bacteria

6.4.1 Sensing of Cold Temperature

Survival at low temperature would depend on the ability of the sensor to perceive the signal and to transduce the signal to the genome that resulted up-regulation of genes, whose products might be involved in cold adaptation (Fig. 6.1). Bacterial temperature sensing appears to be located in the ability of bacterial cell to define and locate those defined changes in its biomolecular constitution that occur as physico-chemical response to temperature changes (Eriksson et al. 2002). Bacterial temperature sensing via alteration in nucleic acid or protein conformation, or changes in membrane lipid behavior, as sensing devices has been reported (Eriksson et al. 2002) (Fig. 6.1).

6.4.1.1 Signal Transduction

Decrease in membrane fluidity due to cold serves as a primary signal for cold perception. A two-component signal transduction system consisting of a membrane based sensory kinase and a cytoplasmic response regulator is reported to constitute the signal transduction reaction in different bacteria (Suzuki et al. 2001; Los et al. 2008). This two-component signal transduction system is known as the phosphor-transfer pathway. In this system, there is a transfer of a phosphate moiety from the sensor to the response regulator. The sensor is normally an integral membrane

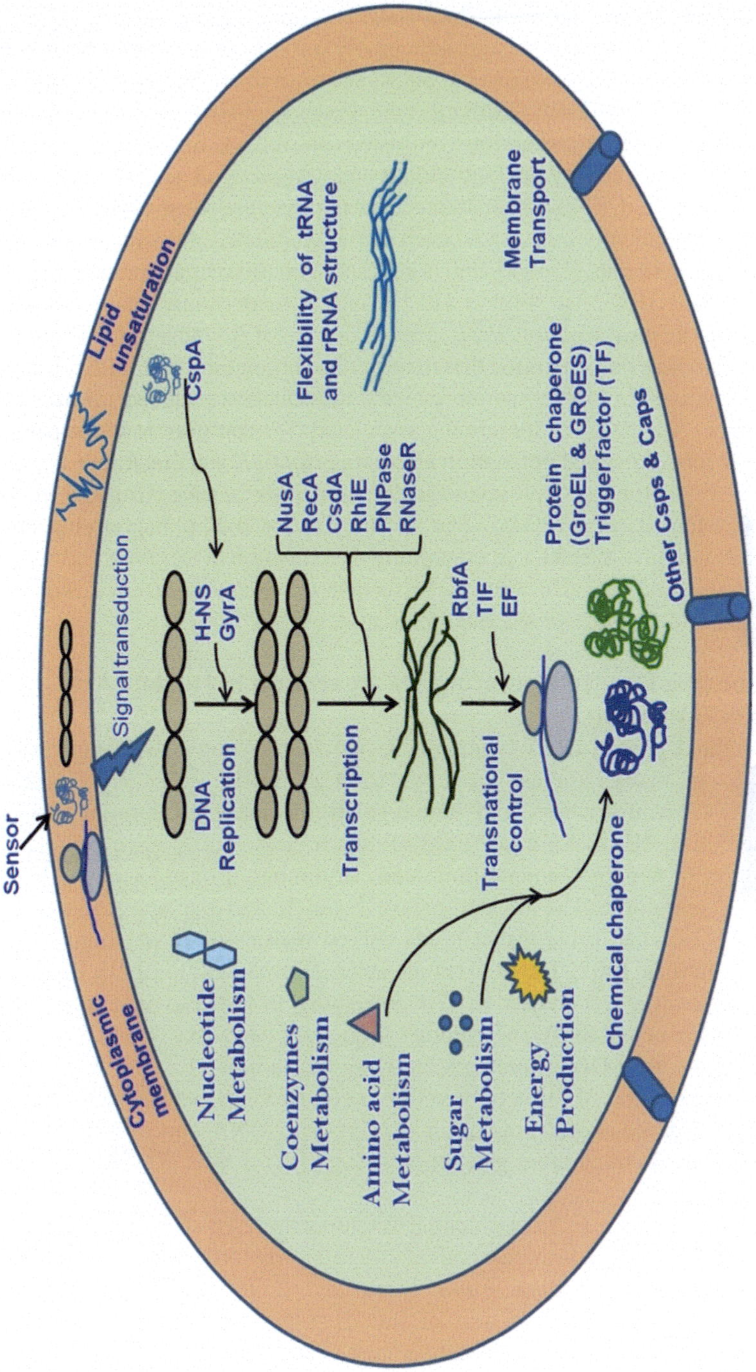

Fig. 6.1 Cellular process that is important for cold adaptation in bacteria

protein consisting of membrane-spanning domains, a signal-recognition domain and a kinase domain (Shivaji and Prakash 2010).

Recognition of signal causes auto-phosphorylation of a histidine by the kinase and the phosphate is then transferred to an aspartate on the response regulator which is present in the cytoplasm (Shivaji and Prakash 2010). In contrast to the two-component signal transduction, multistep phosphorelay systems have been investigated in both prokaryotes and eukaryotes (Appleby et al. 1996; Zhang and Shi 2005). Suzuki et al. (2000) investigated low temperature sensor in cyanobacterial cells. They systematically inactivated each of the 43 putative histidine kinase genes (*hik*) in *Synchocystis* PCC 6803 and identified two histidine kinases, namely a membrane-bound Hik33, a soluble Hik19, and a response regulator, Rer1 that affected the inducibility of the *desB* gene (coding for 15 acyl lipid desaturase). Subsequently, it was demonstrated that the response regulator 26 (Rer 26) along with *Hik33* was involved in low temperature signal transduction pathways (Suzuki et al. 2001; Mikami et al. 2002). A membrane associated two-component signal transduction pathway for cold signal perception consisting of *DesK* and *DesR*, the sensor and the response regulator has been investigated in *Bacillus subtilis* (Kunst et al. 1997; Hoch 2000; Mansilla et al. 2005). The similar CpxA-CpxR phosphorelay system, where CpxA is a histidine kinase containing transmembrane regions and CpxR as response regulators has also been investigated in *Escherichia coli, Salmonella typhimurium,* and *Yersinia pestis* (De-Wult et al. 2000).

6.4.1.2 Sensing Low Temperature via Alteration in Nucleic Acid Conformation

In bacteria, the degree of DNA super helicity varies in response to changes with the temperature. In many studies, the expression of many genes is temperature-dependent DNA conformation, and gene regulation is mastered through changes in DNA supercoiling (Eriksson et al. 2002). Prakash et al. (2009) have demonstrated that inhibition of negative supercoiling leads to inhibition of cold-inducible genes and DNA supercoiling. The topoisomerase I and II and the nucleoid-associated protein H-NS (a small protein that binds curved regions of DNA) regulate DNA supercoiling (Drlica 1992; Tse-Dinh et al. 1997; Dorman et al. 1999). In *Shigella*, when the temperature is increased to 37°C, the ability of H-NS to bind cooperatively to its target sequence at the *VirF* promoter sequences decreases due to a conformational shift in the local DNA topology, allowing transcription of *VirF* (Falconi et al. 1998). Similar function of StpA protein in *E. coli* and many other bacteria has been studied and observed these proteins in concert with DNA appear to serve as an additional bacterial temperature perception system (Dorman et al. 1999; Sonnefield et al. 2001).

RNA molecules have a strong potential as temperature sensor to form pronounced secondary and tertiary structures and ability to form intramolecular RNA:RNA hybrids (Andersen and Delihas 1990; Lease and Belfort 2000). In *E. coli* and *Salmonella typhimurium* expression of the *dsrA* gene encoding a small RpoS regulation is dependent on the cold temperature. The temperature-depended expression of these small regulatory RNAs can modify the activity of proteins and stability and translation of mRNAs and thus play a critical role in regulation of cold genes (Sledjeski et al. 1996; Majdalani et al. 2005).

6.4.1.3 Sensing Low Temperature via Alteration in Protein Conformation

At low temperature, changes in protein conformation as a means for temperature sensing have observed in many bacteria. TlpA, a protein (371 amino acids) in *Salmonella typhimurium* has the ability to sense temperature and to regulate gene expression (Gulig et al. 1993; Hurme et al. 1997). As temperature increases, the proportion of DNA-binding oligomers decreases, leading to repression of the target gene. The sensory capacity of TlpA is dependent on the coiled-coil structure of TlpA, which illustrates sensing of temperature through changes in protein conformation.

Previous studies had suggested that membrane proteins that undergo temperature-dependent phosphorylation–dephosphorylation in bacteria might act as sensors. Changes in membrane proteins due to temperature-depended phosphorylation–dephosphorylation in bacteria act as thermometer (Ray et al. 1994a). In psychrophilic *Pseudomonas syringae* phosphorylation and dephosphorylation of a set of membrane proteins in response to upshift and downshift of temperature were observed. Response to low temperature in *P. syringae* Lz4w, phosphorylation of a cytosolic 66 kDa protein, and differential phosphorylation of lipopolysaccharide were observed (Ray et al. 1994a, c).

6.5 Exopolysaccharide Production

Cold-adapted microorganisms are accustomed to get frozen within their habitats. Such organisms are also expected to have evolved adaptations to survive repeated freezing and thawing, as these processes tend to damage living cells and attenuate cell viability (Mazur 1966). Several diatoms, cyanobacteria and bacteria generate abundant quantities of exopolysaccharide (Palmisano and Sullivan 1985; Cooksey and Cooksey 1995; Costerton et al. 1995; Stoderegger and Herndl 1998), which is stored as a thick gel surrounding the cells. The primary ecological significant characteristic of exopolysaccharide is that it can form and maintain protective microhabitats around microorganisms in aquatic and cold environments (Decho 1990). The physical, rheological, and chemical properties of exopolysaccharide are affected by the length of the polymer chain, which is the principal determinant of the molecular weight (Christensen 1999). As the length of the polymer increases, a greater opportunity for complex entanglement of polymer chains and intramolecular associations occurs and these interactions contribute to the tertiary structure and physical behavior of the polymer (Sutherland 1994).

The role of extracellular polysaccharides (EPS) in the desiccation and freeze tolerance was studied in cyanobacterium *Nostoc commune* (Tamaru et al. 2005). The cells embedded in EPS were highly desiccation tolerant, freeze–thaw cycles and the O_2 evolution was not damaged by air drying. When the cells were completely scrapped off their surrounding EPS (EPS-depleted cells) their O_2 evolution was significantly reduced on desiccation treatment. Similar to the EPS-depleted cells, another *Nostoc commune* KU002 had only small amount of EPS naturally present on cells, which was found to be sensitive to desiccation stress. These findings suggest the role of EPS in stress tolerance during desiccation and freeze thawing.

6.6 Cell Membrane Associated Changes

Low temperature exposure primarily affects the fluidity of lipid bilayer of the bacterial cell membrane, which upon cold shock becomes rigid and impairs membrane associated functions such as transport, energy generation, and cell division (Shivaji and Prakash 2010). At low temperature, the fluidity of bacterial cell membrane decreases, and the maintenance of optimum membrane fluidity becomes crucial for survival. To adapt at low temperature, it is crucial for the bacterium to restore membrane function by increasing the fluidity of the membrane. Bacteria modulate membrane fluidity using various strategies such as by altering the size and charge of the polar head groups, by changing the proportion of short and long chain fatty acids, by changing the extent of fatty acid desaturation, by changing the proportion of *cis* and *trans* fatty acids, and by changing the composition of carotenoids (Shivaji and Prakash 2010).

A predominance of $C_{15:0}$ was found in the fatty acid profile of food-borne pathogen *Listeria monocytogenes* when cells were grown at 5°C, while two cold-sensitive mutants were found to be deficient to synthesize $C_{15:0}$ and another branched chain fatty acid $C_{17:0}$. It is known that a switchover in the synthesis from iso to anteiso fatty acid in bacteria is the CoA ester of 2-methylbutyric acid that is derived from isoleucine. It was postulated that cold sensitivity of mutants might be due to inability to produce 2-methylbutyric acid (Annous et al. 1997). In an Antarctic strain of *Pseudomonas syringae*, decrease in membrane fluidity with concomitant increase in the amount of saturated and *trans* monounsaturated fatty acids was evidenced. However, the *cti* gene was found to be constitutively expressed in the same organism irrespective of growth temperature implying that production of *cis-trans* isomerase in this organism was post-transcriptionally regulated (Kiran et al. 2005).

Polyunsaturated fatty acid and fatty acyl lipid desaturases are essential for the acclimation of cyanobacteria to low temperature (Murata and Wada 1995; Wada and Murata 1990; Chintalapati et al. 2004). Cyanobacteria respond to low temperature by increasing the level of the polyunsaturated fatty acid ($C_{18:3\ (9,\ 12,\ 15)}$) at the expense of mono- and di-unsaturated fatty acids ($C_{18:1(9)}$) and C18:2 (9, 12), respectively (Shivaji and Prakash 2010). In Gram-negative bacteria little is known about the impact of low temperature on outer membrane. Outer membrane major component of Gram-negative bacteria is lipopolysaccharide. This has three components: a polysaccharide that acts as an antigen, a hydrophobic membrane anchor known as lipid A, and a core oligosaccharide that connects the antigen polymer to lipid A. The lipid A moiety of LPS is of interest since cold shock alters the de novo synthesis of the lipid A, such that laurate is replaced by the unsaturated fatty acid palmitoleate in LPS so as to readjust the outer membrane fluidity after cold shock (Carty et al. 1999).

During fractionation of the subcellular components of Antarctic bacteria *Micrococcus roseus* and *Sphingobacterium antarcticus*, carotenoids were always found to be associated with membranes. Therefore, a role of the carotenoids in the regulation of membrane fluidity was postulated. A trend of increase in the amount of non-polar carotenoids was also observed in both the organisms when they were grown at low temperature, compared to ambient temperature (Chattopadhyay and Jogannadham 2001). These researchers also found an increasing trend in the amount of polar carotenoids and decrease in the amount of non-polar carotenoid in both the organisms when they were grown at low temperature compared to their production profile obtained by growth at room temperature. Therefore, their finding suggested that in response to the increase in the synthesis of membrane-fluidizing fatty acids, synthesis of membrane-rigidifying polar carotenoids was also enhanced to counterbalance the effects of fatty acids in the Antarctic bacteria (Chattopadhyay and Jagannadham 2001).

Denarie et al. (1992) emphasized that maintenance of membrane fluidity is a major mode of survival of cold-adapted rhizobia since the symbiotic proteins (p Sym Nod) which are major determinants of nodule competitiveness are membrane associated. However, as per findings of Geiger et al. (1993) the induction of *nod* FE gene in cold-adapted *R. leguminosarum* bv. *viciae* was found to result in the *de novo* synthesis of phospholipids with specific polyunsaturated fatty acids. Drouin et al. (2000) observed that low temperature conditions affected fatty acid composition of all rhizobial strains, regardless of their cold adaptation level. The proportion of unsaturated fatty acids also increased significantly with the decrease in the growth temperature from 25 to 5°C. A specific fatty acid (cis-12 octadecenoic acid) was detected in arctic rhizobial strains during growth at 5°C.

6.7 Cold-Active Enzymes

The most important selective pressure of low temperatures is exerted on chemical reaction rates causing it to drop exponentially. Enzymes are biological catalyst and involved in most of the chemical reactions in the cell, those are necessary for the cell survival. Adaptation of the cells to the low temperature requires the presence of intracellular enzymes, which are active at low temperature. These cold-active enzymes have high catalytic efficiency (K_{cat}/K_m) at low and moderate temperatures (0–30°C) at which homologous enzymes produced by microorganism from other thermal classes are poorly active or not active at all. In addition, these enzymes are generally thermolabile; their activity is shifted toward low temperature (Margesin et al. 2007). The commonly accepted hypothesis for this cold adaptation is the activity–stability–flexibility relationship, which suggests that psychrophilic enzymes increase the flexibility of their structure to cope freezing effect of cold habitats (Somero 2004).

A wide range of molecular determinants confer conformation flexibility of enzymes. These determinants involved in protein stability are either reduced in the number or modified to increase flexibility and to reduce rigidity in protein of cold-adapted microorganisms. These determinants include changes in the frequency of particular molecular bonds (fewer ion pairs, arginine-mediated hydrogen bond, and aromatic interactions) and amino acid side chains (more polar and less hydrophobic residues, a decrease in proline residues in loops, a reduction in arginine residues, or low arginine/lysine ratio), increased interaction with solvent (water and associated ions), reduced hydrophobic interactions between subunit and lose anchoring of *N* and *C* termini (Feller and Gerday 1997; Russell 2000; Sheridan et al. 2000; Marx et al. 2004; Margesin et al. 2005a). Obviously, no cold-active enzymes display all these features; the strategy can be differerent from enzyme to enzyme (Margesin et al. 2007). Characterization of β-galactosidase obtained from psychrotolerant strain of *Arthrobacter* from Antarctic dry-valley soil showed temperature optima near 18°C and it remained 50% active at 0°C. It was also found 2.1 and 5 times more active than β-galactosidase from *E. coli* at 20 and 10°C, respectively. Comparison between β-galactosidase of this *Arthrobacter* strain with another β-galactosidase (temperature optima around 40°C) from psychrotolerant *Arthrobacter psychrolactophilus* revealed that except the decrease in proline residues, most of the criteria for structural features believed so far to confer cold stability and cold-active nature of the enzyme were not satisfied. Again, most of the trends suggested for cold-active enzymes were not found, when the amino acid composition of the cold-active β-galactosidase was compared to *E. coli* β-galactosidase. The thermolability of the enzymes was explained by the fact that it was a tetramer, which dissociated at 25°C into the inactive monomers (Coker et al. 2003). Therefore, the above study indicates that cold stability of enzyme differed from bacterial strain to strain.

Structural feature contributing to gross thermostability structure of glycerol hydrolases are distribution of hydrogen bonds, ion pairs, and amino acid composition (Panasik et al. 2000). These structural features do not always help in generalizing the structural basis of adaptation of enzyme activities at different extremes of temperature. Thermostability or cold-active nature of enzymes could be explained by synergistic and co-operative intramolecular interaction with compatible solutes, viz. sugars and amino acid (Wintrode et al. 2000; Zartler et al. 2001). Recently, Sundareswaran et al. (2010) reported that CSM2, a cold-sensitive mutant of psychrophilic *Pseudomonas syringae*, grows like wild-type cells when cultured at 22 and 28°C. But the growth is retarded at 4°C. In CSM2, *AAT* (aspartate aminotransferase) is identified as the mutated gene. The expression of *AAT* in *Pseudomonas syringae* was transiently enhanced when cells were shifted from 22 to 4°C indicating that *AAT* is cold-inducible. Complementation of the mutated *AAT* transformed CSM2 from a cold-sensitive phenotype to a cold-resistant phenotype like the wild-type cells. This finding indicates the importance of aspartate aminotransferase enzyme in the growth of psychrophilic bacteria at low temperatures.

6.8 Antioxidant Enzymes

Formation of reactive oxygen species (ROS) has been associated with many types of stress. It produces mainly superoxide radicals, generated from oxygen and electrons that have leaked from the electron transport chain. ROS cause cellular damage, such as protein inactivation, membrane damage due to lipid peroxidation, and damage to DNA (Santoro and Thiele 1999). Increase in oxidative stress in cells grown at low temperature was evidenced by increase in the activity of an enzyme and also in the amount of free radicals generated, in the cold-grown cells. The association between cold stress and oxidative stress demonstrated in this investigation bolsters the concept of interlinked stress response in bacteria (Chattopadhyay et al. 2011). The higher level of ROS in microbes nicely correlated as a result of freezing and thawing. The activity of superoxide dismutase and other ROS detoxifying system has likewise proven to be important for freeze tolerance in bacteria (Stead and Park 2000).

6.9 RNA Degradosomes

To protect from cold effect bacteria, synthesize RNA degradosome. This is a protein-complex of several ribonucleases that serve as a major determinant factor for stability of cellular RNA. The degradosome of an Antarctic bacterium *Pseudomonas syringae* has been found to contain an endoribonuclease RNAse E and an RNA helicase (Purusharth et al. 2005). But instead of polynucleotide phosphorylase, which is the exoribonuclease found in mesophilic *E. coli*, the degradosome of the Antarctic bacterium contains another exoribonuclease, called RNAse R. In *E. coli* this enzyme is known to play an important role in ensuring the quality control of rRNA. The significance of the association of this enzyme with RNAse E in the Antarctic bacterium is not definitely known. But it is believed that RNAse R can degrade RNA molecules with extensive secondary structures. This eliminates the necessity of ATP, required by helicase, thereby helping the cell conserve energy at low temperatures (Purusharth et al. 2005). This signifies that RNA metabolism is highly influenced by the RNA secondary structures at low temperature.

6.10 Cold Shock Proteins (Csps)

A sudden drop of environmental temperature causes a number of physico-chemical changes in bacteria that severely affect cellular function. The "cold shock" response in microorganisms is a transient phenomenon that affects growth rate of cell, membrane structure, and function and rates of DNA, RNA, and protein synthesis (Herbraud and Potier 1999). Cold shock response is evidently not confined to psychrophilic (cold loving) and psychrotrophic (cold tolerant) microorganisms but constitutes the beginning of cold adaptation in all microbes. However, it has been noticed that the responses are similar in both groups of organisms, except that the actual temperature which induces them is much lower (0–4°C) in case of psychrophiles than seen in mesophiles (15–20°C). Cold shock response involves the induction and synthesis of cold shock proteins (Csps), for the regulation of protein synthesis and m-RNA folding (Ray 2006).

The first identified member of the family is CspA from *E. coli*, whose homologues were subsequently identified from several bacteria (Graumann et al. 1996; Thieringer et al. 1998; Herbraud and Potier 1999). CspA homologues, in fact, are the most conspicuous group of protein at low temperature in any bacteria. In the case of *E. coli* upon cold shock these homologues constitute more than 10% of total cellular proteins. Among the nine homologues, only four (CspA, CspB, CspG, and CspI) are cold inducible in *E. coli* (Yamanaka 1999; Phadtare and Inouye 2004). In Gram-positive bacterium *B. subtilis*, CspB of the three homologues (CspB, CspC, and CspD) is cold inducible (Graumann et al. 1996, Kaan et al. 2002). This suggests that there is a functional redundancy, as well as probable division of labor among the members of the group. In fact, *E. coli* with only quadruple deletion but not double or triple deletions of *cspA*, *cspB*, *cspG*, and *cspI* genes are cold sensitive (Phadtare and Inouye 2004).

Csp-like proteins were found in more than 50 other bacterial species (Graumann and Marahiel 1998). CspA-like proteins have also been identified in psychrotrophic bacteria: *Pseudomonas fragi* (Hebraud et al. 1994), *Arthrobacter globiformis* (Berger et al. 1996), and *Yersinia enterocolitica* (Neuhaus et al. 2000). In addition, an FKBP family protein was involved in cold adaptation of psychrotrophic bacteria *Shewanella* sp. SIB1 (Suzuki et al. 2004). The cold shock proteins identified in psychrotrophic bacteria and their properties are listed in Table 6.1.

In the Antarctic psychrotrophic *P. fluorescens*, the *cspA* is induced within 30 min when the cultures are shifted from 22°C (optimum temperature) to 4°C (cold stress temperature) (Ray et al. 1998). Similar results were obtained with another psychrotrophic strain *P. fragi*, a refrigerated food spoiling bacterium (Hebraud et al. 1994). The transfer of the food-borne pathogen *Listeria monocytogenes* from 30 to 5°C was characterized by the sharp induction of 18 kDa molecular mass protein that shared a complete sequence identity with a *Listeria innocua* non-heme iron-binding ferritin protein. The purification of these ferritin-like proteins (Flp) revealed a native molecular mass of about 100–110 kDa and a polypeptide composed of six 18 kDa-subunits indicated that these polypeptides might be responsible for cold adaptation (Hébraud and Guzzo 2000).

In another study, cold-adapted strains of *R. leguminosarum* bv. *viciae* were compared with a poorly adapted strain and a cold-sensitive strain for freezing survival, protein induction, and fatty acid composition following a cold shock from 25°C to 10, 5, and 0°C. A common 6.1 kDa cold shock protein was induced in all the strains, but the total number of cold shock proteins synthesized at 0 °C was higher in the cold-adapted strains than in the cold-sensitive strains (Drouin et al. 2000). Transgenic *E. coli* cells expressing the cold-adapted chaperones (Cpn 60 and Cpn 10) from the psychrophilic bacterium *Oleispira antarctica* were found to grow at 4°C. By co-immunoprecipitation of Cpn 60, Northern blot, and in vitro refolding, it was systematically identified that protein–chaperone interactions are the key determinants of protein function at low temperatures (Strocchi et al. 2006).

Table 6.1 Cold shock/Cold acclimation proteins identified in psychrotrophic bacteria

Organisms	Cold shock/Cold acclimation protein	Molecular weight (kDa)	Reference
Pseudomonas fragi	C 7.0 C 8.0	7.0 8.0	Hebraud et al. (1994)
Arthrobacter protophormiae	cspA	7.0	Ray et al. (1994e)
Pseudomonas fluorescens	cspA	7.0	Ray et al. (1994e)
Arthrobacter globiformis	A9	9.0	Berger et al. (1996)
Bacillus cereus WSBC 10201	Csp A Other Csps	7.5 30 35	Mayr et al. (1996)
Bacillus subtilis	CspB CspC CspD	7.365 8.0 13.0	Graumann et al. (1996)
Shewanella violacea DSS12	CspA CspG	7.538 7.614	Fujii et al. (1999)
Listeria monocytogenes	Csp (Flps)	18	Hébraud and Guzzo (2000)
R. leguminosarum bv. viciae	Csp	6.1	Drouin et al. (2000)
Salmonella typhimurium LT2	CspA	7.4	Horton et al. (2000)
Pseudomonas spp. 30-3	CapB	8	Panicker et al. (2002)
Shewanella sp. SIB1	FKBP family	28	Suzuki et al. (2004)
Yersinia enterocolitica	CspA1 CspA2	7.0 7.0	Annamalai and Venkitanarayanan (2005)
Streptococus thermophilus	Clp L	75	Varcamonti et al. (2006)
Colwellia sp. NJ341	CspA GST	8.1 21	Wang et al. (2006)
Pseudomonas fluorescens MTCC 103 *Pseudomonas fluorescens* $CRPF_2$	Csps CRP	14 35 kDa	Khan et al. (2003, 2007)
Lactococcus piscium CNCM I-403	Csp	7-kDa	Garnier et al. (2010)

Functionally, the low molecular-mass Csp group of proteins bind ssDNA and RNA and have a closed β-barrel structure comprising five antiparallel β-strands (Jiang et al. 1997). Bacterial cold shock proteins consist of a single nucleic acid-binding domain, called the cold shock domain (CSD). The CSD is considered to be an ancient molecule present even prior to the advent of single cell life and is the most evolutionarily conserved nucleic acid-binding domain within prokaryotes and eukaryotes (Graumann and Marahiel 1998). The conserved aromatic amino acid residues on the β-strands are involved in binding RNA, with specificity toward "CCAAT" or "ATTGG" sequence. Owing to the design of prokaryotic transcriptional machinery, the cold-induced RNA secondary structure may impose premature transcription termination. The functional significance of bacterial Csps is therefore directly related to the formation of stable secondary RNA structures in response to low temperature stress (Polissi et al. 2003).

6.11 Cold Acclimation Proteins (Caps) or Cold Resistance Protein (CRP)

Cold tolerant bacteria produce permanently one set of proteins called cold acclimation proteins (Caps) during continuous growth at low temperature. Caps may be fundamental to life in the cold and ensure improved protein synthesis at low temperature (Margesin et al. 2007). Cold acclimation proteins are comparable to Csps and are continuously synthesized during prolonged growth at low temperatures and differentiate psychrotrophs from mesophiles. In cold-adapted bacteria (in mesophiles), some of the caps were identified as cold shock proteins, and a typical example being the RNA chaperone CspA (Ray 2006). It has been proposed that cold acclimatization proteins are essential for the maintenance of both growth and cell cycle at low temperatures, but their function is still poorly understood.

Nucleotide sequence of CapA, a gene encoding a protein of the CspA family in psychrotrophic bacterium *A. globiformis* SI55, when compared with that of other CspA related protein from various sources showed that the cold shock response in *A. globiformis* SI55 is an adaptive process that enables cells to protect themselves from deleterious effects of cold (Berger et al. 1996). Radiolabelling of total cellular proteins of *Pseudomonas* spp. 30-3 revealed elevated expression of an 8 kDa protein at 4°C, which suggests that the protein CapB plays a pivotal role in survival and tolerance at cold and subzero temperatures (Panicker et al. 2002). Similarly a 248 bp DNA fragment in *Pseudomonas* spp. 30-3 that was amplified using CapB gene specific primers showed a 98% amino acid sequence homology with CapB of *Pseudomonas fragi* and 62% homology with CspA of *E.coli* (Michel et al. 1997; Panicker et al. 2002). Cold shock proteins (Csps) from *Pseudomonas fluorescens* (MTCC 103) and cold resistance proteins from its mutant $CRPF_2$ of 14 and 35 kDa, respectively, were purified and expression level was checked at different temperatures, i.e., 4, 10, 20, 30, and 37°C. The expression of Csps and CRP increases with decrease in temperature and the cell wall thickness of mutant exhibited two-fold increase, thus facilitating low temperature survival (Khan et al. 2003).

Antarctic sea ice microorganism *Colwellia* sp. NJ341 could grow at temperatures −5 to 22°C, which provided an excellent model system to study microbial adaptation to cold temperatures. To examine such a cold-adaptation mechanism at the protein level, *Colwellia* sp. NJ341 was grown at 0, 8, and 15°C. The results of SDS-PAGE analysis revealed that *Colwellia* sp. NJ341 responded to the low temperature by inducing the synthesis of a set of proteins. Several bands increased in intensity from 15 to 0°C, but the most evident was two proteins CspA and glutathione S-transferase (GST), which indicated that these proteins were responsible for cold adaptation at near-freezing temperature (Wang et al. 2006).

In recent study, adaptation of *Salmonella enterica* sv. typhimurium at refrigerated temperatures involves induction of a multigenic cold shock response (CSR) where gene expression is co-ordinately modified, to express cold shock proteins (Csps). Characterization of CspA instigated the identification of other CspA paralogues, which are highly conserved and widespread across species. Six CspA paralogues have previously been identified in *S. typhimurium* with comparing a csp null strain that lacking all CspA paralogues. This csp null strain is unable to grow following cold shock, demonstrating that the CspA paralogues play an essential role during low temperature adaptation. The individual CspA paralogues exhibit distinct expression profiles, including expression of CspC and CspE at optimal temperature and CspA and CspB following cold shock. This work investigates the transcriptional changes of *S. typhimurium* during cold shock and the role of the CspA paralogues under both optimal and cold shock conditions (Woodall et al. 2011).

6.12 Regulation of Major Cold Shock and Cold Acclimation Genes

Responses to cold shock might be controlled at the transcriptional or translational level, though these two possibilities are not mutually exclusive. Two kinds of proteins are mainly synthesized after downshift of temperature: the so-called cold shock proteins (CSPs) or class I protein, whose level increases sharply for short period within the lag phase and second one is cold-acclimation proteins (CAPs) or class II protein, the level of which increases gradually to a moderate level and does not fall as fast as CSPs (Ray 2006). Examples of the class I protein include cold shock proteins CspA, CspB, CspG, CspI, RNA helicase CsdA ribosome binding protein RbfA, transcription factor NusA, and exoribonuclease PNPase. The class II includes histone-like protein H-NS, recombination protein RecA, DNA gyrase subunit A, translation initiation factor IF2α, trigger factor (TF), pyruvate dehydrogenase subunit E1, and dihydrolipoamide dehydrogenase (Thieringer et al. 1998; Yamanaka 1999; Jones et al. 1987). Thus, there is a distinct change in protein profiles of bacteria during acclimatization to lower temperature. After the temperature downshift the cold shocked cells resume multiplication, transcription and translation of housekeeping genes and continue the synthesis of Caps for some more time before reaching to the steady state level of growth. The growth rate at low temperature is slower and hence all the metabolic activities are adjusted to the new rate. In fact, global analysis of protein profiles (proteome) in few bacteria suggests a change in large number of proteins in cell at low temperature, and the pattern of change is more complex than the induction of 1–2 dozen of new proteins, the understanding of which in relation to low temperature biology is only beginning to unfold (Phadtare and Inouye 2004, Kaan et al. 2002).

The Csps and Caps that have been identified in *E. coli* and their putative role in cell functions would appear from the list that the proteins are of diverse type and involved in diverse cellular functions. Among them, the CspA family of proteins constitute the most common type of "cold shock" proteins in bacterial species. They are small acidic proteins with molecular mass ~7.0 kDa (Polissi et al. 2003). The regulation of *cspA* gene induction after cold shock is rather complicated and as yet not fully understood. The *cspA* gene shows an unusually long leader sequence. The major transcription starts +1 is located 159 bp upstream from the translational starting point. The promoter seems to be σ-70 dependent, since the −35 region (TTGCAT) and the −10 region (CTTAAT) are similar to an σ-70 consensus sequence (TTGACA) for the −35 and TATAAT for the −10 (Qoronfleh et al. 1992; Tanabe et al. 1992). There are other regulatory elements in the gene-sequence of CspA, like the cold box (Fang et al. 1998). The 5′ end of the Csp A m-RNA contains a regulatory sequence (cold box), which may be responsible for cold shock induction. The consensus sequence (5′ UGACGUACAGA) is found in CspA, Csp B, and Csd A. However, if the 5′-end of CspA containing this cold box is overproduced, the expression of cold shock genes is no longer transient, and the synthesis of bulk protein is impaired. Also, the cessation of regrowth after cold shock is prolonged. This fits nicely with the observation that CspA m-RNA in excess is poisonous to the cell. Since overproduction of CspA together with that of the 5′-end restores normal cold shock response, it is likely that CspA itself interacts with the cold box (Jiang et al. 1997). Moreover, besides CspA-mediated autoregulation, a repressor for CspA was found to be CspE, which is abundantly produced at 37°C, and in a CspE mutant, it is derepressed (Fang et al. 1998; Feng et al. 2001). In vitro CspE and CspA cause transcriptional pausing just behind the cold box of CspA, and CspA production is inhibited by addition of CspE to the translating ribosomes (Bae et al. 1999). The induction of CspA is at least partly due to an increase in m-RNA stability. Its half-life is 12 s at 37°C but between 15 and 30 min at 15°C (Tanabe et al. 1992). If the coding region of CspA is fused to the constitutive promoter lpp, it is still cold inducible. This observation is explained by a strong vulnerability of the transcript to RNase E degradation at 37°C. Even if CspA promoter is turned on constitutively, CspA can only be synthesized if the transcript is stabilized, perhaps by CspE (Feng et al. 2001). The transcripts CspB and CspC of *Bacillus subtilis* are also dramatically stabilized, with a half-life of one minute at 37°C and more than 30 min at 15°C (Kaan et al. 1999).

Some metabolic genes are also known to influence gene expression in cold tolerant bacteria. The PP4695 (*cbrA*) and PP4696 (*cbrB)* genes encode the sensory box histidine kinase and a response regulator, respectively. The very similar orthologs of this two-component system in *Pseudomonas aeruginosa* designated *cbrA* and *B* were found to control utilization of carbon and nitrogen sources (Nishijyo et al. 2001). Mutants of *Pseudomonas aeruginosa* deficient in *cbrA* and *B* grew poorly on carbon sources glucose, citrate, or pyruvate and were unable to utilize several amino acids and polyamines (Nishijyo et al. 2001). The latter compounds play an important role in enhancing the translational efficiency (Delcher et al. 2002) and ribosome assembly (Kakegawa et al. 1986). Considering the paramount role of *cbrA/B* in central metabolism it is possible that these genes do have a role to play in regulation of genes responsible for cold tolerance. Cold shock proteins (Csp) and Cold acclimation proteins (Cap) include many diverse kind of proteins, which suggest that the genes are involved in all major aspects of cellular metabolism that are affected by low temperatures.

6.13 Freeze Tolerance in Bacteria

In bacteria, several cellular constituents synthesize to survive in freezing temperature. Most of these freeze protection mechanisms include synthesis of cryoprotectants, ice nucleators, and antifreeze proteins.

6.13.1 Cryoprotectants

Cryoprotectants are chemical substances, which are known to accumulate in the cellular fluid of bacterial cell during cold stress. These substances include both sugar and amino acids. Such cryoprotectants are thought to act as chemical chaperones at cold temperatures (Margesin and Schinner 1999a, b; Russell 1998). These cryoprotectants have been repeatedly shown to protect proteins and other macromolecules against different types of stresses (Chattopadhyay 2006).

6.13.1.1 Sugar Cryoprotectants

Cold tolerant bacteria are endowed with the ability to synthesize several sugars (glucose, fructose, sucrose, and trehalose) and sugar alcohols (glycerol, mannitol, sorbitol, erythritol, threitol) cryoprotectants (Chattopadhyay 2006). The cryoprotective role of glucose was reported by Koda et al. (2002) in *Pantoea ananas* KUIN-3. They found higher glucose in cells at 10°C when cells were subjected to cold stress from 30°C (optimum temperature) to 10°C. The cryotolerance of this bacterium reached 80% after cold acclimation at 10°C. This high level of freezing tolerance in bacteria was correlated with the cryoprotective role of glucose and high activity of glucose-6-phosphatase during cold acclimation.

Trehalose is non-reducing disaccharide (α-*O*-glucopyranosyl-1, 1-2 D glucopyranoside) found in many prokaryotic and eukaryotic organisms, known to be important protectants against stress condition (Kandror et al. 2002). Trehalose plays a vital role in protecting cell against adverse environmental condition. The function of trehalose is stabilizing the membrane and proteins by replacing water and preservation of intracellular water structure (Sano et al. 1999). Exogenous trehalose helps to protect a variety of organisms against freezing and the maximum protection is found when trehalose is present on both sides of the cell membrane, while the exogenous trehalose enhances viability of bacteria during freezing temperature (Herbraud and Potier 1999). Trehalose synthesis is regulated by the genes *otsA* and *otsB* that encode the enzymes, trehalose-6-phosphate synthases and trehalose-6-phosphatase, respectively (Kaasen et al. 1992). A mutant of *E. coli* strain unable to produce trehalose died much faster than the wild type at 4°C due to lack of these two genes *otsA* and *otsB*. But when the mutant strain of *E. coli* was transformed with *otsA* and *otsB genes,* the cells acquired the trehalose synthesizing capability with its viability restored at 4°C (Kaasen et al. 1992). *E. coli* under cold shock conditions and as a consequence of the accumulation of trehalose increase the cells viability as temperature declines to near freezing. Therefore, it may be assumed that increase of trehalose accumulation in *E. coli* is probably responsible for higher viability at low temperature (Kaasen et al. 1992; Mitta et al. 1997).

6.13.1.2 Amino Acid Cryoprotectants

Glycine betaine is a cryoprotectant of bacterial origin and was first demonstrated in the food-borne pathogen *Listeria monocytogenes* which survives at low temperatures and high osmolarity (Angelidis and Smith 2003). Angelidis and Smith (2003) observed that more than 100 colonies appeared on culture plates with betaine after 32 days of incubation at 7°C, while there was no growth without betaine. The exact mechanism of action of glycine betaine is not yet clear. However, it is thought to function as a chemical chaperone, which prevents the aggregation of cellular proteins during stress conditions. The possible function of glycine betaine is to regulate the fluidity of membrane at lower temperatures (Chattopadhyay 2002).

In bacteria proline is an important natural compatible solute during osmotic stress (Yoshiba et al. 1997; Kempf and Bremer 1998). Higher level of intracellular proline has been reported in microorganisms to improve freeze tolerance (Morita et al. 2003; Nanjo et al. 1999). This indicates that proline accumulation might be a general protective mechanism against freeze stress. In addition, intracellular accumulation of charged amino acid arginine and glutamate also seems to enhance microbial freeze tolerance (Shima et al. 2003).

6.13.2 Role of Ice Nucleation in Freeze Tolerance

The term ice nucleation describes the initiation of the phase transition of water from a liquid to a solid state. When a water sample of moderate size is cooled, it will normally not freeze at 0°C. If the water is pure, it can be cooled to temperatures near to −40°C before it freezes. Liquid water at temperatures lower than 0°C is termed supercooled water and this supercooled state is metastable. To enable ice formation to take place, water molecules must cluster in an ice-like pattern and this cluster must reach a critical size. If the initial aggregation of water molecules takes place on a foreign structure, the process is termed *heterogeneous* ice nucleation. If the water molecules aggregate without the help of another structure, the nucleation is termed *homogenous* (Lundheim 2002). A structure that organizes water into an ice-like pattern so that nucleation takes place is called an *ice nucleator* and responsible for heterogeneous ice nucleation (Fig. 6.2). However, the term is usually not used for substances inducing ice nucleation at temperatures lower than −10°C. Many bacteria have ability to minimize freezing injury due to extracellular ice formation and impose an ice like arrangement on the water molecule in contact with their surface

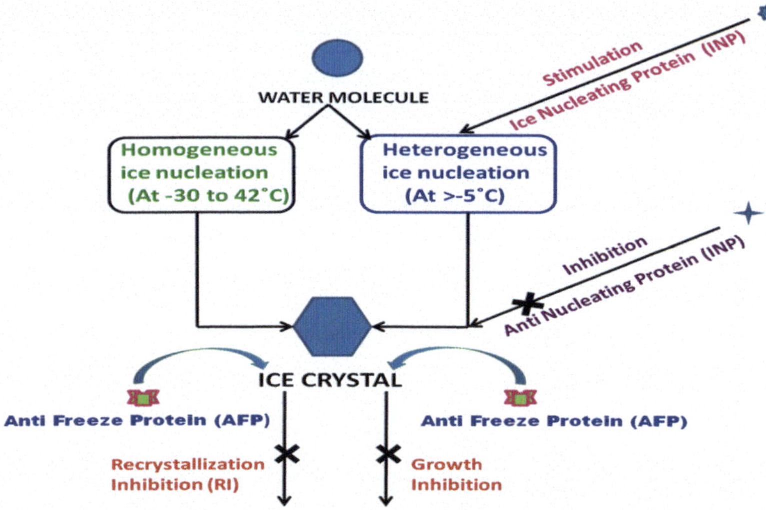

Fig. 6.2 Interaction among ice crystal controlling proteins

that lowers the energy necessary for the initiation of ice formation (Fig. 6.2). The activity of ice nuclei has been classified by the range of temperature in which they initiate freezing: type 1 ice nuclei are active between −2 to −5°C, type 2 are active between −5 and −7°C, and type 3 are active between −7 and −10°C (Yankofsky et al. 1981). Very potent ice nucleators, active at high subfreezing temperature, are produced by bacteria such as *Erwinia herbicola* (Kozloff et al. 1983). Other bacterial genera, *viz. Pseudomonas, Pantoea (Erwinia)*, and *Xanthomonas* can nucleate the crystallization of ice from supercooled water (Lindow et al. 1978; Maki et al. 1974; Obata et al. 1990). The ice nucleation activity of bacteria provides cold protection from the released heat of fusion and establishes protective extracellular freezing instead of lethal intracellular freezing. The "ice plus" bacteria possess INA protein (ice nucleation-active protein) found on the outer bacterial wall that acts as the nucleating center for ice crystals. This protein located on the outer membrane of these bacteria is responsible for ice nucleation (Lee et al. 1995).

The ice nuclei activity has been classified by the range of temperature in which they initiate freezing: type 1 ice nuclei are active between −2 and −5°C, type 2 are active between −5 and −7°C, and type 3 are active between −7 and −10°C (Yankofsky et al. 1981). Very potent ice nucleators, active at high subfreezing temperature, are produced by bacteria such as *Erwinia herbicola* (Kozloff et al. 1983). Other bacterial genera, *viz. Pseudomonas, Pantoea (Erwinia)*, and *Xanthomonas* can nucleate the crystallization of ice from supercooled water (Lindow et al. 1978; Maki et al. 1974; Obata et al. 1990).

Genes conferring ice nucleation activity have been sequenced from many bacterial strains coding for ice nucleating protein (INPs) of 120–180 kDa, with similar primary structure. This protein is a lipoglycoprotein complex that forms large membrane-bound aggregates (Kozloff et al. 1991; Muryoi et al. 2003). INPs comprise of continuous repeat of a consensus octapeptide (Ala-Gly-Thr-Gly-Ser-Thr-Leu-Thr) and function as templates for the formation of small ice crystal seeds termed "ice nuclei." This facilitates ice formation at high subzero temperature, while "ice minus" bacteria do not possess Ina proteins and lower the ice nucleation temperature. Turner et al. (1991) have classified ice nucleating protein into three chemically distinct classes depending on A, B, and C structure. The class C structure was composed of aggregates of ice-nucleating protein (INP). The class B structure was a glycoprotein with sugar residue including glucose, mannose, etc., attached to the protein core and the class A structure was a lipoglycoprotein that was covalently anchored to the cell surface via a mannose-PI (phosphatidylinositol) that is similar to the anchoring of many proteins to cell membranes of eukaryotic cells (Kozloff et al. 1991).

6.13.3 Antifreeze Proteins

In contrast, another strategy used by organism to survive freezing temperature is the production of anti-nucleating proteins or antifreeze proteins (AFPs). These are structurally diverse group of proteins that have the ability to modify ice crystal structure. Unlike the widely used automotive antifreeze (ethylene glycol), AFPs do not lower freezing point in proportion to concentration. Rather, they work in a non-colligative manner. This allows them to act as antifreeze at concentrations 300–500 times lower than the other dissolved solutes. This minimizes their effect on osmotic pressure (Fletcher et al. 2001). The unusual capabilities of AFPs are attributed to their binding ability at specific ice crystal surfaces (Jorov et al. 2004). AFPs create a difference between the melting point and freezing point known as thermal hysteresis (TH), which is a measure of the antifreeze activity. The addition of AFPs at the interface between solid ice and liquid water inhibits the thermodynamically favored growth of the ice crystal. Ice growth is kinetically inhibited by the AFPs covering the water-accessible surfaces of ice (Jorov et al. 2004). This mechanism is also known as freeze avoidance. Besides causing TH, the AFPs act by another mechanism called freeze tolerance. In most of the freezing conditions, formation of ice takes place as a multicrystalline mass. Growth of large ice crystals occurs at the expense of smaller crystals, a phenomenon termed ice recrystallization. AFPs limit the growth of ice crystals at subzero temperatures by being adsorbed on the ice surface (Fig. 6.2). This mechanism is known as recrystallization inhibition (RI) (Duman 2001).

Several bacteria have been reported to exhibit antifreeze activity. These include *Pseudomonas putida* GR12-2, which was originally isolated from the high arctic Canadian soil, *Rhodococcus erythropolis*, isolated from the midguts of beetle larvae, *Micrococcus cryophilus*, isolated from chilled sausages, a *Moraxella* sp. isolated from Antarctic soils, an additional 11 γ- and α-proteobacteria isolated from Antarctic lakes and Antarctic strain *Flavobacterium xanthum* (Muryoi et al. 2004; Kawahara et al. 2007).

The Arctic plant growth promoting rhizobacteria *Pseudomonas putida* GR12-2 secretes an antifreeze protein (AFP) that promotes survival at subzero temperature. Expression of *afpA* in *E. coli* yielded an intracellular 72 kDa protein modified with both sugar and lipid that exhibited lower level of antifreeze and ice nucleation activities. The AfpA sequence was most similar to cell wall associated proteins and less similar to ice nucleation proteins (INPs). Hydropathy plots revealed that the amino acid sequence of AfpA was more hydrophobic than those of the INPs in the domain that forms the ice template, thus suggesting that AFPs and INPs interact differently with ice (Muryoi et al. 2004). The Antarctic *Moraxella* sp. produce 52-kDa antifreeze protein (AFP) and confirmed by formation of hexagonal ice crystal. The purified protein was found to have a TH value of 0.104°C. This was lipoglycoprotein and showed no N-terminal amino acid sequence similarity with *Pseudomonas putida* GR12-2, but it had high sequence similarity with outer membrane proteins of *Moraxella catarrhalis* (Yamashita et al. 2002). This was confirmed that AFP induced in *Moraxella* sp. by low temperature had adaptive strategy to survive at subzero temperature.

An Antarctic sea ice bacterium of the Gram-negative genus *Colwellia* strain SLW05 produces an extracellular substance that changes the morphology of growing ice. The full gene sequence was determined and was found to encode a 253-amino acid protein with a calculated molecular mass of 26,350 Dalton. The predicted amino acid sequence is similar to predicted sequences of ice-binding proteins recently found in two species of sea ice diatoms and a species of snow mold. The function of the protein is unknown, but it may act as an ice recrystallization inhibitor to protect membranes at frozen state (Raymond et al. 2007). Table 6.2 represents characteristics of AFP identified in psychrotolerant bacteria.

Two reports reveal a breach in this widely believed notion. In one investigation, concentrated supernatant of the cell lysate of a bacterium *Marinomonas primoryensis*, isolated from a saline and permanently cold Antarctic lake was found to cause over 2°C freezing point depression, which is higher than the TH value of most of the AFPs isolated from fishes. The activity was reduced in the presence of EDTA but could be restored by saturation of EDTA with calcium chloride. Thus, the antifreeze activity was Ca_2^+ dependent, a feature known so far to be associated with fish AFPs but not with bacterial AFPs (Garnham et al. 2008). Subsequently, a cell-free extract of an Antarctic strain *Flavobacterium xanthum* IAM12026 has been found to have both freezing point depression and ice recrystallization-inhibiting activities. The purified protein was found to have a TH value of 1.19°C (Kawahara et al. 2007). These reports indicate that like fish and insect AFPs, bacterial AFPs can also act by the mechanism of freeze avoidance.

6.14 Biotechnological Applications

Over the past two decades, understanding of the microbial cold tolerance has much increased considerably, which could be attributed by several factors, such as the awareness of accelerated environmental changes in polar regions, evaluation of the

Table 6.2 List of characterized antifreeze protein (AFP) in bacteria

Organisms	Molecular weight[a]	Nature of protein[b]	TH value (°C)	References
Moraxella sp.	52 kDa	LP	0.104	Yamashita et al. (2002)
Pseudomonas putida GR 12-2	164 kDa	LP	UC	Muryoi et al. (2004)
Colwellia strain SLW05	26.35 kDa	UC	UC	Raymond et al. 2007
Flavobacterium xanthum IAM12026	59 kDa	UC	1.19	Kawahara et al. (2007)
Marinomonas primoryensis	>1 MDa	UC	2	Garnham et al. (2008)
Pseudomonas fluorescens	80 kDa	UC	UC	Kawahara et al. (2004)

[a]*kDa* kilo dalton; *MDa* mega dalton;
[b]*LP* lipoglycoprotein; *UN* uncharacterized; *TH* thermal hysteresis

possibility of life on distance frozen planet (astrobiology), genetic tools to create transgenic, and a realization of the considerable biotechnological potential of these organisms (Chattopadhyay 2002; Margesin et al. 2007). By contrast, the number of known or proven current applications remains modest. It should be stressed that confidentiality accompanying commercial products frequently obscures the possible psychrophilic origin of compounds and, accordingly, some current applications are summarized below.

6.14.1 Biotechnological Application of Cold-Active Enzymes

High catalytic efficiency at low and moderate temperatures of cold-active enzyme offers a number of advantages for biotechnology processes, such as the shortening of process times, saving of energy costs, prevention of the loss of volatile compounds, performance of reactions that involve thermosensitive compounds, and reduced risk of contamination (Margesin et al. 2007). The first cold-adapted enzymes from Antarctic bacteria that have been cloned, sequenced, and expressed in a recombinant form were lipases, subtilisins, and α-amylase, i.e. well-known representatives of industrial enzymes. This illustrates besides the fundamental research on biocatalysis in the cold the immense biotechnological potential of psychrophilic enzymes (Fig. 6.3).

6.14.1.1 Detergents Industry

The market for enzymes used in detergents represents 30–40% of all enzymes produced worldwide. Among these enzymatic cleaning agents, subtilisin (an alkaline serine protease predominantly produced by *Bacillus* species) largely dominates in detergents market (Margesin and Schinner 1999b). At the domestic level, the current trend is however to use detergents at lower washing temperatures

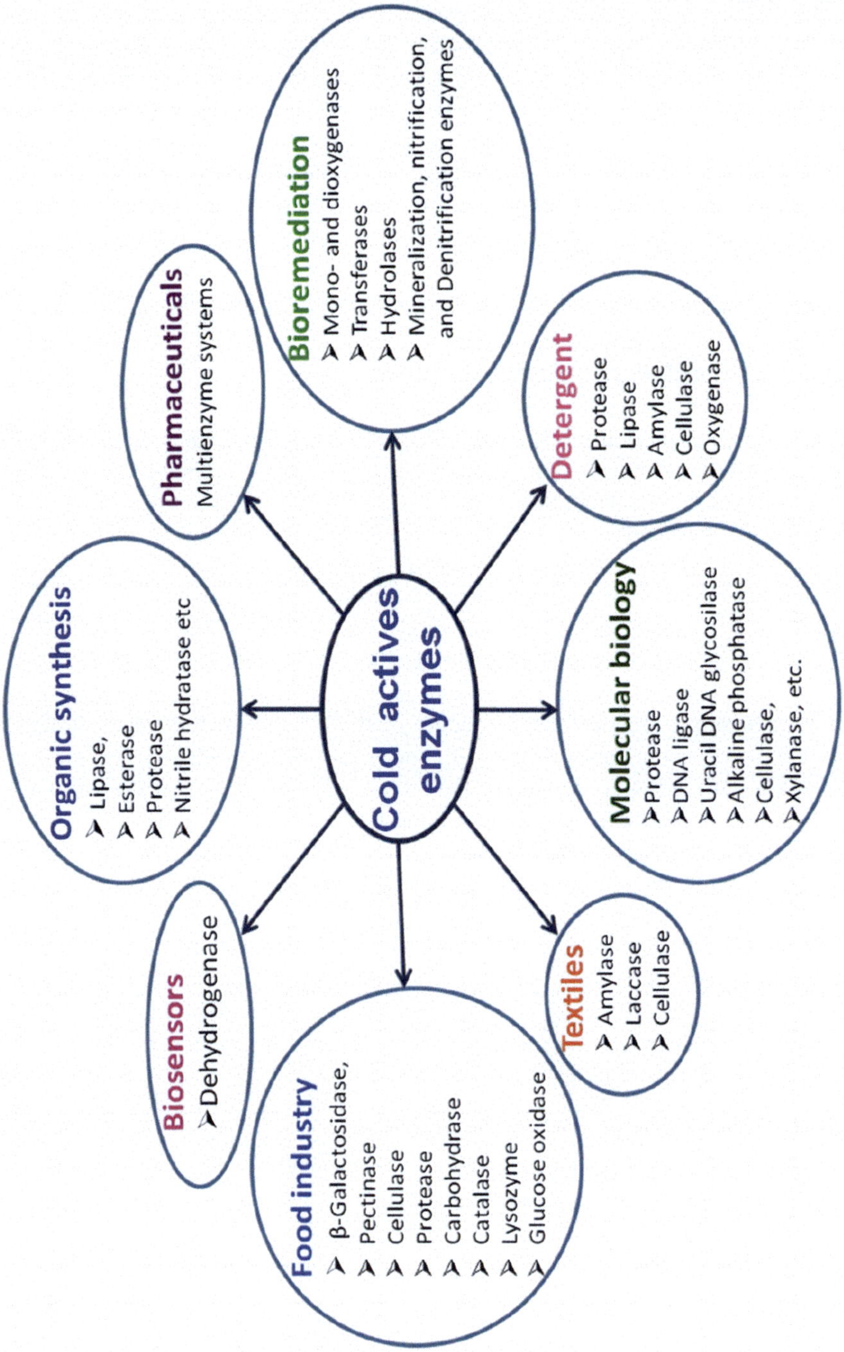

Fig. 6.3 Biotechnological applications of cold-active enzymes

because of the associated reductions in energy consumption and costs as well as to protect texture and colors of the fabrics (Davail et al. 1994; Mohammed et al. 2012). Therefore, subtilisins currently incorporated in cold-active detergents are engineered enzymes that combine storage stability, alkaline stability and activity, and cold activity. Although psychrophilic subtilisins are not components of cold-active detergents, they can largely contribute to the advancement of this economically attractive concept. Besides this cold-active microbial α-amylases indicates the potential of this enzymes as a detergent additive for cold washing that may be useful for domestic processes (Mohammed et al. 2012). Therefore, lowering the wash temperature by using cold-active enzymes can reduce energy consumption and may be used to protect environment because these are biodegradable.

6.14.1.2 Food and Pharmaceutical Industry

The cold-active xylanase from the Antarctic bacterium *Pseudoalteromonas haloplanktis* is a nice example of the successful biotechnological transfer from academic research to industry. Xylanases are glycoside hydrolases that degrade the polysaccharide beta-1, 4-xylan, thus breaking down hemicellulose, one of the major components of plant cell walls. Xylanases are also a key ingredient of industrial dough conditioners used to improve bread quality (De-Vos et al. 2006). Furthermore, baking trials have revealed that the psychrophilic xylanase was very effective in improving the dough properties and final bread quality with, for instance, a positive effect on loaf volume (Collins et al. 2006). This efficiency appears to be related to the high activity of the psychrophilic xylanase at cool temperatures required for dough resting and to its specific mode of xylan hydrolysis. Following careful production optimization of this peculiar xylanase, the product is now sold by Puratos (Belgium). This is apparently the psychrophilic enzyme produced at the highest amounts at present time (Margesin and Feller 2010).

Beta-galactosidase, or lactase, is also a glycoside hydrolase that specifically hydrolyzes the milk sugar lactose into galactose and glucose. It should be stressed that 75% of the world population suffers from lactose intolerance arising from deficient synthesis of intestinal lactase in adults and resulting in digestive disorders due to fermentation of lactose by enteric bacteria. In this context, a cold-active lactase from an Antarctic bacterium has been patented (WO 01/04276A1) for its capacity to hydrolyze lactose during milk storage at low temperatures (Hoyoux et al. 2001). It is worth mentioning that commercially available lactases require milk heating to become active. This heating step has, however, detrimental effects on milk quality as it alters the aspect, the taste, and texture (Maillard reactions, activation of proteases, coagulation, and so on). Although the psychrophilic lactase is apparently not used for this specific application, it is expected that many industries will be produced soon in large quantities to hydrolyze lactose (a by-product of the dairy industry) in the process of the high value sweetener D-tagatose, a natural monosaccharide with low caloric value and glycemic index (Margesin and Feller 2010).

Antarticine-NF3 is a glycoprotein with antifreeze properties produced by the bacterium *Pseudoalteromonas antarctica* that has been patented by Spanish researchers (Parente et al. 2006). It was found that Antarcticine is effective for scar treatments and re-epithelialization of wounds. This glycoprotein is now included in some cosmetic regeneration creams (sometimes under the name Antarctilyne). It is also proposed in association with edelweiss extract: this is of course reminiscent of the peculiar resistance to harsh conditions of both the Antarctic bacterium and the Alp flower. The extracts of the Antarctic algae *Durvillaea antarctica* are included in cosmetic creams to improve skin vitality such as in the Extra-Firming Day Cream, a top seller of Clarins (France) (Margesin and Feller 2010). The successful business endeavor has been the introduction of AFPs into ice cream and yogurt products. AFPs allow the production of very creamy, dense, reduced fat ice cream with fewer additives. They control ice crystal growth brought on by thawing on the loading dock or kitchen table which drastically reduces texture quality (Regand and Goff 2006).

Commercial uses of bacterial INAs for energy and cost saving applications include the production of artificial snow (Snomax®; the addition of INA to water in snow-making machines raises the critical temperature for artificial snow making by several degrees), the production of ice as a construction material for installations in the Arctic and Antarctica, the manufacture of ice-cream and other frozen food (Yin et al. 2005), and the substitution for silver iodide in cloud seeding (Lundheim 2002).

6.14.1.3 Molecular Biology Research

Alkaline phosphatases are mainly used in molecular biology for the dephosphorylation of DNA vectors prior to cloning to prevent recircularization, for the dephosphorylation of 5′-nucleic acid termini before 5′-end labelling by polynucleotide kinase or for removal of dNTPs and pyrophosphate from PCR. However, the phosphatase has to be carefully removed after dephosphorylation to avoid interferences with the subsequent steps. Furthermore, *E. coli* and calf intestinal alkaline phosphatase (that was the preferred enzyme for these applications) are heat-stable and require detergent addition for inactivation. It follows that heat-labile alkaline phosphatases are excellent alternatives as they are inactivated by moderate heat treatment allowing one to perform the subsequent steps in the same test tube and minimizing nucleic acid losses. The heat-labile alkaline phosphatase from Antarctic bacterium is a new tool in molecular biology, this interesting finding is now well established and expressed in *E. coli*. This heat-labile alkaline phosphatase sold as Antarctic phosphatase and now proposed to market by New England Biolabs (USA) (Wang et al. 2007). In the same context, the heat-labile alkaline phosphatase from the Arctic shrimp *Pandalus borealis* is also available, for instance, from Biotec Pharmacon ASA (Norway) or GE Healthcare Life Sciences (UK) (Margesin and Feller 2010).

Two other psychrophilic enzymes are also marketed for molecular biology applications taking advantage of the heat-labile property. Shrimp nuclease selectively degrades double stranded DNA: for instance, it is used for the removal of carry-over contaminants in PCR mixtures, and then it is heat-inactivated prior to addition of the template. This enzyme is produced in recombinant form in *Pichia pastoris* and is available from Biotec Pharmacon ASA (Norway), USB Corporation (USA), or Thermo Scientific (UK). Heat-labile uracil-DNA N-glycosylase from Atlantic cod (*Gadus morhua*) that presents typical cold adaptation features is also used to remove DNA contaminants in sequential PCR (Leiros et al. 2003). When PCR is performed with dUTP instead of dTTP, PCR products become distinguishable from target DNA, and can be selectively degraded by uracil-DNA N-glycosylase. Following degradation of contaminants, the enzyme is completely and irreversibly inactivated after heat treatment. Heat-labile uracil-DNA N-glycosylase, produced in recombinant form in *E. coli*, is available from Biotec Pharmacon ASA (Norway) (Margesin and Feller 2010).

6.14.1.4 Bioremediation

Bioremediation is a process to accelerate the natural biodegradation rates through the optimization of limiting environmental conditions and is an ecologically and economically effective method. Low-temperature biodegradation of organic contaminants in cold ecosystems is a result of the degradation capacity of the indigenous psychrophilic microbial population. They transform or mineralize organic pollutants into less harmful, non-hazardous substances, which are then integrated into natural biogeochemical cycles.

Several schemes have been implemented successfully at the Arctic and Antarctic regions to remediate petroleum-contaminated sites (Aislabie and Foght 2008). Successful on-site treatments include biopiles and land farming, which is now well-developed for cold regions and offers low-cost treatment of petroleum-contaminated soils (Thomassin-Lacroix et al. 2002). The most widely used bioremediation procedure in cold soils is biostimulation of the indigenous microorganisms by supplementation of appropriate nutrients (and optimization of other limiting factors, such as oxygen content, pH, and temperature control); however, care has to be taken to avoid inhibition of biodegradation due to over-fertilization (Walworth et al. 2007). Bioaugmentation by inoculating allochthonous hydrocarbon degraders has been used as a bioremediation option to treat petroleum-contaminated sites in Alaska, Canada, Greenland, and Norway. Bioaugmentation with non-indigenous or genetically modified/engineered microorganisms is banned in Antarctica, Norway, Iceland, and Sweden (Filler et al. 2009). The psychrophiles with specific degradative capabilities based on the transfer of the TOL plasmid from the mesophile *Pseudomonas putida* by conjugation to a psychrophile of the same species; the transconjugant degraded toluene at temperatures as low as 0°C (Kolenc et al. 1988). Recently, the gene coding for a monooxygenase involved in the degradation of aromatic hydrocarbons from the mesophile *Pseudomonas stutzeri* was recombinantly expressed in the Antarctic *Pseudoalteromonas haloplanktis and* performance of such strains has still to be proven (Siani et al. 2006).

Bioleaching is the extraction of specific valuable metals from their ores through the use of bacteria. Several mines worldwide operate at average temperatures of 8–10°C with satisfactory bioleaching performance. Cold-adapted strains of *Acidithiobacillus ferrooxidans* mediate the bioleaching of metal sulfides at such temperatures (Rossi 1999).

Cold-adapted microbial communities able to degrade high amounts of organic compounds within a short time at low temperatures represent a promising source as inocula for low energy wastewater treatment leads to a significant decrease in operational costs. For example, a cold-adapted *Arthrobacter psychrolactophilus* strain displayed all the features necessary for its use as microbial starter, both from the viewpoint of biosafety and production. At 10°C, the strain induced a complete clarification of a synthetic wastewater turbid medium, it hydrolyzed proteins, starch, and lipids, and improved the biodegradability of organic compounds in the wastewater (Gratia et al. 2009). Another example is low-temperature degradation of phenol, which is the most common representative of aromatic toxic pollutants in a wide variety of wastewaters. Psychrophilic *Rhodococcus* spp. able to fully degrade up to 12.5 mM phenol at 10°C under fed-batch cultivation; with some strains phenol degradation occurred even at temperatures as low as 1°C (Margesin et al. 2005b). These studies indicated cold-adapted bacteria inocula as a promising source for accelerated wastewater treatment and also for the construction of biosensors for the rapid monitoring or in situ analysis of pollution (Margesin et al. 2007).

6.14.2 Biological Cryoprotectants

Microbial cryoprotectants like trehalose have immense biotechnological potential and can be used as biological cryoprotectants in a wide range of applications (Lillford and Holt 2002). Similarly, the cold-active enzymes and Anti-Freeze Proteins (Afps) from bacteria can be used in a wide variety of ways. Commercially, there appear to be an infinite number of applications for antifreeze proteins. AFPs appear to be useful in cryosurgery and also in the cryopreservation of whole organisms, isolated organs, cell lines, and tissues (Tange et al. 2003).

The cryoprotective exopolysaccharides producing *Pseudoalteromonas arctica* were isolated from sediment in King George Island, Antarctica. The presence of 0.1% (w/v) purified exopolysaccharide of bacterium showed the survival ratio of *E. coli* cells, which was as high as 82.6% over three repeated freeze–thaw cycles. In addition, at much lower concentrations (0.1–1.0%), purified exopolysaccharide (P-21653) resulted in survival ratios was 83.1–98.4% similar to those of two commercially available cryoprotectants (VEG plus X-1000, 92.9% and VM3, 95.3%), which were utilized at the recommended concentrations (90%). Thus, biochemical characteristics of EPS reflect that this compound may be used as bio-cryoprotectant in medical applications and in the food industry (Kim and Yim 2007).

6.14.3 Microbial Cells as Production Factories

The production level of cold-active (heat-labile) proteins by wild-type strains is usually too low for the production on an industrial scale. To facilitate biotechnological applications of psychrophiles and of their products, a recombinant protein secretion system is a way to produce large-scale production. Therefore, genes encoding for cold-active (heat-labile) proteins have been cloned and expressed in host bacteria, such as *Escherichia coli*, for which efficient expression systems have been designed to obtain high enzyme yields.

Tutino et al. (2001a) described the first recombinant production of a cold-active enzyme (α-amylase from Antarctic *Pseudoalteromonas haloplanktis*) in an Antarctic host bacterium of the same species. The cold gene-expression system was further developed and optimized for the recombinant extracellular secretion of heterologous proteins in *P. Haloplanktis*, with enzymes originating from various Antarctic *P. haloplanktis* strains and a mesophilic yeast (Cusano et al. 2006; Papa et al. 2007). The simultaneous secretion of proteolytic enzymes that degraded the recombinant products could be considerably reduced by inactivating the secretion system with the use of a gene insertion strategy; the mutant strain still secreted the cold-active enzyme (α-amylase) as efficiently as the wildtype and in a stable form (Parrilli et al. 2008). Another recombinant protein expression system working at low temperature was developed by using an Antarctic *Shewanella* sp. strain and was based on the selection of a suitable promoter and a broad host-range plasmid. High yields of ß-lactamase were produced in the *Shewanella* sp. strain at 4°C; the enzyme yield produced at 4°C was 64% of that obtained at 18°C. The efficiency of the system was demonstrated by the production of foreign proteins (putative peptidases and a glucosidase) from the psychrophile *Desulfotalea psychrophila* (Miyake et al. 2007).

Cold-active chaperones have also found very useful application in the production of recombinant proteins. High-level expression of heterologous proteins in *E. coli* can result in the production of large amounts of incorrectly folded proteins, generating aggregates of inactive protein generally in the form of inclusion bodies. To circumvent this insolubility problem, low temperature cultivation of *E. coli* represents a classical strategy and co-expression of chaperones also frequently improves the recovery of soluble proteins. Chaperones are a ubiquitous class of proteins that assist the folding of nascent polypeptides, preventing misfolding or even repairing misfolding. In this context, the chaperonins Cpn10 and Cpn60 (homologous to GroES and GroEL in *E. coli*) from the Antarctic bacterium *Oleispira antarctica* were shown to improve the growth of *E. coli* at low temperatures and to remain optimally active as folding catalysts at these low temperatures (Ferrer et al. 2003). Taking advantage of these properties, the Arctic Express *E. coli* cells from Stratagene (USA) have been engineered to co-express the cold-active chaperonins with the recombinant protein of interest, therefore improving protein processing at low temperatures and increasing the yield of active, soluble recombinant protein.

6.15 Agriculture

Microorganisms (PGPRs) play a major role in sustaining the production and productivity of any agro-ecosystem through a myriad of roles that extend form nitrogen fixation, nutrient solubilization, nutrient mobilization, plant growth promotion, and the suppression of harmful pathogens and insects. Biocontrol of plant diseases using antibiotics produced by cold-adapted bacteria, cell-wall digestive enzymes, and toxins, or induced host resistance at low temperatures (0–5°C) are commercially available as biocontrol agents (Bio-Green©, Plant-Helper©). They are an alternative to chemical pesticides for the control of diseases and pests in cold climates, of winter crops, and during cold storage (Wong and McBeath 1999). Antibiotics produced by *Pseudomonas fluorescens* have been commercialized for the biological control of fire blight in pears and apples (Blightban®) (Lindow and Leveau 2002). Arctic rhizobia increased the production of legumes by 30% through improved nitrogen fixation and are more efficient than commercial rhizobia (Prevost et al. 2003). More recently, psychrotolerant *Pseudomonas* sp. PPERs23 inoculation increased 13.4% grain yield of wheat under rainfed field condition of Indian Himalaya (Bisht et al. 2009). In another study, coinoculation of cold tolerant *Pseudomonas* spp. (PGERs17 & NARs9) with *Rhizobium leguminosarum* PR1 enhanced nodulation, acquisition, nutrient uptake and growth of lentil at 10°C verifying effectiveness and specific functional compatibility relationships between cold tolerant microbial inoculants (Mishra et al. 2010a). Application of psychrotolerant bacterial inoculums and their agricultural importance is recently reviewed by our groups (Mishra et al. 2010b, 2011, 2012).

Freezing injury in plants is particularly complex because of the non-uniform behavior of different plant parts, e.g. stem, leaf, bud, flowers, etc. Also, ice nucleation in plants is frequently not endogenous, but is induced by catalytic sites present in microbial parasites, which can be found on leaves, fruits, or stems (Lindow 1983). A number of bacteria, such as *Pseudomonas syringae* and *Erwinia herbicola*, cause frost injury to plants by triggering ice crystal formation through the action of INAs at subzero temperatures. Reducing the number of such Ice^+ bacteria with naturally occurring or genetically modified "ice-minus" mutants is claimed to be an effective and environmentally safe method of controlling frost damage in plants. A commercial product (Frostban®) consisting of a mixture of three bacterial strains (*Pseudomonas fluorescens* and *Pseudomonas syringae*) can be sprayed on crops to protect plants from frost (Skirvin et al. 2000; Lindow and Leveau 2002).

The development of transgenic plants with increased frost tolerance is another exciting application. The introduction of genes from microorganisms or even whole biosynthetic pathways in plants has already been shown to improve freeze tolerance. *Arabidopsis thaliana* plants transformed with the *codA* gene encoding choline oxidase and accumulating glycine betaine in the chloroplast showed a significant improvement in freeze tolerance (Sakamota et al. 2002). Recently, Castiglioni et al. (2008) demonstrate that bacterial Csp can confer improved stress adaptation to multiple plant species. Transgenic rice (*Oryza sativa*) plants expressing CspA and CspB manifest improved stress tolerance for a number of abiotic stresses, including cold, heat, and water deficits. Improved tolerance was documented by demonstrating improved plant growth rates of transgenic plants relative to their nontransgenic controls, as measured by plant height. Similar observations were reported with transgenic maize plants expressing CspA under greenhouse and field conditions. An across-event analysis demonstrates that the CspA transgenic entries provide a yield increase of 4.6% under water stress, with the two best performing events demonstrated advantage of 30.8 and 18.3%. Yield averages of CspB-positive plants as a group were significantly greater than controls by 7.5%. A number of individual events exhibited significant yield advantages as well; the best two performing events, CspB-Zm event 1 and event 2, demonstrated yield improvements of 20.4 and 10.9%, respectively. The action of Csps in plants through a conserved stress adaptation mechanism common in plants and bacteria is supported by showing a functional RNA binding motif required for the improved stress tolerance in both bacteria and plants. Breadth of tolerance across environments and germplasms is the key element in establishing the value of transgenic strategies for crop stress tolerance improvement and require indicated the potential benefits of bacterial cold stress gene in transgenic development.

6.16 Conclusions

Cold tolerant bacteria are widely distributed in the earth and playing a variety of ecological roles. The most commonly encountered cold tolerance mechanisms include modification of the cell membrane constituents to maintain cell membrane fluidity, induction of cold shock proteins (that act as molecular chaperones), synthesis of cold acclimation protein, cryoprotectant-mediated protection, production of ice nucleation factors or antifreeze proteins and RNA degradosomes. Therefore, complexities of cold adaptation physiology of microorganisms depend on preponderant temperature of the ecosystem and have the ability to survive and grow in a “range of temperature” where the lowest, highest, and the optimal points are different for different organisms. In the past, lot of studies have been carried out mainly with mesophilic organisms, and now it is the turn of “natives” which has been given attention to find out the adaptive mechanisms of psychrotrophic bacteria when subjected to cold temperatures. This would allow one to identify the common mechanisms and the molecules that are necessary for growth of psychrophiles or psychrotrophs, which would in turn led to identify the regulatory elements and cold-stress response mechanisms that organisms have developed for producing common molecules/strategies for the respective cold temperatures in the course of evolution. This would also lead to identification of the unique adaptive features of different groups of organisms for growing at lower temperatures. However, recent developments based on cold-adapted organisms and on their biomolecules, such as those mentioned here, have clearly demonstrated the huge biotechnological potential. This potential appears to be even larger than other extremophiles when considering both the broader psychrophilic biodiversity that encompasses microorganisms, plants, and animals and the broader fields of application. Most biotechnological applications of psychrophiles or psychrotrophs are environmentally friendly and contribute to energy saving, which considered being almost important in current scenario.

Acknowledgements We acknowledge the generous supprot of Indian Council of Agriculrural Research through the project "Application of Microorganisms in Agriculture and Allied Sectors (AMAAS)".

References

Aislabie J, Foght JM (2008) Hydrocarbon-degrading bacteria in contaminated soils. In: Filler DM, Snape I, Barnes DL (eds) Bioremediation of petroleum hydrocarbons in cold regions. Cambridge University Press, Cambridge, pp 69–83

Andersen J, Delihas N (1990) *micF* RNA binds to the 5' end of *ompF* mRNA and to a protein from *Escherichia coli*. Biochem 29:9249–9256

Angelidis AS, Smith GM (2003) Role of glycine betaine and carnitine transporters in adaptation of *Listeria monocytogenes* to chill stress in defined medium. Appl Environ Microbiol 69:7492–7498

Annamalai T, Venkitanarayanan K (2005) Expression of major cold shock proteins and genes by *Yersinia enterocolitica* in synthetic medium and foods. J Food Prot 68:2454–2458

Annous BA, Becker LA, Bayles DO, Labeda DP, Wilkinson BJ (1997) Critical role of anteiso-$C_{15:o}$ fatty acid in the growth of *Listeria monocytogenes* at low temperatures. Appl Environ Microbiol 63:3887–3894

Appleby JL, Parkinson JS, Bourret RB (1996) Signal transduction via the multi-step phosphorelay: not necessarily a road less traveled. Cell 86:845–848

Bae W, Phadtare S, Severinov K, Inouye M (1999) Characterization of *Escherichia coli cspE*, whose product negatively regulates transcription of *cspA*, the gene for the major cold shock protein. Mol Microbiol 31:1429–1441

Berger F, Morellet N, Menu F, Potier P (1996) Cold shock and cold acclimation proteins in the psychrotrophic bacterium *Arthrobacter globiformis* SI55. J Bacteriol 178:2999–3007

Bisht SC, Mishra PK, Ruwari P, Joshi P, Suyal P, Selvakumar G, Bisht JK (2009) Enhancement of chilling tolerance and productivity of inoculated wheat with cold tolerant Plant Growth Promoting *Pseudomonas* sp. strain PPERs23. 4th Annual Conference UCOST, G.B Pant University of Ag. & Tech. Pantnagar, 15

Carpenter EJ, Lin S, Capone DG (2000) Bacterial activity in South Pole snow. Appl Environ Microbiol 66:4514–4517

Carty SM, Sreekumar KR, Raetz CR (1999) Effect of cold shock on lipid A biosynthesis in *Escherichia coli*. Induction at 12 degrees C of an acyltransferase specific for palmitoleoyl-acyl carrier protein. J Biol Chem 274:9677–9685

Castiglioni P, Warner D, Bensen RJ, Anstrom DC, Harrison J, Stoecker M, Abad M, Kumar G, Salvador S, D'Ordine R, Navarro S, Back S, Fernandes M, Targolli J, Dasgupta S, Bonin C, Luethy MH, Heard JE (2008) Bacterial RNA Chaperones confer abiotic stress tolerance in plants and improved grain yield in maize under water limited conditions. Plant Physiol 147:446–455

Cavicchioli R (2006) Cold adapted archaea. Nat Rev Microbiol 4:331–343

Chattopadhyay MK (2002) Bacterial cryoprotectants. Resonance 7:59–63

Chattopadhyay MK (2006) Mechanism of bacterial adaptation to low temperature. J Biosci 31:157–165

Chattopadhyay MK, Jagannadham MV (2001) Maintenance of membrane fluidity in Antarctic bacteria. Polar Biol 24:386–388

Chattopadhyay MK, Raghu G, Sharma YV, Biju AR, Rajasekharan MV, Shivaji S (2011) Increase in oxidative stress at low temperature in an Antarctic bacterium. Curr Microbiol 62:544–546

Chintalapati S, Kiran MD, Shivaji S (2004) Role of membrane lipid fatty acids in cold adaptation. Cell Mol Biol 50:631–642

Christner BC (2002) Incorporation of DNA and protein precursors into macromolecules by bacteria at −15°C. Appl Environ Microbiol 68:6435–6438

Christensen BE (1999) Physical and chemical properties of extracellular polysaccharides associated with biofilms and related substances. In: Wingender J, Neu T, Flemming HC (eds) Microbial extracellular substances: characterization, structure and function. Springer, New York, pp 144–154

Coker JA, Sheridan PP, Loveland-Curtze J, Gutshall KR, Auman AJ, Brenchley JE (2003) Biochemical characterization of a β-galactosidase with a low temperature optimum obtained from an Antarctic *Arthrobacter* isolate. J Bacteriol 185:5473–5482

Collins T, Hoyoux A, Dutron A, Georis J, Genot B, Dauvrin T, Arnaut F, Gerday C, Feller G (2006) Use of glycoside hydrolase family 8 xylanases in baking. J Cereal Sci 43:79–84

Cooksey KE, Cooksey B (1995) Adhesion of bacteria and diatom to surfaces in the sea: a review. Aquat Microb Ecol 9:87–96

Costerton JW, Lewandowski Z, Caldwell DE, Korber DR, Lappin-Scott HM (1995) Microbial biofilms. Ann Rev Microbiol 49:711–745

Cummings SP, Black GW (1999) Polymer hydrolysis in a cold climate. Extremophiles 3:81–87

Cusano AM, Parrilli E, Duilio A, Sannia G, Marino G, Tutino ML (2006) Secretion of psychrophilic alpha-amylase deletion mutants in *Pseudoalteromonas haloplanktis* TAC125. FEMS Microbiol Lett 258:67–71

D'Amico S, Colins T, Marx JC, Feller G, Gerday C (2006) Psychrophilic microorganisms: a challenge for life. EMBO Rep 7:385–389

Davail S, Feller G, Narinx E, Gerday C (1994) Cold adaptation of proteins. Purification, characterization, and sequence of the heat-labile subtilisin from the Antarctic psychrophile *Bacillus* TA41. J Biol Chem 269:17448–17453

DeAngelis M, Gobbetti M (2004) Environmental stress response in *Lactobacillus*: a review. Proteomics 4:106–122

Decho AW (1990) Microbial exopolymer secretions in ocean environments: their role(s) in food webs and marine processes. Oceanogr Mar Biol Ann Rev 28:73–153

Delcher AL, Phillippy A, Carlton J, Salzberg SL (2002) Fast algorithms for large-scale genome alignment and comparison. Nucleic Acids Res 30:2478–2483

Deming JW (2002) Psychrophiles and polar regions. Curr Opin Microbiol 5:301–309

Denarie J, Debelle F, Rosenberg C (1992) Signalling and host range variation in nodulation. Annu Rev Microbiol 46:497–531

De-Vos D, Collins T, Nerinckx W, Savvides SN, Claeyssens M, Gerday C, Feller G, Van-Beeumen J (2006) Oligosaccharide binding in family 8 glycosidases: crystal structures of active-site mutants of the beta-1, 4- xylanase pXyl from *Pseudoaltermonas haloplanktis* TAH3a in complex with substrate and product. Biochem 45:4797–4807

De-Wult P, Akerley BJ, Lin EC (2000) Presence of the Cpx system in bacteria. Microbiol 146:247–248

Dorman CJ, Hinton JCD, Free A (1999) Domain organization and oligomerization among H-NS-like nucleoid-associated proteins in bacteria. Trends Microbiol 7:124–128

Drlica K (1992) Control of bacterial DNA supercoiling. Mol Microbiol 6:425–433

Drouin P, Prevost D, Antoun H (2000) Physiological adaptation to low temperatures of strains of *Rhizobium leguminosarum* bv. viciae associated with *Lathyrus* spp. FEMS Microbiol Ecol 32:111–120

Dufrenne J, Delfgou E, Ritmeester W, Notermans S (1997) The effect of previous growth conditions on the lag phase time of some foodborne pathogenic microorganisms. Int J Food Microbiol 34:89–94

Duman JG (2001) Antifreeze and Ice Nucleator Proteins in Terrestrial Arthropods. Annu Rev Physiol 63:327–357

Eriksson S, Hurme R, Rhen M (2002) Low temperature sensors in bacteria. Philos Trans R Soc Lond B 357:887–893

Falconi M, Colonna B, Prosseda G, Micheli G, Gualerzi CO (1998) Thermoregulation of *Shigella* and *Escherichia coli* EIEC pathogenicity. A temperature-dependent structural transition of DNA modulates accessibility of *virF* promoter to transcriptional repressor H-NS. EMBO J 17:7033–7043

Fang L, Hou Y, Inouye M (1998) Role of the cold box region in the 5′- untranslated region of the *cspA* m-RNA in its transient expression at low temperature in *Escherichia coli.* J Bacteriol 180:90–95

Feller G, Gerday C (1997) Psychrophilic enzymes: molecular basis of cold adaptation. Cell Mol Life Sci 53:830–841

Feng Y, Huang H, Liao J, Cohen SN (2001) *Escherichia coli* poly (A)-binding proteins that interact with components of degradosomes or impede RNA decay mediated by polynucleotide phosphorylase and RNaseE. J Biol Chem 274:31651–31656

Ferrer M, Chernikova TN, Yakimov MM, Golyshin PN, Timmis KN (2003) Chaperonins govern growth of Escherichia coli at low temperatures. Nat Biotechnol 21:1266–1267

Filler DM, Van-Stempvoort DR, Leigh MB (2009) Remediation of frozen ground contaminants with petroleum hydrocarbons: feasibility and limits. In: Margesin R (ed) Environmental technology 843 permafrost soils. Springer-Verlag, Berlin, pp 279–301

Fletcher GL, Hew CL, Davies PL (2001) Antifreeze proteins of teleost fishes. Annu Rev Physiol 63:359–390

Fujii S, Nakasone K, Horikoshi K (1999) Cloning of two cold shock genes, cspA and cspG from the deep sea psycrophilic bacterium Shewanella vioacea strain DSS12. FEMS Microbial Lett 178:123–128

Garnham CP, Gilbert JA, Hartman CP, Campbell RL, Laybourn-Parry J, Davies PL (**2008**) A Ca^{2+}-dependent bacterial antifreeze protein domain has a novel b-helical ice-binding fold. Biochem J 411:171–180

Garnier M, Sebastien M, Didier C, Marie-France P, Francoise L, Odile T (2010) Adaptation to cold and proteomic responses of the psychrotrophic biopreservative *Lactococcus piscium* strain CNCM I-403. Appl Envion Microbiol 76:8011–8018

Geiger O, Spaink HP, Lugtenberg BJJ (1993) Biosynthesis of lipo-oligosaccharides: phospholipids of *Rhizobium* contain nod E- determined highly unsaturated fatty acid moieties. In: Palacios R, Mora J, Newton WE (eds) New horizons in nitrogen fixation. Kluver Academics Publisher, Dordrecht, p 233

Glansdorff N, Xu J (2002) Microbial life at low temperatures: mechanisms of adaptation and extreme biotopes. Implications for exobiology and the origin of life. Recent Res Devl 6:1–21

Gounot AM, Russell NJ (1999) Physiology of cold-adapted microorganisms. In: Margesin R, Schinner F (eds) Cold adapted organisms. Springer, Berlin, pp 33–55

Gratia E, Weekers F, Margesin R, D'Amico S, Thonart P, Feller G (2009) Selection of a cold-adapted bacterium for bioremediation of wastewater at low temperatures. Extremophiles 13:763–768

Graumann P, Marahiel MA (1998) A superfamily of proteins containing the cold shock domain. Trends Biochem Sci 23:286–290

Graumann P, Schroder K, Schmid R, Marahiel MA (1996) Cold shock stress induced proteins in *Bacillus subtilis.* J Bacteriol 178:4611–4619

Gulig PA, Danbara H, Guiney DG, Lax AJ, Norel F, Rhen M (1993) Molecular analysis of *spv* virulence genes of the *Salmonella* virulence plasmid. Mol Microbiol 7:825–830

Hébraud M, Guzzo J (2000) The main cold shock protein of *Listeria monocytogenes* belongs to the family of ferritin-like proteins. FEMS Microbiol Lett 190:29–34

Hebraud M, Dubois E, Potier P, Labadie J (1994) Effect of growth temperature on the protein levels in the psychrotrophic bacterium, *Pseudomonas fragi.* J Bacteriol 176:4017–4024

Herbert RA (1986) The ecology and physiology of psychrotrophic micro-organism. In: Herbert RA, Codd GA (eds) Society for gen microbiology. Academic Press, Cambridge, pp 1–24

Herbraud M, Potier P (1999) Cold shock response and low temperature adaptation in psycrophilic bacteria. J Mol Microbiol Biotechnol 1:211–219

Hoang LC, Dumomt F, Marechal PA, Thanh ML, Gervais P (2007) Rates of chilling to 0°C: implication for the survival of microorganisms and relationship with membrane fluidity modification. Appl Microbiol Biotechnol 77:1379–1387

Hoch JA (2000) Two component and phosphorelay signal transduction. Curr Opin Microbiol 3:165–170

Horton AJ, Hak KM, Steffan RJ, Foster JW, Bej AK (2000) Adaptive response to cold temperatures and characterization of *cspA* in *Salmonella typhimurium* LT2. Antonie van Leeuwenhoek 77:13–20

Hoyoux A, Jennes I, Dubois P, Genicot S, Dubail F, Francois JM, Baise E, Feller G, Gerday C (2001) Cold-adapted beta-galactosidase from the Antarctic psychrophile *Pseudoalteromonas haloplanktis*. Appl Environ Microbiol 67:1529–1535

Hurme R, Berndt K, Normark SJ, Rhen M (1997) A proteinaceous gene regulatory thermometer in *Salmonella*. Cell 90:55–64

Jiang W, Hou Y, Inouye M (1997) CspA, the major cold-shock protein of *Escherichia coli*, is an RNA chaperone. J Biol Chem 272:196–202

Jones PG, Vanbogelen RA, Neidhart FC (1987) Induction of proteins in response to low temperature in *Escherichia coli*. J Bacteriol 69:2092–2095

Jorov A, Zhorov BS, Yang DSC (2004) Theoretical study of interaction of winter flounder antifreeze protein with ice. Protein Sci 13:1524–1537

Junge K, Eicken H, Deming JW (2003) Motility of *Colwellia psychrerythraea* strain 34H at subzero temperatures. Appl Environ Microbiol 69:4282–4284

Kaan T, Jurgen B, Schweder T (1999) Regulation of the expression of the cold shock proteins CspB and CspC in *Bacillus subtilis*. Mol Gen Genom 262:351–354

Kaan T, Mader U, Bandow J, Schweder T (2002) Genome-wide transcriptional profiling of the *Bacillus subtilis* cold-shock response. Microbiol 148:3441–3455

Kaasen I, Falkenberg P, Styrvold OB, Stroem AR (1992) Molecular cloning and physical mapping of the *otsBA* genes, which encode the osmoregulatory trehalose pathway of *Escherichia coli*: evidence that transcription is activated by Kat F(AppR). J Bacteriol 174:889–898

Kakegawa TS, Hirose S, Kashiwagi K, Igarshi K (1986) Effect of polyamines on *in vitro* reconstitution of ribosomal sub units. Eur J Biochem 158:265–269

Kandror O, DeLeon A, Goldberg AL (2002) Trehalose synthesis is induced upon exposure of *Escherichia coli* to cold and is essential for viability at low temperatures. Proc Natl Acad Sc 99:9727–9732

Karner MB, Delong EF, Karl DM (2001) Archeal dominance in the mesophilic zone of Pacific Ocean. Nature 409:507–510

Kawahara H, Li J, Griffth M, Glick BR (2001) Relationship between Antifreeze protein and Freezing Resistance in *Pseudomonas putida* GR 12-2. Curr Microbiol 43:365–370

Kawahara H, Nakano Y, Omiya K, Muryoi N, Nishikawa J, Obata H (2004) Production of two types of ice crystal controlling proteins in Antarctic bacterium. J Biosci Bioeng 98:220–223

Kawahara H, Iwanaka Y, Higa S, Muryoi N, Sato M, Honda M, Omura H, Obata H (2007) A novel, intracellular antifreeze protein in an Antarctic bacterium, *Flavobacterium xanthum*. Cryo Letters 28:39–49

Kempf B, Bremer E (1998) Uptake and synthesis of compatible solutes as microbial stress response to high-osmolality environments. Arch Microbiol 170:319–330

Khan M, Kumar V, Goel R (2003) Characterization and localization of fluorescent *Pseudomonas* cold shock protein(S) by Monospecific polyclonal Antibodies. Microbial Immunol 47:895–901

Khan M, Kumar S, Goel R (2007) Development of immunoassay for the identification of cold shock proteins from diversified micro flora. Afr J Biotechnol 6:252–257

Kim SJ, Yim JH (2007) Cryoprotective Properties of Exopolysaccharide (P-21653) Produced by the Antarctic Bacterium, *Pseudoalteromonas arctica* KOPRI 21653. J Microbiol 45:510–514
Kiran MD, Annapoorni S, Suzuki I, Murata N, Shivaji S (2005) *Cistrans* isomerase gene in psychrophilic *Pseudomonas syringae* is constitutively expressed during growth and under conditions of temperature and solvent stress. Extremophiles 9:117–125
Koda N, Inada Y, Nakayama S, Kawahara H, Obata H (2002) Response of the Ice-nucleating bacterium *Pantoea ananas* KUIN-3 during cold acclimation. Biosci Biochnol Biochem 66:866–868
Kolenc RJ, Inniss WE, Glick BR, Robinson CW, Mayfield CI (1988) Transfer and expression of mesophilic plasmid-mediated degradative capacity in a psychrotrophic bacterium. Appl Environ Microbiol 54:638–641
Kozloff LM, Schofield MA, Lute M (1983) Ice-nucleating activity of *Pseudomonas syringae* and *Erwinia herbicola*. J Bacteriol 153:222–231
Kozloff LM, Turner MA, Arellano F (1991) Formation of bacterial membrane ice-nucleating lipoglycoprotein complexes. J Bacteriol 173:6528–6536
Kunst F, Ogasawara N, Moszer I, Albertini AM, Alloni G, Azevedo V, Bertero MG, Bessieres P, Bolotin A, Borchert S, Borriss R, Boursier L, Brans A, Brauwn M, Brignell SC, Born S, Brouillet S, Bruschi SV, Caldwell B, Capuano V, Carter NM, Choi SK, Codani SJ, Connerton IF (1997) The complete genome sequence of the gram-positive bacterium *Bacillus subtilis*. Nature 390:249–256
Lease RA, Belfort M (2000) A trans-acting RNA as a control switch in *Escherichia coli*: DsrA modulates function by forming alternative structures. Proc Natl Acad Sci USA 97:9919–9924
Lee RE, Warren GJ, Gusta LV (1995) Biochemistry of bacterial ice nuclei. In: Ray F, Paul WK (eds) Biological ice nucleation and its application. APS Press, St Paul, pp 63–83
Leiros I, Moe E, Lanes O, Smalas AO, Willassen NP (2003) The structure of uracil-DNA glycosylase from Atlantic cod (Gadus morhua) reveals cold-adaptation features. Acta Crystallogra D Biol 59:1357–1365
Lillford PJ, Holt CB (2002) *In-vitro* uses of biological cryoprotectants. Philos Trans R Soc Lond Ser B Biol Sci 357:945–951
Lindow SE (1983) The role of bacterial ice nucleation in frost injury to plants. Rev Phytopathol 21:363–384
Lindow SE, Leveau JH (2002) Phyllosphere microbiology. Curr Opin Biotechnol 13:238–243
Lindow SE, Arny DC, Upper CD (1978) *Erwinia herbicola*: a bacterial ice nucleus active in increasing frost injury to corn. Phytopathol 68:523–527
Los DA, Suzuki I, Zinchenko VV, Murata N (2008) Stress responses in *Synechocystis*: regulated genes and regulatory systems. In: Herrero A, Flores E (eds) The cyanobacteria: molecular biology, genetics and evolution. Horizon ScientiWc Press, Wymondham, pp 117–158
Lundheim R (2002) Physiological and ecological significance of biological ice nucleators. Philos Trans R Soc Lond B Biol Sci 357:937–943
Majdalani N, Vanderpool CK, Gottesman S (2005) Bacterial small RNA regulators. Crit Rev Biochem Mol Biol 40:93–113
Maki IR, Galyon EL, Chang-Chien M, Cald WDR (1974) Ice nucleation induced by *Pseudomonas syringae*. Appl Microbiol 28:456–460
Mansilla MC, Albanesi D, Cybulski LE, de Mendoza D (2005) Molecular mechanisms of low temperature sensing bacteria. Ann Hepatol 4:216–217
Margesin R, Feller G (2010) Biotechnological applications of psychrophiles. Environ Technol 31:835–844
Margesin R, Schinner F (1999a) Cold-adapted organisms. In: Ecology, physiology, enzymology and molecular biology. Springer, Berlin
Margesin R, Schinner F (1999b) Biotechnological applications of cold-adapted organisms. Springer, Berlin
Margesin R, Zacke G, Schinner F (2002) Characterization of heterotrophic microorganisms in alpine glacier cryoconite. Arct Antarct Alp Res 34:88–93

Margesin R, Fauster V, Fonteyne PA (2005a) Characterization of cold active pectate lyases from psychrophilic *Mrakia frigida*. Lett Appl Microbiol 40:453–459

Margesin R, Fonteyne PA, Redl B (2005b) Low-temperature biodegradation of high amounts of phenol by *Rhodococcus* spp. and Basidiomycetous yeasts. Res Microbiol 156:68–75

Margesin R, Neuner G, Storey KB (2007) Cold- loving microbes, plants, and animals-fundamental and applied aspects. Naturewisenschaften 94:77–99

Marx JC, Blaise V, Collins T, D'Amico S, Delille D, Gratia E, Hoyoux A, Huston AL, Sonan G, Feller G, Gerday C (2004) A perspective on cold enzymes: current knowledge and frequently asked questions. Cell Mol Biol 50:643–655

Mayr B, Kaplan T, Lechner S, Scherer S (1996) Identification and purification of a family of dimeric major cold shock protein homologs from the psychrotrophic *Bacillus cereus* WSBC 10201. J Bacteriol 178:2916–2925

Mazur P (1966) Physical and chemical basis of injury in single-celled microorganisms subjected to freezing and thawing. In: Merman HT (ed) Cryobiology. Academic Press, New York, pp 214–315

Michel V, Lehoux I, Hebraud P (1997) The cold shock response of the psychrotrophic bacterium *Pseudomonas fragi*. Curr Microbiol 33:16–25

Mikami K, Kanesaki Y, Suzuki I, Murata N (2002) The histidine kinase Hik33 perceives osmotic stress and low-temperature stress in *Synechocystis* sp. PCC 6803. Mol Microbiol 46:905–915

Mishra PK, Bisht SC, Ruwari P, Joshi GK, Singh G, Bisht JK, Bhatt JC (2010a) Bioassociative effect of cold tolerant *Pseudomonas* spp. and *Rhizobium leguminosarum*-PR1 on iron acquisition, nutrient uptake and growth of lentil (*Lens culinaris* L.). Eur J Soil Biol 47:35–43

Mishra PK, Joshi P, Bisht SC, Bisht JK, Selvakumar G, Gupta HS (2010b) Cold-tolerant agriculturally important microorganisms. In: Maheshwari DK (ed) Plant growth and health promoting bacteria. microbiology monographs (Springer series), vol 18. Springer, Berlin, pp 273–296

Mishra PK, Bisht SC, Bisht JK, Bhatt JC (2011) Cold tolerant PGPRs as bioinoculant for stress management. In: Maheshwari DK (ed) Bacteria in agrobiology: stress management. Microbiology monographs. Springer, Berlin, pp 95–118

Mishra PK, Bisht SC, Bisht JK, Bhatt JC (2012) Cold tolerant PGPRs as bioinoculant for stress management. In: Maheshwari DK (ed) Bacteria in agrobiology: stress management. Microbiology Monographs. Springer, Berlin, pp 95–118

Miteva VI, Sheridan PP, Brenchley JE (2004) Phylogenetic and physiological diversity of microorganisms isolated from a deep Greenland glacier ice core. Appl Environ Microbiol 70:202–213

Mitta M, Fang L, Inouye M (1997) Deletion analysis of *cspA* of *Escherichia coli*: requirement of the AT-rich UP element for *cspA* transcription and the downstream box in the coding region for its cold shock induction. Mol Microbiol 26:321–335

Miyake R, Kawamoto J, Wei YL, Kitagawa M, Kato I, Kurihara T, Esaki N (2007) Construction of a low temperature protein expression system using a cold adapted bacterium, *Shewanella* sp. strain Ac10, as the host. Appl Environ Microbiol 73:4849–4856

Mohammed K, Roohi JMA, Pramod W (2012) Structural adaptation and biocatalytic prospective of microbial cold-active α-amylase. African J Microbiol Res 6:206–213

Morita RY (1975) Psychrophilic bacteria. Bacteriol Rev 39:144–167

Morita Y, Nakamori S, Takagi H (2003) L-proline accumulation and freeze tolerance in *Saccharomyces cerevisiae* are caused by a mutation in the PRO1 gene encoding gamma-glutamyl kinase. Appl Environ Microbiol 69:212–219

Murata N, Wada H (1995) Acyl lipid desaturases and their importance in the tolerance and acclimatization to cold of cyanobacteria. Biochem J 308:1–8

Muryoi N, Matsukawa K, Yamade K, Kawahara H, Obata H (2003) Purification and properties of an Ice-nucleating protein from an Ice nucleating bacterium *Pantoea ananatis* KUIN-3. J Biosci Bioeng 95:157–163

Muryoi N, Sato M, Kaneko S, Kawaahara H, Obata H, Yaish MWF, Griffth M, Glick BR (2004) Cloning and expression of afpA, a gene encoding an antifreeze protein from the Arctic plant growth-promoting rhizobacterium *Pseudomonas putida* GR12-2. J Bacteriol 186:5661–5671
Nanjo T, Kobayashi M, Yoshida Y, Kakubari Y, Yamaguchi-Shinozaki K, Shinozaki K (1999) Antisense suspension of proline degradation improves tolerance to freezing and salinity in *Arabidopsis thaliana*. FEBS Lett 461:205–210
Neuhaus K, Rappsch S, Francis KP, Scherer S (2000) Restart of exponential growth of cold-shocked *Yersinia enterocolitica* occurs after down-regulation of cspA1/A2 mRNA. J Bacteriol 182:3285–3288
Nishijyo T, Haas D, Itoh Y (2001) The Cbr A, Cbr B two- component regulatory system controls the utilization of multiple carbon and nitrogen sources in *Pseudomonas aeruginosa*. Mol Microbiol 40:917–931
Obata H, Kakinami K, Tanishita J, Hasegawa Y (1990) Identification of new Ice-nucleating bacterium and its ice nucleation properties. Agri Bio Chem 54:725–730
Olson JC, Nottingham PM (1980) Temperature in microbial ecology of foods volume 1: factors affecting life and death of microorganisms. In: International commission on microbiological specifications for foods. Academic Press, London, pp 1–37
Palmisano AC, Sullivan CW (1985) Growth, metabolism and dark survival in sea ice microalgae. In: Horner RA (ed) Sea ice biota. CRC Press, Boca Raton, FL, pp 131–146
Panasik N, Brenchley JE, Farber GK (2000) Distributions of structural features contributing to thermostability in mesophilic and thermophilic alpha/beta barrel glycosyl hydrolases. Biochim Biophys Acta 1543:189–201
Panicker G, Aislabie SD, Bej AK (2002) Cold tolerance of *Pseudomonas* sp. 30-3 isolated from oil contaminated soil. Antarctica. Polar Biol 25:5–11
Papa R, Rippa V, Sannia G, MarinGO DA (2007) An effective cold inducible expression system developed in *Pseudoalteromonas haloplanktis* TAC125. J Biotechnol 127:199–210
Parente AD, Garces JG, Guinea JS, Garcia JMA, Casaroli RM, Reina MDP, Vilaro SC (2006) Use of a glycoprotein for the treatments and Re-Epithelialisation of wounds. EP20020738185, US7022668
Parrilli E, De Vizio D, Cirulli C, Tutino ML (2008) Development of an improved *Pseudoalteromonas haloplanktis* TAC125 strain for recombinant protein secretion at low temperature. Microb Cell Fact 7:2. https://doi.org/10.1186/1475-2859-7-2
Phadtare S, Inouye M (2004) Genome-wide transcriptional analysis of the cold shock response in wild-type and cold-sensitive, quadruple-csp-deletion strains of *Escherichia coli*. J Bacteriol 186:7007–7014
Polissi A, De Laurentis W, Zangrossi S, Briani F, Loghi V, Pesole G, Deho G (2003) Changes in *Escherichia coli* transcriptome during acclimatization at low temperature. Res Microbiol 154:573–580
Ponder MA, Gilmour SJ, Bergholz PW, Mindock CA, Hollingsworth R, Thomashow MW, Tiedje JM (2005) Characterization of potential stress response in ancient Siberian permafrost psychroactive bacteria. FEMS Microbiol Ecol 53:103–115
Prakash JSS, Zorina A, Kupriyanova E, Sinetova M, Suzuki I, Murata N, Los DA (2009) DNA supercoiling regulates the stress-inducible expression of genes in *Synechocystis* sp. PCC 6803. Mol Biosyst 5:1904–1912
Prevost D, Drouin P, Laberge S, Bertrand A, Cloutier J, Levesque G (2003) Cold-adapted rhizobia for nitrogen fixation in temperate regions. Can J Bot 81:1153–1161
Price PB (2004) Life in solid ice on earth and other planetary bodies. In: Norris R, Stootman F (eds) Bioastronomy 2002: life among the stars, proceedings of IAU symposium 213. Astronomical Society of the Pacific, San Francisco, pp 363–366
Purusharth RI, Klein F, Sulthana S, Jager S, Jagannadham MV, Hackenberg EE, Ray MK, Klug G (2005) Exoribonuclease R interacts with endoribonuclease E and RNA helicase in the psychrotrophic bacterium *Pseudomonas syringae* Lz4W. J Biol Chem 280:14572–14578

Qoronfleh MW, Debouck C, Keller J (1992) Identification and characterization of novel low temperature inducible promoters of *Escherichia coli*. J Bacteriol 174:7902–7909
Ray MK (2006) Cold-stress response of low temperature adapted bacteria. In:Sreedhar AS, Srinivas UK (eds) Stress response: A molecular biology approach. Research Signpost, India. 2006. pp. 1–23
Ray MK, Seshu Kumar G, Shivaji S (1994a) Phosphorylation of Lipopolysaccharides in the Antarctic Psychrotroph *Pseudomonas syringae*: a Possible Role in Temperature Adaptation. J Bacteriol 176:4243–4243
Ray MK, Seshu Kumar G, Shivaji S (1994b) Phosphorylation of membrane proteins in response to temperature in an Antarctic *Pseudomonas syringae*. Microbiol 140:3217–3223
Ray MK, Seshu Kumar G, Shivaji S (1994c) Tyrosine phosphorylation of a cytosolic protein from the Antarctic psychrotrophic bacterium *Pseudomonas syringae*. FEMS Microbiol Lett 122:49–54
Ray MK, Seshu Kumar G, Shivaji S (1994d) Phosphorylation of lipopolysaccharides in the Antarctic psychrotroph *Pseudomonas syringae*: a possible role in temperature adaptation. J Bacteriol 176:4243–4249
Ray MK, Sitaramamma T, Ghandhi S, Shivaji S (1994e) Occurrence and expression of cspA, a cold shock gene, in Antarctic psychrotrophic bacteria. FEMS Microbiol Lett 116:55–60
Ray MK, Kumar GS, Janiyani K, Kannan K, Jagtap P, Basu M, Sivaji S (1998) Adaptation to low temperature and regulation of gene expression in Antarctic psychrotrophic bacteria. J Biosci 23:423–435
Raymond JA, Fritsen C, Shen K (2007) An ice-binding protein from an Antarctic sea ice bacterium. FEMS Microbiol Ecol 61:214–221
Regand A, Goff HD (2006) Ice recrystallization inhibition in ice cream as affected by ice structuring proteins from winter wheat grass. J Dairy Sci 89(1):49–57
Rivkina EM, Friedmann EI, McKay CP, Gilichinsky DA (2000) Metabolic activity of permafrost bacteria below the freezing point. Appl Environ Microbiol 66:3230–3233
Rossi G (1999) Biohydrometallurgical processes and temperature. In: Margesin R, Schinner F (eds) Biotechnological applications of cold-adapted organisms. Springer, Berlin, pp 291–308
Russell NJ (1990) Cold adaptation of microorganisms. Philos Trans R Soc Lond B Biol Sci 329:595–611
Russell NJ (1998) Molecular adaptations in psychrophilic bacteria: potential for biotechnological applications. Adv Biochem Eng Biotechnol 61:1–21
Russell NJ (2000) Toward a molecular understanding of cold activity of enzymes from psychrophiles. Extremophiles 4:83–90
Russell NJ (2006) Antarctic micro-organisms: coming in from the cold. Culture 27:1–7
Sakamoto O, Kitoh T, Ohura T, Ohya N, Iinuma K (2002) Novel missense mutation (R94S) in the TAZ (G4.5) gene in a Japanese patient with Barth syndrome. J Hum Genet 47:229–231
Sano F, Asakawa N, Inouye Y, Sakurai M (1999) A dual role for intracellular trehalose in the resistance of yeast cells to water stress. Cryobiology 39:80–87
Santoro N, Thiele DJ (1999) Oxidative stress responses in the yeast Saccharomyces cerevisiae. In: Hohmann S, Mager WH (eds) Yeast stress response. R.G. Landes Co., Austin TX
Selvakumar G, Gupta AD, Samaresh K, Gupta HS (2009) Cold tolerance mechanism in microorganisms and their agricultural importance. In: Arora DK, Rajendra TP, Srivastava AK (eds) Agriculturally important microorganisms, vol 1, pp 73–93
Sheridan PP, Panasik N, Coombs JM, Brenchely JE (2000) Approaches for deciphering the structural basis of low temperature enzyme activity. Biochim Biophys Acta 1543:417–433
Shima J, Sakata-Tsuda Y, Suzuki Y, Nakajima R, Watanable H, Kawamoto S, Takano H (2003) Disruption of the CARI gene encoding arginase enhances freeze tolerance of the commercial baker's yeast Saccharomyces cerevisiae. Appl Envion Microbiol 69:715–718
Shivaji S, Prakash JSS (2010) How do bacteria sense and respond to low temperature? Arch Microbiol 192:85–95

Siani L, Papa R, Di Donato A, Sannia G (2006) Recombinant expression of Toluene o-Xylene Monooxygenase (ToMO) from *Pseudomonas* stutzeri OX1 in the marine Antarctic bacterium *Pseudoalteromonas haloplanktis* TAC125. J Biotechnol 126:334–341

Skirvin RM, Kohler E, Steiner H, Ayers D, Laughnan A, Norton MA, Warmund M (2000) The use of genetically engineered bacteria to control frost on strawberries and potatoes. Whatever happened to all of that research? Sci Hortic 84:179–189

Sledjeski DD, Gupta A, Gottesman S (1996) The small RNA, DsrA, is essential for the low temperature expression of RpoS during exponential growth in *Escherichia coli.* EMBO J 15:3993–4000

Somero GN (2004) Adaptation of enzymes to temperature: searching for basic "strategies". Comp Biochem Physiol 139:321–333

Sonnefield JM, Burns CM, Higgins CF, Hinton J (2001) The nucleoid- associated protein StpA binds curved DNA, has a greater DNA-binding affinity than H-NS and is present in significant levels in *hns* mutants. Biochimie 83:243–249

Stead D, Park SF (2000) Roles of Fe superoxide dismutase and catalase in resistance of *Campylobacter coli* to freeze-thaw stress. Appl Envion Microbiol 66:3110–3112

Stoderegger K, Herndl G (1998) Production and release of bacterial capsular material and its subsequent utilization by marine bacterioplankton. Limnol Oceanol 43:877–884

Strocchi M, Ferrer M, Timmis KN, Golyshin PN (2006) Low temperature induced systems failure in *Escherichia coli*: Insights from rescue by cold-adapted chaperones. Proteomics 6:193–206

Sundareswaran VR, Singh AK, Dube S, Shivaji S (2010) Aspartate aminotransferase is involved in cold adaptation in psychrophilic *Pseudomonas syringae.* Arch Microbiol 92:663–672

Sutherland IW (1994) Structure function relationships in microbial exopolysaccharides. Biotechnol Adv 12:393–448

Suzuki I, Los DA, Kanesaki Y, Mikami Y, Murata N (2000) The pathway for perception and transduction of low-temperature signals in *Synechocystis.* EMBO J 19:1327–1334

Suzuki I, Kanasaki Y, Mikami K, Kanehisa M, Murata N (2001) Cold regulated genes under the control of cold sensor *hik33* in *Synechocystis.* Mol Microbiol 40:235–245

Suzuki Y, Haruki M, Takano K, Morikawa M, Kanaya S (2004) Possible involvement of an FKBP family member protein from a psychrotrophic bacterium *Shewanella* sp. SIB1 in cold adaptation. Eur J Biochem 271:1372–1381

Tamaru Y, Yoshida T, Sakamoto T (2005) Crucial role of extracellular polysaccharides in desiccation and freezing tolerance in the terrestrial cyanobacterium. Appl Envion Microbiol 71:7327–7333

Tanabe H, Goldstein J, Yang M, Inouye M (1992) Identification of the promoter region of the *Escherichia coli* major cold shock gene, *cspA.* J Bacteriol 174:3867–3873

Tange TQ, Shibuya T, Jurica M, Moare MJ (2003) Biochemical analysis of the EJC reveals two new factors and a stable tetrameric protein core. RNA Society 11:1869–1883

Thieringer HA, Jones PG, Inouye M (1998) Cold shock and adaptation. BioEssays 20:49–57

Thomassin-Lacroix EJ, Eriksson M, Reimer KJ, Mohn WW (2002) Biostimulation and bioaugmentation for on-site treatment of weathered diesel fuel in Arctic soil. Appl Microbiol Biotechnol 59:551–556

Trotsenko YA, Khmelenina VN (2005) Aerobic methanotrophic bacteria of cold ecosystems. FEMS Microbiol Ecol 53:15–26

Tse-Dinh YC, Qi H, Menzel H (1997) DNA supercoiling and bacterial adaptation: thermotolerance and thermoresistance. Trends Microbiol 5:323–326

Turner MA, Arellano F, Kozloff LM (1991) Components of ice nucleation structure of bacteria. J. Bacteriol 173:6515–6527

Tutino M, Duilio A, Parrilli E, Remaut E, Sannia G, Marino G (2001a) A novel replication element from an Antarctic plasmid as a tool for the expression of proteins at low temperature. Extremophiles 5:257–264

Yi T, Takani Y, Yoshida T, Sakamoto T (2005) Crucial role of extracellular polysaccharides in desiccation and freezing tolerance in the terrestrial cyanobacterium *Nostoc commune*. App Envion Microbiol 71:7327–7333

Tutino M, Duilio A, Parrilli E, Remaut E, Sannia G, Marino G (2001b) A novel replication element from an Antarctic plasmid as a tool for the expression of proteins at low temperature. Extremophiles 5:257–264

Varcamonti M, Arsenijevic S, Martirani L, Fusco D, Naclerio G, Felice MD (2006) Expression of the heat shock gene *clpL* of *Streptococcus thermophilus* is induced by both heat and cold shock. Microbial Cell Factories. https://doi.org/10.1186/1475-2859-5-6

Wada H, Murata N (1990) Temperature-induced changes in the fatty acid composition of the cyanobacterium, *Synechocystis* PCC6803. Plant Physiol 92:1062–1069

Walworth JL, Pond A, Snape I, Rayner J, Ferguson S, Harvey P (2007) Nitrogen requirements for maximizing petroleum bioremediation in a sub-Antarctic soil Cold Regions. Sci Tech 48:84–91

Wang E, Koutsioulis D, Leiros HK, Andersen OA, Bouriotis V, Hough E, Heikinheimo P (2007) Crystal structure of alkaline phosphatase from the Antarctic bacterium TAB5. J Mol Biol 366:1318–1331

Wang QF, Miao JL, Hou YH, Ding Y, You LG (2006) Expression of CspA and GST by an Antarctic psychrophilic bacterium *Colwellia* sp. NJ341 at near-freezing temperature. World J Microbiol Biotechnol 22:311–316

Wintrode PL, Miyazaki K, Arnold FH (2000) Cold-adaptation of a mesophilic subtilisin-like protease by laboratory evolution. J Biol Chem 275:31635–31640

Witter LD, Campbell MF, Azuma Y (1966) Formation of bacterial pigments at low temperature by psychrophillic pseudomonads. Dev Ind Microbiol 7:231–239

Wong PTW, McBeath JH (1999) Plant protection by cold-adapted fungi. In: Margesin R, Schinner F (eds) Biotechnological applications of cold-adapted organisms. Springer, Berlin, pp 177–190

Woodall KA, Gallagher M, Blakely G, Ferguson G (2011) Cold shock response of *Salmonella enterica* serovar *typhimurium*; the involvement of the CspA paralogues. PhD thesis, http://www.era.lib.ed.ac.uk/bitstream/1842/5652/1/Woodall2011.pdf. Accessed on 16 July 2012

Xu H, Griffith M, Patten CL, Glick BR (1998) Isolation and characterization of an antifreeze protein with ice nucleation activity from the plant growth promoting rhizobacterium *Pseudomonas putida* GR12-2. Can J Microbiol 44:64–73

Yamanaka K (1999) Cold shock response in *Escherichia coli*. J Mol Microbiol Biotechnol 34:193–202

Yamashita Y, Nakamura N, Omiya K, Nishikawa J, Kawahara H, Obata H (2002) Identification of an antifreeze lipoprotein from *Moraxella* sp. of Antarctic origin. Biosci. Biotechnol Biochem 66:239–247

Yankofsky SA, Levin Z, Bertold T, Sandlerman N (1981) Some basic characteristics of bacterial freezing nuclei. J Appl Meterol 20:1013–1019

Yoshiba Y, Kiyosue T, Nakashima K, Yamaguchi-Shinozaki K, Shinozaki K (1997) Regulation of levels of proline as an osmolyte in plants under water stress. Plant cell Physiol 38:1095–1102

Zartler ER, Jenney FE, Terrell M, Eidsness MK, Adams MW, Prestegard JH (2001) Structural basis for thermostability in aporubredoxins from *Pyrococcus furiosus* and *Clostridium pasteurianum*. Biochem 40:7279–7290

Zhang W, Shi L (2005) Distribution and evolution of multiple-step phosphorelay in prokaryotes: lateral domain recruitment involved in the formation of hybrid-type histidine kinases. Microbiol 151:2159–2173

Zhao G, Zhang G (2005) Effect of protective agents, freezing temperature, rehydration media on viability of malolactic bacteria subjected to freeze-drying. J Appl Microbiol 99:333–338

Microbe-Mediated Mitigation of Abiotic Stress in Plants

7

Maddur Puttaswamy Raghavendra

Abstract

The application of microorganisms is considered to be economically feasible in agricultural practices and moreover eco-friendly because of its natural origin. A wide range of microorganisms are already in field applications as plant growth promoting rhizobacteria (PGPR). They are of rhizosphere origin involved in plant growth promotion and recent reports reveal its application in alleviating plant associated stress, whether it is abiotic or biotic. The diversity of these organisms is huge and even mechanisms employed to mitigate the plant stress are also different and are specific to individual strains. In this chapter, abiotic stress with special reference to high salt concentrations and drought is discussed. Along with these agriculture is facing another peculiar problem of plastic pollution. The recent decade has witnessed incessant and indiscriminate use of plastics and now the microplastics are distributed everywhere and have become an insepara-ble component of the agricultural practices. This chapter hence intends to provide information on the present status of plastic pollution and the role of soil micro-flora in degrading plastics in agricultural fields.

Keywords

Stress · Microorganisms · Plant growth promotion

M. P. Raghavendra (✉)
Postgraduate Department of Microbiology, Maharani's Science College for Women, Mysuru, Karnataka, India

B. Giri, M. P. Sharma (eds.), *Plant Stress Biology*,
https://doi.org/10.1007/978-981-15-9380-2_7

7.1 Introduction

Sustainable agriculture is the main theme of the next green revolution. Scientists over the globe are planning to develop a strategy to use the natural resources instead of pouring too much of chemical fertilizers and pesticides to the agricultural fields. Hence next green revolution should boost the productivity of the plant in an environmentally sustainable manner. Mostly in the previous green revolution plant associated environment and its natural component such as microorganisms were not given much importance (Morrissey et al. 2004). Hence present-day agriculture is facing several challenges in terms of varied abiotic stress such as a change in temperature, salinity, drought, and recently added plastic pollution in agricultural fields. These stresses are the major constraints for the next green revolution and addressing them naturally is really a challenge. They alter the growth pattern, affect plant physiology, alter gene expression and breakdown of macromolecules (Fig. 7.1).

Abiotic and biotic stress is associated with plants naturally in the field conditions. The amount of stress in the agricultural field is comparatively high compared to plants grown in natural habitats. Plants cultivated in the agricultural fields has to face both natural and conventional agriculture practices induced stresses which is more human centric from their planting up to its harvest in the agricultural field. Compared to the wild plants observed in nature, crops cultivated in the agriculture field are more vulnerable to these stresses. Because the wild plants due to its continuous interaction with the prevailing extreme environment show adaptive mechanism,

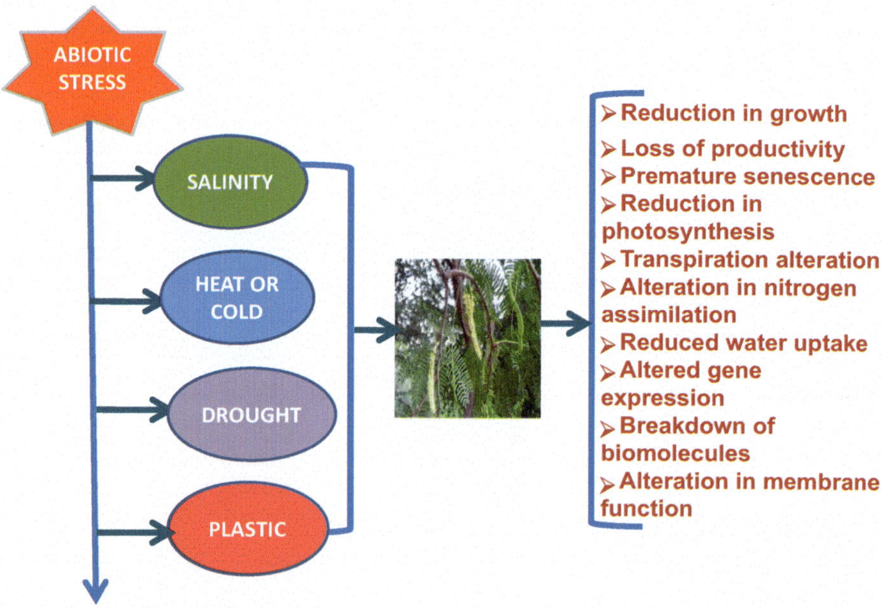

Fig. 7.1 Abiotic stresses associated with plants and its impact on its metabolism

which favors those plants even to survive and multiply. The natural environments tend to pose a given set of stress in particular environment, whereas the plants grown in conventional agricultural systems have to face stress which is specific to agricultural practices and availability of the natural resources in a continuous cropping system.

The list of stresses associated with growing plants is increasing day by day. To overcome these stresses understanding genetics and applying different breeding techniques were widely exploited to obtain stress tolerant varieties. In addition, recent advances in biotechnological and genome sequencing applications are providing a wider platform to develop genetically modified organisms in general and plants in particular. There is a hue and cry on the use of genetically modified plants in several parts of the world especially India and even methods employed questions its suitability and stability in nature. Hence microorganisms being ubiquitous and potentially being gifted with an array of genes which are under simple regulatory systems seem to be one of the best options to induce resistance to these stresses naturally in association with the genes which are inherently associated with stress in plants (Sirari et al. 2016).

Environmental stress and adapting to these changes are not new to plants. They are naturally gifted with a multitude of stress related genes. These genes in response to abiotic or biotic stresses, at the molecular level alter gene expression leading to the synthesis of various proteins or enzymes required to overcome the specific stress (Bagati et al. 2018). Thermal stress is considered to be a major challenge these days with the change in environmental conditions due to global warming which directly relates to the availability of water and increase in heat stress (Godfray et al. 2010; Ferguson 2019). It is reported to directly hamper the productivity of several valuable crops and even its ecological fitness. Along with this high salt concentration is known to hamper several processes of plants at different stages of its growth and recent studies also indicate that plastic pollution is also playing a vital role in agricultural production. To overcome all these problems and to facilitate plant adaptation along with its inherent capacity, microbial inoculants are also reported to support the plants to alleviate the effects of these stresses significantly (Orhan 2016).

Arthrobacter, *Achromobacter*, *Azospirillum*, *Azotobacter*, *Burkholderia*, *Bacillus*, *Enterobacter*, *Paenibacillus*, *Klebsiella*, *Pantoea*, *Microbacterium*, *Streptomyces*, *Pseudomonas* and *Serratia* are the common candidate microorganisms of plant growth promoting rhizobacteria isolated from rhizosphere. The use of these bacteria as biofertilizers, biopesticides, and biostimulants is already proven valuable in the agriculture sector. Numana et al. (2018) are of the opinion that these PGPR can also be exploited as cheap and easily available sources to mitigate the stress associated with plants. It is involved in stress alleviation directly or indirectly by triggering the production of additional growth hormones, supporting through the acquisition of the limiting factors by producing special proteins such as siderophores, providing protection from free radicals through the activation of antioxidant enzymes, and improving plant system through overall nutrition balance (Kumar and Verma 2017).

7.2 Salt Stress Tolerance

Several countries all over the world still depend on agriculture for its economic growth and even per capita income depends on agriculture produce. In such situation, several issues are bothering agriculture which hampers the economy of the nation. One such bothering issue in agriculture is the soil salinity. It is considered as a major threat to agriculture and Ashraf (1994) and Vinocur and Altman (2005) report that by 2050, 50% of all agriculture soils will be affected by increased salinity.

The stress induced in the soil due to the presence of different salt ions soluble in water in excess creates salt stress. It may be because of carbonate (CO_3^{2-}), bicarbonate (HCO_3^-), calcium (Ca^{2+}), sodium (Na^+), magnesium (Mg^{2+}), chloride (Cl^-), sulfate (SO_4^{2-}), and potassium (K^+) ions which leads to salinization. Even among these, sodium chloride is considered to be a major contributor to salinity and the chloride ions associated with it are toxic to many living organisms including plants. Accumulation of these ions may also be a result of poor drainage of these ions from the agricultural field through irrigation. They finally enter plant cells and lead to bioaccumulation. High amount of these ions are detrimental for plant growth (Liu and Zhu 1998; Zhu et al. 2005; Hasegawa 2013).

The vigor of the plant depends on several biochemical and physiological and molecular approaches of the plant to the environment. High salt concentration is known to disturb all these processes at various stages of its growth. The major parameters altered during high salt concentration are membrane integrity, protein synthesis, stomatal function, glandular trichome density, mineral nutrition, volatile exudation, rate of photosynthesis, antioxidant capacity, and metabolism of lipids and carbon (Parida and Das 2005; Zhou et al. 2018).

In an attempt to create a natural soil management strategy in sustainable agriculture, it is imperative to search for new strains of microorganisms which involve along with the plants to alleviate the salinity induced stress, finally leading to plant growth promotion and yield (Grover et al. 2011; Singh et al. 2011). To prove the efficiency of microorganisms in the field, several researchers isolated the microorganisms from the saline environment and attempted to prove their ability in plant growth promotion. Goswamia et al. (2014) reported the isolation of 85 isolates from the rhizosphere of the halotolerant plant *Suaeda fruticosa* located in the saline desert of the Little Rann of Kutch, Gujarat, India. Along with its ability to survive and support plant growth in a saline environment, these isolates were also screened for their additional ability to solubilize phosphate and production of indole acetic acid (IAA). Among 85 isolates, 23 were found phosphate solubilizers, 11 were reported to produce IAA, and 7 isolates were found to have both the activities. The research also revealed the growth promoting activity of *Bacillus licheniformis* even in the soil supplemented with 50 mM NaCl concentration used for testing the treated plants. Highly promising result on significant increase in fresh biomass, root length, and total length compared to control was also observed. These parameters of growth enhancement are found even more significant in treated plants compared to control in absence of sodium chloride.

In another experiment conducted with plants *Hordeum secalinum* and *Plantago winteri* which grow in natural salt meadow, Cardinale et al. (2015) isolated plant growth promoting *Curtobacterium flaccumfaciens* from the rhizosphere of these plants. They observed significant growth promoting activity to the tune of 300% compared to control in barley seeds treated with isolated strain. It was shown to possess nitrogen fixing ability, auxin synthesis, ACC deaminase activity, and calcium and phosphate mobilization required for plant growth promotion.

Another interesting conclusion these scientists claim is while isolating the strains with a specific growth promoting activities mentioned earlier, we may land up in only a few strains with these abilities. Hence, they reported that the screening of organisms in pure culture may not be suitable with selected criteria of plant growth promotions, in turn it is important to screen them thoroughly in consortia and it will lead to understanding of indirect approach these strains use to promote plant growth. Further these mechanisms may be novel and will add more insights into our understanding on plant–microbe interaction in depth.

Electron transport chain in mitochondria and chloroplast is severely affected due to the accumulation of these salt ions. Since two components required for the development of superoxides that is molecular oxygen and electrons both are involved in the electron transport chain, there is more possibility of generation of free radicals due to its abnormal regulation of electron flow. Oxygen in presence of electrons as an electron acceptor generates reactive oxygen species (ROS). This in turn facilitates the formation of other free radicals such as hydroxyl ions, peroxynitrile ions, and hydrogen peroxide, which are strong oxidizing agents and affect plant metabolism severely even compared to ROS (Grob et al. 2013). To overcome this problem plant responds in terms of several enzymes such as glutathione reductase, glutathione and ascorbate peroxidase, superoxide dismutase, and proline catalase. These enzymes involve in scavenging free radicals (Asada 1999; Noreen et al. 2010). Sharma and Sharma (2017) reported the involvement of microbes in scavenging these radicals. They observed an increase in phenol content and also other defensive enzymes in the plants treated with PGPR in response to high salt and drought. Another interesting observation reported by these scientists is the parallel increase in these defensive systems with the addition of silicon compared to control.

The above observation is in line with the need of metal ions as a cofactor for the activation of several enzymes. Different types of superoxide dismutases (SOD) are available in plant systems; they are Mn-SOD, Fe-SOD, and Cu/Zn-SOD (Alscher et al. 2002). The metal ions mentioned along with SOD need to be supplied to make the enzyme active. This indirectly connects the rhizosphere microflora with the stress responses. There are several microorganisms involved in mineralizing the metals ions and supporting its uptake by the plant system through ascent of sap. It means if Mn, Fe, Cu, and Zn are not supplied in sufficient quantity plants fail to respond to stresses due to the nonavailability of active defensive enzymes. Conversely, PGPR are reported to improve physiology of plants and antioxidant potential in providing nutrients which are not capable of direct assimilation by plants from the environment. They are also involved in increasing the activity of several hormones such as ethylene, pyrroloquinoline quinone, cytokinins, abscisic acid, indole acetic acid, and gibberellins (Perrig et al. 2007). Out of these hormones IAA was reported to play important role in resistance against salt along with its plant growth promotion (Ali and Abbas 2003; Kaya et al. 2013; Khalid et al. 2013; Kang et al. 2014). PGPR therefore through augmentation of this hormone favor plant growth instead of uptake of the ions responsible for high salinity (Zhang et al. 2008). Along with this Alavi et al. (2013) reported another molecule spermidine which serves as a protector molecule in saline stress.

ACC deaminase serves as another key factor involved in supporting the plants in high salt concentrations. Several rhizospheres inhabiting pseudomonads are known to have the ability to produce this enzyme along with *Bacillus* sp. and *Brevibacterium* sp. (Siddikee et al. 2010, 2011), *Burkholderia* sp. (Shaharoona et al. 2007), and *Achromobacter* sp. (Mayak et al. 2004; Karthikeyan et al. 2012). Along with this several root colonizers revealed promising results in lettuce (Kohler et al. 2009), cucumber (Egamberdieva et al. 2011), sunflowers (Shilev et al. 2012), rice (Jha et al. 2011), mung bean (Ahmad et al. 2011), wheat (Zahir et al. 2009; Nadeem et al. 2010), and tomato (Tank and Saraf 2010).

The list of microorganisms associated with salt tolerance directly or indirectly with plants is many. The reports reveal that the diversity of these organisms varies with the plant species and the soil or rhizosphere environment. Different organisms are associated with different plants and the possible mechanism associated with salt tolerance is mentioned in Table 7.1.

7.3 Drought Resistance

Growing demand for safe drinking water due to overexplosion of population is creating water scarcity for human consumption itself. Agricultural practices and even the varieties grown in agricultural fields are reported to be resource consuming including demand for more water supplies. Along with these there is a drastic change in the environmental patterns all over the world. Global warming is taking the front seat in making agricultural systems failure in connection with crop loss due to photoperiodism or circadian rhythm.

Lack of water is known to induce stomatal closure through which it directly impacts the rate of photosynthesis and the ability of the plants to acquire nutrients. Several alterations in response to high salt concentration in plants hold good for altered temperature. It is reported that the hormonal variation and production of reactive oxygen species due to drought condition are the main reason for plant suffering. In this situation, PGPR are also known to provide additional protective measures such as water retention and soil aggregation near the root zone along with this PGPR is known to promote the production of enzymes such as 1-aminocyclopropane-1-carboxylic acid (ACC) deaminase, phosphatase to help in the assimilation of more phosphorus, siderophore production, and also growth hormones such as indole-3-acetic acid. ACC deaminase serves as a best candidate to reduce the impact of high salt concentration. It is reported that it removes the

Table 7.1 Microorganisms associated with salt stress tolerance in plants

Microorganisms	Source and crop plant	Mechanism/Effect	References
Klebsiella sp.	Wheat rhizosphere soil/*Avena sativa*	Increased root and shoot length, shoot and root dry weight, and relative water content	Sapre et al. (2018)
Thalassobacillus spp., *Bacillus* sp., *Halomonas* sp., *Oceanobacillus* sp., *Zhihengliuella* sp., and *Staph. succinus*	Isolated from salt-affected soils of the East Anatolian region	Plant growth in a hydroponic culture *Triticum aestivum* with varied sodium chloride concentration *by* ammonia, indole-3-acetic acid, and 1-aminocyclopropane-1-carboxylate-deaminase production, phosphate solubilization, and nitrogen fixation activities	Orhan (2016)
Bacillus amyloliquifaciens	Rhizosphere	Biofilm formation in H. vulgare L.	Kasim et al. (2016)
Achromobacter piechaudii	*Lycopersicum esculentum* Mill. (Tomato)	Plant growth promotion and reduction in ethylene production	Mayak et al. (2004)
Azospirillum	*Zea mays* L. (Maize)	Increase in proline production, nitrogenase and nitrate reductase activity along with regulation of Na^+, K^+, and Ca^{2+} uptake	Hamdia et al. (2004)
Aeromonas hydrophila, *Bacillus* sp., and *Bacillus insolitus*	*Triticum aestivum* L. (Wheat)	Production of exopolysaccharide	Ashraf et al. (2004)
Enterobacter aerogenes, *Pseudomonas fluorescens*, and *Ps. syringae*	*Zea mays* L. (Maize)	Increased ACC deaminase activity	Nadeem et al. (2010)
Ps. fluorescens	*Arachis hypogea* L. (Groundnut)	Increased ACC deaminase activity	Saravanakumar and Samiyappan (2007)
Bacillus subtilis	*Arabidopsis thaliana* (L.) Heynh.	Regulation of tissue specific HKT1 transporter	Zhang et al. (2008)
Ps. mendocina	*Lactuca sativa* L. (Lettuce)	Increased uptake of essential nutrients and ACC deaminase activity	Kohler et al. (2009)
Rhizobium sp. and *Pseudomonas* sp.	Maize	Regulation of water content and electrolyte	Bano and Fatima (2009)

(continued)

Table 7.1 (continued)

Microorganisms	Source and crop plant	Mechanism/Effect	References
		with increased production of proline and uptake of K^+ ions	
Bacillus pumilus and *Ps. pseudoalcaligenes*	*Oryza sativa* L. (Rice)	More production of compatible solute glycine betaine	Jha et al. (2011)
Ps. putida	Cotton	Regulation of absorption of different ions such as Mg^{2+}, K^+, Ca^{2+}, and Na^{2+}	Yao et al. (2010)
Rhizobium phaseoli strains M1, M6, and M9, *Ps. fluorescens* Mk25, and biotype G strains, *Ps. syringae*	Mung bean	Increase in nodulation ability and growth along with ACC deaminase activity	Ahmad et al. (2011)
Raoultella planticola	Cotton	Increase in ACC deaminase activity	Wu et al. (2012)
Haererohalobacter, *Brevibacterium casei*, and *Brachybacterium saurashtrense*	*Arachis hypogea* L. (Groundnut)	Maintenance of high K^+/Na^+ ratio along with Ca^{2+}	Shukla et al. (2012)
Ps. fluorescens Mk25, Mk20 and biotype G strains, *Ps. syringae*, *Rhizobium phaseoli*	*Vigna radiata* L. (Mung bean)	Increase in water maintenance and ACC deaminase	Ahmad et al. (2012)
Pseudomonas and *Rhizobium*	*Vigna radiata* L.	Increased production of ACC deaminase and IAA	Ahmad et al. (2013)
Ps. fluorescens, *Ps. putida*, *Ent. cloacae*, and *Serratia ficaria*	Wheat	Increase in germination, rate, index with improved nutrient uptake	Nadeem et al. (2013)
Bacillus pumilus and *Ps. pseudoalcaligenes*	Salt sensitive rice variety GJ-17	Reduction in superoxide dismutase and lipid peroxidation activity	Jha and Subramanian (2014)
Pseudomonas spp. and *Acinetobacter* spp.	Oats and barley	ACC deaminase and IAA production	Chang et al. (2014)
Streptomyces sp.	Tomato	Phosphate solubilization, IAA, and ACC deaminase activity	Palaniyandi et al. (2014)
Piriformospora indica	Barley	Increased antioxidant activity	Waller et al. (2005)
Ps. syringae, *Bacillus insolitus*, *B. amyloliquifaciens*, *Microbacterium* sp.,	Wheat	Regulating Na^+ influx	Ashraf et al. (2004)

(continued)

Table 7.1 (continued)

Microorganisms	Source and crop plant	Mechanism/Effect	References
Acinetobacter sp., *Ps. putida*, *Curtobacterium* sp.	*Sulla carnosa*	Increased antioxidant enzyme activity, total soluble sugars, chlorophyll leaf content, dry biomass, and photosynthesis	Hmaeid et al. (2019)
Bradyrhizobium sp. and *Paenibacillus graminis*	Cowpea	Increased superoxide dismutase, catalase and phenol peroxidase activities, and lipid peroxidation	Santos et al. (2018)
Streptomyces sp. and *Bacillus* sp.	*Mesembryanthemum crystallinum* L. (Common ice plant)	Production of siderophore, indole acetic acid, and 1-aminocyclopropane-1-carboxylate deaminase along with increased root growth	Mahmood et al. (2019)
Ps. chlororaphis and *Ps. extremorientalis*	Tomato	Enhanced antioxidant enzymes and proline production in plants along with control of foot and root rot of tomatoes caused by *Fusarium solani*	Egamberdieva et al. (2017)

precursor required for ethylene, which is generally involved in inducing the plant response to several abiotic stresses (Dimkpa et al. 2009; Yang et al. 2009).

Along with these the rhizosphere associated microorganisms such as *Aeromonas* sp., *Serratia* sp., and *Bacillus* sp. are known to produce exopolysaccharides (Ashraf et al. 2004), volatile compounds, antioxidants, accumulation of osmolytes, regulation of genes associated with drought stress and very importantly alters root morphology to promote more water acquisition from the soil. These alterations help the plant to develop induced systemic tolerance and even exopolysaccharide produced by microorganisms is a component of drought stress tolerance (Vurukonda et al. 2016).

The list of microorganisms associated with drought tolerance of plants is listed in Table 7.2 and it also provides brief information on the mechanisms favored in the presence of specific microorganisms against drought condition.

7.4 Plastic Pollution

Plastic is a growing monster and its impact on living world is not yet completely documented and so is the case with agriculture. Especially microplastics distributed all over the world are a hidden problem need to be evaluated in detail on its health

Table 7.2 Microorganisms involved in plant drought stress tolerance

Microorganisms	Source and crop plant	Mechanism/Effect	References
Pantoea agglomerans	Wheat	Aggregation of rhizosphere soil by EPS production	Amellal et al. 1998
Paenibacillus polymyxa	*Arabidopsis*	Activation of gene ERD15 responsible for stress resistance	Timmusk and Wagner (1999)
Rhizobium sp.	Sunflower	Aggregation of rhizosphere soil by EPS production	Alami et al. (2000)
Azospirillum sp.	Wheat	Regulating water relation	Creus et al. (2004)
Achromobacter piechaudii	Tomato	ACC deaminase activity	Mayak et al. (2004)
Variovorax paradoxus	Pea	ACC deaminase activity	Dodd et al. (2005)
VAM fungi	Sorghum	Regulating water relation	Cho et al. (2006)
Paraphaeosphaeria quadriseptata	Arabidopsis	HSP induction	McLellan et al. (2007)
Rhizobium tropici and *Ps. polymyxa*	Common bean	Regulation of stomatal conductance and hormones	Figueiredo et al. (2008)
Pseudomonas sp.	Pea	Reduced ethylene production	Arshad et al. (2008)
Glomus intraradices and *Ps. mendocina*	Lettuce	Increase in antioxidants	Kohler et al. (2008)
Ps. putida P45	Sunflower	Aggregation of rhizosphere soil by EPS production	Sandhya et al. (2009)
Azotobacter chroococcum, *Azotobacter salinestris*	Zea mays	Increase in phosphate and potassium solubilization associated with increased shoot dry weight, plant height, chlorophyll content	Shirinbayan et al. (2019)
Azospirillum brasilense	Maize	Increase in biomass and water content	Casanovas et al. (2002)
Ps. koreensis	*Helianthus annuus*	Increased leaf area, water content, and IAA production by bacteria	Macleod et al. (2015)

hazard in animal system and also plants especially in agriculture. Microplastics are the small plastic particles with 1 mm to 5 mm size. These are created by the natural degeneration of several plastics such as tires, domestic goods, packing materials, materials coated with plastics, and other plastic household things used on daily basis. Hence it is observed more in urban environment especially in urban run-off water finally ending up in treatment plants, river, and the final sink ocean. Marine plastics refer to pollution of marine ecosystem with microplastics (Carpenter and Smith 1972).

Microplastics which travel this way through water will definitely reach agricultural fields and less information is available on soil as a microplastic sink (Blasing and Amelung 2018). This in turn pollutes the terrestrial ecosystem. Even today the scientific methodology to evaluate the load of microplastics in soil in general and agricultural fields in particular has not been standardized. Since quantification methodology itself is not standardized, assessing its actual impact on the ecosystem is therefore still at large. Even though few techniques are available they are time consuming and data from different agricultural fields is seldom collected (Corrandini et al. 2019). There are few reports available in recent times to evaluate the effect of components of plastic on soil microflora or accumulation of microplastics in soil conditions.

7.4.1 Microplastics in the Soil Environment

Terrestrial environment is known to receive microplastics from several sources. The main contributor is the human activities associated with environmental sources which are in circulation (Machado et al. 2018a). Atmospheric precipitation (Dris et al. 2016), circulating natural water (Nizzetto et al. 2016), plastic mulches (Ng et al. 2018; Zhou et al. 2020; Huang et al. 2020), and even the compost obtained from the domestic and industrial solid and liquid waste used as an agricultural amendment (Weithmann et al. 2018) contribute to plastic pollution in soil. Scheurer and Bigalke (2018) reported the presence of microplastics even in mountainous and inhabited area up to 0.002% of soil dry weight. After reaching the soil, the process of bioturbation provides microplastics an opportunity to get into the soil matrix (Huerta Lwanga et al. 2017; Rillig et al. 2017; Maass et al. 2017) in association with soil management practices (Steinmetz et al. 2016) and percolating water (Zubris and Richards 2005) and other agriculture field related activities. Once it is done it is not known scientifically what happens to these plastics (Machado et al. 2018b) and hence in this connection plastic degrading rhizosphere microflora need to be evaluated. If it is not effectively degraded it leads to accumulation in the environment and present data reveals that already 7% of microplastic weight is present in the soil (Fuller and Gautam 2016).

Every year the amount of plastic added to soil is constantly increasing leading to more accumulation, even traditional tillage practices increase its distribution. As the amount of plastic is increasing even different size plastics such as macro and mega are added to soil along with microplastic to the agricultural soils (Changrong et al. 2014; Liu et al. 2014; Rillig et al. 2017a; Steinmetz et al. 2016).

7.4.2 Quantification of Microplastic in Soil

Few techniques available to evaluate the microplastics in soil include visual sorting of the plastics (Lots et al. 2017), use of portable spectroradiometer with range 350–2500 nm, Raman and Fourier-transform infrared spectroscopy (FT-IR) (Crawford and Quinn 2017), pyrolysis-gas chromatography-mass spectrometry (Pyr-GC-MS) (Ziajahromi et al. 2017), and thermal desorption GC-MS (Dumichen et al. 2015, 2017). Even though visual sorting is considered as cheap and simpler option to evaluate plastic pollution in soil (Ziajahromi et al. 2017), advanced techniques provide more accurate results. But they are comparatively costly and demand collection of soil from different conditions and carrying them safely to the laboratory for accurate evaluation of the results is also important.

7.4.3 Impact of Plastics on Soil Environment

In recent times, as mentioned earlier effects of microplastics on terrestrial ecosystems are moving into focus from the aquatic ecosystem. Once it reaches the soil, it will impact physical properties of the soil such as structure and texture and also its associated microbial components. Once it is altered, definitely it will impact on the plant growth, whether in good or bad way. Hence Rillig et al. (2019) are of the opinion that there is a necessity to develop mechanistic pathways through which impact of microplastics on plant growth whether positively or negatively can be evaluated. They are of the opinion that the effect will definitely depend on the plant species and the type of plastics.

The research on microplastic pollution in agriculture fields begun mostly from the last few years and it was related to its impact on soil biota (Huerta Lwanga et al. 2016, 2017; Maass et al. 2017; Rillig et al. 2017). Availability of the standard methods in the beginning to quantify the microplastics might have hindered the thorough research (Machado et al. 2018a), but at present few reliable methods are available and hence the research related to this is gaining momentum.

The current research on microplastics highlights this as a physical contaminant of the soil (Machado et al. 2018a) and initial reports revealed that the microfibrils lead to lowered bulk density (Machado et al. 2018b). Earlier this was reported to favor the increased plant root penetration and better soil aeration leading to increased plant growth (Zimmermann and Kardos 1961). But recent findings by Wan et al. (2019) reveal that microplastics will create a channel through which water movement is favored leading to increased loss of water through evaporation, which in turn lead to soil drying. It is a negative consequence; hence, these two reports highlight the need for thorough research on possible impact of plastics on soil environment directly and plant growth indirectly. Rillig et al. (2019) highlighted another impact, if it alters the soil structure, then naturally the microbial flora has to alter, it will also impact microbial interaction with plant roots, its positive influence on plant growth and this shift may alter mycorrhizal association and even nitrogen fixers.

In this connection Machado et al. (2018b) is of the opinion that changes in overall structure of the soil will have significant effect on soil aggregation. If the microfibers facilitate more soil aggregation which in turn favors significant soil aeration and hydration, then it will have a positive influence on root growth. If by chance the soil aggregate fails to accommodate air and water in an altered cycle, then it will definitely impact negatively for the growth of the plant.

It is a proven fact that the soil microflora is important component of the plant performance in the field (Wagg et al. 2014; Powell and Rillig 2018). If soil structure is altered due to microplastics accumulation it will definitely impact microbial communities of the soil (Lowery and Ursell 2019). If it gets altered, biogeochemical cycle connected with mineralization of several trace elements required for plant growth will also alter. It is Feng et al. (2013) who proved that the nanoplastics will impact as a toxic physical component in mycorrhizal functioning.

There are several reports to prove the impact of microplastics on soil biodiversity and function. Machado et al. (2018a, 2018b) conducted a thorough research for 5-weeks in a garden experiment where in loamy sand soil is exposed to polyethylene fragments, polyacrylic fibers, polyester fibers, and polyamide beads up to 2% concentration. After the exposure the bulk density, hydraulic conductivity, water holding capacity, soil aggregation, and overall microbial activity were recorded. They observed that microplastic affected the bulk density, water holding capacity and even they observed the alteration in functional relationship between water stable aggregates and microbial activity. They have a final saying to conclude that these changes in due course of time will alter the terrestrial ecosystems homeostasis and plastic will serve as a long term anthropogenic stressors.

Even though plastic pollution is increasing day by day, its impact on terrestrial ecosystem is not completely worked out. Earthworms are used as a test organism to assess the plastic pollution in soil and its adverse effect on soil fauna, similar extended study on other organisms including human beings, invertebrates, plant associated microflora, and insects needs to be thoroughly reviewed. This will provide the information on overall effect of plastic in agriculture related ecosystem (Chae and An 2018).

7.5 Plastic Degrading Soil Microflora

Microorganisms show inherent capacity to adapt to the environment they come across during its multiplication. Since they are naturally exposed to the diverse environmental conditions it is obvious that they should possess mechanism to use one or the other carbon or energy source available in such situations. Due to this they have the potential to transform variety of raw materials and if it is continuously exposed to plastic rich environment, they do adjust its physiology to make use of these plastics as source of nutrition through complete mineralization or partial degradation. Hence in connection with this, numerous scientific studies are carried out to isolate microorganisms associated with plastic polluted environment especially soil and water. Microbial flora associated with utilization of plastic polymers and its degradation remains as a better alternative to clean up the plastic polluted agricultural soil.

Research carried out on biodegradation associated plastic clean up using microorganisms (Premraj and Doble 2005; Raaman et al. 2012; Alshehrei 2017; Ru et al. 2020), it is observed that in the beginning microbes adhere to the polymer surface leading to its further colonization. After successful colonization microbes are known to produce array of enzymes which are generally extracellular, these enzymes lead to hydrolytic breakdown of the surface. Dang et al. (2018) reported that *Bacillus* sp. produced several hydrolytic enzymes such as CMCase, lipase, chitinase,

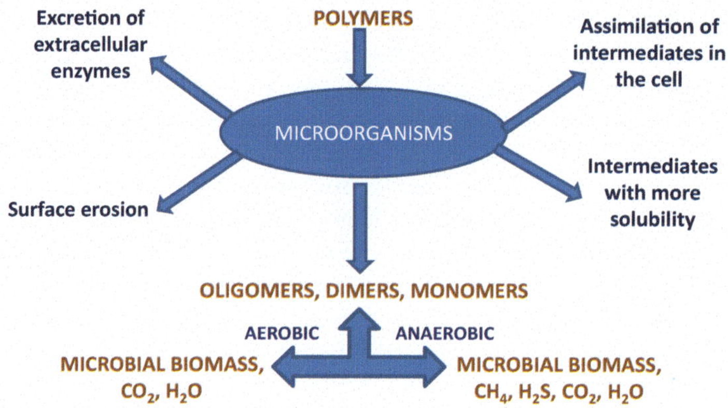

Fig. 7.2 Role of microorganisms in polymer degradation

xylanase, and protease at different levels during degradation of plastic bags, while polyhydroxyalkanoate (PHA) degrading microorganisms are known to produce PHA depolymerases (Jendrossek et al. 1996; Jendrossek 2001).

After successful enzyme action the polymers are degraded to low weight polymers, volatile compounds (Lwanga et al. 2018), water soluble products which may be used as energy or carbon source (Jendrossek et al. 1996; Jendrossek 2001) and these water soluble low weight polymers are further completely mineralized into carbon dioxide and water (Tokiwa et al. 2009). Conversely, these polymer particles enter the cell where they are exposed to cellular enzymes which provide further connect with the pathways involved in its further degradation (Gewert et al. 2015). The role of microorganisms in degradation of plastic in general and polymers in particular is represented in Fig. 7.2.

Understanding the pathways connected with the plastic degradation and their regulations needs to be completely evaluated in future. In this connection use of nuclear magnetic resonance and other spectral analysis will provide more insights into the intermediate compounds or polymers obtained during degradation process. At present regular screening of the plastic or its associated polymer degrading microorganisms is gaining momentum in present-day research.

Identifying a potential microbe having the ability to degrade plastics will help in developing a natural bioremediation strategy to clean up the ecosystem in general and agricultural soil in particular without much adverse effects on the ecological processes (Bharadwaj et al. 2012). According to Mahdiyah and Mukti (2013), bacteria, fungi, and actinomycetes belonging to 90 different genera were reported to date and at present the list is increasing (Ru et al. 2020). These microbes are generally isolated from the plastic or its polymer contaminated sites, soil dumping areas and to have microbes for plastic degrading bacteria in agricultural fields, it should be from rhizosphere soil (Kale et al. 2015). Presently they are observed from different families including pseudonocardiaceae, thermonosporaceae, micromonosporaceae, streptomycetaceae, and streptosporangiaceae. Table 7.3

Table 7.3 Soil microflora associated with partial or complete degradation of plastic

Microorganisms	Source	Component of plastic tested	References
Arthrobacter sp., *E. coli*, *Micrococcus* sp., and *Pseudomonas* sp.	Soil bacteria	Polystyrene film	Lyklema et al. (1989)
Streptomyces strains, *Mucor rouxii*, and *Aspergillus flavus*	Soil microflora	Polyethylene bags	El-Shafei et al. (1998)
Vibrio and *Pseudomonas* sp.	Marine bacteria	Polypropylene	Flemming (1998)
Aspergillus sp. and *Streptomyces* strains	Soil microflora	Poly (3-hydroxybutyrate-co-valerate)	Tansengco and Tokiwa (1998)
White rot fungi	Plant	PVC	Kirbaş et al. (1999)
White rot fungi	Plant	PVC	Kirbaş et al. (1999)
Penicillium simplicissimum	Soil	Polyethylene	Yamada-Onodera et al. (2001)
Rhizopus arrhizus	Soil	Polypropylene	Ramis et al. (2004)
Brevibacillus borstelensis and *Rhodococcus ruber*		Polyethylene	Hadad et al. (2005)
Pseudomonas sp., *Vibrio* sp., and *Aspergillus niger*	Marine bacteria	Polypropylene	Arutchelvi et al. (2008)
Actinobacteria spp.		Low molecular weight PE oligomers	Ghosh et al. (2013a, 2013b)
Kocuria palustris, *Bacillus pumilus*, and *Bacillus subtilis*	Marine bacteria	PE	Harshvardhan and Jha (2013)
Bacillus cereus		PE	Sowmya et al. (2014)
Gut microflora	Gut of the larvae *Plodia interpunctella* (meal moth)	Polyethylene	Yang et al. (2014)
Pseudomonas stutzeri, *Alcaligenes faecalis*, *Pseudomonas putida*, *Streptomyces* sp., *Brevibacillus borstelensis*, and *Staphylococcus* sp.		Polymers	Ghosh et al. (2013a, b); Caruso (2015)
Pseudomonas alcaligenes and *Desulfotomaculum nigrificans*	Plastic contaminated soil	Polyethylene bag	Begum et al. (2015)
Bacillus subtilis		Polyethylene films of thickness 18 μ	Vimala and Mathew (2016)

(continued)

Table 7.3 (continued)

Microorganisms	Source	Component of plastic tested	References
Pseudomonas PL-01 and *Bacillus* PL-01	Indigenous mangrove soil bacteria	Plastic	Shovitri et al. (2016a, 2016b)
Ideonella sakaiensis		PET	Yoshida et al. (2016)
Pseudomonas sp. and *Bacillus* sp.		PS	Mohan et al. (2016)
Bacillus cereus and *Sporosarcina globispora*	Mangrove sediments in Peninsular Malaysia	Polypropylene	Helen et al. (2017)
Bacillus cereus and *Bacillus gottheilii*	Mangrove sediments in Peninsular Malaysia	Polyethylene (PE), polyethylene terephthalate (PET), polypropylene, and polystyrene (PS)	Auta et al. (2017)
Zalerion maritimum		PE	Paco et al. (2017)
Actinobacteria and Firmicutes	Gut of earthworm *Lumbricus terrestris*	Low density polyethylene (LDPE)	Lwanga et al. (2018)
Pseudomonas spp.	Earthworm	LDPE-microplastic	Kyaw et al. (2012); Lwanga et al. (2018)
Bacillus sp.	Compost	Plastic bags	Dang et al. (2018);
Thermophilic microbes		Plastic	Bharadwaj et al. (2012); Bhardwaj et al. (2013); Restrepo-Flórez et al. (2014); Kale et al. (2015); Skariyachan et al. (2016); Raziyafathima et al. (2016); Pathak and Navneet (2017); Ganesh et al. (2017)
Bacillus sp.	Soil bacteria	Polyethylene terephthalate synthetic plastics	Rustinia et al. (2020)

provides brief information on microorganisms associated with the degradation of plastics or its polymers.

7.6 Conclusions and Outlook

Disasters are increasing every day impacting environment directly and human and other living world indirectly. Agriculture being a main source of food supply and also source of economy is in transition phase. It is becoming inevitable in agriculture that newer technologies or the ecofriendly approaches need to be implemented to make it a viable system for food production. Microorganisms being the connecting link between bio and geo need to be evaluated regularly from different environmental conditions so as to identify a potential candidate which will take care of the stress problems associated with plants. In future, it may become compulsory to develop bioinoculants associated with fertility enrichment of the soil along with plastic degrading microorganisms as a soil amendment.

References

Ahmad M, Zahir ZA, Asghar HN, Asghar M (2011) Inducing salt tolerance in mung bean through coinoculation with rhiobia and plant growth promoting rhizobacteria containing 1-aminocycloepropane-1-carboxylate deaminase. Can J Microbiol 57:578–589

Ahmad M, Zahir ZA, Asghar HN, Asghar M (2012) The combined application of rhizobial strains and plant growth promoting rhizobacteria improves growth and productivity of mung bean (*Vigna radiata* L.) under salt-stressed conditions. Ann Microbiol 62:1321–1330

Ahmad M, Zahir ZA, Farheen N, Fareeha A, Arshad M, Khalid M (2013) Effectiveness of halotolerant, auxin producing *Pseudomonas* and *Rhizobium* strains to improve osmotic stress tolerance in mung bean (*Vigna radiata* L.). Braz J Microbiol 44:1341–1348

Alami Y, Achouak W, Marol C, Heulin T (2000) Rhizosphere soil aggregation and plant growth promotion of sunflowers by exopolysaccharide producing *Rhizobium* sp. strain isolated from sunflower roots. Appl Environ Microbiol 66:3393–3398

Alavi P, Starcher M, Zachow C, Muller H, Berg G (2013) Root-microbe systems: the effect and mode of interaction of stress protecting agent (SPA) *Stenotrophomonas rhizophila* DSM14405T. Front Plant Sci 4:141

Ali R, Abbas H (2003) Response of salt stressed barley seedlings to phenylurea. Plant Soil Environ 49:158–162

Alscher RG, Erturk N, heath LS (2002) Role of superoxide dismutases (SODs) in controlling oxidative stress in plants. J Exp Bot 53:1331–1341

Alshehrei F (2017) Biodegradation of synthetic and natural plastic by microorganisms. J Appl Environ Microbiol 5:8–19

Amellal N, Burtin G, Bartoli F, Heulin T (1998) Colonization of wheat rhizosphere by EPS producing Pantoea agglomerans and its effect on soil aggregation. Appl Environ Microbiol 64:3740–3747

Arshad M, Sharoona B, Mahmood T (2008) Inoculation with *Pseudomonas* spp. containing ACC deaminase partially eliminate the effects of drought stress on growth, yield and ripening of pea (*Pisum sativum* L.). Pedosphere 18:611–620

Arutchelvi J, Sudhakar M, Arkatkar A, Doble M, Bhandari S, Uppara PV (2008) Biodegradation of polyethylene and polypropylene. Indian J Biotechnol 7:9–22

Asada K (1999) The water-water cycle in chloroplasts: scavenging of active oxygen and dissipation of excess photons. Annu Rev Plant Biol 50:601.639

Ashraf M (1994) Breeding for salinity tolerance in plants. Crit Rev Plant Sci 13:17–42

Ashraf M, Berge SH, Mahmood OT (2004) Inoculating wheat seedling with exopolysaccharide-producing bacteria restricts sodium uptake and stimulates plant growth under salt stress. Biol Fertil Soils 40:157–162

Auta HS, Emenike CU, Fauziah SH (2017) Screening of *Bacillus* strains isolated from mangrove ecosystems in peninsular Malaysia for microplastic degradation. Environ Pollut 12:1–8

Bagati S, Mahajan R, Nazir M, Dar AA, Zargar SM (2018) "Omics" A gateway towards abiotic. In: Zargar SM, Zargar MY (eds) Abiotic stress mediated sensing and signaling in plants: an omics perspective. Springer, Singapore, pp 1–45

Bano A, Fatima M (2009) Salt tolerance in *Zea mays* (L.) following inoculation with *Rhizobium* and *Pseudomonas*. Biol Fertil Soils 45:405–413

Begum MA, Varalakshmi B, Umamagheswari K (2015) Biodegradation of polythene bag using bacteria isolated from soil. Int J Curr Microbiol App Sci 4:674–680

Bharadwaj H, Gupta R, Tiwari A (2012) Microbial population associated with plastic degradation. Sci Rep 5:272–274

Bhardwaj H, Gupta R, Tiwari A (2013) Communities of microbial enzymes associated with biodegradation of plastics. J Polym Environ 21:575–579

Blasing M, Amelung W (2018) Plastics in soil: analytical methods and possible sources. Sci Total Environ 612:422–435

Cardinale M, Rtering S, Suarez C, Montoya AMZ, Geissler-Plaum R, Schnell S (2015) Paradox of plant growth promotion potential of rhizobacteria and their actual promotion effect on growth of barley (*Hordeum vulgare* L.) under salt stress. Microbiol Res 181:22–32

Carpenter EJ, Smith KL (1972) Plastics on the Sargasso Sea surface. Science 175:1240

Caruso G (2015) Plastic degrading microorganisms as a tool for the bioremediation of plastic contamination in aquatic environments. J Pollut Eff Control 3:112–115

Casanovas EM, Barassi CA, Sueldo RJ (2002) *Azospirillum* inoculation mitigates water stress effects in maize seedlings. Cereal Res Commun 30:343–350

Chae Y, An YJ (2018) Current research trends on plastic pollution and ecological impacts on the soil ecosystem: a review. Environ Pollut 240:387–395

Chang P, Gerhardt KE, Xiao-Dong H, Xiao-Ming Y, Glick BR, Grewing PD, Greenber (2014) Plant growth promoting bacteria facilitates the growth of barley and oats in salt impacted soil: implications for phytoremediation of saline soils. Int J Phytorem 16: 1133–1147

Changrong Y, Wenquing H, Neil C (2014) Plastic film mulch in Chinese agriculture: importance and problems. World Agric 4:32–36

Cho K, Toler H, Lee J, Ownley B, Stutz JC, Moore JL, Auge RM (2006) Mycorrhizal symbiosis and response of sorghum plants to combined drought and salinity stresses. J Plant Physiol 163:517–528

Corrandini F, Bartholomeus H, Lwanga EH, Gertsen H, Geissen V (2019) Predicting soil microplastic concentration using vis-NIR spectroscopy. Sci Total Environ 650:922–932

Crawford CB, Quinn B (2017) Microplastic identification techniques. In: Microplastic pollutants. Elsevier, Amsterdam, pp 219–267

Creus CM, Sueldo RJ, Barassi CA (2004) Water relations and yield in *Azospirillum*-inoculated wheat exposed to drought in the field. Can J Bot 82:273–281

Dang TCH, Nguyen DT, Thai H, Nguyen TC et al (2018) Plastic degradation by thermophilic Bacillus sp. BCBT21 isolated from composting agricultural residual in Vietnam. Adv Nat Sci Nanosci Nanotechnol 9:015014–015025

Dimkpa C, Weinand T, Asch F (2009) Plant–rhizobacteria interactions alleviate abiotic stress conditions. Plant Cell Environ 32:1682–1694

Dodd IC, Belimov AA, Sobeih WY, Safronova VI, Grierson D, Davies WJ (2005) Will modifying plant ethylene status improve plant productivity in water limited environments? In: 4th international crop science congress. http://www.cropscience.org.au/

Dris R, Gasperi J, SAad M, Mirande C, Tassin B (2016) Synthetic fibers in atmospheric fallout: a source of microplastics in the environment? Mar Pollut Bull 104:290–293

Dumichen E et al (2015) Analysis of polyethylene microplastics in environmental samples, using a thermal decomposition method. Water Res 85:451–457

Dumichen E et al (2017) Fast identification of microplastics in complex environmental samples by a thermal degradation method. Chemosphere 174:572–584

Egamberdieva D, Kucharova Z, Davranov K, Berg G, Makarova N, Azarova T, Chebotar V, Tikhonovich I, Kamilova F, Validov S, Lugtenberg B (2011) Bacteria able to control foot and root rot and to promote growth of cucumber in salinated soils. Biol Fertil Soils 47:197–205

Egamberdieva D, Davranov K, Wirth S, Hashen A, Fathi E, Allah A (2017) Impact of soil salinity on the plant-growth-promoting and biological control abilities of root associated bacteria. Saudi J Biol Sci 24:1601–1608

El-Shafei HA, Abd El-Nasser NH, Kansoh AL, Ali AM (1998) Biodegradation of disposable polyethylene by fungi and *Streptomyces* species. Polym Degrad Stab 62:361–365

Feng Y, Cui X, He S, Dong G, Chen M, Wang J, Lin X (2013) The role of metal nanoparticles in influencing arbuscular mycorrhizal fungi effects on plant growth. Environ Sci Technol 47:9496–9504

Ferguson JN (2019) Climate change and abiotic stress mechanisms in plants. Emer Top Life Sci 3:165–181

Figueiredo MVB, Burity HA, Martinez CR, Chanway CP (2008) Alleviation of drought stress in common bean (*Phaseolus vulgaris* L.) by co-inoculation with *Paenibacillus polymyxa* and *Rhizobium tropici*. Appl Soil Ecol 40:182–188

Flemming HC (1998) Relevance of biofilms for the biodeterioration of surfaces of polymeric materials polymer. Degrad Stab 59:309–315

Fuller S, Gautam A (2016) A procedure measuring microplastics using pressurized fluid extraction. Environ Sci Technol 50:5774–5780

Ganesh P, Dineshraj D, Yoganathan K (2017) Production and screening of depolymerising enzymes by potential bacteria and fungi isolated from plastic waste dump yard site. Int J Appl Res 3:693–695

Gewert B, Plassmann MM, MacLeod M (2015) Pathways for degradation of plastic polymers floating in the marine environment. Environ Sci: Processes Impacts 17:1513

Ghosh SK, Pal S, Ray S (2013a) Study of microbes having potentiality for biodegradation of plastics. Environ Sci Pollut Res 20:4339–4355

Ghosh SK, Pal S, Ray S (2013b) Study of microbes having potentiality for biodegradation of plastics. Environ Sci Pollut Res 20:4339–4355

Godfray HCJ, Beddington JR, Crute IR, Haddad L, Lawrence D, Muir JF, Pretty J, Robinson S, Thomas SM, Toulmin C (2010) Food security: the challenge of feeding 9 billion people. Science 327:812–818

Goswamia D, Dhandhukiab P, Patela P, Thakker JN (2014) Screening of PGPR from saline desert of Kutch: growth promotion in *Arachis hypogea* by *Bacillus licheniformis* A2. Microbiol Res 169:66–75

Grob F, Durner J, Gaupels F (2013) Nitric oxide, antioxidants and prooxidants in plant defense responses. Front Plant Sci 4:419

Grover M, Ali SZ, Sandhya V, Rasul A, Venkateswarlu B (2011) Role of microorganisms in adaptation of agriculture crops to abiotic stresses. World J Microbiol Biotechnol 27:1231–1240

Hadad D, Garesh S, Sivan A (2005) Biodegradation of polyethylene by the thermophilic bacterium *Brevibacillus borstelensis*. J Appl Microbiol 98:1093–1100

Harshvardhan K, Jha B (2013) Biodegradation of low-density polyethylene by marine bacteria from pelagic waters, Arabian Sea, India. Mar Pollut Bull 77(1–2):100–106

Hamdia MA, Shaddad MAK, Doaa MM (2004) Mechanism of salt tolerance and interactive effect of *Azospirillum brasilense* inoculation on maize cultivars grown under salt stress conditions. Plant Growth Regul 44:165–174

Hasegawa PM (2013) Sodium (Na^+) homeostasis and salt tolerance of plants. Environ Exp Bot 92:19–31

Helen AS, Uche EC, Hamid FS (2017) Screening for polypropylene degradation potential of bacteria isolated from mangrove ecosystems in peninsular Malaysia. Int J Biosci Biochem Bioinform 7:245–251

Hmaeid N, Walia M, Mahmoud AMB, Pueyo JJ, Ghnaya T, Abdelly C (2019) Efficient rhizobacteria promote growth and alleviate NaCl-induced stress in the plant species *Sulla carnosa*. Appl Soil Ecol 133:104–113

Huang Y, Liu Q, Jia W, Yan C (2020) Agricultural plastic mulching as a source of microplastics in the terrestrial environment. Environ Pollut 260:114096

Huerta Lwanga E, Gertsen H, Gooren H, Peers P, Salanki T, Ploeg VDM, Besseling E, Koelmans AA, Geissen V (2016) Microplastics in the terrestrial ecosystem: implications for *Lumbricus terrestris* (Oligochaeta, Lumbricidae). Environ Sci Technol 50:2685–2691

Huerta Lwanga E, Gertsen H, Gooren H, Peers P, Salanki T, Ploeg VDM, Besseling E, Koelmans AA, Geissen V (2017) Incorporation of microplastics from litter into burrows of *Lumbricus terrestris*. Environ Pollut 220:532–531

Jendrossek D (2001) Microbial degradation of polyesters. In: Biopolyesters. Springer, Berlin, pp 293–325

Jendrossek D, Schirmer A, Schlegel HG (1996) Biodegradation of polyhydroxyalkanoic acids. Appl Microbiol Biotechnol 46:451–463

Jha Y, Subramanian RB (2014) PGPR regulate caspase-like activity, programmed cell death and antioxidant enzyme activity in paddy under salinity. Physiol Mol Biol Plants 20:201–207

Jha Y, Subramanian RB, Patel S (2011) Combination of endophytic and rhizospheric plant growth promoting rhizobacteria in *Oryza sativa* shows higher accumulation of osmoprotectant against saline stress. Acta Physiol Plant 33:797–802

Kale SK, Deshmukh AG, Dudhare MS, Patil VB (2015) Microbial degradation of plastic; a review. J Biochem Technol 6:952

Kang SM, Khan AL, Waqas M, You Y-H, Kim J-H, Kim J-G, Hamayun M, Lee L-J (2014) Plant growth promoting rhizobacteria reduce adverse effects of salinity ad osmotic stress by regulating phytohormones and antioxidants in *Cucumis sativus*. J Plant Interact 9:673–682

Karthikeyan B, Joe MM, Islam MR, Sa T (2012) ACC deaminase containing diazotrophic endophytic bacteria ameliorate alt stress in *Catharanthus roseus* through reduced ethylene levels and induction of antioxidative defense systems. Symbiosis 56:77–86

Kasim WA, Gaffar RM, Abou-Ali RM, Omar MN, Hewait HM (2016) Effect of biofilm forming plant growth promoting rhizobacteria on salinity tolerance in barley. Ann Agric Sci 61:217–227

Kaya C, Ashraf M, Dikilitas M, Tuna AL (2013) Alleviation of salt stress-induced adverse effects on maize plants by exogenous application of indole acetic acid (IAA) and inorganic nutrients- a field trial. Aust J Crop Sci 7:249

Khalid S, Parvaiz M, Nawaz K, Hussain H, Arshad A, Shawaka S, Sarfaraz Z, Waheed T (2013) Effect of Indole acetic acid (IAA) on morphological, biochemical and chemical attributes of two varieties of maize (*Zea mays* L.) under salt stress. World Appl Sci J 26:1150–1159

Kirbaş Z, Keskin N, Güner A (1999) Biodegradation of polyvinylchloride (PVC) by white rot fungi. Bull Environ Contam Toxicol 63:335–342

Kohler J, Hernandez JA, Caravaca F, Roldan A (2008) Plant growth promoting rhizobacteria and arbuscular mycorrhizal fungi modify alleviation biochemical mechanisms in water stressed plants. Funct Plant Biol 35:141–151

Kohler J, Hernandez JA, Caravaca F, Roldan A (2009) Induction of antioxidant enzymes involved in the greater effectiveness of a PGPR versus AM fungi with respect to increasing the tolerance of lettuce to sever salt stress. Environ Exp Bot 65:245–252

Kumar A, Verma JP (2017) Does plant-microbe interaction confer stress tolerance in plants: a review. Microbiol Res 207:41–52

Kyaw MB, Champakalakshmi R, Sakharkar MK, Lim CS, Sakharkar KR (2012) Biodegradation of low density polyethylene (LDPE) by *Pseudomonas* species. Indian J Microbiol 52:411–419

Liu J, Zhu J-K (1998) A calcium sensor homolog required for plant salt tolerance. Science 280:1943–1945

Liu E, He W, Yanc C (2014) "White revolution" to "white pollution"-agricultural plastic film mulch in China. Environ Res Lett 9:091001

Lots FAE, Behrens P, Vijver MG, Horton AA, Bosker T (2017) A large scale investigation of microplastic contamination: abundance and characteristics of microplastics in European beach sediment. Mar Pollut Bull 123:219–226

Lowery VN, Ursell T (2019) Structured environments fundamentally alter dynamics and stability of ecological communities. Proc Natl Acad Sci USA 116:379–388

Lwanga EH, Thapa B, Yang X, Gertsen H, Salanki T, Geissen V, Garbeva P (2018) Decay of low-density polyethylene by bacteria extracted from earthworm's gut: a potential for soil restoration. Sci Total Environ 624:753–757

Lyklema J, Norde W, van Loosdrecht MCM, Zehnder AJB (1989) Adhesion of bacteria to polystyrene surfaces. Colloids Surf 39:175–187

Maass S, Daphi D, Lehmann A, Rillig MC (2017) Transport of microplastics by two *Collembolan* species. Environ Pollut 225:456–459

Machado AAS, Kloas W, Zarfl C, Hempel S, Rillig MC (2018a) Microplastics as an emerging threat to terrestrial ecosystems. Glob Chang Biol 24:1405–1416

Machado AAS, Lau CW, Till J, Kloas W, Lehmann A, Becker R, Rillig MC (2018b) Impacts of microplastics on the soil biophysical environment. Environ Sci Technol 52:9656–9665

Macleod K, Rumbold K, Padayachee K (2015) A systems approach to uncover the effects of the PGPR *Pseudomonas koreensis* on the level of drought stress tolerance in *Helianthus annuus*. Procedia Environ Sci 29:262–263

Mahdiyah D, Mukti BH (2013) Isolation of polyethylene plastic degrading bacteria. Biosci Int 2:29–32

Mahmood A, Amaya R, Turgay OC, Yaprak AE, Taniguchi T, Kataoka R (2019) High salt tolerant plant growth promoting rhizobacteria from the common ice-plant *Mesembryanthemum crystallinum* L. Rhizosphere 9:10–17

Mayak S, Tirosh T, Glick BR (2004) Plant growth promoting bacteria confer resistance in tomato plants to salt stress. Plant Physiol Biochem 42:565–572

McLellan CA, Turbyville TJ, Wijeratne K, Kerschen A, Vierling E, Queitsch C, Whiteshell L, Gunatilaka AAL (2007) A rhizosphere fungus enhances *Arabidopsis* thermotolerance through production of an HSP90 inhibitor. Plant Physiol 145:174–182

Mohan AJ, Sekhar VC, Bhaskar T, Nampoothiri KM (2016) Microbial assisted high impact polystyrene (HIPS) degradation. Bioresour Technol 213:204–207

Morrissey JP, Dow JM, Mark GL, Gara FO (2004) Are microbes at the root of a solution to world food production? EMBO Rep 5:922–926

Nadeem SM, Zaheer ZA, Naveed M, Asghar HN, Arshad M (2010) Rhizobacteria capable of producing ACC-deaminase may mitigate salt stress in wheat. Soil Sci Soc Am 74:533–542

Nadeem SM, Zaheer ZA, Naveed M, Nawaz S (2013) Mitigation of salinity induced negative impact on the growth and yield of wheat by plant growth promoting rhizobacteria in naturally saline conditions. Ann Microbiol 63:225–232

Ng EL, Huerta Lwanga E, Eldrige SM, Johnston P, Hu HW, Geissen V, Chen D (2018) An overview of microplastic and nanoplastic pollution in agroecosystems. Sci Total Environ 627:1377–1388

Nizzetto L, Bussi G, Futte MN, Butterfield D, Whitehead PG (2016) A theoretical assessment of microplastic transport in river catchments and their retention by soils and river sediments. Environ Sci Process Impacts 18:1051–1059

Noreen A, Ashraf M, Akram N (2010) Salt induced regulation of some key antioxidant enzymes and physio-biochemical phenomenon in five diverse cultivars of turnip (*Brassica rapa* L.). J Agron Crop Sci 196:273–285

Numana M, Bashir S, Khan Y, Mumtaz R, Shinwari ZK, Khan AL, Khan A, AL-Harrasi A (2018) Plant growth promoting bacteria as an alternative strategy for salt tolerance in plants: a review. Microbiol Res 209:21–32

Orhan F (2016) Alleviation of salt stress by halotolerant and halophilic plant growth promoting bacteria in wheat (*Triticum aestivum*). Braz J Microbiol 47:621–627

Paco A, Duarte K, da Costa JP, Santos PSM, Pereira R, Pereira ME, Freitas AC, Duarte AC, Rocha-Santos TAP (2017) Biodegradation of polyethylene microplastics by the marine fungus *Zalerion maritimum*. Sci Total Environ 586:10–15

Palaniyandi SA, Damodharan K, Yang SH, Suh JW (2014) *Streptomyces* sp. strain PFPA39 alleviates salt stress and promotes growth of "micro tom" tomato plants. J Appl Microbiol 117:766–773

Parida A, Das A (2005) Salt tolerance and salinity effects on plants: a review. Ecotoxicol Environ Saf 60:324–349

Pathak VM, Navneet XX (2017) Review on the current status of polymeric degradation: a microbial approach. Bioresour Bioprocess 4:15–46

Perrig D, Boiero M, Masciarelli O, Penna C, Ruiz O, Cassan F, Luna M (2007) Plant growth promoting compounds produced by two agronomically important strains of *Azospirillum brasilense* and implications for inoculants formulation. Appl Microbiol Biotechnol 75:1143–1150

Powell JR, Rillig MC (2018) Biodiversity of arbuscular mycorrhizal fungi and ecosystem function. New Phytol 220:1059–1075

Premraj R, Doble M (2005) Biodegradation of polymers. Indian J Biotechnol 4:186–193

Raaman N, Rajitha N, Jayshree A, Jegadeesh R (2012) Biodegradation of plastic by *Aspergillus* spp. isolated from polythene polluted sites around Chennai. J Acad Ind Res 1:313–316

Ramis X, Cadenato A, Salla JM, morancho JM, Valles A, Contat L, Ribes A (2004) Thermal degradation of polypropylene/starch-based materials with enhanced biodegradability. Polym Degrad Stab 86:483–491

Raziyafathima M, Praseetha PK, Rimal Isaac RS (2016) Microbial degradation of plastic waste: a review. J Pharm Chem Biol Sci 4:231–242

Restrepo-Flórez JM, Bassi A, Thompson MR (2014) Microbial degradation and deterioration of polyethylene: a review. Int Biodeterior Biodegrad 88:83–90

Rillig MC, Ziersch L, Hempel S (2017) Microplastic transport in soil by earthworms. Sci Rep 7:1362

Rillig MC, Ingraffia R, Machado AAD (2017a) Microplastic incorporation into soil in agroecosystems. Front Plant Sci 8:1805–1809

Rillig MC, Lehmann A, Machado AADS, Yang G (2019) Microplastic effects on plants. New Phytol. https://orcid.org/0000-0003-3541-7853

Ru J, Huo Y, Yang Y (2020) Microbial degradation and valorization of plastic wastes. Front Microbiol 11:442

Rustinia KSC, Jannah M, Dewi AP, Djamaana A (2020) Polyethylene terephthalate plastic degradation using soil bacteria from mount Jayawijaya, Papua, Indonesia. J Agric Vet Sci 13:2319–2372

Sandhya V, Ali SZ, Grover M, Reddy G, Venkateswarlu B (2009) Alleviation of drought stress effects in sunflower seedlings by exopolysaccharides producing *Pseudomonas putida* strain P45. Biol Fertil Soils 46:17–26

Santos ADA, Silveira JADA, Bonifacio A, Rodrigues AC, Figueiredo MDVB (2018) Antioxidant response of cowpea co-inoculate with plant growth-promoting bacteria under salt stress. Braz J Microbiol 49:513–521

Sapre S, Gontia-Mishra I, Tiwari S (2018) *Klebsiella* sp. confers enhanced tolerance to salinity and plant growth promotion in oat seedlings (*Avena sativa*). Microbiol Res 206:25–32

Saravanakumar D, Samiyappan R (2007) ACC deaminase from *Pseudomonas fluorescens* mediated saline resistance in groundnut (*Arachis hypogea*) plants. J Appl Microbiol 102:1283–1292

Scheurer M, Bigalke M (2018) Microplastics in Swiss floodplain soils. Environ Sci Technol 52:3591–3598

Shaharoona B, Jamro GM, Zahir ZA, Arshad M, Memon KS (2007) Effectiveness of various *Pseudomonas* spp. and *Burkholderia caryophylli* containing ACC-Deaminase for improving growth and yield of wheat (*Triticum aestivum* L.). J Microbiol Biotechnol 17:1300–1307

Sharma IP, Sharma A (2017) Physiological and biochemical changes in tomato cultivar PT-3 with dual inoculation of mycorrhiza and PGPR against root-knot nematode. Symbiosis 71:175–183

Shilev S, Sancho ED, Benlloch-Gonzalez M (2012) Rhizospheric bacteria alleviate salt-produced stress in sunflower. J Environ Manag 95:S37–S41

Shirinbayan S, Khosravi H, Malakouti MJ (2019) Alleviation of drought stress in maize (*Zea mays*) by inoculation with *Azotobacter* strains isolated from semi-arid regions. Appl Soil Ecol 133:138–145

Shovitri M, Nafiah R, Antika TR, Alami NH, Kuswytasai ND, Zulaikha E (2016a) Proceedings of international biology conference. AIP Publishing, pp 020035–020039

Shovitri M, Nafiah R, Antika TR, Alami NH, Kuswytasai ND, Zulaikha E (2016b) Proceedings of international biology conference. AIP Publishing, pp 020035–020035-9

Shukla PS, Agarwal PK, Jha B (2012) Improved salinity tolerance of *Arachis hypogaea* (L.) by the interaction of halotolerant plant growth- promoting rhizobacteria. J Plant Growth Regul 31:195–206

Siddikee MA, Chauhan PS, Anandham R, Han G-H, Sa T (2010) Isolation, characterization, and use for plant growth promotion under salt stress, of ACC deaminase producing halotolerant bacteria derived from coastal soil. J Microbiol Biotechnol 20:1577–1584

Siddikee MA, Glick BR, Chauhan PS, Yim WJ, Sa T (2011) Enhancement of growth and salt tolerance of red pepper seedlings (*Capsicum annuum* L.) by regulating stress ethylene synthesis with halotolerant bacteria containing 1-aminocyclopropane-1-carboxylic acid deaminase activity. Plant Physiol Biochem 49:427–434

Singh JS, Pandey VC, Singh DP (2011) Efficient soil microorganisms: a new dimension for sustainable agriculture and environmental development. Agric Ecosyst Environ 140:339–353

Sirari K, Kashyap L, Mehta CM (2016) Stress management practices in plants by microbes. In: Singh DP, Singh HB, Prabha R (eds) Microbial inoculants in sustainable agricultural productivity, Functional applications, vol 2. Springer, New Delhi, pp 85–99

Skariyachan S, Manjunatha V, Sultana S, Jois C, Bai V, Vasist KS (2016) Novel bacterial consortia isolated from plastic garbage processing areas demonstrated enhanced degradation for low density polyethylene. Environ Sci Pollut Res 23:18307–18319

Sowmya HV, Ramalingappa KM, Thippeswamy B (2014) Biodegradation of polyethylene by *Bacillus cereus*. Adv Polym Sci Technol Int J 4:28–32

Steinmetz Z, Wollmann C, Schaefer M, Buchmann C, David J, Troger J, Munoz K, Fror O, Schaumann GE (2016) Plastic mulching in agriculture. Trading short term agronomic benefits for long term soil degradation? Sci Total Environ 550:690–705

Tank N, Saraf M (2010) Salinity-resistant plant growth promoting rhizobacteria ameliorates sodium chloride stress on tomato plants. J Plant Interact 5:51–58

Tansengco M, Tokiwa Y (1998) Thermophilic microbial degradation of polyethylene succinate. World J Microbiol Biotechnol 14:133–138

Timmusk S, Wagner EGH (1999) The plant growth promoting rhizobacterium *Paenibacillus polymyxa* induces changes in *Arabidopsis thaliana* gene expression: a possible connection between biotic and abiotic stress responses. Mol Plant-Microbe Interact 12:951–959

Tokiwa Y, Calabia BP, Ugwu CU, Aiba S (2009) Biodegradability of plastics. Int J Mol Sci 10:3722–3742

Vimala PP, Mathew L (2016) Biodegradation of polyethylene using *Bacillus subtilis*. Proc Technol 24:232–239

Vinocur B, Altman A (2005) Recent advances in engineering plant tolerance to abiotic stress: achievements and limitations. Curr Opin Biotechnol 16:123–132

Vurukonda SSKP, Vardharajula S, Shrivastava M, SkZ A (2016) Enhancement of drought stress tolerance in crops by plant growth promoting rhizobacteria. Microbiol Res 184:13–24

Wagg C, Bender SF, Widmer F, van der Heijden MGA (2014) Soil biodiversity and soil community composition determine ecosystem multifunctionality. Proc Nat Acad Sci USA 111:5266–5270

Waller F, Achatz B, Baltruschat H, Fodor J, Becker K, Fischer M, Heier T, Hückelhoven R, Neumann C, Wettstein D (2005) The endophytic fungus *Piriformospora indica* reprograms barley to salt-stress tolerance, disease resistance, and higher yield. Proc Natl Acad Sci U S A 102:13386–13391

Wan Y, Wu C, CXue Q, Hui X (2019) Effects of plastic contamination on water evaporation and desiccation cracking in soil. Sci Total Environ 654:576–582

Weithmann N, Möller JN, Löder MGJ, Piehl S, Laforsch C, Freitag R (2018) Organic fertilizer as a vehicle for the entry of microplastic into the environment. Sci Adv 4:8060–8067

Wu Z, Yue H, Lu J, Li C (2012) Characterization of rhizobacterial strain Rs-2 with ACC deaminase activity and its performance in promoting cotton growth under salinity stress. World J Microbiol Biotechnol 28:2383–2393

Yamada-Onodera K, Mukumoto H, Katsuyaya Y, Saiganji A, Tani Y (2001) Degradation of polyethylene by a fungus, *Penicillium simplicissimum* YK. Polym Degrad Stab 72:323–327

Yang J, Kloepper JW, Ryu CM (2009) Rhizosphere bacteria help plants tolerate abiotic stress. Trends Plant Sci 14:1–4

Yang J, Yang Y, Wu WM, Zhao J, Jiang L (2014) Evidence of polyethylene biodegradation by bacterial strains from the guts of plastic-eating waxworms. Environ Sci Technol 48:13776–13784

Yao L, Wu Z, Zheng Y, Kaleem I, Li C (2010) Growth promotion and protection against salt stress by *Pseudomonas putida* Rs-198 on cotton. Eur J Soil Biol 46:49–54

Yoshida S, Hiraga K, Takehana T, Taniguchi I, Yamaji H, Maeda Y, Toyohara K, Miyamoto K, Kimura Y, Oda K (2016) A bacterium that degrades and assimilates poly ethylene terephthalate. Science 351:1196–1199

Zahir ZA, Ghani U, Naveed M, Nadeem SM, Asghar HN (2009) Comparative effectiveness of *Pseudomonas* and *Serratia* sp. containing ACC-deaminase for improving growth and yield of wheat (*Triticum aestivum* L.) under salt-stressed conditions. Arch Microbiol 191:415–424

Zhang H, Kim MS, Sun Y, Dowd SE, Shi H, Paré PW (2008) Soil bacteria confer plant salt tolerance by tissue-specific regulation of the sodium transporter HKT1. Mol Plant-Microbe Interact 21:737–744

Zhou Y, Tang N, Huang L, Zhao Y, Tang X, Wang K (2018) Effects of salt stress on plant growth, antioxidant capacity, glandular trichome density, and volatile exudates of *Schizonepeta tenuifolia* Briq. Int J Mol Sci 19:252

Zhou B, Wnag J, Zhang H, Shi H, Fei Y, Huang S, Tong Y, Wen D, Luo Y, Barcelo D (2020) Microplastics in agricultural soils on the coastal plain of Hangzhou Bay, East China: multiple sources other than plastic mulching film. J Hazard Mater 388:121814

Zhu JK, Bressan RA, Hasegawa PM, Pardo JM, Bohnert HJ (2005) Salt and crops: salinity tolerance. In: Success stories in agriculture, vol 32. Council for Agricultural Science and Technology, p 13

Ziajahromi S, Neale PA, Rintoul L, Leusch FDL (2017) Wastewater treatment plants as a pathway for microplastics: development of a new approach to sample waste water based microplastics. Water Res 112:93–99

Zimmermann RP, Kardos LT (1961) Effect of bulk density on root growth. Soil Sci 91:280–288

Zubris KAV, Richards BK (2005) Synthetic fibers as an indicator of land application of sludge. Environ Pollut 138:201–211

Orchestration of MicroRNAs and Transcription Factors in the Regulation of Plant Abiotic Stress Response

8

Sombir Rao, Sonia Balyan, Sarita Jha, Chandni Bansal, Jaishri Rubina Das, Apoorva Gupta, and Saloni Mathur

Abstract

Plants have to tackle daily vagaries of the environment like temperature extremes, drought, excess salt, etc. Together with nutritional imbalances present in the soil, these abiotic stresses are major obstacles in achieving sustainable food production. To mitigate the effects of harsh environmental conditions, an array of regulatory pathways has evolved in plants to coordinate gene expression and cellular signaling to maintain cellular homeostasis. This dynamic reprogramming of gene expression is majorly governed by two *trans*-regulators: the transcription factors (TFs) and microRNAs (miRNAs). While TFs regulate transcription by binding to the *cis*-regulatory elements in the promoters of various genes, the miRNAs regulate gene expression by silencing their target genes through mRNA cleavage, translational repression, or DNA methylation. Recent studies have elucidated the intricate stress responsive regulatory circuits involving TFs and miRNAs as focal nodes. Several TF:miRNA modules have been shown to be active under various environmental stress conditions like HSFA1b/A7b:miR398 (heat stress); SPL7:miR398 and HY5-SPL7:miR408 (copper starvation); PHR1: miR399 (phosphate starvation); and SLIM1:miR395 (sulfate deficiency). Since many of the miRNA targets are TFs themselves, several of these modules work as critical sub-networks in abiotic stress response including the miR156:*SPL*, miR159:*MYB*, miR169:*NF-YA*, and miR166:*HD-ZIP*. This control of plant gene expression at both transcriptional and post-transcriptional levels regulated by TFs and miRNAs, respectively, leads to the establishment of complex regulatory networks which are involved in abiotic stress response in plants. In this chapter, we have reviewed and discussed the detailed information available on the

S. Rao · S. Balyan · S. Jha · C. Bansal · J. R. Das · A. Gupta · S. Mathur (✉)
National Institute of Plant Genome Research, New Delhi, India
e-mail: saloni@nipgr.ac.in

B. Giri, M. P. Sharma (eds.), *Plant Stress Biology*,
https://doi.org/10.1007/978-981-15-9380-2_8

roles of TF:miRNA:target modules in different abiotic stress responses in various plant species including model and crop plants. This information will aid in understanding the mechanisms of action and interactions between these regulators in plant growth and development under normal as well as environmental stress conditions and pave the path to engineer more resilient plants in the future.

Keywords

Transcription factors · microRNA · Abiotic stress · Heat · Cold · Salinity and heavy metal stress · Homeostasis

8.1 Introduction

Environmental factors that negatively affect plant growth and development are collectively recognized as various forms of abiotic stress. Plants are constantly exposed to adverse conditions, such as high temperature, drought, high salinity, cold, heavy metals, and nutrient deficiency. These factors limit the geographical distribution of plants and also affect crop yield. Being sessile in nature, plants have evolved sophisticated and robust mechanisms to cope with the myriad stresses they encounter (Zhang 2015; Zhu 2016; Bielach et al. 2017; Nolan et al. 2017; Zarattini and Forlani 2017). At the forefront of these are regulatory responses to maintain physiological homeostasis through specific and well-defined reprogramming of the transcriptional and post-transcriptional activities by transcription factors (TFs) and a class of small 20–22 nucleotides long non-coding endogenous RNAs called microRNAs (miRNAs), respectively.

Transcription factors are the key regulators of gene expression belonging to multigene families in plants (Salih et al. 2016). TFs act by recognizing the *cis*-elements in the promoter region of different stress responsive genes including various miRNAs (Yant et al. 2010; Wang and Perry 2013; Biłas et al. 2016). TF genes are regulated at both transcriptional and post-transcriptional levels in plants (Mitsuda and Ohme-Takagi 2009; Payne and Wagner 2015; Chen et al. 2017; Hernando et al. 2017). MicroRNAs act via direct mRNA cleavage, translational repression, and DNA methylation of its targets based on sequence complementarity. *MIRNA* genes are first transcribed into primary miRNAs (pri-miRNAs) by RNA polymerase II (Pol II) (Xie et al. 2005; Kim et al. 2011). Besides the TATA box core promoter element, *MIR* promoters are enriched in various *cis*-regulatory elements that regulate the transcription of *MIRs* in different developmental stages and/or varied environmental cues by recruiting different TFs. In contrast to animals, only a small portion of genes have been validated as true targets of plant miRNAs, i.e., less than 1% of protein coding genes (Addo-Quaye et al. 2009; Li et al. 2010). However, the overall impact of miRNA-mediated gene regulation in plants cannot be underestimated as most of the target genes are TFs (Jones-Rhoades et al. 2006) that regulate several developmental and plant stress responses. The identification and functional characterization of several genes including TFs and miRNAs that are altered in response to stress suggest their involvement in the maintenance of stress tolerance (Nelson et al. 2007; Sunkar et al. 2007; Petroni et al. 2012; Yan et al. 2013; Baxter et al. 2014; Kumar et al. 2014; Lee et al. 2014; Golldack et al. 2014; Weng et al. 2016; Zanetti et al. 2017; Zhou and Tang 2019; Liebsch and Palatnik 2020; Millar 2020). Moreover, detailed analysis of several miRNA biogenesis mutants like *hyponastic leaves 1* (*hyl1*), *cap-binding protein 80/aba hypersensitive 1* (*cbp80/abh1*), and *sickle* (sic) under different abiotic stress conditions clearly suggests the critical role of miRNAs in plant stress regulation (Lu and Fedoroff 2000; Kim et al. 2008; Zhang et al. 2008; Zhan et al. 2012). These alterations in TFs and miRNAs levels in response to plant stress responses have revealed the existence of complex regulatory networks that control the onset of different gene networks. Elucidation of such networks is pivotal for understanding the molecular mechanisms of plants stress response. Exploring the interplay between TFs and miRNAs will help in understanding the organization of several stress responsive networks in plants, some of which form feedback loop circuits. This chapter will provide deeper insights into TF:miRNA:target mediated gene regulation and their crosstalk during plant abiotic stress responses.

8.2 TFs-miRNAs: Regulating Plant Heat Stress Response

Global mean temperatures have seen an upsurge in the past few decades. Heat stress (HS) is a major limiting factor in plant growth, development, and productivity. Previously, heat stress transcription factors (HSFs) were considered as the key players of plant heat stress response (HSR) regulating the expression of heat-responsive genes like heat shock proteins (HSPs) (Wang et al. 2004). However, many non-coding RNAs including miRNAs have now been found to mediate plant HSR also (Fig. 8.1 and Table 8.1). In Arabidopsis, miR398 has four target genes, viz. *copper/zinc superoxide dismutases 1* and *2* (*CSD1* and *CSD2*), mitochondrial *cytochrome c oxidase* (*Cox5b-1*), and *CCS1* [*copper chaperone for superoxide dismutase* (*SOD*)] (Sunkar and Zhu 2004). CSD proteins are important scavengers of reactive oxygen species (ROS) and CSD/CCS negatively regulates the accumulation of different ROS species (Mittler et al. 2004; Sunkar et al. 2006). It has been shown that under HS conditions, there is a rapid rise in the miR398 levels via direct HSF-mediated (HSFA1b and HSFA7b) transcriptional regulation of *MIR398* precursors (Guan et al. 2013). Downregulation of its targets *CSD1*, *CSD2*, and *CCS* modifies the redox status of cells that is sensed by HSFs which in turn regulate not only miR398 expression but also other HSR genes. The loss of function *csd1*, *csd2*, and *ccs* Arabidopsis mutants are more heat tolerant and maintain higher HSF and HSP levels while the miR398-resistant forms of target genes that avoid cleavage by miRNAs are more sensitive to HS (Guan et al. 2013; Lu et al. 2013). This miR398:*CSDs*/*CCS* module is also functional in the HS responses in *Brassica rapa* and *Populus tomentosa* (Kotak et al. 2007; Yu et al. 2011), indicating that the HSF:miR398:*CSD*/*CCS* pathway is widely conserved in the HS response in

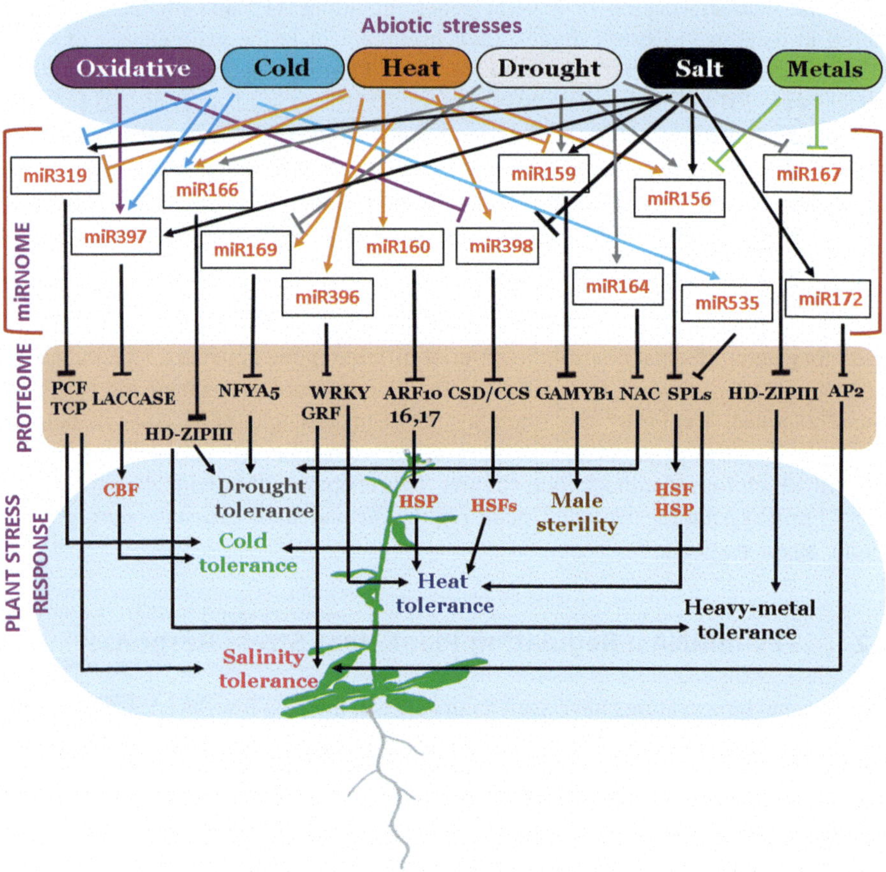

Fig. 8.1 Schematic representation of various miRNA:TF modules orchestrated in response to different environmental stresses in plants. The miRNAs miR319, miR397, miR166, miR169, miR396, miR160, miR398, miR159, miR164, miR156, miR535, miR167, and miR172 respond to different abiotic stresses and exhibit stress mediated differential regulation in expression levels. MiRNAs that are upregulated in a particular stress are shown by pointed arrows, while downregulated miRNAs are shown by blunt arrows. These stress responsive miRNAs attenuate the expression of their target transcription factor transcripts like *TCP/PCF, HD-ZIPIII, NF-YA5, WRKY/GRF, ARF, HSF, MYB, NAC, SPL,* and *AP2* (depicted by blunt arrows connecting the miRNA with their respective targets). These miRNA:target modules fine-tune the abiotic stress signaling via altering different hormonal, metabolic, and molecular pathways to confer heat/cold/salt/drought/oxidative and heavy metal stress tolerance to the plants

plants. Stief et al. (2014) have shown HS-mediated strong induction of miR156 that is maintained for several days even when HS ceases, signifying that miR156 is required for HS memory. Elevated miR156 levels enhance and prolong HS memory by the repression of *SPL2* and *SPL11*, the *SQUAMOSA-PROMOTER BINDING PROTEIN-LIKE* transcription factor genes in Arabidopsis. Higher expression of miR156 maintains sustained expression of HSR genes that lead to the improved

Table 8.1 Table summarizing the TF:miRNA:target modules involved in different biological functions and environment interactions

TF	miRNA family	Target	Biological functions	Species	References
	miR156	SPL	Heat stress memory	Arabidopsis	Stief et al. (2014) and Matthews (2018)
			Drought tolerance	Medicago	Arshad et al. (2017)
			Phosphate-deficiency	Arabidopsis	Lei et al. (2016)
			Shade-avoidance	Arabidopsis	Xie et al. (2017)
	miR159	MYB	Sensitivity to heat stress	Rice	Wang et al. (2012b)
			ABA hypersensitivity	Arabidopsis	Roy (2016)
	miR160	ARF	Drought tolerance	Arabidopsis	Kinoshita et al. (2012)
			Root responses to nitrate	Arabidopsis	Gifford et al. (2008)
			Heat stress tolerance	Arabidopsis, cotton	Lin et al. (2018) and Ding et al. (2016)
	miR164	NAC	Drought tolerance	Arabidopsis	Fang et al. (2014)
	miR165/ miR166	HD-ZIPIII/CLP-1/ RDD1	Drought/cold tolerance	Arabidopsis, rice	Ramachandran et al. (2016), Yan et al. (2016), and Zhang et al. (2018)
			Ion uptake and accumulation	Rice	Iwamoto and Tagiri (2016)
HYH	miR169	NF-YA/HAP2	Drought tolerance	Arabidopsis	Du et al. (2017), Li et al. (2008), and Ni et al. (2013)
			Nutrient-deficiency response	Wheat	Qu et al. (2015)
			Salinity, high temperature	Arabidopsis	Kong et al. (2014) and Serivichyaswat et al. (2017)
			Abscisic acid response	Arabidopsis	Contreras-Cubas et al. (2012) and Cheng et al. (2016)

(continued)

Table 8.1 (continued)

TF	miRNA family	Target	Biological functions	Species	References
SVP	miR172	AP2	Thermo-sensory flowering	Arabidopsis	Lee et al. (2010)
	miR319	TCP/PCF	Cold tolerance	Sugarcane, rice	Thiebaut et al. (2012), Yang et al. (2013), and Wang et al. (2014)
			Drought/salt tolerance	Creeping bentgrass/ Switchgrass	Zhou et al. (2013) and Liu et al. (2020a)
SLIM1	miR395	APS/SULTR2:1	Sulfate-deficiency response	Arabidopsis, rice	Jagadeeswaran et al. (2014), Kawashima et al. (2011), and Yuan et al. (2016)
	miR396	GRF	Leaf growth upon UV-B irradiation	Arabidopsis	Casadevall et al. (2013)
		WRKY	Response to arsenic treatment	Rice	Liu and Zhang (2012)
			Response to high temperature	Sunflower	Giacomelli et al. (2012)
HSFA1b, HSFA7b	miR398	CSD1/CSD2/ CCS1	Heat stress tolerance	Arabidopsis	Guan et al. (2013)
			Oxidative-stress tolerance	Arabidopsis	Sunkar et al. (2006)
SPL7			Copper-deficient response	Arabidopsis	Beauclair et al. (2010) and Yamasaki et al. (2007, 2009)
PHR1 WRKY74 MYB2P-1	miR399	PHO2/UBC24	Phosphate-deficiency response	Arabidopsis, rice, common bean, barley	Franco-Zorrilla et al. (2007), Hu et al. (2011), Dai et al. (2012, 2016), Liu et al. (2010), and Hackenberg et al. (2013)
HY5, SPL7	miR408	LAC/ PLASTACYANIN	Copper-deficiency response	Arabidopsis	Abdel-Ghany and Pilon (2008) and Jung et al. (2012)
			Salt/drought/cold/oxidative/ osmotic-stress response	Arabidopsis, wheat	Feng et al. (2013) and Ma et al. (2015)
	miR444	MADS	Nitrate/phosphate-deficiency responses	Rice	Yan et al. (2014)

OsSPL9	miR528	AO/LAC	Salt and nitrate-deficient response	Creeping bentgrass	Yuan et al. (2015)
			Lodging resistance under nitrogen-luxury conditions	Maize	Sun et al. (2018)
	miR529	SPL	Oxidative-stress tolerance	Rice	Yue et al. (2017a)
HSFA1 HSFA2	miR824	AGL	Post-stress development	Arabidopsis	Szaker et al. (2019)
PHR2	miR827	NLA/SPX-MFS	Phosphate-deficient response	Barley, rice, Arabidopssis	Guan et al. (2013), Kant et al. (2011), Lin et al. (2013), and Yue et al. (2017b)
	miR828, miR858	MYB	Response to high temperature	Cotton	Wang et al. (2016)

AFB auxin signaling F-box, AGO1 argonaute 1, AO L-ascorbate oxidase, APS ATP sulfurylase, ARF auxin response factor, CCS1 copper chaperone of CSD1, CLP-1 cysteine protease-1, CSD copper/zinc superoxide dismutase, GRFs growth-regulating factors, HAP2 apetala2 protein homolog, HD-ZIP III homeodomain-leucine zipper III, LAC laccase, MYB myeloblastosis, NF-YA nuclear transcription factor Y subunit alpha, SVP short vegetative phase, NLA nitrogen limitation adaptation, PHO2 phosphate 2, PPR pentatricopeptide repeat, SULTR2:1 sulfate transporter 2;1, SPL squamosa-promoter binding protein-like, SPX-MFS S YG1/P ho81/X PR1-major facilitator superfamily, TCP teosinte branched1, cycloidea proliferating cell nuclear antigen binding factor, TIR1 transport inhibitor response1, UBC24 ubiquitin-conjugating enzyme 24, SLIM1 sulfur limitation 1, HSF heat stress transciption factor; HY5 elongated hypocotyl 5, PHR1 Phosphate Starvation Response 1 acquired thermo-tolerance. In *Medicago sativa*, miR156 overexpression and its target *SPL13* RNAi plants show enhanced tolerance to HS (Matthews 2018). Contrary to the roles of miR398 and miR156, miR159 acts as a negative regulator of HS response by targeting gibberellic acid (GA) regulated *MYB* (*GAMYB*)*-like* family TFs in plants. Overexpression of wheat miR159 leads to enhanced sensitivity to HS in rice (Wang et al. 2012b). Similarly, heat-mediated downregulation of csa-miR159b is also reported in cucumber (*Cucumis sativa*) plants (Li et al. 2016). Overexpression of csa-miR159b leads to decreased heat tolerance in Arabidopsis, further confirming the negative role of miR159 in heat tolerance. Giacomelli et al. (2012) have identified a WRKY transcription factor, *HaWRKY6* as a novel miR396 target in sunflower (*Helianthus annuus*) that is unique to the Asteraceae family. Transgenic plants expressing the miR396-resistant version of *HaWRKY6* are sensitive to HS indicating the positive role of this recently evolved miR396:*HaWRKY6* miRNA:target module has in tolerance to high temperatures. These results also show how a miRNA (miR396) that is normally known for regulating plant development via its canonical targets the *GROWTH-REGULATING FACTOR* (*GRF*) TF genes (Rodriguez et al. 2010), can also be recruited for HS tolerance. Similarly, miR172 known for regulating the juvenile to adult phase transition has been reported to be downregulated in heat stress in Arabidopsis, wheat, and sunflower, while its target TF genes *TOE1* and *TOE2* (*Target of early activation tagged 1* and *2*) are upregulated (May et al. 2013; Li et al. 2014). In addition, other developmentally regulated miRNA:TF target modules such as miR164:*NAC1* (*NAM, ATAF1/2,* and *CUC2*), miR166:*PHV/REV/HOX9* (*PHAVOLUTA/REVOLUTA/HOMEOBOX A9*), miR169:*NF-YA* (*Nuclear factor-YA*), and miR171:*SCL6-III* (*Scarecrow-like protein 6-III*) have also been shown to be heat stress responsive in genome-wide large data studies (Barku et al. 2013; May et al. 2013; Li et al. 2014; Kumar et al. 2014: Kruszka et al. 2014; Rao et al. 2020). Lin et al. (2018) have reported that overexpression of miR160 improves seed germination and seedling survival under heat stress. miR160 overexpression and its target mimic MIM160 plants show heat tolerance and heat sensitivity, respectively, via regulating the expression of target *Auxin Response Factor* (*ARF*) TFs, namely *ARF10*, *ARF16*, and *ARF17*. The authors have shown that miR160 alters the expression of heat shock proteins as well as plant development to allow plants to survive heat stress. Furthermore, a study has delineated a Brassicaceae-specific heat-mediated regulation of miR824 in integrating the HS signals to modulate the MADS-box TF, *AGAMOUS LIKE 16* (*AGL16*), leading to the derepression of *FLOWERING LOCUS T* (*FT*) gene (Szaker et al. 2019). During HS the HSFA1 and HSFA2 mediated transcriptional induction of miR824 regulates *AGL16* post-transcriptionally, which alters the *FT* levels to fine-tune the post-stress development of the plants (Szaker et al. 2019).

8.3 miRNA-TFs: Regulating Drought and Salinity

Drought and salinity are two major constraints to agricultural productivity worldwide. Over the past two decades the molecular basis of plant tolerance to these stresses has identified numerous drought and salt stress responsive genes (Bartels and Sunkar 2005). In addition to the protein coding genes, stress conditions also alter miRNAs expression in plants (Fig. 8.1 and Table 8.1) (Sunkar and Zhu 2004; Zhao et al. 2007; Li et al. 2008; Trindade et al. 2010; Kulcheski et al. 2011). In response to drought stress, plants modify their root structure architecture (RSA) by inhibiting primary root growth and increasing lateral root formation for maximum water assimilation (Gilbert and Medina 2016). Accumulation of bioactive auxin indole-3-acetic acid (IAA) is a known morphogenic trigger for lateral root formation and contributes to the plastic changes in RSA (Benková et al. 2009). In Arabidopsis, drought-mediated downregulation of miR167a that targets the TF *HD-ZIP* (*Homeodomain-leucine zipper*) leads to accumulation of IAA-Ala RESISTANT 3 that promotes IAA accumulation and development of lateral roots (Kinoshita et al. 2012). The miR165/166:*HD-ZIP III* module is negatively regulated by drought and confers enhanced drought tolerance in Arabidopsis and rice, by elevated abscisic acid (ABA) levels via enhanced *HD-ZIP III* activity (Yan et al. 2016; Zhang et al. 2018). Furthermore, miR164:*NAC* module also plays an important role in regulating drought tolerance in rice, where overexpression of miR164 and concomitant depression of target *NAC* TFs leads to susceptibility to drought stress (Fang et al. 2014). In another study by Jiang et al. (2019) it is shown that rice plants overexpressing the miR164b-resistant form of *OsNAC2* have higher levels of drought and salt tolerance than the wild-type plants, and that the ABA content is increased in the transgenic plants. In Arabidopsis, drought-mediated downregulation of miR169 allows the induction of *NF-YA5*. Overexpression of *NF-YA5* TF enhances drought tolerance, while *nf-ya5* mutant and miR169 overexpressing plants are hypersensitive to drought (Li et al. 2008). In soybean, the miR169:*GmNF-YA3* module also confers drought tolerance, establishing conserved regulatory roles for miR169:*NF-YA* modules in plant drought tolerance (Ni et al. 2013). In addition, plants have evolved another mechanism to ensure the induction of *NF-YA5* upon drought. In Arabidopsis, an *NF-YA5 cis*-natural antisense gene called *NERF* (*NF-YA5 ENHANCING RING FINGER*) can produce short interfering RNAs (siRNAs) having sequence similar to miR169 that cannot direct the cleavage of *NF-YA5* transcripts. Both miRNA and siRNA compete for the *NF-YA5* binding preventing miR169-mediated repression of *NF-YA5* expression resulting in high accumulation of *NF-YA5* and enhanced drought tolerance in *NERF* overexpression lines (Gao et al. 2015b). Du et al. (2017) have demonstrated that miR169i and miR169l positively regulate *NF-YA5* expression via translational activation in response to dehydration shock in Arabidopsis. Another report in soybean suggests that gma-miR169c negatively regulates the drought stress response by inhibiting the expression of the targets *AtNF-YA1* and *AtNF-YA5* and reducing the transcript levels of the stress response genes *AtRD29A*, *AtRD22*, *AtGSTU25*, and *AtCOR15A* (Yu et al. 2019). Feyissa et al. (2019) have shown that low to moderate levels of miR156 expression are adequate to stimulate drought tolerance in alfalfa. Moderate levels of miR156 expression silence *SPL13* induce WD40–1 expression to adjust *DFR* (*DIHYDROFLAVONOL-4-REDUCTASE*) expression for the biosynthesis of anthocyanin and regulate various developmental, physiological, and biochemical processes in alfalfa leading to improved drought resilience.

Salt stress causes both osmotic stress and ion toxicity in plants (Huang et al. 2019). Many miRNAs have been reported as potential candidates for salt stress tolerance in plants having targets that are TFs as well as genes which function in electron-transfer shuttles between proteins, laccase, K^+ transporter gene *HAK5*, and others (Ding et al. 2009; Li et al. 2013; Wang et al. 2013; Sun et al. 2015a; Yuan et al. 2015). In perennial creeping bentgrass (*Agrostis stolonifera*), overexpression of rice osa-miR396c and subsequent downregulation of its target TF *GRF* provides salt tolerance by improving water retention, enhanced chlorophyll content, cell membrane integrity, and Na^+ exclusion during high salinity exposure (Yuan et al. 2019). Another study in bentgrass shows salt stress induced accumulation of mature miR319 (Zhou and Luo 2014). Increased miR319 levels cause the downregulation of targets *AsPCF5*, *AsPCF6*, *AsPCF8* and the TF *AsTCP14* (*TEOSINTE BRANCHED1, CYCLOIDEA, PROLIFERATING CELL NUCLEAR ANTIGEN BINDING FACTOR*) and positively contribute towards salinity tolerance in bentgrass. Liu et al. (2020a) have shown that overexpressing osa-miR319b and downregulating the targets positively regulate ethylene synthesis and salt tolerance in switchgrass (*Panicum virgatum*). Salt stress enhances the expression of gma-miR172 with a strong peak at 6 h (Song et al. 2011). Furthermore, transient overexpression of gma-miR172a in soybean significantly enhances the survival rate than that of the vector control plants (Pan et al. 2016). Detailed molecular analysis has revealed that miR172a promotes salt tolerance mainly through cleaving the *AP2/EREBP*-type TF gene *SSAC1* to relieve its protein inhibition on thiamine biosynthesis gene *THI1* that encodes a positive regulator of salt tolerance (Pan et al. 2016).

8.4 miRNAs-TFs: Regulating Cold Stress

Low temperature includes chilling (0–10 °C) and freezing (<4 °C) and is known to impact the longevity and geographical distribution of plants (Josine et al. 2011). Transcriptional control of the expression of cold responsive genes is well known (Chinnusamy et al. 2010). miRNAs have also been added to the suite of cold responsive gene regulatory networks. Many cold stress responsive miRNAs, including miR396, miR397, and miR319 have been identified in various plant species, such as wheat, rice, Arabidopsis, tomato, and *Brachypodium distachyon* (Fig. 8.1 and Table 8.1) (Zhou et al. 2008; Barrera-figueroa et al. 2012; Tang et al. 2012; Cao et al. 2014; Zhang et al. 2014b). In plants, cold stress induces a different set of responses depending on the species. For example, during cold stress, miR172 is induced in *Brachypodium* and *Prunus persica* (Zhang et al. 2009; Barakat et al. 2012) but is repressed in grapevine and wheat (Tang et al. 2012; Sun et al. 2015b). In sugarcane and rice, cold stress leads to the induction of evolutionarily conserved miR319. Overexpression of miR319 downregulates *TCP* transcription factor genes and enhances cold tolerance in both species (Thiebaut et al. 2012; Yang et al. 2013), suggesting that miR319 acts as a positive regulator of cold tolerance. Plants overexpressing miR397 are tolerant to chilling stress of 4 °C for 2 months (Dong and Pei 2014). Higher transcript levels of cold induced *CBF* (*C-repeat/dehydration binding factor*) TF genes and downstream cold responsive genes in miR397a overexpression plants indicate possible role of miR397a in the CBF-regulon (Dong and Pei 2014). Overexpression of miR394a and its target *F-box* gene *LCR* (*LEAF CURLING RESPONSIVENESS*) in Arabidopsis has demonstrated the positive role of this miRNA-target pair in response to low temperature stress (Song et al. 2016). *MIR394a* overexpressing plants have higher levels of *CBF1*, *CBF2,* and *CBF3* TF transcripts relative to the wild-type plants, suggesting the involvement of miR394 in CBF-dependent cold tolerance pathway. In rice, miR535 negatively regulates cold tolerance by aggravating cell death and ROS accumulation. Overexpression of osa-miR535 downregulates the expression of *SPL* TF genes during cold conditions in rice, thereby negatively regulating cold tolerance (Sun et al. 2020) (Table 8.1).

8.5 miRNA-TFs: Regulating Heavy Metal Stress

Plants acquire essential heavy-metal elements like iron (Fe), zinc (Zn), copper (Cu), and manganese (Mn) from the soil. These are required for many physiological and biochemical processes of the plant in small concentrations. However, when present in higher concentrations, they become harmful for the plant, often leading to oxidative stress and stunted growth (Noman et al. 2019). Certain non-essential heavy metals like lead (Pb), mercury (Hg), and cadmium (Cd) are also present in the heavy metal contaminated soils and can be taken up by the plants, thus leading to heavy metal toxicity (Gupta et al. 2014). Plants have devised various mechanisms to prevent the accumulation of these heavy metal ions at harmful concentrations. miRNAs play an important role in protecting the plant from heavy metal toxicity by targeting specific TFs involved in pathways which ultimately lead to translocation and compartmentalization of heavy metals, heavy metal chelation or control of oxidative damage (Noman and Aqeel 2017). Gao et al. (2019) have shown that miR156, miR166, miR167, and miR171, all of which target different TFs, are downregulated in response to cadmium stress in maize. The downregulation of miR171 (and thus accumulation of its target *GRAS* transcription factor transcripts) shows the highest fold-difference by qPCR and is also confirmed by in situ hybridization. Furthermore, overexpression of miR166 in rice improves Cd tolerance by reducing the Cd-induced oxidative burst in transgenic rice plants (Ding et al. 2018). Overexpression of miR166 reduces both Cd translocation from roots to shoots and Cd accumulation in the grains by targeting class-III HD-Zip transcription factor *HOMEODOMAIN CONTAINING PROTEIN4* (*OsHB4*) transcripts in rice (Ding et al. 2018). In *Medicago truncatula,* high Cd, Hg, and Al induce miR319 expression that leads to the degradation of the *TCP* TF genes that in turn affect leaf growth and senescence (Mendoza-Soto et al. 2012) (Fig. 8.1).

8.6 TFs-miRNA: Regulating Phosphate Homeostasis

Phosphate is a very crucial macroelement for all life forms. It is a major structural component of DNA/RNA and phospholipids that is also involved in energy transfer, protein activation, and metabolic regulation. Inorganic phosphate (Pi) is required for the growth and development of plants (Liu et al. 2014; Kuo and Chiou 2011; Nilsson et al. 2010). Plants acquire Pi from the soil by roots. But the primary challenge is low availability of Pi in soil as the majority of Pi gets converted into organic matter (Marschner 1995). Consequently, plants have evolved counter strategies to optimize Pi acquisition and assimilation. Phosphate starvation response 1 (PHR1) and PHL1 (PHR1-like 1) are MYB TFs that play a critical role in Pi deficiency signaling by transcriptionally regulating a wide range of phosphate starvation genes (Rubio et al. 2001; Bustos et al. 2010). PHR1 acts upstream to the miR399 by controlling the transcription of all six forms of miR399 (Bari et al. 2006). Pi deficiency induced miR399 reduces the expression of its target gene *PHO2* (*phosphate over-accumulator* 2), an *E2 conjugase* gene (Fujii et al. 2005; Chiou et al. 2006; Zhu et al. 2020). Under Pi-sufficient condition, the overexpression of miR399 increases the uptake of Pi and allocation to shoots which eventually causes Pi toxicity in Arabidopsis (Fujii et al. 2005; Aung et al. 2006; Bari et al. 2006; Chiou et al. 2006). In constitutively overexpressing osa-miR399f or osa-miR399j transgenic rice, a similar phenotype is observed (Hu et al. 2011). The Pi content in shoot and root increases in heterologous overexpression of Arabidopsis miR399d in tomato due to enhanced expression of Pi transporter genes and improves proton exudation from roots (Gao et al. 2015a). Moreover, the expression of osa-miR399a and osa-miR399j is also regulated by an R2R3 MYB TF, OsMYB2P-1 in response to Pi deficiency (Dai et al. 2012). The transcriptional activity of osa-miR399a, osa-miR399f, and osa-miR399j is regulated by OsWRKY74 TF in Pi deficient conditions. Plants overexpressing *OsWRKY74* have enhanced expression of osa-miR399a, osa-miR399f, and osa-miR399j (Dai et al. 2016). The activity of miR399 is also controlled by a long non-coding RNA *IPS1* that contains a near-perfect non-cleavable binding site for miR399 that efficiently sequesters miR399 (Franco-Zorrilla et al. 2007) (Fig. 8.2). This results in reduction of miR399 levels leading to protection of *PHO2* transcripts from cleavage (Franco-Zorrilla et al. 2007). Notably, the regulatory module *IPS1*:miR399:*PHO2* is conserved within different plant species including Arabidopsis, soybean, common bean, *Medicago,* and tomato (Valdés-López et al. 2008; Branscheid et al. 2010; Liu et al. 2010; Hu et al. 2011; Huang et al. 2011; Wang et al. 2020). In rice, Pi deficiency response leads to the activation of OsPHR2 TF-regulated miRNA827:*OsSPX-MFS1,2* (*SYG1/Pho81/XPR1-MAJOR FACILITATOR SUPERFAMILY*) module that in turn maintains phosphate homeostasis (Wang et al. 2012a).

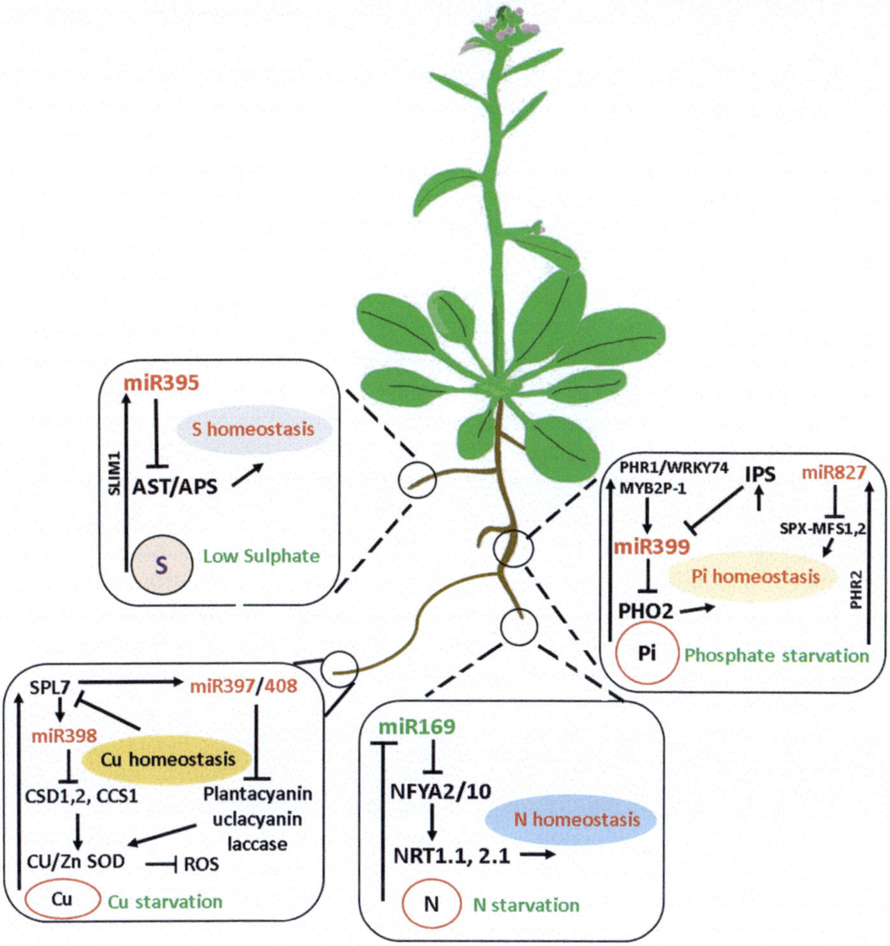

Fig. 8.2 TF:miRNA:target modules mediated signaling network for regulating nutrient homeostasis in plants. Low nutrients in soil upregulate signaling molecules like transcription factors (TFs) which can act as an activator of nutrient stress responsive miRNA transcription. Activation of miRNA attenuates the expression of target TFs. These miRNAs:TF modules stabilize the transport of nutrients and maintain the nutrient homeostasis by altering the expression of different transporter genes. Different transcription factors like SLIM1, SPL7, and PHR1 sense the nutrient deficiency and activate the transcription of *MIR395*, *MIR398*, and *MIR399*, respectively (positive regulation depicted by pointed arrowheads). These miRNAs further regulate their downstream targets (*AST/CSD/PHO2*; negative regulation depicted by blunt headed arrows) and maintain the nutrient homeostasis (sulfate/Cu/Pi/N) by altering the expression of different transporters

8.7 miRNAs-TFs: Regulating Nitrogen Homeostasis

Nitrogen (N) plays an essential role in growth and development of plants. It is a major component of nucleic acids, amino acids, co-enzymes, and a myriad of plant secondary metabolites. In soil, N is available in different forms, plants predominantly take up N in the form of nitrate and ammonium. The availability of N to plant roots is often an important limitation for plant growth and crop yield (Richardson et al. 2009; McAllister and Beatty 2012). To counter this, plants have developed multiple strategies, including physiological, morphological, and biochemical adaptations (Schachtman and Shin 2007; Kant et al. 2011; Kraiser et al. 2011). Plants can adapt in N-limiting soil conditions by up- or downregulating a specific group of nitrogen exporter or importer proteins. Several miRNAs regulating transporters have been reported in the literature. In Arabidopsis, nitrogen responsive miRNAs are classified into two groups, the first is named as N-starvation-induced (NSI) miRNA families that include miR156, miR160, miR169, miR171, miR319, miR826, miR829, miR839, and miR846, whereas miR167, miR172, miR399, miR395, miR850, miR857, miR863, and miR827 are grouped into N-starvation-suppressed (NSS) miRNAs (Liang et al. 2012). Two members of the NSS group. Viz. miR167 and miR393 regulate root growth in response to N (Gifford et al. 2008; Vidal et al. 2010). Nitrate deficiency mediated regulation of miRNA expression is reported in multiple plants (Pant et al. 2009; Jeong et al. 2011; Fischer et al. 2013; Liu et al. 2020b; Hou et al. 2020; Vakilian 2020). About 15 and 14 miRNA families have been identified to be responsive in N-limiting conditions in rice and maize, respectively (Xu et al. 2011). In Arabidopsis, miR156 family has been found at the highest abundance and miR156h is thought to be the most important among the three members of the miR156 family under N-limited conditions (Liang et al. 2012). N-starvation mediated induction of miR160 inhibits lateral root development, whereas miR170 mediated cleavage of *AUXIN RESPONSE FACTOR* (ARF16/17) TF and *SCL6* regulatory protein transcripts hastens the primary root growth, respectively (Liang et al. 2012). In contrast, the perturbation of miR167 biogenesis in N-limiting condition attenuates the expression of *ARF6/8*, which in turn facilitates the development of lateral and adventitious roots (Jones-Rhoades and Bartel 2004; Gifford et al. 2008). Furthermore, nitrate deficiency leads to downregulation of miR169a and upregulation of its target *NF-YA* TF family members in Arabidopsis. The miR169-NF-YA module likely regulates the adaptive response of nitrate uptake systems as is evident by overexpression of miR169. Enhanced miR169 levels reduce expression of multiple nitrate transporter genes and cause an early senescence phenotype (Zhao et al. 2011). This regulatory role of miR169:*NF-YA* module in response to nitrate starvation is also conserved in wheat where overexpression of *TaNF-YAB1* significantly enhances the uptake of both nitrate and phosphate and leads to enhanced grain yield in a field experiment with different levels of nutrient supply (Qu et al. 2015).

8.8 TFs-miRNA: Regulating Copper Homeostasis

In plants, the micronutrient, copper (Cu) plays indispensable role in plant growth and development by serving as an essential cofactor for many proteins operative in biological processes like photosynthesis, ethylene perception, ROS detoxification, and cell wall metabolism (Marschner 1995; Burkhead et al. 2009). The protein classes that require Cu as a cofactor mainly belong to plastocyanin, copper/zinc (Cu/Zn) superoxide dismutase (CSD), cytochrome c oxidase, plantacyanin (plastocyanin-like copper containing proteins), ethylene receptors, salicylic acid receptor NPR1, ascorbate oxidases, amine oxidases, polyphenol oxidase, and laccases (Carr and Winge 2003; Weigel et al. 2003; Kuper et al. 2004; Mayer 2006; Puig et al. 2007; Burkhead et al. 2009; Wu et al. 2012). Deep investigation on copper homeostasis in diverse plant species has delineated the conserved regulatory TF:Cu-miRNAs:target modules. Under limited Cu availability, plants activate the Cu-economy mode wherein the activation of Cu-sensing transcription results in the transcriptional activation of several Cu-miRNAs (Yamasaki et al. 2007). This leads to the mRNA cleavage of the non-essential copper requiring proteins to save copper for the most crucial and abundant protein plastocyanin and cytochrome oxidase, which is required for plant autotrophic growth. Among the known Cu-miRNAs miR408, miR398, and miR397 are conserved in plants, while few miRNAs are species-specific including miR528 (monocot specific), miR508 (grapevine), miR1444 (popular), miR1073 (mosses), and miR857 (Arabidopsis) (Sunkar et al. 2006; Yamasaki et al. 2007; Abdel-Ghany and Pilon 2008; Zhang and Li 2013; Balyan et al. 2017). The transcriptional regulation of the above Cu-miRNAs in response to Cu requires the conserved TF AtSPL7 in Arabidopsis (Yamasaki et al. 2009) and OsSPL9 (Balyan et al. 2017) in rice. The above SPL members share similarity with *Chlamydomonas reinhardtii* CRR1 (copper response regulator), which is the transcriptional regulator of Cu-homeostasis (Kropat et al. 2005). AtSPL7, CRR1, and OsSPL9 bind to the Cu-response element (CuRE) having GTAC core *cis*-regulatory motif in the promoter regions of their target *MIR* genes and other protein genes like *copper transporters* (*COPTs*) (Kropat et al. 2005; Yamasaki et al. 2009; Balyan et al. 2017). In *Arabidopsis,* SPL7 regulates the transcription of *MIR398, MIR408, MIR857, MIR397, MIR159, COPT1, COPT2, ZIP2, FRO3, COPT6,* and *CCH*, as well as a large number of genes involved in the photosynthesis process (Yamasaki et al. 2009; Jung et al. 2012; Zhang et al. 2014a). While in rice, OsSPL9 regulates miR408 and miR528 expression (Balyan et al. 2017; Yao et al. 2019; Yang et al. 2019). In addition, elongated Hypocotyl 5 (HY5, a bzip TF), the master regulator of light signaling regulates the transcription of *MIR408* in Arabidopsis (Zhang et al. 2011, 2014a). The transcription of *MIR408* is co-regulated by SPL7 and HY5 in *Arabidopsis* under changing light and copper regimens (Zhang et al. 2011, 2014a; Zhang and Li 2013). The target prediction, experimental evidences, and degradome sequencing provide evidence that the majority of the target genes of the above described Cu-miRNA encode for proteins that require 1–4 Cu molecules as cofactors. The very first Cu-responsive miRNA, miR398 targets the *CSD1*, *CSD2*, *CCS1*, *blue copper-binding protein* (*BCBP*) and a subunit of mitochondrial *cytochrome oxidase*, *COX5b*-1 mRNAs in Arabidopsis (Sunkar et al. 2006; Yamasaki et al. 2007; Beauclair et al. 2010; Brousse et al. 2014). miR408 targets *plantacyanin* gene family members in rice (Li et al. 2010; Zhou et al. 2010; Mutum et al. 2016). miR397 is involved in the downregulation of the *laccase* gene family members (Abdel-Ghany and Pilon 2008; Lu et al. 2013; Wang et al. 2014). The above described copper Cu-miRNA:targets are organized into intertwined feedback loops that regulate the cellular Cu levels (Pilon 2017). The SPL7 TF governs the abundance of Cu-miRNA:target modules, by sensing any change in the available pool of Cu. Lower levels of Cu lead to activation of SPL7 which enhances the transcription of Cu-miRNAs, which, in turn, lead to the reduction in target gene expression causing a re-adjustment of the system by diminishing the Cu utilization by target genes (Pilon 2017). The above regulatory module helps plants to regulate the local and systemic distribution of Cu making the plant to modulate the Cu protein expression and development (Pilon 2017). The Cu-miRNA shows a wide range of stress response and developmental roles suggesting the critical involvement of their regulatory functions across diverse plant species (Pilon 2017).

8.9 TFs-miR395: Regulating Sulfate Homeostasis

Sulfur is another essential macronutrient available in soil in the form of sulfate. On absorption, by plants, sulfate is initially assimilated into the amino acid cysteine and subsequently into glutathione, phytoalexins, and glucosinolates required by plant for optimum growth under normal and stress conditions (Rausch and Wachter 2005). During sulfate deprivation, plants tightly regulate sulfur uptake and metabolism at different levels and miRNAs are also involved in this regulatory network (Lewandowska and Sirko 2008). In sulfate deprived conditions, the *SLIM1* (SULFUR LIMITATION 1) TF gets accumulated which induces the expression of miR395 (Liang et al. 2010). Enhanced miR395 levels further target the *APS* (*ATP SULFURYLASE*) genes and thus regulate the accumulation of sulfate in shoot. miR395 also cleaves the transporter *SULTR2;1* (*SULFATE TRANSPORTER 2;1*) gene which finally affects the translocation of sulfate between the leaves. Besides miR395, levels of miR160, miR164, miR167, miR168, miR156, and miR394 are also altered during sulfate deprivation in *Brassica rapa* (Huang et al. 2010), suggesting possible roles of these miRNAs in modulating necessary growth and developmental adjustments during sulfate deprived conditions (Fig. 8.2 and Table 8.1).

8.10 Conclusion and Future Perspective

Plant stress response is a complex trait involving some mechanisms that are conserved across plant species while other processes that are specific to only a few plants. The role of TFs in various stress responses is well worked out in many plant species, highlighting their fundamental role and placing them at the top of the regulatory hierarchy. Conversely, since the discovery of miRNAs, the progress on the roles of miRNAs as critical regulators of transcriptome has gained momentum defining various abiotic responses. This is in part owing to the fact that most of the targets of miRNAs are TFs themselves. Thus, many TF and miRNA regulatory modules have emerged as key nodes in the plant abiotic stress networks that can be categorized into two genetic sub-networks: one where miRNA controls the post-transcriptional regulation of its target TF mRNAs directly by transcript cleavage or by translational repression and second where the TF drives the miRNA gene transcription. Several studies form the basis for the first sub-network through utilizing the NGS and degradome technologies. However, only a few have been functionally validated *in-planta* for stress tolerance. Some such examples where miRNAs regulate various TF families include miR159:*MYB*, miR169:*NF-YA*, miR396:*WRKY*, miR164:*NAC*, miR395:*GRF,* and miR156:*SPL.* These modules have been validated by different wet lab experiments for their specific roles in drought tolerance, heat tolerance, nutrient homeostasis, salinity, and cold tolerance by analyzing transgenic plants with miRNA overexpression, target overexpression, resistant targets that bypass miRNA cleavage and miRNA target-mimics that sponge up the mature miRNA. Several of these miRNA:TF modules exhibit similar expression in different stress conditions which suggest that they are part of a canonical response that is required to counter stress. However, there are some modules that show contrasting regulation in response to different stress regimens suggesting a possible functional diversification of mature miRNAs to target novel forms or these miRNA genes have acquired new regulatory regions in their promoters that respond to different stresses by recruiting stress-specific TFs. In nature, plants always encounter a combination of different stress. Therefore, it is important to study the miRNA-TF sub-networks in combination of multiple stress conditions to not only understand their mechanism of action but also to be able to identify modules that would give broad-spectrum tolerance.

One hurdle that needs attention is how specific are these miRNA:TF modules? This question stems from the fact that a single miRNA can functionally regulate several targets at a time depending upon tissue and stress regimens. This requires specific and sophisticated strategies to modulate miRNA and target levels *in-planta* followed by careful phenotyping and assessing stress response. It is also the need of the hour to revisit, compile, and reanalyze the previously published large amount of next-generation sequence data on different tissues and molecular levels and uncover novel aspects about not just for the TF:miRNA but also about plant stress regulatory networks. Still, very few reports are available in literature on the TFs as upstream regulator of miRNA transcription. Thus, continued inputs are required for the experimental mapping of transcriptional networks using various methods like chromatin immunoprecipitation (ChIP) and yeast one hybrid (Y1H) assays. Regulatory interactions of TFs and miRNAs determine the expression of specific genes in a spatio-temporal manner that in turn governs the implementation of particular cellular or developmental processes. Current research focuses majorly on the networks operating at tissue or organ level, however to get detailed and significant insights of particular process, future research needs to explore the networks operating at cellular levels.

Till now, our understanding about the stress regulatory mechanisms has primarily been confined to model plant species, thus, research efforts need to be channelized to gain similar level of in-depth knowledge for crop plants. Moreover, analyzing the function of the above sub-networks in contrasting cultivars exhibiting either tolerance or sensitivity could narrow down regulatory pathways governing tolerance to a particular stress. These established miRNAs and TFs networks could be pivotal in guiding future research to engineer or develop plants with desired phenotypes and stress tolerance ability. These stress resilient crop plants will be able to grow in adverse climatic conditions, thereby offering a solution to food security to an ever-increasing world population.

Acknowledgement Grants from DBT-NIPGR, New Delhi, India, DST-SERB India grant (EMR/2016/006229), CSIR, DST-ISPIRE and University Grants Commission, Government of India are acknowledged for providing fellowships. The authors are thankful to DBT-eLibrary Consortium (DeLCON) for providing access to e-resources.

References

Abdel-Ghany SE, Pilon M (2008) MicroRNA-mediated systemic down-regulation of copper protein expression in response to low copper availability in Arabidopsis. J Biol Chem 283 (23):15932–15945

Addo-Quaye C, Snyder JA, Park YB, Li YF, Sunkar R, Axtell MJ (2009) Sliced microRNA targets and precise loop-first processing of MIR319 hairpins revealed by analysis of the Physcomitrella patens degradome. RNA 15(12):2112–2121

Arshad M, Feyissa BA, Amyot L, Aung B, Hannoufa A (2017) MicroRNA156 improves drought stress tolerance in alfalfa (Medicago sativa) by silencing SPL13. Plant Sci 258:122–136

Aung K, Lin SI, Wu CC, Huang YT, Su CL, Chiou TJ (2006) Pho2, a phosphate overaccumulator, is caused by a nonsense mutation in a microRNA399 target gene. Plant Physiol 141 (3):1000–1011

Balyan S, Kumar M, Mutum RD, Raghuvanshi U, Agarwzal P, Mathur S, Raghuvanshi S (2017) Identification of miRNA-mediated drought responsive multi-tiered regulatory network in drought tolerant rice, Nagina 22. Sci Rep 7(1):1–17

Barakat A, Sriram A, Park J, Zhebentyayeva T, Main D, Abbott A (2012) Genome wide identification of chilling responsive microRNAs in Prunus persica. BMC Genomics 13:481

Bari R, Datt Pant B, Stitt M, Scheible WR (2006) PHO2, microRNA399, and PHR1 define a phosphate signaling pathway in plants. Plant Physiol 141:988–999

Barku MM, Bashasab F, Sudipta G, Krishnaraj PU (2013) LNA mediated in situ hybridization of miR171 and miR397a in leaf and ambient root tissues revealed expressional homogeneity in response to shoot heat shock in Arabidopsis thaliana. J Plant Biochem Biotechnol 23:93–103

Barrera-Figueroa BE, Gao L, Wu Z, Zhou X, Zhu J, Jin H, Liu R, Zhu JK (2012) High throughput sequencing reveals novel and abiotic stress-regulated microRNAs in the inflorescences of rice. BMC Plant Biol 12(1):132

Bartels D, Sunkar R (2005) Drought and salt tolerance in plants. Crit Rev Plant Sci 24(1):23–58

Baxter A, Mittler R, Suzuki N (2014) ROS as key players in plant stress signaling. J Exp Bot 65:1229–1240

Beauclair L, Yu A, Bouché N (2010) microRNA-directed cleavage and translational repression of the copper chaperone for superoxide dismutase mRNA in Arabidopsis. Plant J 62(3):454–462

Benková E, Ivanchenko MG, Friml J, Shishkova S, Dubrovsky JG (2009) A morphogenetic trigger: is there an emerging concept in plant developmental biology? Trends Plant Sci 14:189–193

Bielach A, Hrtyan M, Tognetti VB (2017) Plants under stress: involvement of auxin and cytokinin. Int J Mol Sci 4(18):7

Biłas R, Szafran K, Hnatuszko-Konka K, Kononowicz AK (2016) Cis-regulatory elements used to control gene expression in plants. Plant Cell Tissue Organ Cult 127(2):269–287

Branscheid A, Sieh D, Pant BD, May P, Devers EA, Elkrog A, Schauser L, Scheible WR, Krajinski F (2010) Expression pattern suggests a role of MiR399 in the regulation of the cellular response to local Pi increase during arbuscular mycorrhizal symbiosis. Mol Plant-Microbe Interact 23:915–926

Brousse C, Liu Q, Beauclair L, Deremetz A, Axtell MJ, Bouche N (2014) A non-canonical plant microRNA target site. Nucleic Acids Res 42(8):5270–5279

Burkhead JL, Gogolin Reynolds KA, Abdel-Ghany SE, Cohu CM, Pilon M (2009) Copper homeostasis. New Phytol 182(4):799–816

Bustos R, Castrillo G, Linhares F, Puga MI, Rubio V, Pérez-Pérez J, Solano R, Leyva A, Paz-Ares J (2010) A central regulatory system largely controls transcriptional activation and repression responses to phosphate starvation in Arabidopsis. PLoS Genet 6:e1001102

Cao X, Wu Z, Jiang F, Zhou R, Yang Z (2014) Identification of chilling stress-responsive tomato microRNAs and their target genes by high-throughput sequencing and degradome analysis. BMC Genomics 15(1):1–16

Carr HS, Winge DR (2003) Assembly of cytochrome c oxidase within the mitochondrion. Acc Chem Res 36(5):309–316

Casadevall R, Rodriguez RE, Debernardi JM, Palatnik JF, Casati P (2013) Repression of growth regulating factors by the microRNA396 inhibits cell proliferation by UV-B radiation in Arabidopsis leaves. Plant Cell 25:3570–3583

Chen Q, Li M, Zhang Z, Tie W, Chen X, Jin L, Zhai N, Zheng Q, Zhang J, Wang R, Xu G (2017) Integrated mRNA and microRNA analysis identifies genes and small miRNA molecules associated with transcriptional and post-transcriptional-level responses to both drought stress and re-watering treatment in tobacco. BMC Genomics 18(1):62

Cheng HY, Wang Y, Tao X, Fan YF, Dai Y, Yang H, Ma XR (2016) Genomic profiling of exogenous abscisic acid-responsive microRNAs in tomato (Solanum lycopersicum). BMC Genomics 17(1):423

Chinnusamy V, Zhu JK, Sunkar R (2010) Gene regulation during cold stress acclimation in plants. In: Plant stress tolerance. Humana press, pp 39–55

Chiou TJ, Aung K, Lin SI, Wu CC, Chiang SF, Su CL (2006) Regulation of phosphate homeostasis by microRNA in Arabidopsis. Plant Cell 18:412–421

Contreras-Cubas C, Palomar M, Arteaga-Vázquez M, Reyes JL, Covarrubias AA (2012) Non-coding RNAs in the plant response to abiotic stress. Planta 236(4):943–958

Dai X, Wang Y, Yang A, Zhang WH (2012) OsMYB2P-1, an R2R3 MYB transcription factor, is involved in the regulation of phosphate-starvation responses and root architecture in rice. Plant Physiol 159(1):169–183

Dai X, Wang Y, Zhang WH (2016) OsWRKY74, a WRKY transcription factor, modulates tolerance to phosphate starvation in rice. J Exp Bot 67(3):947–960

Ding D, Zhang L, Wang H, Liu Z, Zhang Z, Zheng Y (2009) Differential expression of miRNAs in response to salt stress in maize roots. Ann Bot 103(1):29–38

Ding Q, Zeng J, He XQ (2016) MiR169 and its target PagHAP2-6 regulated by ABA are involved in poplar cambium dormancy. J Plant Physiol 198:1–9

Ding Y, Gong S, Wang Y, Wang F, Bao H, Sun J, Cai C, Yi K, Chen Z, Zhu C (2018) MicroRNA166 modulates cadmium tolerance and accumulation in rice. Plant Physiol 177 (4):1691–1703

Dong CH, Pei H (2014) Over-expression of miR397 improves plant tolerance to cold stress in Arabidopsis thaliana. J Plant Biol 57:209–217

Du Q, Zhao M, Gao W, Sun S, Li WX (2017) microRNA/microRNA* complementarity is important for the regulation pattern of NFYA5 by miR169 under dehydration shock in Arabidopsis. Plant J 91(1):22–33

Fang Y, Xie K, Xiong L (2014) Conserved miR164-targeted NAC genes negatively regulate drought resistance in rice. J Exp Bot 65(8):2119–2135

Feng H, Zhang Q, Wang Q, Wang X, Liu J, Li M, Huang L, Kang Z (2013) Target of tae-miR408, a chemocyanin-like protein gene (TaCLP1), plays positive roles in wheat response to high-salinity, heavy cupric stress and stripe rust. Plant Mol Biol 83:433–443

Feyissa BA, Arshad M, Gruber MY, Kohalmi SE, Hannoufa A (2019) The interplay between miR156/SPL13 and DFR/WD40–1 regulate drought tolerance in alfalfa. BMC Plant Biol 19 (1):1–19

Fischer JJ, Beatt PH, Good AG, Muench DG (2013) Manipulation of microRNA expression to improve nitrogen use efficiency. Plant Sci 210:70–81

Franco-Zorrilla JM, Valli A, Todesco M, Mateos I, Puga MI, Rubio-Somoza I, Leyva A, Weigel D, Garcia JA, PazAres J (2007) Target mimicry provides a new mechanism for regulation of microRNA activity. Nat Genet 39(8):1033

Fujii H, Chiou TJ, Lin SI, Aung K, Zhu JK (2005) A miRNA involved in phosphate-starvation response in Arabidopsis. Curr Biol 15:2038–2043

Gao N, Qiang XM, Zhai BN, Min J, Shi WM (2015a) Transgenic tomato overexpressing ath-miR399d improves growth under abiotic stress conditions. Russian journal of plant physiology 62(3):360–366

Gao W, Liu W, Zhao M, Li WX (2015b) NERF encodes a RING E3 ligase important for drought resistance and enhances the expression of its antisense gene NFYA5 in Arabidopsis. Nucl Acid Res 43:607–617

Gao J, Luo M, Peng H, Chen F, Li W (2019) Characterization of cadmium-responsive MicroRNAs and their target genes in maize (Zea mays) roots. BMC Mol Biol 20(1):14

Giacomelli JI, Weigel D, Chan RL, Manavella PA (2012) Role of recently evolved miRNA regulation of sunflower HaWRKY6 in response to temperature damage. New Phytol 195:766–773

Gifford ML, Dean A, Gutierrez RA, Coruzzi GM, Birnbaum KD (2008) Cell-specific nitrogen responses mediate developmental plasticity. Proc Natl Acad Sci U S A 105(2):803–808

Gilbert ME, Medina V (2016) Drought adaptation mechanisms should guide experimental design. Trends Plant Sci 21(8):639–647

Golldack D, Li C, Mohan H, Probst N (2014) Tolerance to drought and salt stress in plants: unraveling the signaling networks. Front Plant Sci 5:151

Guan Q, Lu X, Zeng H, Zhang Y, Zhu J (2013) Heat stress induction of miR398 triggers a regulatory loop that is critical for thermotolerance in Arabidopsis. Plant J 74:840–851

Gupta OP, Sharma P, Gupta RK, Sharma I (2014) MicroRNA mediated regulation of metal toxicity in plants: present status and future perspectives. Plant Mol Biol 84(1–2):1–18

Hackenberg M, Shi BJ, Gustafson P, Langridge P (2013) Characterization of phosphorus-regulated miR399 and miR827 and their isomiRs in barley under phosphorus-sufficient and phosphorus-deficient conditions. BMC Plant Biol 13:214

Hernando CE, Romanowski A, Yanovsky MJ (2017) Transcriptional and post-transcriptional control of the plant circadian gene regulatory network. Biochim Biophy Acta (BBA) Gene Regul Mech 1860(1):84–94

Hou G, Du C, Gao H, Liu S, Sun W, Lu H, Kang J, Xie Y, Ma D, Wang C (2020) Identification of microRNAs in developing wheat grain that are potentially involved in regulating grain characteristics and the response to nitrogen levels. BMC Plant Biol 20(1):1–21

Hu B, Zhu C, Li F, Tang J, Wang Y, Lin A, Liu L, Che R, Chu C (2011) LEAF TIP NECROSIS1 plays a pivotal role in the regulation of multiple phosphate starvation responses in rice. Plant Physiol 156:1101–1015

Huang SQ, Xiang AL, Che LL, Chen S, Li H, Song JB, Yang ZM (2010) A set of miRNAs from Brassica napus in response to sulphate deficiency and cadmium stress. Plant Biotechnol J 8 (8):887–899

Huang CY, Shirley N, Genc Y, Shi B, Langridge P (2011) Phosphate utilization efficiency correlates with expression of low-affinity phosphate transporters and noncoding RNA, IPS1, in barley. Plant Physiol 156(3):1217–1229

Huang X, Feng J, Wang R, Zhang H, Huang J (2019) Comparative analysis of microRNAs and their targets in the roots of two cultivars with contrasting salt tolerance in rice (Oryza sativa L.). Plant Growth Regul 87(1):139–148

Iwamoto M, Tagiri A (2016) MicroRNA-targeted transcription factor gene RDD1 promotes nutrient ion uptake and accumulation in rice. Plant J 85:466–477

Jagadeeswaran G, Li YF, Sunkar R (2014) Redox signaling mediates the expression of a sulfate deprivation-inducible microRNA395 in Arabidopsis. Plant J 77:85–96

Jeong DH, Park S, Zhai J, Gurazada SG, De Paoli E, Meyers BC, Green PJ (2011) Massive analysis of rice small RNAs: mechanistic implications of regulated microRNAs and variants for differential target RNA cleavage. Plant Cell 23:4185–4207

Jiang D, Zhou L, Chen W, Ye N, Xia J, Zhuang C (2019) Overexpression of a microRNA-targeted NAC transcription factor improves drought and salt tolerance in Rice via ABA-mediated pathways. Rice 12(1):76

Jones-Rhoades MW, Bartel DP (2004) Computational identification of plant microRNAs and their targets, including a stress-induced miRNA. Mol Cell 14:787–799

Jones-Rhoades MW, Barte DP, Bartel B (2006) MicroRNAs and their regulatory roles in plants. Annu Rev Plant Biol 57:19–53

Josine TL, Ji J, Wang G, Guan CF (2011) Advances in genetic engineering for plants abiotic stress control. Afr J Biotechnol 10:5402–5413

Jung HI, Gayomba SR, Rutzke MA, Craft E, Kochian LV, Vatamaniuk OK (2012) COPT6 is a plasma membrane transporter that functions in copper homeostasis in Arabidopsis and is a novel target of SQUAMOSA promoter-binding protein-like 7. J Biol Chem 287(40):33252–33267

Kant S, Peng M, Rothstein SJ (2011) Genetic regulation by NLA and microRNA827 for maintaining nitrate-dependent phosphate homeostasis in Arabidopsis. PLoS Genet 7:e1002021

Kawashima CG, Matthewman CA, Huang S, Lee BR, Yoshimoto N, Koprivova A, Rubio-Somoza-I, Todesco M, Rathjen T, Saito K, Takahashi H (2011) Interplay of SLIM1and miR395 in the regulation of sulfate assimilation in Arabidopsis. Plant J 66:863–876

Kim S, Yang JY, Xu J, Jang IC, Prigge MJ, Chua NH (2008) Two cap-binding proteins CBP20 and CBP80 are involved in processing primary microRNAs. Plant Cell Physiol 49:1634–1644

Kim YJ, Zheng B, Yu Y, Won SY, Mo B, Chen X (2011) The role of mediator in small and long noncoding RNA production in Arabidopsis thaliana. EMBO J 30:814–822

Kinoshita N, Wang H, Kasahara H, Liu J, MacPherson C, Machida Y, Kamiya Y, Hannah MA, Chua NH (2012) IAA-Ala Resistant3, an evolutionarily conserved target of miR167, mediates Arabidopsis root architecture changes during high osmotic stress. Plant Cell 24:3590–3602

Kong X, Zhang M, Xu X, Li X, Li C, Ding Z (2014) System analysis of microRNAs in the development and aluminium stress responses of the maize root system. Plant Biotechnol J 12:1108–1121

Kotak S, Vierling E, Bäumlein H, von Koskull-Döring P (2007) A novel transcriptional cascade regulating expression of heat stress proteins during seed development of Arabidopsis. Plant Cell 19(1):182–195

Kraiser T, Gras DE, Gutiérrez AG, González B, Gutiérrez RA (2011) A holistic view of nitrogen acquisition in plants. J Exp Bot 62(4):1455–1466

Kropat J, Tottey S, Birkenbihl RP, Depege N, Huijser P, Merchant S (2005) A regulator of nutritional copper signaling in Chlamydomonas is an SBP domain protein that recognizes the GTAC core of copper response element. Proc Natl Acad Sci 102(51):18730–18735

Kruszka K, Pacak A, Swida-Barteczka A, Nuc P, Alaba S, Wroblewska Z, Karlowski W, Jarmolowski A, Szweykowska-Kulinska Z (2014) Transcriptionally and post-transcriptionally regulated microRNAs in heat stress response in barley. J Exp Bot 65(20):6123–6135

Kulcheski FR, de Oliveira LF, Molina LG, Almerão MP, Rodrigues FA, Marcolino J, Barbosa JF, Stolf-Moreira R, Nepomuceno AL, Marcelino-Guimarães FC, Abdelnoor R (2011) Identification of novel soybean microRNAs involved in abiotic and biotic stresses. BMC Genomics 12 (1):307

Kumar RR, Pathak H, Sharma SK, Kale YK, Nirjal MK, Singh GP (2014) Novel and conserved heat-responsive microRNAs in wheat (Triticum aestivum L.). Funct Integr Genomics 15:1–26

Kuo HF, Chiou TJ (2011) The role of microRNAs in phosphorus deficiency signalling. Plant Physiol 156(3):1016–10124

Kuper J, Llamas A, Hecht HJ, Mendel RR, Schwarz G (2004) Structure of the molybdopterin-bound Cnx1G domain links molybdenum and copper metabolism. Nature 430(7001):803–806

Lee H, Yoo SJ, Lee JH, Kim W, Yoo SK, Fitzgerald H, Carrington JC, Ahn JH (2010) Genetic framework for flowering-time regulation by ambient temperature-responsive miRNAs in Arabidopsis. Nucleic Acids Res 38(9):3081–3093

Lee SH, Li HW, Koh KW, Chuang HY, Chen YR, Lin CS, Chan MT (2014) MSRB7 reverses oxidation of GSTF2/3 to confer tolerance of Arabidopsis thaliana to oxidative stress. J Exp Bot 65:5049–5062

Lei KJ, Lin YM, An GY (2016) miR156 modulates rhizosphere acidification in response to phosphate limitation in Arabidopsis. J Plant Res 129(2):275–284

Lewandowska M, Sirko A (2008) Recent advances in understanding plant response to sulfur-deficiency stress. Acta Biochim Pol 55(3):457–471

Li WX, Oono Y, Zhu J, He XJ, Wu JM, Iida K, Lu XY, Cui X, Jin H, Zhu JK (2008) The Arabidopsis NFYA5 transcription factor is regulated transcriptionally and posttranscriptionally to promote drought resistance. Plant Cell 20:2238–2251

Li YF, Zheng Y, Addo-Quaye C, Zhang L, Saini A, Jagadeeswaran G, Axtell MJ, Zhang W, Sunkar R (2010) Transcriptome-wide identification of microRNA targets in rice. Plant J 62(5):742–759

Li B, Duan H, Li J, Deng XW, Yin W, Xia X (2013) Global identification of miRNAs and targets in Populus euphratica under salt stress. Plant Mol Biol 81(6):525–539

Li MY, Wang F, Xu ZS, Jiang Q, Ma J, Tan GF, Xiong AS (2014) High throughput sequencing of two celery varieties small RNAs identifies microRNAs involved in temperature stress response. BMC Genomics 15(1):242

Li H, Wang Y, Wang Z, Guo X, Wang F, Xia XJ, Zhou J, Shi K, Yu JQ, Zhou YH (2016) Microarray and genetic analysis reveal that csa-miR159b plays a critical role in abscisic acid mediated heat tolerance in grafted cucumber plants. Plant Cell Environ 39(8):1790–1804

Liang G, Yang F, Yu D (2010) MicroRNA395 mediates regulation of sulfate accumulation and allocation in Arabidopsis thaliana. Plant J 62(6):1046–1057

Liang G, He H, Yu D (2012) Identification of nitrogen starvation-responsive microRNAs in Arabidopsis thaliana. PLoS One 7(11):e48951

Liebsch D, Palatnik JF (2020) MicroRNA miR396, GRF transcription factors and GIF co-regulators: a conserved plant growth regulatory module with potential for breeding and biotechnology. Curr Opin Plant Biol 53:31–42

Lin WY, Huang TK, Chiou TJ (2013) Nitrogen limitation adaptation, a target of microRNA827, mediates degradation of plasma membrane-localized phosphate transporters to maintain phosphate homeostasis in *Arabidopsis*. Plant Cell 25:4061–4074

Lin JS, Kuo CC, Yang IC, Tsai WA, Shen YH, Lin CC, Liang YC, Li YC, Kuo YW, King YC, Lai HM, Jeng ST (2018) MicroRNA160 modulates plant development and heat shock protein gene expression to mediate heat tolerance in Arabidopsis. Front Plant Sci 9:68

Liu Q, Zhang H (2012) Molecular identification and analysis of arsenite stress-responsive miRNAs in rice. J Agric Food Chem 60(26):6524–6536

Liu JQ, Allan DL, Vance CP (2010) Systemic signaling and local sensing of phosphate in common bean: cross-talk between photosynthate and microRNA399. Mol Plant 3:428–443

Liu TY, Lin WY, Huang TK, Chiou TJ (2014) MicroRNA-mediated surveillance of phosphate transporters on the move. Trends Plant Sci 19(10):647–655

Liu Y, Yan J, Wang K, Li D, Han Y, Zhang W (2020a) Heteroexpression of Osa-miR319b improved switchgrass biomass yield and feedstock quality by repression of PvPCF5. Biotechnol Biofuels 13(1):1–13

Liu ZW, Li H, Liu JX, Wang Y, Zhuang J (2020b) Integrative transcriptome, proteome, and microRNA analysis reveals the effects of nitrogen sufficiency and deficiency conditions on theanine metabolism in the tea plant (Camellia sinensis). Horticult Res 7(1):1–13

Lu C, Fedoroff N (2000) A mutation in the Arabidopsis HYL1 gene encoding a dsRNA binding protein affects responses to abscisic acid, auxin and cytokinin. Plant Cell 12:2351–2365

Lu X, Guan Q, Zhu J (2013) Downregulation of CSD2 by a heat-inducible miR398 is required for thermotolerance in Arabidopsis. Plant Signaling Behav 8:e24952

Ma C, Burd S, Lers A (2015) miR408 is involved in abiotic stress responses in Arabidopsis. Plant J 84:169–187

Marschner H (1995) Mineral nutrition of higher plants, 2nd edn. Academic, Great Britain

Matthews C (2018) Characterizing the role of the miR156-SPL network in heat stress response in Medicago sativa

May P, Liao W, Wu Y, Shuai B, McCombie WR, Zhang MQ, Liu QA (2013) The effects of carbon dioxide and temperature on microRNA expression in Arabidopsis development. Nat Commun 4 (1):1–11

Mayer AM (2006) Polyphenol oxidases in plants and fungi: going places? A review. Phytochemistry 67(21):2318–2331

McAllister PH, Beatty AG (2012) Good engineering nitrogen use efficient crop plants: the current status. Plant Biotechnol J 10:1011–1025

Mendoza-Soto AB, Sánchez F, Hernández G (2012) MicroRNAs as regulators in plant metal toxicity response. Front Plant Sci 3:105

Millar AA (2020) The function of miRNAs in plants. Plants 9(2):198

Mitsuda N, Ohme-Takagi M (2009) Functional analysis of transcription factors in Arabidopsis. Plant Cell Physiol 50(7):1232–1248

Mittler R, Vanderauwera S, Gollery M, Van Breusegem F (2004) Reactive oxygen gene network of plants. Trends Plant Sci 9:490–498

Mutum RD, Kumar S, Balyan S, Kansal S, Mathur S, Raghuvanshi S (2016) Identification of novel miRNAs from drought tolerant rice variety Nagina 22. Sci Rep 6(1):1–5

Nelson DE, Repetti PP, Adams TR, Creelman RA, Wu J, Warner DC, Anstrom DC, Bensen RJ, Castiglioni PP, Donnarummo MG, Hinchey BS (2007) Plant nuclear factor Y (NF-Y) B subunits confer drought tolerance and lead to improved corn yields on water-limited acres. Proc Natl Acad Sci U S A 104(42):16450–16455

Ni Z, Hu Z, Jiang Q, Zhang H (2013) GmNFYA3, a target gene of miR169, is a positive regulator of plant tolerance to drought stress. Plant Mol Biol 82:113–129

Nilsson L, Müller R, Nielsen TH (2010) Dissecting the plant transcriptome and the regulatory responses to phosphate deprivation. Physiol Plant 139(2):129–143

Nolan T, Chen J, Yin Y (2017) Cross-talk of Brassinosteroid signaling in controlling growth and stress responses. Biochem J 474:2641–2661

Noman A, Aqeel M (2017) miRNA-based heavy metal homeostasis and plant growth. Environ Sci Pollut Res 24(11):10068–10082

Noman A et al (2019) Crosstalk between plant miRNA and heavy metal toxicity. In: Sablok G (ed) Plant metallomics and functional omics. Springer, Cham

Pan WJ, Tao JJ, Cheng T, Bian XH, Wei W, Zhang WK, Ma B, Chen SY, Zhang JS (2016) Soybean miR172a improves salt tolerance and can function as a long-distance signal. Mol Plant 9 (9):1337–1340

Pant BD, Musialak-Lange M, Nuc P, May P, Buhtz A, Kehr J, Walther D, Scheible WR (2009) Identification of nutrient-responsive Arabidopsis and rapeseed microRNAs by comprehensive real-time polymerase chain reaction profiling and small RNA sequencing. Plant Physiol 150:1541–1555

Payne JL, Wagner A (2015) Mechanisms of mutational robustness in transcriptional regulation. Front Genet 6:322

Petroni K, Kumimoto RW, Gnesutta N, Calvenzani V, Fornari M, Tonelli C, Holt BF, Mantovani R (2012) The promiscuous life of plant nuclear factor Y transcription factors. Plant Cell 24 (12):4777–4792

Pilon M (2017) The copper microRNAs. New Phytol 213(3):1030–1035

Puig S, Andrés-Colás NU, García-Molina AN, Penarrubia L (2007) Copper and iron homeostasis in Arabidopsis: responses to metal deficiencies, interactions and biotechnological applications. Plant Cell Environ 30(3):271–290

Qu B, He X, Wang J, Zhao Y, Teng W, Shao A, Zhao X, Ma W, Wang J, Li B, Li Z (2015) A wheat CCAAT box-binding transcription factor increases the grain yield of wheat with less fertilizer input. Plant Physiol 167(2):411–423

Ramachandran P, Carlsbecker A, Etchells JP (2016) Class III HD-ZIPs govern vascular cell fate: an HD view on patterning and differentiation. J Exp Bot 68:55–69

Rao S, Balyan S, Jha S, Mathur S (2020) Novel insights into expansion and functional diversification of MIR169 family in tomato. Planta 251(2):55

Rausch T, Wachter A (2005) Sulfur metabolism: a versatile platform for launching defence operations. Trends Plant Sci 10(10):503–509

Richardson AE, Barea JM, McNeill AM, Prigent-Combaret C (2009) Acquisition of phosphorus and nitrogen in the rhizosphere and plant growth promotion by microorganisms. Plant Soil 321:305–339

Rodriguez RE, Mecchia MA, Debernardi JM, Schommer C, Weigel D, Palatnik JF (2010) Control of cell proliferation in Arabidopsis thaliana by microRNA miR396. Development 137 (1):103–112

Roy S (2016) Function of MYB domain transcription factors in abiotic stress and epigenetic control of stress response in plant genome. Plant Signaling Behav 11:e1117723

Rubio V, Linhares F, Solano R, Martín AC, Iglesias J, Leyva A, Paz-Ares J (2001) A conserved MYB transcription factor involved in phosphate starvation signaling both in vascular plants and in unicellular algae. Genes Dev 15:2122–2133

Salih H, Gong W, He S, Sun G, Sun J, Du X (2016) Genome-wide characterization and expression analysis of MYB transcription factors in Gossypium hirsutum. BMC Genet 17:129

Schachtman DP, Shin R (2007) Nutrient sensing and signaling: NPKS. Annu Rev Plant Biol 58:47–69

Serivichyaswat PT, Susila H, Ahn JH (2017) Elongated hypocotyl 5-homolog (HYH) negatively regulates expression of the ambient temperature-responsive microRNA gene MIR169. Front Plant Sci 8:2087

Song QX, Liu YF, Hu XY, Zhang WK, Ma B, Chen SY, Zhang JS (2011) Identification of miRNAs and their target genes in developing soybean seeds by deep sequencing. BMC Plant Biol 11(1):5

Song JB, Gao S, Wang Y, Li BW, Zhang YL, Yang ZM (2016) miR394 and its target gene LCR are involved in cold stress response in Arabidopsis. Plant Gene 5:56–64

Stief A, Altmann S, Hoffmann K, Pant BD, Scheible W, Bäurle I (2014) Arabidopsis miR156 regulates tolerance to recurring environmental stress through SPL transcription factors. Plant Cell 26:1792–1807

Sun X, Xu L, Wang Y, Yu R, Zhu X, Luo X, Gong Y, Wang R, Limera C, Zhang K, Liu L (2015a) Identification of novel and salt-responsive miRNAs to explore miRNA-mediated regulatory network of salt stress response in radish (Raphanus sativus L.). BMC Genom 16(1):197

Sun XM, Fan GT, Su LY, Wang WJ, Liang ZC, Li SH, Xin HP (2015b) Identification of cold-inducible microRNAs in grapevine. Front Plant Sci 6:595

Sun Q, Liu X, Yang J, Liu W, Du Q, Wang H, Fu C, Li WX (2018) MicroRNA528 affects lodging resistance of maize by regulating lignin biosynthesis under nitrogen-luxury conditions. Mol Plant 11(6):806–814

Sun M, Shen Y, Yang J, Cai X, Li H, Zhu Y, Jia B, Sun X (2020) miR535 negatively regulates cold tolerance in rice. Mol Breed 40(1):1–2

Sunkar R, Zhu JK (2004) Novel and stress-regulated microRNAs and other small RNAs from Arabidopsis. Plant Cell 16:2001–2019

Sunkar R, Kapoor A, Zhu JK (2006) Posttranscriptional induction of two Cu/Zn superoxide dismutase genes in Arabidopsis is mediated by downregulation of miR398 and important for oxidative stress tolerance. Plant Cell 18:2051–2065

Sunkar R, Chinnusamy V, Zhu J, Zhu JH (2007) Small RNAs as big players in plant abiotic stress responses and nutrient deprivation. Trends Plant Sci 12:301–309

Szaker HM, Darkó É, Medzhiradszky AR, Janda T, Liu HC, Charng YY, Csorba T (2019) miR824/AGAMOUS-LIKE16 module integrates recurring environmental heat stress changes to fine-tune post-stress development. Front Plant Sci 10:1454

Tang Z, Zhang L, Xu C, Yuan S, Zhang F, Zheng Y, Zhao C (2012) Uncovering small RNA-mediated responses to cold stress in a wheat thermosensitive genic male-sterile line by deep sequencing. Plant Physiol 159:721–738

Thiebaut F, Rojas CA, Almeida KL, Grativol C, Domiciano GC, Lamb CR, DE Almeida Engler JA, Hemerly AS, Ferreira PC (2012) Regulation of miR319 during cold stress in sugarcane. Plant Cell Environ 35:502–512

Trindade I, Capitão C, Dalmay T, Fevereiro MP, Dos Santos DM (2010) miR398 and miR408 are up-regulated in response to water deficit in Medicago truncatula. Planta 231(3):705–716

Vakilian KA (2020) Determination of nitrogen deficiency-related microRNAs in plants using fluorescence quenching of graphene oxide nanosheets. Mol Cell Probes:101576

Valdés-López O, Arenas-Huertero C, Ramírez M, Girard L, Sánchez F, Vance CP, Luis Reyes J, Hernández G (2008) Essential role of MYB transcription factor: PvPHR1 and microRNA: PvmiR399 in phosphorus-deficiency signalling in common bean roots. Plant Cell Environ 31:1834–1843

Vidal EA, Araus V, Lu C, Parry G, Green PJ, Coruzzi GM, Gutiérrez RA (2010) Nitrate-responsive miR393/AFB3 regulatory module controls root system architecture in Arabidopsis thaliana. Proc Natl Acad Sci U S A 107(9):4477–4482

Wang F, Perry SE (2013) Identification of direct targets of FUSCA3, a key regulator of Arabidopsis seed development. Plant Physiol 161:1251–1264

Wang W, Vinocur B, Shoseyov O, Altman A (2004) Role of plant heat-shock proteins and molecular chaperones in the abiotic stress response. Trends Plant Sci 9:244–252

Wang C, Huang W, Ying Y, Li S, Secco D, Tyerman S, Whelan J, Shou H (2012a) Functional characterization of the rice SPX-MFS family reveals a key role of OsSPX-MFS1 in controlling phosphate homeostasis in leaves. New Phytol 196(1):139–148

Wang Y, Sun F, Cao H, Peng H, Ni Z, Sun Q, Yao Y (2012b) TamiR159 directed wheat TaGAMYB cleavage and its involvement in anther development and heat response. PLoS One 7:e48445

Wang M, Wang Q, Zhang B (2013) Response of miRNAs and their targets to salt and drought stresses in cotton (Gossypium hirsutum L.). Gene 530(1):26–32

Wang J, Sun N, Deng T, Zhang L, Zuo K (2014) Genome-wide cloning, identification, classification and functional analysis of cotton heat shock transcription factors in cotton (Gossypium hirsutum). BMC Genomics 15(1):961

Wang Q, Liu N, Yang X, Tu L, Zhang X (2016) Small RNA-mediated responses to low-and high-temperature stresses in cotton. Sci Rep 6:35558

Wang R, Fang YN, Wu XM, Qing M, Li CC, Xie KD, Deng X, Guo WW (2020) The miR399-CsUBC24 module regulates reproductive development and male fertility in citrus. Plant Physiol. https://doi.org/10.1104/pp.20.00129

Weigel M, Varotto C, Pesaresi P, Finazzi G, Rappaport F, Salamini F, Leister D (2003) Plastocyanin is indispensable for photosynthetic electron flow in Arabidopsis thaliana. J Biol Chem 278 (33):31286–31289

Weng L, Bai X, Zhao F, Li R, Xiao H (2016) Manipulation of flowering time and branching by overexpression of the tomato transcription factor SlZFP2. Plant Biotechnol J 14(12):2310–2321
Wu Y, Zhang D, Chu JY, Boyle P, Wang Y, Brindle ID, De Luca V, Després C (2012) The Arabidopsis NPR1 protein is a receptor for the plant defense hormone salicylic acid. Cell Rep 1 (6):639–647
Xie Z, Allen E, Fahlgren N, Calamar A, Givan SA, Carrington JC (2005) Expression of Arabidopsis MIRNA genes. Plant Physiol 138:2145–2154
Xie Y, Liu Y, Wang H, Ma X, Wang B, Wu G, Wang H (2017) Phytochrome-interacting factors directly suppress MIR156 expression to enhance shade-avoidance syndrome in Arabidopsis. Nat Commun 8(1):1–11
Xu Z, Zhong S, Li X, Li W, Rothstein SJ, Zhang S, Bi Y, Xie C (2011) Genome-wide identification of microRNAs in response to low nitrate availability in maize leaves and roots. PLoS One 6(11): e28009
Yamasaki H, Abdel-Ghany SE, Cohu CM, Kobayashi Y, Shikanai T, Pilon M (2007) Regulation of copper homeostasis by micro-RNA in Arabidopsis. J Biol Chem 282(22):16369–16378
Yamasaki H, Hayashi M, Fukazawa M, Kobayashi Y, Shikanai T (2009) SQUAMOSA promoter binding protein–like7 is a central regulator for copper homeostasis in Arabidopsis. Plant Cell 21 (1):347–361
Yan X, Dong C, Yu J, Liu W, Jiang C, Liu J, Hu Q, Fang X, Wei W (2013) Transcriptome profile analysis of young floral buds of fertile and sterile plants from the self-pollinated offspring of the hybrid between novel restorer line NR1 and Nsa CMS line in Brassica napus. BMC Genomics 14(1):26
Yan Y, Wang H, Hamera S, Chen X, Fang R (2014) miR444a has multiple functions in the rice nitrate-signaling pathway. Plant J 78(1):44–55
Yan J, Zhao C, Zhou J, Yang Y, Wang P, Zhu X, Tang G, Bressan RA, Zhu JK (2016) The miR165/166 mediated regulatory module plays critical roles in ABA homeostasis and response in Arabidopsis thaliana. PLoS Genet 12:e1006416
Yang C, Li D, Mao D, Liu XU, Ji C, Li X, Zhao X, Cheng Z, Chen C, Zhu L (2013) Overexpression of microRNA319 impacts leaf morphogenesis and leads to enhanced cold tolerance in rice (Oryza sativa L.). Plant Cell Environ 36:2207–2218
Yang R, Li P, Mei H, Wang D, Sun J, Yang C, Hao L, Cao S, Chu C, Hu S, Song X (2019) Fine-tuning of MiR528 accumulation modulates flowering time in rice. Mol Plant 12(8):1103–1113
Yant L, Mathieu J, Dinh TT, Ott F, Lanz C (2010) Orchestration of the floral transition and floral development in Arabidopsis by the bifunctional transcription factor APETALA2. Plant Cell 22:2156–2170
Yao S, Yang Z, Yang R, Huang Y, Guo G, Kong X, Lan Y, Zhou T, Wang H, Wang W, Cao X (2019) Transcriptional regulation of miR528 by OsSPL9 orchestrates antiviral response in rice. Mol Plant 12(8):1114–1122
Yu X, Wang H, Lu Y, de Ruiter M, Cariaso M, Prins M, He Y (2011) Identification of conserved and novel microRNAs that are responsive to heat stress in Brassica rapa. J Exp Bot 63 (2):1025–1038
Yu Y, Ni Z, Wang Y, Wan H, Hu Z, Jiang Q, Sun X, Zhang H (2019) Overexpression of soybean miR169c confers increased drought stress sensitivity in transgenic Arabidopsis thaliana. Plant Sci 285:68–78
Yuan S, Li Z, Li D, Yuan N, Hu Q, Luo H (2015) Constitutive expression of rice microRNA 528 alters plant development and enhances tolerance to salinity stress and nitrogen starvation in creeping bentgrass. Plant Physiol 169:576–579
Yuan N, Yuan S, Li Z, Li D, Hu Q, Luo H (2016) Heterologous expression of a rice miR395 gene in Nicotiana tabacum impairs sulfate homeostasis. Sci Rep 6:28791
Yuan S, Zhao J, Li Z, Hu Q, Yuan N, Zhou M, Xia X, Noorai R, Saski C, Li S, Luo H (2019) MicroRNA396-mediated alteration in plant development and salinity stress response in creeping bentgrass. Horticult Res 6(1):48
Yue E, Liu Z, Li C, Li Y, Liu Q, Xu JH (2017a) Overexpression of miR529a confers enhanced resistance to oxidative stress in rice (Oryza sativa L.). Plant Cell Rep 36:1171–1182

Yue W, Ying Y, Wang C, Zhao Y, Dong C, Whelan J, Shou H (2017b) Os NLA 1, a RING-type ubiquitin ligase, maintains phosphate homeostasis in Oryza sativa via degradation of phosphate transporters. Plant J 90(6):1040–1051

Zanetti ME, Rípodas C, Niebel A (2017) Plant NF-Y transcription factors: key players in plant-microbe interactions, root development and adaptation to stress. Biochim Biophys Acta 1860 (5):645–654

Zarattini M, Forlani G (2017) Toward unveiling the mechanisms for transcriptional regulation of proline biosynthesis in the plant cell response to biotic and abiotic stress conditions. Front Plant Sci 2(8):927

Zhan X, Wang B, Li H, Liu R, Kalia RK, Zhu JK, Chinnusamy V (2012) Arabidopsis proline-rich protein important for development and abiotic stress tolerance is involved in microRNA biogenesis. Proc Natl Acad Sci 109(44):18198–18203

Zhang B (2015) MicroRNAs: a new target for improving plant tolerance to abiotic stress. J Exp Bot 66:1749–1761

Zhang H, Li L (2013) SQUAMOSA promoter binding protein-like7 regulated microRNA408 is required for vegetative development in Arabidopsis. Plant J 74(1):98–109

Zhang JF, Yuan LJ, Shao Y, Du W, Yan DW, Lu YT (2008) The disturbance of small RNA pathways enhanced abscisic acid response and multiple stress responses in *Arabidopsis*. Plant Cell Environ 31:562–574

Zhang J, Xu Y, Huan Q, Chong K (2009) Deep sequencing of Brachypodium small RNAs at the global genome level identifies microRNAs involved in cold stress response. BMC Genomics 10:449

Zhang H, He H, Wang X, Wang X, Yang X, Li L, Deng XW (2011) Genome-wide mapping of the HY5-mediated gene networks in Arabidopsis that involve both transcriptional and post-transcriptional regulation. Plant J 65(3):3463–3458

Zhang H, Zhao X, Li J, Cai H, Deng XW, Li L (2014a) MicroRNA408 is critical for the HY5-SPL7 gene network that mediates the coordinated response to light and copper. Plant Cell 26 (12):4933–4953

Zhang Y, Zhu X, Chen X, Song C, Zou Z, Wang Y, Wang M, Fang W, Li X (2014b) Identification and characterization of cold-responsive microRNAs in tea plant (Camellia sinensis) and their targets using high-throughput sequencing and degradome analysis. BMC Plant Biol 14(1):271

Zhang J, Zhang H, Srivastava AK, Pan Y, Bai J, Fang J, Shi H, Zhu JK (2018) Knockdown of rice microRNA166 confers drought resistance by causing leaf rolling and altering stem xylem development. Plant Physiol 176(3):2082–2094

Zhao B, Liang R, Ge L, Li W, Xiao H, Lin H, Ruan K, Jin Y (2007) Identification of drought-induced microRNAs in rice. Biochem Biophys Res Commun 354:585–590

Zhao M, Ding H, Zhu JK, Zhang F, Li WX (2011) Involvement of miR169 in the nitrogen-starvation responses in Arabidopsis. New Phytol 190:906–915

Zhou M, Luo H (2014) Role of microRNA319 in creeping bentgrass salinity and drought stress response. Plant Signaling Behav 9(4):1375–1391

Zhou M, Tang W (2019) MicroRNA156 amplifies transcription factor-associated cold stress tolerance in plant cells. Mol Gen Genomics 294(2):379–393

Zhou X, Wang G, Sutoh K, Zhu JK, Zhang W (2008). Identification of cold-inducible microRNAs in plants by transcriptome analysis. Biochim Biophys Acta (BBA) Gene Regul Mech 11:780–788

Zhou M, Gu L, Li P, Song X, Wei L, CHen Z, Cao X (2010) Degradome sequencing reveals endogenous small RNA targets in rice (Oryza sativa L. ssp. indica). Front Biol 5(1):67–90

Zhou M, Li D, Li Z, Hu Q, Yang C, Zhu L, Luo H (2013) Constitutive expression of a miR319 gene alters plant development and enhances salt and drought tolerance in transgenic creeping bentgrass. Plant Physiol 161:1375–1391

Zhu JK (2016) Abiotic stress signaling and responses in plants. Cell 167:313–324

Zhu FR, Qiu ZB, Zhang YM, Zhang XR, Wang WL (2020) Characterization and functional analysis of microRNA399 in Cunninghamia lanceolata. Biologia Plantarum 64:193–199

Phytohormones: A Promising Alternative in Boosting Salinity Stress Tolerance in Plants

9

Ayushi Mahajan, Meenakshi Raina, Sameena Ejaz Tanwir, Neha Katru, Akanksha Sharma, and Deepak Kumar

Abstract

In order to meet the food requirements of our growing population, the agricultural land has been expanded worldwide but still certain constraints (stress factors-abiotic and biotic) exist that limit the yield of many crops. These constraints/stress factors cause severe damage to the crop. These abiotic stresses including drought, salinity, cold and extreme temperature, etc. led to a drastic reduction in the yield of the number of crops and among all of these, salinity is becoming more severe problem day by day. Salinity hampers the growth and development of plants by inducing several changes like osmotic stress, ion toxicity, oxidative stress (ROS), membrane disorganisation, etc. and thus poses a major threat to agricultural yield. Several approaches have been applied to develop salt-tolerant plants to reduce the destruction of crops caused by salinity. In this regard, conventional breeding proved very tedious while in vitro techniques proved useful as they provide keen

A. Mahajan · N. Katru · A. Sharma
Department of Botany, Central University of Jammu, Rahya-Suchani (Bagla), Jammu and Kashmir, India

M. Raina
Department of Botany, Central University of Jammu, Rahya-Suchani (Bagla), Jammu and Kashmir, India

Department of Biotechnology, Government Degree College (Boys), Udhampur, Jammu and Kashmir, India

S. E. Tanwir
Citrus Research and Education Centre, University of Florida, Lake Alfred, FL, USA

D. Kumar (✉)
Department of Botany, Institute of Science, Banaras Hindu University, Varanasi, Uttar Pradesh, India
e-mail: deepakkumar@bhu.ac.in

B. Giri, M. P. Sharma (eds.), *Plant Stress Biology*,
https://doi.org/10.1007/978-981-15-9380-2_9

insight into plant's physiology growing under stress conditions and thereby helps to develop stress-tolerant plants. In recent years, several salt-tolerant plants have also been developed using a transgenic approach involving genes involved in the biosynthesis of individual phytohormones. This chapter reviews in brief about the approaches used in the development of salt-stress-tolerant plants.

Keywords

Jasmonic acid · Phytohormones · Reactive oxygen species · Salt stress · Salicylic acid · Transgenic

9.1 Introduction

Agriculture plays a vital role in the Indian economy and will continue to do so for a long time. It meets the food requirements of a large population of our country. However, it is observed that the yield of various crop plants is greatly affected by stress factors, both biotic (pests, rodents, etc.) and abiotic factors (salinity, temperature, drought, etc.), prevailing in their environment. A large-scale agricultural loss has been observed due to these abiotic stress factors (Boyer 1982). Among all these stress factors, salinity is a major factor that limits the growth and productivity of plants. About 6.74 million hectares of land is salinity affected in India and this proportion is estimated to increase to 16.2 million hectares by 2050 (CSSRI 2015). Salinity affects the growth and productivity of plants greatly as it leads to osmotic stress, ionic imbalance, ionic toxicity, reduction in gaseous exchanges, etc. and increases susceptibility to other stresses. Osmotic stress degrades the ability of a plant under stress to detoxify reactive oxygen species (ROS). ROS cause immense damage to proteins, lipids, etc. that leads to membrane disintegration and even cell death (Gill and Tuteja 2010). In order to develop salinity-tolerant plants, it is essential to have detailed information regarding the metabolism, functioning and responses of plants under salinity. Different approaches have been adopted within the past few years to develop plants that are tolerant to the various environmental stresses (Jisha et al. 2013; Tran et al. 2010). Due to the complexity associated with salinity tolerance, limited success is achieved to develop salt-tolerant plants through breeding but with genetic engineering, it is possible to do so. Scientists also have tested the potential of phytohormones in alleviating the harmful effects of salinity by regulating various plants' processes. Having a complete understanding of molecular mechanisms and genes associated with salt tolerance, it is possible to develop plants with salt tolerance using transgenic technology. To sum up, the different techniques being used to alleviate the harmful effects of salt stress are transgenic approach/ genetic engineering, invitro-strategies, phytohormone application, etc. In the present book chapter, we emphasise on different strategies that have been used to alleviate the harmful effects of salinity which thereby results in the production of salinity-tolerant plants.

9.2 Environmental Constraints Limiting Agricultural Production

Food is the basic necessity of life and to fulfil this need, agriculture plays a very crucial role. Several strategies have been applied in the agricultural field to increase the yield of crops but still the crop plants come across various constraints (stress factors) that hinder their growth and yield. These factors are majorly categorised as biotic factors (pests, rodents, etc.) and abiotic factors (salinity, drought, heavy metal, etc.) as depicted in Fig. 9.1. All these stress factors hamper the normal growth and development of plants and thereby, hinder sustained food production. More than 50% loss in productivity in most of the crop plants has been observed due to abiotic stresses including drought, salinity, cold and extreme temperature, etc. (Bray et al. 2000; Wang et al. 2003).

9.2.1 Salinity as Major Abiotic Stress

The different abiotic stresses limiting plant growth and metabolism include salinity, drought, heat, cold, metal stress, etc. Among all these stress factors, salinity has a major negative impact on plants and pose a major threat to global food security. The proportion of salinity affected cultivated land worldwide is estimated to be 20% and this figure is increasing day by day. Salinity of soil can be defined as existence of high concentration of soluble salts in the soil. Soils having an electrical conductivity

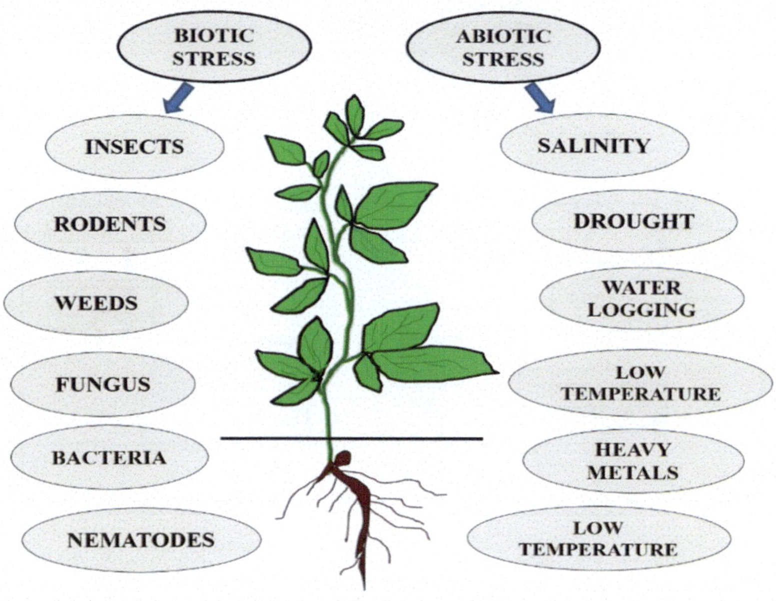

Fig. 9.1 List of stress factors affecting plant growth and metabolism

of 4 dS/m (generating an osmotic pressure of approximately 0.2 MPa) or more are categorised as saline soils (USDA-ARS 2008). Poor drainage, improper irrigation, irrigation water with high levels of salts, excessive use of fertilisers, etc. are some of the factors that degrade the land towards salinity (Kijne 2016). Currently salt-affected area in India is about 6.74 million hectare and it is expected to increase to 16.2 million hectares by 2050 (CSSRI 2015). Reduction in biomass production is one of the major effects of salinity. Salinity hampers the growth and metabolism of plants by inducing various changes like osmotic stress, ion toxicity, oxidative stress (ROS), membrane disorganisation (Parihar et al. 2015; Machado and Serralheiro 2017).

9.2.2 Impact of Salinity on Plants

Salinity has a major negative impact on plant growth and poses a major threat to global food security (Fig. 9.2). Salinity hampers the growth and metabolism of plants by inducing a number of changes like osmotic stress, ion toxicity, oxidative stress (ROS), membrane disorganisation, etc. The osmotic stress makes the plant unable to absorb water and thus leads to reduction in growth rate. Reduced expansion of leaves, closure of stomatal aperture, reduced photosynthesis, etc. are some of the consequences observed in salinity due to water deficit condition created by

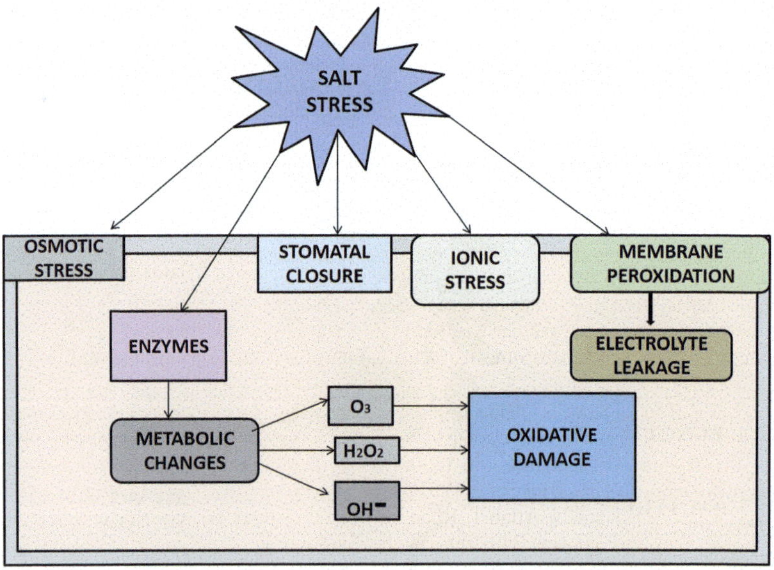

Fig. 9.2 Impact of salinity stress in plants include changes like ion toxicity, osmotic stress, browning of leaves and death; closing of stomata and reduced photosynthesis process, membrane peroxidation, production of reactive oxygen species, oxidative damage, decreased water uptake efficiency and poor root growth, etc.

osmotic stress (Rahnama et al. 2010). Damage to photosynthetic pigments, stomatal aperture, etc. under salinity leads to overproduction of reactive oxygen species like H_2O_2, O_2^- (Saed-Moucheshi et al. 2014). ROS leads to the degradation of proteins, inactivation of enzymes, etc., thereby leading to cellular damages and even leads to cell death. Reduced shoot and root weight and yield due to decline in photosynthesis under salinity have been well documented in many plants like strawberry (Yaghubi et al. 2016), turfgrasses (Sekar 2016), tomato (Rivero et al. 2014), etc. Under ionic stress, more influx of Na^+ occurs in place of K^+ ion and Cl^- decreases NO_3^- acceptance that interrupts the normal ion balance and hampers the major functions performed by the plant, majorly photosynthesis. The expulsion of Na^+ from leaves results in salinity tolerance as reported in rice (Haq et al. 2010), barley (Shavrukov et al. 2010), etc. Plants differ greatly in showing resistance to salinity due to the difference in their internal organisations. After understanding the mechanism underlying salinity tolerance, researches have started exploring the solution to this problem. In this regard, they identified phytohormones as a major signalling molecule in plants playing a crucial role in stress responses (Sharma et al. 2005; Shaterian et al. 2005). Javid et al. (2011) presented an extensive review describing the role of different phytohormones in alleviating salinity stress. The phytohormones reported by them include abscisic acid, indoleacetic acid, cytokinins, gibberellic acid, brassinosteroids, jasmonates, salicylic acid, etc. After going through the literature, we found salicylic acid and jasmonic acid to be more significant in alleviating salinity stress. A lot of reports are available where jasmonic acid (Qiu et al. 2014; Ahmad et al. 2018) and salicylic acid (Idrees et al. 2012; Liu et al. 2016) when exogenously supplied to plants results in reduced damage by salinity.

9.3 Role of Phytohormones in Plants

Plants have the inherent capability to sense and respond against these adverse stress conditions usually by modulating their internal cellular metabolic processes. Such modulation in metabolic processes is possible due to the presence of signalling molecules in plants that are able to sense the stress factors prevailing in the plant's environment and activate the signal transduction pathway that signals the plant to modify its physiological and biochemical processes in order to cope up with the stress conditions. In view of this, phytohormone acts as a major signalling molecule that perceives signals from the plant's niche and produces required modifications in cellular processes such as changes in the activity of ion-channels, protein modifications, protein degradation and gene expression. Auxin (IAA), Cytokinins (CKs), Abscisic acid (ABA), Ethylene (ET) and Gibberellins (GAs) are the major classical phytohormones and recently, Salicylic acid (SA) and Jasmonates (JAs) are also included in the list of phytohormones (Fig. 9.3). Recent researches have shown the potential of salicylic acid (SA), jasmonic acid (JA) and its derivative methyl jasmonate (MeJ) to reduce the negative impact of abiotic stress in plants and enhances the stress tolerance capacity of a plant under abiotic stresses (Walia et al.

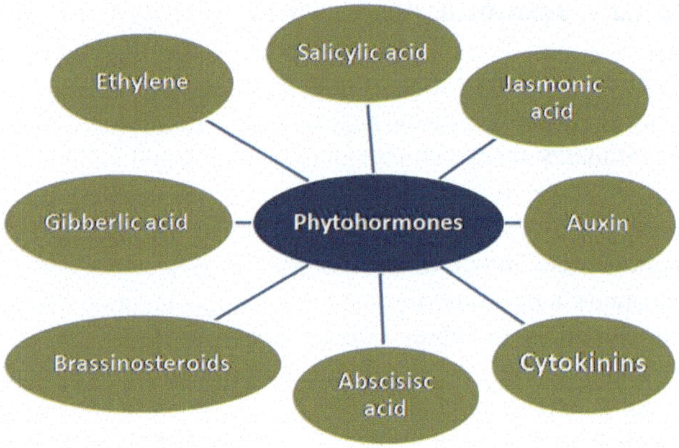

Fig. 9.3 List of major phytohormones found in plants

2007; Du et al. 2013; Qiu et al. 2014; Salimi et al. 2016; Fayez and Bazaid 2014; Singh et al. 2015) (Fig. 9.3).

The exact pathway of action of these phytohormones for conferring tolerance in plants is yet to be revealed and for this, extensive research has been going on. These phytohormones are thought to be involved in up regulation of the antioxidant system (Ascorbate peroxidase, Catalase, etc.), regulation of osmolytes (proline) which lead to enhanced water relation and gaseous exchange attributes, etc. under stress conditions. Hence, phytohormones work by adjusting the physiological and biochemical processes of plants under stress conditions which helps them to cope up with the stress conditions.

9.3.1 Salicylic Acid

Salicylic acid (SA) (ortho-hydroxybenzoic acid), a phenolic compound, is ubiquitously present in the plant kingdom and is synthesised via two different pathways. In one pathway, ICS (isochorismate synthase) converts chorismate to isochorismate which is then catalysed to SA by IPL (isochorismate pyruvate lyase) (Wildermuth et al. 2001; Strawn et al. 2007). Another pathway for SA biosynthesis starts from phenylalanine which is first converted to cinnamic acid by PAL (phenylalanine ammonia lyase) and finally to SA via O-coumaric acid or benzoic acid. Garcion and Métraux 2006). It has an evident role in seed germination, flowering, ion uptake and transport, stomatal conductance, transpiration, etc. It has been observed that it has a positive impact on plants when applied exogenously both in stressed and non-stressed conditions. Although it is widely known for its role in biotic resistance in plants, its role in abiotic stress tolerance is also evident (Fayez and Bazaid 2014; Singh et al. 2015). Several reports have indicated that SA on exogenous application at an optimum concentration confers tolerance in plants against several types of stresses (abiotic and biotic).

9.3.2 Jasmonic Acid

Jasmonic acid and its conjugates are collectively known as jasmonates and the most widely used are jasmonoyl isoleucine (JA-Ile) and methyl jasmonate (MeJA). Jasmonates are small lipid-derived carotenoid molecules (Fig. 9.4). In plants, JA is found to play role in the growth and developmental processes like seed germination, callus growth, primary root growth, flowering, formation of gum and bulb and senescence (Huang et al. 2017). MeJA is reported to enhance abiotic stress tolerance in plants at optimum levels by regulation of the antioxidant system and by regulation of secondary metabolism (Maserti et al. 2014). Several studies have indicated the role of JA in conferring abiotic stress tolerance in plants, although the exact mechanism is not clearly understood (Wasternack 2014; Wasternack and Strnad 2016; Sharma and Laxmi 2016) (Fig. 9.4).

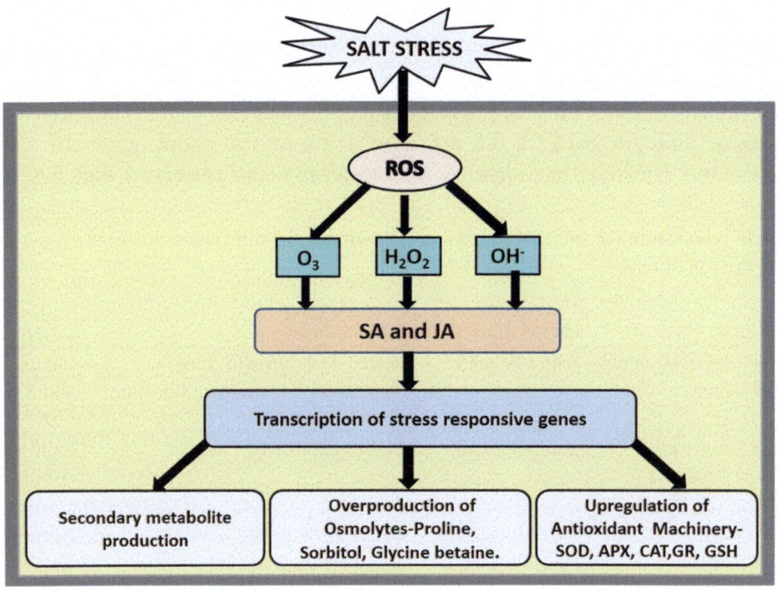

Fig. 9.4 Overview of the proposed salinity tolerance mechanism in plants by action of Salicylic acid (SA) and Jasmonic acid (JA). Upon salt stress, Reactive oxygen species (ROS) are generated that leads to oxidative damage in plants. ROS activate SA and JA signalling pathway which initiate the transcription of stress responsive genes [Including secondary metabolites, osmolytes, antioxidants like Superoxide dismutase(SOD), Ascorbate peroxidase (APX), Catalase (CAT), Glutathione reductase (GR) and Glutathione (GSH) resulting in alleviation of salinity stress

9.4 Approaches to Combat Salinity Stress

Salinity is one of the major problems affecting the plants and an immediate solution to this is required. Phytohormone application and genetic engineering are the two most effective strategies in this regard.

9.4.1 Phytohormone Treatment to Enhance Salinity Stress Tolerance

Phytohormones are proposed to be effective signalling molecules in plants that play a crucial role in salinity stress alleviation. Several reports have been published where jasmonic acid (Qiu et al. 2014; Ahmad et al. 2018) and salicylic acid (Idrees et al. 2012; Liu et al. 2016) when exogenously supplied to plants resulted in reducing damage by salinity.

9.4.1.1 Application of Phytohormones for Salinity Tolerance in Plants Under In-Vitro Conditions

Several approaches have been adopted for producing salt-tolerant plants. One such approach is by using phytohormones (Salicylic acid and Jasmonic acid) under in vitro conditions for salinity tolerance in plants (Fig. 9.4). Some such studies are listed in Table 9.1. Sakhanokho and Kelley (2009) reported enhancement in salinity tolerance in *Hibiscus* plants using Salicylic acid under in vitro conditions. They induced salinity stress with NaCl treatment (0, 175 and 200 mM). They investigated the effect of Salicylic acid (0, 0.5 and 1 mM) on invitro shoot apices of *Hibiscus acetosella* and *Hibiscus moscheutos* under salinity and observed that SA treated

Table 9.1 Application of phytohormones for salinity stress alleviation in some crop's species under invitro conditions

Crop species	Phytohormone/ chemical used	Observations	References
Hibiscus acetosella and *Hibiscus moscheutos*	Salicylic acid	Better shoot growth, root elongation and increased proline accumulation	Sakhanokho and Kelley (2009)
Solanum tuberosum cv. i.e. Cardinal and Desiree	Salicylic acid	Improved growth	Sajid and Aftab (2012)
Phoenix dactylifera cv. Nersy	Salicylic acid and ascorbic acid	Better growth, and increased antioxidant activity of SOD and APX	Almayahi (2016)
Fragaria × ananassa Duch.) cv. Queen Elisa	Salicylic acid and iron nanoparticle	Better growth, increased pigment content, increased relative water content.	Mozafari et al. (2018)
Solanum melongena cv. Mardin Kiziltepe and Kemmer	Jasmonic acid	Better growth	Gunalp et al. (2011)

plants showed better shoot growth, root elongation and more proline accumulation as compared to plants treated with only NaCl.

Almayahi (2016) also investigated the effect of Salicylic acid (50 mgl^{-1}) and Ascorbic acid (100 mgl^{-1}) in micro-propagated shoots of *Phoenix dactylifera* cv. Nersy under salinity stress. Their investigation revealed salinity tolerance and improved growth with increased antioxidant activity of SOD and APX in *Phoenix dactylifera* cv. Nersy on application of Salicylic acid (50 mgl^{-1}) and Ascorbic acid (100 mgl^{-1}). Studies carried out by Mozafari et al. (2018) showed similar expression of Salicylic acid (0.0, 0.01, 0.05 mM) conferring salinity stress alleviation on strawberry explants in better growth, increased pigment content, increased relative water content, etc. and thereby mitigate harmful effects of salinity. Another tissue culture-based study also reported salinity stress alleviation in two potato cultivars, i.e. Cardinal and Desiree under in vitro conditions by application of Salicylic acid (0.125, 0.25, 0.50 and 0.75 mM (Sajid and Aftab 2012).

Gunalp et al. (2011) investigated the effect of Jasmonic acid (10 and 20μM JA) on embryos of eggplant grown invitro conditions under salinity stress and their observation revealed salinity tolerance with better growth in JA (10μM) treated plants as compared to controls under salinity. Hence, phytohormones can be effectively employed to develop salt-tolerant plants (Table 9.1).

9.4.1.2 Exogenous Application of Phytohormone to Enhance Salinity Stress Tolerance in Plants

Salinity is one of the major constraints that limit plant growth and development and to overcome this problem, several approaches have been applied; one is exogenous phytohormone application at the optimum concentration. Several researches have been done where the exogenous application of phytohormones at optimum concentration conferred tolerance against salinity stress. Some of them are depicted in Table 9.2. Several studies are available which reported the positive role of salicylic and jasmonic acid in salinity stress alleviation in crop plants. Khan et al. (2010) reported the positive role of SA (0.1, 0.5, and 1.0 mM) in tolerance of salinity (50 mM NaCl) in *Vigna radiata L.* (Wilczek) cultivar Pusa Vishal. The salt stress-induced high K^+/Na^+ ratio in plants while the SA treatment alleviated the effect of salinity with a reduction in Na^+, Cl^-, H_2O_2 content. Also, SA treated plants exhibited increased N, P and K contents, increased antioxidant activity and increased photosynthesis.

Hussein and co-workers (Hussein et al. 2007) examined the effect of salicylic acid (200 ppm) on growth parameters in maize plants cv. Single Hybrid 10 under salinity (2000 and 4000 ppm NaCl) and observed improved growth parameters including plant height, the number of green leaves, the diameter of stem and dry weight on SA application under salinity. In another study by Idrees et al. (2012), they observed reduction in growth in two varieties of *Cymbopogon* (Krishna and Neema) exposed to salinity (50, 100 and 150 mM of NaCl) and reported that SA treatment (10^{-5} M) resulted in mitigation of salinity stress along with improvement in the activities of carbonic anhydrase and nitrate reductase enzymes of salicylic acid (1 mM) and nitric oxide (100μM) on *Vigna angularis*. SA and NO application

Table 9.2 Exogenous application of phytohormones (Salicylic acid and Jasmonic acid) for salinity stress alleviation in some crops

Crop species	Phytohormone/ chemical used	Observations	References
Vigna angularis	Salicylic acid and nitric oxide	Enhanced growth, upregulation of antioxidant system, increased proline content, etc.	Ahanger et al. (2019)
Dianthus superbus	Salicylic acid	Increased photosynthetic activity and antioxidant system activity	Ma et al. (2017)
Solanum tuberosum cv. N-Y LARA and 720-110 NARC	Salicylic acid	Increased photosynthetic activity and antioxidant activity	Faried et al. (2016)
Vigna radiata L.	Salicylic acid	Improved growth	Khan et al. (2014)
Cymbopogon flexuosus var. Neema and Krishna	Salicylic acid	Improved growth and better tolerance.	Idrees et al. (2012)
Vigna radiata L. (Wilczek) cultivar Pusa Vishal	Salicylic acid	Increased N, Pand K contents, increased antioxidant activity and increased photosynthesis.	Khan et al. (2010)
Zea mays c.v. single hybrid 1	Salicylic acid	Improved growth and salinity tolerance	Hussein et al. (2007)
Carthamus tinctorius var. IL111 and Isfahan	Methyl Jasmonate	Improved growth with increase in chlorophyll content, proline content, etc.	Maryam et al. (2019)
Vigna unguiculate L.	Methyl Jasmonate	Increased chlorophyll content, stomatal conductance, proline accumulation and relative water content	Sadeghipour (2017)
Brassica napus L. variety GSC-6	Jasmonic acid	Better growth	Kaur and Sirhindhi (2017)
Glycine max L.	Salicylic acid, Jasmonic acid and 24-Epibrassinolide	Improved growth with increase in photosynthetic pigments and proline content	Sheokand et al. (2018)
Glycine max	Salicylic acid and Jasmonic acid	Improved growth, better tolerance during SA + JA treatment as compared to only SA or JA treatment	Golezani and Salar (2018)
Fragaria × ananassa cv. 'Camarosa'	Salicylic acid (SA) and methyl Jasmonate (MeJ)	Better tolerance to salinity	Faghih et al. (2018)

were reported to significantly enhance plant growth and metabolism with upregulation of antioxidant system including superoxide dismutase (SOD), catalase (CAT), ascorbate peroxidase (APX), dehydroascorbate reductase (DHAR), and glutathione reductase (GR), increased proline content in SA + NO treated seedlings as compared to control.

Impact of SA application on biochemical characteristics of plants under salinity stress was also thoroughly studied (Ahanger et al. 2019; Khan et al. 2014). Ahanger et al. (2019) examined the effect of exogenously applied salicylic acid (SA) and nitric oxide (NO) on *Vigna angularis* exposed to salinity and concluded that SA and NO application significantly enhanced plant growth and metabolism with upregulation of the antioxidant system including superoxide dismutase (SOD), catalase (CAT), ascorbate peroxidase (APX), dehydroascorbate reductase (DHAR), and glutathione reductase (GR) and with increased osmolyte (proline) accumulation. Foliar application of SA (0.5 mM) in mung bean (*Vigna radiata* L.) under salinity resulted in alleviation of salt stress with increased glycinebetaine (GB) and methionine (Met) accumulation in SA treated plants (Khan et al. 2014). Ma et al. (2017) reported enhancement in salinity tolerance in *Dianthus superbus* with improved photosynthetic activity under salinity stress (0.3, 0.6, and 0.9% NaCl). This was achieved by exogenous application of salicylic acid (0.5 mM) with significant improvement in photosynthetic activity and antioxidant system activity in SA treated plants under salinity as compared to untreated plants. Consistent with the results of Ma et al. (2017) in *Dianthus superbus*, foliar application of SA in two potato cultivars N-Y LARA and 720-110 NARC grown under salt stress alleviates the harmful effects of salinity and resulted in increased photosynthetic activity and antioxidant system activity. Faried et al. (2016) also observed similar mitigation of salinity stress on exogenous application of salicylic acid (0.5 mM) on potato cultivars N-Y LARA and 720–110 NARC under salt stress (50 mmol L^{-1}) with increased photosynthetic activity and antioxidant activity.

The exogenous application of jasmonic acid (JA) also proved useful in alleviating the adverse effects of salinity as reported in various studies (Maryam et al. 2019; Sadeghipour 2017). Maryam et al. (2019) observed that methyl jasmonate (0.1 and 0.5 mMMeJ) on the exogenous application on *Carthamus tinctorius* varieties IL111 and Isfahan exposed to salinity (6 and 12 ds m^{-1}) mitigated the negative effects of salinity and improved plant growth and resulted in increased chlorophyll content, proline content, etc.

Sadeghipour (2017) also investigated the effect of exogenously applied methyl jasmonate (0, 25 and 50μM MeJA) for salinity tolerance in *Vigna unguiculata* L. seedlings and observed that plants treated with MeJ showed improved growth, increased chlorophyll content, stomatal conductance, proline accumulation and relative water content (RWC) under salinity stress (50 and 100 mM) as compared to controls.

Studies carried out by Kaur and Sirhindhi (2017) showed similar effects of jasmonic acid (JA) in salinity stress alleviation in seedlings of *Brassica napus* L. exposed to salinity (0, 140, 160, 180 mM NaCl) with jasmonic acid studied under salinity stress. JA treatment (0, 6, 9, 12 M) resulted in reduced toxicity of salt stress on seedling growth with increased proline content and decreased electrolyte leakage and lipid peroxidation. Simultaneous application of both SA and JA was found to result in better mitigation of salinity stress than their individual treatments (Sheokand et al. 2018). Exogenous application of SA (10^{-6} M), JA (0.5μM) and 24-Epibrassinolide (10^{-7} M) on soybean plants exposed to salinity mitigate the negative effect of salinity with a decline in MDA content and increase in photosynthetic pigments and proline content as compared to controls.

Another finding describing amelioration of salinity tolerance by use of Salicylic acid and jasmonic acid was provided by Golezani and Salar (2018) where they observed that soybean plants when applied SA + JA treatment displayed much better tolerance, increased relative water content, etc. than the plants getting individual phytohormone treatment under salinity.

9.4.2 Transgenic Approach for Generation of Salinity-Tolerant Plants

Apart from phytohormone exogenous application, another effective strategy for the generation of salt-tolerant plants is genetic engineering where a gene of interest from a specific source is inserted in the host plant to develop transgenic plants. This approach has been applied for the development of many transgenic plants resistant to salt stress. Transcription factors, signal transduction genes, water channel proteins, ion transporters, detoxifying genes, molecular chaperones, dehydrins and osmoprotectants are the most widely used genes for the development of stress-tolerant plants by genetic engineering.

Table 9.3 represents some transgenic plants for salinity tolerance where transgene induces changes in expression of genes coding for phytohormone (SA or JA) or intermediate and impart salinity tolerance.

Ye et al. (2009) reported transgenic rice overexpressing the repressor gene OsJAZ9 with the accumulation of a higher level of proline than wild plants under salinity conditions. Zhao et al. (2014) reported that transgenic *Arabidopsis* developed by inserting bread wheat gene TaAOC1 (encoding an allene oxide cyclase involved in the α-linolenic acid metabolism pathway) showed enhanced salinity tolerance with higher accumulation of jasmonic acid. This was the first evidence

Table 9.3 Development of some salinity-tolerant plants using transgenic approach

Transgene	Source species	Transformed species	Observations	References
OsJAZ9	*Oryza sativa L.* sp. Indica	*Oryza sativa L.* sp. japonica	Increased proline content, improved salinity tolerance	Ye et al. (2009)
TaAOC1	*Triticum aestivum*	*Arabidopsis thaliana*	Transgenic plants with improved salt tolerance	Zhao et al. (2014)
OsCYP94C2b	*Oryza sativa*	*Oryza sativa*	Transgenic plants with better salinity tolerance and better growth	Kurotani et al. (2015)
GarWRKY5	*Gossypium aridum*	*Arabidopsis thaliana*	Improved salinity tolerance in transgenic lines	Guo et al. (2019)
NahG	*Nicotiana tabacum*	*Arabidopsis thaliana*	Tolerance to moderate stress but not to severe stress	Cao et al. (2009)

indicating the positive role of JA in the plant salinity response and that the α-linolenic acid metabolism pathway governs this response.

Contrary to these findings, there are some reports which proposed that salinity tolerance is due to reduction in jasmonic acid level or its inactivation in plants. Kurotani et al. (2015) in their finding observed that transgenic rice expressing transgene OsCYP94C2b resulted in decreased sensitivity to salinity and enhanced the survival rate. This gene encodes an enzyme that catalyses the conversion of jasmonoyl isoleucine to inactive form. Guo et al. (2019) reported that transcription factor GarWRKY5 regulates salinity stress in *Arabidopsis* on its overexpression by jasmonic or salicylic acid-mediated signalling pathway.

Some reports suggested that SA may enhance the formation of ROS which leads to increased oxidative damage during salinity. Cao et al. (2009) reported transgenic *Arabidopsis* expressing NahG showed better tolerance to moderate salinity and have higher glutathione/oxidised glutathione ratio and the ascorbate/dehydroascorbate ratio than wild type plants but not under severe stress.

9.4.3 Genome Editing for Salinity Stress Alleviation

Genomics has emerged progressively over the last decade and has a tremendous contribution to the agricultural field particularly by providing crucial information for crop improvement. With its help, we can identify novel genes involved in salinity tolerance and have been able to use them for enhancing tolerance of our crops. Genome editing is targeted mutagenesis of genomes that allows us to introduce specific changes at specific sites in the genome. It utilises DNA cleavage reagents and cellular DNA repair pathways (Orellana et al. 2010). The DNA cleavage reagents are mostly nucleases (engineered) that cleave the target DNA site at a specific site and the double-stranded breaks so generated are repaired by non–homologous end joining, NHEJ or homologous recombination, HR (Carroll 2014) Zinc finger nucleases (ZFNs), the Transcription activator-like effector nucleases (TALENs), the Meganucleases are used to create breaks in targeted double-stranded DNA at or close to place of the target gene and DNA can be exploited to make specific changes like insertion or deletion and repaired by non-homologous end joining or homologous recombination (Curtin et al. 2012; Carroll 2014). The CRISPR/Cas RNA-guided system is recently identified which has provided newer and faster means for the production of precisely engineered crops.

This newly emerging technology CRISPR/Cas has immense potential in improving the stress tolerance ability of our crop plants and several researchers are working on developing stress-tolerant plant species. Recently, Zhang et al. (2019a) succeeded in enhancing salinity tolerance in rice via CRISPR/Cas9 targeted mutagenesis of OsRR22 (*Oryza sativa* response regulator 22) gene. Bo et al. (2019) observed that targeted mutagenesis of NAC (NAM, ATAF and CUC) transcription factor coding gene, OsNAC041 resulted in salt sensitivity in rice which pointed out the role of OsNAC041 gene in imparting salinity tolerance in rice. Zhang et al. (2019b) in their study produced knockout of the SUMO protease, *OsOTS1* (Rice OVERLY

TOLERANT TO SALT 1) gene in the *Oryza sativa* L. sp. *japonica cv.* Kitaake) rice cultivar Kitaake using the CRISPR/Cas9 gene-editing system and observed that *Os*OTS1 gene plays a crucial role in imparting salt stress tolerance in rice.

9.5 Conclusion and Future Outlook

The salinity stress poses a major threat to plants and influences their physiology, metabolism, viability, yield, etc. The extent of damage caused by salinity depends on various factors like severity and duration of stress, genotype of the affected plant, plant's growth stage exposed to stress, etc. The strategies employed to alleviate salt stress include the exogenous application of phytohormones and genetic manipulation using targeted genes. The phytohormones, salicylic acid and jasmonic acid have shown their potential role for protecting plants from saline conditions, probably by modifying physiological and biochemical processes (by promoting the accumulation of antioxidant molecules and osmolytes, etc.) in plants. The effects of SA and JA are dependent not only on the concentration applied but also on the method of application. The utility of transgenic technology and genome editing can be further enhanced through the discovery and exploitation of stress-inducible promoters, genes of biosynthetic pathways of these phytohormones, etc. which could enhance salt tolerance. Still, several questions regarding these phytohormones remain unanswered. It is unclear if exogenous SA and JA application directly or indirectly increase endogenous SA and JA levels under stress. Further clarification of these questions could lead to a better understanding of the exact role of SA and JA in salinity stress adaptation to plants. The applications of these phytohormones and their metabolic engineering hold great promise as a management tool for enhancing productivity and protecting our agricultural crops against the aforesaid constraints ultimately aiding to increase potential crop yield in near future.

Acknowledgements Dr. Deepak Kumar is thankful to the University Grant Commission for providing financial support to the laboratory as Startup research grant (No. F.30-352/2017) (BSR), New Delhi, Government of India.

References

Ahanger MA, Aziz U, Alsahli AA, Alyemeni MN, Ahmad P (2019) Influence of exogenous salicylic acid and nitric oxide on growth, photosynthesis, and ascorbate-glutathione cycle in salt stressed *Vigna angularis*. Biomol Ther 10(1):42. https://doi.org/10.3390/biom10010042

Ahmad P, Ahanger MA, Alyemeni MN, Wijaya L, Alam P, Ashraf M (2018) Mitigation of sodium chloride toxicity in *Solanum lycopersicum* L. by supplementation of jasmonic acid and nitric oxide. J Plant Interact 13:64–72

Almayahi AMW (2016) Influence of salicylic acid (SA) and ascorbic acid (ASA) on in vitro propagation and salt tolerance of date palm (Phoenix dactylifera L.) cv. 'Nersy'. Aust J Crop Sci 10:969–976. https://doi.org/10.21475/ajcs.2016.10.07.p7640

Bo WA, Zhaohui ZH, Huanhuan ZH, Xia WA, Bingli LI, Lijia YA, Xiangyan HA, Deshui Y, Xuelia Z, Chunguo W, Wenqin S, Chengbin C, Yon Z (2019) Targeted mutagenesis of NAC transcription factor gene, OsNAC041, leading to salt sensitivity in rice. Rice Sci 26(2):98–108. https://doi.org/10.1016/j.rsci.2018.12.005

Boyer JS (1982) Plant productivity and environment. Science 218:443–448. https://doi.org/10.1126/science.218.4571.443

Bray EA, Bailey SJ, Weretilnyk E (2000) Responses to abiotic stresses. In: Buchanan BB, Gruissem W, Jones RL (eds) Biochemistry and molecular biology of plants. American Society of Plant Physiologists, Rockville, MA, pp 1158–1203

Cao Y, Zhang ZW, Xue LW, Du JB, Shang J, Xu F, Yuan S, Lin HH (2009) Lack of salicylic acid in Arabidopsis protects plants against moderate salt stress. Z Naturforsch C 64:231–238. https://doi.org/10.1515/znc-2009-3-414

Carroll D (2014) Genome engineering with targetable nucleases. Annu Rev Biochem 83:409–439. https://doi.org/10.1146/annurev-biochem-060713-035418

CSSRI (2015) Vision 2050. ICAR-Central Soil Salinity Research Institute, Karnal, Haryana, India

Curtin SJ, Voytas DF, Stupar RM (2012) Genome engineering of crops with designer nucleases. Plant Genome J 5(42):12. https://doi.org/10.3835/plantgenome2012.06.0008

Du H, Liu H, Xiong L (2013) Endogenous auxin and jasmonic acid levels are differentially modulated by abiotic stresses in rice. Front Plant Sci 4:397. https://doi.org/10.3389/fpls.2013.00397

Faghih S, Zarei A, Ghobadi C (2018) Positive effects of plant growth regulators on physiology responses of Fragaria × ananassa cv. 'Camarosa' under salt stress. Int J Fruit Sci. https://doi.org/10.1080/15538362.2018.1462291

Faried HN, Ayyub CM, Amjad M, Ahmed R (2016) Salinity impairs ionic, physiological, and biochemical attributes in potato. Pak J Agric Sci 53:17–25. https://doi.org/10.21162/PAKJAS/16.4766

Fayez KA, Bazaid SA (2014) Improving drought and salinity tolerance in barley by application of salicylic acid and potassium nitrate. J Saudi Soc Agric Sci 13:45–55. https://doi.org/10.1016/j.jssas.2013.01.001

Garcion C, Métraux JP (2006) Salicylic acid. In: Hedden P, Thomas SG (eds) Plant hormone signaling, vol 24. Blackwell Press, Oxford, pp 229–255

Gill SS, Tuteja N (2010) Reactive oxygen species and antioxidant machinery in abiotic stress tolerance in crop plants. Plant Physiol Biochem 48:909–930. https://doi.org/10.1016/j.plaphy.2010.08.016

Golezani KG, Salar FA (2018) Foliar sprays of salicylic acid and jasmonic acid stimulate H+-ATPase activity of tonoplast, nutrient uptake, and salt tolerance of soybean. Ecotoxicol Environ Saf 166:18–25. https://doi.org/10.1016/j.ecoenv.2018.09.059

Gunalp B, Yasar F, Horasan O, Kusvuran S, Tipirdamaz R, Ellialtioglu S (2011) Effects of jasmonic acid and salt applications on antioxidative enzyme activities of the eggplant seedlings grown in vitro culture. Curr Opin Biotechnol 22:S152. https://doi.org/10.1016/j.copbio.2011.05.465

Guo Q, Zhao L, Fan X, Peng X, Zhenzhen X, Zhang X, Shan M, Shen X (2019) Transcription factor GarWRKY5 is involved in salt stress response in diploid cotton species (*Gossypium aridum* L.). Int J Mol Sci 20(21):5244. https://doi.org/10.3390/ijms20215244

Haq TU, Gorham J, Akhtar J, Akhtar N, Steele KA (2010) Dynamic quantitative trait loci for salt stress components on chromosome 1 of Rice. Funct Plant Biol 37:634–645. https://doi.org/10.1071/FP09247

Huang H, Liu B, Liu L, Song S (2017) Jasmonate action in plant growth and development. J Exp Bot 68(6):1349–1359. https://doi.org/10.1093/jxb/erw495

Hussein MM, Balbaa LK, Gaballah MS (2007) Salicylic acid and salinity effects on growth of maize plants. Res J Agric Biol Sci 3(4):312–312

Idrees M, Naeem M, Khan MN, Aftab T, Khan MM (2012) Alleviation of salt stress in lemongrass by salicylic acid. Protoplasma 249:709–720. https://doi.org/10.1007/s00709-011-0314-1

Javid MG, Sorooshzadeh A, Moradi F, Sanavy SA, Allahadadi I (2011) The role of phytohormones in alleviating salt stress in crop plants. Aust J Crop Sci 5(6):726–734

Jisha KC, Vijayakumari K, Puthur JT (2013) Seed priming for abiotic stress: an overview. Acta Physiol Plant 35:1381–1396. https://doi.org/10.1007/s11738-012-1186-5
Kaur H, Sirhindhi G (2017) Exogenous application of jasmonic acid offers tolerance to salinity by altering stress responses in *Brassica napus* L. variety GSC 6. Int J Sci Res 6(2):1930–1934
Khan N, Syeed S, Masood A, Nazar R, Iqbal N (2010) Application of salicylic acid increases contents of nutrients and antioxidative metabolism in mungbean and alleviates adverse effects of salinity stress. Int J Plant Biol 1:1–8. https://doi.org/10.4081/pb.2010.e1
Khan MIR, Asgher M, Khan NA (2014) Alleviation of salt-induced photosynthesis and growth inhibition by salicylic acid involves glycine betaine and ethylene in mungbean (*Vigna radiata* L.). Plant Physiol Biochem 80:67–74. https://doi.org/10.1016/j.plaphy.2014.03.026
Kijne JW (2016) Abiotic stress and water scarcity: identifying and resolving conflicts from plant level to global level. Field Crops Res 97:3–18. https://doi.org/10.1016/j.fcr.2005.08.011
Kurotani K, Hayashi K, Hatanaka S, Toda Y, Ogawa D, Ichikawa H, Ishimaru Y, Tashita R, Suzuki T, Ueda M, Hattori T (2015) Elevated levels of CYP94 family gene expression alleviate the jasmonate response and enhance salt tolerance in rice. Plant Cell Physiol 56(4):779–789. https://doi.org/10.1093/pcp/pcv006
Liu W, Zhang Y, Yuan X, Xuan Y, Gao Y, Yan Y (2016) Exogenous salicylic acid improves salinity tolerance of *Nitraria tangutorum*. Russ J Plant Physiol 63:132–142. https://doi.org/10.1134/S1021443716010118
Ma X, Zheng J, Xule Zhang X, Qingdi Hu O, Qian R (2017) Salicylic acid alleviates the adverse effects of salt stress on *Dianthus superbus* (Caryophyllaceae) by activating photosynthesis, protecting morphological structure, and enhancing the antioxidant system. Front Plant Sci 8 (200). https://doi.org/10.3389/fpls.2017.00600
Machado R, Serralheiro RP (2017) Soil salinity: effect on vegetable crop growth. Management practices to prevent and mitigate soil salinization. Horticulturae 3(2):30
Maryam CK, Kalantai MK, Arvin MJ (2019) Effect of salinity stress and exogenously applied methyl jasmonate on growth and physiological traits of two *Carthamus tinctorius* varieties, 2019. Int J Hortic Sci Technol 6(1):39–49. https://doi.org/10.22059/ijhst.2019.277800.283
Maserti BE, Del Carratore R, Della Croce CM, Podda A, Migheli Q, Froelicher Y, Luro F, Morillon R, Ollitrault P, Talon M, Rossignol M (2014) Comparative analysis of proteome changes induced by the two spotted spider mite Tetranychus urticae and methyl jasmonate in citrus leaves. J Plant Physiol 168:392–402. https://doi.org/10.1016/j.jplph.2010.07.026
Mozafari AA, Dedejani S, Ghaderi N (2018) Positive responses of strawberry (Fragaria × ananassa Duch.) explants to salicylic and iron nanoparticle application under salinity conditions. Plant Cell Tissue Org Cult 134(2):267–275
Orellana S, Yañez M, Espinoza A, Verdugo I, González E, Ruiz-Lara S, Casaretto JA (2010) The transcription factor SlAREB1 confers drought, salt stress tolerance and regulates biotic and abiotic stress-related genes in tomato. Plant Cell Environ 33:2191–2208
Parihar P, Singh S, Singh R, Singh VP, Prasad SM (2015) Effect of salinity stress on plants and its tolerance strategies: a review. Environ Sci Pollut Res 22(6):4056–4075
Qiu Z, Guo J, Zhu AJ, Zhang L, Zhang M (2014) Exogenous jasmonic acid can enhance tolerance of wheat seedlings to salt stress. Ecotoxicol Environ Saf 104:202–208
Rahnama A, Poustini K, Munns R, James RA (2010) Stomatal conductance as a screen for osmotic stress tolerance in durum wheat growing in saline soil. Funct Plant Biol 37:255–263
Rivero RM, Mestre TC, Mittler R, Rubio F, Garcia-sanchez F, Martinez V (2014) The combined effect of salinity and heat reveals a specific physiological, biochemical and molecular response in tomato plants. Plant Cell Environ 37:1059–1073. https://doi.org/10.1111/pce.12199
Sadeghipour O (2017) Amelioration of salinity tolerance in cowpea plants by seed treatment with methyl jasmonates. Legum Res 40:1100–1106. https://doi.org/10.18805/lr.v0i0.8394
Saed-Moucheshi A, Pakniyat H, Pirasteh-Anosheh H, Azooz M (2014) Role of ROS as signalling molecules in plants. In: Ahmad P (ed) Oxidative damage to plants: antioxidant networks and signaling. Elsevier, Amsterdam, pp 585–620. https://doi.org/10.1016/B978-0-12-799963-0.00001-0

Sajid Z, Aftab F (2012) Role of salicylic acid in amelioration of salt tolerance in potato (solanum tuberosum l.) under in vitro conditions. Pak J Bot 44:37–42

Sakhanokho HF, Kelley RY (2009) Influence of salicylic acid on in vitro propagation and salt tolerance in Hibiscus acetosella and Hibiscus moscheutos (cv 'Luna Red'). Afr J Biotechnol 8:1474–1481

Salimi F, Shekari F, Hamzei J (2016) Methyl jasmonate improves salinity resistance in German chamomile (Matricaria chamomilla L.) by increasing activity of antioxidant enzymes. Acta Physiol Plant 38:1–14. https://doi.org/10.1007/s11738-015-2023-4

Sekar SK (2016) Efficacy of salinity mitigation on warm season turfgrasses. Cornell University, New York

Sharma M, Laxmi A (2016) Jasmonates: emerging players in controlling temperature stress tolerance. Front Plant Sci 6:1129. https://doi.org/10.3389/fpls.2015.01129

Sharma N, Abrams SR, Waterer DR (2005) Uptake, movement, activity, and persistence of an abscisic acid analog (80 acetylene ABA methyl ester) in marigold and tomato. J Plant Growth Regul 24:28–35. https://doi.org/10.1007/s00344-004-0438-z

Shaterian J, Waterer D, Jong DH, Tanino KK (2005) Differential stress responses to NaCl salt application in early- and late maturing diploid potato (Solanum sp.) clones. Environ Exp Bot 54 (3):202–212. https://doi.org/10.1016/j.envexpbot.2004.07.005

Shavrukov Y, Gupta NK, Miyazaki J, Baho MN, Chalmers KJ, Tester M, Langridge P, Collins NC (2010) HvNax3 -a locus controlling shoot sodium exclusion derived from wild barley (*Hordeum vulgare* ssp. spontaneum). Funct Integr Genomics 10:277–291. https://doi.org/10.1007/s10142-009-0153-8

Sheokand M, Jakhar S, Singh V, Sikerwal V (2018) Effect of salicylic acid, 24-Epibrassinolide and jasmonic acid in modulating morpho-physiological and biochemical constituents in *Glycine max* L. merill under salt stress. Int J Res Anal Rev 5(4):i434–i442

Singh PK, Shahi SK, Singh AP (2015) Effects of salt stress on physico-chemical changes in maize (*Zea mays* l.) plants in response to salicylic acid. Indian J Plant Sci 4:69–77

Strawn MA, Marr SK, Inoue K, Inaba N, Zubieta C, Wildermuth MC (2007) *Arabidopsis* isochorismate synthase functional in pathogen-induced salicylate biosynthesis exhibits properties consistent with a role in diverse stress responses. J Biol Chem 282:5919–5933

Tran LS, Nishiyama R, Yamaguchi-Shinozaki K, Shinozaki K (2010) Potential utilization of NAC transcription factors to enhance abiotic stress tolerance in plants by biotechnological approach. GM crops 1(1):32–39. https://doi.org/10.4161/gmcr.1.1.10569

USDA-ARS (2008) Research databases. Bibliography on salt tolerance. George E. Brown, Jr Salinity Lab. US Dep. Agric., Agric. Res. Serv. Riverside, CA. http://www.ars.usda.gov/Services/docs.htm?docid=8908

Walia H, Wilson C, Condamine P, Liu X, Ismail AM, Close TJ (2007) Large-scale expression profiling and physiological characterization of jasmonic acid-mediated adaptation of barley to salinity stress. Plant Cell Environ 30:410–421

Wang W, Vinocur B, Altman A (2003) Plant responses to drought; salinity and extreme temperatures: towards genetic engineering for stress tolerance. Planta 218:1–14

Wasternack C (2014) Action of jasmonates in plant stress responses and development applied aspects. Biotechnol Adv 32:31–39. https://doi.org/10.1016/j.biotechadv.2013.09.009

Wasternack C, Strnad M (2016) Jasmonatesignaling in plant stress responses and development-active and inactive compounds. New Biotechnol 33:604–613. https://doi.org/10.1016/j.nbt.2015.11.001

Wildermuth MC, Dewdney J, Wu G, Ausubel FM (2001) Isochorismate synthase is required to synthesize salicylic acid for plant defense. Nature 414:562–565

Yaghubi K, Ghaderi N, Vafaee Y, Javadi T (2016) Potassium silicate alleviates deleterious effects of salinity on two strawberry cultivars grown under soilless pot culture. Sci Hortic 213:87–95. https://doi.org/10.1016/j.scienta.2016.10.012

Ye H, Du H, TangN LX, Li X (2009) Identification and expression profiling analysis of TIFY family genes involved in stress and phytohormone responses in rice. Plant Mol Biol 71:291–305. https://doi.org/10.1007/s11103-009-9524-8

Zhang A, Liu Y, Wang F, Li T, Chen Z, Kong D, Bi J, Zhang F, Luo X, Wang J, Tang J, Yu X, Liu G, Luo L (2019a) Enhanced rice salinity tolerance via CRISPR/Cas9-targeted mutagenesis of the *OsRR22* gene. Mol Breed 39:47. https://doi.org/10.1007/s11032-019-0954-y

Zhang C, Srivastava AK, Sadanandom A (2019b) Targeted mutagenesis of the SUMO protease, *Overly Tolerant to Salt1* in rice through CRISPR/Cas9-mediated genome editing reveals a major role of this SUMO protease in salt tolerance. BioRxiv:555706. https://doi.org/10.1101/555706

Zhao Y, Dong W, Naibo Z, Xinghui A, Huang MWZ, Langtao X, Xia G (2014) A wheat allene oxide cyclase gene enhances salinity tolerance via jasmonate signaling. Plant Physiol 164 (2):1068–1076. https://doi.org/10.1104/pp.113.227595

Microbe-Mediated Biotic Stress Signaling and Resistance Mechanisms in Plants

10

Lebin Thomas and Ishwar Singh

Abstract

Modern agriculture is heavily dependent on agrochemicals for management of biotic and abiotic stresses faced by crop plants. However, the exclusive dependence on these chemicals has caused an elevated concern about environment, deleterious effects on non-target organisms, and resistance in target organisms against synthetic pesticides. Plants respond to numerous biotic and abiotic stresses by morphological, biochemical, and molecular mechanisms with interacting signaling pathways involving the membrane-bound or intracellular receptors that perceive different elicitors such as pathogen-associated molecular patterns or herbivore-associated molecular patterns (PAMPs/MAMPs/HAMPs) or effectors, thereby causing a PAMP-triggered immunity (PTI) or pathogens/insect pest effector-triggered immunity (ETI). One of the recommendations to overcome the biotic stress concerns includes the development and implementation of biopesticides and biofertilizers, containing the beneficial plant growth promoting microorganisms (PGPM). These PGPM enhance the growth, yield, and nutrient uptake of plants, and further, exhibit biological control of plant diseases. Under natural habitats, the plant–microbe interactions can be crucial for proper plant nutrient mobilization, growth, development, and protection against pathogens. Colonization of roots by specific beneficial microbes may cause induced resistance locally and systemically in plants, which is characterized by the activation of concealed defense mechanisms that is hormonally regulated by interconnected signaling pathways. The biotic or abiotic elicitors induce systemic acquired resistance (SAR) in plant tissues via salicylic acid (SA) signaling which results in an accumulation of pathogenesis-related proteins (PR proteins). Whereas, the exposure of roots to PGPM under influence of

L. Thomas · I. Singh (✉)
Department of Botany, Hansraj College, University of Delhi, Delhi, India

B. Giri, M. P. Sharma (eds.), *Plant Stress Biology*,
https://doi.org/10.1007/978-981-15-9380-2_10

jasmonic acid (JA) and ethylene (ET) signaling, activation of NPR1 gene, transcription factors (MYB72 and MYC2) and callose formation cause induction of induced systemic resistance (ISR). Both SAR and ISR cause the plants to acquire a special condition of priming, thus making them more tolerant or resistant to existent and subsequent infections by broad spectrum of pathogens and insects. The rhizo-microflora is extremely diverse and implicated in elicitation of ISR. Moreover, the biocontrol of soil borne diseases by PGPM can be a significant contributor to crop yield under various biotic stress conditions. These beneficial microbes by various initiative mechanisms, including production and release of different substances, trigger physiological and biochemical changes for biotic stress tolerance in plants. Because conventional applications against pathogens are inefficient and disease management is highly challenging, the use of beneficial microorganisms has been suggested as a biocontrol solution, for providing an eco-friendly and cost-effective alternative for sustainable crop production involving the disease resistance by ISR and promotion of growth in plants.

Keywords

Biotic stresses · PAMP-triggered immunity (PTI) · Effector-triggered immunity (ETI) · Plant growth promoting microorganisms (PGPM) · Systemic acquired resistance (SAR) · Induced systemic resistance (ISR)

10.1 Introduction

Plants face various kinds of biotic stresses in the form of damage, disease, and competition due to their exposure to pathogens, insects, and weeds competing for resources, which is a major reason for quantitative and qualitative losses in terms of crop yield in agriculture. It has been reported that about 15% of total global food production is lost due to different diseases caused by pathogens (Onaga and Wydra 2016). Climate change may reduce food production further as it may cause evolution of aggressive phytopathogens and expansion of disease or insect pest outbreaks to newer areas (Anderson et al. 2004; Ijaz and Khan 2012). Biotic stress divests the plants of their nutrients and causes reduced robustness and mortality in severe circumstances. Several factors such as unfavorable weather conditions, poor crop-management and cultivation practices, and vulnerable or less resistant crop varieties can aggravate biotic stress (Das and Rakshit 2016). Therefore, management of stress promoting factors is paramount for sustainable crop productivity.

Exploration of resistance mechanisms used by plants to combat stress-associated biotic factors is one of the approaches that helps in development of resistant cultivars using the diverse disease and pest resistance alleles prevailing in gene pools of cultivated crops and their wild relatives (Islam et al. 2016). Further, many microbes, especially those belong to rhizomicrobiome, can suppress several diseases and ameliorate the harmful impact of the biotic and abiotic stresses by stimulation of various cellular components including resistance mechanisms, thereby enhancing growth attributes and reducing disease susceptibility of plants (Bari and Jones 2009; Dohroo and Sharma 2012). This microbial activity has the potential to be used as an eco-friendly and cost-effective alternative to chemical pesticides and fertilizers, towards a sustainable approach for promoting plant growth, development, and yield. This chapter is intended to describe microbe-activated resistance mechanisms related to alleviation of biotic stresses in plants.

10.2 Impact of Biotic Stresses in Plants

The biotic stresses, primarily responsible for damages and diseases in plants, are caused by pathogens (fungi—necrotrophs and biotrophs, bacteria, and viruses), nematodes, and pests (insects and mites). The pathogens infect different parts of plants to derive nutrients for their own growth and development and disturb hormonal balance, nutrient regulation, and physiology of the host plant; all such changes once taken place in the host body are reflected through various diseases in the form of vascular wilts, leaf spots, cankers, blight, nutritional imbalances, root rot, scab, damping-off, blister, mildew, rust, smut, among other symptoms (Ramegowda and Senthil-Kumar 2015).

Worldwide, pathogens are one of the foremost prolonged threats to food production and ecosystem stability as many crops are extremely prone to biotic stress in the form of soil-borne and root diseases that cause huge losses in both quality and yield (Sharma et al. 2004). For instance, the bacterial wilt disease of potato caused by *Ralstonia solanacearum* causes a global loss of more than USD 950 million per annum by affecting about 1.7 million hectares of crop in around 80 countries (Champoiseau et al. 2009; Nion and Toyota 2015). Whereas, the epidemic bacterial leaf blight disease of rice has caused severe crop losses of 50% in Asia (Ranjani et al. 2018). At least 350 different plant diseases caused by *Xanthomonas* spp. affect agricultural crops (especially, rice, citrus, cassava, tomato, sugarcane, passion fruit, and brassicas) of great importance in the world (Marin et al. 2019).

Fusarium spp. are amongst the most diverse and widely dispersed phytopathogenic fungi that cause economically important blights, rots, wilts, and cankers of numerous field, forest, horticultural, and ornamental crops in agricultural commodities and natural ecosystems. Amongst various species of *Fusarium*, *F. graminearum* and *F. verticillioides* mainly infect cereals, and *F. oxysporum* affects a wide host range of monocotyledonous and dicotyledonous plants (Sharma et al. 2018). The fungus *Hemileia vastatrix* that causes a rust disease on coffee has affected every coffee-producing region of the world (Melchor et al. 2018). Grapevine (*Vitis vinifera*) is a major fruit crop worldwide, which is susceptible to downy mildew (caused by *Plasmopara viticola*), powdery mildew (caused by *Erysiphe necator*), and gray mold (caused by *Botrytis cinerea*) diseases, all of which can cause severe economic losses in both wine and table grape production (El-Sharkawy et al. 2018a).

Many highly contagious plant viruses can drastically reduce crop yield and quality depending on the cultivar involved. Viruses affect different parts of the plant and cause local lesions and systemic damage due to stunting, leaf curl, mosaic, chlorosis, and malformations. Furthermore, parasitic nematodes feed on all parts of the plant including roots and cause wilting or stunting. The root-knot nematodes, *Meloidogyne* spp., are amongst the important plant fauna that limit the productivity of many susceptible crops (Mostafa et al. 2014). In addition, insects and mites also damage plants by their feeding and egg laying activities, and as vectors of various pathogens (Schumann and D'Arcy 2006).

Various biotic threats especially diseases caused by pathogens are managed with the help of protectant and systemic pesticides, which are applied routinely to suppress the spread of the causative agents of these stresses. However, majority of pesticides in vogue are non-biodegradable and have unwanted side effects such as pathogen resistance towards pesticides, environmental pollution, ground and surface water contamination, and other non-target deleterious effects on beneficial soil microorganisms, humans, insects, birds, and fishes (Savci 2012; Muñoz-Leoz et al. 2013). Therefore, alternative approaches to handle this problem of agri-sector are utmost.

10.3 How Do Plants Manage Biotic Stress?

Plants have evolved an array of constitutive or inducible morphological, genetic, biochemical, and molecular mechanisms of resistance to various biotic stresses caused by pathogens and insect pests as a part of their immune system (Howe and Jander 2008; Nurnberger and Kemmerling 2009). The passive defense that prevents pathogens or insect herbivores from getting access into plant-body has physical barriers such as waxes, thick cuticles, specialized trichomes, and production of antimicrobial compounds (Nejat and Mantri 2017; Singh 2017). Additionally, as a second line of defense, plants also possess an inducible defense mechanism against pathogens that break primary constitutive defense system. This includes the steps of recognition of invaders and actual defense reaction.

10.3.1 Recognition of Biotic Stress

During invasion, pathogens release a plethora of chemicals in the plant-body, called as elicitors, which pertain to various classes of biomolecules such as proteins, lipids, oligosaccharides, and nucleotides. Having conserved structures and essential roles in their producers, elicitors are capable of provoking the defense response of the host on their recognition. Further, elicitors may be general elicitors, also termed as pathogen- or microbe-associated molecular patterns (PAMPs/MAMPs), if produced by a number of microorganisms belonging to a group, and specific elicitors, also termed as effectors, if produced by a particular microorganism after successful invasion (Yu et al. 2017). Further, insects also release herbivore-associated elicitors

(HAEs) comprising of herbivore-associated molecular patterns (HAMPs) and herbivore effectors during plant–insect interaction (Santamaria et al. 2013). Apart from this, plants also as a consequence of infection or injury release some chemicals, called as danger- or damage-associated molecular patterns (DAMPs) (Yu et al. 2017). Some commonly reported PAMPs/MAMPs are bacterial flagellin and elongation factor Tu (EF-Tu), fungal chitin, yeast mannans, and Oomycete xylanase and heptaglucan (Dodds and Rathjen 2010; Newman et al. 2013). HAMPs are present in oral secretions released by chewing insects containing proteins, fatty acid-amino acid conjugates (FACs), sulfur-containing fatty acids, plant-derived degradation products of ATP synthase and cell walls generated following insect herbivory, and insect egg ovipositional fluids (Wu and Baldwin 2009; Foyer et al. 2015). Similarly, plant derived chemicals such as proteins (*At*PEPs, PIPs, and HMGB3), oligogalacturonides and extracellular ATP may act as DAMPs (Yu et al. 2017). Plants detect various elicitors with the help of certain receptors. PAMPs/MAMPs are recognized by plasma membrane bound extracellular receptors, called as pattern recognition receptors (PRRs), of either receptor-like kinase or receptor-like protein families (Nurnberger and Kemmerling 2009). The structure of PRRs contain an extracellular ligand-specific domain and a transmembrane domain. Examples of extracellular domains include leucine-rich repeats (LRRs) that help in peptide ligand perception and signaling, lysine motifs (LysMs) that help in GlcNAc containing ligand perception and signaling, and lectin motifs that help in lipopolysaccharide perception (Couto and Zipfel 2016; Yu et al. 2017). Similarly, HAMPs and DAMPs are also recognized by PRRs (Heil et al. 2012; Santamaria et al. 2013).

In contrast to PAMPs/MAMPs, effectors are recognized intracellularly by resistance (R) proteins, encoded by R genes that directly or indirectly recognize specific virulence effectors (Avr proteins) from pathogens or pests (Cui et al. 2015). Most R proteins contain specific domains, namely, a variable N-terminal effector, a conserved central Nucleotide Binding Site (NBS), and a C-terminal Leucine-Rich Repeat (LRR). With other accessory proteins, the R proteins form a nucleotide-binding-leucine rich repeat (NB-LRR) receptor protein complex that can recognize many specific pathogen effector molecules (Meyers et al. 2003; Dodds and Rathjen 2010).

10.3.2 Overcoming Biotic Stress

Subsequent to recognition of PAMPs/MAMPs and effectors, plants develop PAMP-triggered immunity (PTI) and effector-triggered immunity (ETI), respectively (Peng et al. 2018). These defense mechanisms induce a cascade of somewhat overlapping events that include rapid ion fluxes across the plasma membrane, oxidative bursts with generation of reactive oxygen species (ROS), activation of mitogen-activated protein kinases (MAPKs) and calcium-dependent protein kinases (CDPKs), localized induction of defense-related genes or pathogen cell wall/cell membrane lysing enzymes/peptides (e.g., chitinases, glucanases, and defensins), production of antimicrobial phytoalexins, plant cell wall modifications (e.g., deposition of papillae, enriched with 1,3-β-glucan cell wall polymer callose), lignin biosynthesis, or changes in cell wall proteins and pectic polysaccharide structures (Wydra and Beri 2006; Boller and Felix 2009; Zipfel 2009; Yu et al. 2017; Peng et al. 2018). The characteristic cellular defense responses observed for virus-derived molecules (nucleic acids, e.g., dsRNAs) are very similar to that for microbial elicitors, which trigger PTI (for innate immunity) and RNA interference (RNAi) (for adaptive immunity) (Nicaise 2014).

The recognition of HAEs and HAMPs by PRRs induces ion imbalances, variations in membrane potential and Ca^{2+} fluxes, and generation of ROS that stimulate downstream signaling activities in plants (Maffei et al. 2007). Besides inducing defense responses, plants can either directly dissuade the attacking insects by producing volatile compounds from the lipoxygenase (LOX) and terpenoid pathways or indirectly apprentice natural enemies of their invaders, both of which are regulated by the interacting signaling pathways of jasmonic acid (JA) with ethylene (ET), salicylic acid (SA), and abscisic acid (ABA) (Pichersky and Gershenzon 2002; Van Oosten et al. 2008; Arimura et al. 2009; War et al. 2012; Gouhier-Darimont et al. 2013). Further, plants have complex locally and systemically induced signaling pathways in response to phytophagous herbivore insects to reduce the capacity of their digestion, which involve JA, systemin, oligogalacturonides (OGAs), hydrogen peroxide, and expression of downstream defense protein inhibitors (amylase inhibitors, lectins, chitinases, and polyphenol oxidases) (Fürstenberg-Hägg et al. 2013).

Besides, the successful pathogens that have evolved by utilizing virulence effector molecules can subdue PTI signaling or preclude host detection (Pel and Pieterse 2013). Consequently, plants have evolved NB-LRR receptor protein complex that activates plant defense mechanisms more effectively. This recognition forms the R-gene-mediated or vertical resistance, constituting the effector-triggered immunity (ETI) for the second line of defense (Dodds and Rathjen 2010). Further, Toll-interleukin 1 receptor (TIR) domain (TNLs) and coiled-coil (CC) domain (CNLs) subgroups of NB-LRR can functionally interact with various proteins for resistance signaling (Griebel et al. 2014). These interactions among pathogen and host plants have resulted in a variety of pathogen effectors and resistance genes, as a manifestation of gene-for-gene resistance, that activates hypersensitive responses (HR) with programmed cell death (PCD) in infected cells and the surrounding plant parts, and RNA silencing for antiviral defense (Soosaar et al. 2005; Huang et al. 2016).

10.4 Microbe-Induced Resistance Against Biotic Stress in Plants

Plants secrete around 5–21% of their total assimilated carbon in the form of low (amino acids, organic acids, sugars, phenolics, and secondary metabolites) and high (polysaccharides and proteins) molecular mass compounds into rhizosphere via roots (Hernández et al. 2015). Chemotactically, these compounds attract and/or repel a complex diverse mixture of microorganisms including beneficial as well as harmful ones and cause their proliferation, establishment, and colonization on/in the plants (Badri and Vivanco 2009; Gaiero et al. 2013). Further, plants have the ability to modulate the composition of the communities of commensal, mutualistic, and pathogenic microorganisms that live in close association with its roots (root microbiome). Secretion of certain specific compounds that selectively stimulate and enrich beneficial microbes while repressing other microbes can provide important fitness benefits to the plants (Berendsen et al. 2012). This constitutes the rhizosphere effect of plant root secretions in the rhizosphere on microbial biomass, activity, and community composition compared with the majority soil. For instance, maize plants secrete antimicrobial compounds of benzoxazinoids that inhibit most microbes, but are selective for *Pseudomonas putida* KT2440 (Neal et al. 2012). The plant-growth promoting microbes (PGPM) that include plant-growth promoting bacteria (PGPB), such as rhizobia, and plant-growth-promoting fungi (PGPF), such as mycorrhizal fungi, establish a mutually beneficial relationship with plants and promote growth and development of plants by making them tolerant to environmental stresses of mineral deficiency, water scarcity, and phytopathogens (Drogue et al. 2013). The PGPF arbuscular mycorrhizal fungi (AMF) develop mutual beneficial relationships with over 90% of terrestrial plants including agricultural and horticultural crops, and improve the acquisition of less soluble or immobile nutrients like phosphate (Singh and Giri 2017). Similarly, the PGPB rhizobia fix atmospheric nitrogen in root nodules of plants belonging to family, Fabaceae (Geurts et al. 2012).

The disease suppressing capabilities of many of the PGPM are because of their antagonistic activities such as antibiosis, competition, parasitism, and induction of host-defense system against different pathogens (Thakur and Singh 2018). PGPM can hormonally modulate and activate defensive reaction mechanisms within plants by the systemic acquired resistance (SAR) and the induced systemic resistance (ISR). Under biotic stress, the SAR comprises accumulation of SA and PR proteins, while ISR involves pathways regulated by JA and ET (Bari and Jones 2009; Salas-Marina et al. 2011). However, the induction of resistance in plants depends on the released non-pathogenic microbial elicitor.

10.5 SAR Signaling

The SAR develops in plants as a fully active defense mechanism, involving recognition of molecular patterns of pathogen followed by detoxification utilizing altered gene expression, and production of hormones and metabolites. There is an increase in the local endogenous SA levels upon elicitation, which then concomitantly generates a mobile signal that after transportation within the plant to distant leaves initiates an additional distal or systemic SA production (Durrent and Dong 2004). The SA provides this SAR against a wide variety of plant pathogens including bacteria, fungi, and viruses (Malamy et al. 1990; Ryals et al. 1996; Durner et al. 1997). This can induce a hypersensitive response (HR), which involves localized cell death at the point of pathogen entry, caused by interactions of a host confrontation gene product with a specific pathogen-produced elicitor (Staskawicz et al. 1995).

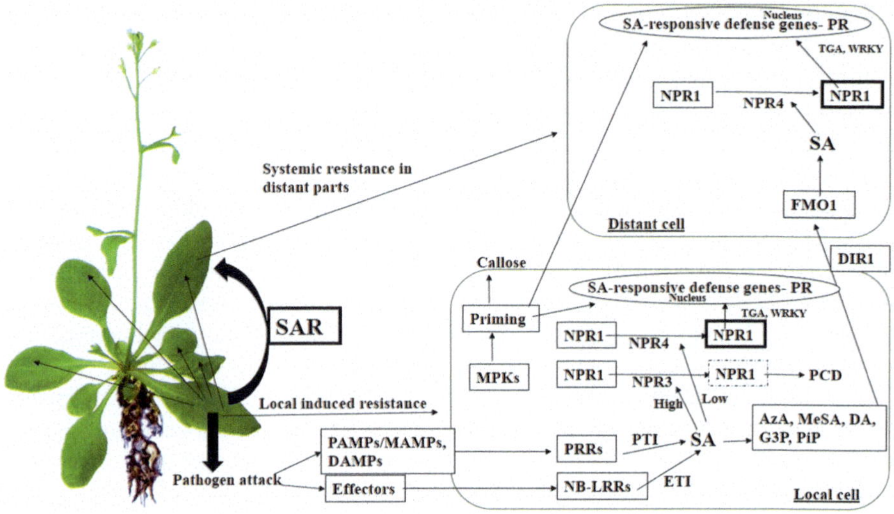

Fig. 10.1 Systemic acquired resistance (SAR) signaling in plants. See text for details

The SAR signaling in plants is schematically represented in Fig. 10.1. In the SAR, any infection in the plant triggers a local activation of a PTI or ETI (Mishina and Zeier 2007). This is characterized by an increased accumulation of SA, accompanied with the activation of SA-responsive defense genes such as PATHOGENESIS-RELATED (PR) genes that encode PR proteins like PR-1, possessing antifungal or antibacterial properties (Vernooij et al. 1994; Van Loon et al. 2006; Vlot et al. 2009). Locally, further downstream of SA in SAR signaling, there is activation of NONEXPRESSOR OF PR GENES1 (NPR1), which is a redox-regulated, ankyrin-repeat family receptor protein that functions as a transcriptional co-activator of many PR genes (Dong 2004; Pieterse et al. 2012). In the cytoplasm of healthy cells NPR1 remains in oligomeric form but in an infected cell SA modifies the cellular redox state that reduces NPR1 to its monomeric form. These NPR1 monomers are translocated to the nucleus and activate SA-responsive defense genes by interacting with their promoters and transcription factors (in the TGA and WRKY family) (Pajerowska-Mukhtar et al. 2013). The NPR1 has paralogs NPR3 and NPR4, which bind to SA with different affinities and function as adaptors of the CULLIN 3 (CUL3) ubiquitin E3 ligase (for mediating NPR1 degradation) and regulators of NPR1 stability and activity (Fu et al. 2012). At lower concentrations of SA (as in PTI or in distal SAR-expressing tissues) NPR4 stabilizes NPR1, and subsequently, activates SA-responsive PR gene expression. Whereas, at higher SA concentrations (as in ETI) NPR3 mediates degradation of NPR1, ensuing in local programmed cell death (PCD).

The initiation of SAR in other distal organs occurs by involving a long-distance signaling cascade in the vascular tissues. The lipid-transfer protein DEFECTIVE IN INDUCED RESISTANCE1 (DIR1) is considered to be a crucial chaperone for the portable SAR signals, though SA itself does not get translocated (Maldonado et al. 2002; Champigny et al. 2011). However, the several metabolites that are putatively considered to be involved in long-distance SAR signaling are methyl ester of SA (MeSA), diterpenoid dehydroabietinal (DA), a glycerol-3-phosphate (G3P)-dependent factor, azelaic acid (AzA), and pipecolic acid (Pip). Then a FLAVIN-DEPENDENT MONOOXYGENASE 1 (FMO1) transduces or amplifies these long-distance signals in the systemic tissues (Mishina and Zeier 2006).

In SAR, other known signaling molecules involved is defense priming, in which mitogen-activated protein kinases (MAPKs) such as MPK3 and MPK6 get accumulated after pathogen infection, resulting in potentiated *PR-1* gene expression, callose formation, and systemic immunity (Beckers et al. 2009). Moreover, SAR priming involves chromatin modifications in the promoters of WRKY transcription factor genes that cause regulation of SA-dependent defenses from pathogens (Jaskiewicz et al. 2011) (Fig. 10.1).

10.6 ISR Signaling

In plants, PGPM induce a resistance against various phytopathogens after an initial infection and for forthcoming attacks. This forms the PGPM escorted ISR, facilitated by production of allelochemicals (such as siderophores and antibiotics that effectively inhibit pathogen growth), and competition for ecotype and nutrient (Choudhary and Johri 2009; Jain et al. 2013). Plants can procure a state of ISR to a broad spectrum of disease causing pathogens with trivial effects on yield and growth, subsequent to an interface with different PGPB including *Pseudomonas fluorescens*, *Pseudomonas putida*, *Bacillus pumilus*, *Serratia marcescens*, *Paenibacillus alvei*, *Acinetobacter lwoffii*, *Chryseobacterium balustinum*, and *Azospirillum brasilense* (Van Hulten et al. 2006; Van Loon 2007). Similarly, the PGPF *Piriformospora indica* that colonizes the roots of many plants confers disease resistance systemically in response to pathogen attack by stimulating the host to synthesize phosphatidic acid and triggering the OXI1 pathway (Camehl et al. 2011).

10.6.1 Elicitors of ISR

ISR-inducing beneficial microbes produce various elicitors that are necessary for the commencement of ISR. Elicitation of ISR shows similarities to certain non-specific plant defense reactions to the common MAMPs such as cell surface components of lipopolysaccharides (LPS) and flagella (Erbs and Newmann 2003). Moreover, diverse ISR occasioning determinants including effectors and volatiles have been identified for certain rhizobacterial strains. For instance, LPS, iron-regulated metabolites pyoverdin and SA, antibiotics, such as 2,4-diacetylphloroglucinol (DAPG) and pyocyanin, flagella, *N*-acyl homoserine lactones, and biosurfactants are considered to be involved in PGPR induced ISR (De Vleesschauwer and Hofte 2009). Furthermore, volatile organic compounds such as 2R- and 3R-butanediol synthesized by *Bacillus subtilis* GB03 and a C13 volatile secreted by *Paenibacillus polymyxa* can elicit ISR (Ryu et al. 2004; Lee et al. 2012). In *Arabidopsis*, the O-antigenic side chain of cell wall LPS, flagella, and the siderophore pyoverdine of *Pseudomonas putida* WCS358 can elicit ISR (Meziane et al. 2005). However, *Pseudomonas putida* WCS358 mutants that lack these are still capable of triggering ISR suggesting multiple bacterial elicitors of ISR in this strain. The LPS, siderophore, and Fe-regulated compounds of *Pseudomonas fluorescens* WCS 417 are known to elicit ISR in *Arabidopsis*, carnation, and radish (Van Peer and Schippers 1992; Leeman et al. 1996). For PGPF, comparative genomics of *Trichoderma* spp. and mycorrhizal fungi has indicated the presence of many genes that encode putative effectors and elicitors (Mukherjee et al. 2013; Tisserant et al. 2013). Several elicitors of ISR identified in PGPF include enzymes (xylanases and cellulases) and specific proteins and peptides (Sm1 from *Trichoderma virens*) (Shoresh et al. 2006; Djonovic et al. 2007).

10.6.2 Root Colonization as an Early Signaling Event in ISR

Initiation of ISR in the plant requires a variety of signals whose generation needs an efficient root-colonization by the beneficial microbes (Lugtenberg and Kamilova 2009; Shoresh et al. 2010; Zamioudis and Pieterse 2012). PGPM respond to the root exudates and are then subsequently involved in chemotaxis, root colonization, and energy metabolism (Fig. 10.2). In mycorrhizal and rhizobial symbioses, plant released strigolactones and flavonoids kindle these microbes to produce symbiotic Sym and Nod factors that activate the needed plant symbiosis (Sym) signaling pathway in roots (Oldroyd et al. 2009). PGPR develops a mutual relationship with plants by having the ability to colonize roots at all stages of plant development, thereby providing benefits to both partners. After colonization, the PGPR can stay epiphytic (by living on the root surface) or endophytic (by penetrating into the root by the main root, lateral roots, or root hair, and systemically spread into the aerial

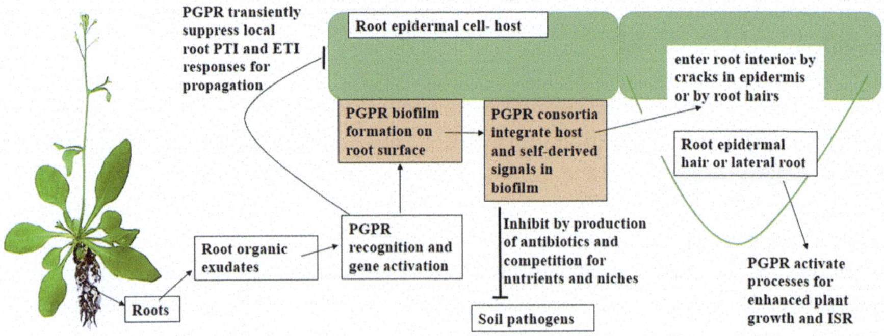

Fig. 10.2 Root colonization with biofilm formation by PGPR as an early signaling event in ISR. See text for details

parts) by utilizing specific mechanisms (Huang et al. 2011; Reinhold-Hurek and Hurek 2011).

The endophytes prior to penetration form a biofilm on the root surface, in which a single microbial cell adheres to the surface, multiplies to form multiple microcolonies, which are linked and embedded in a matrix of extracellular exopolysaccharides (Fig. 10.2). These biofilms provide protection from external stress, decrease microbial competition, and promote the host-plant growth (Ramey et al. 2004). The formation of biofilms further involves chemical signals among bacteria by quorum sensing, which allows microbial communities to respond fervently as a synchronized single unit to inhibit hostile organisms, enhance nutrient uptake, adapt to altering environmental conditions, and controls bacterial size and population. These diffusible signals can be N-acyl-homoserine lactones (AHLs), 2-heptyl-3-hydroxy-4-quinoline and autoinducer-2 in *Proteobacteria*, gamma-butyrolactones in *Streptomyces*, cis-11-methyl-2-dodecanoic acid in *Xanthomonas,* and oligopeptides in Gram-positive bacteria (Danhorn and Fuqua 2007). Further, swarming, a common property of motile PGPR such as strains of genera, *Bacillus* and *Pseudomonas* occurs, in which the bacterial cells translocate on a surface-linked with neighboring bacteria by extensive flagella in a coordinated manner that facilitates in root-colonization (Tremblay et al. 2007; Oura et al. 2015). Besides, *Trichoderma* spp. form certain structures analogous to the appressorium of plant-pathogenic fungi for colonizing the root hairs (Mukherjee et al. 2013). In *Bacillus subtilis* for root colonization, there is a stimulation and expression of genes involved in biofilm matrix production, in response to the polysaccharides exuded from host plant cell walls that function as signaling molecules (Beauregard et al. 2013). This biofilm matrix is considered as the mutualistic interface in which bacterial cells integrate both self-derived and host solutes and chemical signals for the coordination of plant growth promotion, nutrition, and ISR (Fig. 10.2).

The further entry of PGPR in the root takes place by root hairs, root apex, or cracks in the newly emerged lateral roots, which is facilitated by cell wall-degrading cellulase and pectinase exo-enzymes (Reinhold-Hurek and Hurek 2011). After colonization, *Pseudomonas*, *Bacillus*, and *Trichoderma* strains can initiate an auxin-dependent plant growth-promoting activity including increased root hair length, abundant lateral root formation, and enhanced plant biomass production, though these can function independently from the triggering of ISR (Zhang et al. 2007; Contreras-Cornejo et al. 2009; Zamioudis et al. 2013).

10.6.3 Suppression of Plant PTI or ETI

Colonization of roots further necessitates local suppression of PTI or ETI to protect the PGPR against MAMP or effector triggered production of antimicrobial compounds (Figs. 10.2 and 10.3). The ISR-inducing beneficial microbes must suppress and evade plant immune responses to establish a prolonged mutualism with the host (Wang et al. 2012; Zamioudis and Pieterse 2012). For suppressing the ETI responses and promoting fungal biotrophy, the AMF *Rhizophagus intraradices*

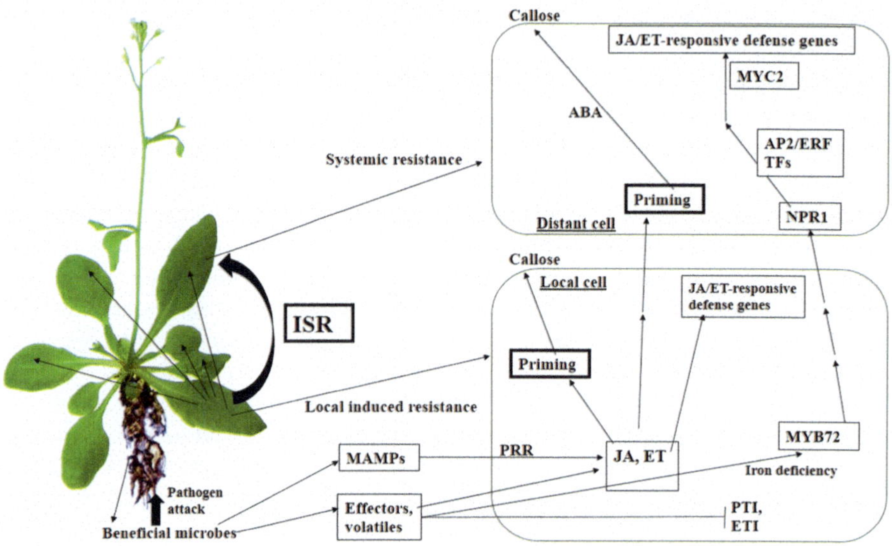

Fig. 10.3 Induction and hormonal regulation of ISR by beneficial microbes in plants. See text for details

utilizes the symbiotic effector SP7 (Kloppholz et al. 2011). In *Arabidopsis* roots, the JA signaling pathway is activated along with secretion of many immune suppressive effector proteins by the PGPF *Piriformospora indica* to suppress both early and late defense responses (Jacobs et al. 2011; Zuccaro et al. 2011). Similar down–regulation of root immune responses, which might also involve a type III secretion system, has also been described for other ISR–inducing PGPM such as *Trichoderma, Bacillus subtilis* FB17 and *Pseudomonas fluorescens* WCS417r (Weller et al. 2012; Brotman et al. 2013; Lakshmanan et al. 2013).

10.6.4 Regulation of ISR

The ISR is consequential of a long-distance signaling mechanism, which is liable for regulating the colonization density of the symbionts. PGPM are known to trigger ISR which is regulated by signaling pathways dissimilar to the pathogen-induced SAR. The plant hormones JA and ET are opined as important regulators of PGPR-mediated ISR, which is shown to be effective against necrotrophic pathogens and insect herbivores that are sensitive to JA/ET-dependent defenses (Fig. 10.3). This JA/ET regulation of ISR is observed in *Arabidopsis*, tomato and rice, from various PGPR (*Pseudomonas fluorescens* WCS417r, *Serratia marcescens* 90–166, *Pseudomonas protegens* CHA0, and *Pseudomonas fluorescens* Q2–87), and PGPF (*Penicillium* sp. GP16–2, *Trichoderma harzianum* T39 and *Piriformospora indica*) (Pieterse et al. 1998; Knoester et al. 1999; Iavicoli et al. 2003; Ryu et al. 2004;

Ahn et al. 2007; De Vleesschauwer et al. 2008; Hossain et al. 2008; Korolev et al. 2008; Van der Ent et al. 2009b; Weller et al. 2012).

In plant-root, initiation of ISR requires certain essential signaling components. For instance, in response to ISR-inducing PGPR *Pseudomonas fluorescens* WCS417r and PGPF *Trichoderma*, there is a significant expression of the MYB transcription factor gene *MYB72* of the R2R3-type that encodes MYB72, a root-specific transcription factor which functions with other signaling components and is necessary for the instigation of ISR in the epiblema and cortex of *Arabidopsis* roots (Van der Ent et al. 2008; Segarra et al. 2009). Typically, during iron-deficiency conditions, roots extrude protons by H^+-ATPase to make Fe(III)/Fe^{3+} more soluble by acidification of the soil environment. This is then reduced to Fe(II)/Fe^{2+} by ferric chelate reductase-FRO2 and is transported into root cells by the iron transporter, IRT1. In *Arabidopsis* roots colonized by ISR-inducing rhizobacterial *Pseudomonas* strains, there is a coordinated up-regulation of *MYB72* with *FRO2*, *IRT1,* and other iron deficiency-regulated genes (Pieterse et al. 2014). Alternatively, MYB72 functions in the production and/or secretion of root semiochemicals that further stimulate ISR by PGPR. Thus, this indicates a link between iron homeostasis and ISR initiation, as MYB72 is induced under iron-limited or distorted iron uptake conditions (Palmer et al. 2013).

The transcriptional factor NPR1 that has essential role as coregulator of SA-dependent *PR* genes in SAR is also known to be required for JA/ET-dependent rhizobacteria-mediated ISR by many PGPR and PGPF without *PR* gene activation (Fig. 10.3). For the SA signaling, NPR1 is associated within the nucleus, while for JA/ET signaling in ISR it is deliberated to have a cytosolic function (Spoel et al. 2003; Dong 2004; Pajerowska-Mukhtar et al. 2013). In *Arabidopsis*, ISR is associated with an enhanced expression of the JA/ET-responsive defense genes *VSP*, *PDF1.2* and *HEL* against the insect herbivore *Spodoptera exigua* (Van Wees et al. 1999). In carnation, ISR is accompanied with increased accumulation of phytoalexins at the site of pathogen infection (Van Peer et al. 1991) (Fig. 10.3).

Moreover, the PGPM mediated ISR is commonly based on priming, in which whole plant is sensitized by an enhanced activation of a combination of cellular defenses including altered gene expression and production of structural barriers (Conrath et al. 2006; Liu et al. 2007). There can be increased frequency of callose depositions at the infection site to effectively block pathogen access that is observed in the ISR of *Pseudomonas fluorescens* WCS417r in *Arabidopsis* against downy mildew pathogen *Hyaloperonospora arabidopsidis* (Fig. 10.3) (Van der Ent et al. 2009a). However, callose deposition is considered to be regulated by plant hormone ABA, as biotrophic pathogen *H. arabidopsidis* is insensitive to JA/ET-dependent defenses. Additional priming can be observed by *Bacillus subtilis* FB17 mediated ISR in *Arabidopsis*, where there is an augmented closure of the stomata in response to infection of its leaves by *Pseudomonas syringae* (Kumar et al. 2012). Thus, the regulation by hormone ABA and the structural barriers of priming provide an auxiliary stratum of protection that extends the effective range of ISR.

In *Arabidopsis*, during the ISR priming condition, transcription factor genes of the AP2/ERF family transcription factors (TFs) are highly expressed (Memelink

2009; Van der Ent et al. 2009b). These TFs are implicated in the regulation of JA- and ET-dependent defenses (Fig. 10.3). Further, an *in-silico* analysis of the promoter sequences of JA-responsive *Arabidopsis* genes with expression pattern in ISR-expressing plants has revealed the presence of a *cis*-acting G-box-like motif within the promoters of the ISR-primed genes, which functions as a binding site for an essential transcriptional regulator, MYC2, of JA-dependent defenses (Pozo et al. 2008). The ISR has been observed to be induced in *Arabidopsis*, by *Bacillus amyloliquefaciens* IN 937a (against *Erwinia carotovora*), *Bacillus pumilus* SE34 and *Bacillus pumilus* (against *Pseudomonas syringae* pv. *maculicola*), *Pseudomonas fluorescens* 89B61 and *Serratia marcescens 90–166* T4 (against *Pseudomonas syringae* pv. *maculicola* and *Pseudomonas syringae* pv. *tomato*), *B. subtilis* GB03 (against *Erwinia carotovora*), *Pseudomonas fluorescens* CHA0 (against *Peronospora parasitica*), and *Pseudomonas fluorescens* WCS417 (against *Pseudomonas syringae* pv. *tomato*) (Iavicoli et al. 2003; Ryu et al. 2003). Whereas, in tobacco, ISR has been observed to be induced by *Bacillus pumilus* SE34 (against *Peronospora tabacina*) and *Pseudomonas chlororaphis* 06 and *Serratia marcescens* 90–166 (against *Pseudomonas syrigae* pv. *tabaci*) (Press et al. 1997; Zhang et al. 2002; Spencer et al. 2003). In tomato, ISR has been observed to be induced by *Pseudomonas aeruginosa* 7NSK2 and *Pseudomonas fluorescens* CHA0 (against *Meloidogyne javanica*) and *Pseudomonas fluorescens* 89B61 (against *Phytophthora infestans*) (Yan et al. 2002; Siddiqui and Saukat 2004). Furthermore, the rhizobacterial strains can do differential ISR for variety of plant species or even for narrow range of plants in a species-specific manner. For instance, *Pseudomonas fluorescens* WCS 417 promotes ISR in *Arabidopsis*, bean, tomato, carnation, and radish, and *Pseudomonas putida* WCS 358 promotes ISR in *Arabidopsis*, bean and tomato, while *Pseudomonas fluorescens* WCS 374 is known to cause ISR in only radish (Gómez-Gómez 2004).

Though many rhizobacteria produce SA, it is frequently not considered as the causative factor of the perceived systemic resistance (Pieterse et al. 1996; Ran et al. 2005; Djavaheri et al. 2012). However, certain PGPR (*Pseudomonas aeruginosa* 7NSK2, *Paenibacillus alvei* K165, *Pseudomonas fluorescens* SS101) and PGPF (*Trichoderma*) are known to prompt an SA-dependent type of ISR resembling the pathogen induced SAR (Tjamos et al. 2005; Mathys et al. 2012). In most of these, ROS accumulate as an important elicitor at the site of tissue colonization.

10.7 Herbivore-Induced Resistance (HIR) Signaling

The initiation of HIR signaling at the site of tissue injury occurs after the release of various HAMPs, plant-derived signals (e.g., DAMPs), and elicitors in insect oral secretions, which are recognized by PRRs for PTI (Wu and Baldwin 2010; Heil et al. 2012). It is schematically represented in Fig. 10.4. Production of insecticidal toxins and other feeding deterrents constitute a direct defense of plants against herbivory. Besides, plants have an indirect defense system against herbivory involving production of volatile compounds that entice natural predators of the attacking herbivores

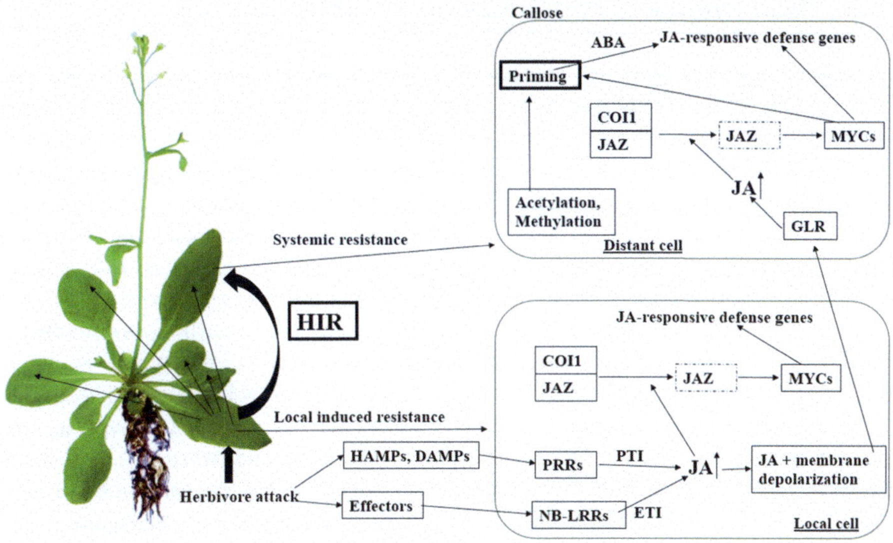

Fig. 10.4 HIR signaling in plants. See text for details

(Dicke and Baldwin 2010). Moreover, plants have evolved *R* genes, involving NB-LRRs for ETI against certain suppressive herbivore effectors. For instance in tomato, wounding caused by insect herbivory elicits a localized and a mediated distant systemic expression, and accumulation of proteinase inhibitors that affect digestive enzymes of insect-gut (Green and Ryan 1972), while the *Mi* gene is involved in conferring resistance against aphid feeding (Rossi et al. 1998). The HIR is known to be associated with JA biosynthesis and signaling. Following perception of HAMPs, there is an increased release of the biologically active signal of jasmonoyl-isoleucine (JA-Ile), which is then perceived by a coreceptor complex containing the F-box protein CORONATINE INSENSITIVE1 (COI1) and JASMONATE ZIM-domain (JAZ) proteins (Howe and Jander 2008). This JAZ protein of the coreceptor complex usually represses positive regulators of JA-mediated defense responses, such as the transcription factors MYC2, 3, and 4, within uninduced cells. However, after JA-Ile perception by the coreceptor complex, there is activation of JA-responsive genes and thus, de-repression of the JA-mediated defense responses (Memelink 2009) (Fig. 10.4).

After herbivory, JA itself is considered as the long-distance transmitted signal, required for the systemic expression of HIR (Sun et al. 2011). In *Arabidopsis*, the herbivory induced wounding stimulates a membrane surface depolarization by ion fluxes, which is then perceived by GLUTAMATE RECEPTORLIKE proteins (GLRs) that further mediates JA biosynthesis and JA-responsive gene expression in distal leaves. This indicates the importance of both electric and JA signaling in wound-induced systemic HIR signaling (Mousavi et al. 2013).

10.8 Microbe-Assisted Mitigation of Biotic Stresses

Several management practices have been suggested and implemented for management of biotic stresses in plants, though none are completely effective because of the wide distribution and the high diversity of pathogens and their host range (Muthoni et al. 2013). Most of these practices are dependent on the use of chemical fertilizers and pesticides to ensure food security however, use of such synthetic agrochemicals causes environmental problems and obliteration of soil structure and soil microorganism, and possible diminishing of food quality (Ward 2016). One of the alternative approaches to have sustainable agriculture along with environment preservation is the development and implementation of biopesticides and biofertilizers containing agro-friendly microorganisms, which enhance the growth and development of plants and suppress different plant diseases (Youssef and Eissa 2014; Thomas and Singh 2019). The dynamic interplay of the diverse rhizospheric microflora through different synergistic and antagonistic interactions within the limits of the available nutrients helps plants in their development and acclimation to variety of stresses (Van Loon and Glick 2004). Studies on plant rhizosphere, mycorrhizosphere, and endorhiza have revealed presence of a varied microbial community of PGPM (Nion and Toyota 2015). The use of rhizospheric, free-living PGPR that colonize plant roots, as biocontrol agents (BCAs) of plant diseases can be an ecological means to manage agricultural disease complications along with other beneficial effects on plant (Bouizgarne 2013). PGPR can specifically or non-specifically suppress plant diseases by either antagonism or inducing a plant systemic resistance against multiple root and foliar pathogens. The biocontrol mechanisms of most BCAs are discoursed to involve biosynthesis of antibiotics, siderophores, surfactants and phytohormones, niche and nutrient competition, mycoparasitism, ISR, phage therapy, and quorum quenching (Diallo et al. 2011; Thakur and Singh 2018).

Several microorganisms that display natural antagonism to pathogens have been identified. This microbe mediated biotic stress tolerance has been reported in many plants, and is thus implemented as BCAs of several pathogens. PGPR produce different secondary metabolites such as lipopeptides and polyketides that antagonize other microorganisms including phytopathogens. For instance, the widely distributed, resistant endospore forming *Bacilli* demonstrate various forms of biocontrol mechanisms against different plant pathogens including antagonism, competition for niche space and nutrients, and induction of host resistance (Stein 2005; Aleti et al. 2015; Villarreal-Delgado et al. 2018). *Bacillus subtilis* is one of the most commercialized BCAs that produces various bioactive compounds, particularly cyclic lipopeptides having antibacterial, antifungal, and antiviral activities that develop an ISR, including the surfactin, iturin, fengycin bacillomycin, bacilysin, lichenysin, and mycobacillin families, against a wide range of pathogens (Deleu et al. 2008; Ongena and Jacques 2008; Jourdan et al. 2009; Falardeau et al. 2013; Cawoy et al. 2014; Farace et al. 2015). Besides, *Bacillus* sp. can promote tolerance in rice plants to a leaf blight (caused by *Xanthomonas oryzae*), mediated by increased accumulation of phenylalanine ammonia lyase, peroxidase and polyphenol oxidase (Udayashankar et al. 2011).

Certain group of PGPR including fluorescent pseudomonads and other organisms is known for biocontrol and protection of a range of crop plants from many pathogens. Irrespective of antibiotic production, these PGPR elicit ISR in the host and allow plants to withstand pathogens attacking leaves/roots (Ongena et al. 2004). The different strains of *Pseudomonas* spp. (*Pseudomonas fluorescens* WCS 417r, *Pseudomonas thivervalensis* and *Pseudomonas fluorescens* CHA0) primed *Arabidopsis thaliana*, in which the plant reacted more rapidly and sturdily to pathogen attack, as part of facilitated ISR with JA/ET inducible defensive pathway (Verhagen et al. 2004). The PGPR, *Pseudomonas fluorescens* GRP3 promotes ISR in rice against sheath blight (Pathak et al. 2004). In *P. fluorescens* GRP3, rhamnolipids are considered an important determinant of biocontrol, with plant growth promoting and anti-mycelial activities (along with lysis of zoospore plasma membrane) against *Pythium* and *Phytophthora* caused damping-off in chili and tomato (Sharma et al. 2007). Molecular characterization of rhamnolipids in strain GRP3 revealed presence of a number of mono- and di-rhamnolipids that include rhamnose (Rha)–C8–C10, Rha–C10–C8, Rha–C10–C10, Rha–C10–C12:1, Rha–C10–C12, Rha–Rha–C8–C10, Rha–Rha C10–C10, Rha–Rha–C10–C10:1, Rha–Rha–C10–C12, Rha–Rha–C10–C12:1, Rha–Rha–C12–C12:1, and Rha–Rha–C12–C12. Furthermore, strain GRP3 effectively increased shoot length and activities of ISR responsive proteins peroxidase and phenylalanine ammonia lyase (PAL) involved in active lignification. Besides, PGPR produce a complex mixture of volatiles that can kindle plant growth, stimulate ISR for disease suppression, or antagonize phytopathogens, nematodes, and insects (Ryu et al. 2004; Vespermann et al. 2007; Kai et al. 2009; Farag et al. 2013).

AMF, another potential group which is ubiquitous in natural and agricultural terrestrial ecosystems, is an economically and ecologically important group of symbiotic fungi that provide varied benefits to plants including enhanced phosphorus nutrition and tolerance towards metal toxicity and drought. Besides, AMF are considered important BCA of pathogens in the natural agriculture systems for *Aphanomyces*, *Fusarium*, *Phytophthora,* and *Sclerotium* (Zambolim and Schenck 1983; Rosendahl 1985; Mark and Cassells 1996; Cordier et al. 1998). The several mechanisms explained for biocontrol by AMF include stress alleviation, along with alterations of rhizosphere, root system, nutrient availability status, other biochemical and anatomical aspects of plants cells, and induction of ISR (Singh and Giri 2017). In addition, isolates of *Trichoderma* have the ability to induce plant growth by direct and indirect mechanisms, enhance photosynthetic activity, and to reduce disease severity in plants by resilient antagonistic and mycoparasitic effects against phytopathogens, and inducing systemic resistance by releasing proteins and secondary metabolites (Keswani et al. 2016; Zachow et al. 2016). Moreover, endophytes produce antimicrobial, insecticidal, antioxidant, anti-tumor, and anti-viral metabolites. For instance, endophytic fungi can produce alkaloid (perfumoid, phomoenamide, joxysporidinone, alantrypinene, alantryleunone, anhydrooxysporidinone, and deoxyoxysporidinone) and other cytotoxic compounds (nidurufin, sterigmatocystin, averantin, 11a-methoxycurvularin

4, 11b-methoxycurvularin, tenellone H, phomopene, and 1-chloro-2,4-dihydroxy-5-methoxy-7-methylanthraquinone) (Segaran and Sathiavelu 2019).

Lytic bacteriophages, with their very narrow host ranges that infect very specific target bacteria are also considered as a tool for biological control (Loc-Carrillo and Abedon 2011). In their lytic cycle, a bacteriophage actively infects host bacteria to multiply inside and kill the host to release progeny (Orlova 2012). This has allowed lytic bacteriophages to be used for phage therapy for controlling many bacterial pathogens caused disease in plants (Buttimer et al. 2017; Doffkay et al. 2015). Effective and eco-friendly methods for the control of the devastating pathogen *Fusarium* spp. are still not available but mycovirus associated hypovirulence has been proposed to be a potential solution for biocontrol of *Fusarium* (Sharma et al. 2018).

Thus, the biocontrol by PGPM can be an effective means of managing biotic stresses in addition to abiotic stresses. Some BCA for different biotic stresses with their biocontrol mechanisms are mentioned in Table 10.1.

10.9 Conclusion and Future Prospective

Biotic stresses affect plant growth attributes, development, productivities and survivability, which stands as a foremost constraint for crop yield, food quality and thus, global food security. One suitable alternative solution of these stresses in plants is the development of microbial tools and techniques involving plant–microbe–soil interaction, which can sustain plants in stress conditions by altering their physiological and biological properties. To combat biotic stress conditions, plants have developed different mechanisms which include various pathogen recognition mechanisms that trigger different defense responses. There are two important types of pathogen recognition mechanisms, first, where the plant pattern recognition receptors (PRRs) perceive pathogen-associated molecular patterns or herbivore-associated molecular patterns (PAMPs/MAMPs/HAMPs), thereby causing a PAMP-triggered immunity (PTI); and second, R proteins perceive effectors, thereby causing an effector-triggered immunity (ETI), which is successful in controlling pathogens that evade PTI.

Plant protection involves the accumulation of defense proteins both at the site of infection and systemically in uninfected tissues and/or plants. The systemic acquired resistance (SAR) provides long-term defense against a broad-spectrum of pathogens and insects. Induced systemic resistance (ISR), a remarkable variety of induced resistance, is potentiated by plant growth promoting rhizobacteria (PGPR), especially *Pseudomonas* spp. Both SAR and ISR convene a fitness advantage to plants in conditions of high disease pressure towards an efficient signaling system capable of interpreting and transporting signals produced at the plant–pathogen interface. The SA pathway stimulated long-term resistance responses to a broad spectrum of biotrophic and hemi-biotrophic pathogens, mediates SAR, involving the expression of pathogenesis-related (PR) genes. Whereas, the jasmonic acid (JA) and ethylene

Table 10.1 The biocontrol mechanisms of some BCA

Biotic stress	Causal agent	BCA	Biocontrol mechanisms	Reference
Flavescence dorée—an epidemic yellow disease of grapevine (*Vitis vinifera*)	Phytoplasma (FDP)	Endophyte *Pseudomonas migulae* 8R6	The 1-aminocyclopropane-1-carboxylate (ACC) deaminase activity regulates the level of the stress-ethylene and reduces the FDP titer under the quantification threshold of 38% in plants	Gamalero et al. (2017)
Tobacco (*Nicotiana* sp.) bacterial wilt (TBW) and tobacco black shank (TBS)	*Ralstonia solanacearum*, *Phytophthora nicotianae*	*Pseudomonas aeruginosa* NXHG29	Suppressive effects on pathogens, by colonization in differentiation and subsequently in the elongation and maturation zones of the primary roots	Ma et al. (2018)
Tomato (*Lycopersicon esculentum*) bacterial wilt	*Ralstonia pseudosolanacearum*	*Ralstonia* sp. TCR112 and *Mitsuaria* sp. TWR114	Disease suppression by production of siderophore, indole-3-acetic acid, protease and polygalacturonase, competing for nutrients, and inducing resistance	Marian et al. (2018)
Bacterial wilt of eggplant (*Solanum melongena* cv. *agassaim*)	*Ralstonia solanacearum*	*Agrobacterium tumefaciens* XB1R, *Enterobacter* sp. XB99R and *Bacillus cereus* XB177R	Varying degree of colonization of endophytic tissues of leaves, rhizoplane, cortex and xylem vessels is considered an important pre-requisite for effective wilt prevention	Achari and Ramesh (2019)
Bacterial wilt of tomato (*Lycopersicon esculentum*)	*Ralstonia solanacearum*	*Trichoderma asperellum* (T4 and T8)	Disease control and yield enhancement with increased peroxidase (POX), phenylalanine ammonium lyase (PAL), polyphenol oxidase	Konappa et al. (2018)

(continued)

Table 10.1 (continued)

Biotic stress	Causal agent	BCA	Biocontrol mechanisms	Reference
			(PPO), β-1,3-glucanase and total phenol activities	
Damping-off disease in cucumber (*Cucumis sativus*)	*Rhizoctonia solani*	*Bacillus pumilus* SQR-N43	Biofilm formation on the root surface, with induction of hyphal deformation, enlargement of cytoplasmic vacuoles and cytoplasmic leakage in *R. solani* mycelia	Huang et al. (2012)
Fire blight disease of apple (*Malus domestica*) and pear (*Pyrus* sp.)	*Erwinia amylovora*	*Pantoea* spp.	Production of one or more antimicrobial compounds of histidine-reversible antibiotics herbicolin O, MccEh252, and pantocin A	Smits et al. (2019)
Fire blight and black shoot blight disease in apple (*Malus domestica*) and pear (*Pyrus* sp.)	*Erwinia amylovora* and *Erwinia pyrifoliae*	Bacteriophage phiEaP-8 of the family Podoviridae	By lytic activity	Park et al. (2018)
Bacterial blight of rice (*Oryza* sp.)	*Xanthomonas oryzae* pv. *oryzae*	*Pseudomonas aeruginosa* CGK-KS-1	Inhibition of quorum sensing signal, xanthan gum secretion, biofilm formation in *Xanthomonas*, by the production of two bioactive extrolites Chumacin-1 and Chumacin-2	Kanugala et al. (2019)
Bacterial blight of rice (*Oryza* sp.)	*Xanthomonas oryzae* pv. *oryzae*	Bacteriophage φXOF4 of the family Siphoviridae	By lytic activity	Ranjani et al. (2018)
Fungal sheath blight of rice (*Oryza* sp.)	*Rhizoctonia solani*	*Bacillus subtilis*	Discretion of *Rhizoctonia* mycelial growth by larger levels and early accretion of phenolics and phytoalexins	Raj et al. (2019)

Bacterial leaf blight in rice (*Oryza* sp.)	*Xanthomonas oryzae* pv. *oryzae*	*Paenibacillus polymyxa* Sx3	Reduced bacterial growth, biofilm formation, disruption of the cell morphology of *Xanthomonas*, by synthesis of the secondary metabolites fusaricidins and polymyxin	Abdallah et al. (2019a)
Crown gall disease of tomato (*Lycopersicon esculentum*), enhanced by nematode caused wounds	*Agrobacterium tumefaciens* and nematode *Meloidogyne incognita*	Rhizobacteria—*Pseudomonas japonica* NBRC 103040, *Bacillus megaterium* CtST3.5, *Pseudomonas* sp. Gamma-81, *P. tolaasii* ATCC 33618, *P. chlororaphis* Lzh-T5, and *P. mosselii* CV25	Inhibition of *Agrobacterium* growth and affect viability of *Meloidogyne* juveniles by producing a volatile secondary metabolite hydrogen cyanide, which forms stable complexes with the essential elements (Cu^{2+}, Fe^{2+} and Mn^{2+}) for the protein function, and inhibits the electron transport that disrupts the energy supply to the cell	El-Rahman et al. (2019)
Crown gall disease of tomato (*Lycopersicon esculentum*)	*Agrobacterium tumefaciens*	*Bacillus amyloliquefaciens* subsp. *plantarum* 32a	Anti-agrobacterium activity involving production of shorter alkyl chains for surfactins (C13 and C14)	Abdallah et al. (2019b)
Basal stem rot of oil palm (*Elaeis guineensis*)	*Ganoderma boninense*	Actinomycete—*Streptomyces* sp. A19	Hyphal damage of *Ganoderma* by production of antimicrobial compounds ribostamycin, benzylmalic acid, landomycin B, and salinomycin	Lim et al. (2018)
Potato (*Solanum tuberosum*) common scab	*Streptomyces scabies*	*Streptomyces violaceusniger* AC12AB	Production of azalomycin RS-22A with plant growth promotion features including production of indole-3-acetic acid and siderophores, nitrogen	Sarwar et al. (2019)

(continued)

Table 10.1 (continued)

Biotic stress	Causal agent	BCA	Biocontrol mechanisms	Reference
			fixation, and phosphate solubilization	
Anthracnose in Andean lupin (*Lupinus mutabilis*)	*Colletotrichum acutatum*	*Bacillus* spp.	Inhibition of *Colletotrichum* mycelial growth and conidial germination by production of fengycin, iturin, and surfactin lipopeptides	Yánez-Mendizábal and Falconí (2018)
Anthracnose in papaya (*Carica papaya*)	*Colletotrichum gloeosporioides*	Hemiascomycetous yeast—*Debaryomyces hansenii*	Production of volatile organic compounds (VOCs), β-1, 3 glucanase and protease, inhibition of spore germination, and competition for saccharose, glucose, fructose, and total carbohydrates	Hernandez-Montiel et al. (2018)
Strawberry (*Fragaria* × *ananassa*) anthracnose	*Colletotrichum nymphaeae*	*Staphylococcus sciuri* MarR44	Production of protease, chitinase, HCN, siderophore, IAA, gibberellin, and biofilm. Production of antifungal volatile compounds (antibiosis), which inhibited *Colletotrichum* mycelial growth and conidial germination	Alijani et al. (2019)
Anthracnose of chili pepper (*Capsicum* sp.) and tomato (*Lycopersicon esculentum*)	*Colletotrichum capsici*	*Bacillus* sp. strain M10	Production of the antifungal protein similar to catalase (KatA), which induced abnormal *Colletotrichum* hyphal elongation and conidial swelling and rupture	Srikhong et al. (2018)

Anthracnose in common bean (*Phaseolus vulgaris*)	*Colletotrichum lindemuthianum*	*Bacillus amylolicefaciens*	Production of the toxic volatiles from 3-methylbutanoic acid and 2-methylbutanoic acid	Martins et al. (2019)
Root-knot of tomatoes (*Lycopersicon esculentum*)	Nematode—*Meloidogyne incognita*	PGPR—*Pseudomonas aeruginosa* and *Burkholderia gladioli*	Production of enzymatic and non-enzymatic antioxidants, accumulation of superoxide anion, H_2O_2 and malondialdehyde, causing nematode nuclear damage and loss of cell viability	Khanna et al. (2019)
Tomato (*Lycopersicon esculentum*) bacterial canker disease	*Clavibacter michiganensis* subsp. *michiganensis*	*Pseudomonas* sp. 23S	Solubilization of inorganic phosphorus, production of siderophores, indole acetic acid and hydrogen cyanide, and by ISR with the involvement of SA signaling pathways	Takishita et al. (2018)
Damping-off and stem rot on cowpea (*Vigna unguiculata*) plants	*Rhizoctonia solani*	Yeasts—*Candida saopaulonensis* C6A, *Cryptococcus laurentii* FVC10 and *Bullera sinensis* FVF10	By competition and induction of resistance in plant involving production of peroxidase, catalase, and ascorbate peroxidase	de Tenório et al. (2019)
Mulberry (*Morus* sp.) powdery mildew	*Phyllactinia* sp.	*Pseudozyma aphidis* CNm2012	Hyphae of *Pseudozyma* grow around *Phyllactinia* sp. conidial surface and utilize the conidia as a nutrient source, thereby causing conidial atrophy, cleavage and collapse	Liu et al. (2018)
Downy mildew of grapevine (*Vitis vinifera*)	The biotrophic oomycete *Plasmopara viticola*	*Bacillus subtilis* GLB191	From both direct effect (by production of the cyclic lipopeptides fengycin and surfactin) against *Plasmopara*	Li et al. (2019)

(continued)

Table 10.1 (continued)

Biotic stress	Causal agent	BCA	Biocontrol mechanisms	Reference
			and stimulation of the plant defenses (induction of defense gene expression and callose production)	
Cucumber (*Cucumis sativus*) downy mildew	*Pseudoperonospora cubensis*	*Streptomyces padanus* PMS-702	Production of a polyene macrolide antibiotic fungichromin which displays antagonistic activities, including reduced sporangial germination and caused cytoplasmic aggregation	Fan et al. (2019)
Stem rust disease of wheat (*Triticum* sp.)	*Puccinia graminis*	*Trichoderma* spp.—*T. harzianum* and *T. viride*	Significant reduction of the disease measures with induction of peroxidase and polyphenol oxidase enzymes, increased total phenol, and enhanced plant growth and yield parameters	El-Sharkawy et al. (2018b)
Groundnut (*Arachis hypogaea*) rust	*Puccinia arachidis*	*Acremonium obclavatum*	Glucan from *Acremonium* increased levels of chitinase and β-1,3-glucanase in the apoplastic fluid, along with an increase in endogenous levels of SA	Sathiyabama and Balasubramanian (2018)
Leaf rust of wheat (*Triticum* sp.)	*Puccinia triticina* f.sp. *tritici*	BCA—*Bacillus subtilis*, *Bacillus chitinospours*, *Bacillus pumilus*, *Trichoderma viride* and *Trichoderma harzianum*	Significant decrease of number of pustules, pustule length and width, final rust severity, along with increase activities of catalase and peroxidase	Omara et al. (2019)
Coffee (*Coffea* sp.) rust	*Hemileia vastatrix*	Yeast—*Pichia membranifaciens*	Effective slowing down the progress of the rust by	Melchor et al. (2018)

			production of long-chain carboxylic acids of ethyl formate, octadecenoic acid, propionic acid, 3-(octadecanoyl)-propionic acid and methyl acetate with fungicide properties	
Brown eye spot of coffee (*Coffea* sp.)	*Cercospora coffeicola*	*Phialomyces macrosporus*	Reduction of the germination of *Cercospora* conidia by production of both volatile and non-volatile compounds that inhibited their growth, sporulation, and viability	Laborde et al. (2019)
Red rot in sugarcane (*Saccharum* sp.)	*Colletotrichum falcatum*	*Trichoderma*—*T. harzianum*, *T. asperullum*	Expression of antifungal genes of cell wall degrading enzymes (chitinase, endochitinase, β-1, 3-glucanase, exochitinase 1, exochitinase 2), proteases (alkaline proteinase, trypsin-like protease, subtilin-like serine protease), and stress and defense proteins (choline dehydrogenase, FKBP-type peptidyl-prolyl cis-trans isomerase, superoxide dismutase) against *Colletotrichum*.	Elamathi et al. (2018)
Crown rot of wheat (*Triticum* sp.)	*Fusarium graminearum*	*Lysobacter antibioticus* HS124	Abnormal hyphal structures including swelling and distortion, causing inhibition toward mycelial growth, by	Kim et al. (2019)

(continued)

Table 10.1 (continued)

Biotic stress	Causal agent	BCA	Biocontrol mechanisms	Reference
			production of volatile compounds	
Verticillium wilt of olive (*Olea europaea*)	*Verticillium dahliae*	*Pseudomonas* spp.	Plant growth promotion and/or biocontrol abilities (e.g., phytase, xylanase, catalase, cellulase, chitinase, glucanase activities, and siderophore and HCN production)	Gómez-Lama Cabanás et al. (2018)
Tomato (*Lycopersicon esculentum*) vascular wilt disease	*Verticillium dahliae*	*Bacillus velezensis* C2	Production of lipopeptides (bacillomycin, fengycin, and surfactin), polyketides (macrolactin, bacillaene, and difficidin), the dipeptide bacilysin, volatile metabolites (tetradecane, benzeneacetic acid, benzaldehyde, 1-decene, and phenylethyl alcohol), lytic enzymes (protease, chitinase, and β-gluacanase), siderophore and indole-3-acetic acid, and solubilization of inorganic phosphate	Dhouib et al. (2019)
Tomato (*Lycopersicon esculentum*) spotted wilt	Tomato spotted wilt virus (TSWV)	*Pseudomonas fluorescens*	Reduced TSWV by increased activity of polyphenol oxidase, β-1,3-glucanase and chitinase	Kandan et al. (2005)
Tomato (*Lycopersicon esculentum*) leaf curl	Tomato leaf curl virus (ToLCV)	*Pseudomonas* sp.	Involvement of chitosan elicitor and ISR	Mishra et al. (2014)

(ET) pathways, which have substantial differences in gene expression than the SA pathway, are induced against necrotrophic pathogens and chewing insects.

Despite the occurrence of several defense mechanisms in plants to overcome biotic stress conditions, biotic stresses cause major economic losses, every year whose prevention is indispensable as world human-population, which demands more food, is increasing. Therefore, amendment of the current approaches as well as design of new methodologies for crop protection is paramount. The application of microbial consortium of plant growth promoting bacteria (PGPR) and plant growth promoting fungi (PGPF) can enhance plant growth under multiple abiotic and biotic stress conditions by maintaining soil health, enhancing nutrition, regulating plant hormones and antioxidant system, siderophore production, and ISR induction. The consideration of the prospective of microbes to resolve food security problems for global economy necessitates future research to identify potential stress tolerant and diverse strains of PGPM that can further be used to formulate effective microbial consortia. Furthermore, biological control by these beneficial microorganisms is among the most eco-friendly and economical disease and pest management strategies. The use of PGPM for biocontrol provides the advantage of being a safe alternative to hazardous chemicals, which minimizes problems associated with environmental pollution, ecosystem disruption, residual chemicals on crops and chemical bioaccumulation in the food chain. Thus, special interest in the development of beneficial microbes as biotic stress biocontrol agents has the potential to limit recurrent crop losses and promote plant growth.

References

Abdallah Y, Yang M, Zhang M, Masum MM, Ogunyemi SO, Hossain A, Li B (2019a) Plant growth promotion and suppression of bacterial leaf blight in rice by *Paenibacillus polymyxa* Sx3. Lett Appl Microbiol 68:423–429

Abdallah DB, Tounsi S, Frikha-Gargouri O (2019b) Inoculum type affect the efficacy of the endophytic *Bacillus amyloliquefaciens* subsp. *plantarum* strain 32a against the plant pathogen *Agrobacterium tumefaciens*. Appl Soil Ecol 134:25–30

Achari GA, Ramesh R (2019) Colonization of eggplant by endophytic bacteria antagonistic to *Ralstonia solanacearum*, the bacterial wilt pathogen. Proc Natl Acad Sci India Sect B Biol Sci 89:585–593

Ahn IP, Lee SW, Suh SC (2007) Rhizobacteria–induced priming in *Arabidopsis* is dependent on ethylene, jasmonic acid, and NPR1. Mol Plant-Microbe Interact 20:759–768

Aleti G, Sessitsch A, Brader G (2015) Genome mining: prediction of lipopeptides and polyketides from *Bacillus* and related firmicutes. Comput Struct Biotechnol J 13:192–203

Alijani Z, Amini J, Ashengroph M, Bahramnejad B (2019) Antifungal activity of volatile compounds produced by *Staphylococcus sciuri* strain MarR44 and its potential for the biocontrol of *Colletotrichum nymphaeae*, causal agent strawberry anthracnose. Int J Food Microbiol 307:108276. https://doi.org/10.1016/j.ijfoodmicro.2019.108276

Anderson PK, Cunningham AA, Patel NG, Morales FJ, Epstein PR, Daszak P (2004) Emerging infectious diseases of plants: pathogen pollution, climate change and agrotechnology drivers. Trends Ecol Evol 19:535–544

Arimura GI, Matsui K, Takabayashi J (2009) Chemical and molecular ecology of herbivore–induced plant volatiles: proximate factors and their ultimate functions. Plant Cell Physiol 50:911–923

Badri DV, Vivanco JM (2009) Regulation and function of root exudates. Plant Cell Environ 32:666–681

Bari R, Jones JD (2009) Role of plant hormones in plant defence responses. Plant Mol Biol 69:473–488

Beauregard PB, Chai YR, Vlamakis H, Losick R, Kolter R (2013) *Bacillus subtilis* biofilm induction by plant polysaccharides. Proc Natl Acad Sci U S A 110:E1621–E1630

Beckers GJM, Jaskiewicz M, Liu Y, Underwood WR, He SY et al (2009) Mitogen–activated protein kinases 3 and 6 are required for full priming of stress responses in *Arabidopsis thaliana*. Plant Cell 21:944–953

Berendsen RL, Pieterse CMJ, Bakker PAHM (2012) The rhizosphere microbiome and plant health. Trends Plant Sci 17:478–486

Boller T, Felix G (2009) A renaissance of elicitors: perception of microbe–associated molecular patterns and danger signals by pattern–recognition receptors. Annu Rev Plant Biol 60:379–406

Bouizgarne B (2013) Bacteria for plant growth promotion and disease management. In: Maheshwari D (ed) Bacteria in agrobiology: disease management. Springer, Berlin, pp 15–47. https://doi.org/10.1007/978-3-642-33639-3_2

Brotman Y, Landau U, Cuadros-Inostroza A, Takayuki T, Fernie AR et al (2013) *Trichoderma*–plant root colonization: escaping early plant defense responses and activation of the antioxidant machinery for saline stress tolerance. PLoS Pathog 9:e1003221

Buttimer C, McAuliffe O, Ross RP, Hill C, O'Mahony J, Coffey A (2017) Bacteriophages and bacterial plant diseases. Front Microbiol 8:34. https://doi.org/10.3389/fmicb.2017.00034

Camehl I, Drzewiecki C, Vadassery Y, Shahollari B, Sherameti I, Forzani C, Munnik T, Hirt H, Oelmüller R (2011) The OXI1 kinase pathway mediates *Piriformospora indica*–induced growth promotion in *Arabidopsis*. PLoS Pathog 7:e1002051

Cawoy H, Mariutto M, Henry G, Fisher C, Vasilyeva N, Thonart P, Dommes J, Ongena M (2014) Plant defense stimulation by natural isolates of *Bacillus* depends on efficient surfactin production. Mol Plant-Microbe Interact 27:87–100

Champigny M, Shearer H, Mohammad A, Haines K, Neumann M, Thilmony R, He SY, Fobert P, Dengler N, Cameron RK (2011) Localization of DIR1at the tissue, cellular and subcellular levels during systemic acquired resistance in *Arabidopsis* using DIR1:GUS and DIR1:EGFP reporters. BMC Plant Biol 11:125. https://doi.org/10.1186/1471-2229-11-125

Champoiseau PG, Jones JB, Allen C (2009) *Ralstonia solanacearum* race 3 biovar 2 causes tropical losses and temperate anxieties. Plant Health Prog 10:1–10

Choudhary DK, Johri BN (2009) Interactions of *Bacillus* spp. and plants–with special reference to induced systemic resistance (ISR). Microbiol Res 164:493–513

Conrath U, Beckers GJM, Flors V, Garcıa-Agustın P, Jakab G et al (2006) Priming: getting ready for battle. Mol Plant-Microbe Interact 19:1062–1071

Contreras-Cornejo HA, Macias-Rodriguez L, Cortes-Penagos C, Lopez-Bucio J (2009) *Trichoderma virens*, a plant beneficial fungus, enhances biomass production and promotes lateral root growth through an auxin–dependent mechanism in *Arabidopsis*. Plant Physiol 149:1579–1592

Cordier C, Pozo MJ, Barea JM, Gianinazzi S, Gianinazzi-Pearson V (1998) Cell defense responses associated with localised and systemic resistance to *Phytophthora parasitica* in tomato by an arbuscular mycorrhizal fungus. Mol Plant-Microbe Interact 11:1017–1028

Couto D, Zipfel C (2016) Regulation of pattern recognition receptor signalling in plants. Nat Rev Immunol 16:537–552

Cui H, Tsuda K, Parker JE (2015) Effector–triggered immunity: from pathogen perception to robust defense. Annu Rev Plant Biol 66:487–511

Danhorn T, Fuqua C (2007) Biofilm formation by plant–associated bacteria. Annu Rev Microbiol 61:401–422

Das IK, Rakshit S (2016) Millets, their importance, and production constraints. In: Biotic stress resistance in millets. Academic Press, Cambridge, MA, pp 3–19
de Tenório DA, de Medeiros EV, Lima CS, da Silva JM, de Barros JA, Neves RP, Laranjeira D (2019) Biological control of *Rhizoctonia solani* in cowpea plants using yeast. Trop Plant Pathol 44:113–119
De Vleesschauwer D, Hofte M (2009) Rhizobacteria–induced systemic resistance. Adv Bot Res 51:223–281
De Vleesschauwer D, Djavaheri M, Bakker PAHM, Hofte M (2008) *Pseudomonas fluorescens* WCS374r–induced systemic resistance in rice against *Magnaporthe oryzae* is based on pseudobactin–mediated priming for a salicylic acid–repressible multifaceted defense response. Plant Physiol 148:1996–2012
Deleu M, Paquot M, Nylander T (2008) Effect of fengycin, a lipopeptide produced by *Bacillus subtilis*, on model biomembranes. Biophys J 94:2667–2679
Dhouib H, Zouari I, Abdallah DB, Belbahri L, Taktak W, Triki MA, Tounsi S (2019) Potential of a novel endophytic *Bacillus velezensis* in tomato growth promotion and protection against Verticillium wilt disease. Biol Control 139:104092. https://doi.org/10.1016/j.biocontrol.2019.104092
Diallo S, Crépin A, Barbey C, Orange N, Burini JF, Latour X (2011) Mechanisms and recent advances in biological control mediated through the potato rhizosphere. FEMS Microbiol Ecol 75:351–364
Dicke M, Baldwin IT (2010) The evolutionary context for herbivore–induced plant volatiles: beyond the "cry for help". Trends Plant Sci 15:167–175
Djavaheri M, Mercado-Blanco J, Versluis C, Meyer JM, Van Loon LC, Bakker PAHM (2012) Iron regulated metabolites produced by *Pseudomonas fluorescens* WCS374r are not required for eliciting induced systemic resistance (ISR) against *Pseudomonas syringae* pv. *tomato* in *Arabidopsis*. Microbiology 1:311–325
Djonovic S, Vargas WA, Kolomiets MV, Horndeski M, Wiest A, Kenerley CM (2007) A proteinaceous elicitor Sm1 from the beneficial fungus *Trichoderma virens* is required for induced systemic resistance in maize. Plant Physiol 145:875–889
Dodds PN, Rathjen JP (2010) Plant immunity: towards an integrated view of plant–pathogen interactions. Nat Rev Genet 11:539–548
Doffkay Z, Dömötör D, Kovács T, Rákhely G (2015) Bacteriophage therapy against plant, animal and human pathogens. Acta Biol Szeged 59:291–302
Dohroo A, Sharma DR (2012) The role of plant growth promoting rhizobacteria, arbuscular mycorrhizal fungi and their helper bacteria on growth parameters and root rot of apple. Walailak J Sci Technol 2:35–38
Dong X (2004) NPR1, all things considered. Curr Opin Plant Biol 7:547–552
Drogue B, Combes-Meynet E, Moënne-Loccoz Y, Wisniewski-Dyé F, Prigent-Comaret C (2013) Control of the cooperation between plant growth–promoting rhizobacteria and crops by rhizosphere signals. In: de Bruijn FJ (ed) Molecular microbial ecology of the rhizosphere, vol 1 and 2. Wiley, Hoboken, pp 281–294
Durner J, Shah J, Klessig DF (1997) Salicylic acid and disease resistance in plants. Trends Plant Sci 2:266–274
Durrent WE, Dong X (2004) Systemic acquired resistance. Annu Rev Phytopathol 42:185–209
Elamathi E, Malathi P, Viswanathan R, Sundar AR (2018) Expression analysis on mycoparasitism related genes during antagonism of *Trichoderma* with *Colletotrichum falcatum* causing red rot in sugarcane. J Plant Biochem Biotechnol 27:351–361
El-Rahman AA, Shaheen HA, El-Aziz RMA, Ibrahim DS (2019) Influence of hydrogen cyanide–producing rhizobacteria in controlling the crown gall and root–knot nematode, *Meloidogyne incognita*. Egypt J Biol Pest Control 29:41
El-Sharkawy HHA, Abo-El-Wafa TSA, Ibrahim SA (2018a) Biological control agents improve the productivity and induce the resistance against downy mildew of grapevine. J Plant Pathol 100:33–42

El-Sharkawy HHA, Ibrahim SA, Rashad YM (2018b) Biocontrol of stem rust disease of wheat using arbuscular mycorrhizal fungi and *Trichoderma* spp. Physiol Mol Plant Pathol 103:84–91

Erbs G, Newmann MA (2003) The role of lipopolysaccharides in induction of plant defense responses. Mol Plant Pathol 4:421–425

Falardeau J, Wise C, Novitsky L, Avis TJ (2013) Ecological and mechanistic insights into the direct and indirect antimicrobial properties of *Bacillus subtilis* lipopeptides on plant pathogens. J Chem Ecol 39:869–878

Fan YT, Chung KR, Huang JW (2019) Fungichromin production by *Streptomyces padanus* PMS–702 for controlling cucumber downy mildew. Plant Pathol J 35:341–350

Farace G, Fernandez O, Jacquens L, Coutte F, Krier F, Jacques P, Clément C, Barka EA, Jacquard C, Dorey S (2015) Cyclic lipopeptides from *Bacillus subtilis* activate distinct patterns of defence responses in grapevine. Mol Plant Pathol 16:177–187

Farag MA, Zhang H, Ryu CM (2013) Dynamic chemical communication between plants and bacteria through airborne signals: induced resistance by bacterial volatiles. J Chem Ecol 39:1007–1018

Foyer CH, Verrall SR, Hancock RD (2015) Systematic analysis of phloem–feeding insect–induced transcriptional reprogramming in *Arabidopsis* highlights common features and reveals distinct responses to specialist and generalist insects. J Exp Bot 66:495–512

Fu ZQ, Yan S, Saleh A, Wang W, Ruble J, Oka N, Mohan R, Spoel SH, Tada Y, Zheng N, Dong X (2012) NPR3 and NPR4 are receptors for the immune signal salicylic acid in plants. Nature 486:228–232

Fürstenberg-Hägg J, Zagrobelny M, Bak S (2013) Plant defense against insect herbivores. Int J Mol Sci 14:10242–10297

Gaiero JR, McCall CA, Thompson KA, Day NJ, Best AS, Dunfield KE (2013) Inside the root microbiome: bacterial root endophytes and plant growth promotion. Am J Bot 100:1738–1750

Gamalero E, Marzachì C, Galetto L, Veratti F, Massa N, Bona E, D'Agostino G (2017) An 1–Aminocyclopropane–1–carboxylate (ACC) deaminase–expressing endophyte increases plant resistance to flavescence dorée phytoplasma infection. Plant Biosyst 151:331–340

Geurts R, Lillo A, Bisseling T (2012) Exploiting an ancient signalling machinery to enjoy a nitrogen fixing symbiosis. Curr Opin Plant Biol 15:438–443

Gómez-Gómez L (2004) Plant perception systems for pathogen recognition and defence. Mol Immunol 41:1055–1062

Gómez-Lama Cabanás C, Legarda G, Ruano-Rosa D, Pizarro-Tobías P, Valverde-Corredor A, Niqui JL, Mercado-Blanco J (2018) Indigenous *Pseudomonas* spp. strains from the olive (*Olea europaea* L.) rhizosphere as effective biocontrol agents against *Verticillium dahliae*: from the host roots to the bacterial genomes. Front Microbiol 9:277. https://doi.org/10.3389/fmicb.2018.00277

Gouhier-Darimont C, Schmiesing A, Bonnet C, Lassueur S, Reymond P (2013) Signalling of *Arabidopsis thaliana* response to *Pieris brassicae* eggs shares similarities with PAMP triggered immunity. J Exp Bot 64:665–674

Green TR, Ryan CA (1972) Wound–induced proteinase inhibitor in plant leaves: a possible defense mechanism against insects. Science 175:776–777

Griebel T, Maekawa T, Parker JE (2014) NOD–like receptor cooperativity in effector–triggered immunity. Trends Immunol 35:562–570

Heil M, Ibarra-Laclette E, Adame-Alvarez RM, Martinez O, Ramirez-Chavez E, Molina-Torres J, Herrera-Estrella L (2012) How plants sense wounds: damaged–self recognition is based on plant–derived elicitors and induces octadecanoid signaling. PLoS One 7:e30537

Hernández M, Dumont MG, Yuan Q, Conrad R (2015) Different bacterial populations associated with the roots and rhizosphere of rice incorporate plant–derived carbon. Appl Environ Microbiol 81:2244–2253

Hernandez-Montiel LG, Gutierrez-Perez ED, Murillo-Amador B, Vero S, Chiquito-Contreras RG, Rincon-Enriquez G (2018) Mechanisms employed by *Debaryomyces hansenii* in biological control of anthracnose disease on papaya fruit. Postharvest Biol Tech 139:31–37

Hossain MM, Sultana F, Kubota M, Hyakumachi M (2008) Differential inducible defense mechanisms against bacterial speck pathogen in Arabidopsis thaliana by plant–growth–promoting–fungus *Penicillium* sp. GP16–2 and its cell free filtrate. Plant Soil 304:227–239

Howe GA, Jander G (2008) Plant immunity to insect herbivores. Annu Rev Plant Biol 59:41–66

Huang B, Lv C, Zhang H, Fan L (2011) Endophytic colonisation of *Bacillus subtilis* in the roots of *Robinia pseudoacacia* L. Plant Biol 13:925–931

Huang X, Zhang N, Yong X, Yang X, Shen Q (2012) Biocontrol of *Rhizoctonia solani* damping–off disease in cucumber with *Bacillus pumilus* SQR–N43. Microbiol Res 167:135–143

Huang J, Yang M, Zhang X (2016) The function of small RNAs in plant biotic stress response. J Integr Plant Biol 58:312–327

Iavicoli A, Boutet E, Buchala A, Métraux J-P (2003) Induced systemic resistance in *Arabidopsis thaliana* in response to root inoculation with *Pseudomonas fluorescens* CHA0. Mol Plant-Microbe Interact 16:851–858

Ijaz S, Khan AI (2012) Genetic pathways of disease resistance and plants–pathogens interactions. Mol Pathog 3:19–26

Islam F, Gill RA, Ali B, Farooq MA, Xu L, Najeeb U, Zhou W (2016) Sesame. In: Breeding oilseed crops for sustainable production. Academic Press, Cambridge, MA, pp 135–147

Jacobs S, Zechmann B, Molitor A, Trujillo M, Petutschnig E et al (2011) Broad–spectrum suppression of innate immunity is required for colonization of *Arabidopsis* roots by the fungus *Piriformospora indica*. Plant Physiol 156:726–740

Jain S, Vaishnav A, Kasotia A, Kumari S, Gaur RK, Choudhary DK (2013) Bacteria–induced systemic resistance and growth promotion in *Glycine max* L. Merrill upon challenge inoculation with *Fusarium exosporium*. Proc Natl Acad Sci India Sect B Biol Sci 83:561–567

Jaskiewicz M, Conrath U, Peterhansel C (2011) Chromatin modification acts as a memory for systemic acquired resistance in the plant stress response. EMBO Rep 12:50–55

Jourdan E, Henry G, Duby F, Dommes J, Barthelemy JP, Thonart P, Ongena MARC (2009) Insights into the defense-related events occurring in plant cells following perception of surfactin-type lipopeptide from *Bacillus subtilis*. Mol Plant-Microbe Interact 22:456–468

Kai M, Haustein M, Molina F, Petri A, Scholz B, Piechulla B (2009) Bacterial volatiles and their action potential. Appl Microbiol Biotechnol 81:1001–1012

Kandan A, Ramiah M, Vasanthi VJ, Radjacommare R, Nandakumar R, Ramanathan A, Samiyappan R (2005) Use of *Pseudomonas fluorescens*–based formulations for management of tomato spotted wilt virus (TSWV) and enhanced yield in tomato. Biocontrol Sci Tech 15:553–569

Kanugala S, Kumar CG, Rachamalla HKR, Palakeeti B, Kallaganti VSR, Nimmu NV, Thipparapu G (2019) Chumacin–1 and Chumacin–2 from *Pseudomonas aeruginosa* strain CGK–KS–1 as novel quorum sensing signaling inhibitors for biocontrol of bacterial blight of rice. Microbiol Res 228:126301. https://doi.org/10.1016/j.micres.2019.126301

Keswani C, Bisen K, Singh V, Sarma BK, Singh HB (2016) Formulation technology of biocontrol agents: present status and future prospects. In: Arora NK, Mehnaz S, Balestrini R (eds) Bioformulations: for sustainable agriculture. Springer, New Delhi, pp 35–52

Khanna K, Jamwal VL, Kohli SK, Gandhi SG, Ohri P, Bhardwaj R, Ahmad P (2019) Role of plant growth promoting Bacteria (PGPRs) as biocontrol agents of *Meloidogyne incognita* through improved plant defense of *Lycopersicon esculentum*. Plant Soil 436:325–345

Kim YT, Monkhung S, Lee YS, Kim KY (2019) Effects of *Lysobacter antibioticus* HS124, an effective biocontrol agent against *Fusarium graminearum*, on crown rot disease and growth promotion of wheat. Can J Microbiol 65:904–912

Kloppholz S, Kuhn H, Requena N (2011) A secreted fungal effector of *Glomus intraradices* promotes symbiotic biotrophy. Curr Biol 21:1204–1209

Knoester M, Pieterse CMJ, Bol JF, Van Loon LC (1999) Systemic resistance in *Arabidopsis* induced by rhizobacteria requires ethylene–dependent signaling at the site of application. Mol Plant-Microbe Interact 12:720–727

Konappa N, Krishnamurthy S, Siddaiah CN, Ramachandrappa NS, Chowdappa S (2018) Evaluation of biological efficacy of *Trichoderma asperellum* against tomato bacterial wilt caused by *Ralstonia solanacearum*. Egypt J Biol Pest Control 28:63. https://doi.org/10.1186/s41938-018-0069-5

Korolev N, David DR, Elad Y (2008) The role of phytohormones in basal resistance and *Trichoderma* induced systemic resistance to *Botrytis cinerea* in *Arabidopsis thaliana*. BioControl 53:667–683

Kumar AS, Lakshmanan V, Caplan JL, Powell D, Czymmek KJ et al (2012) Rhizobacteria *Bacillus subtilis* restricts foliar pathogen entry through stomata. Plant J 72:694–706

Laborde MCF, Botelho DMDS, Rodriguez GAA, Resende MLVD, Queiroz MVD, Batista AD, Medeiros FHVD (2019) *Phialomyces macrosporus* reduces *Cercospora coffeicola* survival on symptomatic coffee leaves. https://doi.org/10.25186/cs.v14i1

Lakshmanan V, Castaneda R, Rudrappa T, Bais HP (2013) Root transcriptome analysis of *Arabidopsis thaliana* exposed to beneficial *Bacillus subtilis* FB17 rhizobacteria revealed genes for bacterial recruitment and plant defense independent of malate efflux. Planta 238:657–668

Lee B, Farag MA, Park HB, Kloepper JW, Lee SH, Ryu CM (2012) Induced resistance by a long chain bacterial volatile: elicitation of plant systemic defense by a C13 volatile produced by *Paenibacillus polymyxa*. PLoS One 7:e48744

Leeman M, Den OFM, van Pelt JA, Dirkx FPM, Steijl H, Bakker PAHM, Schippers B (1996) Iron availability affects induction of SR against *Fusarium* wilt of radish *P. fluorescens*. Phytopathology 86:149–155

Li Y, Héloir MC, Zhang X, Geissler M, Trouvelot S, Jacquens L, Adrian M (2019) Surfactin and fengycin contribute to the protection of a *Bacillus subtilis* strain against grape downy mildew by both direct effect and defence stimulation. Mol Plant Pathol 20:1037–1050

Lim P, Gansau JA, Chong KP (2018) *Streptomyces* spp. a potential biocontrol agent against *Ganoderma boninense* of basal stem rot. J Oil Palm Res 30:265–275

Liu J, Maldonado-Mendoza I, Lopez-Meyer M, Cheung F, Town CD, Harrison MJ (2007) Arbuscular mycorrhizal symbiosis is accompanied by local and systemic alterations in gene expression and an increase in disease resistance in the shoots. Plant J 50:529–544

Liu X, Qiu X, Duan Z, Ping D, Zhou X, Yang J, Wan Y (2018) A novel strain of *Pseudozyma aphidis* from mulberry parasitises the conidia of mulberry powdery mildew fungus *Phyllactinia* sp. and its biocontrol effect in the fields. Biocontrol Sci Tech 28:62–76

Loc-Carrillo C, Abedon ST (2011) Pros and cons of phage therapy. Bacteriophage 1:111–114

Lugtenberg B, Kamilova F (2009) Plant–growth–promoting rhizobacteria. Annu Rev Microbiol 63:541–556

Ma L, Zhang HY, Zhou XK, Yang CG, Zheng SC, Duo JL, Mo MH (2018) Biological control tobacco bacterial wilt and black shank and root colonization by bio–organic fertilizer containing bacterium *Pseudomonas aeruginosa* NXHG29. Appl Soil Ecol 129:136–144

Maffei ME, Mithöfer A, Boland W (2007) Before gene expression: early events in plant–insect interaction. Trends Plant Sci 12:310–316

Malamy J, Carr JP, Klessig DF, Raskin I (1990) Salicylic acid: a likely endogenous signal in the resistance of tobacco to viral infection. Science 250:1002–1004

Maldonado AM, Doerner P, Dixon RA, Lamb CJ, Cameron RK (2002) A putative lipid transfer protein involved in systemic resistance signaling in *Arabidopsis*. Nature 419:399–403

Marian M, Nishioka T, Koyama H, Suga H, Shimizu M (2018) Biocontrol potential of *Ralstonia* sp. TCR112 and *Mitsuaria* sp. TWR114 against tomato bacterial wilt. Appl Soil Ecol 128:71–80

Marin VR, Ferrarezi JH, Vieira G, Sass DC (2019) Recent advances in the biocontrol of *Xanthomonas* spp. World J Microbiol Biotechnol 35:72. https://doi.org/10.1007/s11274-019-2646-5

Mark GL, Cassells AC (1996) Genotype–dependence in the interaction between *Glomus fistulosum*, *Phytophthora fragariae* and the wild strawberry (*Fragaria vesca*). Plant Soil 185:233–238

Martins SJ, Faria AF, Pedroso MP, Cunha MG, Rocha MR, Medeiros FHV (2019) Microbial volatiles organic compounds control anthracnose (*Colletotrichum lindemuthianum*) in common bean (*Phaseolus vulgaris* L.). Biol Control 131:36–42

Mathys J, De Cremer K, Timmermans P, Van Kerckhove S, Lievens B et al (2012) Genome–wide characterization of ISR induced in *Arabidopsis thaliana* by *Trichoderma hamatum* T382 against *Botrytis cinerea* infection. Front Plant Sci 3:108. https://doi.org/10.3389/fpls.2012.00108

Melchor RL, Rosales VG, Pérez MC, Fernández SP, Álvarez GO, Mastache JM (2018) Effectiveness of carboxylic acids from *Pichia membranifaciens* against coffee rust. Ciênc Agrotecnol 42:42–50

Memelink J (2009) Regulation of gene expression by jasmonate hormones. Phytochemistry 70:1560–1570

Meyers BC, Kozik A, Griego A, Kuang H, Michelmore RW (2003) Genome–wide analysis of NBS–LRR–encoding genes in *Arabidopsis*. Plant Cell 15:809–834

Meziane H, Vander SI, van Loon LC, Höfte M, Bakker PAHM (2005) Determinants of *P. putida* WCS 358 involved in induced systemic resistance in plants. Mol Plant Pathol 6:177–185

Mishina TE, Zeier J (2006) The *Arabidopsis* flavin–dependent monooxygenase FMO1 is an essential component of biologically induced systemic acquired resistance. Plant Physiol 141:1666–1675

Mishina TE, Zeier J (2007) Pathogen–associated molecular pattern recognition rather than development of tissue necrosis contributes to bacterial induction of systemic acquired resistance in *Arabidopsis*. Plant J 50:500–513

Mishra S, Jagadeesh KS, Krishnaraj PU, Prem S (2014) Biocontrol of tomato leaf curl virus (ToLCV) in tomato with chitosan supplemented formulations of *Pseudomonas* sp. under field conditions. Aust J Crop Sci 8:347–355

Mostafa FAM, Khalil AE, Nour El Deen AH, Ibrahim DS (2014) Induction of systemic resistance in sugar–beet against root–knot nematode with commercial products. J Plant Pathol Microbiol 5:236–243

Mousavi SAR, Chauvin A, Pascaud F, Kellenberger S, Farmer EE (2013) GLUTAMATE RECEPTORLIKE genes mediate leaf–to–leaf wound signaling. Nature 500:422–426

Mukherjee PK, Horwitz BA, Herrera-Estrella A, Schmoll M, Kenerley CM (2013) *Trichoderma* research in the genome era. Annu Rev Phytopathol 51:105–129

Muñoz-Leoz B, Garbisu C, Charcosset JY, Sánchez-Pérez JM, Antigüedad I, Ruiz-Romera E (2013) Non–target effects of three formulated pesticides on microbially–mediated processes in a clay–loam soil. Sci Total Environ 449:345–354

Muthoni JJ, Shimelis H, Melis R (2013) Potato production in Kenya: farming systems and production constraints. J Agric Sci 5:182–197

Neal AL, Ahmad S, Gordon-Weeks R, Ton J (2012) Benzoxazinoids in root exudates of maize attract *Pseudomonas putida* to the rhizosphere. PLoS One 7:e35498

Nejat N, Mantri N (2017) Plant immune system: crosstalk between responses to biotic and abiotic stresses the missing link in understanding plant defense. Curr Issues Mol Biol 23:1–16

Newman M-A, Sundelin T, Nielsen JT, Erbs G (2013) MAMP (microbe–associated molecular pattern) triggered immunity in plants. Front Plant Sci 4:1–10

Nicaise V (2014) Crop immunity against viruses: outcomes and future challenges. Front Plant Sci 5:660. https://doi.org/10.3389/fpls.2014.00660

Nion YA, Toyota K (2015) Recent trends in control methods for bacterial wilt diseases caused by *Ralstonia solanacearum*. Microbes Environ 30:1–11

Nurnberger T, Kemmerling B (2009) Pathogen–associated molecular patterns (PAMP) and PAMP–triggered immunity. In: Parker J (ed) Molecular aspects of plant disease resistance. Annual plant reviews, vol 34. Wiley, Oxford, pp 16–47. https://doi.org/10.1002/9781444301441.ch2

Oldroyd GED, Harrison MJ, Paszkowski U (2009) Reprogramming plant cells for endosymbiosis. Science 324:753–754

Omara RI, El-Kot GA, Fadel FM, Abdelaal KA, Saleh EM (2019) Efficacy of certain bioagents on patho–physiological characters of wheat plants under wheat leaf rust stress. Physiol Mol Plant P 106:102–108

Onaga G, Wydra K (2016) Advances in plant tolerance to biotic stresses. INTECH Plant Genomics:229–272. https://doi.org/10.5772/64351

Ongena M, Jacques P (2008) *Bacillus* lipopeptides: versatile weapons for plant disease biocontrol. Trends Microbiol 16:115–125

Ongena M, Duby F, Rossignol F, Fauconnier ML, Dommes J, Thonart P (2004) Stimulation of the lipooxygenase pathway is associated with systemic resistance induced in bean by a nonpathogenic *Pseudomonas* strain. Mol Plant-Microbe Interact 17:1009–1018

Orlova EV (2012) Bacteriophages and their structural organisation. In: Bacteriophages. IntechOpen, Rijeka, pp 3–30

Oura H, Tashiro Y, Toyofuku M, Ueda K, Kiyokawa T, Ito S, Takahashi Y, Lee S, Nojiri H, Nakajima-Kambe T, Uchiyama H, Futamata H, Nomura N (2015) Inhibition of *Pseudomonas aeruginosa* swarming motility through 1–naphthol and other bicyclic compounds bearing hydroxyl groups. Appl Environ Microbiol 81:2808–2818

Pajerowska-Mukhtar KM, Emerine DK, Mukhtar MS (2013) Tell me more: roles of NPRs in plant immunity. Trends Plant Sci 18:402–411

Palmer CM, Hindt MN, Schmidt H, Clemens S, Guerinot ML (2013) MYB10 and MYB72 are required for growth under iron–limiting conditions. PLoS Genet 9:e1003953

Park J, Lee GM, Kim D, Park DH, Oh CS (2018) Characterization of the lytic bacteriophage phiEaP–8 effective against both *Erwinia amylovora* and *Erwinia pyrifoliae* causing severe diseases in apple and pear. Plant Pathol J 34:445–450

Pathak A, Sharma A, Johri BN (2004) *Pseudomonas* strain GRP3 induces systemic resistance to sheath blight in rice. Int Rice Res Notes 29:35–36

Pel MJC, Pieterse CMJ (2013) Microbial recognition and evasion of host immunity. J Exp Bot 64:1237–1248

Peng Y, Van Wersch R, Zhang Y (2018) Convergent and divergent signaling in PAMP–triggered immunity and effector–triggered immunity. MPMI 31:403–409

Pichersky E, Gershenzon J (2002) The formation and function of plant volatiles: perfumes for pollinator attraction and defense. Curr Opin Plant Biol 5:237–243

Pieterse CMJ, Van Wees SC, Hoffland E, Van Pelt JA, Van Loon LC (1996) Systemic resistance in *Arabidopsis* induced by biocontrol bacteria is independent of salicylic acid accumulation and pathogenesis related gene expression. Plant Cell 8:1225–1237

Pieterse CMJ, Van Wees SC, Van Pelt JA, Knoester M, Laan R, Gerrits H, Weisbeek PJ, van Loon LC (1998) A novel signaling pathway controlling induced systemic resistance in *Arabidopsis*. Plant Cell 10:1571–1580

Pieterse CMJ, Van der Does D, Zamioudis C, Leon-Reyes A, Van Wees SC (2012) Hormonal modulation of plant immunity. Annu Rev Cell Dev Biol 28:489–521

Pieterse CMJ, Zamioudis C, Berendsen RL, Weller DM, Van Wees SC, Bakker PA (2014) Induced systemic resistance by beneficial microbes. Annu Rev Phytopathol 52:347–375

Pozo MJ, Van Der Ent S, Van Loon LC, Pieterse CMJ (2008) Transcription factor MYC2 is involved in priming for enhanced defense during rhizobacteria–induced systemic resistance in *Arabidopsis thaliana*. New Phytol 180:511–523

Press CM, Wilson M, Tuzun S, Kloepper JW (1997) Salicylic acid produced by *Serratia marcescens* 91–166 is not the primary determinant of induced systemic resistance in cucumber or tobacco. Mol Plant-Microbe Interact 10:761–768

Raj TS, Muthukumar A, Renganathan P, Kumar RSR, Ann H (2019) Biological control of sheath blight of rice caused by *Rhizoctonia solani* Kuhn using marine associated *Bacillus subtilis*. Int Arch Appl Sci Technol 10:148–153

Ramegowda V, Senthil-Kumar M (2015) The interactive effects of simultaneous biotic and abiotic stresses on plants: mechanistic understanding from drought and pathogen combination. J Plant Physiol 176:47–54

Ramey BE, Matthysse AG, Fuqua C (2004) The FNR–type transcriptional regulator SinR controls maturation of *Agrobacterium tumefaciens* biofilms. Mol Microbiol 52:1495–1511
Ran LX, Van Loon LC, Bakker PAHM (2005) No role for bacterially produced salicylic acid in rhizobacterial induction of systemic resistance in *Arabidopsis*. Phytopathology 95:1349–1355
Ranjani P, Gowthami Y, Gnanamanickam SS, Palani P (2018) Bacteriophages: a new weapon for the control of bacterial blight disease in Rice caused by *Xanthomonas oryzae*. MBL 46:346–359
Reinhold-Hurek B, Hurek T (2011) Living inside plants: bacterial endophytes. Curr Opin Plant Biol 14:435–443
Rosendahl S (1985) Interactions between the vesicular–arbuscular mycorrhizal fungus *Glomus fasciculatum* and *Aphanomyces eutieches* root rot of peas. Phytopathol Z 114:31–41
Rossi M, Goggin FL, Milligan SB, Kaloshian I, Ullman DE, Williamson VM (1998) The nematode resistance gene Mi of tomato confers resistance against the potato aphid. Proc Natl Acad Sci U S A 95:9750–9754
Ryals JA, Neuenschwander UH, Willits MG, Molina A, Steiner HY, Hunt MD (1996) Systemic acquired resistance. Plant Cell 8:1809–1819
Ryu CM, Hu CH, Reddy MS, Kloepper JW (2003) Different signaling pathways of induced resistance by rhizobacteria in *Arabidopsis thaliana* against two pathovars of *Pseudomonas syringae*. New Phytol 160:413–420
Ryu CM, Farag MA, Hu CH, Reddy MS, Kloepper JW, Paré PW (2004) Bacterial volatiles induce systemic resistance in *Arabidopsis*. Plant Physiol 134:1017–1026
Salas-Marina MA, Silva-Flores MA, Uresti-Rivera EE, Castro-Longoria E, Herrera-Estrella A, Casas-Flores S (2011) Colonization of *Arabidopsis* roots by *Trichoderma atroviride* promotes growth and enhances systemic disease resistance through jasmonic acid/ethylene and salicylic acid pathways. Eur J Plant Pathol 131:15–26
Santamaria ME, Martínez M, Cambra I, Grbic V, Diaz I (2013) Understanding plant defence responses against herbivore attacks: an essential first step towards the development of sustainable resistance against pests. Transgenic Res 22:697–708
Sarwar A, Latif Z, Zhang S, Hao J, Bechthold A (2019) A potential biocontrol agent *Streptomyces violaceusniger* AC12AB for managing potato common scab. Front Microbiol 10:202. https://doi.org/10.3389/fmicb.2019.00202
Sathiyabama M, Balasubramanian R (2018) Protection of groundnut plants from rust disease by application of glucan isolated from a biocontrol agent *Acremonium obclavatum*. Int J Biol Macromol 116:316–319
Savci S (2012) An agricultural pollutant: chemical fertilizer. Int J Environ Sci Technol 3:73–80
Schumann GL, D'Arcy CJ (2006) Essential plant pathology. APS Press, St. Paul
Segaran G, Sathiavelu M (2019) Fungal endophytes: a potent biocontrol agent and a bioactive metabolites reservoir. Biocatal Agric Biotechnol 21:101284
Segarra G, Van der Ent S, Trillas I, Pieterse CMJ (2009) MYB72, a node of convergence in induced systemic resistance triggered by a fungal and a bacterial beneficial microbe. Plant Biol 11:90–96
Sharma MP, Gaur A, Tanu U, Sharma OP (2004) Prospects of arbuscular mycorrhiza in sustainable management of root and soil borne diseases of vegetable crops. In: Mukerji KG (ed) Disease management of fruits and vegetables, fruit and vegetable diseases. Kluwer Academic Publishers, Dordrecht, pp 501–539
Sharma A, Wray V, Johri BN (2007) Molecular characterization of plant growth promoting rhizobacteria that enhance peroxidase and phenylalanine ammonia lyase activities in Chile (*Capsicum annuum* L.) and tomato (*Lycopersicon esculentum* Mill.). Arch Microbiol 188:483–494
Sharma M, Guleria S, Singh K, Chauhan A, Kulshrestha S (2018) Mycovirus associated hypovirulence, a potential method for biological control of *Fusarium* species. Virus 29:134–140
Shoresh M, Gal-On A, Leibman D, Chet I (2006) Characterization of a mitogen–activated protein kinase gene from cucumber required for *Trichoderma*–conferred plant resistance. Plant Physiol 142:1169–1179

Shoresh M, Harman GE, Mastouri F (2010) Induced systemic resistance and plant responses to fungal biocontrol agents. Annu Rev Phytopathol 48:21–43
Siddiqui IA, Saukat SS (2004) Systemic resistance in tomato induced by biocontrol bacteria against the root–knot nematode, *Meloidogyne javanica* is independent of SA production. J Phytopathol 152:48–54
Singh I (2017) Antimicrobials in higher plants: classification, mode of action and bioactivities. Chem Biol Lett 4:48–62
Singh I, Giri B (2017) Arbuscular mycorrhiza–mediated control of plant pathogens. In: Varma A, Prasad R, Tuteja N (eds) Mycorrhiza–nutrient uptake, biocontrol, ecorestoration. Springer, Basel, pp 131–160. https://doi.org/10.1007/978-3-319-68867-1_7
Smits TH, Duffy B, Blom J, Ishimaru CA, Stockwell VO (2019) Pantocin A, a peptide–derived antibiotic involved in biological control by plant–associated *Pantoea* species. Arch Microbiol 201:713–722
Soosaar JL, Burch-Smith TM, Dinesh-Kumar SP (2005) Mechanisms of plant resistance to viruses. Nat Rev Microbiol 3:789–798
Spencer M, Ryu CM, Yang KY, Kim YC, Kloepper JW, Anderson AJ (2003) Induced defense in tobacco by *P. chlororaphis* strain O6 involves at least the ethylene pathway. Physiol Mol Plant Pathol 63:27–34
Spoel SH, Koornneef A, Claessens SMC, Korzelius JP, Van Pelt JA et al (2003) NPR1 modulates crosstalk between salicylate– and jasmonate–dependent defense pathways through a novel function in the cytosol. Plant Cell 15:760–770
Srikhong P, Lertmongkonthum K, Sowanpreecha R, Rerngsamran P (2018) *Bacillus* sp. strain M10 as a potential biocontrol agent protecting chili pepper and tomato fruits from anthracnose disease caused by *Colletotrichum capsici*. BioControl 63:833–842
Staskawicz BJ, Ausubel FM, Baker BJ, Ellis JG, Jones JDG (1995) Molecular genetics of plant disease resistance. Science 268:661–667
Stein T (2005) *Bacillus subtilis* antibiotics: structures, syntheses and specific functions. Mol Microbiol 56:845–857
Sun JQ, Jiang HL, Li CY (2011) Systemin/jasmonate–mediated systemic defense signaling in tomato. Mol Plant 4:607–615
Takishita Y, Charron JB, Smith DL (2018) Biocontrol rhizobacterium *Pseudomonas* sp. 23S induces systemic resistance in tomato (*Solanum lycopersicum* L.) against bacterial canker *Clavibacter michiganensis* subsp. *michiganensis*. Front Microbiol 9:2119. https://doi.org/10.3389/fmicb.2018.02119
Thakur P, Singh I (2018) Biocontrol of soilborne root pathogens: an overview. In: Giri B, Prasad R, Varma A (eds) Root biology, Soil biology, vol 52. Springer, Cham, pp 181–220
Thomas L, Singh I (2019) Microbial biofertilizers: types and applications. In: Giri B, Prasad R, Wu Q-S, Varma A (eds) Biofertilizers for sustainable agriculture and environment. Soil biology, vol 55. Springer, Cham, pp 1–19
Tisserant E, Malbreil M, Kuo A, Kohler A, Symeonidi A et al (2013) Genome of an arbuscular mycorrhizal fungus provides insight into the oldest plant symbiosis. Proc Natl Acad Sci U S A 110:20117–20122
Tjamos SE, Flemetakis E, Paplomatas EJ, Katinakis P (2005) Induction of resistance to *Verticillium dahlia* in *Arabidopsis thaliana* by the biocontrol agent K–165 and pathogenesis–related proteins gene expression. Mol Plant-Microbe Interact 18:555–561
Tremblay J, Richardson AP, Lépine F, Déziel E (2007) Self–produced extracellular stimuli modulate the *Pseudomonas aeruginosa* swarming motility behaviour. Environ Microbiol 9:2622–2630
Udayashankar AC, Nayaka SC, Reddy MS, Srinivas C (2011) Plant growth–promoting rhizobacteria mediate induced systemic resistance in rice against bacterial leaf blight caused by *Xanthomonas oryzae* pv. *oryzae*. Biol Control 59:114–122
Van der Ent S, Verhagen BW, Van Doorn R, Bakker D, Verlaan MG, Pel MJ, Joosten RG, Proveniers MC, Van Loon LC, Ton J, Pieterse CMJ (2008) MYB72 is required in early

signaling steps of rhizobacteria–induced systemic resistance in *Arabidopsis*. Plant Physiol 146:1293–1304

Van der Ent S, Van Hulten MHA, Pozo MJ, Czechowski T, Udvardi MK, Pieterse CMJ, Ton J (2009a) Priming of plant innate immunity by rhizobacteria and β–aminobutyric acid: differences and similarities in regulation. New Phytol 183:419–431

Van der Ent S, Van Wees SC, Pieterse CMJ (2009b) Jasmonate signaling in plant interactions with resistance–inducing beneficial microbes. Phytochemistry 70:1581–1588

Van Hulten M, Pelser M, Van Loon LC, Pieterse CMJ, Ton J (2006) Costs and benefits of priming for defense in Arabidopsis. Proc Natl Acad Sci U S A 103:5602–5607

Van Loon LC (2007) Plant responses to plant growth–promoting rhizobacteria. Eur J Plant Pathol 119:243–254

Van Loon LC, Glick BR (2004) Increased plant fitness by rhizobacteria. In: Sandermann H (ed) Molecular ecotoxicology of plants. Springer–Verlag, Berlin, pp 177–205

Van Loon LC, Rep M, Pieterse CMJ (2006) Significance of inducible defense–related proteins in infected plants. Annu Rev Phytopathol 44:135–162

Van Oosten VR, Bodenhausen N, Reymond P, Van Pelt JA, Van Loon LC, Dicke M, Pieterse CMJ (2008) Differential effectiveness of microbially induced resistance against herbivorous insects in *Arabidopsis*. Mol Plant-Microbe Interact 21:919–930

Van Peer R, Schippers B (1992) LPS of plant growth–promoting *Pseudomonas* sp. strain WCS417r induce resistance in carnation to *Fusarium* wilt. Neth Plant Pathol 98:129–139

Van Peer R, Niemann GJ, Schippers B (1991) Induced resistance and phytoalexin accumulation in biological control of *Fusarium* wilt of carnation by *Pseudomonas* sp. strain WCS417r. Phytopathology 81:728–734

Van Wees SC, Luijendijk M, Smoorenburg I, Van Loon LC, Pieterse CMJ (1999) Rhizobacteria mediated induced systemic resistance (ISR) in *Arabidopsis* is not associated with a direct effect on expression of known defense–related genes but stimulates the expression of the jasmonate–inducible gene Atvsp upon challenge. Plant Mol Biol 41:537–549

Verhagen BWM, Glazebrook J, Zhu T, Chang H-S, Van Loon LC, Pieterse CMJ (2004) The transcriptome of rhizobacteria–induced systemic resistance in *Arabidopsis*. Mol Plant-Microbe Interact 17:895–908

Vernooij B, Friedrich L, Morse A, Reist R, Kolditz-Jawhar R et al (1994) Salicylic acid is not the translocated signal responsible for inducing systemic acquired resistance but is required in signal transduction. Plant Cell 6:959–965

Vespermann A, Kai M, Piechulla B (2007) Rhizobacterial volatiles affect the growth of fungi and *Arabidopsis thaliana*. Appl Environ Microbiol 73:5639–5641

Villarreal-Delgado MF, Villa-Rodríguez ED, Cira-Chávez LA, Estrada-Alvarado MI, Parra-Cota F, de los Santos-Villalobos S (2018) The genus *Bacillus* as a biological control agent and its implications in the agricultural biosecurity. Mex J Phytopathol 36:95–130

Vlot AC, Dempsey DA, Klessig DF (2009) Salicylic acid, a multifaceted hormone to combat disease. Annu Rev Phytopathol 47:177–206

Wang E, Schornack S, Marsh JF, Gobbato E, Schwessinger B et al (2012) A common signaling process that promotes mycorrhizal and oomycete colonization of plants. Curr Biol 22:2242–2246

War AR, Paulraj MG, Ahmad T, Buhroo AA, Hussain B, Ignacimuthu S, Sharma HC (2012) Mechanisms of plant defense against insect herbivores. Plant Signal Behav 7:1306–1320

Ward MG (2016) The regulatory landscape for biological control agents. EPPO Bull 46:249–253

Weller DM, Mavrodi DV, Van Pelt JA, Pieterse CMJ, Van Loon LC, Bakker PAHM (2012) Induced systemic resistance (ISR) in *Arabidopsis thaliana* against *Pseudomonas syringae* pv. Tomato by 2,4–diacetylphloroglucinol–producing *Pseudomonas fluorescens*. Phytopathology 102:403–412

Wu J, Baldwin IT (2009) Herbivory–induced signaling in plants: perception and action. Plant Cell Environ 32:1161–1174

Wu J, Baldwin IT (2010) New insights into plant responses to the attack from insect herbivores. Annu Rev Genet 44:1–24

Wydra K, Beri H (2006) Structural changes of homogalacturonan, rhamnogalacturonan I and arabinogalactan protein in xylem cell walls of tomato genotypes in reaction to *Ralstonia solanacearum*. Physiol Mol Plant P 68:41–50

Yan Z, Reddy MS, Ryu CM, McInroy JA, Wilson M, Kloepper JW (2002) Induced systemic protection against tomato late blight elicited by PGPR. Phytopathology 92:1329–1333

Yánez-Mendizábal V, Falconí CE (2018) Efficacy of *Bacillus* spp. to biocontrol of anthracnose and enhance plant growth on Andean lupin seeds by lipopeptide production. Biol Control 122:67–75

Youssef MMA, Eissa MFM (2014) Biofertilizers and their role in management of plant parasitic nematodes. A review. E3 J Biotechnol Pharm Res 5:1–6

Yu X, Feng B, He P, Shan L (2017) From chaos to harmony: responses and signaling upon microbial pattern recognition. Annu Rev Phytopathol 55:109–137

Zachow C, Berg C, Muller H, Monk J, Berg G (2016) Endemic plants harbor specific *Trichoderma* communities with an exceptional potential for biocontrol of phytopathogens. J Biotechnol 235:162–170

Zambolim L, Schenck NC (1983) Reduction of the effects of pathogenic root rot infecting fungi on soybean by the mycorrhizal fungus *Glomus mosseae*. Phytopathology 73:1402–1405

Zamioudis C, Pieterse CMJ (2012) Modulation of host immunity by beneficial microbes. Mol Plant-Microbe Interact 25:139–150

Zamioudis C, Mastranesti P, Dhonukshe P, Blilou I, Pieterse CMJ (2013) Unraveling root developmental programs initiated by beneficial *Pseudomonas* spp. bacteria. Plant Physiol 162:304–318

Zhang S, Moyne AL, Reddy MS, Kloepper JW (2002) The role of salicylic acid in induced systemic resistance elicited by plant growth promoting rhizobacteria against blue mold of tobacco. Biol Control 25:288–296

Zhang H, Kim MS, Krishnamachari V, Payton P, Sun Y, Grimson M, Farag MA, Ryu CM, Allen R, Melo IS, Pare PW (2007) Rhizobacterial volatiles emission regulate auxin homeostasis and cell expansion in *Arabidopsis*. Planta 226:839–851

Zipfel C (2009) Early molecular events in PAMP–triggered immunity. Curr Opin Plant Biol 12:414–420

Zuccaro A, Lahrmann U, Guldener U, Langen G, Pfiffi S et al (2011) Endophytic life strategies decoded by genome and transcriptome analyses of the mutualistic root symbiont *Piriformospora indica*. PLoS Pathog 7:e1002290

Role of WRKY Transcription Factor Superfamily in Plant Disease Management

11

Laxman Singh Rajput, Sumit Kumar Aggarwal, Sahil Mehta, Sanjeev Kumar, Vennampally Nataraj, M. Shivakumar, Hemant Singh Maheshwari, Sanjay Yadav, and Divyanshu Goswami

Abstract

In their natural habitat, the plants face numerous challenges from various plant pathogens simultaneously. In response, they activate the stress-response related machinery through regulation of a complex system of genes, miRNAs, siRNAs, and most importantly, transcriptional factors (TFs). Among the various activated TFs, WRKY TFs superfamily encodes maximum number of regulatory proteins. The WRKY TFs regulate downstream genes by both direct (auto- and cross-regulation) and indirect mechanisms (physical interactions within themselves or other TFs, proteins, and small RNAs). All the WRKY TF members possess a conserved WRKY domain consisting of nearly 60 amino acids with a specific heptapeptide sequence (WRKYGQK), with a Zn2+-finger motif that binds to specific cis-regulatory elements of defense gene called as W-box (TTGAC[C/T]). This W-box has been reported to be contemporary in the genes promoter region related to plants' innate immunity including PAMP triggered immunity (PTI), effectors triggered immunity (ETI), basal defense, and systemic acquired resistance (SAR). Because of this specific molecular orchestration primarily in plant immunity, this WRKY TFs superfamily has been established as a good target in

L. S. Rajput (✉) · S. Kumar · V. Nataraj · M. Shivakumar · H. S. Maheshwari · S. Yadav · D. Goswami
ICAR-Indian Institute of Soybean Research, Indore, Madhya Pradesh, India

S. K. Aggarwal
ICAR-Indian Institute of Maize Research, Ludhiana, Punjab, India

S. Mehta
Crop Improvement Group, International Centre for Genetic Engineering and Biotechnology, New Delhi, India

Department of Microbiology, Faculty of Life sciences, Institute of Advanced Management and Research, Ghaziabad, Uttar Pradesh, India

B. Giri, M. P. Sharma (eds.), *Plant Stress Biology*,
https://doi.org/10.1007/978-981-15-9380-2_11

plant disease management. However, in certain cases, along with beneficial effects, overexpression or repression of various WRKY TFs leads to genetic drag which needs to be identified and eliminated for most advantageous plant growth and development. This chapter will be fully focused on the regulation of the WRKY TFs and its role in plant disease management specifically.

Keywords

Plants · WRKY · Disease · Pathogens · Defense · Bacteria · Virus · Fungi

11.1 Introduction

Every day, the plants face various environmental stresses under the changing climatic scenario. Among all of the stresses, biotic stresses incited by plant pathogens especially affect the growth of various plants such as rice, barley, sugarcane, lentil, faba bean, chickpea (Mehta et al. 2019; Rajput et al. 2017a). Upon being challenged by phytopathogens, the plants change the growth rate and pathogenicity of phytopathogen, which leads to the modification of host–pathogen interaction (Madhusudhan et al. 2019; Rajput et al. 2017b). Because of continuous changes in climatic conditions, new virulent races or pathovars develop that can infect the crop resistant to pathogens previously (Rahman et al. 2019; Singh et al. 2019). So, therefore in order to combat these new entities, there is always a pressure/emphasis to enhance the current or develop new management strategies for phytopathogens.

However, in order to improve the strategies, the special focus is also given in understanding changes occurring at the genic, transcriptional, protein, metabolic as well as the cellular level (Chen et al. 2019; Singh et al. 2018a). The converging point of all these changes can be traced back to the transcriptional regulation controlled by transcription factors (TFs). The various TFs related to stress-responsive genes are WRKY, AP2/ERF, bZIP, MYC, MYB, MADS, NAM, ATAF, CUC, and NAC (Chen et al. 2019). Each TFs consist of a specific polypeptide binding domain which binds to a particular sequence/stretch of DNA bases together known as *cis*-regulatory elements (CREs) in response to specific stress (Mittal et al. 2018).

Among all of the TFs, WRKY TF superfamily is of meticulous importance as they are reported to be involved in a diverse range of plant development, metabolism, senescence, wounding, etc. (Chen et al. 2019). The WRKY TFs are reported to be found in higher plants as well as a lower plant with few exceptions (Bakshi and Oelmüller 2014). The members interact among themselves or with other TFs, proteins, and small RNAs to regulate the target genes (Chen et al. 2019). With regard to the discussion in here, WRKY TFs modulation in complex defense gene network is a key step for signal transduction pathways related to plant immunity. This results in the PTI, as well as a specific plant pathogen, triggered ETI (Zhang et al. 2019; Singh et al. 2018b). In both of the processes, salicylic acid (SA) and jasmonic acid (JA) get modulated at the downstream level. This eventually induces

NPR1 gene that leads to the induction of various defense genes such as PR proteins, phytoalexins, reactive oxygen species, peroxidase, and other related enzymes (Zhang et al. 2019). As a result, in this chapter, the structure and classification of WRKY as well as downstream regulation/interaction with the same or other TFs, proteins, and small RNAs to regulate the defense gene for various plant pathogens such as fungi, bacteria, and viruses have been summarized. We have also provided important highlights of understanding the regulation and advancement in crop implement strategies by using WRKY TFs for the management of phytopathogens.

11.2 WRKY TFs: Structure

A specific DNA-binding domain is the distinguishing character of WRKY TFs which consists of a greatly conserved partly protruding region of almost 60 amino acids in plants (Duan et al. 2015). In this region, there is an almost invariable sequence at N terminal of protein, WRKYGQK (W; Tryptophan, R; Arginine, K; Lysine, Y; Tyrosine, G; Glycine Q; Glutamine, K; Lysine).WRKY TFs generally bind to a considerably conserved region of DNA identified as the W-box elements having the conserved motif TGACC/Tsingaly or in tandem repeats of the promoter region of defense genes (Chen et al. 2019). Recently, the WRKY domain structure with W-box as a binding site was identified and showed four-stranded β-sheet makes grove of DNA in an unusual manner where the plane of β-sheet is almost at right angles to the helical axis of the DNA (Yamasaki et al. 2012). Whereas C-terminal of protein consists of 4–5 stranded anti-parallel β-sheets of zinc-finger-like motif (Cx4-HxC or 5Cx22-23HxH or Cx7Cx23) (Alves et al. 2014). A tryptophan amino acid makes the core structure of the conserved WRKYGQK sequence, whereas the rest of amino acids were bind to DNA. The glycine amino acid helps in creating protrudes that make grove for binding to W-box by hydrophobic interaction with the methyl groups of thymine nitrogenous bases of the DNA. Mutation in the thymine base or Zn^{2+}-binding site drastically reduced the DNA-binding activity by a disorder of the active structure of DNA-binding domain protein (Yamasaki et al. 2013). Only in very few crops like *Arabidopsis* (WRIY), soybean (WHQY), potato (WHKC and WRKC), black cottonwood (FRKY), *tomato (*WRKR, WIKY, WSKY, and WQKY), *and* French bean (ARKM, WWKN, and WRMY), WRKY proteins have been reported to have changes in the conserved sequence of WRKY (Mohanta et al. 2016).

11.3 Classification of WRKY TFs

Based on the total number of WRKY domains and presence/absence of zinc-finger-like motif, WRKY protein is classified into three groups, i.e., group I, group II, and group III (Chen et al. 2019). Group I members consist of two WRKY domains, whereas group II members have one WRKY domain along with Cys2-His2-type of zinc-finger motif. Group II members are further divided into five subgroups IIa to IIe based on the presence of amino acid motifs additional in the WRKY domain. Group III members also contain one WRKY domain with Cys2-His/Cysor Cys2-His2 type of zinc-finger motif (Eulgem et al. 2000). The criteria of classification method are exclusively based on protein structure; however, this classification does not concern about evolution, origin, and duplications of the gene for WRKY TFs. So, in the year 2005, Zhang and Wang (2005) again reclassified WRKY TFs into five groups, i.e., group 1, group 2_a + 2_b, groups 2_c, group 2_d + 2_e, and group 3, based on phylogenetic analysis, domains conservation, and intron position in WRKY domain. However, based on the insertion position in an intron, WRKY TFs were again classified into two groups where group 1 includes R-type of the intron in the WRKY domain, whereas group 2 members include V-type of an intron (Zhang and Wang 2005).

Initially, WRKY TF was identified for the regulation of sporamin and β-amylase production from sweet potato as SPF1 (Ishiguro and Nakamura 1994). However, later, Rushton and group identified three different WRKY TFs (WRKY1, WRKY2, and WRKY3) from parsley in a stress response against elicitor Pep25 of *Phytophthora parasitica* and given the name of "worky" in 1996 (Rushton et al. 1996). This opened a way for identification of stress-responsive WRKY TFs, which has resulted now in a superfamily. They formerly considered to be reported from plants only, but they have also found in protists (*Giardia lamblia*) and Metazoa (*Dictyostelium discoideum*) (Finatto et al. 2018).

11.4 Regulation of WRKY TFs

11.4.1 Kinases

MAP kinases can regulate most of WRKY TF by upregulation of various defense genes against several pathogens (Aamir et al. 2018). After recognition of PAMP or MAMP molecules, plants trigger a series of events in MAPKKK signaling which leads to activation of *AtWRKY33* (Qiu et al. 2008). Then *AtWRKY33* activates phytoalexin producing defense gene naming PAD3 (phytoalexin deficient 3) that produces camalexin and provides defense against *Pseudomonas syringae* by disruption of bacterial membranes (Rogers et al. 1996). *AtWRKY33* could be activated by two other MAP kinases, i.e., MPK3 and MPK6, which provide defense against *Botrytis cinerea* by producing camalexin (Mao et al. 2011). Other kinases like calcium-dependent protein kinases like CPK11 and CPK4 modulate *AtWRKY28* against *P. syringae* pv. *tomato*. It induces PKS2 (SOS2-like protein kinase 5) gene which phosphorylates NPR1 (non-expressor of pathogenesis-related gene 1) gene that in turn induces systemic resistance (Gao and He 2013). This interaction again induces a couple of WRKY genes like *AtWRKY38* and *AtWRKY62,* which ultimately induces plant defense genes (Xie et al. 2010). Similarly, in rice, *OsWRKY45* WRKY TF regulated by MAPK provides resistance against *Magnaporthe oryzae* and *Xanthomonas oryzae pv. oryzae* by induction of SAR pathway (Nakayama et al. 2013). Likewise, *Sl*WRKY33 interacts with MAPK5 and provides resistance against

F. oxysporum f. sp. *lycopersici* in tomato plants (Aamir et al. 2018) Therefore, various kinases are key regulator in controlling or modulating defense genes by phosphorylation and activation of various WRKY TFs.

11.4.2 Autoregulation and Cross Regulation

Various external and internal stimuli induce numerous signaling pathways that form "WRKY web." Signaling of this WRKY web is autoregulated when WRKY TFs interact with their own promoters while cross regulated when WRKY TFs interact with promoters of other WRKY TFs (Eulgem and Somssich 2007). Chromatin immunoprecipitation (ChIP) analysis on parsley showed that *PcWRKY1* TF binds to own promoter and defense gene *Pc*PR10, whereas *PcWRKY1* also binds to the promoter of *Pc*WRKY3 and defense gene *Pc*PR1 (Turck et al. 2004). The members of the same family contribute related and superfluous roles in controlling signaling pathways by regulating each other transcription level. *WRKY18, WRKY40*, and *WRKY60* in A. thaliana cross regulate abscisic acid signaling by controlling each other transcription (Yan et al. 2013). WRKY TFs regulated the defense process by auto- and cross-regulation mechanisms with the formation of homocomplexes and heterocomplexes (Liu et al. 2016a). WRKY web provides resistance through the cross-regulation mechanism. Likewise, *SlWRKY23* provides resistance against *Oidium neolycopersici,* a causal agent of tomato powdery mildew under alkaline soil only (Kissoudis et al. 2016).

11.4.3 Positive and Negative Regulation

In some cases, same WRKY TFs provide reistance to one pathogen and susceptibility to another pathogen. For example, *OsWRKY45-1* and *OsWRKY45-2* found in rice encode protein differing in a total of 10 amino acids. Upon overexpression, both the alleles showed resistance to the rice blast fungal pathogen *M. oryzae* positively. While *OsWRKY45-1* regulates negative resistance towards rice bacterial blight pathogen *X. oryzae* subsp. *Oryzae*, *OsWRKY45-2* regulates positive resistance towards the same pathogens. This is due to OsWRKY 45-1 that regulates both SA and JA pathway genes, while *OsWRKY45-1* regulates only JA pathway genes (Tao et al. 2009). After the challenge of *Botrytis cinerea, AtWRKY33* induces camalexin biosynthetic with a positive feedback regulatory loop which provides defense against *B. cinerea* (Mao et al. 2011). In contrast to that *OsWRKY62* and *OsWRKY76* increase susceptibility against *M. oryzae* and *X. oryzae* subsp. *Oryzae*, respectively, through an alternative splicing mechanism, these WRKY TFs regulated with a negative feedback regulation mechanism (Liu et al. 2016a, 2016b). WRKY TFs have another interesting property that is to show the opposite effects on both types of stress tolerance while complex communications among various signaling networks that lead to may be positive and negative effects on the regulation of different stresses (Bai et al. 2018). For example, *OsWRKY45* provides resistance against

M. oryzae positively while altering abiotic stress tolerance (Tao et al. 2009), and on the other hand, *OsWRKY75* enhances *M. oryzae* susceptibility same time increase adaptation of rice towards cold stress (Yokotani et al. 2013).

11.4.4 Epigenetic Regulation

Epigenetic regulation is based on the regulation of histone proteins. *AtWRKY70 was expressed epigenetically after* histone protein H3K4 gets trimethylated due to trithorax binding (ATX1). *AtWRKY70* activates defense genes such as PR-1 and THI2.1 (Alvarez-Venegas et al. 2007). *AtWRKY38* and *AtWRKY62* get repressed due to the removal of acetyl groups from histone protein by histone deacetylase 19 (HDA19) and lead to enhanced susceptibility towards *P. syringae* (Kim et al. 2008). The promoter region of *AtWRKY40 was activated due to* histone methylations and provides systemic acquired resistance (Alvarez et al. 2010). Through histone modification at promoter regions of *At*WRKY29and *At*WRKY6 two histone protein H3K4 (histone H3 lysine 4) and H3K14 increase and provide SAR against *P. syringae* pv. tomato (Singh et al. 2014). Some proteins such as VQ protein also induce histone modification and provide defense against pathogens. Two VQ proteins sigma factor binding protein 1 (SIB1) and SIB2 interacted with *At*WRKY33 and provided resistance towards *Botrytis cinerea (*Lai et al. 2011). Interestingly, DNA demethylases in Arabidopsis enhance resistance against *Fusarium oxysporum* by altering the modulation of genes involving pathogenicity (Le et al. 2014). Similarly, demethylation in the promoter region of *AtWRKY22* provides resistance against *P. syringae* pv. *tomato* (Yu et al. 2013a).

11.4.5 Proteasome Regulation

In normal conditions, the production of WRKY TFs is checked by proteasome-mediated degradation. *Os*WRKY45 has a major role in SAR against rice blast pathogen *M. oryzae which is regulated by the nuclear ubiquitin-proteasome system (UPS). After the M. oryzae attack on rice, polyubiquitination allows the accumulation of Os*WRKY45 to induce defense against *M. oryzae* (Matsushita et al. 2013). Pathogen resistance is positively regulated by *At*WRKY53, while leaf senescence is negatively regulated. Ubiquitin protein ligase 5 (UPL5) interacts with the leucine zipper domain of *At*WRKY53 for degradation and accumulation of UPS. The expression of *At*WRKY53 is highly regulated by inducing pathogen response and UPS production (Miao and Zentgraf 2010). UPS mediated degradation of transcription factor also regulates other WRKY TFs, for example, in Chinese wild grapevine (*Vitis pseudoreticulata*) *VpWRKY11* provides defense against pathogen *Golovinomyces cichoracearum. VpWRKY11* TF production was regulated by interaction with E3 ubiquitin ligase Erysiphe necator induced RING finger protein 1 (EIRP1) through degradation by the 26S proteasome (Yu et al. 2013a, 2013b).

11.4.6 Small RNA Regulation

Small RNAs (smRNA) regulate WRKY TFs under pathogen response by RNA interference, RNA silencing, or post-transcriptional gene silencing (PTGS) (Voinnet 2009). After several phytohormone treatments in rice, several miRNAs were induced; out of which, miR167f encoded an NBS-LRR disease resistance protein (Liu et al. 2009). Double mutants of *At*wrky18 *At*wrky40 provide resistance to *G. orontii* by enhaning the expression of RCD One5 (SRO5) which allows siRNA production and leads to suppression of WRKY TFs (Borsani et al. 2005). Similarly, AvrPtoB an effector secreted by *P. syringae* protein suppresses host miRNAs which lead to suppression of AtWRKY30 (Navarro et al. 2008). Production of secondary metabolites can be regulated by miRNA which provides resistance against fungi. Expression profiling of apple (Cv. Golden Delicious) infected with leaf spot fungus (*A. alternata* f. sp. *mali.*) confirmed that both Md-miR156ab and MdmiR395 target *MdWRKYN1* and *MdWRKY26*, respectively, and regulate resistance against the pathogen (Zhang et al. 2017). After the inoculation of *F. oxysporum* in *Persicaria minor* (a herb), the miRNAs get upregulated and downregulated by the WRKY TFs and terpenoid biosynthesis was reduced (Samad et al. 2019).

11.5 Role of WRKY TFs Against Phytopathogens

11.5.1 Role of Host Plant WRKY Against Viral Diseases

As described earlier, the WRKY TFs are one of the highly studied regulatory protein family which play both positive and negative role in plant immunity (Chen et al. 2019). One of the major challenges to plant immunity is the onset of viral diseases (Honjo et al. 2020). There are already many reports in the literature by various research groups for the bolstering support of the role of WRKY TFs against plant viruses. Out of all published articles, the first-ever preliminary report was by Yoda and colleagues from Nara Institute of Science and Technology, Japan. In their pioneer work, they screened a set of defense reaction genes upregulated during the hypersensitive response (HR) in wild tobacco (Cv. Xanthi nc) upon *Tobacco mosaic virus* (TMV) infection using fluorescent differential display (Yoda et al. 2002). The full-length deduced TIZZ protein contained a single WRKY domain which showed high similarity to one of the WRKY family members, namely WIZZ. Their results indicated the presence of a novel type of WRKY protein(s) that might play a critical role in HR signal activation. The more profound support for the role of WRKY genes against viral pathogens was given by Liu and group article published in The Plant Journal (Liu et al. 2004). In their article, they used a candidate gene approach to identify defense genes that play a role in immunity against TMV. TRV-VIGS based downregulation of several candidate genes including *WRKY1-WRKY3* confirmed the compromised N-mediated resistance to TMV (Liu et al. 2004). In another study, a WRKY gene, namely CaWRKY-a from *Capsicum annuum* (Cv. Bugang) was isolated (Park et al. 2006). The overexpression of *CaWRKY-a* in transgenic plants confirmed that *CaWRKY-a* might be involved as a transcription factor in defense-related signal transduction pathway against TMV (P1.2 pathotype). The positive regulatory roles of *CaWRKY1* and *CaWRKY2* against the deadly viruses, namely TMV and Pepper mild mottle virus were first put up forwarded by Oh and colleagues in their series of publications, respectively (Oh et al. 2006). In another published report, Dardick from Appalachian Fruit Research Station focused on identifying the statistically significant changes in gene expression concomitant with *Plum pox potyvirus* (PPV), *Tomato ringspot nepovirus* (ToRSV), and *Prunus necrotic ringspot ilarvirus* (PNRSV) symptoms in *N. benthamiana* leaves. The expression levels of identified WRKY transcription factors are in accordance with the severity of the observed symptoms in all three viruses (Dardick 2007). In another report, overexpression of four *Medicago truncatula MtWRKY* genes in the tobacco confirmed their regulatory roles in upregulating PR genes expression as well as lignin deposition against TMV (Naoumkina et al. 2008). Similarly, Babu et al. (2008) focused on the global gene expression using *Arabidopsis* Affymetrix ATH1 array after PPV infection in protoplasts of Arabidopsis accession Col-0. In this study, the 263 genes were upregulated including family members of WRKY TF, MADS-box protein, TIR-NBS-LRR, and zinc-finger family protein. In a similar study, McGregor and colleagues focused on one of the most devastating *Ipomoea batatas* disease, namely sweet potato virus disease. They used a similar global gene expression approach in two different sweet potato cultivars NASPOT1 (resistant) and Beauregard (susceptible). Their group found that cell expansion genes, as well as chloroplastic genes, were suppressed, while stress-related and various transcription family genes (WRKY, homeodomain proteins, and NAC-like proteins) were induced highly. After virus infection, the protein synthesis-related genes induction was in co-relation with virus accumulation in susceptible plants. This switch in the expression of all these specific host-encoded genes was established as a reason to cause developmental defects in susceptible plants (McGregor et al. 2009). In another publication, Alfenas-Zerbini and his Brazilian colleagues inoculated susceptible plants of tomato (Cv. Moneymaker) with *Pepper yellow mosaic virus* and further constructed a subtractive library from inoculated leaves at 72 h after inoculation. The upregulated genes were identified to be related to stress/defense response, cell cycle regulation, signal transduction, and transcriptional regulation (e.g., WRKY22 and SCARECROW TF). Few of the differentially expressed genes (DEGs) including WRKY22 were validated using macroarray analysis (Alfenas-Zerbini et al. 2009). In the year 2009, an Italian group headed by Catoni elucidated the organ-specific responses upon *Tomato spotted wilt virus* (TSWV) infection at the transcriptional level. They reported a total of 17 WRKY TFs were found differentially expressed in shoots and roots (Catoni et al. 2009). Similarly, the antiviral role of tobacco *NtWRKY4* in viral stress tolerance was affirmed by using RNAi technology. Upon TMV inoculation, the mosaic pattern leaves were highly twisted in the transgenic plants as compared to the uninoculated wild-type (Ren et al. 2010). Chen and Yeh (2010) used a microarray assay to analyze the effect of TMV infection on *A. thaliana* protoplasts. They affirmatively reported that approximately eight transcriptional regulation genes showed greater than threefold changed expression. Furthermore, the overexpression, as well as knockout of *AtWRKY6* in transgenic plants, helped in concluding *AtWRKY6* has a dual role in supporting and inhibiting virus infection. Inoculation of *Rice stripe virus* causes phenotypic symptoms such as stunting, necrosis, chlorosis, and weakness. In addition, it also covers temporal changes at the transcriptomic level of protein-synthesis machinery, energy production, cell-structure component synthesis as well as WRKY, AP2, and NAC TF genes (Satoh et al. 2010). The same trend of results was observed for the defense systems regulated by WRKY45 upon *Rice dwarf virus* infection (all three strains, namely O, D84, and S) (Satoh et al. 2011). Similarly, in another report, it was confirmed *CaWRKY30* is upregulated upon application of TMV, *Ralstonia solanacearum,* and *P. capsici* (Jingyuan et al. 2011). It was also supported by the work of Naqvi and the group who confirmed the infection of *Tomato leaf curl virus* (ToLCNDV) increase, the *LeWRKY30* and LePR-1 in tomato (Naqvi et al. 2011). With WRKY domain-specific differential display protocol, the *CaWRKYb* and *CaWRKYd* genes were identified which get rapidly induced upon TMV inoculation. Later, the overexpression and knockout studies revealed the positive roles of *CaWRKYb* and *CaWRKYd* in the hypersensitive response between hot pepper (Cv. Bugang) and TMV (pathotype P0) (Huh et al. 2012).

Using *Bean pod mottle virus*-based VIGS technology, the role of *WRKY6*, as well as *WRKY30* in Rsv1-mediated resistance, was elucidated in soybean (Zhang et al. 2012). In another article, mild infection of *Citrus tristeza virus* isolate in sweet *Citrus aurantifolia* also upregulated the WRKY TFs and ethylene-responsive element binding factors (ERFs) profiles confirmed with suppression subtractive hybridization (Liu et al. 2012). Similarly, the overexpression of cotton *GhWRKY15* and *GhWRKY11* in transgenic tobacco plants activated the expression of several PR, POD, and APX genes, therefore, triggering systemic acquired resistance (SAR) to protect the plant against viral pathogens such as TMV and *cucumber mosaic virus* (CMV) as compared with the wild-type (Sun et al. 2012). The similar role of WRKY TF (*AtWRKY8*) also has been elucidated in the defense response to crucifer-infecting TMV. The reason lies in the mediatory role of *AtWRKY* in ET and ABA signaling crosstalk (Chen et al. 2013a). In another published article, a total of 16 WRKY genes were downregulated even up to 7.9-fold in the *Tomato yellow leaf curl virus* (TYLCV)-susceptible tomato line (TMXA48-4-0). However, there were nearly 7 WRKY genes upregulated in the TYLCV-resistant CLN2777A line (Chen et al. 2013b). Inoculation of *Rice tungro spherical virus* (RTSV) on susceptible rice (Cv. TN1) changed the transcripts levels of multiple stress-related genes including multiple members of the WRKY gene family (*OsWRKY1.V2, OsWRKY5, OsWRKY9, OsWRKY28, OsWRKY29*, and *OsWRKY45*) (Satoh et al. 2013).

In their extensive work, the expression profiles of 13 selected papayas genotypes, *CpWRKY* TF genes under both two biotic and three abiotic stresses were evaluated through qRT-PCR. The expression levels of TF12.199, TF807.3, TF21.156, and TF18.51 were notably induced by the *Papaya ringspot virus* (Pan and Jiang 2014). Later, focus was also done on finding the common WRKY TFs coding genes in induced plant expression profiles by multiple viruses like Potato virus Y, TMV, and CMV through using databases, online literature, and quantitative analyses. Data suggested that the Tobacco 06G gene is a member of WRKY subfamily II that was common in all virus infection-induced plant expression profiles and can be utilized to enhance virus resistance in many plants (Zhou et al. 2014). Using techniques like the Y2H system, co-immunopurification assay, and mutant lines of A. thaliana, the binding of *WRKY70* with RCY1-encoded CC-NB domain fragment was affirmed, which directly confirmed the function of *WRKY70* in resistance to a yellow strain of CMV (Ando et al. 2014).

In another instance, the chrysanthemum *WRKY11*, a homolog of *AtWRKY11* gets upregulated after infection by all three viruses, namely CMV, TSWV, and Potato virus X (Choi et al. 2015). Mandal et al. (2015) analyzed the promoter region of SlTORNADO1, a gene important for the cell expansion, vein formation as well as symptoms development during ToLCNDV infection. Their team confirmed the *SlWRKY16* active interaction with W-boxes present in the SlTORNADO1 promoter region. Kundu et al. (2015) correlated the role of *Vigna mungo* WRKY in imparting resistance to the *mungbean yellow mosaic India virus* (MYMIV) while working on VMR84 (MYMIV-resistant) and T9 (susceptible) cultivars. Huh and colleagues functionally characterized the control of MAPK-1 and -2 in *CaWRKYa*-based L-mediated transcriptional reprogramming of PR gene expression during TMV infection in hot pepper (Huh et al. 2015).

In the year 2016, the role of six tomato WRKYs (*WRKY41, WRKY42, WRKY53, WRKY54, WRKY80,* and *WRKY81*) in TYLCV infection was elucidated using subcellular localization analysis, interaction network analysis, and TRV-VIGS (Huang et al. 2016). The overexpression and knockout studies of *AtWRKY61* in A. thaliana transgenic plants confirmed the negative correlation to Turnip crinkle virus accumulation and symptoms which elucidated its role in plant immunity (Gao et al. 2016).

Time-course analysis of the effect of *Ugandan cassava brown streak virus* grafting on resistant and susceptible cassava varieties transcriptome revealed the upregulation of differentially expressed-defense genes response genes including LRR-containing, NBARC-containing, PR, LEA, WRKY, GATA, NAC, and HSPs (Amuge et al. 2017). A similar kind of RNA-Seq study was conducted for *cucumber green mottle mosaic virus* (CGMMV), a member of the Tobamovirus genus which induces fruit decay in watermelon plants (*Citrullus lanatus*). Out of all 1621 DEGs, various members of WRKY family such as *WRKY13, WRKY31, WRKY46, WRKY48, WRKY53,* and *WRKY70* were highly upregulated (Li et al. 2017). *NtWRKY12* gene expression is induced upon TMV infection and PAMP elicitation (Gullner et al. 2017). Similarly, microarray analysis coupled with GO and MapMan analysis of two different ToLCNDV-resistant and susceptible potato cultivars also revealed there were more than 3500 genes differentially regulated including multiple WRKY family members (log2FC $>$ 2) following the ToLCNDV infection (Jeevalatha et al. 2017). All these published work together opened a list of multiple biomarkers and candidate genes to be used in resistance breeding against viruses.

Recently, in an article published in Plant Cell Reports, Madronero and colleagues conducted the global gene expression analysis on *Papaya meleira virus* complex induced changes in infected papaya at pre-and post-flowering stages. At the pre-flowering stage, a total of 633 DEGs were observed including multiple salicylic acid, ethylene-pathway genes, PR genes, ROS genes, and WRKY TF encoding genes (Madronero et al. 2018). In another published article, in-depth RNA-Seq was carried out to unravel the transcriptional changes associated with the symptoms of TYLCV and *tomato chlorosis virus*. Comparative analysis of DEGs for both viruses revealed that *WRKY6* levels were highly altered among the WRKY TF family which positively regulates the crosstalk between the expressions of phytopathogens-related as well as senescence-associated genes (Seo et al. 2018). The in-depth promoter region analysis for *Cestrum yellow leaf curling virus* was done to elucidate the interacting factors responsible for NPR1-dependent SA signaling induction which leads to synergistic identification of *WRKY53* TF as a partner (Sarkar et al. 2018).

The positive role of *AtWRKY30* was confirmed in resistance against CMV using overexpression and mutant studies in addition to oxidative stress, fungus, SA, and ABA (Zou et al. 2016). In CGMMV-infected cucumbers, expression profiling data confirmed both miRNA854 and miRNA5658 target *WRKY21* and LRR receptor kinase as well as FLS2-like protein which is together known in the literature to play role in defense responses/phytopathogens resistance (Liang et al. 2019). The combinatorial effect of low light intensity/shading and *Soybean mosaic virus* on the transcriptome level of soybean plants was assessed. Among all the 24 DEGs related to plant–pathogen interaction, a total of two WRKY genes (*WRKY33* and *WRKY62*) were differentially expressed under both light conditions (Zhang et al. 2019a).

More recently, the effect of infection by both rice tungro viruses (*Rice tungro bacilliform virus* and RTSV) at the rice transcriptomic landscape was deduced by using global gene expression changes using Illumina Hiseq 2500 platform followed by qRT-PCR. About 959 DEGs were related to stress-responsive pathways and hormonal homeostasis. Among all DEGs, the reported WRKY transcription factors were LOC_Os05g25770, LOC_Os08g38990, LOC_Os09g25060, and LOC_Os11g02520 (Kumar and Dasgupta 2020). Taking together all these studies at a point, we can conclude that the WRKY TFs regulate host defenses against viruses at various levels directly or indirectly. The various methods cover direct modulation of viral target/plant defense genes (downstream), repression as well as activation of additional TFs through feed-forward or -backward regulation.

11.5.2 Role of Host Plant WRKY Against Bacterial Diseases

Unlike viruses, the bacteria grow in the spaces between plant cells and cause multiple symptoms including cankers, wilts, soft rots, blights, scabs, galls, and leaf spots (APS 2020). The change in host plant WRKY TFs levels in response to the bacterial diseases has been reported in multiple reports of the literature. The first report in the literature regarding the elicitor-induced nature of WRKYs in response to bacteria was by the group of Dellagi et al. (2000). Their group first-time isolated an upregulated potato *StWRKY1* protein using the SSH technique upon inoculation of *Erwinia carotovora* subsp. *atroseptica* culture filtrate. *P. syringae* pv. *tomato*

(Strain DC300) infiltration on mature leaves resulted in *AtWRKY6* induced expression which implicated the role of *AtWRKY6* in regulating a few aspects of host defense response (Robatzek and Somssich 2001).

In one of their article in series, Robatzek and Somssich studied the targets of senescence- and defense-associated *AtWRKY6* factors. Their study revealed the *WRKY6* negative regulation on its promoter activity as well as promoters of *AtPR1, AtSIRK*, and other closely related WRKY family members (Robatzek and Somssics 2002). The *AtWRKY18* was overexpressed under the control of CaMV35S promoter in *Arabidopsis* which results in stunting, increased levels of PR genes as well as resistance against *P. syringae in* the transgenic plants (Chen and Chen 2002). Dong and colleagues identified a list of common *AtWRKYs* induced under treatments such as SA application as well as inoculation of an avirulent strain of *P. syringae* (Dong et al. 2003).

The *OsWRKY03* transcript profile in rice seedlings specifically varies with treatment such as hormones, fungicide, rice bacterial blight pathogen, and mechanical wounding (Liu et al. 2005). In *X. oryzae* pv. *oryzae* infected rice plants, systematic expression analysis revealed that 12 *Os*WRKYs-encoding genes were differentially regulated including *OsWRKY-7, OsWRKY-10, OsWRKY-11, OsWRKY-30, OsWRKY-32, OsWRKY-67, OsWRKY-70, OsWRKY-83,* and *OsWRKY-85* (Ryu et al. 2006). Similarly, using a domain-specific differential display procedure, Park et al. (2006) isolated a rapidly induced WRKY gene, namely *CaWRKY-a* during *X. campestris* pv. *vesicatoria* infection (Park et al. 2006). Similarly, the positive role of nuclear-localized *AtWRKY7* gene in *P. syringae* susceptibility was confirmed using loss- and gain-of-function studies in *A. thaliana* (Kim et al. 2006). Similarly, the negative role of the *At*WRKY25 in plant defense against the bacterial pathogen *P. syringae* was also deduced using T-DNA insertion mutants and overexpression studies (Zheng et al. 2007).

The complexity of the *CaWRKY1* networking in chili pepper leaves was studied upon inoculation of *X. axonopodis* pv. *vesicatoria and P. syringe* pv. *tabaci* using VIGS and overexpression technology. Their work suggested that *CaWRKY1* got strongly induced upon bacterial inoculation (Oh et al. 2008). In another study, the rice *OsWRKY45* expression was highly upregulated upon treatments such as ABA, NaCl, PEG, heat stress, cold stress, blast pathogen as well as rice bacterial blight pathogen. Furthermore, *OsWRKY45* over expressed plants showed increase in *PR* genes' expression as well as enhaced tolerance/resistance to *P. syringae*, salt and drought stresses (Qiu and Yu 2009). In another instance, the AtWRKY8 induction was observed upon various treatments such as maggot infestation, ABA, H2O2, wounding as well as P.syringae infection. Furthermore, T-DNA insertion mutants showed increased resistance to *P. syringae,* whereas OE *AtWRKY8* transgenic plants displayed an increase in susceptibility to the might bacteria *P. syringae* infection. Combined, the study suggested that *AtWRKY8* acts as a negative regulator to *P. syringae* resistance (Chen et al. 2010).

Hwang and colleagues firstly isolated a group II *OsWRKY6* from rice samples infected with *X. oryzae* pv. *oryzae*. Furthermore, their heterologous overexpression study in Arabidopsis confirmed *OsWRKY6* acts as a transcriptional regulator of the plant defense response against *X. campestris* pv. *campestris* (Hwang et al. 2011). The genome-wide profiling of WRKY TFs in the red tomato resulted in an arrangement of a total of 81 *SlWRKY* genes into 3 main groups. Furthermore, the qRT-PCR analysis of *Sl*WRKYs showed spatial expression patterns in response to various treatments including four WRKYs, namely *SlWRKY23*, *SlWRKY39*, *SlWRKY80,* and *SlWRKY81* in response to *P. syringae* invasion (Huang et al. 2016). In a report published in famous Plant, Cell & Environment, Dang et al. (2013) clarified the role of pepper *Ca*WRKY40 in imparting resistance against *R. solanacearum* infection. In another report published in Molecular Plant Pathology, Wang and colleagues concluded the negative role of *CaWRKY58* in imparting resistance to *R. solanacearum* infection (Wang et al. 2013). Similarly, the overexpression of *Gh*WRKY40 *in N. benthamiana* enhanced the *R. solanacearum* susceptibility as compared to the wild-type plants (Wang et al. 2014). In another report published in the Physiologia Plantarum, the overexpression of *CaWRKY27* enhanced the resistance of tobacco transgenic plants to *R. solanacearum* (Dang et al. 2014). Wu and colleagues compared the transcriptome-scale changes in two maize NILs upon inoculation of bacterial brown spot pathogen. They observed WRKY-encoding genes such as *WRKY33, WRKY53,* and *WRKY71* were pronouncedly activated in both resistant and susceptible NIL (Wu et al. 2015). In another article, the heterologous overexpression of rapeseed *Br*WRKY7 enhanced the resistance in *Arabidopsis* transgenic plants against *Pectobacterium carotovorum,* the causal organism of bacterial soft rot (Ko et al. 2015).

In another study, constitutive overexpression of the poplar *PtrWRKY89* in *Arabidopsis plants* resulted in enhanced susceptibility to *P. syringae* as compared to the wild-type plants. This was further confirmed by the qRT-PCR study which confirmed the downregulation of marker genes related to SA as well as JA pathways at the molecular level (Jiang et al. 2016). On the other instance, Hwang and colleagues published their article in *Plant Cell Reports* which elucidated the positive role of *OsWRKY51* in defense against *X. oryzae* pv. Oryzae (Hwang et al. 2016). In the year of 2017, Nemchinov et al. first selected and inoculated bacterial stem blight-resistant and susceptible alfalfa (*Medicago sativa* L) plants and then performed their temporal transcript profiling. Their analysis revealed that there were plenty of DEGs in two contrasting genotypes at the molecular level. The reason for resistance appeared to be mediated primarily by 20 WRKY family transcription factors and other function-related genes (Nemchinov et al. 2017). On the other instance, Liu and colleagues published their article in Frontiers in Plant Science which elucidated the positive role of *NtWRKY50* in imparting resistance to *R. solanacearum* by altering both SA and JA production (Liu et al. 2017). In the next year, Liu and colleagues published their article in *BMC Plant Biology* which elucidated about the positive role of *OsWRKY67* in regulating bacteria blight resistance in rice (Liu et al. 2018). Their work was further validated by the report published by Vo and group in the journal Frontiers in Plant Science (Vo et al. 2018).

Recently, Sureshkumar and other scientists from ICAR-IARI generated a database "RiceMetaSysB." This database contained a collaborative as well as a curated list of bacterial blight responsive genes in rice as well as opened a channel to utilize already identified key WRKY genes for developing blast- and bacterial blight-tolerant plants in the future (Sureshkumar et al. 2019). The constitutive overexpression of wild grapevine *VdWRKY53* in Arabidopsis resulted in multi-fold enhancement in resistance to multiple pathogens including *P. syringae* pv. *tomato* (DC3000) (Zhang et al. 2019). More recently, Gao and group from Southwest University of Science and Technology (China) characterized the role of *SlWRKY8* in the resistance to *P. syringae* pv. tomato DC3000 (Pst DC3000) along with other abiotic stresses. The constitutive overexpression in the tomato plants (Cv. Ailsa Craig) resulted in increased resistance to Pst DC3000 by enhancing expression levels of PR genes, namely *Sl*PR1a1 as well as *Sl*PR7. Overall, their report suggested the role of *Sl*WRKY8 in plant immunity against bacterial pathogen and other prominent abiotic stresses (Gao et al. 2019).

11.5.3 Role of Host Plant WRKY Against Fungal Diseases

As described earlier, the WRKY TFs are one of a highly studied regulatory superfamily that plays a role in plant immunity (Singh et al. 2018b). Next to plant viruses, the next major challenge to plant immunity is the onset of fungal diseases (Rahman et al. 2019). There are already many reports in the literature by various research groups for the bolstering support of the role of WRKY TFs against fungal diseases. Out of all published articles, the first-ever preliminary report by Rushton et al. (1996) used both gain- and loss-of-function experiments in parsley (*Petroselinum crispum*) and identified the presence of *WRKY1*, *WRKY2,* and *WRKY3* TFs binding W-box in the promoters of *PR1-1* and *PR1-2* genes. Furthermore, they confirmed the Pep25 elicitor treatment in parsley cells induced a rapid increase in the mRNA levels of only *WRKY1* and *WRKY3*. Their work suggested that WRKY TFs might play a role in the signal transduction pathway. The next published report came after a period of long 3 years, i.e., in 1996, when the Euglem and colleagues published their in vivo work in The EMBO Journal (Eulgem et al. 2000). They confirmed the parsley *WRKY1* acts as a transcriptional activator that mediates the *Phytophthora sojae* elicitor-induced gene expression in parsley leaf tissue using in situ RNA hybridization. In the same year, Suzuki from the National Institute of Bioscience and Human-Technology (Japan) studied the signal transduction by inoculating two elicitors, namely purified xylanase of *Trichoderma viride* and *P. infestans* cell wall extract in the wild tobacco (Cv. Xanthi) cell suspension (Suzuki 1999). They reported many regions including a putative EIRE that contained conserved motif of W-boxes which might get activated by *NtWRKY1, NtWRKY2, NtWRKY3*, and *NtWRKY4* homologous to earlier reported parsley WRKYs. The effect of rice blast fungus race KJ301-derived elicitor on the gene responsiveness in Asian rice (Cv. Milyang 117) cell suspension culture cells was evaluated using cDNA library screening and mRNA differential display. The results revealed the increase in the cDNA levels of *OsERG1, OsERG2, OsEREBP1, OsHin1, OsCPX1, OsLPL1, OsMEK1* along with *OsWRKY1* (Kim et al. 2008). Using suppression subtractive hybridization (SSH), a putative StWRKY1 protein-encoding gene was identified in potato after inoculation of *P. infestans* as well as *E. carotovora* subsp. *atroseptica* filtrate (Dellagi et al. 2000). The same results were evidentially supported by the work of Beyer et al. (2001) upon compatible interaction of *P. infestans* with the potato. They identified multiple induced potato genes including *S1-9D* (WRKY-box transcription factor-like) during the fungus colonization using SSH, inverse northern analysis, and RNA blot analysis, For more detailed information, the readers can refer to the same publication. In another study, Cormack et al. (2002) identified two new WRKY TFs from parsley, namely *WRKY4* and *WRKY5* using the Y1-hybrid system. Furthermore, Pep25 elicitor treatment increased the transient expression of the *WRKY5* gene, a group III family member. Additionally, their results for other WRKY TFs such as *WRKY1* and *WRKY3* were following the results of Rushton et al. (1996). In *Arabidopsis thaliana* (Ecotype Columbia and Landsberg erecta), the transient expression time-course kinetics for a total of 13 WRKY group members were analyzed upon *Peronospora parasitica* and *Blumeria graminis* f. sp. *hordei* in both compatible and incompatible interactions. Their findings indicated that the WRKY TFs such as *WRKY38, WRKY54, WRKY55, WRKY66, WRKY67*, and *WRKY70* upregulated in susceptible Ler-1 plants (Kalde et al. 2003). Similarly, the *CaWRKY31* gene was found to be upregulated in coffee upon coffee rust fungus (*Hemileia vastatrix*) infection (Fernandez et al. 2004). In another work, they confirmed the upregulation in the transcripts level of *CaWRKY1* upon inoculation of coffee rust fungus (Kenyan isolate 1427) (Ganesh et al. 2006).

The arbuscular mycorrhiza colonization-induced changes in the whole rice transcriptome were compared upon with changes induced by two devastating pathogens, *M. grisea* and *Fusarium moniliforme,* and reported over 40% changes. Among the total mycorrhiza-regulated genes, about 30 genes including *OsAM205*, a WRKY-encoding gene was differentially expressed in a similar way upon AM and pathogen colonization reflects a general plant response to fungi colonization (Guimil et al. 2005). Ryu and colleagues in 2006 confirmed the changes in the expression level of total of 15 host WRKY genes upon inoculation of *M. grisea* (Philippines isolate PO6-6). Their extensive profiling analysis work revealed that the transcript levels of *OsWRKY7, OsWRKY10, OsWRKY11, OsWRKY30, OsWRKY45, OsWRKY62, OsWRKY76, OsWRKY82*, and *OsWRKY85* were significantly increased by 6–48 h. The positive role of *AtWRKY33* in plant immunity against necrotrophic fungi such as *Alternaria brassicicola* and *B. cinerea* was confirmed with gain- and loss-of-function studies (Zheng et al. 2006). Using the model plant *A. thaliana*, the physical interaction and complex regulatory role of *AtWRKY18, AtWRKY40*, and *AtWRKY60* in *B. cinerea* resistance was deduced (Xu et al. 2006). The role of the so-called SA activator, namely benzothiadiazole is well depicted in the literature. Using the group of techniques like microarray screening, RNAi, and transient overexpression system, the role of BTH-inducible *WRKY45* was identified in providing resistance against rice blast disease (Shimono et al. 2007). The same type of results upon overexpression of *AtWRKY70* in *A. thaliana* confirmed the RPP4-based resistance against *Hyaloperonospora parasitica* (Knoth et al. 2007). The positive role of nuclear-localized, rapidly inducing *AtWRKY3* and *AtWRKY4* in PR1-based immunity against the necrotrophic fungus *B. cinerea* was revealed using

T-DNA insertion mutants and gain-of-function mutants (Lai et al. 2008). Similarly, the constitutive overexpression of rice blast-induced *OsWRKY31* in Japanese rice cultivar, namely Zhonghua 17 leads to enhanced shoot length, root length, and resistance against two blast fungus strain P131 and MS220 of *M. oryzae* (Zhang et al. 2008). Levee and colleagues isolated a poplar *PtWRKY23*, an *AtWRKY23* ortholog primarily. Later, they examined the response of *PtWRKY23*-misexpressing plants for resistance to *Melampsora* rust using histological techniques, qRT-PCR, and Affymetrix GeneChip. The data revealed the role of *PtWRKY23* in resistance, redox homeostasis as well as cell wall-related metabolism (Levee et al. 2009). Similarly, the role of OsWRKY45 in multiple stresses such as salt, osmotic, drought, cold, fungus (*M. oryzae* Cav.), and bacteria (*X. oryzae* pv. *oryzae* and *P. syringae*) was established using various physiological parameters, gain- and loss-of-function studies (Qiu and Yu 2009). Yang et al. (2009) studied the expression of a total of 46 WRKY TFs encoding genes in the canola infected with two devastating fungal pathogens, namely *Sclerotinia sclerotiorum* and *A. brassicae* using quantitative real time-PCR (qRT-PCR). Their study revealed that about 13 BnWRKYs transcript abundance changed significantly following the fungal challenge. In another instance, using the same protocol by Fernandez et al. (2004), 6-month-old *C. arabica* plants (Var. Caturra and Tupi) were challenged with two different coffee rust (*H. vastatrix*) isolates. In comparison, the second isolate elicited more number of WRKY TFs (from twofold to sevenfold) (Ramiro et al. 2010).

The role of *GhWRKY3* upon infection with mighty fungi such as *Fusarium oxysporum* f. sp. *vasinfectum*, *Colletotrichum gossypii*, and *Rhizoctonia solani* was elucidated (Guo et al. 2011). Fan et al. (2011) confirmed the positive role of nuclear-encoded Chinese "Qinguan" apple in resistance against *A. alternata* f. sp. *mali.* Furthermore, the *MdWRKY1* overexpression tobacco plants showed enhanced resistance against the deadly oomycete, *P. parasitica* var. *nicotianae*. Similarly, the constitutive *TaWRKY45* overexpression lines also elicited the resistance against three fungal pathogens such as *F. graminearum*, *B. graminis*, and *Puccinia triticina* (Bahrini et al. 2011). Overexpression of nuclear-localized *OsWRKY30* gene in rice plants depicted the enhanced *R. solani* and *M. grisea* resistance. This occurred due to the activated expression of JA- and PR- synthesis-related genes (Peng et al. 2012). The similar role of *OsWRKY2* (Abbruscato et al. 2012), *OsWRKY45* (Shimono et al. 2012), *ClWRKY70* (Cho et al. 2012), *GbWRKY1* (Shu-Ling et al. 2012), *OsWRKY28* (Chujo et al. 2013), *OsWRKY76* (Yokotani et al. 2013), *AtWRKY28* (Chen et al. 2013a), *AtWRK75* (Chen et al. 2013c), *GhWRKY39-1* (Shi et al. 2014a) *GhWRKY39* (Shi et al. 2014b) was also elucidated against fungal pathogens such as *M. grisea, M. oryzae, S. sclerotiorum, Cladosporium cucumerinum, R. solani*, and *Verticillium dahlia*.

In the year 2015, an SA pathway-inducible poplar *PtrWRKY73* was isolated, characterized, and overexpressed in *A. thaliana* to increase the resistance against *B. cinerea* (Duan et al. 2015). Similarly, Cheng et al. (2015) clarified the interrelationship between various OsWRKYs in the rice blast resistance using gain- and loss-of-function studies. Their data revealed that *OsWRKY13, WRKY42*, and *WRKY45-2* interact with each other in a sequential transcriptional regulatory cascade to provide resistance against blast pathogen. In another study, the cotton *GhWRKY44* was isolated, characterized, localized, and constitutively overexpressed in *N. benthamiana*. The overexpression plants exhibited enhanced resistance to *R. solani* as compared with the wild-type plants. The resistance increased importantly due to the upregulation of several defense-related genes belonging to PR- and SA-signaling (Li et al. 2015).

In the next year, i.e., 2016, another poplar *PtrWRKY89* was characterized as a transcription activator in the nucleus for increasing the susceptibility to bacteria *P. syringae* as well as fungus *B. cinerea* as compared to the wild-type plants using constitutive overexpression (Jiang et al. 2016). In contrast, Jiang and colleagues overexpressed BoWRKY6 in the broccoli plants and reported enhanced downy mildew resistance (Jiang et al. 2016). In another published article, conidial suspension of *B. cinerea* (strain BCT 314) inoculated in 10-week-old OE *GhWRKY25* tobacco lines resulted in enhanced sensitivity by downregulating the SA- as well as ET-signaling related genes. Their paper indicated that this *GhWRKY25* plays a negative role in response to fungal pathogen resistance (Liu et al. 2016a, 2016b).

Recently, Lui and colleagues from Nanjing Agricultural University (China) classified various WRKY family members into three main groups based on the conserved domains structure, collinearity, exon–intron structures, and duplications. Furthermore, they illustrated the expression pattern of MdWRKY-encoding genes in response to *A. alternate* (Lui et al. 2017). Expression profiling of apple (Cv. Golden Delicious) infected with leaf spot fungus (*A. alternata* f. sp. *mali.*) confirmed that both Md-miR156ab and MdmiR395 target MdWRKYN1 and MdWRKY26, respectively, and regulate resistance against the pathogen (Zhang et al. 2017). Genome-wide identification and expression analysis of the WRKY TF family in *Moniliophthora perniciosa* infected cacoa (*Theobroma cacao*) revealed that Tc01_p014750, Tc04_p016130, Tc06_p013130, Tc06_p004420, Tc09_p001530, and Tc10_p016570 have shown promising changes in their transcript levels (de Almeida et al. 2017). In another published report, *TaWRKY49* and *TaWRKY62* silencing in wheat (Cv. Xiaoyan6) lead to a change in resistance against *P. striiformis* f. sp. *tritici* (Wang et al. 2017). In another published article, grape *VlWRKY48* overexpression *A. thaliana* lines showed the enhanced disease resistance against *Golovinomyces cichoracearum* as well as drought conditions in comparison to the control-plants (Zhao et al. 2018). The reason for the enhancement was the increased defense-related genes expression. Similarly, the melon (*Cucumis melo*) showed upregulation of *CmWRKY6, CmWRKY19*, and *CmWRKY48* even up to sevenfold after powdery mildew fungus infection (Jiao et al. 2018). Aamir and colleagues investigated the differential tissue-specific expression of WRKY genes in tomato plants after challenging *F. oxysporum* f. sp. *lycopersici*. Their qRT-PCR work revealed that *SlWRKY4, SlWRKY33*, and *SlWRKY37* showed a clear-cut difference (even up to fivefold) in gene expression in both leaf and root tissues. Besides, they also showed that the *SlWRKY33* interact with other proteins such as *WRKY1, WRKY40, WRKY70, MAPK5*, and *SIB1* (Aamir et al. 2018).

Recently, Dong et al. (2019) studied the transcriptome level changes in soybean upon inoculation with the causal organism of soybean downy mildew, *Peronospora manshurica* in resistant and susceptible genotypes. In total, 16 WRKY TFs were differentially expressed in response to fungal inoculation. Furthermore, they revealed that the *GmWRKY31* bind to the W-box element present in the *GmSAGT1* gene promoter region using the Y1H assay. Similarly, the negative role of *GhWRKY70* in resistance against *V. dahlia* was also elucidated (Xiong et al. 2019). More recently, barley *HvWRKY6* and *HvWRKY70* were validated to be a key regulator between BTH- and NPR1-mediated acquired resistances by generating heterologous overexpressed wheat transgenic lines followed by challenging with *P. striiformis* f. sp. *tritici* and *B. graminis* f. sp. *tritici* (Li et al. 2020). The similar results were obtained by overexpressing *TaWRKY142* in transgenic Arabidopsis plants against *C. higginsianum* (Kuki et al. 2020).

11.6 Conclusion

Plants are facing challenges from various phytopathogens for accurate plant growth and development. In this chapter, we summarized and highlighted the regulation and role of WRKY TFs in plant defense mechanism. Advancement in molecular technologies in the last two decades enlightens the understanding in regulation and role of WRKY TFs, this provides economic advantages to plant. Present information of WRKY TFs can be implemented in the improvement of plant ideotypes, have residence or tolerant against various phytopathogens. Overexpression or knockdown of WRKY TFs in the early life of plant provides an opportunity for good growth and development of plant with either resistance or tolerance against various phytopathogens. Recent finding enlightens that regulation of various plant defense genes by WRKY TFs is by regulating itself positively or negatively, auto- or cross-regulation, by activation or suppression through other WRKY TFs or other mechanisms regulated downstream plant defense genes. Regulation of WRKY TFs is very complex phenomenon sine it requires to maintain harmony among entire physiological and biochemical events within the plant. One WRKY TF can regulate several defense or development process at a time, this cross-talk mechanism is needed to be investigated. In the genomic era, study of entire genome of various crops through transcriptomics and genomics help in better understanding of regulation of WRKY TFs against various plant diseases for strategic disease management. Clustered regularly interspaced short palindromic repeats and their associated genes can also allow manipulating the negatively controlled WRKY TFs genes functional characterization of these WRKY TFs in defense. In addition, response of WRKY TFs for various climatic challenges is to be studied and synchronization among environmental factors and defense should be done.

References

Aamir M, Singh VK, Dubey MK, Kashyap SP, Zehra A, Upadhyay RS, Singh S (2018) Structural and functional dissection of differentially expressed tomato WRKY transcripts in host defense response against the vascular wilt pathogen (*Fusarium oxysporum* f. sp. *lycopersici*). PLoS One 13(4). https://doi.org/10.1371/journal.pone.0193922

Abbruscato P, Nepusz T, Mizzi L, Del Corvo M, Morandini P, Fumasoni I, Michel C, Paccanaro A, Guiderdoni E, Schaffrath U, Morel JB (2012) OsWRKY22, a monocot WRKY gene, plays a role in the resistance response to blast. Mol Plant Pathol 13(8):828–841

Alfenas-Zerbini P, Maia IG, Fávaro RD, Cascardo JC, Brommonschenkel SH, Zerbini FM (2009) Genome-wide analysis of differentially expressed genes during the early stages of tomato infection by a potyvirus. Mol Plant Microbe Inter 22(3):352–361

Alvarez ME, Nota F, Cambiagno DA (2010) Epigenetic control of plant immunity. Mol Plant Pathol 11(4):563–576

Alvarez-Venegas R, Abdallat AA, Guo M, Alfano JR, Avramova Z (2007) Epigenetic control of a transcription factor at the cross section of two antagonistic pathways. Epigenetics 2(2):106–113

Alves MS, Dadalto SP, Gonçalves AB, De Souza GB, Barros VA, Fietto LG (2014) Transcription factor functional protein-protein interactions in plant defense responses. Proteomes 2(1):85–106

Amuge T, Berger DK, Katari MS, Myburg AA, Goldman SL, Ferguson ME (2017) A time series transcriptome analysis of cassava (*Manihot esculenta* Crantz) varieties challenged with Ugandan cassava brown streak virus. Sci Rep 7(1):1–21

Ando S, Obinata A, Takahashi H (2014) WRKY70 interacting with RCY1 disease resistance protein is required for resistance to cucumber mosaic virus in *Arabidopsis thaliana*. Physiol Mol Plant Pathol 85:8–14

Babu M, Griffiths JS, Huang TS, Wang A (2008) Altered gene expression changes in Arabidopsis leaf tissues and protoplasts in response to plum pox virus infection. BMC Genomics 9(1):325

Bahrini I, Ogawa T, Kobayashi F, Kawahigashi H, Handa H (2011) Overexpression of the pathogen-inducible wheat TaWRKY45 gene confers disease resistance to multiple fungi in transgenic wheat plants. Breed Sci 61(4):319–326

Bai Y, Kissoudis C, Yan Z, Visser RG, van der Linden G (2018) Plant behaviour under combined stress: tomato responses to combined salinity and pathogen stress. Plant J 93(4):781–793

Bakshi M, Oelmüller R (2014) WRKY transcription factors: Jack of many trades in plants. Plant Signal Behav 9(2):e27700

Beyer K, Binder A, Boller T, Collinge M (2001) Identification of potato genes induced during colonization by Phytophthora infestans. Mol Plant Pathol 2(3):125–134

Borsani O, Zhu J, Verslues PE, Sunkar R, Zhu JK (2005) Endogenous siRNAs derived from a pair of natural cis-antisense transcripts regulate salt tolerance in Arabidopsis. Cell 123(7):1279–1291

Catoni M, Miozzi L, Fiorilli V, Lanfranco L, Accotto GP (2009) Comparative analysis of expression profiles in shoots and roots of tomato systemically infected by tomato spotted wilt virus reveals organ-specific transcriptional responses. Mol Plant-Microbe Inter 22(12):1504–1513

Chen C, Chen Z (2002) Potentiation of developmentally regulated plant defense response by AtWRKY18 a pathogen-induced Arabidopsis transcription factor. Plant Physiol 129 (2):706–716

Chen CE, Yeh HH (2010) Plant defense-related transcription factor WRKY 6 plays both supportive and inhibitory roles in tobacco mosaic virus infection. Plant Pathol Bull 19(1):31–40

Chen L, Zhang L, Yu D (2010) Wounding-induced WRKY8 is involved in basal defense in Arabidopsis. Mol Plant-Microbe Inter 23(5):558–565

Chen L, Zhang L, Li D, Wang F, Yu D (2013a) WRKY8 transcription factor functions in the TMV-cg defense response by mediating both abscisic acid and ethylene signaling in Arabidopsis. Proceed Nat Acad Sci 110(21):1963–1971

Chen T, Lv Y, Zhao T, Li N, Yang Y, Yu W, He X, Liu T, Zhang B (2013b) Comparative transcriptome profiling of a resistant vs. susceptible tomato (*Solanum lycopersicum*) cultivar in response to infection by tomato yellow leaf curl virus. PLoS One 8(11). https://doi.org/10.1371/journal.pone.0080816

Chen X, Liu J, Lin G, Wang A, Wang Z, Lu G (2013c) Overexpression of AtWRKY28 and AtWRKY75 in Arabidopsis enhances resistance to oxalic acid and *Sclerotinia sclerotiorum*. Plant Cell Rep 32(10):1589–1599

Chen X, Li C, Wang H, Guo Z (2019) WRKY transcription factors: evolution, binding, and action. Phytopathol Res 1(1):13

Cheng H, Liu H, Deng Y, Xiao J, Li X, Wang S (2015) The WRKY45-2 WRKY13 WRKY42 transcriptional regulatory cascade is required for rice resistance to fungal pathogen. Plant Physiol 167(3):1087–1099

Cho SM, Kang EY, Min KH, Lee YK, Kim YC, Yang KY, Kim KS, Choi Y, Cho BH (2012) A positive regulatory role of the watermelon ClWRKY70 gene for disease resistance in transgenic *Arabidopsis thaliana*. Biol Plant 56(3):560–565

Choi H, Jo Y, Lian S, Jo KM, Chu H, Yoon JY, Choi SK, Kim KH, Cho WK (2015) Comparative analysis of chrysanthemum transcriptome in response to three RNA viruses: cucumber mosaic virus, tomato spotted wilt virus and potato virus X. Plant Mol Biol 88(3):233–248

Chujo T, Miyamoto K, Shimogawa T, Shimizu T, Otake Y, Yokotani N, Nishizawa Y, Shibuya N, Nojiri H, Yamane H, Minami E (2013) OsWRKY28, a PAMP-responsive transrepressor, negatively regulates innate immune responses in rice against rice blast fungus. Plant Mol Biol 82(1–2):23–37

Cormack RS, Eulgem T, Rushton PJ, Kochner P, Hahlbrock K, Somssich IE (2002) Leucine zipper-containing WRKY proteins widen the spectrum of immediate early elicitor-induced WRKY transcription factors in parsley. Biochim Biophys Acta Gene Struct Exp 1576(1–2):92–100

Dang FF, Wang YN, Yu LU, Eulgem T, Lai YAN, Liu ZQ, Wang XU, Qiu AL, Zhang TX, Lin J, Chen YS (2013) CaWRKY40, a WRKY protein of pepper, plays an important role in the regulation of tolerance to heat stress and resistance to *Ralstonia solanacearum* infection. Plant Cell Environ 36(4):757–774

Dang F, Wang Y, She J, Lei Y, Liu Z, Eulgem T, Lai Y, Lin J, Yu L, Lei D, Guan D (2014) Overexpression of CaWRKY27, a subgroup IIe WRKY transcription factor of *Capsicum annuum*, positively regulates tobacco resistance to *Ralstonia solanacearum* infection. Physiol Plant 150(3):397–411

Dardick C (2007) Comparative expression profiling of *Nicotiana benthamiana* leaves systemically infected with three fruit tree viruses. Mol Plant Microbe Inter 20(8):1004–1017

de Almeida DSM, Do Amaral DOJ, Del-Bem LE, dos Santos EB, Silva RJS, Gramacho KP, Vincentz M, Micheli F (2017) Genome-wide identification and characterization of cacao WRKY transcription factors and analysis of their expression in response to witches' broom disease. PLoS One 12(10). https://doi.org/10.1371/journal.pone.0187346

Dellagi A, Heilbronn J, Avrova AO, Montesano M, Palva ET, Stewart HE, Toth IK, Cooke DE, Lyon GD, Birch PR (2000) A potato gene encoding a WRKY-like transcription factor is induced in interactions with *Erwinia carotovora* subsp. *atroseptica* and *Phytophthora infestans* and is coregulated with class I endochitinase expression. Mol Plant Microbe Inter 13(10):1092–1101

Dong J, Chen C, Chen Z (2003) Expression profiles of the Arabidopsis WRKY gene superfamily during plant defense response. Plant Mol Biol 51(1):21–37

Dong H, Tan J, Li M, Yu Y, Jia S, Zhang C, Wu Y, Liu Y (2019) Transcriptome analysis of soybean WRKY TFs in response to *Peronospora manshurica* infection. Genomics 111(6):1412–1422

Duan Y, Jiang Y, Ye S, Karim A, Ling Z, He Y, Yang S, Luo K (2015) *PtrWRKY73*, a salicylic acid-inducible poplar WRKY transcription factor, is involved in disease resistance in *Arabidopsis thaliana*. Plant Cell Rep 34(5):831–841

Eulgem T, Somssich IE (2007) Networks of WRKY transcription factors in defense signaling. Curr Opin Plant Biol 10(4):366–371

Eulgem T, Rushton PJ, Robatzek S, Somssich IE (2000) The WRKY superfamily of plant transcription factors. Trends Plant Sci 5(5):199–206

Fan H, Wang F, Gao H, Wang L, Xu J, Zhao Z (2011) Pathogen-induced MdWRKY1 in 'Qinguan' apple enhances disease resistance. J Plant Biol 54(3):150–158

Fernandez D, Santos P, Agostini C, Bon MC, Petitot AS, Silva C, Guimaraes MG, Leonor R, Ana AX, Nicole M (2004) Coffee (*Coffea arabica* L) genes early expressed during infection by the rust fungus (*Hemileia vastatrix*). Mol Plant Pathol 5(6):527–536

Finatto T, Viana VE, Woyann LG, Busanello C, Maia LCD, Oliveira AC (2018) Can WRKY transcription factors help plants to overcome environmental challenges? Gen Mol Biol 41 (3):533–544

Ganesh D, Petitot AS, Silva MC, Alary R, Lecouls AC, Fernandez D (2006) Monitoring of the early molecular resistance responses of coffee (*Coffea arabica* L.) to the rust fungus (*Hemileia vastatrix*) using real-time quantitative RT-PCR. Plant Sci 170(6):1045–1051

Gao X, He P (2013) Nuclear dynamics of Arabidopsis calcium-dependent protein kinases in effector-triggered immunity. Plant Signal Behav 8(4):e23868

Gao R, Liu P, Yong Y, Wong SM (2016) Genome-wide transcriptomic analysis reveals correlation between higher WRKY61 expression and reduced symptom severity in turnip crinkle virus infected *Arabidopsis thaliana*. Sci Rep 6:24604

Gao YF, Liu JK, Yang FM, Zhang GY, Wang D, Zhang L, Ou YB, Yao YA (2019) The WRKY transcription factor WRKY8 promotes resistance to pathogen infection and mediates drought and salt stress tolerance in *Solanum lycopersicum*. Physiol Plant 168(1):98–117

Guimil S, Chang HS, Zhu T, Sesma A, Osbourn A, Roux C, Ioannidis V, Oakeley EJ, Docquier M, Descombes P, Briggs SP (2005) Comparative transcriptomics of rice reveals an ancient pattern of response to microbial colonization. Proc Nat Acad Sci 102(22):8066–8070

Gullner G, Juhasz C, Nemeth A, Barna B (2017) Reactions of tobacco genotypes with different antioxidant capacities to powdery mildew and tobacco mosaic virus infections. Plant Physiol Biochem 119:232–239

Guo R, Yu F, Gao Z, An H, Cao X, Guo X (2011) GhWRKY3, a novel cotton (*Gossypium hirsutum* L.) WRKY gene, is involved in diverse stress responses. Mol Biol Rep 38(1):49–58

Honjo MN, Emura N, Kawagoe T, Sugisaka J, Kamitani M, Nagano AJ, Kudoh H (2020) Seasonality of interactions between a plant virus and its host during persistent infection in a natural environment. ISME J 14(2):506–518

Huang Y, Li MY, Wu P, Xu ZS, Que F, Wang F, Xiong AS (2016) Members of WRKY group III transcription factors are important in TYLCV defense signaling pathway in tomato (*Solanum lycopersicum*). BMC Genomics 17(1):788

Huh SU, Lee GJ, Kim YJ, Paek KH (2012) *Capsicum annuum* WRKY transcription factor d (CaWRKYd) regulates hypersensitive response and defense response upon tobacco mosaic virus infection. Plant Sci 197:50–58

Huh SU, Lee GJ, Jung JH, Kim Y, Kim YJ, Paek KH (2015) *Capsicum annuum* transcription factor WRKYa positively regulates defense response upon TMV infection and is a substrate of CaMK1 and CaMK2. Sci Rep 5:7981

Hwang SH, Yie SW, Hwang DJ (2011) Heterologous expression of OsWRKY6 gene in Arabidopsis activates the expression of defense related genes and enhances resistance to pathogens. Plant Sci 181(3):316–323

Hwang SH, Kwon SI, Jang JY, Fang IL, Lee H, Choi C, Park S, Ah I, Bae SC, Hwang DJ (2016) OsWRKY51 a rice transcription factor functions as a positive regulator in defense response against *Xanthomonas oryzae* pv. *oryzae*. Plant Cell Rep 35(9):1975–1985

Ishiguro S, Nakamura K (1994) Characterization of a cDNA encoding a novel DNA-binding protein SPF1 that recognizes SP8 sequences in the 5′ upstream regions of genes coding for sporamin and β-amylase from sweet potato. Mol Gen Genet 244(6):563–571

Jeevalatha A, Siddappa S, Kumar A, Kaundal P, Guleri A, Sharma S, Nagesh M, Singh BP (2017) An insight into differentially regulated genes in resistant and susceptible genotypes of potato in response to tomato leaf curl New Delhi virus-potato infection. Virus Res 232:22–33

Jiang Y, Guo L, Liu R, Jiao B, Zhao X, Ling Z, Luo K (2016) Overexpression of poplar PtrWRKY89 in transgenic Arabidopsis leads to a reduction of disease resistance by regulating defense-related genes in salicylate-and jasmonate-dependent signaling. PLoS One 11(3). https://doi.org/10.1371/journal.pone.0149137

Jiao Z, Sun J, Wang C, Dong Y, Xiao S, Gao X, Cao Q, Li L, Li W, Gao C (2018) Genome-wide characterization, evolutionary analysis of WRKY genes in Cucurbitaceae species and assessment of its roles in resisting to powdery mildew disease. PLoS One 13(12). https://doi.org/10.1371/journal.pone.0199851

Jingyuan Z, Xuexiao Z, Zhenchuan M, Bingyan X (2011) A novel pepper (*Capsicum annuum* L.) WRKY gene CaWRKY30 is involved in pathogen stress responses. J Plant Biol 54(5):329

Kalde M, Barth M, Somssich IE, Lippok B (2003) Members of the Arabidopsis WRKY group III transcription factors are part of different plant defense signaling pathways. Mol Plant Microbe Inter 16(4):295–305

Kim KC, Fan B, Chen Z (2006) Pathogen-induced Arabidopsis WRKY7 is a transcriptional repressor and enhances plant susceptibility to *Pseudomonas syringae*. Plant Physiol 142 (3):1180–1192

Kim KC, Lai Z, Fan B, Chen Z (2008) Arabidopsis WRKY38 and WRKY62 transcription factors interact with histone deacetylase 19 in basal defense. Plant Cell 20(9):2357–2371

Kissoudis C, Sunarti S, Van De Wiel C, Visser RG, van der Linden CG, Bai Y (2016) Responses to combined abiotic and biotic stress in tomato are governed by stress intensity and resistance mechanism. J Exp Bot 67(17):5119–5132

Knoth C, Ringler J, Dangl JL, Eulgem T (2007) Arabidopsis WRKY70 is required for full RPP4-mediated disease resistance and basal defense against *Hyaloperonospora parasitica*. Mol Plant Microbe Inter 20(2):120–128

Ko YJ, Lee S, Song K, Park SY, Ahn I, Bae SC, Lee YH, Hwang DJ (2015) Heterologous expression of the Brassica rapa transcription factor BrWRKY7 enhances resistance against bacterial soft rot caused by *Pectobacterium carotovorum* in Arabidopsis. Plant Biotechnol Rep 9(4):179–186

Kuki Y, Ohno R, Yoshida K, Takumi S (2020) Heterologous expression of wheat WRKY transcription factor genes transcriptionally activated in hybrid necrosis strains alters abiotic and biotic stress tolerance in transgenic Arabidopsis. Plant Physiol Biochem 150:71–79

Kumar G, Dasgupta I (2020) Comprehensive molecular insights into the stress response dynamics of rice (*Oryza sativa* L.) during rice tungro disease by RNA-seq-based comparative whole transcriptome analysis. J Biosci 45(1):27

Kundu A, Patel A, Paul S, Pal A (2015) Transcript dynamics at early stages of molecular interactions of MYMIV with resistant and susceptible genotypes of the leguminous host *Vigna mungo*. PLoS One 10(4). https://doi.org/10.1371/journal.pone.0124687

Lai Z, Vinod KM, Zheng Z, Fan B, Chen Z (2008) Roles of ArabidopsisWRKY3 and WRKY4 transcription factors in plant responses to pathogens. BMC Plant Biol 8(1):68

Lai Z, Li Y, Wang F, Cheng Y, Fan B, Yu JQ, Chen Z (2011) Arabidopsis sigma factor binding proteins are activators of the WRKY33 transcription factor in plant defense. Plant Cell 23 (10):3824–3841

Le TN, Schumann U, Smith NA, Tiwari S, Au PCK, Zhu QH, Taylor JM, Kazan K, Llewellyn DJ, Zhang R, Dennis ES (2014) DNA demethylases target promoter transposable elements to positively regulate stress responsive genes in Arabidopsis. Genome Biol 15(9):458

Levee V, Major I, Levasseur C, Tremblay L, MacKay J, Seguin A (2009) Expression profiling and functional analysis of populus WRKY23 reveals a regulatory role in defense. New Phytol 184 (1):48–70

Li J, Wang J, Wang N, Guo X, Gao Z (2015) GhWRKY44 a WRKY transcription factor of cotton, mediates defense responses to pathogen infection in transgenic *Nicotiana benthamiana*. Plant Cell Tiss Org 121(1):127–140

Li X, An M, Xia Z, Bai X, Wu Y (2017) Transcriptome analysis of watermelon (*Citrullus lanatus*) fruits in response to cucumber green mottle mosaic virus (CGMMV) infection. Sci Rep 7 (1):1–12

Li H, Wu J, Shang X, Geng M, Gao J, Zhao S, Yu X, Li D, Kang Z, Wang X, Wang X (2020) WRKY transcription factors shared by BTH-induced resistance and NPR1-mediated acquired resistance improve broad-spectrum disease resistance in wheat. Mol Plant Microbe Int 33 (3):433–443

Liang C, Liu H, Hao J, Li J, Luo L (2019) Expression profiling and regulatory network of cucumber microRNAs and their putative target genes in response to cucumber green mottle mosaic virus infection. Archiv Virol 164(4):1121–1134

Liu Y, Schiff M, DineshKumar SP (2004) Involvement of MEK1 MAPKK NTF6 MAPK WRKY/MYB transcription factors, COI1 and CTR1 in N-mediated resistance to tobacco mosaic virus. Plant J 38(5):800–809

Liu XQ, Bai XQ, Qian Q, Wang XJ, Chen MS, Chu CC (2005) OsWRKY03 a rice transcriptional activator that functions in defense signaling pathway upstream of OsNPR1. Cell Res 15 (8):593–603

Liu Q, Zhang YC, Wang CY, Luo YC, Huang QJ, Chen SY, Zhou H, Qu LH, Chen YQ (2009) Expression analysis of phytohormone-regulated microRNAs in rice, implying their regulation roles in plant hormone signaling. FEBS Lett 583(4):723–728

Liu Y, Wang G, Wang Z, Yang F, Wu G, Hong N (2012) Identification of differentially expressed genes in response to infection of a mild Citrus tristeza virus isolate in *Citrus aurantifolia* by suppression subtractive hybridization. Sci Hortic 134:144–149

Liu J, Chen X, Liang X, Zhou X, Yang F, Liu J, He SY, Guo Z (2016a) Alternative splicing of rice WRKY62 and WRKY76 transcription factor genes in pathogen defense. Plant Physiol 171 (2):1427–1442

Liu X, Song Y, Xing F, Wang N, Wen F, Zhu C (2016b) GhWRKY25, a group I WRKY gene from cotton, confers differential tolerance to abiotic and biotic stresses in transgenic *Nicotiana benthamiana*. Protoplasma 253(5):1265–1281

Liu Q, Liu Y, Tang Y, Chen J, Ding W (2017) Overexpression of NtWRKY50 increases resistance to *Ralstonia solanacearum* and alters salicylic acid and jasmonic acid production in tobacco. Front Plant Sci 8:1710

Liu Q, Li X, Yan S, Yu T, Yang J, Dong J, Zhang S, Zhao J, Yang T, Mao X, Zhu X (2018) OsWRKY67 positively regulates blast and bacteria blight resistance by direct activation of PR genes in rice. BMC Plant Biol 18(1):1–13

Lui S, Luo C, Zhu L, Sha R, Qu S, Cai B, Wang S (2017) Identification and expression analysis of WRKY transcription factor genes in response to fungal pathogen and hormone treatments in apple (*Malus domestica*). J Plant Biol 60(2):215–230

Madhusudhan P, Sinha P, Rajput LS, Bhattacharya M, Sharma T, Bhuvaneshwari V, Gaikwad K, Krishnan SG, Singh AK (2019) Effect of temperature on Pi54-mediated leaf blast resistance in rice. World J Microbiol Biotechnol 35(10):148

Madronero J, Rodrigues SP, Antunes TFS, Abreu PMV, Ventura JA, Fernandes AAR, Fernandes PMB (2018) Transcriptome analysis provides insights into the delayed sticky disease symptoms in *Carica papaya*. Plant Cell Rep 3:967–980

Mandal A, Sarkar D, Kundu S, Kundu P (2015) Mechanism of regulation of tomato TRN1 gene expression in late infection with tomato leaf curl New Delhi virus (ToLCNDV). Plant Sci 241:221–237

Mao G, Meng X, Liu Y, Zheng Z, Chen Z, Zhang S (2011) Phosphorylation of a WRKY transcription factor by two pathogen-responsive MAPKs drives phytoalexin biosynthesis in Arabidopsis. Plant Cell 23(4):1639–1653

Matsushita A, Inoue H, Goto S, Nakayama A, Sugano S, Hayashi N, Takatsuji H (2013) Nuclear ubiquitin proteasome degradation affects WRKY 45 function in the rice defense program. Plant J 73(2):302–313

McGregor CE, Miano DW, LaBonte DR, Hoy M, Clark CA, Rosa GJ (2009) Differential gene expression of resistant and susceptible sweet potato plants after infection with the causal agents of sweet potato virus disease. J Am Soc Hortic Sci 134(6):658–666

Mehta S, Singh B, Dhakate P, Rahman M, Islam MA (2019) Rice marker-assisted breeding, and disease resistance. In: Disease resistance in crop plants. Springer, Cham, pp 83–111

Miao Y, Zentgraf U (2010) A HECT E3 ubiquitin ligase negatively regulates Arabidopsis leaf senescence through degradation of the transcription factor WRKY53. Plant J 63(2):179–188

Mittal S, Banduni P, Mallikarjuna MG, Rao AR, Jain PA, Dash PK, Thirunavukkarasu N (2018) Structural, functional, and evolutionary characterization of major drought transcription factors families in maize. Front Chem 6:177

Mohanta TK, Park YH, Bae H (2016) Novel genomic and evolutionary insight of WRKY transcription factors in plant lineage. Sci Rep 6(1):1–22

Nakayama A, Fukushima S, Goto S, Matsushita A, Shimono M, Sugano S, Jiang CJ, Akagi A, Yamazaki M, Inoue H, Takatsuji H (2013) Genome-wide identification of WRKY45-regulated genes that mediate benzothiadiazole-induced defense responses in rice. BMC Plant Biol 13 (1):150

Naoumkina MA, He X, Dixon RA (2008) Elicitor-induced transcription factors for metabolic reprogramming of secondary metabolism in *Medicago truncatula*. BMC Plant Biol 8(1):132

Naqvi AR, Sarwat M, Pradhan B, Choudhury NR, Haq QMR, Mukherjee SK (2011) Differential expression analyses of host genes involved in systemic infection of tomato leaf curl New Delhi virus (ToLCNDV). Virus Res 160(1–2):395–399

Navarro L, Jay F, Nomura K, He SY, Voinnet O (2008) Suppression of the microRNA pathway by bacterial effector proteins. Science 321(5891):964–967

Nemchinov LG, Shao J, Lee MN, Postnikova OA, Samac DA (2017) Resistant and susceptible responses in alfalfa (*Medicago sativa*) to bacterial stem blight caused by *Pseudomonas syringae* pv. *syringae*. PLoS One 12(12). https://doi.org/10.1371/journal.pone.0189781

Oh SK, Yi SY, Yu SH, Moon JS, Park JM, Choi D (2006) CaWRKY2, a chili pepper transcription factor, is rapidly induced by incompatible plant pathogens. Mol Cells 22(1):58–64

Oh SK, Baek KH, Park JM, Yi SY, Yu SH, Kamoun S, Choi D (2008) *Capsicum annuum* WRKY protein CaWRKY1 is a negative regulator of pathogen defense. New Phytol 177(4):977–989

Pan LJ, Jiang L (2014) Identification and expression of the WRKY transcription factors of *Carica papaya* in response to abiotic and biotic stresses. Mol Biol Rep 41(3):1215–1225

Park CJ, Shin YC, Lee BJ, Kim KJ, Kim JK, Paek KH (2006) A hot pepper gene encoding WRKY transcription factor is induced during hypersensitive response to tobacco mosaic virus and *Xanthomonas campestris*. Planta 223(2):168–179

Peng X, Hu Y, Tang X, Zhou P, Deng X, Wang H, Gu Z (2012) Constitutive expression of rice WRKY30 gene increases the endogenous jasmonic acid accumulation, PR gene expression and resistance to fungal pathogens in rice. Planta 236(5):1485–1498

Qiu Y, Yu D (2009) Over-expression of the stress-induced OsWRKY45 enhances disease resistance and drought tolerance in Arabidopsis. Environ Exp Bot 65(1):35–47

Qiu JL, Fiil BK, Petersen K, Nielsen HB, Botanga CJ, Thorgrimsen S, Palma K, Suarez-Rodriguez MC, Sandbech-Clausen S, Lichota J, Brodersen P (2008) Arabidopsis MAP kinase 4 regulates gene expression through transcription factor release in the nucleus. EMBO J 27(16):2214–2221

Rahman M, Sultana S, Nath D, Kalita S, Chakravarty D, Mehta S, Wani SH, Islam MA (2019) Molecular breeding approaches for disease resistance in sugarcane. In: Disease resistance in crop plants. Springer, Cham, pp 131–155

Rajput LS, Sharma T, Madhusudhan P, Sinha P (2017a) Effect of temperature on growth and sporulation of rice leaf blast pathogen *Magnaporthe oryzae*. Int J Curr Microbiol App Sci 6 (3):394–401

Rajput LS, Sharma T, Madhusudhan P, Sinha P (2017b) Effect of temperature on rice blast infection process with emphasis on appressoria formation *by Magnaporthe oryzae*. Int J Curr Microbiol App Sci 6(4):1931–1939

Ramiro D, Jalloul A, Petitot AS, De Sá MFG, Maluf MP, Fernandez D (2010) Identification of coffee WRKY transcription factor genes and expression profiling in resistance responses to pathogens. Tree Genet Genom 6(5):767–781

Ren XJ, Huang WD, Li WZ, Yu DQ (2010) Tobacco transcription factor WRKY4 is a modulator of leaf development and disease resistance. Biol Plant 54(4):684–690

Robatzek S, Somssich IE (2001) A new member of the Arabidopsis WRKY transcription factor family, AtWRKY6 is associated with both senescence and defence related processes. Plant J 28 (2):123–133

Robatzek S, Somssics IE (2002) Targets of AtWRKY6 regulation during plant senescence and pathogen defense. Gen Dev 16(9):1139–1149
Rogers EE, Glazebrook J, Ausubel FM (1996) Mode of action of the Arabidopsis thaliana phytoalexin camalexin and its role in Arabidopsis-pathogen interactions. Mol Plant Microbe Inter 9:748–757
Rushton PJ, Torres JT, Parniske M, Wernert P, Hahlbrock K, Somssich IE (1996) Interaction of elicitor-induced DNA-binding proteins with elicitor response elements in the promoters of parsley PR1 genes. EMBO J 15(20):5690–5700
Ryu HS, Han M, Lee SK, Cho JI, Ryoo N, Heu S, Lee YH, Bhoo SH, Wang GL, Hahn TR, Jeon JS (2006) A comprehensive expression analysis of the WRKY gene superfamily in rice plants during defense response. Plant Cell Rep 25(8):836–847
Samad AFA, Rahnamaie-Tajadod R, Sajad M, Jani J, Murad AMA, Noor NM, Ismail I (2019) Regulation of terpenoid biosynthesis by miRNA in *Persicaria minor* induced by *Fusarium oxysporum*. BMC Genomics 20(1):586
Sarkar S, Das A, Khandagale P, Maiti IB, Chattopadhyay S, Dey N (2018) Interaction of Arabidopsis TGA3 and WRKY53 transcription factors on Cestrum yellow leaf curling virus (CmYLCV) promoter mediates salicylic acid-dependent gene expression in planta. Planta 247 (1):181–199
Satoh K, Kondoh H, Sasaya T, Shimizu T, Choi IR, Omura T, Kikuchi S (2010) Selective modification of rice (Oryza sativa) gene expression by rice stripe virus infection. J Gen Virol 91(1):294–305
Satoh K, Shimizu T, Kondoh H, Hiraguri A, Sasaya T, Choi IR, Omura T, Kikuchi S (2011) Relationship between symptoms and gene expression induced by the infection of three strains of Rice dwarf virus. PLoS One 6(3). https://doi.org/10.1371/journal.pone.0018094
Satoh K, Kondoh H, De Leon TB, Macalalad RJA, Cabunagan RC, Cabauatan PQ, Mauleon R, Kikuchi S, Choi IR (2013) Gene expression responses to Rice tungro spherical virus in susceptible and resistant near-isogenic rice plants. Virus Res 171(1):111–120
Seo JK, Kim MK, Kwak HR, Choi HS, Nam M, Choe J, Choi B, Han SJ, Kang JH, Jung C (2018) Molecular dissection of distinct symptoms induced by tomato chlorosis virus and tomato yellow leaf curl virus based on comparative transcriptome analysis. Virol 516:1–20
Shi W, Hao L, Li J, Liu D, Guo X, Li H (2014a) The *Gossypium hirsutum* WRKY gene GhWRKY39-1 promotes pathogen infection defense responses and mediates salt stress tolerance in transgenic *Nicotiana benthamiana*. Plant Cell Rep 33(3):483–498
Shi W, Liu D, Hao L, Wu CA, Guo X, Li H (2014b) GhWRKY39 a member of the WRKY transcription factor family in cotton has a positive role in disease resistance and salt stress tolerance. Plant Cell Tiss Org 118(1):17–32
Shimono M, Sugano S, Nakayama A, Jiang CJ, Ono K, Toki S, Takatsuji H (2007) Rice WRKY45 plays a crucial role in benzothidiazole-inducible blast resistance. Plant Cell 19(6):2064–2076
Shimono M, Koga H, Akagi AYA, Hayashi N, Goto S, Sawada M, Kurihara T, Matsushita A, Sugano S, Jiang CJ, Kaku H (2012) Rice WRKY45 plays important roles in fungal and bacterial disease resistance. Mol Plant Pathol 13(1):83–94
Shu-Ling Z, XingFen W, Yan Z, Jian-Feng L, Li-Zhu W, Dong-Mei Z, Zhi-Ying M (2012) GbWRKY1 a novel cotton (*Gossypium barbadense*) WRKY gene isolated from a bacteriophage full-length cDNA library, is induced by infection with *Verticillium dahliae*. IJBB 49 (6):405–413
Singh V, Roy S, Singh D, Nandi AK (2014) Arabidopsis flowering locus D influences systemic-acquired-resistance-induced expression and histone modifications of WRKY genes. J Biosci 39 (1):19–126
Singh B, Mehta S, Tiwari M, Bhatia S (2018a) Legume breeding for fungal resistance: a lesson to learn. In: Molecular approaches for plant improvement. Kalpaz Publication, New Delhi, pp 159–180

Singh P, Shekhar S, Rustagi A, Sharma V, Kumar D (2018b) Insights into the role of WRKY superfamily of protein transcription factor in defense response. In: Molecular aspects of plant-pathogen interaction. Springer, Singapore, pp 185–202

Singh B, Mehta S, Aggarwal SK, Tiwari M, Bhuyan SI, Bhatia S, Islam MA (2019) Barley disease resistance, and molecular breeding approaches. In: Disease resistance in crop plant. Springer, Cham, pp 261–299

Sun J, An H, Shi W, Guo X, Li H (2012) Molecular cloning and characterization of GhWRKY11, a gene implicated in pathogen responses from cotton. S Afr J Bot 81:113–123

Sureshkumar V, Dutta B, Kumar V, Prakash G, Mishra DC, Chaturvedi KK, Rai A, Sevanthi AM, Solanke AU (2019) RiceMetaSysB a database of blast and bacterial blight responsive genes in rice and its utilization in identifying key blast-resistant WRKY genes. Database 2019:baz015. https://doi.org/10.1093/database/baz015

Suzuki K (1999) Elicitor signal transduction that leads to hypersensitive reaction in cultured tobacco cells. Plant Biotechnol 16(5):343–351

Tao Z, Liu H, Qiu D, Zhou Y, Li X, Xu C, Wang S (2009) A pair of allelic WRKY genes play opposite roles in rice-bacteria interactions. Plant Physiol 151(2):936–948

Turck F, Zhou A, Somssich IE (2004) Stimulus-dependent promoter-specific binding of transcription factor WRKY1 to its native promoter and the defense-related gene PcPR1-1 in parsley. Plant Cell 16(10):2573–2585

Vo KT, Kim CY, Hoang TV, Lee SK, Shirsekar G, Seo YS, Lee SW, Wang GL, Jeon JS (2018) OsWRKY67 plays a positive role in basal and Xa21-mediated resistance in rice. Front Plant Sci 8:2220

Voinnet O (2009) Origin biogenesis and activity of plant microRNAs. Cell 136(4):669–687

Wang Y, Dang F, Liu Z, Wang X, Eulgem T, Lai Y, Yu L, She J, Shi Y, Lin J, Chen C (2013) CaWRKY58 encoding a group I WRKY transcription factor of *Capsicum annuum*, negatively regulates resistance to *Ralstonia solanacearum* infection. Mol Plant Pathol 14(2):131–144

Wang X, Yan Y, Li Y, Chu X, Wu C, Guo X (2014) GhWRKY40 a multiple stress-responsive cotton WRKY gene plays an important role in the wounding response and enhances susceptibility to *Ralstonia solanacearum* infection in transgenic *Nicotiana benthamiana*. PLoS One 9 (4). https://doi.org/10.1371/journal.pone.0093577

Wang J, Tao F, Tian W, Guo Z, Chen X, Xu X, Shang H, Hu X (2017) The wheat WRKY transcription factors TaWRKY49 and TaWRKY62 confer differential high-temperature seedling-plant resistance to *Puccinia striiformis* f. sp. *tritici*. PLoS One 12(7). https://doi.org/10.1371/journal.pone.0181963

Wu XJ, Li X, Zhao PF, Li N, Wu L, He Y, Wang SC (2015) Comparative transcriptome profiling of two maize near-isogenic lines differing in the allelic state for bacterial brown spot disease resistance. J Integr Agric 14(4):610–621

Xie C, Zhou X, Deng X, Guo Y (2010) PKS5 a SNF1-related kinase, interacts with and phosphorylates NPR1, and modulates expression of WRKY38 and WRKY62. J Genet Genom 37(6):359–369

Xiong X, Sun S, Li Y, Zhang X, Sun J, Xue F (2019) The cotton WRKY transcription factor GhWRKY70 negatively regulates the defense response against *Verticillium dahliae*. Crop J 7 (3):393–402

Xu X, Chen C, Fan B, Chen Z (2006) Physical, functional interactions between pathogen-induced Arabidopsis WRKY18 WRKY40, WRKY60 transcription factors. Plant Cell 18(5):1310–1326

Yamasaki K, Kigawa T, Watanabe S, Inoue M, Yamasaki T, Seki M, Shinozaki K, Yokoyama S (2012) Structural basis for sequence-specific DNA recognition by an Arabidopsis WRKY transcription factor. J Biol Chem 287(10):7683–7691

Yamasaki K, KigawaT SM, Shinozaki K, Yokoyama S (2013) DNA-binding domains of plant-specific transcription factors: structure, function, and evolution. Trends Plant Sci 18(5):267–276

Yan L, Liu ZQ, Xu YH, Lu K, Wang XF, Zhang DP (2013) Auto-and cross-repression of three Arabidopsis WRKY transcription factors WRKY18 WRKY40, WRKY60 negatively involved in ABA signaling. J Plant Growth Regul 32(2):399–416

Yang B, Rahman MH, Deyholos MK, Kav NN, Jiang Y (2009) Identification and expression analysis of WRKY transcription factor genes in *Brassica napus* (canola) in response to fungal pathogens and hormone treatments. BMC Plant Biol 9(1):68

Yoda H, Ogawa M, Yamaguchi Y, Koizumi N, Kusano T, Sano H (2002) Identification of early-responsive genes associated with the hypersensitive response to tobacco mosaic virus and characterization of a WRKY-type transcription factor in tobacco plants. Mol Genet Genomics 267(2):154–161

Yokotani N, Sato Y, Tanabe S, Chujo T, Shimizu T, Okada K, Yamane H, Shimono M, Sugano S, Takatsuji H, Kaku H (2013) WRKY76 is a rice transcriptional repressor playing opposite roles in blast disease resistance and cold stress tolerance. J Exp Bot 64(16):5085–5097

Yu A, Lepere G, Jay F, Wang J, Bapaume L, Wang Y, Abraham AL, Penterman J, Fischer RL, Voinnet O, Navarro L (2013a) Dynamics and biological relevance of DNA demethylation in Arabidopsis antibacterial defense. Proc Natl Acad Sci 110(6):2389–2394

Yu Y, Xu W, Wang J, Wang L, Yao W, Yang Y, Xu Y, Ma F, Du Y, Wang Y (2013b) The Chinese wild grapevine (*Vitis pseudoreticulata*) E3 ubiquitin ligase Erysiphe necator induced RING finger protein 1 (EIRP1) activates plant defense responses by inducing proteolysis of the VpWRKY11 transcription factor. New Phytol 200(3):834–846

Zhang Y, Wang L (2005) The WRKY transcription factor superfamily: its origin in eukaryotes and expansion in plants. BMC Evol Biol 5(1):1

Zhang J, Peng Y, Guo Z (2008) Constitutive expression of pathogen-inducible OsWRKY31 enhances disease resistance and affects root growth and auxin response in transgenic rice plants. Cell Res 18(4):508–521

Zhang C, Grosic S, Whitham SA, Hill JH (2012) The requirement of multiple defense genes in soybean Rsv1–mediated extreme resistance to soybean mosaic virus. Mol Plant-Microbe Inter 25(10):1307–1313

Zhang Q, Li Y, Zhang Y, Wu C, Wang S, Hao L, Wang S, Li T (2017) Md-miR156ab and Md-miR395 target WRKY transcription factors to influence apple resistance to leaf spot disease. Front Plant Sci 8:526

Zhang Y, Yao JL, Feng H, Jiang J, Fan X, Jia YF, Wang R, Liu C (2019) Identification of the defense-related gene VdWRKY53 from the wild grapevine *Vitis davidii* using RNA sequencing and ectopic expression analysis in Arabidopsis. Hereditas 156(1):14

Zhang L, Shang J, Wang W, Du J, Li K, Wu X, Yu L, Liu C, Khaskheli MI, Yang W (2019a) Comparison of transcriptome differences in soybean response to soybean mosaic virus under normal light and in the shade. Viruses 11(9):793

Zhao J, Zhang X, Guo R, Wang Y, Guo C, Li Z, Chen Z, Gao H, Wang X (2018) Over-expression of a grape WRKY transcription factor gene VlWRKY48 in *Arabidopsis thaliana* increases disease resistance and drought stress tolerance. Plant Cell Tiss Org 132(2):359–370

Zheng Z, Qamar SA, Chen Z, Mengiste T (2006) Arabidopsis WRKY33 transcription factor is required for resistance to necrotrophic fungal pathogens. Plant J 48(4):592–605

Zheng Z, Mosher SL, Fan B, Klessig DF, Chen Z (2007) Functional analysis of Arabidopsis WRKY25 transcription factor in plant defense against *Pseudomonas syringae*. BMC Plant Biol 7(1):2

Zhou H, Xi Y, Wei S (2014) Expression of a novel WRKY gene in response to three RNA viruses in tobacco. J Anhui Agric Univ 41(6):1020–1026

Zou LJ, Deng XG, Han XY, Tan WR, Zhu LJ, Xi DJ, Zhang DW, Lin HH (2016) Role of transcription factor HAT1 in modulating *Arabidopsis thaliana* response to cucumber mosaic virus. Plant Cell Physiol 57:1879–1889. https://doi.org/10.1093/pcp/pcw109

Unraveling the Molecular Mechanism of *Magnaporthe oryzae* Induced Signaling Cascade in Rice

12

Nisha Khatri, Qingfeng Meng, Sun Tae Kim, and Ravi Gupta

Abstract

Rice is one of the most important staple food crops, which feeds more than half of the world's population. Rice blast disease, caused by *Magnaporthe oryzae*, is one of the major factors affecting rice yield negatively. Several epidemics of rice blast disease have been reported across the globe which results in the loss of up to 100% of the total rice production in the region. Development and use of disease-resistant cultivars is the most effective and environment-friendly method for the control of this deadly disease. Therefore, several lines of studies have been carried out to understand the molecular mechanism of rice-*M. oryzae* interaction. Being the first line of defense, numerous studies were performed on pathogen-associated molecular patterns (PAMPs) and effectors in *M. oryzae*, as well as plasma membrane-localized and cytosolic receptors, which contribute to rice blast resistance by activation of PAMP-triggered immunity (PTI) and effector-triggered immunity (ETI). In this chapter, we briefly summarize the progress made so far in the PTI and ETI signaling networks including downstream cellular and physiological responses in rice–*M. oryzae* pathosystem.

N. Khatri
Department of Botany, Dyal Singh College, University of Delhi, New Delhi, India

Q. Meng
Department of Instrument Science and Engineering, School of Electronic Information and Electrical Engineering, Shanghai Jiao Tong University, Shanghai, China

S. T. Kim (✉)
Department of Plant Bioscience, Pusan National University, Miryang, South Korea
e-mail: stkim71@pusan.ac.kr

R. Gupta (✉)
Department of Botany, School of Chemical and Life Science, New Delhi, India
e-mail: dr.ravigupta@jamiahamdard.ac.in

B. Giri, M. P. Sharma (eds.), *Plant Stress Biology*,
https://doi.org/10.1007/978-981-15-9380-2_12

Keywords

PAMP-triggered immunity · Effector-triggered immunity (ETI) · Cellular and physiological responses · MAPK cascade · Biotic stress

12.1 Introduction

Rice is one of the most cultivated staple crops across the globe with an annual production of approximately 480 million tons worldwide (Skorbiansky 2018). However, similar to the other crops, rice productivity is constrained by various factors including biotic and abiotic stresses (Wang et al. 2019). Among the biotic stresses, rice blast disease, caused by a hemibiotrophic pathogen, *Magnaporthe oryzae* is the most devastating rice disease limiting up to 30% of the global rice productivity (Meng et al. 2019b). Because food security has become a global issue in recent years, especially for staple crops such as rice, efforts have been put in the past few decades to control this deadly disease of rice. The development and cultivation of resistant crop plants is the most effective and environmental-friendly approach for disease control (Liu et al. 2013).

During the course of evolution, plants have developed a two-layered immune system to fight against the invading pathogens (Gupta et al. 2015a). During plant–pathogen interaction, both the organisms secrete various small proteins and other small molecules including lipids, nucleic acids, and carbohydrates in the host apoplast, where these proteins interact with each other (Jones and Dangl 2006). This interaction among the pathogen and host-derived proteins determines the outcome of their relationship. While pathogen-derived proteins facilitate the pathogenicity for infecting the plants, host-derived proteins are involved in the recognition of these proteins to activate the defense signaling (Liu et al. 2014). During incompatible interactions, the molecular signatures in the pathogen-derived molecules, termed as pathogen-associated molecular patterns (PAMPs), are recognized by plants plasma membrane-localized pattern recognition receptors (PRRs) to activate the first line of defense, termed as PAMP-triggered immunity (PTI) (Jones and Dangl 2006). PTI responses are not host specific and are relatively weak and function to restrict the pathogen colonization (Miller et al. 2017). To overcome the PTI, pathogens secrete effector proteins directly inside the host cells, which are recognized by the intracellular receptors of the plants to activate the second line of defense which is known as effector-triggered immunity (ETI). As these pathogen secreted proteins trigger the ETI responses in plants upon recognition by the intracellular receptors, these effectors are termed as avirulence (Avr) proteins (Stotz et al. 2014). Intracellular receptors of the plants are the products of resistance (R)-genes and are associated with the well-known gene-for-gene hypothesis, to directly or indirectly recognize cytosolic pathogen effectors to activate the ETI (Jones and Dangl 2006). ETI responses are host specific and are more rapid and robust than PTI. ETI responses often culminate into the hypersensitive response at the site of infection (Gupta et al. 2015a).

During the past few decades, several lines of studies have been conducted to understand rice–*M. oryzae* interactions in greater detail (Meng et al. 2018b, 2019a). Moreover, the development of high-throughput omics-based approaches has facilitated the identification of novel genes, proteins, and metabolites from both the involved in the plant–pathogen interaction (Gupta et al. 2015b). Results obtained from these studies have been successfully mapped to the biological pathways involved in pathogen infection, plant response, and disease progression (Wang et al. 2017a). Quantitative multi-omics datasets are mapped to known metabolic networks to identify pathways that are up- or downregulated upon pathogen attack (Meng et al. 2018a). Alternatively, combined with phenotype data, multi-omics datasets are used to construct correlation networks during pathogen attack (Gupta et al. 2018a, 2020).

12.2 PTI Responses in Rice–*M. oryzae* Interaction

12.2.1 PRRs and PAMPs Identified So Far

As of today, three PAMPs have been identified from the *M. oryzae* including chitin (Kuchitsu et al. 1993), MSP1 (Jeong et al. 2007; Wang et al. 2016), and *M. oryzae* hypersensitive response-inducing protein 1 (MoHRIP1) (Chen et al. 2012) (Fig. 12.1). However, the number of identified PRRs from the rice is much higher as compared to the number of identified PAMPs from *M. oryzae*. In general, PRRs include two classes of proteins including receptor-like kinases (RLKs) and receptor-like proteins (RLPs). While RLKs are composed of three domains including an ectodomain (ECD), a transmembrane domain, and a cytoplasmic kinase domain, RLPs lack a cytoplasmic kinase domain. Based on the domains or motifs in ECDs, PRRs are classified into different subfamilies including leucine-rich repeat (LRR) domain, lysine motifs (LysM), lectin domain, or epidermal growth factor (EGF)-like domain. LysM-RLPs and LysM-RLKs represent a major class of receptors for the perception of microbial N-acetyl glucosamine-containing glycans, including fungal chitin and bacterial peptidoglycan (PGN). The completion of the rice genome sequencing project has led to the identification of 1131 RLK and 90 RLP genes, which may be involved in cellular signaling and developmental events. Although, the genome sequencing project of *M. oryzae* was also completed almost two decades ago (Dean et al. 2005), the exact number of PAMPs present could not be predicted because of the absence of any specific domain or conserved sequence in them. PAMPs, therefore, can only be identified based on their molecular characterization. Although several rice PRRs have been predicted, the PRRs that perceive conserved *M. oryzae*-derived PAMPs including MSP1, and MoHrip1 could not be identified so far.

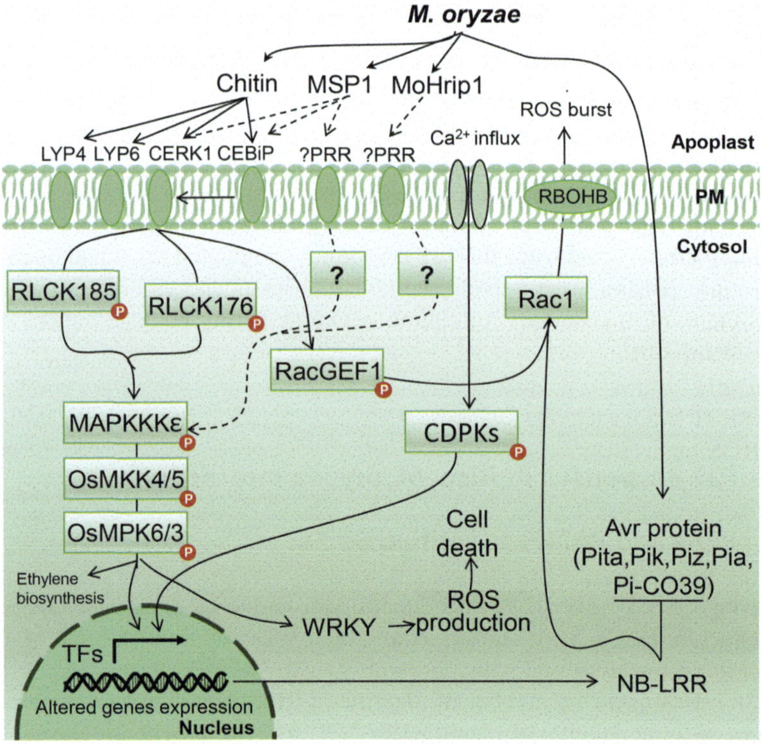

Fig. 12.1 A putative model explaining the rice immune responses to *Magnaporthe oryzae*. To date, three PRRs have been identified from *M. oryzae* that are identified by PM localized PRRs. Avr proteins of the *M. oryzae* are directly injected inside the rice cells that are identified by the NB-LRR proteins to activate the downstream responses. Further details of the model are in the text. Solid lines represent confirmed responses while dashed lines represent predicted responses

12.2.2 Chitin–LysM Domain Protein-Mediated Immunity

Chitin, a polymer of β-1,4-linked N-acetyl glucosamine, is a constituent of fungal cell walls that acts as a PAMP in plants including rice. Among the three identified PAMPs in *M. oryzae*, chitin is the best characterized one. In rice, at least four PRRs have been identified that directly or indirectly recognize chitin fragments. These chitin recognizing PRRs include CEBiP (chitin oligosaccharide elicitor-binding protein), CERK1 (chitin elicitor receptor kinase 1), LYP4, and LYP6 (LysM domain-containing proteins 4 and 6) (Kaku et al. 2006; Liu et al. 2014) (Fig. 12.1). It was shown that knockdown of any of these four genes expression remarkably reduces rice resistance against *M. oryzae* by suppressing the chitin-triggered immunity. Among these, CEBiP, an RLP containing a transmembrane domain and two LysM motifs as ECDs, is responsible for chitin binding, however, as it lacks the additional kinase, it is not able to transduce chitin-triggered signals downstream. Therefore, an additional LysM RLK protein OsCERK1 cooperates with CEBiP and functions as a crucial component for chitin-triggered immunity in rice. OsCERK1 and CEBiP heterodimers to form a plasma membrane receptor complex (Shimizu et al. 2010). LYP4 and LYP6 also bind to chitin. As all four chitin-binding proteins including CEBiP, OsCERK1, LYP4, and LYP6 contain at least one LysM domain, it can be speculated that this domain is critical to the perception of chitin oligosaccharides in rice (Liu et al. 2014). CERK1 probably functions either as a shared coreceptor or signaling partner for LysM domain-containing PRRs.

After the perception of chitin by CEBiP, CERK1 gets activated and phosphorylates the guanine nucleotide exchange factor RacGEF1 which is involved in the activation of the small GTPase OsRac1 (Akamatsu et al. 2013) (Fig. 12.1). Subsequently, OsRac1 positively regulates RBOHB (RESPIRATORY BURST OXIDASE HOMOLOGUE PROTEIN B), which is the NADPH oxidase responsible for ROS bursts in rice, showing a direct link between PRR complex activation and ROS production (Torres et al. 2006; Wong et al. 2007). In addition to RacGEF1, OsRac1 is also activated through directly interacting with the rice NB-LRR-type R protein Pit and is necessary for Pit-mediated immunity to the rice blast fungus, suggesting that OsRac1 plays a pivotal role in both PTI and ETI in rice (Kawano et al. 2010).

In addition to OsRacGEF1, RLCK176, and RLCK185, which are members of the rice RLCK family VII, both function downstream of OsCERK1 in chitin- and peptidoglycan-induced plant immunity (Yamaguchi et al. 2013). Loss of RLCK185 in rice specifically impaired MAPK activation in response to chitin. Moreover, RLCK185 is not required for chitin-triggered ROS burst, suggesting another branch downstream signaling of PRR complex independent of ROS production (Meng et al. 2019b; Yamaguchi et al. 2013).

12.2.3 MSP1-Triggered Immunity in Rice

In addition to the chitin, two additional PAMPs identified include an MSP1 and MoHRIP1. Of these, MSP1 is a fungal secreted cerato-platanin family protein with four conserved cysteine residues and is required for pathogen virulence (Wang et al. 2016). At first, it was shown that the deletion of the MSP1 gene in *M. oryzae* did not show any phenotypic differences and developed normal appressoria, however, the virulence of the mutants was greatly reduced as compared to the wild type (Jeong et al. 2007). Further, it was observed that MSP1 is a secreted protein and is not associated with the fungal cell wall (Jeong et al. 2007). Ectopic expression of MSP1 (also called as MoSM1) in Arabidopsis induced broad-spectrum disease resistance against *Botrytis cinerea, Alternaria brassicicola,* and *Pseudomonas syringae* pv. *tomato* (Yang et al. 2009). Further, the accumulation of reactive oxygen species (ROS) was observed in the transgenic plants expressing MSP1 (Yang et al. 2009). Exogenous application of recombinant MSP1 protein resulted in autophagic cell death and the production of H_2O_2 in both rice suspension-cultured cells and rice leaves (Wang et al. 2016). Besides, it was also speculated that MSP1-induced signaling in rice leaves is mediated by the activity of protein kinase(s). Furthermore, it was shown that the phytohormones jasmonic acid and abscisic acid positively regulate the MSP1 induced signaling while salicylic acid was involved in the suppression of MSP1 induced signaling in rice (Wang et al. 2016). In contrast, a recent report showed the accumulation of both JA and SA in MSP1 overexpression lines of rice together with the upregulation of SA and JA signaling genes. Further, MSP1 overexpression lines conferred broad-spectrum resistance to rice blast and bacterial blight diseases, however, no effects on the resistance against sheath blight and drought and salt stress tolerance were observed in the MSP1 overexpressing lines as compared to the wild type (Hong et al. 2017). Moreover, MSP1 overexpression did not affect the grain yield, suggesting that MSP1 overexpression lines of rice can be used in the future to combat the two deadliest rice pathogens including *M. oryzae* and *X. oryzae* without compromising the overall yield.

In order to understand the molecular mechanism of MSP1 induced signaling in rice, transcriptomic, high-throughput proteomic, and phosphoproteomic analyses of rice leaves have been carried out upon exogenous treatment of recombinant MSP1 (Gupta et al. 2019; Meng et al. 2018a, b, 2019a). It was shown that exogenous treatment of recombinant MSP1 protein resulted in the oxidative burst, MAPK3/6 phosphorylation, and upregulation of pathogenesis-related genes such as DUF26, PBZ, and PR-10 in rice leaves. Transcriptomic analysis showed that the proteins related to photosynthesis, secondary metabolism, lipid synthesis, and protein synthesis were specifically downregulated in response to MSP1 treatment. In contrast, the upregulated proteins were found to be related to the protein and lipid degradation, posttranslational modifications, and signaling (Meng et al. 2018a). These results were further supplemented by the proteome analysis using a label-free quantitative proteomics approach (Meng et al. 2019a). In addition, proteomics results also showed an increased abundance of various peroxidases and receptor kinases and proteins related to jasmonic acid biosynthesis, redox signaling, and MAP kinase signaling upon MSP1 perception, suggesting key functions of these in MSP1 induced signaling in rice (Gupta et al. 2019; Meng et al. 2018b, 2019a). Phosphoproteome analysis showed MSP1 induced phosphorylation of some of the key proteins associated with the PTI response, suggesting the function of MSP1 as a PAMP (Gupta et al. 2019).

12.2.4 MoHRIP1-Induced Signaling in Rice

MoHrip1 is a 14.32 kDa Alt A 1 (AA1) family protein, which was found to be secreted out from the *M. oryzae* as well as associated with the fungal cell wall. Downstream responses of MoHrip1 are similar to that of MSP1 as exogenous treatment of recombinant MoHrip1 resulted in H_2O_2 production, callose deposition, and induction of hypersensitive response in tobacco (Chen et al. 2012). Further, it was shown that rice seedlings treated with MoHrip1 showed upregulation of PR-proteins, and enhanced systemic resistance against *M. oryzae* (Chen et al.

2012). It was shown that ectopic expression of MoHrip1 in rice showed higher resistance of transgenic plants to *M. oryzae* and enhanced tolerance to drought stress than wild type (Wang et al. 2017b). Moreover, increased expression of SA and ABA related genes was observed in the MoHrip1 expressing plants as compared to the wild types, suggesting that MoHrip1 induced signaling is mediated by the functioning of these two phytohormones (Wang et al. 2017b). Recently, it was shown that MoHrip1 can bind to the tobacco and rice plant plasma membrane, indicating that MoHrip1 is a PAMP perceived by the plant immune system (Zhang et al. 2017). The expression of MoHrip1 was found to be increased during penetration and colonization of *M. oryzae* (Nie et al. 2019). Moreover, it was shown that MoHrip1 is required for fungal virulence as MoHrip1 deletion mutants showed significantly reduced virulence on rice (Nie et al. 2019). However, as no or very fewer efforts have been dedicated to understanding the molecular mechanism of MoHrip1-induced signaling in rice, the exact mechanism of MoHrip1-induced PTI responses is still to be deciphered. Moreover, the PRRs for both MSP1 and MoHrip1 is yet to be identified.

12.3 Downstream Responses of PTI Signaling

PTI signaling orchestrates a number of events including activation of MAP kinase cascade, production of antimicrobial compounds, commonly termed as phytoalexins, synthesis, production of ROS, and secretion of PR-proteins, among others (Thomma et al. 2011). In addition, closure of stomata also takes place as a downstream signaling event of PTI, however, it is majorly effective against bacterial pathogens because fungal pathogens generally use mechanical pressure to rupture the leaf surfaces (Bigeard et al. 2015). The majority of the PTI events are common with the ETI events and there is no sharp distinction between these two signaling cascades (Liu et al. 2013).

12.3.1 Activation of MAPK Cascade

Mitogen-activated protein kinase (MAPK) cascades are well established, highly conserved signaling modules that play pivotal roles in regulating PTI. It is well established that rapid and transient activation of MAPKs occurs during the activation of PTI responses. MAPK cascades are activated as one of the earliest signaling events after recognition of PAMPs by plant PRRs, which consist of a MAPKKK-MEK-MPK module. For instance, Group A MAPKs including rice MPK3 and MPK6 are involved in plant responses to biotic and abiotic stresses and are also involved in growth and development. Similar phosphorylation of MPK3/6 was observed upon MSP1 treatment in rice leaves (Gupta et al. 2019; Meng et al. 2019a). OsMPK5, also known as OsBIMK1 and OsMPK3, has been reported to be involved in the disease resistance responses, positive regulation of the JA signaling pathway, and plant resistance to a chewing herbivore in rice. Moreover, MAPKKK1 is involved in the activation of ethylene biosynthesis and thereby positively and negatively regulates fungal and bacterial disease resistance, respectively. In parallel, phosphorylation of CDPKs, which are unique Ca^{2+} sensor protein kinases, and Calmodulin has also been observed upon MSP1 and other PAMP perception (Gupta et al. 2019).

12.3.2 Transcription Factor (TFs)-Mediated Downstream Responses

PTI-activated defense responses include physical cell wall reinforcement, generation, and secretion of antimicrobial chemicals (such as secondary metabolite phytoalexin accumulation), and expression of TFs (Katagiri and Tsuda 2010). A number of rice TFs have been identified that directly or indirectly play roles in the biotic and abiotic stresses. These TFs include WRKY, MADS (MCM1, AGAMOUS, DEFICIENS, and SRF) box, and NAC (NAM, ATAF1,2, CUC2), and of these WRKY TFs are most characterized during rice blast infection (Ramamoorthy et al. 2008). WRKY TFs function downstream of MAPK cascades and as many as 11 WRKY TFs have been identified playing crucial roles in the rice resistance against rice blast diseases. These TFs include OsWRKY23, OsWRKY24, OsWRKY28, OsWRKY45, OsWRKY51, OsWRKY53, OsWRKY62, OsWRKY70, OsWRKY71, OsWRKY72, and OsWRKY76. Using transcriptome analysis, increased expressions of OsWRKY45, OsWRKY47, OsWRKY53, OsWRKY55, OsWRKY62, and OsWRKY71 have been identified in response to *M. oryzae* infection (Chujo et al. 2007; Ryu et al. 2006; Shimono et al. 2007; Wei et al. 2013). Further, it was shown that overexpression of some of these genes enhanced resistance to rice blast infection. In addition, some of the WRKY TFs regulate rice immune response both in response to multiple pathogens. As an example, the involvement of OsWRKY45 was shown in rice resistance against both *M. oryzae* and *X. oryzae* pv. *oryzae* via SA hormone signaling and negatively modulates resistance against the brown planthopper, *Nilaparvata lugens* (Lee et al. 2017; Shimono et al. 2012). Moreover, OsWRKY45 and Pb1 interaction contribute to blast resistance through the protection of OsWRKY45 from ubiquitin proteasome system degradation (Matsushita et al. 2013).

12.3.3 Apoplastic Reactive Oxygen Species Burst

A transient and rapid generation of apoplastic ROS is one of the earliest events in the PTI signaling. Typically, a ROS burst is initiated within ~4–6 min, reaches its peak ~30–45 min, then gradually declines to the resting state ~60 min after PAMP treatments in various plant species. Apoplastic ROS burst is mediated by the activity of a plasma membrane-localized RBOHB and several proteomics studies have shown increased abundance and even phosphorylation of this protein in response to *M. oryzae* or exogenous MSP1 treatment (Gupta et al. 2019; Kim et al. 2013). Apoplastic ROS can act as a toxin barrier against subsequent pathogen infections and also involved in plant cell walls strengthening by forming oxidative cross-linking of polymers (Torres et al. 2006). However, the virulent strains of rice blast can suppress the ROS generated by rice cells by the secretion AvrPii and AvrPiz-t effectors (Kou et al. 2019). Moreover, virulent strains of *M. oryzae* utilize the activity of a variety of proteins including thioredoxin 2, glutathione peroxidase, glutathione reductase, and nitronate monooxygenases, among others, for the successful detoxification of rice generated ROS to facilitate the infection (Kou et al. 2019).

12.3.4 Production of Antimicrobial Compounds and Phytohormones

It has long been observed that pathogen infection or treatment of chitin results in an accumulation of antimicrobial compounds in plants, commonly referred to as phytoalexins. In monocots including rice, these phytoalexins include various terpenoids, and phenolic compounds, such as phenylamides (Schmelz et al. 2014). In particular, the accumulation of two phytoalexins, sakuranetin, and momilactone A was measured in *M. oryzae*-infected leaves using HPLC–MS/MS. Among them, the accumulation of Momilactone A was found to be higher in incompatible fungal-infected leaves compared to compatible ones, and Sakuranetin was specifically accumulated in incompatible fungal-infected samples (Wang et al. 2014).

SA (salicylic acid), JA (jasmonate acid), and ET (ethylene) are three major phytohormones that play important roles in plant immunity. In dicots, it has been well established that SA regulates immunity against biotrophic pathogens while JA regulates growth development and stress responses, especially defense responses to herbivores and necrotrophic pathogens (Browse 2009). However, there is no such distinction in the case of monocots including rice (Meng et al. 2019b). SA measurements in response to *M. oryzae* inoculation showed no significant change in SA concentrations before and after inoculation (Silverman et al. 1995). However, rice plants indeed respond to exogenous SA treatment, indicating that the involvement of SA in rice defense responses is more dependent on the SA signaling, rather than the endogenous SA level (Silverman et al. 1995). In the case of JA signaling, it has been observed that JA is involved in rice basal defense against bacterial and fungal pathogens (Tamaoki et al. 2013; Yamada et al. 2012). JA together with ABA positively regulate MSP1 induced cell death, while overexpression of MSP1 in rice confers broad-spectrum resistance through modulation of the SA- and JA-mediated signaling pathways (Wang et al. 2016). MoHrip1 activates both the SA signaling pathway and the gibberellin (GA) pathway and suppresses JA signaling (Hong et al. 2017). In the case of ET, multiple reports have been published suggesting that ET confers broad-spectrum resistance, especially against the fungal infections (Gupta et al. 2018b).

12.3.5 Callose Deposition

Callose is a β-1,3 glucan polymer with a high molecular weight that strengthens plant cell walls (Luna et al. 2011). Callose deposits in a timely manner at the site of infections, forming a prominent physical barrier against pathogen attacks. Various PAMPs or DAMPs, including flg22, elf18, chitin, PGN (peptidoglycan), and OG (oligogalacturonide), induce callose deposits in plant roots, cotyledons, and leaves (Luna et al. 2011). Callose deposition seems to be regulated at multiple levels and requires the activity of multiple proteins. For instance, function loss of RBOHD, a key enzyme involved in apoplastic ROS formation, results in compromised flg22- and OG-induced callose deposits (Clay et al. 2009).

12.4 ETI in Rice–*M. oryzae* Interaction

ETI signaling events have been well characterized by the gene-for-gene concept where identification of a pathogen secreted Avr protein by the R-gene product of the plants results in the activation of defense responses and culminating into resistance (Stotz et al. 2014). However, a growing body of evidence suggests that direct interaction between R-gene and Avr gene products is relatively rare and the recognition of an Avr protein by corresponding R-gene product is major because of the indirect interactions between these two, employing one or more additional components (Jia et al. 2000).

12.4.1 Effector Suppression of PTI

Bacterial and fungal pathogens suppress plant immunity through the secretion of numerous effector proteins that either hinders the plant defense signaling or increase their susceptibility. The number and types of effector proteins secreted by the pathogens determine their virulence level and host range. For instance, XOO1488 targets RLCK185 to suppress the OsCERK1-mediated defense (Yamaguchi et al. 2013). Moreover, when host LysM immune receptor proteins detect fungal-derived chitin to activate immunity, fungi employ LysM effectors to prevent the recognition of chitin by host immune receptors. Slp1 (secreted LysM protein 1), such one secreted protein with two LysM domains in *M. oryzae*, prevents chitin recognition by CEBiP via direct binding to chitin oligosaccharides released from the fungal cell wall (Mentlak et al. 2012).

12.4.2 R-Genes and Avr Effectors

To date, more than 100 rice R-genes conferring rice blast resistance have been identified, of which 23 were cloned, while a total of 13 Avr effector genes have been cloned from *M. oryzae* (Sharma et al. 2012). Five cloned rice R-gene and

M. oryzae Avr gene pairs (*Pita/AvrPita, Pik/Avr-Pik, Piz-t/AvrPiz-t, Pia/Avr-Pia, and Pi-CO39/Avr1-CO39*) are well studied for rice blast resistance. Most of the 23 cloned *M. oryzae* R genes are dominant NB-LRR (nucleotide-binding (NB) domain, leucine-rich repeat (LRR)-containing receptors) genes (Chen et al. 2010).

A single Avr gene may be identified by one or more R-gene products or conversely one R-gene may be involved in the recognition of multiple Avr proteins. Pita/AvrPita and Piz-t/AvrPiz-t are two common and well-studied examples of recognition of a single Avr gene by a single dominant R gene (Jia et al. 2000). While Pik-, Pi5-, and Pia/Pi-CO39 require two NB-LRR-type R-gene including Pik-1 and Pik-2, Pi5-1 and Pi5-2, a locus called Pia or Pi-CO39 consisting of RGA4 and RGA5, respectively (Ashikawa et al. 2008; Cesari et al. 2013; Lee et al. 2009; Okuyama et al. 2011). In contrast, identification of two Avr proteins including Avr1-CO39 and Avr-Pia, which share no sequence similarity, by RGA5-A is an example of recognition of two different Avr proteins by a single R-gene product. A total of four atypical R genes (Pi21, Ptr, BSR-D1, and BSR-K1) have been isolated from rice involved in the non-race-specific resistance (Zhao et al. 2018; Zhou et al. 2018).

12.5 Rice Blast Resistance Breeding

The development of disease resistance cultivar has been proven to be the most effective and economical method for disease control. There can be either complete or true resistance or partial or field resistance in rice (Parlevliet 1979). While complete resistance is race-specific, partial resistance is not. In addition, complete resistance is controlled by a single dominant or recessive R gene and partial resistance is controlled by Quantitative Trait Locus (QTLs), which implies more general mechanisms and is thought to be more durable (Skamnioti and Gurr 2009). Molecular marker technology offers the opportunity to improve the efficiency and resolution of genetic analysis of resistance genes as well as to select the lines harboring proper resistance genes for durable resistance breeding. Moreover, gene and individual QTL pyramiding should be considered for durable resistance to blast fungus. QTLs for the rice resistance against *M. oryzae* are co-localized with Pi loci or with other QTLs involved in the rice resistance against other pathogens (Paterson et al. 1991). Some SSR (simple sequence repeat markers) (RM168, RM8225, RM1233, RM6836, RM5961, and RM413) have been found by Ashkani et al. that could be used in MAS (marker-assisted selection) programs (Ashkani et al. 2011). MAS is used for screening of selected populations to track introgression of resistance genes Pib, Pik, Pii, Piz, and Pita (Jia et al. 2002). Also, it is possible to pyramid Pi-ta with either of these major resistance genes to achieve broad-spectrum resistance in the improved germplasm. Pyramiding three blast R genes, Pi1, Piz-5, and Pita-2, into cultivars provides broad-spectrum resistance to many isolates of *M. oryzae* (Jia et al. 2002). It is mandatory to facilitate the discovery and transfer of new DNA markers to breeders for the development of disease-resistant cultivars. In rice, GWAS (genome-wide association study) has been used to identify genes and QTLs associated with traits related to agronomic performance, grain quality, abiotic stress tolerance, and domestication, however, very fewer reports have been published to date on the use of GWAS for the identification of loci associated with disease resistance (Kang et al. 2016).

Acknowledgments This work was supported by grants from the Basic Science Research Program through the National Research Foundation of Korea (NRF) funded by the Ministry of Education, Science and Technology (NRF-2018R1A4A1025158) provided by STK. RG was supported by the Ramalingaswami re-entry fellowship (BT/HRD/35/02/2006) from the Department of Biotechnology, Ministry of Science and Technology, Government of India.

References

Akamatsu A, Wong HL, Fujiwara M, Okuda J, Nishide K, Uno K et al (2013) An OsCEBiP/OsCERK1-OsRacGEF1-OsRac1 module is an essential early component of chitin-induced rice immunity. Cell Host Microbe 13:465–476

Ashikawa I, Hayashi N, Yamane H, Kanamori H, Wu J, Matsumoto T et al (2008) Two adjacent nucleotide-binding site–leucine-rich repeat class genes are required to confer Pikm-specific rice blast resistance. Genetics 180:2267–2276

Ashkani S, Rafii MY, Sariah M, Akmar ASN, Rusli I, Rahim HA et al (2011) Analysis of simple sequence repeat markers linked with blast disease resistance genes in a segregating population of rice (Oryza sativa). Genet Mol Res 10:1345–1355

Bigeard J, Colcombet J, Hirt H (2015) Signaling mechanisms in pattern-triggered immunity (PTI). Mol Plant 8:521–539

Browse J (2009) Jasmonate passes muster: a receptor and targets for the defense hormone. Annu Rev Plant Biol 60:183–205

Cesari S, Thilliez G, Ribot C, Chalvon V, Michel C, Jauneau A et al (2013) The rice resistance protein pair RGA4/RGA5 recognizes the Magnaporthe oryzae effectors AVR-Pia and AVR1-CO39 by direct binding. Plant Cell 25:1463–1481

Chen L, Hamada S, Fujiwara M, Zhu T, Thao NP, Wong HL et al (2010) The Hop/Sti1-Hsp90 chaperone complex facilitates the maturation and transport of a PAMP receptor in rice innate immunity. Cell Host Microbe 7:185–196

Chen M, Zeng H, Qiu D, Guo L, Yang X, Shi H et al (2012) Purification and characterization of a novel hypersensitive response-inducing elicitor from Magnaporthe oryzae that triggers defense response in rice. PLoS One 7:e37654

Chujo T, Takai R, Akimoto-Tomiyama C, Ando S, Minami E, Nagamura Y et al (2007) Involvement of the elicitor-induced gene OsWRKY53 in the expression of defense-related genes in rice. Biochim Biophys Acta (BBA)-Gene Struct Exp 1769:497–505

Clay NK, Adio AM, Denoux C, Jander G, Ausubel FM (2009) Glucosinolate metabolites required for an Arabidopsis innate immune response. Science 323:95–101

Dean RA, Talbot NJ, Ebbole DJ, Farman ML, Mitchell TK, Orbach MJ et al (2005) The genome sequence of the rice blast fungus Magnaporthe grisea. Nature 434:980–986

Gupta R, Lee SE, Agrawal GK, Rakwal R, Park S, Wang Y et al (2015a) Understanding the plant-pathogen interactions in the context of proteomics-generated apoplastic proteins inventory. Front Plant Sci 6:352

Gupta R, Wang Y, Agrawal GK, Rakwal R, Jo IH, Bang KH et al (2015b) Time to dig deep into the plant proteome: a hunt for low-abundance proteins. Front Plant Sci 6:1–3. https://doi.org/10.3389/fpls.2015.00022

Gupta R, Kwon S-Y, Kim ST (2018a) An insight into the tomato spotted wilt virus (TSWV), tomato and thrips interaction. Plant Biotechnol Rep 12:157–163

Gupta R, Min CW, Kramer K, Agrawal GK, Rakwal R, Park KH et al (2018b) A multi-omics analysis of Glycine max leaves reveals alteration in flavonoid and isoflavonoid metabolism upon ethylene and Abscisic acid treatment. Proteomics 18. https://doi.org/10.1002/pmic.201700366

Gupta R, Min CW, Kim YJ, Kim ST (2019) Identification of Msp1-induced signaling components in rice leaves by integrated proteomic and phosphoproteomic analysis. Int J Mol Sci 20. https://doi.org/10.3390/ijms20174135

Gupta R, Min CW, Kim SW, Yoo JS, Moon A-R, Shin A-Y et al (2020) A TMT-based quantitative proteome analysis to elucidate the TSWV induced signaling cascade in susceptible and resistant cultivars of Solanum lycopersicum. Plan Theory 9:290

Hong Y, Yang Y, Zhang H, Huang L, Li D, Song F (2017) Overexpression of MoSM1, encoding for an immunity-inducing protein from Magnaporthe oryzae, in rice confers broad-spectrum resistance against fungal and bacterial diseases. Sci Rep 7:41037

Jeong JS, Mitchell TK, Dean RA (2007) The Magnaporthe grisea snodprot1 homolog, MSP1, is required for virulence. FEMS Microbiol Lett 273:157–165

Jia Y, McAdams SA, Bryan GT, Hershey HP, Valent B (2000) Direct interaction of resistance gene and avirulence gene products confers rice blast resistance. EMBO J 19:4004–4014

Jia Y, Wang Z, Singh P (2002) Development of dominant rice blast Pi-ta resistance gene markers. Crop Sci 42:2145–2149

Jones JDG, Dangl JL (2006) The plant immune system. Nature 444:323

Kaku H, Nishizawa Y, Ishii-Minami N, Akimoto-Tomiyama C, Dohmae N, Takio K et al (2006) Plant cells recognize chitin fragments for defense signaling through a plasma membrane receptor. Proc Natl Acad Sci 103:11086–11091

Kang H, Wang Y, Peng S, Zhang Y, Xiao Y, Wang D et al (2016) Dissection of the genetic architecture of rice resistance to the blast fungus Magnaporthe oryzae. Mol Plant Pathol 17:959–972

Katagiri F, Tsuda K (2010) Understanding the plant immune system. Mol Plant-Microbe Interact 23:1531–1536

Kawano Y, Akamatsu A, Hayashi K, Housen Y, Okuda J, Yao A et al (2010) Activation of a Rac GTPase by the NLR family disease resistance protein Pit plays a critical role in rice innate immunity. Cell Host Microbe 7:362–375

Kim SG, Wang Y, Lee KH, Park Z-Y, Park J, Wu J et al (2013) In-depth insight into in vivo apoplastic secretome of rice-Magnaporthe oryzae interaction. J Proteome 78:58–71

Kou Y, Qiu J, Tao Z (2019) Every coin has two sides: reactive oxygen species during rice–Magnaporthe oryzae interaction. Int J Mol Sci 20:1191

Kuchitsu K, Kikuyama M, Shibuya N (1993) N-acetylchitooligosaccharides, biotic elicitor for phytoalexin production, induce transient membrane depolarization in suspension-cultured rice cells. Protoplasma 174:79–81

Lee SE, Gupta R, Jayaramaiah RH, Lee SH, Wang Y, Park SR et al (2017) Global transcriptome profiling of xanthomonas oryzae pv. Oryzae under in planta growth and in vitro culture conditions. Plant Pathol J 33:458–466. https://doi.org/10.5423/PPJ.OA.04.2017.0076

Lee S-K, Song M-Y, Seo Y-S, Kim H-K, Ko S, Cao P-J et al (2009) Rice Pi5-mediated resistance to Magnaporthe oryzae requires the presence of two coiled-coil–nucleotide-binding–leucine-rich repeat genes. Genetics 181:1627–1638

Liu W, Liu J, Ning Y, Ding B, Wang X, Wang Z et al (2013) Recent progress in understanding PAMP-and effector-triggered immunity against the rice blast fungus Magnaporthe oryzae. Mol Plant 6:605–620

Liu W, Liu J, Triplett L, Leach JE, Wang G-L (2014) Novel insights into Rice innate immunity against bacterial and fungal pathogens. Annu Rev Phytopathol 52:213–241. https://doi.org/10.1146/annurev-phyto-102313-045926

Luna E, Pastor V, Robert J, Flors V, Mauch-Mani B, Ton J (2011) Callose deposition: a multifaceted plant defense response. Mol Plant-Microbe Interact 24:183–193

Matsushita A, Inoue H, Goto S, Nakayama A, Sugano S, Hayashi N et al (2013) Nuclear ubiquitin proteasome degradation affects WRKY 45 function in the rice defense program. Plant J 73:302–313

Meng Q, Gupta R, Kwon SJ, Wang Y, Agrawal GK, Rakwal R et al (2018a) Transcriptomic analysis of Oryza sativa leaves reveals key changes in response to Magnaporthe oryzae MSP1. Plant Pathol J 34:257

Meng Q, Gupta R, Min CW, Kim J, Kramer K, Wang Y et al (2018b) Label-free quantitative proteome data associated with MSP1 and flg22 induced signaling in rice leaves. Data Brief 20:204–209

Meng Q, Gupta R, Min CW, Kim J, Kramer K, Wang Y et al (2019a) A proteomic insight into the MSP1 and flg22 induced signaling in Oryza sativa leaves. J Proteome 196:120–130. https://doi.org/10.1016/j.jprot.2018.04.015

Meng Q, Gupta R, Min CW, Kwon SW, Wang Y, Je BI et al (2019b) Proteomics of rice—Magnaporthe oryzae interaction: what have we learned so far? Front Plant Sci 10. https://doi.org/10.3389/fpls.2019.01383

Mentlak TA, Kombrink A, Shinya T, Ryder LS, Otomo I, Saitoh H et al (2012) Effector-mediated suppression of chitin-triggered immunity by Magnaporthe oryzae is necessary for rice blast disease. Plant Cell 24:322–335

Miller RNG, Costa Alves GS, Van Sluys M-A (2017) Plant immunity: unravelling the complexity of plant responses to biotic stresses. Ann Bot 119:681–687

Nie H-Z, Zhang L, Zhuang H-Q, Shi W-J, Yang X-F, Qiu D-W et al (2019) The secreted protein MoHrip1 is necessary for the virulence of Magnaporthe oryzae. Int J Mol Sci 20:1643

Okuyama Y, Kanzaki H, Abe A, Yoshida K, Tamiru M, Saitoh H et al (2011) A multifaceted genomics approach allows the isolation of the rice Pia-blast resistance gene consisting of two adjacent NBS-LRR protein genes. Plant J 66:467–479

Parlevliet JE (1979) Components of resistance that reduce the rate of epidemic development. Annu Rev Phytopathol 17:203–222

Paterson AH, Damon S, Hewitt JD, Zamir D, Rabinowitch HD, Lincoln SE et al (1991) Mendelian factors underlying quantitative traits in tomato: comparison across species, generations, and environments. Genetics 127:181–197

Ramamoorthy R, Jiang S-Y, Kumar N, Venkatesh PN, Ramachandran S (2008) A comprehensive transcriptional profiling of the WRKY gene family in rice under various abiotic and phytohormone treatments. Plant Cell Physiol 49:865–879

Ryu H-S, Han M, Lee S-K, Cho J-I, Ryoo N, Heu S et al (2006) A comprehensive expression analysis of the WRKY gene superfamily in rice plants during defense response. Plant Cell Rep 25:836–847

Schmelz EA, Huffaker A, Sims JW, Christensen SA, Lu X, Okada K et al (2014) Biosynthesis, elicitation and roles of monocot terpenoid phytoalexins. Plant J 79:659–678

Sharma TR, Rai AK, Gupta SK, Vijayan J, Devanna BN, Ray S (2012) Rice blast management through host-plant resistance: retrospect and prospects. Agric Res 1:37–52

Shimizu T, Nakano T, Takamizawa D, Desaki Y, Ishii-Minami N, Nishizawa Y et al (2010) Two LysM receptor molecules, CEBiP and OsCERK1, cooperatively regulate chitin elicitor signaling in rice. Plant J 64:204–214

Shimono M, Sugano S, Nakayama A, Jiang C-J, Ono K, Toki S et al (2007) Rice WRKY45 plays a crucial role in benzothiadiazole-inducible blast resistance. Plant Cell 19:2064–2076

Shimono M, Koga H, Akagi AYA, Hayashi N, Goto S, Sawada M et al (2012) Rice WRKY45 plays important roles in fungal and bacterial disease resistance. Mol Plant Pathol 13:83–94

Silverman P, Seskar M, Kanter D, Schweizer P, Metraux J-P, Raskin I (1995) Salicylic acid in rice (biosynthesis, conjugation, and possible role). Plant Physiol 108:633–639

Skamnioti P, Gurr SJ (2009) Against the grain: safeguarding rice from rice blast disease. Trends Biotechnol 27:141–150

Skorbiansky SR (2018) Rice outlook. Econ Res Serv:1–33

Stotz HU, Mitrousia GK, de Wit PJGM, Fitt BDL (2014) Effector-triggered defence against apoplastic fungal pathogens. Trends Plant Sci 19:491–500

Tamaoki D, Seo S, Yamada S, Kano A, Miyamoto A, Shishido H et al (2013) Jasmonic acid and salicylic acid activate a common defense system in rice. Plant Signal Behav 8. https://doi.org/10.4161/psb.24260

Thomma BPHJ, Nürnberger T, Joosten MHAJ (2011) Of PAMPs and effectors: the blurred PTI-ETI dichotomy. Plant Cell 23:4–15

Torres MA, Jones JDG, Dangl JL (2006) Reactive oxygen species signaling in response to pathogens. Plant Physiol 141:373–378

Wang Y, Kwon SJ, Wu J, Choi J, Lee YH, Agrawal GK et al (2014) Transcriptome analysis of early responsive genes in rice during Magnaporthe oryzae infection. Plant Pathol J 30:343–354. https://doi.org/10.5423/PPJ.OA.06.2014.0055

Wang Y, Wu J, Kim SG, Tsuda K, Gupta R, Park S-Y et al (2016) Magnaporthe oryzae-secreted protein MSP1 induces cell death and elicits defense responses in rice. Mol Plant-Microbe Interact 29. https://doi.org/10.1094/MPMI-12-15-0266-R

Wang Y, Gupta R, Song W, Huh H-H, Lee SE, Wu J et al (2017a) Label-free quantitative secretome analysis of Xanthomonas oryzae pv. oryzae highlights the involvement of a novel cysteine protease in its pathogenicity. J Proteome 169. https://doi.org/10.1016/j.jprot.2017.02.012

Wang Y, Wu J, Kim SG, Yoo JS, Gupta R, Je BI et al (2019) A modified transient gene expression protocol for subcellular protein localization analysis in rice. Springer

Wang Z, Han Q, Zi Q, Lv S, Qiu D, Zeng H (2017b) Enhanced disease resistance and drought tolerance in transgenic rice plants overexpressing protein elicitors from Magnaporthe oryzae. PLoS One 12

Wei T, Ou B, Li J, Zhao Y, Guo D, Zhu Y et al (2013) Transcriptional profiling of rice early response to Magnaporthe oryzae identified OsWRKYs as important regulators in rice blast resistance. PLoS One 8

Wong HL, Pinontoan R, Hayashi K, Tabata R, Yaeno T, Hasegawa K et al (2007) Regulation of rice NADPH oxidase by binding of Rac GTPase to its N-terminal extension. Plant Cell 19:4022–4034

Yamada S, Kano A, Tamaoki D, Miyamoto A, Shishido H, Miyoshi S et al (2012) Involvement of OsJAZ8 in jasmonate-induced resistance to bacterial blight in rice. Plant Cell Physiol 53:2060–2072

Yamaguchi K, Yamada K, Ishikawa K, Yoshimura S, Hayashi N, Uchihashi K et al (2013) A receptor-like cytoplasmic kinase targeted by a plant pathogen effector is directly phosphorylated by the chitin receptor and mediates rice immunity. Cell Host Microbe 13:347–357

Yang Y, Zhang H, Li G, Li W, Wang X, Song F (2009) Ectopic expression of MgSM1, a Cerato-platanin family protein from Magnaporthe grisea, confers broad-spectrum disease resistance in Arabidopsis. Plant Biotechnol J 7:763–777

Zhang Y, Liang Y, Dong Y, Gao Y, Yang X, Yuan J et al (2017) The Magnaporthe oryzae Alt A 1-like protein MoHrip1 binds to the plant plasma membrane. Biochem Biophys Res Commun 492:55–60

Zhao H, Wang X, Jia Y, Minkenberg B, Wheatley M, Fan J et al (2018) The rice blast resistance gene Ptr encodes an atypical protein required for broad-spectrum disease resistance. Nat Commun 9:1–12

Zhou X, Liao H, Chern M, Yin J, Chen Y, Wang J et al (2018) Loss of function of a rice TPR-domain RNA-binding protein confers broad-spectrum disease resistance. Proc Natl Acad Sci 115:3174–3179

The Role of Endophytic Insect-Pathogenic Fungi in Biotic Stress Management

13

Imtiaz Ahmad, María del Mar Jiménez-Gasco, and Mary E. Barbercheck

Abstract

A diverse beneficial microbial community inhabits soil, with fungi comprising a major component. Fungi can establish in plants as endophytes and as a result, these plant–fungal interactions are associated with different trade-offs. Endophytic insect-pathogenic fungi (EIPF) establish a unique association with plants that can promote growth of host plants and can have detrimental effects on pest insects. The mechanisms of plant growth promotion are diverse and include nutrient exchange, plant defense modulation, and protection from pests and diseases. There is increasing interest in understanding the potential for exploiting EIPF to enhance plant productivity and tolerance to arthropod pests in agricultural systems, especially under changing climate conditions. EIPF mediate plant defense signaling through crosstalk in phytohormone-based defense systems and fine-tuning signaling pathways in the presence of stress. Secondary metabolites and other compounds in root exudates can affect the fate of EIPF–plant–stress interactions. Through these mechanisms, plants can fine-tune growth and defense in a balanced manner. Developing a better understanding of the ecology and biology of EIPF and their interactions with plants and arthropods will contribute to the achievement of agricultural sustainability.

I. Ahmad (✉) · M. E. Barbercheck
Department of Entomology, The Pennsylvania State University, University Park, PA, USA
e-mail: ixa12@psu.edu

M. M. Jiménez-Gasco
Department of Plant Pathology and Environmental Microbiology, The Pennsylvania State University, University Park, PA, USA

B. Giri, M. P. Sharma (eds.), *Plant Stress Biology*,
https://doi.org/10.1007/978-981-15-9380-2_13

Keywords

Fungal endophytes · Insect-pathogenic fungi · Plant defense · Plant growth promotion · Stress · Sustainable agriculture

13.1 Introduction

The soil microbiome comprises diverse and complex microbial communities that provide many beneficial services to agroecosystems (Spence and Bais 2013; Zolla et al. 2013). Several biotic and abiotic factors affect the recruitment and maintenance of microbial communities in the rhizosphere, the thin layer of soil in close proximity to the plant roots, and in plants (Rosier et al. 2016). For example, plant species, root morphology, and soil characteristics give rise to distinct microbial communities in the rhizosphere (Berg and Smalla 2009). Rhizodeposition, the organic compounds released from plant roots into the surrounding environment, contains a large amount of photosynthetically-fixed carbon and is a major resource that supports the soil microbial community (Rosier et al. 2016).

The relationship between plants and soil microbes may be beneficial, harmful, or neutral for plants. Hence, understanding the diversity in community structure and key functions of the microbiome can provide valuable information for improving agricultural sustainability (Ahmad and Zaib 2020). Endophytes are plant symbionts, often bacteria or fungi, that can live within a plant for at least part of their life cycle without causing apparent disease symptoms and can inhabit the tissues of most terrestrial plants (reviewed in Vega 2018). Endophytes can affect plant physiology and growth by conferring abiotic and biotic stress tolerance, increasing growth and biomass, reducing water consumption, improving drought tolerance, and altering resource allocation (Dara 2019; Hu and Bidochka 2019; Vega 2018; Yan et al. 2019). Other benefits of endophytes include enhanced mineral uptake (Ren et al. 2011), improved N use efficiency (Behie et al. 2012, 2017), protection against plant diseases (Dara 2019), and defense against herbivory (Ahmad et al. 2020b; Akutse et al. 2013; Lopez and Sword 2015; Sullivan et al. 2007; Zhang et al. 2009). However, in some cases, endophytic colonization can improve the performance of plant-feeding insects, e.g. aphids (Clifton et al. 2018).

Endophytic insect pathogens are a ubiquitous but often-overlooked group of natural enemies, and fungi are among the most common insect pathogens. There are almost 700 species of fungal insect pathogens belonging to more than 100 genera, the majority of them belonging to the Hypocreales. Endophytic insect-pathogenic fungi (EIPF) have multifunctional lifestyles. As direct pathogens of insects, they can infect and kill susceptible insect by the germination of spores on the host insect's cuticle, and formation of specialized infection structures that penetrate the insect cuticle that result in fatal infections in the insect host. *Metarhizium* (Clavicipitaceae) and *Beauveria* (Cordycipitaceae) genera are hypocrealean EIPF that are ubiquitously found in agricultural soils and can endophytically colonize a broad range of plant hosts (Vega 2018). As endophytes, they can promote plant growth, translocate nutrients, and enhance tolerance of colonized plants against environmental stresses (reviewed in (Bamisile et al. 2018; Hu and Bidochka 2019; Vega 2018). Tritrophic interactions involving EIPF in which the growth or fecundity of herbivorous insects and plant pathogens that attack endophyte-colonized plants is suppressed could have important implications for biological control in agroecosystems.

13.2 Role of Fungal Endophytes in Plant Growth Promotion and Protection

Fungal endophytes in plants are not randomly distributed in plant tissues or in the environment. Both soil and plant factors exert strong influences (Chaparro et al. 2012, 2014; Hacquard et al. 2015) and opinions are divided on which factors predominate in the recruitment of endophytes by plants (Berg and Smalla 2009; Lareen et al. 2016). Soil-borne EIPF are influenced initially in the rhizosphere by many biotic and abiotic factors, e.g. plant species, plant growth stage, fungal species, other microbes, soil properties, climate, and geographical characteristics (Glynou et al. 2016; Lê Van et al. 2017). The factors and interactions in soil that drive changes in the rhizosphere community are complex, dynamic, and not fully understood.

Plants can modify soil microbial communities directly, presumably through differences in root exudates (Chaparro et al. 2014; Hartmann et al. 2009; Oldroyd 2013), or indirectly via their influence on the abiotic environment (Bulgarelli et al. 2013). Qualitatively and quantitatively diverse root exudates play a role in modulating the microbial composition and assembly in the bulk soil microbiome (Bruck 2009; Busby et al. 2017; Sasse et al. 2018). Root exudates contain highly diverse low-molecular weight compounds such as phenolics, metabolites, amino acids, organic acids and sugars and relatively less diverse high-molecular weight compounds such as proteins and mucilage (Bais et al. 2006). The same chemical in root exudates can vary in their function as signals that can attract or deter specific microbes, thus influencing their interaction with plants (Bais et al. 2006). Fungal endophytes can modulate the composition of root exudates by altering the level of phenolics and other metabolites that may contribute to plant growth promotion and tolerance against stresses (Gargallo-Garriga et al. 2018; Guo et al. 2015). The composition of root exudates and their interaction with specific microbes is a signature of chemical communication of the plant and rhizosphere microbiome (Bais et al. 2006). However, the actual mechanism and the multitude of factors by which quality and composition of root exudation is controlled are still unclear (Hu et al. 2018).

The mediation of plant–microbe interactions through root exudates is associated with significant carbon costs to plants (Uren 2000). Plants release carbon-rich rhizodeposits that are used as chemical signals that stimulate microbial growth. For example, the concentration of carbohydrates and organic carbon in the root exudates of tall fescue was greater in the presence of the endophyte *Neotyphodium coenophialum,* compared with endophyte-free plants (Van Hecke et al. 2005). The change in the composition of root exudates may be due to the presence of endophytes or may be a strategy by plants to attract endophytes to cope with nutrient deficiency stress. Physico-chemical properties of soils can directly select for specific microbes by creating conditions that benefit them and influence the availability of plant root exudates affecting microbial recruitment by the plant. For example, soil pH and nutrient availability (e.g., C, N, P) can affect the abundance of both pest and beneficial soil biota in crops (Dumbrell et al. 2010; Garbeva et al. 2004).

13.3 EIPF and Their Role in Plant Growth Promotion

EIPF can live as saprophytes on soil organic matter, as plant symbionts, and as pathogens of insects (Moonjely et al. 2016; Vega 2018). The relationship between EIPF and plants appears to be facultative as they can survive and reproduce solely on insect hosts. Many genera are classified as EIPF but among them the most well-studied are fungi in the genera *Beauveria* and *Metarhizium* (Bamisile et al. 2018; Vega 2018; Zhang et al. 2018).

EIPF in the genus *Metarhizium* occur primarily in soil, have a broad arthropod host range, and are well-adapted to agricultural systems (Meyling and Eilenberg 2007; Steinwender et al. 2011; Tiago et al. 2014). *Metarhizium* spp. can infect more than 200 arthropod host species from 17 families of Insecta and Acari (Roberts and St. Leger 2004; Zimmermann 2007b). In agricultural soils, the prevalence of *Metarhizium* spp. can reach 10^6 colony forming units (CFU) g^{-1} soil and are among the most abundant soil-borne entomopathogenic fungi (Lomer et al. 2001). Even though the insect host range at the generic level is very broad, at the species level, *Metarhizium* contains both specialist and generalist insect pathogens (Gao et al. 2011). Some level of plant specificity of *Metarhizium* spp. has been observed in nature. For example, at two experimental locations in Ontario, Canada, the association of *M. brunneum*, *M. robertsii*, and *M. guizhouense* with different plant species (shrubs, grasses, trees, and wildflowers) was recorded (Wyrebek et al. 2011). When co-occurring, *M. robertsii* was found associated with grass roots, whereas *M. guizhouense* and *M. brunneum* were exclusively associated with the rhizosphere of wildflowers. *M. guizhouense* was found associated with the rhizosphere of trees, whereas *M. brunneum* was associated with the rhizosphere of trees and shrubs (Wyrebek et al. 2011). Fisher et al. (2011) reported varying degree of association and diversity of four *Metarhizium* spp. with the rhizosphere of blueberry (*Vaccinium corymbosum*), grapevines (*Vitis vinifera*), strawberry (*Fragaria ananassa*), and Christmas tree spp. (*Picea engelmannii, Abies procera*, and *Pseudotsuga menziesii* (Pinales: Pinaceae) in the USA. Strawberry and Christmas trees had greater species richness and were colonized by *M. robertsii*, *M. brunneum*, *M. guizhouense*, and *M. flavoviride*. Blueberry and grape vines had lower species richness and diversity. Blueberry was colonized only by *M. brunneum* and *M. guizhouense*, whereas grapes were colonized by *M. robertsii*, *M. brunneum*, and *M. guizhouense* (Fisher et al. 2011). In another study in Japan, there was fungal diversity in the rhizosphere soil but there was no evidence of specificity of *M. robertsii*, *M. lepidiotae*, *M. pemphigi*,

M. guizhouense, and *M. pingshaense* with plant species from Poaceae and Asteraceae (Nishi and Sato 2019). Different soil properties, composition of the microbial community and root exudates may have contributed to inconsistency in results (Mendes et al. 2013; Wyrebek et al. 2011).

Plant growth promotion associated with endophytic colonization by *Metarhizium* spp. has been observed for multiple plant species after endophytic colonization of various tissues via seed, soil, or foliar inoculation (Ahmad et al. 2020a; Batta 2013; Behie et al. 2012, 2015; Jaber and Enkerli 2016; Krell et al. 2018; Vega 2018). For example, three isolates of *M. anisopliae* significantly enhanced root length and dry biomass of root and shoot tissues after soil inoculation of 14-day-old tomato seedlings, although the effect depended on the isolate and inoculation concentration (Elena et al. 2011). In field experiments, seed inoculation of maize with *M. robertsii* and *M. anisopliae* enhanced leaf collar formation, stand density, stalk length, stalk and foliage fresh weight, and biomass of stalk, ear, and foliage compared to control plants (Kabaluk and Ericsson 2007; Liao et al. 2014). Seed inoculation of *M. robertsii* enhanced height and above-ground biomass of maize at V4 stage and the growth promotion effects were correlated with the frequency of detection of the fungus in plant tissues under greenhouse conditions (Ahmad et al. 2020b). These results provide evidence that *Metarhizium* may as a group have a wide host range, which could facilitate their development and use as plant growth promoters for agricultural applications.

EIPF in the genus *Beauveria* endophytically colonized above- and below-ground plant tissues of various plant species and conferred plant growth promotion and protection (Bamisile et al. 2018; Hu and Bidochka 2019; Russo et al. 2018; Vega 2018). For example, after foliar application of conidia of *Beauveria bassiana* to maize (*Zea mays* L.), endophytic hyphal growth was observed between epidermal cells, in the air spaces between parenchymal cells, and in the vascular elements of the xylem (Wagner and Lewis 2000). *B. bassiana* was detected in the root as well as aerial parts of haricot beans (*Phaseolus vulgaris*) and tomato (*Solanum lycopersicum*) (Behie et al. 2015; Russo et al. 2018) suggesting systemic colonization of EIPF in the plant. *B. bassiana* colonized various tissues of strawberry plants (Dara and Dara 2015) and improved their growth (Dara 2013). *B. bassiana* colonized palm (*Phoenix dactylifera*) and a proteomic analysis revealed that endophytic colonization induced stress and defense-related proteins, and improved photosynthesis and energy metabolism (Gómez-Vidal et al. 2009). Seed treatment of broad bean (*Vicia faba*) with *B. bassiana* increased plant growth as a result of systemic colonization (Jaber and Enkerli 2016). After foliar application, endophytic colonization of maize with *B. bassiana* resulted in greater plant height, grain weight, number of leaves, and yield compared with uninoculated control plants (Russo et al. 2019). Foliar application of soybean with *B. bassiana* resulted in plant growth promotion under field conditions (Russo et al. 2019). Such growth promotion effects by endophytic *Beauveria* suggest that they may be suitable for development as inoculants to promote growth in crop plants.

Growth promotion effects associated with endophytic colonization by EIPF in other genera have also been reported. For example, *Lecanicillium dimorphum* and *L.*

cf. *psalliotae* colonized palm plants and induced stress and defense-related proteins, improved photosynthesis and energy metabolism (Gómez-Vidal et al. 2009). *L. psalliotae* enhanced the growth of cardamom (*Elettaria cardamomum*) through production of siderophores, indole acetic acid (IAA), and increased chlorophyll content (Kumar et al. 2018). *Isaria javanica* pf185 promoted the growth of tobacco (*Nicotiana tabacum*) by enhanced root branching and root hair formation and the authors attributed these effects to fungal volatiles (Lee and Kim 2019).

Multiple mechanisms have been described that contribute to plant growth promotion effects, including protection from pest insects and phytopathogens; nutrient translocation and assimilation, plant defense modulation; production of phytohormones and biologically-active metabolites, and improved photosynthesis and energy metabolism (Ahmad et al. 2020b; Behie et al. 2012, 2017; Cherry et al. 2004; Gómez-Vidal et al. 2009; Khan et al. 2012; Kumar et al. 2018; Liao et al. 2017). Several studies reported that EIPF colonization promoted plant growth through the suppression of herbivory and plant diseases. For example, under laboratory conditions, *M. brunneum* colonization of tomato effectively controlled the larval growth of the elaterid click beetle, *Agriotes obscurus* (Mayerhofer et al. 2017). Inoculation of maize with *B. bassiana* resulted in suppression of the noctuid stem-borer (*Sesamia calamistis*) and looper, *Rachiplusia nu* (Cherry et al. 2004; Russo et al. 2019). *M. brunneum* acted antagonistically against the olive phytopathogens, *Verticillium dahliae* and *Phytophthora megasperma* (Lozano-Tovar et al. 2017). *B. bassiana* produced IAA that promoted plant growth by inducing changes in root architecture that may facilitate a symbiotic relationship with the host plant (Liao et al. 2017).

Another mechanism by which EIPF may benefit plants is nutrient transfers from infected insects in the soil, linking the pathogenic and endophytic functions of EIPF. Using radioisotopic labeling, Behie et al. (2012) discovered that *M. robertsii* transferred nitrogen from a fungus-colonized cadaver to a plant through a mycelial association with roots. In return, *M. robertsii* received sugar as a source of carbon from the plant (Behie et al. 2017). If this is a common phenomenon in nature, improved plant nutrient status associated with EIPF may positively impact plant energy metabolism and photosynthetic efficiency under nutrient deficient conditions (Gómez-Vidal et al. 2009; Krell et al. 2018) (Fig. 13.1).

13.4 The Role of EIPF in Plant Defense

Plant responses to stress have been well-studied and include morphological, physiological, cellular, and molecular changes (Dastogeer 2018; Gray and Brady 2016). Abiotic and biotic stresses can impact microbes directly, altering microbial population, community composition, and the processes they mediate (Frey et al. 2013; Hagerty et al. 2014; Karhu et al. 2014). The outcome of plant–endophyte interactions depends on the identity of the plant and fungal symbionts (Dastogeer 2018). Mutualistic plant–microbe interactions can confer plants with molecular

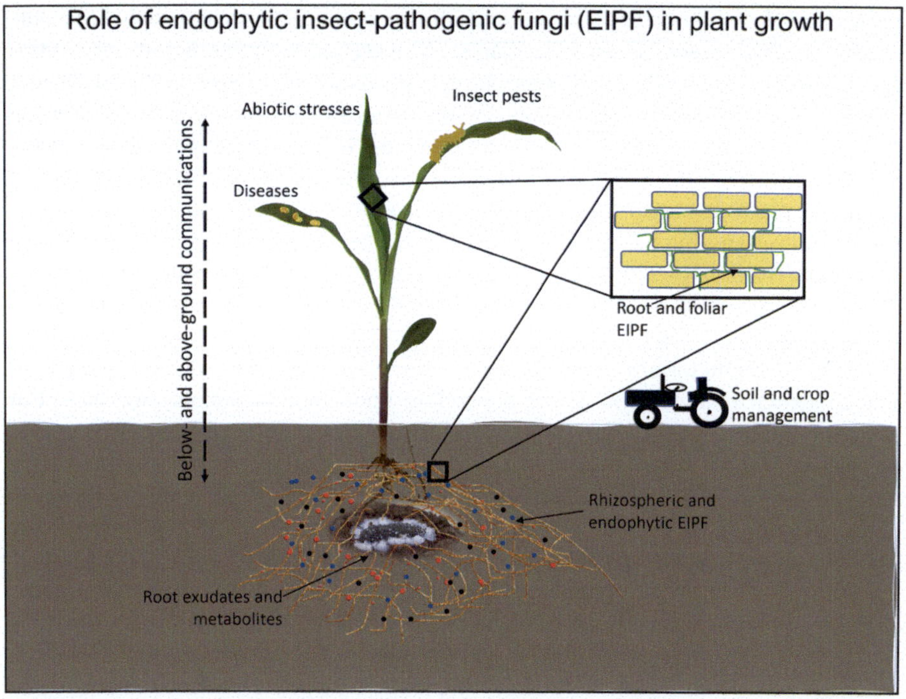

Fig. 13.1 Role of EIPF in plant growth promotion and protection from biotic and abiotic stress. EIPF are direct pathogens of a wide range of arthropods and as endophytes can suppress herbivory. Soil-borne EIPF can also live as endophytes or in the rhizosphere. As endophytes, they can promote plant growth by mediating plant defense against stress, nutrient translocation and production of metabolites, and suppression of phytopathogens and herbivory by arthropods

defense mechanisms against challenging conditions through multifaceted strategies (Koricheva et al. 2009; Pineda et al. 2013; Saikkonen et al. 2013).

Plants can perceive microbe-associated molecular patterns (MAMPs) and the reaction of the plant to these signals can determine the fate of plant–microbe interactions (Acevedo et al. 2015). When phytopathogens or mutualistic symbionts, such as arbuscular mycorrhizae (AM), colonize plants, the plant and fungus recognize each other, undergo morphological and physiological changes, and events including signal perception, signal transduction, and defense gene activation occur (Reinhardt 2007). As with plant–pathogen or AM–plant interactions, the induction or suppression of plant defense mechanisms may play a key role in endophytic colonization by EIPF. Our understanding of the impacts of mycorrhizal fungi-induced plant defenses has increased rapidly, but relatively little is known about the defense modulation of host plants induced by EIPF. EIPF may face critical challenges when plants activate their defense against foreign organisms upon early events during association (Hu and Bidochka 2019).

Specific signaling molecules play key roles in determining whether the foreign organism is a friend or foe. Based on molecular signals, plants perceive their association as symbiotic or pathogenic (Zeilinger et al. 2016). Scientists believe that, similar to AM association with plants, N-acetylglucosamine-based oligomers, lipochito-oligosaccharides, and other chitin signals play a key role in differentiating beneficial fungi from phytopathogens (Genre et al. 2013; Maillet et al. 2011). Similar molecules may be involved in differentiating EIPF from pathogens (Heinz et al. 2018; Razinger et al. 2018). Strigolactones are phytohormones used as chemical signals by plants to attract and facilitate establishment of symbiotic association with AM by promoting hyphal branching and increasing the probability of contact with plant roots (López-Ráez et al. 2017; Massalha et al. 2017; Mori et al. 2016). When EIPF come into contact with host plants, the early events of association may involve similar molecules for recognition (Hu and Bidochka 2019).

Different phytohormones are involved in specific defense tasks and plants are able to switch pathways on or off depending upon the challenges they face. For example, abscisic acid (ABA) is involved in the modulation of plant growth and defense against abiotic stress (Herrera-Medina et al. 2007). Jasmonic acid (JA) and salicylic acid (SA) defense pathways respond in many plant systems where biotrophic pathogens and phloem-feeding insects induce the SA-dependent pathway and necrotrophic pathogens and chewing insects induce the JA-dependent pathway (De Vos et al. 2005; Glazebrook 2005; Pieterse et al. 2012, 2014). However, eliciting a plant defense response can be associated with a trade-off between growth and defense (Karasov et al. 2017) and the antagonistic crosstalk of SA and JA pathways can be an adaptive strategy by plants to improve resource allocation to integrate plant growth and defense (Thaler et al. 2012).

EIPF in the genus *Metarhizium* play a role in plant growth and defense modulation under stress conditions through phytohormone-mediated pathways. For successful establishment of EIPF as endophytes, signaling cascades involving different defense pathways may become activated. For example, Liao et al. (2017) detected IAA in the filtrates of *M. robertsii, M. brunneum,* and *M. acridum* that promoted root proliferation, possibly through auxin-dependent mechanisms in early signaling events of root colonization. In plants, IAA plays a critical role in cell division, differentiation, and elongation (Woodward and Bartel 2005). Changes in the concentration of signaling molecules and phytohormones in plant–EIPF interactions can indicate changes in the status of plant defense. *M. anisopliae*-colonized plants showed a decreased concentration of SA compared with uninoculated plants, suggesting a suppression of plant defense against biotrophic phytopathogens (Hao et al. 2017). Ahmad et al. (2020b) reported down-regulation of gene expression of pathogenesis-related protein 5 (*pr5*) from the SA pathway in *M. robertsii*-colonized maize plants that suggests plants may have suppressed their defense in the absence of stress. Compared with uninoculated plants, *Metarhizium*-inoculated soybean plants showed a decrease in plant ABA and increase in JA levels under salinity stress suggesting a stress ameliorating effect (Khan et al. 2012). *M. anisopliae*-inoculated peanut plants differentially expressed genes responsible for various transcription factors, signaling effectors, and binding proteins compared to uninoculated plants (Hao et al. 2017).

Beauveria spp. modulate defense reactions under stress conditions through defense-related pathways. The strategy of *B. bassiana* to colonize roots and evade plant defense may be similar to other fungal phytopathogens that use LysM effector molecules in modulating host immunity (Cen et al. 2017). Artificial wounding of above-ground maize tissues enhanced the persistence of *B. bassiana* inoculum in the rhizosphere suggesting signal transduction and a strategy by the plant to cope with herbivory (McKinnon et al. 2018). Raad et al. (2019) reported the upregulation of molecular pattern recognition receptors and changes in transcriptional factors of plant defense pathway in *B. bassiana*-inoculated *Arabidopsis thaliana*, suggesting MAMP-triggered immunity. These reports provide evidence that suggests *Beauveria* spp. may play a role in plant growth and defense modulation under stress conditions. Field-based studies will be necessary to confirm that observations from laboratory and greenhouse experiments occur in the field.

Plants can regulate the production of secondary metabolites for enhanced defense. For example, endophytic colonization by *B. bassiana* enhanced the biosynthesis of terpenes in tomato leaves that may lead to the production of biologically-active metabolites (Shrivastava et al. 2015). Terpenes are considered to be chemical defense molecules that plants use to alert other plants of the presence of herbivory (Ahern and Whitney 2014; Reid and Purcell 2011).

EIPF from other genera have been reported to induce defense in plant hosts under various stress conditions. For example, proteomic analysis revealed that date palm colonized with *L. dimorphum* and *L.* cf. *psalliotae* showed differential induction of stress and defense-related proteins (Gómez-Vidal et al. 2009).

13.5 EIPF in Nutrient Exchange

One of the strategies that some plant-associated microbes have evolved is nutrient exchange, in which the endophyte provides some benefit to the plant host, such as plant growth promotion, in exchange for nutrients from the host plant. Some endophytes aid in nutrient absorption and assimilation possibly by affecting root morphology directly or indirectly. Accessible nitrogen is one of the most important nutrients for plants and accessible carbon is the limiting nutrient for soil fungi (Smith et al. 2009). Genes associated with metabolic pathways of nitrogen and carbon in plants and EIPF may affect root colonization by EIPF (Liao et al. 2013). Paungfoo-Lonhienne et al. (2015) demonstrated that nitrogen concentration in growth media can strongly affect the abundance of EIPF in the fungal community of the rhizosphere. The source and fertility of the plant growth medium are a major growth determining factor and plants can shift their association with EIPF according to their needs. For example, the detection of *M. anisopliae* and *B. bassiana* was less variable in sterile vermiculite compared with sterile soil:sand:peat growth medium suggesting an effect due to variability in nutrients due to greater competition in a non-sterile environment (Parsa et al. 2018). Non-sterile growth environments may also reduce endophytic colonization efficiency of EIPF possibly due to competition with other microbes (Parsa et al. 2018; Tefera and Vidal 2009).

Some *Metarhizium* spp. appear to promote plant growth through nutrient transfer (reviewed in Vega 2018). Behie et al. (2017) and Behie et al. (2012) reported that

M. robertsii can translocate nitrogen from a colonized insect cadaver to a host plant and in return receive sugar compounds from the plant. EIPF may enhance growth promotion of plants by affecting photosynthetic efficiency. Endophytic colonization of potato by *M. brunneum* promoted host plant growth by alleviating nutrient deficiency effects (Krell et al. 2018). Additionally, Krasnoff et al. (2014) showed that *M. robertsii* can produce siderophores (iron-chelating molecules) under iron-deficient conditions which improved photosynthetic efficiency. *M. brunneum* increased the availability of iron in calcareous and non-calcareous growth media that resulted in an increase in root length and number of roots of sorghum plants (Raya-Díaz et al. 2017a, 2017b). Endophytic colonization of sorghum plants by seed and soil inoculation with *M. brunneum* increased chlorophyll content, even under iron stress conditions, compared to the control (Raya-Díaz et al. 2017a, 2017b). Endophytic colonization of *M. anisopliae* resulted in increased chlorophyll content compared to control soybean plants (Khan et al. 2012).

The potential of *Beauveria* spp. to promote plant growth through nutrient supplementation and translocation has been reported in multiple studies. For example, endophytic colonization of date palm by *B. bassiana* resulted in improved photosynthesis and energy metabolism (Gómez-Vidal et al. 2009). Endophytic *B. bassiana* enhanced iron availability in calcareous and non-calcareous soil and resulted in improved growth of sorghum (Raya-Díaz et al. 2017a, 2017b). Soil fertility status, sources of nutrient, and crops management can also affect the association of EIPF with plants. For example, *B. bassiana* detection was greater in organically-managed fields where fertilization sources were compost, green and animal manures compared to conventional fields where fertilization sources were synthetic chemicals (Ramos et al. 2017). EIPF can help convert heavy metals to easily-assimilated nutrients in the soil. For example, *B. caledonica* showed tolerance to heavy metals and converted them to oxalates (Fomina et al. 2004) but it was not evident whether those effects were due to direct or indirect effects of EIPF colonization. The effects of EIPF on plant nutrition and growth promotion vary among EIPF species and strain or isolate, plants species and variety, and the composition of the rhizosphere microbiome. In an experiment by Raad et al. (2019), one strain of *B. bassiana* had a positive effect on the growth of *A. thaliana*, while the other strain did not affect plant growth.

13.6 EIPF in Plant Protection Against Pests and Diseases

In addition to plant growth promotion and induction of plant defenses, EIPF have been associated with indirect suppressive effects on pest insects and diseases. Soil drenching with *M. brunneum* resulted in the improvement of plant growth parameters at 7- and 35-days post-inoculation, and delayed the onset of reproduction, reduced the birth rate, and prolonged development time of the green peach aphid (*Myzus persicae*) (Jaber and Araj 2018). Maize seed inoculated with *M. anisopliae* showed increased stand density, stalk and foliage fresh weight, and yield (Kabaluk and Ericsson 2007). The authors attributed these effects to the control of wireworm (*Agriotes obscurus*) resulting in an increase in stand density and yield of corn. *M. anisopliae* contributed to the control of the rice leafroller (*Cnaphalocrocis medinalis)* after foliar spray application under field conditions (Hong et al. 2017). Suppressive effects of endophytic *Metarhizium* spp. have also been observed for larval *Plutella xylostella* (Batta 2013), aphids (Castrillo et al. 2011; de Faria and Wraight 2007; Dutta et al. 2015), wireworms (Reddy et al. 2014), cabbage maggot (*Delia radicum*) (Razinger et al. 2014), and the western corn rootworm (*Diabrotica v. virgifera*) (Rudeen et al. 2013). Under laboratory conditions, Mayerhofer et al. (2017) observed that *M. brunneum* colonization of tomato effectively controlled the growth of *Agriotes obscurus*. Endophytic colonization by *M. brunneum* helped strawberry plants tolerate an infestation by two-spotted spider mite (Dara and Dara 2015). In a detached leaf assay, colonization of maize after seed inoculation resulted in suppressed growth of black cutworm (*Agrotis ipsilon*) larvae compared to uninoculated plants (Ahmad et al. 2020b). Pest suppressive effects have been attributed to the systemic response of plants against herbivory (Lopez and Sword 2015; Lopez et al. 2014). Such herbivore growth suppressive effects could also be due to accumulation of anti-herbivory fungal metabolites and mycotoxins in plant tissues (Gurulingappa et al. 2010).

Although most reports on the interaction of EIPF and insects indicate negative effects on insect pest performance, in some cases, EIPF may increase the susceptibility of plants to insect pests and phytopathogens or have neutral effects. For example, Clifton et al. (2018) reported that seed inoculation of soybean with *M. brunneum* increased aphids but the effects of *B. bassiana* were neutral. Increased plant susceptibility to insects could possibly be due to decreased plant defenses to allow EIPF colonization, thus resulting in increased plant susceptibility.

The capability of *Metarhizium* spp. as an antagonist of various phytopathogens could contribute to crop protection from plant diseases (Lacey et al. 2015). *M. robertsii* protected haricot bean from root rot caused by *Fusarium solani* f. sp. *phaseoli* (Sasan and Bidochka 2013). In vitro, *M. brunneum* produced secondary metabolites with antagonistic activity against *Verticillium dahliae* and *Phytophthora megasperma* pathogens of olive, Olea europaea (Lozano-Tovar et al. 2017). *M. anisopliae* inoculation protected banana plants against *F. oxysporum* and the authors attributed these effects to the production of hydrolytic antifungal compounds (Picardal et al. 2019). In a laboratory bioassay, conidia and the cultural filtrate of *M. anisopliae* caused mortality of green peach aphids and showed antifungal properties against the phytopathogen, *Botrytis cinerea* (Yun et al. 2017).

Endophytic *Beauveria* has also been shown to suppress phytopathogens (Ownley et al. 2008). Endophytic colonization of cotton and tomato by *B. bassiana* exerted antagonistic effects against disease caused by *Pythium myriotylum* and *Rhizoctonia solani* (Ownley et al. 2008). Raad et al. (2019) reported that *B. bassiana* protected *A. thaliana* from *Sclerotinia sclerotiorum.* Other less-studied EIPF have been reported to suppress herbivores and phytopathogens. For example, endophytic colonization of pepper plants by *Isaria javanica* protected the pepper, *Capsicum* sp., plants from green peach aphids and from the fungal phytopathogens, *Colletotrichum gloeosporioides* and *Phytophthora capsici* (Kang et al. 2018).

13.7 Mechanisms of Plant Protection

EIPF employ various strategies to bypass plant host immunity, and the plant protective effects mediated by EIPF colonization may partially be due to priming of phytohormone-based defenses (Pieterse et al. 2012), enhanced production of secondary metabolites (Gibson et al. 2014; Khan et al. 2012; Shrivastava et al. 2015), organic compounds (Bitas et al. 2013), and accumulation of anti-herbivory fungal metabolites and mycotoxins in plant tissues (Gurulingappa et al. 2010). EIPF-mediated plant defense activation or priming effects may be elicited due to endophytic colonization, thus resulting in enhanced plant protection against pests and pathogens (Pieterse et al. 2012; Raad et al. 2019). Microbe-mediated priming activates the plant defenses slightly or transiently in the absence of stress. Later, when a plant perceives a challenging signal, primed plants can activate defense more quickly, with greater strength and in a more sustained manner compared to non-primed plants (Martinez-Medina et al. 2016).

Metarhizium spp. can produce secondary metabolites such as destruxins, cytochalasin, and swainsonine (Gibson et al. 2014; Golo et al. 2014; Pedras et al. 2002; Singh and Kaur 2014). *M. robertsii* produced multiple destruxins in artificial medium and endophytically colonized cowpea (*Vigna unguiculata*) but not in cucumber (*Cucumis sativus*) (Golo et al. 2014). Foliar inoculation of seedlings with *M. brunneum* resulted in the production of destruxins in potato and melon leaves (Resquín-Romero et al. 2016). Destruxins were detected in insects that were fed on *M. brunneum*-colonized tomato and melon (Garrido-Jurado et al. 2017; Resquín-Romero et al. 2016). Endophytic *M. brunneum* in olive produced secondary metabolites with antagonistic activity against the phytopathogens *V. dahliae* and *P. megasperma* (Lozano-Tovar et al. 2017). Chromatographic analysis showed that tyrosine betaine was detected from the conidial extract of *M. anisopliae* (Carollo et al. 2010).

Beauveria spp. produce various secondary metabolites, including beauvericin, beauveriolides, bassianolides, oxalic acid, and oosporein (Scharf et al. 2014; Strasser et al. 2000). Culture filtrates and conidial extracts of *B. bassiana* produced secondary metabolites that reduced the survival and fecundity of *Aphis gossypii* (Gurulingappa et al. 2010). Oosporein was detected in potato tubers following the application of a commercial formulation of *B. brongniartii* to potato fields (Seger et al. 2005).

Secondary metabolites are also produced by other EIPF genera such as *Lecanicillium*, *Isaria*, and *Fusarium* (Scharf et al. 2014; Zimmermann 2007a, 2007b, 2008). Bassianolides produced by *Lecanicillium* spp. have antifungal, insecticidal, antibacterial, and anti-tumor properties (Scharf et al. 2014). Mycelial extracts and culture filtrates of *L. lecanii* reduced the survival and fecundity of *Aphis gossypii* and authors attributed the effects to fungal secondary metabolites (Gurulingappa et al. 2010). Farinosone was detected in plants that were endophytically colonized by *I. farinosa* (Zimmermann 2008).

Plants can also produce secondary metabolites and their concentration and composition can vary in plant–EIPF interactions through effects on root exudates (Chaparro et al. 2014; Hartmann et al. 2009; Oldroyd 2013) and by modifying the abiotic environment (Bulgarelli et al. 2013). Root exudates contain different metabolites that play a role in modulating the composition of the root microbiome (Busby et al. 2017; Sasse et al. 2018). One of the functions of production and induction of plant secondary metabolites is to deter the herbivores. Plant secondary metabolites can affect the growth and development of EIPF in vitro and herbivores directly; however, some herbivores can utilize plant metabolites for their own benefit (Joshi et al. 2018). There is a lack of data about the *in planta* production of secondary metabolites by EIPF even if the metabolites have been produced by EIPF in vitro. Lack of accuracy in the detection of secondary metabolites produced by EIPF may in some cases be because the metabolites may be transient or in levels below our ability to detect them (Fan et al. 2017). Additionally, it may be difficult to differentiate whether the concentration and composition of metabolites is produced by the plant or EIPF in multitrophic associations.

13.8 Applications of EIPF in Sustainable Agriculture and Biotechnology

There is increasing interest in understanding the role of and potential for exploiting the soil microbiome, and of EIPF, specifically, to enhance plant productivity and tolerance to biotic and abiotic stress in agricultural systems. Plants can establish diverse associations with a variety of microbes, and developing multi-species microbiological consortium may help achieve effective pest management and better crop production. EIPF have the potential to provide beneficial ecosystem services where synthetic fertilizers and pesticides are not available, are too expensive, or are not allowed, such as in organic production systems (Kang et al. 2018). Indeed, various species of *Metarhizium*, *Beauveria* and *Isaria* have been developed commercially as biopesticides (Bing and Lewis 1991; Castrillo et al. 2011). In the United States, examples of commercial products registered for direct arthropod pest control in various settings include Met52® (*Metarhizium brunneum*); BotaniGard®, Mycotrol® and Aprehend® (*Beauveria bassiana*), and NoFly® and Pfr-97® (*Isaria fumosorosea*). The discovery of the multiple beneficial effects of EIPF, for example, as a biofertilizer, could be used to promote greater market demand for current and development of new EIPF-based products to improve the sustainability and productivity of crop production (de Faria and Wraight 2007; Mascarin and Jaronski 2016).

Many EIPF species have shown promising results in the laboratory in increasing tolerance of plants to insect pests and phytopathogens (Jaber and Ownley 2018). The impacts of EIPF as endophytes in tritrophic interactions are under-explored although their role in plant growth promotion and insect pathogenicity is now well-described for many plant species (Liao et al. 2017; Sasan and Bidochka 2012). Even so, little is known about how to consistently exploit these beneficial interactions in the field. Research focused on improvement of environmental persistence, competitiveness with other rhizosphere microorganisms, and improvements in direct and indirect virulence against target pests and efficiency of endophytic colonization of plants by EIPF must be expanded (Moonjely et al. 2016).

EIPF are commonly found in agricultural soils. In addition to research to understand underlying mechanisms of action for development of commercial products, there are opportunities to develop strategies to conserve endemic EIPF. Future research to understand the relevance of laboratory experiments to field settings where the environment is more variable are needed to more fully understand the role of EIPF in plant protection, growth promotion, and nutrient exchange in agricultural production systems, and if they can be used to benefit crop production in agroecosystems. We know relatively little about how EIPF–plant–stress interactions determine the frequency, specificity, and extent of EIPF–plant associations, their effects on plant growth promotion, and the persistence of these effects on crop plants in the field. Research on the influence of abiotic and biotic factors on the prevalence and recruitment of endophytes in general, and EIPF, in particular is needed because plant and soil characteristics vary in their ability to be managed. Farming practices that support beneficial rhizosphere and endophytic fungi could benefit crop yields and suppress pests. For example, if factors related to crop species or variety drive recruitment or conservation of EIPF, then plants with those characteristics could be used or bred into crops. Even though EIPF appear to confer many benefits in crop production, more research is needed to assess the risks for EIPF to cause damage to crops, including yield loss, and the environment, including risk of non-target effects (Seiedy et al. 2015). EIPF produce secondary metabolites that may be helpful in enhancing insect pathogenicity and in facilitating endophytic interaction with plants (Fan et al. 2017), but there is potential for development of EIPF beyond agriculture; for example, in medicine. The ecological roles of EIPF remain elusive and need further research in the field. Better understanding of the ecology and biology of EIPF and their association with plants will help achieve sustainable agriculture goals.

Acknowledgements This work was funded by USDA National Institute of Food and Agriculture, Organic Transitions (ORG) (No. 2016-51106-25715 and No. 2019-51106-30198), and Organic Farming Research Foundation (No. 232490).

Conflict of Interests The authors have no conflict of interests.

References

Acevedo FE, Rivera-Vega LJ, Chung SH, Ray S, Felton GW (2015) Cues from chewing insects—the intersection of DAMPs, HAMPs, MAMPs and effectors. Curr Opin Plant Biol 26:80–86

Ahern JR, Whitney KD (2014) Stereochemistry affects sesquiterpene lactone bioactivity against an herbivorous grasshopper. Chemoecology 24:35–39

Ahmad I, Zaib S (2020) Mighty microbes: plant growth promoting microbes in soil health and sustainable agriculture. In: Giri B, Varma A (eds) Soil health. Springer, Berlin, pp 243–264

Ahmad I, Jiménez-Gasco MM, Luthe DS, Barbercheck ME (2020a) Systemic colonization by *Metarhizium robertsii* enhances cover crop growth. J Fungi 6:64

Ahmad I, Jiménez-Gasco MM, Luthe DS, Shakeel SN, Barbercheck ME (2020b) Endophytic *Metarhizium robertsii* promotes maize growth, suppresses insect growth, and alters plant defense gene expression. Biol Control:104167

Akutse KS, Maniania NK, Fiaboe KKM, Van den Berg J, Ekesi S (2013) Endophytic colonization of *Vicia faba* and *Phaseolus vulgaris* (Fabaceae) by fungal pathogens and their effects on the life-history parameters of *Liriomyza huidobrensis* (Diptera: Agromyzidae). Fungal Ecol 6:293–301

Bais HP, Weir TL, Perry LG, Gilroy S, Vivanco JM (2006) The role of root exudates in rhizosphere interactions with plants and other organisms. Annu Rev Plant Biol 57:233–266

Bamisile BS, Dash CK, Akutse KS, Keppanan R, Wang L (2018) Fungal endophytes: beyond herbivore management. Front Microbiol 9:544

Batta YA (2013) Efficacy of endophytic and applied *Metarhizium anisopliae* (Metch.) Sorokin (Ascomycota: Hypocreales) against larvae of *Plutella xylostella* L. (Yponomeutidae: Lepidoptera) infesting *Brassica napus* plants. Crop Protect 44:128–134

Behie SW, Zelisko PM, Bidochka MJ (2012) Endophytic insect-parasitic fungi translocate nitrogen directly from insects to plants. Science 336:1576–1577

Behie SW, Jones SJ, Bidochka MJ (2015) Plant tissue localization of the endophytic insect pathogenic fungi *Metarhizium* and *Beauveria*. Fungal Ecol 13:112–119

Behie SW, Moreira CC, Sementchoukova I, Barelli L, Zelisko PM, Bidochka MJ (2017) Carbon translocation from a plant to an insect-pathogenic endophytic fungus. Nat Commun 8:14245

Berg G, Smalla K (2009) Plant species and soil type cooperatively shape the structure and function of microbial communities in the rhizosphere. FEMS Microbiol Ecol 68:1–13

Bing LA, Lewis LC (1991) Suppression of *Ostrinia nubilalis* (Hübner) (Lepidoptera: Pyralidae) by endophytic *Beauveria bassiana* (Balsamo) Vuillemin. Environ Entomol 20:1207–1211

Bitas V, Kim H-S, Bennett JW, Kang S (2013) Sniffing on microbes: diverse roles of microbial volatile organic compounds in plant health. Mol Plant-Microbe Interact 26:835–843

Bruck DJ (2009) Fungal entomopathogens in the rhizosphere. In: Roy HE, Vega FE, Chandler D, Goettel MS, Pell J, Wajnberg E (eds) The ecology of fungal entomopathogens. Springer, Dordrecht, pp 103–112

Bulgarelli D, Schlaeppi K, Spaepen S, Ver Loren van Themaat E, Schulze-Lefert P (2013) Structure and functions of the bacterial microbiota of plants. Annu Rev Plant Biol 64:807–838

Busby PE, Soman C, Wagner MR, Friesen ML, Kremer J, Bennett A, Morsy M, Eisen JA, Leach JE, Dangl JL (2017) Research priorities for harnessing plant microbiomes in sustainable agriculture. PLoS Biol 15:e2001793

Carollo CA, Calil ALA, Schiave LA, Guaratini T, Roberts DW, Lopes NP, Braga GU (2010) Fungal tyrosine betaine, a novel secondary metabolite from conidia of entomopathogenic *Metarhizium* spp. fungi. Fungal Biol 114:473–480

Castrillo LA, Griggs MH, Ranger CM, Reding ME, Vandenberg JD (2011) Virulence of commercial strains of *Beauveria bassiana* and *Metarhizium brunneum* (Ascomycota: Hypocreales) against adult *Xylosandrus germanus* (Coleoptera: Curculionidae) and impact on brood. Biol Control 58:121–126

Cen K, Li B, Lu Y, Zhang S, Wang C (2017) Divergent LysM effectors contribute to the virulence of *Beauveria bassiana* by evasion of insect immune defenses. PLoS Path 13:e1006604

Chaparro JM, Sheflin AM, Manter DK, Vivanco JM (2012) Manipulating the soil microbiome to increase soil health and plant fertility. Biol Fertil Soils 48:489–499

Chaparro JM, Badri DV, Vivanco JM (2014) Rhizosphere microbiome assemblage is affected by plant development. ISME J 8:790–803

Cherry AJ, Banito A, Djegui D, Lomer C (2004) Suppression of the stem-borer *Sesamia calamistis* (Lepidoptera; Noctuidae) in maize following seed dressing, topical application and stem injection with African isolates of *Beauveria bassiana*. Int J Pest Manage 50:67–73

Clifton EH, Jaronski ST, Coates BS, Hodgson EW, Gassmann AJ (2018) Effects of endophytic entomopathogenic fungi on soybean aphid and identification of *Metarhizium* isolates from agricultural fields. PLoS One 13:e0194815

Dara SK (2013) Entomopathogenic fungus *Beauveria bassiana* promotes strawberry plant growth and health. UCANR eJ Strawberries Vegetables 30

Dara SK (2019) Non-entomopathogenic roles of entomopathogenic fungi in promoting plant health and growth. Insects 10:277

Dara SK, Dara SR (2015) Entomopathogenic fungus *Beauveria bassiana* endophytically colonizes strawberry plants. UCANR eJ Strawberries Vegetables 17

Dastogeer KMG (2018) Influence of fungal endophytes on plant physiology is more pronounced under stress than well-watered conditions: a meta-analysis. Planta 248:1403–1416

de Faria MR, Wraight SP (2007) Mycoinsecticides and mycoacaricides: a comprehensive list with worldwide coverage and international classification of formulation types. Biol Control 43:237–256

De Vos M, Van Oosten VR, Van Poecke RM, Van Pelt JA, Pozo MJ, Mueller MJ, Buchala AJ, Metraux JP, Van Loon LC, Dicke M, Pieterse CM (2005) Signal signature and transcriptome changes of *Arabidopsis* during pathogen and insect attack. Mol Plant-Microbe Interact 18:923–937

Dumbrell AJ, Nelson M, Helgason T, Dytham C, Fitter AH (2010) Relative roles of niche and neutral processes in structuring a soil microbial community. ISME J 4:337–345

Dutta P, Kaushik H, Bhawmick P, Puzari K, Hazarika G (2015) *Metarhizium anisopliae* as endophyte has the ability of plant growth enhancement. Int J Curr Res 7:14300–14304

Elena GJ, Beatriz PJ, Alejandro P, Lecuona R (2011) *Metarhizium anisopliae* (Metschnikoff) Sorokin promotes growth and has endophytic activity in tomato plants. Adv Biol Res 5:22–27

Fan Y, Liu X, Keyhani NO, Tang G, Pei Y, Zhang W, Tong S (2017) Regulatory cascade and biological activity of *Beauveria bassiana* oosporein that limits bacterial growth after host death. Proc Natl Acad Sci 114:E1578–E1586

Fisher JJ, Rehner SA, Bruck DJ (2011) Diversity of rhizosphere associated entomopathogenic fungi of perennial herbs, shrubs and coniferous trees. J Invertebr Pathol 106:289–295

Fomina M, Alexander IJ, Hillier S, Gadd G (2004) Zinc phosphate and pyromorphite solubilization by soil plant-symbiotic fungi. Geomicrobiol J 21:351–366

Frey SD, Lee J, Melillo JM, Six J (2013) The temperature response of soil microbial efficiency and its feedback to climate. Nat Clim Chang 3:395–398

Gao Q, Jin K, Ying S-H, Zhang Y, Xiao G, Shang Y, Duan Z, Hu X, Xie X-Q, Zhou G (2011) Genome sequencing and comparative transcriptomics of the model entomopathogenic fungi *Metarhizium anisopliae* and *M. acridum*. PLoS Genet 7:e1001264

Garbeva P, van Veen JA, van Elsas JD (2004) Microbial diversity in soil: selection microbial populations by plant and soil type and implications for disease suppressiveness. Annu Rev Phytopathol 42:243–270

Gargallo-Garriga A, Preece C, Sardans J, Oravec M, Urban O, Peñuelas J (2018) Root exudate metabolomes change under drought and show limited capacity for recovery. Sci Rep 8:1–15

Garrido-Jurado I, Resquín-Romero G, Amarilla S, Ríos-Moreno A, Carrasco L, Quesada-Moraga E (2017) Transient endophytic colonization of melon plants by entomopathogenic fungi after foliar application for the control of *Bemisia tabaci Gennadius* (Hemiptera: Aleyrodidae). J Pest Sci 90:319–330

Genre A, Chabaud M, Balzergue C, Puech-Pagès V, Novero M, Rey T, Fournier J, Rochange S, Bécard G, Bonfante P (2013) Short-chain chitin oligomers from arbuscular mycorrhizal fungi trigger nuclear C a2+ spiking in *Medicago truncatula* roots and their production is enhanced by strigolactone. New Phytol 198:190–202

Gibson DM, Donzelli BG, Krasnoff SB, Keyhani NO (2014) Discovering the secondary metabolite potential encoded within entomopathogenic fungi. Nat Prod Rep 31:1287–1305

Glazebrook J (2005) Contrasting mechanisms of defense against biotrophic and necrotrophic pathogens. Annu Rev Phytopathol 43:205–227

Glynou K, Ali T, Buch AK, Haghi Kia S, Ploch S, Xia X, Çelik A, Thines M, Maciá-Vicente JG (2016) The local environment determines the assembly of root endophytic fungi at a continental scale. Environ Microbiol 18:2418–2434

Golo PS, Gardner DR, Grilley MM, Takemoto JY, Krasnoff SB, Pires MS, Fernandes EK, Bittencourt VR, Roberts DW (2014) Production of destruxins from *Metarhizium* spp. fungi in artificial medium and in endophytically colonized cowpea plants. PLoS One 9:e104946

Gómez-Vidal S, Salinas J, Tena M, Lopez-Llorca LV (2009) Proteomic analysis of date palm (*Phoenix dactylifera* L.) responses to endophytic colonization by entomopathogenic fungi. Electrophoresis 30:2996–3005

Gray SB, Brady SM (2016) Plant developmental responses to climate change. Dev Biol 419:64–77

Guo J, McCulley RL, McNear DH Jr (2015) Tall fescue cultivar and fungal endophyte combinations influence plant growth and root exudate composition. Front Plant Sci 6:183

Gurulingappa P, Sword GA, Murdoch G, McGee PA (2010) Colonization of crop plants by fungal entomopathogens and their effects on two insect pests when in planta. Biol Control 55:34–41

Hacquard S, Garrido-Oter R, González A, Spaepen S, Ackermann G, Lebeis S, McHardy AC, Dangl JL, Knight R, Ley R (2015) Microbiota and host nutrition across plant and animal kingdoms. Cell Host Microbe 17:603–616

Hagerty SB, van Groenigen KJ, Allison SD, Hungate BA, Schwartz E, Koch GW, Kolka RK, Dijkstra P (2014) Accelerated microbial turnover but constant growth efficiency with warming in soil. Nat Clim Chang 4:903–906

Hao K, Wang F, Nong X, McNeill MR, Liu S, Wang G, Cao G, Zhang Z (2017) Response of peanut *Arachis hypogaea* roots to the presence of beneficial and pathogenic fungi by transcriptome analysis. Sci Rep 7:1–15

Hartmann A, Schmid M, van Tuinen D, Berg G (2009) Plant-driven selection of microbes. Plant Soil 321:235–257

Heinz KM, Harding PA, Ek-Ramos MJ, Hernandez H, Krauter PC, Sword GA (2018) Fungal endophytes in knock out® rose and performance effects of entomopathogens on marigold and zinnia. HortScience 53:1791–1798

Herrera-Medina MJ, Steinkellner S, Vierheilig H, Ocampo Bote JA, Garcia Garrido J (2007) Abscisic acid determines arbuscule development and functionality in the tomato arbuscular mycorrhiza. New Phytol 175:554–564

Hong M, Peng G, Keyhani NO, Xia Y (2017) Application of the entomogenous fungus, *Metarhizium anisopliae*, for leafroller (*Cnaphalocrocis medinalis*) control and its effect on rice phyllosphere microbial diversity. Appl Microbiol Biotechnol 101:6793–6807

Hu S, Bidochka MJ (2019) Root colonization by endophytic insect-pathogenic fungi. J Appl Microbiol

Hu L, Robert CA, Cadot S, Zhang X, Ye M, Li B, Manzo D, Chervet N, Steinger T, Van Der Heijden MG (2018) Root exudate metabolites drive plant-soil feedbacks on growth and defense by shaping the rhizosphere microbiota. Nat Commun 9:1–13

Jaber LR, Araj SE (2018) Interactions among endophytic fungal entomopathogens (Ascomycota: Hypocreales), the green peach aphid *Myzus persicae Sulzer* (Homoptera: Aphididae), and the aphid endoparasitoid *Aphidius colemani Viereck* (Hymenoptera: Braconidae). Biol Control 116:53–61

Jaber LR, Enkerli J (2016) Effect of seed treatment duration on growth and colonization of *Vicia faba* by endophytic *Beauveria bassiana* and *Metarhizium brunneum*. Biol Control 103:187–195

Jaber LR, Ownley BH (2018) Can we use entomopathogenic fungi as endophytes for dual biological control of insect pests and plant pathogens? Biol Control 116:36–45

Joshi S, Sahgal M, Sahu S, Prakash A (2018) Fungal endophytes and their secondary metabolites: role in sustainable agriculture. In: Gehlot P, Singh J (eds) Fungi and their role in sustainable development: current perspectives. Springer, Berlin, pp 121–146

Kabaluk JT, Ericsson JD (2007) *Metarhizium anisopliae* seed treatment increases yield of field corn when applied for wireworm control. Agron J 99:1377–1381

Kang BR, Han JH, Kim JJ, Kim YC (2018) Dual biocontrol potential of the entomopathogenic fungus, Isaria javanica, for both aphids and plant fungal pathogens. Mycobiology 46:440–447

Karasov TL, Chae E, Herman JJ, Bergelson J (2017) Mechanisms to mitigate the trade-off between growth and defense. Plant Cell 29:666–680

Karhu K, Auffret MD, Dungait JA, Hopkins DW, Prosser JI, Singh BK, Subke JA, Wookey PA, Agren GI, Sebastia MT, Gouriveau F, Bergkvist G, Meir P, Nottingham AT, Salinas N, Hartley IP (2014) Temperature sensitivity of soil respiration rates enhanced by microbial community response. Nature 513:81–84

Khan AL, Hamayun M, Khan SA, Kang SM, Shinwari ZK, Kamran M, Ur Rehman S, Kim JG, Lee IJ (2012) Pure culture of *Metarhizium anisopliae* LHL07 reprograms soybean to higher growth and mitigates salt stress. World J Microbiol Biotechnol 28:1483–1494

Koricheva J, Gange AC, Jones T (2009) Effects of mycorrhizal fungi on insect herbivores: a meta-analysis. Ecology 90:2088–2097

Krasnoff SB, Keresztes I, Donzelli BG, Gibson DM (2014) Metachelins, mannosylated and N-oxidized coprogen-type siderophores from *Metarhizium robertsii*. J Nat Prod 77:1685–1692

Krell V, Jakobs-Schoenwandt D, Vidal S, Patel AV (2018) Encapsulation of *Metarhizium brunneum* enhances endophytism in tomato plants. Biol Control 116:62–73

Kumar CS, Jacob T, Devasahayam S, Thomas S, Geethu C (2018) Multifarious plant growth promotion by an entomopathogenic fungus *Lecanicillium psalliotae*. Microbiol Res 207:153–160

Lacey L, Grzywacz D, Shapiro-Ilan D, Frutos R, Brownbridge M, Goettel M (2015) Insect pathogens as biological control agents: back to the future. J Invertebr Pathol 132:1–41

Lareen A, Burton F, Schafer P (2016) Plant root-microbe communication in shaping root microbiomes. Plant Mol Biol 90:575–587

Lê Van A, Quaiser A, Duhamel M, Michon-Coudouel S, Dufresne A, Vandenkoornhuyse P (2017) Ecophylogeny of the endospheric root fungal microbiome of co-occurring Agrostis stolonifera. PeerJ 5:e3454

Lee Y-S, Kim YC (2019) Tobacco growth promotion by the Entomopathogenic fungus, *Isaria javanica* pf185. Mycobiology 47:126–133

Liao X, Fang W, Lin L, Lu H-L, St. Leger RJ (2013) *Metarhizium robertsii* produces an extracellular invertase (MrINV) that plays a pivotal role in rhizospheric interactions and root colonization. PLoS One 8:e78118

Liao XG, O'Brien TR, Fang WG, St. Leger RJ (2014) The plant beneficial effects of *Metarhizium* species correlate with their association with roots. Appl Microbiol Biotechnol 98:7089–7096

Liao XG, Lovett B, Fang WG, St. Leger RJ (2017) *Metarhizium robertsii* produces indole-3-acetic acid, which promotes root growth in *Arabidopsis* and enhances virulence to insects. Microbiology-Sgm 163:980–991

Lomer CJ, Bateman RP, Johnson DL, Langewald J, Thomas M (2001) Biological control of locusts and grasshoppers. Annu Rev Entomol 46:667–702

Lopez DC, Sword GA (2015) The endophytic fungal entomopathogens *Beauveria bassiana* and *Purpureocillium lilacinum* enhance the growth of cultivated cotton (*Gossypium hirsutum*) and negatively affect survival of the cotton bollworm (*Helicoverpa zea*). Biol Control 89:53–60

Lopez DC, Zhu-Salzman K, Ek-Ramos MJ, Sword GA (2014) The entomopathogenic fungal endophytes *Purpureocillium lilacinum* (formerly *Paecilomyces lilacinus*) and *Beauveria bassiana* negatively affect cotton aphid reproduction under both greenhouse and field conditions. PLoS One 9:e103891

López-Ráez JA, Shirasu K, Foo E (2017) Strigolactones in plant interactions with beneficial and detrimental organisms: the Yin and Yang. Trends Plant Sci 22:527–537

Lozano-Tovar M, Garrido-Jurado I, Quesada-Moraga E, Raya-Ortega M, Trapero-Casas A (2017) *Metarhizium brunneum* and *Beauveria bassiana* release secondary metabolites with antagonistic activity against *Verticillium dahliae* and *Phytophthora megasperma* olive pathogens. Crop Protect 100:186–195

Maillet F, Poinsot V, André O, Puech-Pagès V, Haouy A, Gueunier M, Cromer L, Giraudet D, Formey D, Niebel A (2011) Fungal lipochitooligosaccharide symbiotic signals in arbuscular mycorrhiza. Nature 469:58–63

Martinez-Medina A, Flors V, Heil M, Mauch-Mani B, Pieterse CM, Pozo MJ, Ton J, van Dam NM, Conrath U (2016) Recognizing plant defense priming. Trends Plant Sci 21:818–822

Mascarin GM, Jaronski ST (2016) The production and uses of *Beauveria bassiana* as a microbial insecticide. World J Microbiol Biotechnol 32:177

Massalha H, Korenblum E, Tholl D, Aharoni A (2017) Small molecules below-ground: the role of specialized metabolites in the rhizosphere. Plant J 90:788–807

Mayerhofer J, Eckard S, Hartmann M, Grabenweger G, Widmer F, Leuchtmann A, Enkerli J (2017) Assessing effects of the entomopathogenic fungus *Metarhizium brunneum* on soil microbial communities in *Agriotes* spp. biological pest control. FEMS Microbiol Ecol 93:fix117

McKinnon AC, Glare TR, Ridgway HJ, Mendoza-Mendoza A, Holyoake A, Godsoe WK, Bufford JL (2018) Detection of the entomopathogenic fungus *Beauveria bassiana* in the rhizosphere of wound-stressed *Zea mays* plants. Front Microbiol 9:1161

Mendes R, Garbeva P, Raaijmakers JM (2013) The rhizosphere microbiome: significance of plant beneficial, plant pathogenic, and human pathogenic microorganisms. FEMS Microbiol Rev 37:634–663

Meyling NV, Eilenberg J (2007) Ecology of the entomopathogenic fungi *Beauveria bassiana* and *Metarhizium anisopliae* in temperate agroecosystems: potential for conservation biological control. Biol Control 43:145–155

Moonjely S, Barelli L, Bidochka M (2016) Insect pathogenic fungi as endophytes. In: Lovett B, St. Leger RJ (eds) Advances in genetics: genetics and molecular biology of entomopathogenic fungi, vol 94. Elsevier, Amsterdam, pp 107–135

Mori N, Nishiuma K, Sugiyama T, Hayashi H, Akiyama K (2016) Carlactone-type strigolactones and their synthetic analogues as inducers of hyphal branching in arbuscular mycorrhizal fungi. Phytochemistry 130:90–98

Nishi O, Sato H (2019) Isolation of *Metarhizium* spp. from rhizosphere soils of wild plants reflects fungal diversity in soil but not plant specificity. Mycology 10:22–31

Oldroyd GED (2013) Speak, friend, and enter: signalling systems that promote beneficial symbiotic associations in plants. Nat Rev Microbiol 11:252–263

Ownley BH, Griffin MR, Klingeman WE, Gwinn KD, Moulton JK, Pereira RM (2008) *Beauveria bassiana*: endophytic colonization and plant disease control. J Invertebr Pathol 98:267–270

Parsa S, Ortiz V, Gomez-Jimenez MI, Kramer M, Vega FE (2018) Root environment is a key determinant of fungal entomopathogen endophytism following seed treatment in the common bean, *Phaseolus vulgaris*. Biol Control 116:74–81

Paungfoo-Lonhienne C, Yeoh YK, Kasinadhuni NRP, Lonhienne TG, Robinson N, Hugenholtz P, Ragan MA, Schmidt S (2015) Nitrogen fertilizer dose alters fungal communities in sugarcane soil and rhizosphere. Sci Rep 5:8678

Pedras MSC, Nycholat CM, Montaut S, Xu Y, Khan AQ (2002) Chemical defenses of crucifers: elicitation and metabolism of phytoalexins and indole-3-acetonitrile in brown mustard and turnip. Phytochemistry 59:611–625

Picardal JP, Tundag ED, Picardal MT, Goc-ong G (2019) Antagonistic activity of *Metarhizium anisopliae* (Metschnikoff) against Phytopathogenic *Fusarium oxysporum* f. sp. cubense (Schlecht.) as a biological control. CNU J Higher Educ 13:25–33

Pieterse CM, Van der Does D, Zamioudis C, Leon-Reyes A, Van Wees SC (2012) Hormonal modulation of plant immunity. Annu Rev Cell Dev Biol 28:489–521

Pieterse CM, Zamioudis C, Berendsen RL, Weller DM, Van Wees SC, Bakker PA (2014) Induced systemic resistance by beneficial microbes. Annu Rev Phytopathol 52:347–375

Pineda A, Dicke M, Pieterse CM, Pozo MJ (2013) Beneficial microbes in a changing environment: are they always helping plants to deal with insects? Funct Ecol 27:574–586

Raad M, Glare TR, Brochero HL, Müller C, Rostás M (2019) Transcriptional reprogramming of *Arabidopsis thaliana* defence pathways by the entomopathogen *Beauveria bassiana* correlates with resistance against a fungal pathogen but not against insects. Front Microbiol 10:615

Ramos Y, Portal O, Lysøe E, Meyling NV, Klingen I (2017) Diversity and abundance of *Beauveria bassiana* in soils, stink bugs and plant tissues of common bean from organic and conventional fields. J Invertebr Pathol 150:114–120

Raya-Díaz S, Quesada-Moraga E, Barrón V, Del Campillo MC, Sánchez-Rodríguez AR (2017a) Redefining the dose of the entomopathogenic fungus *Metarhizium brunneum* (Ascomycota, Hypocreales) to increase Fe bioavailability and promote plant growth in calcareous and sandy soils. Plant Soil 418:387–404

Raya-Díaz S, Sanchez-Rodriguez AR, Segura-Fernandez JM, del Campillo MD, Quesada-Moraga E (2017b) Entomopathogenic fungi-based mechanisms for improved Fe nutrition in sorghum plants grown on calcareous substrates. PLoS One 12:e0185903

Razinger J, Lutz M, Schroers HJ, Urek G, Grunder J (2014) Evaluation of insect associated and plant growth promoting fungi in the control of cabbage root flies. J Econ Entomol 107:1348–1354

Razinger J, Lutz M, Grunder J, Urek G (2018) Laboratory investigation of cauliflower–fungus–insect interactions for cabbage maggot control. J Econ Entomol 111:2578–2584

Reddy GV, Tangtrakulwanich K, Wu S, Miller JH, Ophus VL, Prewett J, Jaronski ST (2014) Evaluation of the effectiveness of entomopathogens for the management of wireworms (Coleoptera: Elateridae) on spring wheat. J Invertebr Pathol 120:43–49

Reid ML, Purcell J (2011) Condition-dependent tolerance of monoterpenes in an insect herbivore. Arthropod-Plant Int 5:331–337

Reinhardt D (2007) Programming good relations—development of the arbuscular mycorrhizal symbiosis. Curr Opin Plant Biol 10:98–105

Ren A, Li C, Gao Y (2011) Endophytic fungus improves growth and metal uptake of *Lolium arundinaceum* Darbyshire ex. Schreb. Int J Phytoremediation 13:233–243

Resquín-Romero G, Garrido-Jurado I, Delso C, Ríos-Moreno A, Quesada-Moraga E (2016) Transient endophytic colonizations of plants improve the outcome of foliar applications of mycoinsecticides against chewing insects. J Invertebr Pathol 136:23–31

Roberts DW, St. Leger RJ (2004) *Metarhizium* spp., cosmopolitan insect-pathogenic fungi: mycological aspects. Adv Appl Microbiol 54:1–70

Rosier A, Bishnoi U, Lakshmanan V, Sherrier DJ, Bais HP (2016) A perspective on inter-kingdom signaling in plant–beneficial microbe interactions. Plant Mol Biol 90:537–548

Rudeen ML, Jaronski ST, Petzold-Maxwell JL, Gassmann AJ (2013) Entomopathogenic fungi in cornfields and their potential to manage larval western corn rootworm *Diabrotica virgifera virgifera*. J Invertebr Pathol 114:329–332

Russo ML, Pelizza S, Vianna M, Allegrucci N, Cabello MN, Toledo AV, Mourelos C, Scorsetti AC (2018) Effect of endophytic entomopathogenic fungi on soybean *Glycine max* (L.) Merr. growth and yield. J King Saud Univ Sci 31:728–736

Russo ML, Scorsetti AC, Vianna MF, Cabello M, Ferreri N, Pelizza S (2019) Endophytic effects of *Beauveria bassiana* on corn (*Zea mays*) and its herbivore, Rachiplusia nu (lepidoptera: Noctuidae). Insects 10:110

Saikkonen K, Gundel PE, Helander M (2013) Chemical ecology mediated by fungal endophytes in grasses. J Chem Ecol 39:962–968

Sasan RK, Bidochka MJ (2012) The insect-pathogenic fungus *Metarhizium robertsii* (Clavicipitaceae) is also an endophyte that stimulates plant root development. Am J Bot 99:101–107

Sasan RK, Bidochka MJ (2013) Antagonism of the endophytic insect pathogenic fungus *Metarhizium robertsii* against the bean plant pathogen *Fusarium solani* f. sp phaseoli. Can J Plant Pathol 35:288–293

Sasse J, Martinoia E, Northen T (2018) Feed your friends: do plant exudates shape the root microbiome? Trends Plant Sci 23:25–41

Scharf DH, Heinekamp T, Brakhage AA (2014) Human and plant fungal pathogens: the role of secondary metabolites. PLoS Pathog 10:e1003859

Seger C, Längle T, Pernfuss B, Stuppner H, Strasser H (2005) High-performance liquid chromatography-diode array detection assay for the detection and quantification of the *Beauveria* metabolite oosporein from potato tubers. J Chromatogr 1092:254–257

Seiedy M, Tork M, Deyhim F (2015) Effect of the entomopathogenic fungus *Beauveria bassiana* on the predatory mite *Amblyseius swirskii* (Acari: Phytoseiidae) as a non-target organism. Syst Appl Acarol 20:241–250

Shrivastava G, Ownley BH, Augé RM, Toler H, Dee M, Vu A, Köllner TG, Chen F (2015) Colonization by arbuscular mycorrhizal and endophytic fungi enhanced terpene production in tomato plants and their defense against a herbivorous insect. Symbiosis 65:65–74

Singh D, Kaur G (2014) The antileukaemic cell cycle regulatory activities of swainsonine purified from *Metarhizium anisopliae* fermentation broth. Nat Prod Res 28:2044–2047

Smith FA, Grace EJ, Smith SE (2009) More than a carbon economy: nutrient trade and ecological sustainability in facultative arbuscular mycorrhizal symbioses. New Phytol 182:347–358

Spence C, Bais H (2013) Probiotics for plants: rhizospheric microbiome and plant fitness. Mol Microb Ecol Rhizosph 1:713–721

Steinwender BM, Enkerli J, Widmer F, Eilenberg J, Meyling NV (2011) Molecular diversity of the *Metarhizium anisopliae* lineage in an agricultural field. IOBC/WPRS Bull 66:113–115

Strasser H, Vey A, Butt TM (2000) Are there any risks in using entomopathogenic fungi for pest control, with particular reference to the bioactive metabolites of *Metarhizium*, *Tolypocladium* and *Beauveria* species? Biocontrol Sci Tech 10:717–735

Sullivan TJ, Rodstrom J, Vandop J, Librizzi J, Graham C, Schardl CL, Bultman TL (2007) Symbiont-mediated changes in *Lolium arundinaceum* inducible defenses: evidence from changes in gene expression and leaf composition. New Phytol 176:673–679

Tefera T, Vidal S (2009) Effect of inoculation method and plant growth medium on endophytic colonization of sorghum by the entomopathogenic fungus *Beauveria bassiana*. BioControl 54:663–669

Thaler JS, Humphrey PT, Whiteman NK (2012) Evolution of jasmonate and salicylate signal crosstalk. Trends Plant Sci 17:260–270

Tiago PV, de Oliveira NT, Lima EADA (2014) Biological insect control using *Metarhizium anisopliae*: morphological, molecular, and ecological aspects. Cienc Rural 44:645–651

Uren NC (2000) Types, amounts and possible functions of compounds released into the rhizosphere by soil grown plants. In: Pinton R, Varani Z, Nanniperi P (eds) The Rhizosphere: biochemistry, and organic substances at the soil interface. Marcel Dekker Inc, New York, pp 19–40

Van Hecke MM, Treonis AM, Kaufman JR (2005) How does the fungal endophyte *Neotyphodium coenophialum* affect tall fescue (*Festuca arundinacea*) rhizodeposition and soil microorganisms? Plant Soil 275:101–109

Vega FE (2018) The use of fungal entomopathogens as endophytes in biological control: a review. Mycologia 110:4–30

Wagner BL, Lewis LC (2000) Colonization of corn, *Zea mays*, by the entomopathogenic fungus *Beauveria bassiana*. Appl Environ Microbiol 66:3468–3473

Woodward AW, Bartel B (2005) Auxin: regulation, action, and interaction. Ann Bot 95:707–735

Wyrebek M, Huber C, Sasan RK, Bidochka MJ (2011) Three sympatrically occurring species of *Metarhizium* show plant rhizosphere specificity. Microbiol SGM 157:2904–2911

Yan L, Zhu J, Zhao X, Shi J, Jiang C, Shao D (2019) Beneficial effects of endophytic fungi colonization on plants. Appl Microbiol Biotechnol 103:3327–3340

Yun H-G, Kim D-J, Gwak W-S, Shin T-Y, Woo S-D (2017) Entomopathogenic fungi as dual control agents against both the pest *Myzus persicae* and phytopathogen *Botrytis cinerea*. Mycobiology 45:192–198

Zeilinger S, Gupta VK, Dahms TE, Silva RN, Singh HB, Upadhyay RS, Gomes EV, Tsui CK-M, Nayak SC (2016) Friends or foes? Emerging insights from fungal interactions with plants. FEMS Microbiol Rev 40:182–207

Zhang D-X, Nagabhyru P, Schardl CL (2009) Regulation of a chemical defense against herbivory produced by symbiotic fungi in grass plants. Plant Physiol 150:1072–1082

Zhang X, Lei L, Lai J, Zhao H, Song W (2018) Effects of drought stress and water recovery on physiological responses and gene expression in maize seedlings. BMC Plant Biol 18:68

Zimmermann G (2007a) Review on safety of the entomopathogenic fungi *Beauveria bassiana* and *Beauveria brongniartii*. Biocontrol Sci Tech 17:553–596

Zimmermann G (2007b) Review on safety of the entomopathogenic fungus *Metarhizium anisopliae*. Biocontrol Sci Tech 17:879–920

Zimmermann G (2008) The entomopathogenic fungi *Isaria farinosa* (formerly *Paecilomyces farinosus*) and the *Isaria fumosorosea* species complex (formerly *Paecilomyces fumosoroseus*): biology, ecology and use in biological control. Biocontrol Sci Tech 18:865–901

Zolla G, Bakker MG, Badri DV, Chaparro JM, Sheflin AM, Manter DK, Vivanco J (2013) Understanding root–microbiome interactions. Mol Microb Ecol Rhizosph 1:743–754

Biological Overview and Adaptability Strategies of Tamarix Plants, *T. articulata* and *T. gallica* to Abiotic Stress

14

Karima Bencherif, Fares Trodi, Mohamed Hamidi, Yolande Dalpè, and Anissa Lounès Hadj-Sahraoui

Abstract

Terrestrial plants are often found in extreme environmental conditions, such as water deficiency, unbalanced temperature, high salinity, and soil pollution. The study of plants under stress conditions can help to find a solution in the context of biodiversity conservation. *Tamarix* genus contains more than 85 species among them *Tamarix articulata* (*aphylla, orientalis*) and *Tamarix gallica* found under natural stresses. The two species represent a great ubiquity in the Algerian area whether under drought, soils salinity, calcareous soils, and polluted soils. They represent typical thermo-xerophytes plants. In order to studies the comportment of *Tamarix articulata* and *Tamarix gallica* in the stressed Algerian area we want to define in this chapter the biological strategies adapted by these two species under stresses areas morphologically, biochemically, and physiologically. Ions, heavy metals, and pollutants can be taken up by *Tamarix* species; this selective

K. Bencherif (✉)
Faculté des Sciences de la Nature et de la Vie. Université de Djelfa, Djelfa, Algeria

Université du Littoral Côte d'Opale, Unité de Chimie Environnementale et Interactions sur le Vivant (UCEIV), Calais Cedex, France

F. Trodi
Faculté des Sciences de la Nature et de la Vie. Université de Djelfa, Djelfa, Algeria

M. Hamidi
Faculté des Sciences de la Nature et de la Vie. Université de Djelfa, Djelfa, Algeria

Laboratoire de chimie organique et de substances naturelles, Université de Djelfa, Djelfa, Algeria

Y. Dalpè
Agriculture et agroalimentaire Canada, Ottawa, ON, Canada

A. L. Hadj-Sahraoui
Université du Littoral Côte d'Opale, Unité de Chimie Environnementale et Interactions sur le Vivant (UCEIV), Calais Cedex, France

B. Giri, M. P. Sharma (eds.), *Plant Stress Biology*,
https://doi.org/10.1007/978-981-15-9380-2_14

absorption strategy was detailed in addition to morphological strategies to counteract erosion and drought stresses. Moreover, chemicals compounds such as polyphenols produced by *Tamarix* itself as a response to abiotic stresses are examined. Moreover, the role of arbuscular mycorrhizal fungi as biological tools to alleviate abiotic stresses was underlined. These responses collectively determine different strategies to overcome abiotic stress and open the ways to exploit the genes responsible for resistance strategies and to transfer them into other agricultural plants to improve production in stressed areas.

Keywords

Abiotic stresses · Adaptive strategies response · Polyphenols · Reactive oxygen species · Catalase activity

14.1 Introduction

Plants growing in field conditions are subjected to variety of environmental stress conditions such as desertification, salinity, toxicity, drought, heat, nutrient limitation, inundation, sandstorm, and frost (Kuzminsky et al. 2014). To counteract these abiotic stresses plant species have distinctly different adaptation strategies (Han et al. 2013). For that, it seems interesting to guide researches towards resistant plants, such as thermo-xerophytes, heliophyles, and halophytic species (Khabtane and Rahmoun 2012) in order to exploit their mechanism of adaptation to obtain ecological balance stability and to enhance economic productivity. Among these plants, *Tamarix* species have been reported to be highly tolerant to many abiotic stresses (Han et al. 2013), which makes exploration of their species diversity and their natural tolerance to some particular stress highly interesting (Kuzminsky et al. 2014). In fact, riparian species from the Tamaricaceae family are little exploited despite their resistance to drought stress and their capacity for stabilization of the banks (Lavaine et al. 2011a). However, *Tamarix* species possesses different mechanisms than those currently known in other plant that allows them to grow and prosper in different abiotic stress conditions (Kuzminsky et al. 2014; Guallar 2019), with some specific genes which are expressed to overcome adverse conditions (Guallar 2019). Moreover, in Mediterranean region, about eleven *Tamarix* species are recorded, occupying coastal dunes (Kuzminsky et al. 2014), Saharan areas (Baum 1978), and steppic regions (Bencherif et al. 2016). The most widespread species are *Tamarix articulata* (*aphylla*), *Tamarix gallica,* and *Tamarix africana* (Kaabech and Benkheira 2000). In Algerian arid and semiarid areas *Tamarix articulata* and *Tamarix gallica* are found on soils generally made of course materials and poor in organic matter such as those found on the gravel-sandy banks of the edge of rivers, *T. gallica* is also fond of clayey and sometimes saline soils (Kaabech and Benkheira 2000). Indeed, arid and semiarid Algerian areas are exposed to combination of abiotic stress, which classified them as harsh ecosystems. Nevertheless, the species which occupy these areas must be adapted to these drastic conditions. *T. articulata* (*aphylla*) and *T. gallica* are classified as heliophylous species with efficient sexual and vegetative reproduction with a large production of seeds and a high success rate in germination and cuttings (Lavaine et al. 2011b). Often in arid and semiarid Algerian areas, their regeneration process is blocked by a harsh environmental condition such salinity (DGF 2017), which require the presence of symbiotic microorganisms such arbuscular mycorrhizal fungi to accomplish their regeneration (Bencherif et al. 2019a). However, *T. articulata* (*aphylla*) and *T. gallica* are cited for their biological adaptation strategy for many abiotic stresses. Such as in salinized areas for salt extraction in South Africa (Newet et al. 2018), in dunes' fixation, and in inhabiting degraded areas, which would otherwise be subjected to desertification (Newete et al. 2020). These ecological roles of *Tamarix* species are still not well known and not exploited (Kuzminsky et al. 2014). Thus the use of Tamarix species to study the biological process adaptation to abiotic stress is rare, for that, in this book chapter, we focused on the biological (morphological, anatomical, and biochemical) strategies adopted by two *Tamarix* species (*T. articulata* and *T. gallica*) to alleviate the most abundant abiotic stresses in Algerian arid and semiarid areas. At first *Tamarix spp* is presented in Algerian areas, before focalizing on the two abundant species *Tamarix articulata* and *Tamarix gallica*, dealing with the ecological, physiological, morphological, and biological strategies employed on abiotic stress adaptation with the aim of generating consciousness who will serve as serious perspective for the degraded Algerian and Mediterranean areas.

14.2 Biology, Ecology, and Phylogeography of *Tamarix* Genus in Algeria

Tamarix genus are the first genus of four one composed Tamaricaeae family, *Hololachna*; *Myricaria*; *Reaumuria,* and *Tamarix.* Angiosperm phylogeny group classification (APG 2016) places this family under Caryophyllales order, with 78 total species whose 60 are placed under *Tamarix* genera (Christenhusz and Bung 2016). The species of *Tamarix* genus are characterized by multibranched riparian trees, which are up to 12 m high (such *T. articulata*), or shrubs with 6–8 m high (*T. gallica*). These species are considered to be tolerant of highly saline soil, hence their common name "Saltcedar" (Brotherson and Field 1987; Marlin et al. 2017). The *Tamarix* species can excrete salt through specialized glands in their leaves which increases the salinity of the soil surface under the tree canopy and may make the surrounding area unusable by other species (Thomson and Liu 1967; Marlin et al. 2017). But this faculty can be useful for phytoremediation of contaminated soils (Manousaki and Kalogerakis 2009; Marlin et al. 2017; Newete et al. 2020). *Tamarix* are long lived (up to 100 years), and are classified as facultative phreatophytes because of their extensive, deep, root systems (up to 30 m), allowing the plants to exploit deep water tables, and tolerate drought stress (Brotherson and Field 1987; Marlin et al. 2017). *Tamarix* species have two forms of growth:

- A normal shape which gives them the appearance of an ordinary tree with the main stem when they develop in ordinary environments, such as *Tamarix articulata* (*aphylla*);
- A second form is characterized by an abundant branching when the feet are in an environment subjected to stress conditions or in the case of accumulation of alluvial sediments: This is the case of *Tamarix gallica*.

14.2.1 *Tamarix* Species Distribution in Algerian Areas

The *Tamarix* genus is native to Middle Asian deserts, where it evolved in riparian areas with saline soils to Mongolia through China into India, across the Middle East towards the eastern Mediterranean and southern Europe (such Spain), across northern Africa (Morocco, Senegal, and southern Algeria) and through eastern and southern Africa (Baum 1978; Marlin et al. 2017). This genus has been described for the first time by Willdenow in 1816 when he considered only 16 species (Willdenow 1816). In 1978, Baum, conducted an extensive revision of the genus, adding the Pakistani species and increasing the total to 55 species (Baum 1978). The genus is characterized by considerable morphological and ecological similarity among species, making it one of the most taxonomically challenging genera in the angiosperms (Baum 1978; Marlin et al. 2017). In addition, *Tamarix* can grow in varied climatic and edaphic conditions, and each species exhibits considerable phenotypic variation under different circumstances (Thomson and Liu 1967; Marlin et al. 2017). In Fig. 14.1 the presence of Tamarix species in Algerian steppic areas is shown with animal exploitation.

a b

Fig. 14.1 (**a**) *Tamarix articulata* (aphylla) forest in Kaser-ElBoukhari, Algeria as TAMARICION AFRICANAE group (Authors, 2013). (**b**) *Tamarix articulata* grazed by a cheptel of sheep and goat in Kaser-ElBoukhari (Medea, Algeria) (Authors, 2017)

Table 14.1 Different *Tamarix* species recorded in Algerian areas

Tamarix species	Statue	Localization	Source of information
T. africana Desf.	Native	Oued Biskra Sebkha de Miserghine Khanechla region (northeast Algeria) Rachgoun, Oued Isser, Oued Zenata et Hammam Bougrara (Tlemcen) Chott Zahrez (Djelfa)	Battandier (1907), Quezel and Santa (1963), Baum (1978), Khabtane and Rahmoun (2012), Hadj Allal (2014), Nedjimi et al. (2012)
T. amplexicaulis (*balansea* J. Gay)	Native	Very rare species, observed in southern of Biskra (Balansa forest) and near to Oud Seggeur (Oasis of Brezina. Willaya d'El-Bayadh), in the great western erg	Battandier (1907), DGF (2017)
T. articulata (*aphylla*, *orientalis*)	Native Introduced	Recorded in oasis, in the Hoggar, Plain of Abadla (Bechar): Tamarix forest Introduced in Tell Algerian area	Battandier (1907), BNEDER (2015), DGF (2017)
T. boveana (*bounoupea*)		Saline saharian chott Marsh and sands in altitude between 0-300m. Chott Zahrez (Djelfa)	Battandier (1907), Nedjimi et al. (2012)
T. canariensis		In sandy slope in altitude between 0-800m	Marlin et al. (2017)
T. gallica (arborea, *tamariscus gallicus*, *narbonensis*, *pentendra*, *anglica*)	Native Naturalized	Saharan area: Ai ElBaida, Ouargla; West of Algeria Rechgoun (Beni Saf); Oum El bouaghi, El Taref Chutt Edhiba (Souf region, Northern Sahara)	Battandier (1907), Koull and Chehma (2014), Hassaine et al. (2014), Fellah et al. 2018 ; Khachkhouch et al. (2020)
T. parviflora (*cretica*, *laxa* var. *subspicata*, *lucronensis*, *rubella*)	Naïve	Batna	Battandier (1907), Marlin et al. (2017)
T. passerinoides Desf. (*tenuifolia*, *Trichaurus pycnocarpus* (unresolved), *passerinoides* var. *macrocarpa* (infraspecific taxa)	Native	Located in the same region with *T. amplexicaulis*, in southern of Biskra and in oued Seggeur (willaya d'El-Bayadh): Great western erg.	Battandier (1907), DGF (2017)

In Algerian areas, eight *Tamarix* species were recorded (Marlin et al. 2017) (Table 14.1). Indeed, the *Tamarix* genus very widespread in Algeria from the edge of the sea at the bottom of the Sahara. They have not yet been sufficiently studied because the flowering of these plants being short duration, so species remained unknown until the studies of Battandier (1907). However, it should be noted that Vahl and Desfontier around 1820 recorded three Tamarix species in Algerian areas: *T. gallica*, *T. Africana,* and *T. articulata* in the south (Ouargla, Biskra, and Bechar, respectively). After that, De Bunge (1852) described *T. boveana* observed in western Algeria, in Tafna near to Arzew. In 1853, Jack Gay described four other spices *T. Balansae*, *T. bounopeae*, *T. brachystilis* and *T. panculata*. But these results were not published until 1907 when Battendier published them in "La flore d'Algérie".

14.2.2 Biology of *Tamarix* Species

A few species of *Tamarix* are employed in traditional medicine in Asia and Africa (Saoud Orfali 2005). They are employed in several treatments, especially as astringent, aperitif, diuretic, and as stimulus of secretion (Gaston 1998). For the Egyptian, *Tamarix* have great importance as medicinal and spiritual plant, the leaves and young branches are cooked to heal edema of the spleen and mixed with ginger for uterus infections, while the bark is used boiled in addition with vinegar against lice (Bulos 1983; Saoud Orfali 2005). In addition, population of Dhofar in South of Oman extracts compounds from Tamarix and use the as antiseptic, they also use the dried leaves for abscesses and lesions. A solution comosed of Tamarix dried leaves bioled in water is used to comfort prolonged women labor. The leaves of a *Tamarix* species are also used by the population of Southwestern Saudi Arabia to relieve headache and fever; they employed them locally, draped on the head (Ghazanfar 1994). *Tamarix* leaves are suitable for leukoderma treatment, irritation, and eye diseases (Sharma and Parmar 1998). Moreover to medicinal value, *Tamarix* species are also used for tanning and persistent staining (Nawwar et al. 1982; Saoud Orfali 2005). Recently, all these traditional biological activities were attributed to chemical constituents of various *Tamarix* species. They are identified as polyphenolics, triterpenes, Flavonoids, tannins, and volatile constituents (Ghazanfar 1994; Saoud Orfali 2005). Polyphenolics represent the most common phytochemically investigated compounds in *Tamarix* species (Saoud Orfali 2005; Bencherif et al. 2019c). These different phenolic compounds include flavonoids, coumarins, hydrolyzable tannins, organic acids (gallic, ellagic, and cinnamic acids), and finally, lignans. It was recently proven that these bioactive compounds act as a natural antioxidant. They can delay the lipids and proteins oxidation by inhibiting the initiation or/and propagation of oxidative chain reaction. Thus, they may prevent or repair cell damage caused by oxygen (Bettaib et al. 2017).

14.2.3 *Tamarix articulata* Vahll (Aphylla (L) Karst, Orientalis)

Tamarix articulata (*aphylla*) is a desertic species, which is natively found in Asia, North Africa, and Southeastern Europe (Jaseim et al. 2019) and spreads from Asia and Arabia to Senegal (Battandier 1907). It was introduced into many areas as ornamentals, windbreaks, or as stabilize stream banks from those areas, such as in North America in 1823 (Jaseim et al. 2019) and on steppic Algerian areas (DGF 2017). It is widely distributed in many arid areas of North Africa (Battandier 1907; Ferlin 1981). *T. articulata* was naturally recorded in Algerian Sahara where it was required for their gall "Takaout" used by local population for staining. For that it was often cultivated in oasis. After 1880, it was more cultivated in the Tell, but it does not produce the gall, because the insect that bites it to produce the gall is absent (Battandier 1907). *T. articulata* (*aphylla*) is a long flowering and spreading trees, which grown up 15 meters high with pendulous multi-divided branches and long lived about 100 years (Jaseim et al. 2019). The old one has different roots lateral and deep to adsorb the water and minerals (Litwak 1957; Jaseim et al. 2019). It rises in very varied soils, since from sandy soils to dirty soils in depressions. It is a particularly interesting species due to its extreme drought resistance, salt tolerance, ease of multiplication by cuttings, and its speed of growth in favorable conditions. It is also one of the rare woody species that can be used both north and south of the Sahara (Ferlin 1981).

In Algerian areas, *Tamarix articulata* formed a natural eco-botanical group "NERIO-TAMARICETEA BRAUN-BLANQUET and BOLOS 1958" (Braun-Blanquet and Bolos 1958). This class colonizes the sandy areas with a little deep of aquifer undergrounded water; it has been identified in the region of Ben Khellil (Nâama) and in Taghit (Bechar). *Tamarix articulata* represent a dominant botanical group in this region named: "*TAMARICETALIA AFRICANAE*" and second name "*TAMARICION AFRICANAE*" by Braun-Blanquet and Blos (1958). *T. articulata* group was identified in: (1) Sandy valleys of great Saharan wadis (Guir, Zousfana); and in (2) Border of depression (Dayet Tiour: Taghit; Oglat Ed-Daïra: Naâma). This group is distributed in saline depressions with superficial underground water with the presence of thick alluvium retaining constant humidity, which qualified them as "Azonal vegetation."

14.2.4 *Tamarix gallica* L.

(*Tamariscus gallicus*, *Tamariscus narbonensis*, *Tamariscus pentandra* (unresolved name), *T. anglica*). *Tamarix gallica* is a species of smaller size than *T. aphylla*, it is not exceeding 6–9 m high, and more often bushy. It can be used for fixing the dunes (Ferlin 1981), as used in Algerian steppic areas for fixing the cordon dune (INRF 1984). But *T. aphylla* is usually preferred due to its greater development (Ferlin 1981). *T. gallica* is a shrub with reddish brown bark and thin hairless twigs, appearing feathery. The leaves are alternate, of a pinkish white green, they are only about 5 mm in diameter, but very numerous, arranged in dense spikes, giving the whole plant a pink coloring. The fruit is a seed topped with a short tuft of hair (Belhadj-Sgheir et al. 2019). *T. gallica* is one of the constituents of Indian traditional herbal recipes, which is hepatotonic, stimulant and has been used traditionally in the treatment of various liver disorders (Bettaib et al. 2017). Moreover, recent study reported that *T. gallica* has a potent chemopreventive effect and may suppress thioacetamide (TAA)-mediated hepatic oxidative stress, toxicity and tumor promotion response in rats (Hatano et al. 1988; Bettaib et al. 2017). *T. gallica* is very widespread in Algeria and even the only one of its group which faces with certainty part of the flora of the country (Battandier 1907; Marlin et al. 2017). Quezel and Santa (1963) recorded this species in Algerian Sahara and in western of Algeria. *T. gallica* is abundant along streams and wadis, on the banks of the rivers in moist sandy soil, with a high salt content. It has long taproots that can penetrate to the deep water tables. It flourishes within a wide range of rainfall (around 600 mm/year in the Mediterranean regions to 100 mm/year in the Sahara) (Benhouhou 2018).

14.3 Morphological and Anatomical Adaptation Strategies of *Tamarix*t to Abiotic Stress

What is abiotic stress? That corresponding to the prevailing force applied by an environmental parameter, disturbing the usual functioning of the plant. Besides, the plant's response depends, on these environmental parameters (type of phenomenon, its concentration, and its period) and genetic parameters relative to plant (species and genotype) (Hopkins 2003). In Algerian areas, drought, salinity, calcareous, wind erosion, hydric erosion, silting and desertification increasing constitute the most important abiotic stress, for that plant adopts special actions to counteract these conditions.

Saline stress is the most important problem that the *Tamarix* species reencounter in Algerian areas. It concerns the accumulation of Na^+ and Cl^- in plant tissues, which are conducted to stress states manifested by: Reduction of water potential; Disturbance in ionic homeostasis and ionic toxicity (Hopkins 2003). According to this abiotic stress, two morphological strategies exist: (1) plants that exclude salt, (2) and plant which accumulate salt. The exclusion of the salt from a plant consists in avoiding or restricting the penetration of this substance in the tissues, especially in buds (shoots and flowers) (Benhouhou 2018). This phenomenon is specific to "*glycophytes*" who do not tolerate large accumulation of inorganic ions in their tissues. In return, salt penetration in plant leads to increase of inorganic ions in the plant tissues; which are the case of "*halophytes*" plants (Hopkins 2003). In the case of *Tamarix* species the osmoregulation strategy is the most important phenomenon, it is applied by "osmoregulatory glands." *Tamarix articulata* and *Tamarix gallica* are considered as halophytic plants.

14.3.1 Adaptation Strategies Employed by *T. articulata* Under Algerian Abiotic Stress Conditions

That concerns all modifications which affect leaves, roots, or flowers either directly or during the growth stages. In Algerian areas, three abiotic stresses are dominant: drought, salinity, and erosion.

14.3.1.1 Anatomical Adaptation Strategies to Erosion Stress

Tamarix articulata (*aphylla*) species are known to be adapted to windy conditions by evolving extremely effective and robust erosion-resistant characteristics (Han et al. 2013). In fact, Han et al. (2013) studied the relationships between *T. aphylla* surface cracks, the internal histology of plants, and biomechanics principals, in order to providing information for its protection strategies against wind–sand erosion stress. They focused on plant histological cracks, their rings, elasticity modulus, and growth stress. The results of their study revealed that the directionally eccentric growth rings of *T. aphylla*, which are attributed to reduced stress and accelerated cell division, stimulate the formation of surface cracks (Fig. 14.2). The rings exposed to winds are more extensive than the protected one. Exposed wind surfaces are more disposed to a crack, which improves erosion resistance.

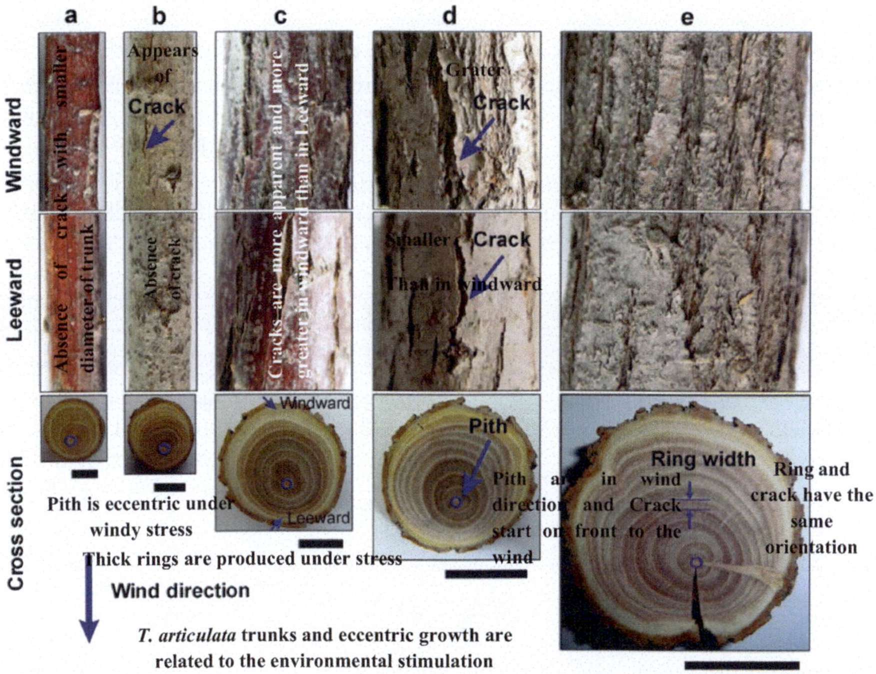

Fig. 14.2 Schematization of *T. aphylla* surface crack and asymmetric radial cross-section to alleviate windy conditions. Panels (**a**)–(**e**) show the surface cracks of the windward and leeward side and the corresponding cross-sectional microscopic structures. The diameters of the samples were approximately 8, 14, 24, 30, and 44 mm, respectively. (**a**) Cracks were absent in both the windward side and the leeward side. (**b**) Cracks started to appear in the windward, but not on the leeward side. The number and size of the cracks gradually increased with increasing trunk diameter. The cracks in the windward side were larger and more numerous than those in the leeward side of the same trunk, respectively (**c**–**e**). The piths of *T. aphylla* with different diameters were far away from wind direction, or windward side (bottom, **a**–**e**). Scale bars, 4 mm (**a**); 7 mm (**b**); 8 mm (**c**); 15 mm (**d**), and 22 mm (**e**). (Han et al. 2013 modified)

14.3.1.2 Anatomical Adaptation Strategies to Drought Stress

The difference between the numbers of flowers per kittens is a form of resistance to abiotic stress. It is decreasing in stressed condition such as high temperature and high soil salinity, which was proven by Khabtane and Rahmoun (2012). The authors studied the morphometric difference between *Tamarix* in three different sites with various pedoclimatic characteristics and when floristic richness in the undergrowth is different according to these abiotic characteristics. They recorded 52 flowers per kitten in the Ouarezne site (Khanechla), which represents a biotope with a permanent drought. In the two other sites with non-permanent stress, they recorded 84 flowers per kitten. That indicated that *T. articulata* reduces flower production as strategies adaptation to drought stress. In fact, it was proven that under drought stress, all growth parameters and biomass production are decreased with increasing the percentage of vessels and the vessels tended to be in groups in the radial direction (Fig. 14.3) (Al-Mefarrej 2013).

14.3.1.3 Anatomical Adaptation Strategies to Saline Stress

In Algeria, *T. articulata* is often found in saline areas, which accommodate this species to adopt a solid strategy to alleviate saline stress. Salt secretion by salt gland is one of the possible avoidance mechanisms of halophytes (Rozema and Gude 1981). The halophyte *Tamarix articulata (aphylla)*, has a root system which absorbs saline water from his habitats, for that he owns in its leaves specified glands in order to purge itself from salt excess: “salt-excreting glands” (Brown and Mies 2012). That

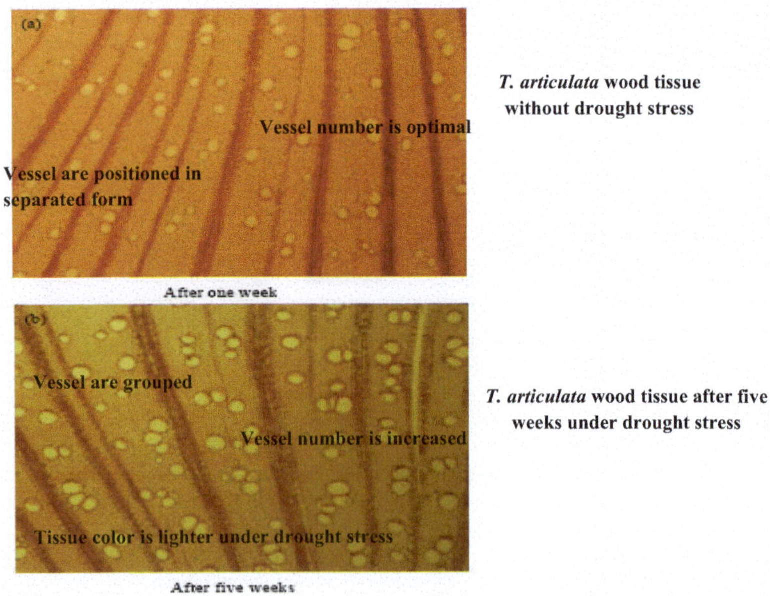

Fig. 14.3 Cross-section of *T. aphylla* wood as affected by drought stress induced by irrigation periods (adapted from Al-Mefarrej 2013)

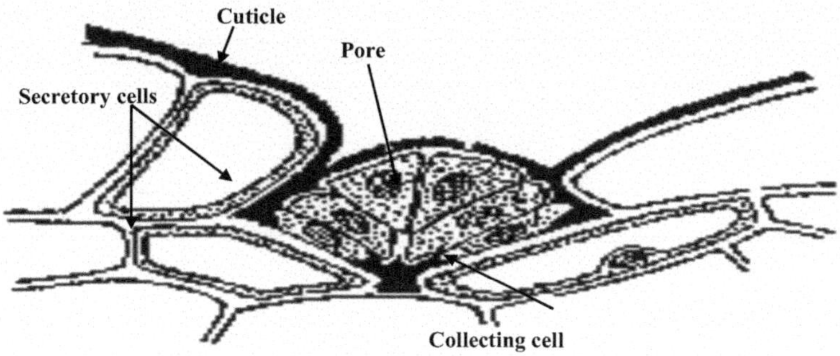

Fig. 14.4 *Tamarix aphylla* osmoregulatory structure according to Bosabalidis and Thomson (1986) description

classified it from the "recretohalophytes" plants (Dassanayake and Larkin 2017). However, leaf primordia of branchlet apices of this halophyte have been found to bear external salt glands at different developmental stages. Further light and electron microscopic observations revealed that all these stages from initiation of salt glands up to their final differentiation also occur in fully-expanded mature leaves. In *T. atriculata* (*aphylla*), unlike the species of other families, external leaf glands are not permanent on leaves, but by the time some glands mature and disintegrate, some others initiate and develop (Bosabalidis 1992). It was recognized that salt glands vary greatly among plants and can be divided into multicellular salt glands, bicellular salt glands, and salt bladders according to their structure (Ma et al. 2011). For *T. aphylla* young branches and leaves have many epidermal salt glands, and each salt gland originates from a single protodermal cell which divides anticlinal to give rise to two equally sized daughter cells. The latter further undergoes two successive asynchronous, asymmetrical divisions resulting in the formation of six, densely cytoplasmic, secretory cells. In the next developmental stage, two inner, vacuolated parenchyma cells adjacent and adherent to the inner secretory cells progressively transform into what have been termed in the literature the collecting cells of the glands. Thus, it appears at maturity that the glands consist of eight cells, and all eight cells may be directly involved in the secretory process (Fig. 14.4) (Bosabalidis and Thomson 1986; Mauseth 1988). Theses salt glands are characterized by their lateral position in epidermis (Rozema and Gude 1981). Bosabalidis (2010) explained that the salt gland of *T. aphylla* is considered as a salt-excreting root and gives a new description of the salt gland and cell functioning on the secretion stage with a new finding consisting of interaction between microvacuoles and wall protuberances as well as the genesis of wall protuberances. The salt gland is divided into three pairs of secretory cells arranged one upon the other. At the stage of secretion, the upper and middle pair of secretory cells develop a "protuberance," which defined as an internal system of anastomosed sticks developed in walls (Fig. 14.5). In the formation of the wall protuberances, Golgi vesicles and microtubules appear to participate (Bosabalidis 2010). The stage of salt secretion is also characterized by the presence of numerous mitochondria and microvacuoles. However, microvacuoles contain the

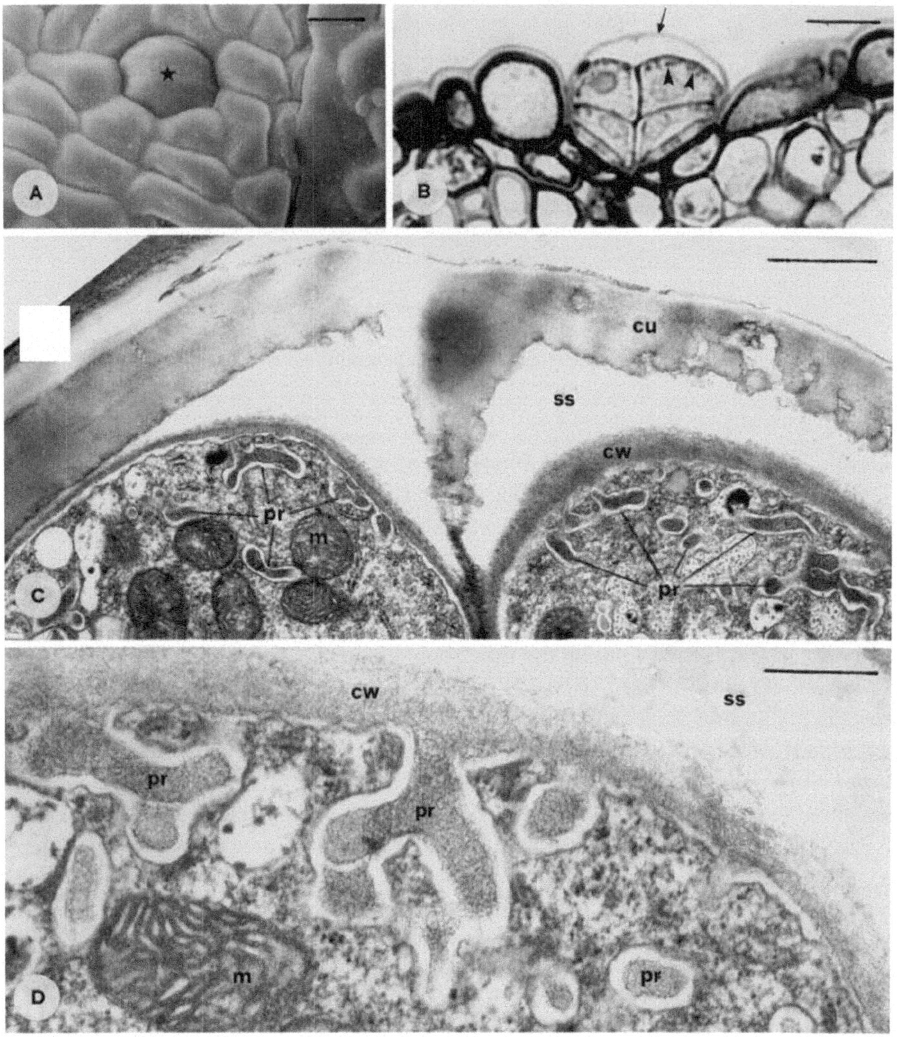

Fig. 14.5 Protuberance formation process in *T. aphylla* leaf Light and micrographs illustration (Bosabalidis 2010). (Light, SEM, and TEM micrographs illustrating salt glands. (**a**) SEM micrograph illustrating a salt gland (asterisk) on the leaf surface; (**b**) LM micrograph of a secreting salt gland. The cuticle (arrow) is detached from the apical cell walls, which bear a system of internal protuberances (arrowheads); (**c**) TEM micrograph taken at the upper part of a salt gland. The cuticle (cu) is raised forming a subcuticular space (ss). The apical walls (3cw) Bear towards the cytoplasm many anastomosed protuberances (pr) m = mitochondrion; (**d**) Higher magnification of the protuberances. They exhibit a fine granular substructure, denser than that of the typical wall. Bars: 12 mm (**a**), 10 mm (**b**), 1 mm (**c**), 0.4 mm (**d**)

secreted solution and accumulate in the region of the wall protuberances. In addition, *T. articulata* secretory cells observed a large number of mitochondria (Thomson and Liu 1967), which indicated that the intercellular transportation of ions in the salt glands was an energy-consuming process (Ma et al. 2011). The process from cell division to the formation of wall protuberance with energy consuming was explained in Fig. 14.6.

14.3.2 Anatomical and Morphological Adaptation Strategies of *T. gallica* to Abiotic Stress Conditions

T. gallica is an intelligent plant that operates many morphological modifications to grow under stressed conditions.

14.3.2.1 Anatomical Adaptation to Drought Stress

T. gallica resistance to drought stress manifests by (i) morphological and physiological modification induced by drought on aerial and root parts; (ii) plant biomasses under drought conditions and (iii) root depth under prolonged drought stress conditions (Lavaine et al. 2011a). Certainly, *T. gallica* is considered as optional phreatophytes, which gives them better efficiency of water extraction in soils. They feed in the tablecloth but also in the unsaturated compartment and can switch from one food to another without significant damage (Smith et al. 1998). This faculty to use multiple water sources minimizes the drought stress effect (Lavaine et al. 2011a). In addition, root architecture in itself is also particularly geared towards the acquisition of resources. *T. gallica* roots are extremely abundant and spread out, their lateral extension exceeding by four to seven meters the diameter of the crown of the tree near the water table, which allows them to overcome frequent drying of the upper layers. In the event of a dropping from the water table, the species quickly emit a large number of roots towards the water table in order to counter water deficiency (Horton and Clark 2001). In addition, Lavaine et al. (2011a) recorded a recapture rate of *T. gallica* under drought stress about 96% with a reducing in aerial biomass compared to root biomass. This suggested that the resources captured are thus mainly allocated to the enlargement of the rhizosphere, and to the reduction of evaporative surface. Indeed, stomata are reduced to overcome drought stress (Abbruzzese et al. 2013) and absorb condensed quantities of water therefore the opening of the ostiole becomes small, which decreases the exchange with the external environment (Abbruzzese et al. 2013; Marius and Toma 2017). That suggested the plantation of *T. gallica* in environments subject to significant variations in the phreatic level (Lavaine et al. 2011b).

14.3.2.2 Anatomical Adaptation to Water Erosion

The aerial architecture of *T. gallica* made up of many flexible and divided branches allows to appropriate fine sediments and to reduce the current. This flexibility lets plants to lie down in the event of an inundation, inducing a carpet effect (soil protection) and a comb effect (reducing the current and capturing sediment) while minimizing the possibilities of scratching by the current (Lavaine et al. 2011a). In fact, *T. gallica* easily survives sediment recovery during floods emitting adventitious roots from buried stems or developing stems from the collar (Muller 1995; Bill et al.

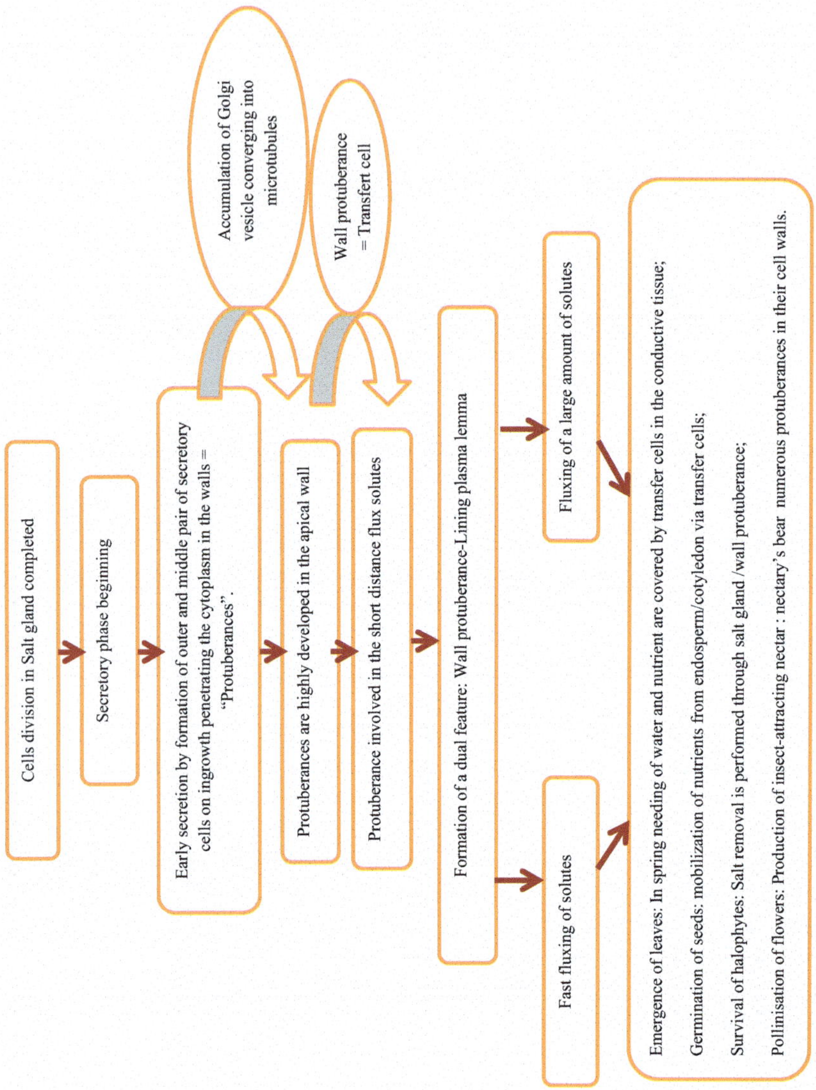

Fig. 14.6 Salt gland of *Tamarix aphylla* in secretory phases adapted from an explication of Bosabalidis (2010)

1997). They thrive well in water-saturated soils; and they support prolonged total submersion (Bill et al. 1997; Lavaine et al. 2011a). In addition, *T. gallica* has a great root density in the first few centimeters, so they can effectively protect the soil (Lavaine et al. 2011b), which suggest the plantation of this species in the riverbank protection program (Lavaine et al. 2011a, b).

14.3.2.3 Anatomical Adaptation to Saline Stress

As same to all Tamaricaceae species *T. gallica* has good salt tolerance (Lavaine et al. 2011b). It was recorded that Na^+ levels was about 23,419 $\mu g.g^{-1}$ *T. gallica* dray weight in saline conditions and can reach 28,199 $\mu g.g^{-1}$ in case of combined stresses (saline + other toxic compounds) while keeping a good growth, which confirms its halophytic character (Belhadj-Sgheir et al. 2019). In addition, high salinity caused marked anatomical in the structure of leaves, such as organization and distance among the vascular bundles, diameter, and number of xylems vessels (Rancic et al. 2019). Under saline stress, *T. gallica* reduces the leave surface (Durate et al. 2013), which is caused by reduction of the diameter of xylem vessels (Rancic et al. 2019) therefore the leaves became thinner and lighter than leaves in non-saline soils (Durate et al. 2013) (Fig. 14.7). As same as *T. articulata*, the *T. gallica* use secretion strategy to overcome high salinity, implying salt glands located at the leaf surface. These salt glands were characterized by a "flower-like" structure, with 5–9 epidermal cells surrounding a central secreting cell, with a density of 36.31 + 0.41 salt $gland.m^{-2}$ of leaf surface (Fig. 14.8) (Abbruzzese et al. 2013). The number of salt gland was less abundant on *T. gallica* leaves than other *Tamarix* species, and it is considered as a salt-excreting leaf (Rancic et al. 2019).

14.3.2.4 *T. gallica* Anatomical Tolerance Mechanisms to Pollutants: Arsenic

It was proven that *T. gallica* is strongly adapted to environmental stress such as metal trace elements (Wang et al. 2016; Belhadj-Sgheir et al. 2019). This adaptation is associated to their salt-tolerance characteristics (Wang et al. 2016). In fact, *T. gallica* occurs in coastal and desertic depression used as sites for accumulation of industrial

Fig. 14.7 *Tamarix gallica* leaves of individuals subjected to 200 and 0 mM NaCl (Durate et al., 2013)

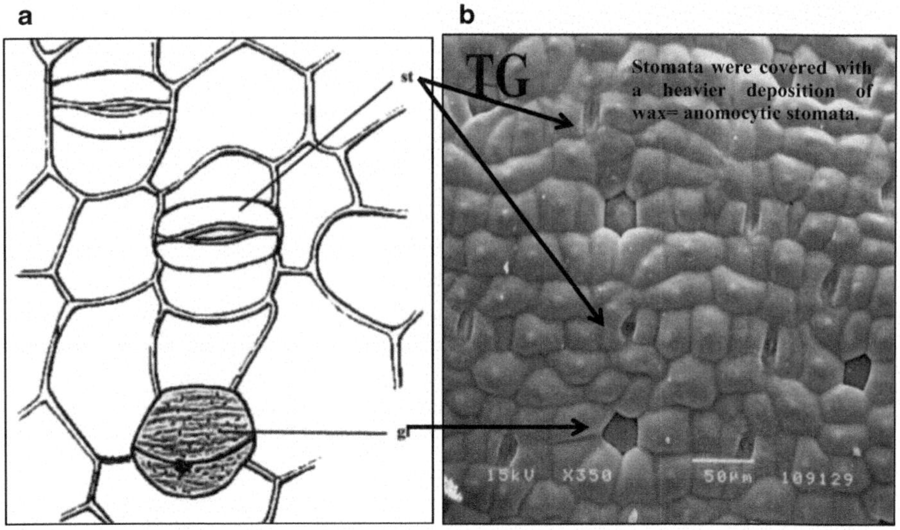

Fig. 14.8 Salt gland (gl) structure and stomata (st) in *Tamarix gallica* leaves (**a**. Vesque, 1883 in Marius and Toma 2017; **b**. Abbruzzese et al. 2013)

and urban effluents contaminated by metals (Ghnaya et al. 2007), where it has the mechanisms to regulate internal and cell wall metal concentration that determine their survival (Belhadj-Sgheir et al. 2019). Sookbirsingh et al. (2010) explained that glandular structures of *T. gallica* are not specific to Na^+, they have the ability to uptake a high concentration of ions and concentrate them in their shoots to be excreted at a later stage. Moreover, exposition to trace element stress leads *T. gallica* to show some morphological modification with a decline in total biomasses. The performance of *T. gallica* and its capacity to accumulate amounts greater than 7834 $\mu g.g^{-1}$ Dry Weight of aluminum toxic metal (Al), leads to classify it among the hyper-accumulator species of Aluminum (Al) (Belhadj-Sgheir et al. 2019). Based on that, Newete et al. (2020) explained that the native genotype of *Tamarix* species such native *T. gallica* is the preferred plant choice for phytoremediation programs in South Africa, which confirm its interest in phytoremediation programs.

14.4 Biochemical Adaptation of *Tamarix* to Abiotic Stress

Biochemical adaptation is reflected by plant excretion of the specific chemical compounds under abiotic stress. However, in addition to the sodium chloride (NaCl) predominantly secreted by most halophytes, *Tamarix* species can transport, sequester, or secrete ions (Weasel 1961; Berry 1970; Sookbirsingh et al. 2010), heavy metals (Conesa et al. 2006; Kadukova and Kalogerakis 2007), and even pollutants (Dreesen and Wangen 1981; Urbansky et al. 2000; Sorensen et al.

2009) what makes Tamarix unique (Meinhardt and Gehring 2013). In addition to these compounds stored and secreted by *Tamarix*, chemical compounds such as polyphenols that are produced by the plant itself as a chemical defense against herbivores and pathogens (Ksouri et al. 2009), and they are considered as the most important product due to abiotic stress (Saïdana et al. 2008; Meinhardt and Gehring 2013). Polyphenols have been shown to have antioxidant, antibacterial, and antifungal activity (Sultanova et al. 2001; Saïdana et al. 2008; Ksouri et al. 2009; Bettaib et al. 2017; Bencherif et al. 2019c). The compositions of polyphenols differ from *Tamarix* species to another one, for that we listed the polyphenolic compounds of *T. gallica* and *T. articulata* (*aphylla*) with their role in the alleviation of abiotic stress. The study of polyphenolic compounds revealed that their secretion is related to induction of some genes, which are induced by abiotic stress, but there is no study proven the efficiency of these genes on the antioxidant properties of polyphenols and their role in the stress resistance process. However, they are considered as a marker of stress because their concentration was higher in summer with high temperatures (Fevreau 2012). On the other hand, the abilities of *Tamarix* species to survive under drought and salt stress have been used to identify abiotic stress tolerance-related genes. That leading to the identification of two groups of genes: Late embryogenesis abundant (LEA) genes and dehydration-responsive element binding (DREB) genes (Swaminathan et al. 2020). Late embryogenesis abundant (LEA) proteins are a type of highly hydrophilic glycine-rich protein with antioxidant, metal ion binding, membrane and protein stabilization, hydration buffering, DNA, and RNA interaction properties. They play an important role in protecting cells from abiotic stress, and in plant normal growth and development. More importantly, LEA expression is often induced by abiotic stresses such as cold, drought, or high salinity (Chen et al. 2019). Many studies were focused on LEA genes of some *Tamarix* species (*T. hispida*, *T. androssowii*, and *T. chinensis*) in order to identify genes responsible in tolerance to salt stress. These genes are used to confer abiotic stress resistance to other species such as rice and poplar species (Gao et al. 2014; Wang et al. 2016; Swaminathan et al. 2020).

14.4.1 Biochemical and Physiological Adaptation Strategies of *T. articulata* to Abiotic Stress

This concerns all changes affecting the internal functionality of the plant.

14.4.1.1 Neutralization of Reactive Oxygen Species Damages

Under drought, salinity, and high calcareous condition *T. articulata* as the same as all plants, increased reactive oxygen species (ROS) that damage plant tissues. Excess of ROS expression induces severe oxidative threats to cell components including: protein, lipid, DNA and RNA associated with cell structural damage, tissue injury, and gene mutation (Bettaib et al. 2017). To counteract the toxicity of ROS, *T. articulata* develops an enzymatic complex and nonenzymatic antioxidant defense systems. Such as superoxide dismutase (SOD) that acts to scavenging superoxide

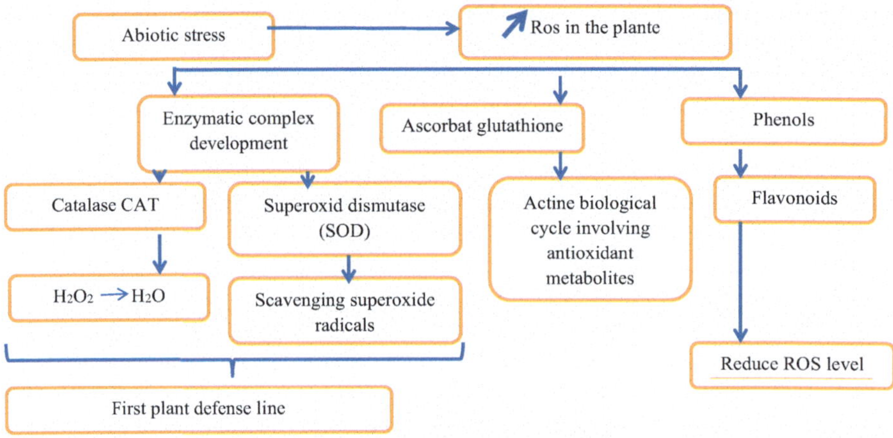

Fig. 14.9 Physiological reaction of *Tamarix* species to abiotic stress

radicals, and catalases (CAT) which transform H_2O_2 to H_2O (Ma et al. 2017). Catalase and superoxide dismutase are the first line defense enzyme against abiotic stress (Ma et al. 2017) (Fig. 14.9). At the second line, we have ascorbate–glutathione (ASC–GSH) (Surowka et al. 2019), which act in *T. articulata* biological cycle involving antioxidant metabolites: ASC and GSH, and the enzymes linking these metabolites such as "Glutathione reductase (GR), monodehydroascorbate reductase (MDHAR), dehydroascorbate reductase (DHAR), and ascorbate peroxidase (APX)" (Li et al. 2013; Ma et al. 2017). Recently, it has been suggested that flavonoids could inhibit the generation of ROS, and reduce levels of ROS once formed in response to environmental stress (Ma et al. 2017). In addition, the enhancement of enzymes antioxidant activity is due to the presence of a large number of mitochondria in salt secretory cells (Thomson and Liu 1967). The majority of antioxidant reactions act in mitochondria; superoxide dismutase, catalase, ascorbate oxidase. . . .,etc. Table 14.2 gives the different catalyzed reactions of enzymatic antioxidant and nonenzymatic ROS scavengers happening in *T. aphylla* mitochondria.

14.4.1.2 Inhibition of Abiotic Stress by Polyphenols

T. articulata (*aphylla*) are known as rich natural resource of polyphenolic compounds, they are used for their pharmacological effect, but they have an interesting role in the plant life to attenuate abiotic stress effect. Polyphenols are an important diverse group of nonenzymatic antioxidant (flavonoids, tannin, anthocyanin, and lignin) those possess antioxidant proprieties (Kumar et al. 2019). Table 14.3 listed the most important phenolic compounds in different part of *T. articulata* and their role on stress alleviation. Phenolics, flavonoids, tannins, and anthocyanidins can delay the lipids and proteins oxidation by inhibiting the initiation and propagation of oxidative chain reaction, and then they may prevent or repair cell damage caused by oxygen (Bettaib et al. 2017). A strong correlation between salinity and polyphenol accumulation was found in both photosynthetic actives tissues and roots

Table 14.2 Different catalyzed reactions in *T. aphylla* mitochondria of enzymatic antioxidant and nonenzymatic ROS scavengers according to Surowka et al. (2019)

Action	Antioxidant phenomenon	Reaction/function
Enzymatic activities	Superoxide dismutase (SOD)	$O^-_2+O^-_2+2H^+ \rightarrow H_2O_2+O_2$
	Catalase (CAT)	$2\ H_2O_2 \rightarrow 2H_2O+O_2$
	Ascorbate oxidase (APX)	H_2O_2+AsC → 2 H_2O+DHA
	Monohydroascorbate reductase (MDHAR)	2MDHA+NADH → 2AsC+NAD
	Dehydroascorbate reductase (DHAR)	DHA+2GSH → AsC+GSSG
	Glutathione reductase (GR)	GSSG+NADPH+H^+ → 2GHS+$NADP^+$
	Glutathione peroxidase (GPX)	2GSH+ H_2O_2 → GSSG+2H_2O
	Glutathione transferase(GST)	RX+GSH → AX+R-S-GSH
	Guaiacol peroxidase (GPX)	2GSH+ H_2O_2 → GSSG+2 H_2O
	Thioredoxin reductase (TrxTr)	NADPH+H^++TrxS2 → $NADP^+$+Trx(SH2)
	Thioredoxin (Trx)	TrxR-S2+NADPH+ H^+ → TrxR-$(SH)_2$+$NADP^+$
	Thioredoxin peroxidase	Trx+ H_2O_2 → Trx-$(SH)_2$+ H_2O
	Peroxiredoxins (Prxs)	2R'-SH+ROOH → R'-S-S-R'+H_2O+ROH
	Glutaredoxins (Grxs)	RSSR'+R"SH → RSSR"+R'SH
Nonenzymatic activities	Ascorbic Acid (AsA reducing agent)	Scavenger O^-_2 and H_2O_2 directly or through Asadda-Halliwell-Foyer pathway recycles the lipid-soluble antioxidant α-tocopherol
	Glutathione(reduced, GSH)	Detoxifying co-substrate for enzymes (GR, GST,...) a component of the cellular redox buffer
	Proline(Pro)	Direct scavenger of O_2 or OH and indirect scavenger of H_2O_2 and O^-_2 osmoprotectants involved in cellular energy and homeostasis; protect photochemical efficiency of PSII, act as electron shuttle to balance redox potential between chloroplast and mitochondria. Used as energy source in abiotic stress. Compatible solute capable of maintaining cell turgor.

(Surowka et al. 2019), which suggests the increasing of polyphenol extraction in saline conditions. In addition, Surowka et al. (2019) reported that *T. aphylla* flavonoids direct scavenger H_2O_2, O_2, and OH^- in plant vacuole and reduced the activity of polyphenol oxidase. Despite, it was recently proven that divers' phenolic compounds are efficient protectors against the oxidative cytotoxicity of hydrogen peroxide by the regulation of the endogenous antioxidant defense system and the modulation of signaling pathways (Bettaib et al. 2017).

Table 14.3 List of phenolic compounds extracted from different parts of *Tamarix articulata* (*aphylla*)

T. articulata Phenolic compounds		Plant part	Action	References
Flavonoids	Tamarixetin (3',3,3,7-tetrahydroxy-4'-methoxyflavone)	Flowers	Direct scavenger H_2O_2, O_2 and OH^- in plant vacuole and reduced the activity of polyphenol oxidase regulation of the endogenous antioxidant defense system and the modulation of Signaling pathways. They have the capacity to regulate CAT, SOD and GPx activities as well as for their ability to restrict JNK and p38 MAPKs phosphorylation induced by H_2O_2 stress	Bettaib et al. (2017), AlHourani et al. (2018), Jaseim et al. (2019)
	Quercetin (3-*O*-isoferulyl-β-glucuronide)			
	Kaempferol 7,4'-dimethyl-ether-3-*O*-sulfates			
	Rhamnetin-3'-glucuronide (Quercetin 7-methylether)			
	Rhamnocitrin3-glucoside	Galls and Barks		
	3-rhamnoside			
	Isoquercetin			
	Tamarixin			
	Taxifolin			
Gallic acid	2,3 galloyl-ester (2,3-di-*O*-galloyl-(α/β) $C^4{}_1$glucopyranose	Galls		Saoud Orfali (2005)
	(isomer) 3,6 galloyl-ester (2,3-di-*O*-galloyl-(α/β) $C^4{}_1$glucopyranose			Saoud Orfali (2005)
	1'-decarboxydehydrodigallic acid (Dehydrodigallic acid)	Bark		AlHourani et al. (2018)
Ellagic acids and their tannins	Tamaricellagic acid	Gall		Saoud Orfali (2005)

(continued)

Table 14.3 (continued)

T. articulata Phenolic compounds		Plant part	Action	References
Lignans	Diaryloxy furanofuran lignin			Saoud Orfali (2005)
	1-isoferulyl-3-pentacosanoyl-glycerol (Niloticol)			Saoud Orfali (2005)
	Triterpenes			AlHourani et al. (2018)
	Myricadial			Saoud Orfali (2005)
	3-ketone, 28-hydroxy-D-friedoolean-14-en-3-one			Saoud Orfali (2005)
	D-friedoolean-14 en-3α,28-diol	Bark		Jaseim et al. (2019)
Aphylin	Glycosylated isoferulic acid	Gall		AlHourani et al. (2018)
Coumarins		Stem bark		AlHourani et al. (2018)
Tannin				Jaseim et al. (2019)
Triterpenes				Jaseim et al. (2019)
Alkaloids				Jaseim et al. (2019)
Steroids, Terpenoids				AlHourani et al. (2018)
Amino acids, protein, carbohydrates, essential oil				Saoud Orfali (2005), Jaseim et al. (2019)
Saponin				Jaseim et al. (2019)
Anthocyanidins	cyanidin 3-O-glycoside	Flower		Jaseim et al. (2019)

14.4.1.3 Biochemical Adaptation Strategies to Saline Conditions

T. articulata tolerate salinity concentration between 200 and 500 mM of NaCl. It is considered as salinity regulators because it possesses specifics osmoregulators, which act in saline environments (Meinhardt and Gehring 2013). *T. articulata* have the ability to absorb chemicals from the soil and deposit them on the surface. This ability allows them to make abiotic changes in soils, such as salinity and electrical conductivity, soil pH, and phosphorus augmentation. Indeed, the salt excreted crystallizes on the surface of the leaves, where it becomes harmless (Hopkins 2003). The expression in the salt glands is very specific. For *T. articulata* Na^+ and Cl^- are extracted against the gradient concentration, while other ions such Ca^{+2} are maintained contrary to their gradient. In addition, the salt gland cations secretion is a preferential function. The cation secretory abilities of the multicellular salt glands of *T. aphylla* are as follows: monovalent cation > divalent cation: $Na^+ > K^+ > Ca^{2+}$ (Ma et al. 2011). In fact, the flow of ions between the eight sub-cells of the salt gland was considered to be driven by hydrostatic pressure (Zhang et al. 2002). However, with the discovery of ion channels and the use of a variety of inhibitors whether the ion secretion is due to active transport has once again become a focus for research on salt secretion mechanisms (Balsamo and Thomson 1996; Kobayachi et al. 2007). But whether *Tamarix* preferentially secretes Na^+ when the concentration of each ion is identical is not yet clarified. In addition, considering that the salt glands in *Tamarix* are multicellular, no reports have answered questions such as whether the ion secretion involves energy consuming active transport? (Ma et al. 2011).

14.4.1.4 Action of *T. articulata* Face to Metals Trace Elements

T. articulate (aphylla) can take up chemical compounds from the soil, transport them upward, and eventually deposit them on the soil surface or in leaf litter by excreting them through salt glands (vesiculated trichomes) (Sookbirsingh et al. 2010; Meinhardt and Gehring 2013) a detoxification mechanism (Marlin et al. 2017). In addition, Lefevre et al. (2009) reported that *T. aphylla* accumulate and excrete Cd and Pb on their leaf surface. They suggest that this plant may be used as phytoremediation plants in Cd and Pb contaminated soils.

14.5 Biochemical and Physiological Adaptation Strategies of *T. gallica* to Abiotic Stress

14.5.1 *T. gallica* Strategy of Neutralization of Reactive Oxygen Species Damages

Similar to *T. articulata* species, the *T. gallica* occurs in natural stressed areas under saline, drought, or calcareous abiotic stress, and in urban areas, it is subject to pollution. So the reactive oxygen species (ROS) is often stimulated by these conditions, conducting to dysfunction of mitochondrial with the enhancement of catalase activity. To contract these phenomenon *T. gallica* increase its secretion of biochemical compound which manifest cytoprotective effect and diminution of

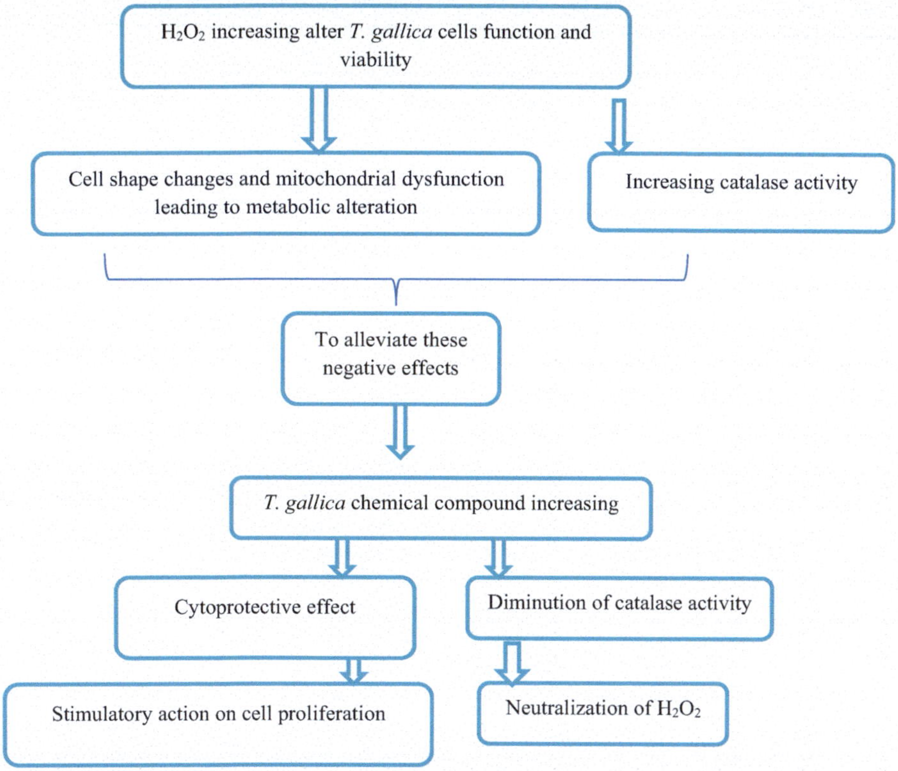

Fig. 14.10 Cytoprotective effect of chemical compounds of *T. gallica*

catalase activity (Fig. 14.10) (Ksouri et al. 2009; Boulaaba et al. 2015; Bettaib et al. 2017). In fact, the abiotic stress increases secretion of H_2O_2 in *T. gallica* cells with serious modification of their structures and reduction of viability. That leads to metabolic alteration with mitochondrial dysfunction activities and increasing in catalase activity. To counteract these effects, *T. gallica* increased secretion of its chemicals compounds that have a cytoprotective effect with the diminution of catalase activity. This process conducts to the reduction of H_2O_2 amount and allows to normal function of cell metabolisms (Ksouri et al. 2009; Boulaaba et al. 2013, 2015; Bettaib et al. 2017).

14.5.2 Action of Polyphenolic on *T. gallica* Abiotic Stress Adaptation

All biochemical and physiological adaptation strategies of *T. gallica* to alleviate abiotic stress are attributed to the high richness on polyphenolic compounds. Table 14.4 listed the most important polyphenolic compounds and their role in mitigation abiotic stress. Indeed, in North Africa, *T. gallica* has a great consideration by population for their therapeutic practices; it was studied by scientists for its important phenolic content; high antioxidant and antimicrobial activities when it

Table 14.4 Most important polyphenolic compound in *T. gallica* and their role in alleviation abiotic stress

T. gallica Phenolic compounds		Plant part	Action	References
Flavonoids	Kaempferide	Flowers	Reducing agents hydrogen donners singlet oxygen quenchers free radical scavengers and chelating agent of pro-oxidants metals Cytoprotective effect against oxidative stress	Saoud Orfali (2005), Bettaib et al. (2017), Kumar et al. (2019)
	Quercetin 4'-methyl ether-3-sulfate (Tamarixetin)			Saoud orfali (2005)
	Kaempferol			Ksouri et al. (2009)
	(Quercetin)		Protecting Ca-co circuit in the plant cell	Saoud Orfali (2005), Ksouri et al. (2009)
	quercetin 3-O-glucuronide		Prevent and repair cell damage caused by oxygen reactive by inhibition and propagation of oxidative chain reaction	Boulaaba et al. (2015)
	Herbarone			Saoud Orfali (2005)
	Kaempferide: Kaempferol-4'-dimethylether-3-sulfate	Bark		Saoud Orfali (2005), Kumar et al. (2019)
Gallic acid	3,4,5-trihydroxybenoic acid (polyhydroxy phenolic compound)			Saoud Orfali (2005), Ksouri et al. (2009)
	Cinnamyl alcohol sulfates: Transconiferyl alcohol-4-O-sulfhate			Ksouri et al. (2009), Boulaaba et al. (2015)
	1,2-Hentriacontranol: Aliphatic compounds			Saoud Orfali (2005)
	Transconiferyl alcohol-4-O-Sulfate			Saoud Orfali (2005)
	12-Hentriacontanol			Saoud Orfali (2005)
Resveratrol		Leaves		Bettaib et al. (2017)
Anthocyanin	Pyrocyanidin	Leaves	Reduce H_2O_2 effect	Lebreton and Bouchez (1967).

even showed efficient antitumoral capacity (Ksouri et al. 2009; Boulaaba et al. 2013, 2015; Bettaib et al. 2017). This species is further investigated for its cytoprotective effect against oxidative stress. In fact, phenolics, flavonoids tannins, and anthocyanidins can delay the lipids and proteins oxidation by inhibiting the initiation and propagation of oxidative chain reaction, than they may prevent or repair cell damage caused by oxygen (Bettaib et al. 2017; Kumar et al. 2019). Moreover, Bettaib et al. (2017) evaluated *T. gallica* total phenolic amount and their antioxidant capacities. They recorded à high phenol level (132.5 mg Gallic Acid Extract. g^{-1} *T. gallica* Dry Weight), with strong total antioxidant activity (107.7 mg Gallic Acid Extract.g^{-1} *T. gallica* Dry Weight). That suggests a close relationship between phenolic amounts and antioxidant capacities in *T. gallica* aerial parts. However, *T. gallica* methanolic extract provides protection against liver carcinogenesis, which might be mediated by multiple actions including restoration of cellular antioxidant enzymes, detoxifying enzymes, ornithine decarboxylase activity, and DNA synthesis (Bettaib et al. 2017). Kaempferide is an important flavonoid in *T. gallica* aerial part (Saoud Orfali 2005), it was proven that it maintains catalase activity in normal level under H_2O_2 elevated concentration, which suggests the biological strategies employed by *T. gallica* to alleviate abiotic stress by exceeding secretion of flavonoids (Boulaaba et al. 2015; Bettaib et al. 2017; Kumar et al. 2019).

14.5.3 Biochemical Adaptation Strategy of *T. gallica* to Metals Elements Pollution

It was reported by Sghaier et al. (2015) that *T. gallica* is a tolerant plant to trace metals elements, which are associated with their salt-tolerance characteristics. In fact, Belhadj-Sgheir et al. (2019) noted that *T. gallica* accumulated 92 and 94 μg.g^{-1} of Arsenic respectively in shoots and roots. For that, he is not considered as an accumulator plants because these one must accumulate at least 1000 μg.g^{-1}. *T. gallica* is considered as an Arsenic-tolerant plant, when the presence of salt gland may play a role in this tolerance. The highest concentration of toxic metals is stored in *T. gallica* shoots, while the root system is considered as the dominant absorption path for metals (Belhadj-Sgheir et al. 2019). Contrary to other halophytes, *T. gallica* manifested no difference in chlorophyll fluorescence parameters when it was exposed to high arsenic concentration. They manifest only an increase in the energy dissipation fluxes (Sghaier et al. 2015). Suggesting the suitable mechanism adopted by *T. gallica* to cope with the excess of trace metals elements (Belhadj-Sgheir et al. 2019). The complexity of *T. gallica* adaptation mechanisms to excess metals elements is schematized in Fig. 14.11. It was proven that *T. gallica* accumulates more than 70% of arsenic in the polysaccharides fraction of leaves cells wall, while a small amount accumulated in the protein fraction and in the intracellular central part of the cells. In addition, arsenic ions were distributed at similar concentration in the different polysaccharide, aqueous and protein fractions (Belhadj-Sgheir et al. 2019). That suggested the existing balance between the different parts of cell wall as strategy to avoid arsenic toxicity in *T. gallica* cell.

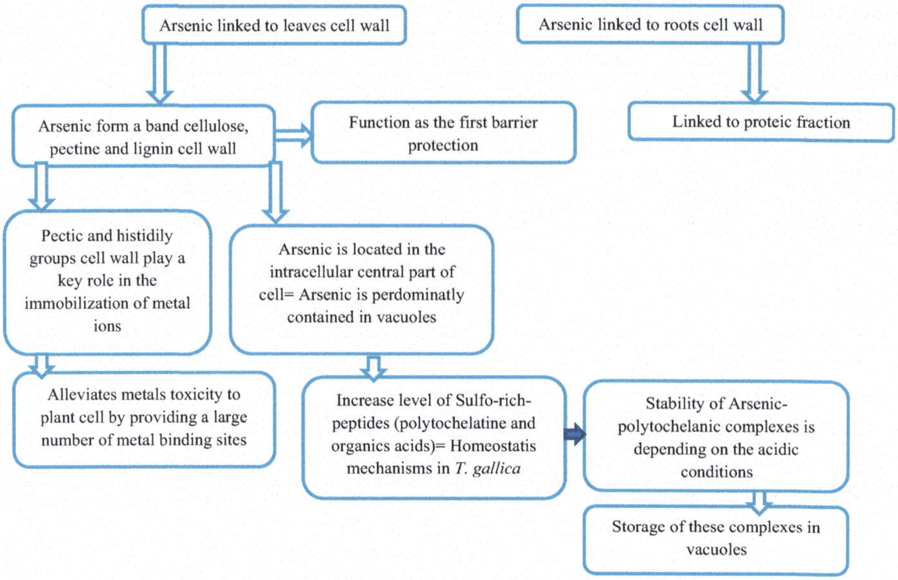

Fig. 14.11 *T. gallica* adaptation strategies to arsenic metal pollution

14.6 AMF Application Strategies

Arbuscular mycorrhizal fungi (AMF) can play a major role in plant response to alleviate abiotic stresses. They can facilitate soil nutrient uptake by the plants, and the fungi benefit from photosynthetic carbon produced by plants (Smith and Read 2008). AMF can improve the performance of individual plants while also having higher order effects on plant community structure, plant productivity, soil stability, and nutrient cycling (Smith and Read 2008; Sellosse 2017). The relationships between *Tamarix* species (*T. articulata*, *T. gallica*) and soil biota have been most extensively studied within plant-fungal mycorrhizal symbioses (Bencherif et al. 2015, 2016, 2019b, c; Stefani et al. 2020). These studies have proven the mycotrophic statue of *T. articulata* and *T. gallica* and the influence of seasonal variation on this symbiosis in arid and semiarid Algerian areas (Bencherif et al. 2016). The impact of soil salinity stress on *T. articulata* AMF and their associated microbiota in the field has been determined (Bencherif et al. 2015). Moreover, the Taxonomic assignment of arbuscular mycorrhizal fungi in an 18S metagenomic dataset of *T. articulata* (*aphylla*) was determined (Stefani et al. 2020). Based on these studies it may be suggested that AMF act by countering stress-induced oxidative damage and increase flavonoids content, which are known to improve plant redox status. In fact, the biochemical phenomenon of abiotic stress alleviation by AMF was studied for some species (*Cicer arietinum* L) (Garg and Singla 2015), (*Ocimum basilicum* L.) (El Hindi et al. 2017); (*Cajanus cajan*) (Garg and Singh 2018) but never for *T. articulata* or *T. gallica*. In fact, the combined complex flavonoids/AMF such as naringenin—*F. mosseae*, increased antioxidants efficiently and attenuated oxidative loads, with the greatest redox stability attained by elevation recycling of reduced glutathione and ascorbate facilitating the higher activity of scavenging antioxidants (Garg and Singla 2015). AMF act by upregulating the antioxidant enzymes as well as by modulating the level of glycine, betaine, proline, and phenols expressed as response to abiotic stress such as salinity (Hashem et al. 2016). The intervention of AMF is very important to maintain physiological plant function at stable level. Moreover, Bencherif et al. (2019b) explained the positive role played by native inoculum and native co-inoculum to increase phosphorus and nitrogen uptake from natural saline soils on *T. articulata*. On the other hand, positive correlation between AMF and polyphenolic compounds was recorded on *T. gallica* studied in natural semiarid soil (Bencherif et al. 2019c). Taken together, these studies suggested that *T. articulata* and *T. gallica* have specific rhizosphere adapted to different abiotic stress and play an important role in plant adaptation to these environments, which allow the creating specific combinations based on native AMF strains in order to alleviate abiotic stress for many agricultural plants (Fig. 14.12).

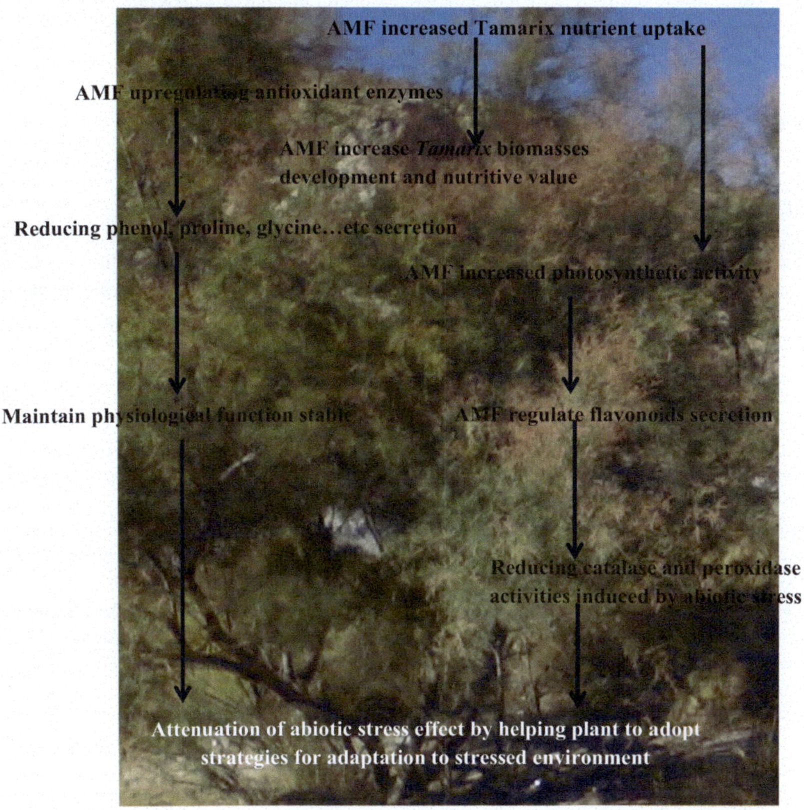

***T. articulata*, *T. gallica* mycotrophic plants**

Fig. 14.12 Role of AMF in plant adaptation to abiotic stresses

14.7 Conclusion

Arid and semiarid Algerian areas are highly dynamic ecosystems subjected to different abiotic stress, which are important factor in plant occurrence. *T. articulata* and *T. gallica* are two interesting plants that occur in these zones and manifest interesting anatomical and physiological adaptation strategies. In fact salt glands present on the leaf surface acts not only as an accumulator of salt but also of heavy metals, which open the way for new phytoremediation technics. Moreover, polyphenolic compounds which are considered as specific elements for *Tamarix* species play an important role in abiotic stress alleviation with interesting strategies. Furthermore, *T. articulata* and *T. gallica* are known to be mycotrophic species, so all the adaptation strategies employed to alleviate abiotic stress may be controlled by arbuscular mycorrhizal fungi. The biological adaptation strategies of *T. articulata* and *T. gallica* to alleviate abiotic stresses give new ideas to develop agriculture and revegetation programs on degraded and desertic lands. It may be interesting to study *T. gallica* and *T. articulata* genes responsible for stress tolerance in arid and semiarid Mediterranean areas and to used their capacities for agricultural plantation to improve their survivability in theses stressed areas such for cereals.

References

Abbruzzese G, Kuzminsky E, Abou Jaoudé V, Angelaccio C, Eshel V, Scoppola A, Valentini R (2013) Leaf epidermis morphological differentiation between Tamarix africana Poir. and Tamarix gallica L. (Tamaricaceae) with ecological remarks. Plant Biosyst Int J Deal Aspects Plant Biol 147(3):573–582. https://doi.org/10.1080/11263504.2012.714805

Alhourani N, Kasabri V, Bustanji Y, Abbassi R, Hudaib M (2018) Potential antiproliferative activity and evaluation of Essential oil composition of the aerial parts of *Tamarix aphylla* (L.) H. Karst.: a wild grown medicinal plant in Jordan. Evidence-Based Complementary and Alternative Medicine 2018:9363868, 7 pages and Industrial Research 57, 873–890 (1998)

Al-Mefarrej AH (2013) Growth characteristics and some wood quality of *Tamarix aphylla* seedlings irrigated with primary treated wastewater under drought stress. Asian J Plant Sci 12:109–118

APG (2016) An update of the angiosperm phylogeny group classification for the orders and families of flowering plants: APG IV. Bot J Linn Soc 181:1–20. https://doi.org/10.1111/boj.12385

Balsamo RA, Thomson WW (1996) Isolation of mesophylle and secretory cell protoplasts of the halophytes Ceratostigina Plumbaginoides (L): a comparison of ATPase concentration and activity. Plant Cell Reprod 15:418–422

Baum BR (1978) The genus Tamarix. Israel Academy of Sciences and Humanities, Jerusalem

Belhadj-Sgheir D, Pedro S, Duarte B, Caçador I, Seleimi N (2019) Arsenic tolerance mechanisms on halophytes the case of Tamarix gallica. In: Hasanuzaman et al. (ed.) Halophytes and climates changes adaptative mechanisms and potentials uses

Bencherif K, Boutekrabt A, Fontaine J, Laruelle F, Dalpè Y, Lounès-Haj Sahraoui A (2016) Soil and seasons affect arbuscular mycorrhizal fungi associated with Tamarix rhizosphere in arid and semi-arid steppes. Appl Soil Ecol 107:182–190

Bencherif K, Dalpé Y, Hadj-Sahraoui A (2019a) Arbuscular mycorrhizal fungi alleviate soil salinity stress in arid and semiarid areas. Soil Biol 56:375–400

Bencherif K, Dalpé Y, Sahraoui A (2019b) Influence of native arbuscular mycorrhizal fungi and Pseudomonas fluorescens on Tamarix shrubs under different salinity levels. Soil Biol 56:265–283

Bencherif K, Djaballah Z, Brahimi F, Boutekrabt A, Dalpè Y, Lounès-Hadj Sahraoui A (2019c) Arbuscular mycorrhizal fungi affect total phenolic content and antimicrobial activity of Tamarix gallica in natural semi-arid Algerian areas. S Afr J Bot 125:39–45

Benhouhou S (2018) A guide to medicinal plants in North Africa. UICNmed.org

Berry WL (1970) Characteristics of salts secreted by *Tamarix aphylla*. Am J Bot 57:1226–1230

Bettaib J, Talarmin H, Droguet M, Magné C, Boulaaba M, Giroux-metges M-A, Ksouri R (2017) Tamarix gallica phenolics protect IEC-6 cells against H 2 O 2 induced stress by restricting oxidative injuries and MAPKs signaling pathways. Biomed Pharmacother 89:490–498. https://doi.org/10.1016/j.biopha.2017.02.047

Bill HC et al (1997) The distribution and patch dynamics of the German tamarisk Myricaria germanica (L) Desv, in the upper river Isar, Bavaria. Zeitschrift fur Okologie und Naturschutz 6(3):137–150

BNEDER (2015) Bulletin d'information trimestriel N°6. Etude de rehabilitation de la plaine d'Abadla, Béchar

Bosabalidis AM (1992) A morphological approach to the question of salt gland lifetime in leaves of Tamarix aphylla L. Israel J Plant Sci 41(3):115–121. https://doi.org/10.1080/0021213X.1992.10677220

Bosabalidis AM (2010) Wall protuberance formation and function in secreting salt gland of *Tamarix aphylla* L. Acta Bot Coatica 69:229–235

Bosabalidis AM, Thomson WW (1986) Light Microscopical studies on salt gland development in Tamarix aphylla L. Ann Bot 54(2):169–174. https://doi.org/10.1093/oxfordjournals.aob.a086780

Boulaaba M, Tsolmon S, Ksouri R, Han J, Kawada K, Smaoui A, Abdelly C, Isoda H (2013) Anticancer effect of Tamarix gallica extracts on human colon cancer cells involves Erk1/2 and p38 action on G2/M cell cycle arrest. Cytotechnology 65:927–936

Boulaaba M, Snoussi M, Saada M, Mkadmini K, Smaoui A, Abdelly C, Ksouri R (2015) Antimicrobial activities and phytochemical analysis of Tamarix gallica extracts. Ind Crop Prod 76:1114–1122. https://doi.org/10.1016/j.indcrop.2015.08.020

Braun-Blanquet J, de Bolòs O (1958) Les groupements végétaux du bassin moyen de l'Ebre et leur dynamisme. Anales Est Exp Aula Dei 5(1–4):1–266

Brotherson JD, Field D (1987) Tamarix: impacts of a successful weed. Rangelands 9:110–112

Brown G, Mies B (2012) Vegetation ecology of Socotra. Springer, Berlin, pp 105–106

Bulos L (1983) Medicinal plant of North Africa, Michigan

Battandier MA (1907) Revision des Tamarix algériens et description de deux espèces nouvelles, Bulletin de la Société Botanique de France, 54:5, 252-257, DOI: 10.1080/00378941.1907.10831266

Chen Y, Li C, Zhang B, Yi J, Yang Y, Kong C, Lei C, Gong M (2019) The role of the late embryogenesis-abundant (LEA) protein family in development and the abiotic stress response: a comprehensive expression analysis of potato (Solanum Tuberosum). Genes 10:148. https://doi.org/10.3390/genes10020148

Christenhusz MJM, Bung JW (2016) The number of known plants species in the world and its annual increase. Phytotaxa 3:201–2017. https://doi.org/10.11646/phytotaxa.261.3.1

Conesa HM, Faz A, Arnaldos R (2006) Heavy metal accumulation and tolerance in plants from mine tailings of the semiarid Cartagena – La Unión mining district (SE Spain). Sci Total Environ 366:1–11

Dassanayake M and Larkin JC (2017) Making Plants Break a Sweat: the Structure, Function, and Evolution of Plant Salt Glands. Front. Plant Sci. 8:406. doi: 10.3389/fpls.2017.00406

DGF (2017) Rapport d'activité de la Direction général des forêts, Evaluation des projets de plantations dans la steppe sud algéroise

Dreesen DR, Wangen LE (1981) Elemental composition of saltcedar (*Tamarix chinensis*) impacted by effluents from a coal-fired power plant. J Environ Qual 10:410–416

Durate B, santos D, Marquez JC, Caçador I (2013) Ecophysiological adaptation of two halophytes to salt stress photosynthesis PSII photochemistry and antioxidant feedback: implication of resilience in climates changes. Plant Physiol Biochem 67:178–188

El Hindi MK, Sharaf El-Din A, Elgorban AM (2017) The impact of arbuscular mycorrhizal fungi in mitigating salt-induced adverse effects in sweet basil (Ocimum basilicum L.). Saudi J Biol Sci 24:170–179

Fellah O, Hameurlaine S, Gherraf N, Zellagui A, Ali T, Abidi A, Altun M. Demirtas I, Yaglioglu AS (2018) Anti-proliferative activity of ethyl acetate extracts of *Tamarix gallica* L. grown at different climatic conditions in Algeria. Acta Sci Nat 5(2):23–31

Ferlin GR (1981) Techniques de reboisement dans les zones désertiques d'Afrique. Centre de Recherches Pour le Développement International, Ottawa, p 46

Fevreau B (2012) Impact du stress hydrique sur l'anatomie et la teneur en polyphénols des feuilles de 2 génotypes d'Eucalyptus au champ. Mémoire de Master en Biologie fonctionnelle des plantes. Université Montpellier 2. 44 p

Gao C, Liu Y, Wang C, Zhang K, Wang Y (2014) Expression profiles of 12 late embryogenesis abundant protein genes from Tamarix hispida in response to abiotic stress. Sci World J:9. https://doi.org/10.1155/2014/868391

Garg N, Singh S (2018) Arbuscular Mycorrhiza Rhizophagus irregularis and silicon modulate growth, Proline biosynthesis and yield in Cajanus cajan L. Millsp. (pigeonpea) genotypes under cadmium and zinc stress. J Plant Growth Regul 37:46–63

Garg N, Singla P (2015) Naringenin- and Funneliformis mosseae-mediated alterations in redox state synchronize antioxidant network to alleviate oxidative stress in Cicer arietinum L. genotypes under salt stress. J Plant Growth Regul 34:595–610. https://doi.org/10.1007/s00344-015-9494-9

Gaston B (1998) La grande flore en couleurs, 1st ed. Belgique et pays voisins, France, Suisse, pp 373

Ghazanfar SA (1994) Handbook of Arabian medicinal plants. CRC Press, Boca Roton, p 203

Ghnaya T, Slama I, Messedi D, Grignon C, Ghorbel MH et al (2007) Cd induced growth reduction in the halophytes Sesuvium portula castrum is significantly improved by NaCl. J Plant Res 120:309–316

Guallar MB (2019) Understanding how metabolism can make or break a plant's ability to adapt to extreme conditions: the Tamarix case study in the Negev. Université de Leida

Hadj Allal FZ (2014) Contribution à l'étude du genre Tamarix: aspects botanique et Phytoécologique dans la région de Tlemcen. Mémoire de magistère. Université de Tlemcen, 170 p

Han Z, Yin W, Zhang J, Niu S, Ren L (2013) Active anti-erosion protection strategy in tamarisk (Tamarix aphylla). Sci Rep 3:3429. https://doi.org/10.1038/srep03429

Hashem A, Abd_Allah EF, Alqarawi A, Al-Huqail ESD (2016) The interaction between arbuscular mycorrhizal fungi and endophytic bacteria enhances plant growth of acacia gerrardii under salt stress. Front Microbiol. https://doi.org/10.3389/fmicb.2016.01089

Hassaine C, Aboura R, Merzouk A, Benmansour D (2014) Study of Halophytes dispersion in the North-West region of Algeria. Open J Ecol 4:628–640

Hatano T, Kagawa H, Yasohara T, Okuda T (1988) Two new flavonoids and other constituents in licorice roots: their relative astringency and radical scavenging effect. Chem Pharm Bull 36:337–341

Hopkins WG (2003) Physiologie végétale. Traduction de la 2ème édition américaine par SERGE R. Ed. De Boeck, pp 66–81

Horton JL, Clark JL (2001) Water table decline alters growth and survival of Salix gooddingii and Tamarix chinensis seedlings. For Ecol Manag 140(2–3):239–247. https://doi.org/10.1155/2018/9363868

INRF (1984) Institut National de Recherche Forestière. Rapport d'activité. Semestre 1. Ministère de l'Agriculture

Jaseim TM, Nasser NM, Al-Bazaz HK (2019) Tamarix aphylla L.: a review. Res J Pharm Tech 12 (7)
Kaabech M, Benkheira A (2000) Guide des habitats arides et sahariens. Typologie phytosociologique de la vegetation d'algérie. DGF. PNUD. FEM
Kadukova J, Kalogerakis N (2007) Lead accumulation from non-saline and saline environment by Tamarixsmyrensis Bunge. Eur J Soil Biol 43:216–223
Khabtane A, Rahmoun C (2012) Effet du biotope sur la diversité floristique et le polymorphisme phénotypique des groupements à *Tamarix africana* Poir., dans les zones arides de la région de Khenchela (Est Algerien). J Agric Environ Int Dev 106(2):123–137
Khachkhouch E, Zeid A, Khezzani B, Guehef ZH (2020) Study of the floristic composition of a wetland (Chott Edhiba) in the Souf region (northern sahara, Algeria). Int J Sci Res 76. https://doi.org/10.21506/j.ponte.2020.3.16
Kobayachi H, Massaoka Y, Takahashi Y (2007) Ability of salt glands in Rhodes grass (Chloris gayana Kunth) to secrete Na+ and K+. Soil Sci Plant Nut 53:764–771
Koull N, Chehma A (2014) Soil-vegetation relationships of saline wetlands in north east of Algerian Sahara. Arid Land Res Manag 29(1):72–84. https://doi.org/10.1080/15324982.2014.898346
Ksouri R, Falleh H, Megdiche W, Trabelsi N, Mhamdi B, Chaieb K, Bakrouf A, Magné C, Abdelly C (2009) Antioxidant and antimicrobial activities of the edible medicinal halophyte Tamarix gallica L. and related polyphenolic constituents. Food Chem Toxicol 47:2083–2091
Kumar N, Lamba S, Kumar A, Kumar P, Mann A, Devi S, Kumari PA, Rani B (2019) Antioxident defencse in halophytes under high salinity. In: Hasanuzaman et al. (ed.) Halophytes and climates changes adaptative mechanisms and potentials uses
Kuzminsky E, De Angelis P, Abou Jaoudé R et al (2014) Biodiversity of Italian *Tamarix* spp. populations: their potential as environmental and productive resources. Rend Fis Acc Lincei 25:439–452. https://doi.org/10.1007/s12210-014-0309-x
Lavaine C, Evette A, Piégay H, Brahic P (2011a) 'Génie végétal contre l'érosion des berges de cours d'eau dans un contexte de changement climatique: quelles nouvelles espèces utiliser?' 1ères Rencontres Interdisciplinaires Doctorales de l'Aménagement Durable (RIDAD 2011)
Lavaine C, Evette A, Piégay H, Lachat B, Brahic P (2011b) Les Tamaricaceae en génie végétal. Sciences eaux et territoires. https://doi.org/10.14758/SET-REVUE.2011.HS.04
Lebreton P, Bouchez M-P (1967) Recherches chimiotaxinomiques sur les plantes vasculaires — V. Phytochemistry 6:1601–1608
Lefevre I, Marchal G, Meerts P, Correal E, Lutts S (2009) Chloride salinity reduces cadmium accumulation by the Mediterranean halophyte species Atriplex halimus L. Environ Exp Bot 65 (1):142–152
Li J, Yu B, Zhao C, Nowak RS, Zhao Z, Sheng Y, Li J (2013) Physiological and morphological responses of tamarix ramosissima and populus euphratica to altered groundwater availability. Tree Physiol 33:57–68
Litwak M (1957) The influence of T. aphylla on soil composition in the northern Negev of Israel. Bull Resour Council Israel 6D:38–45
Ma HY, Tian CY, Feng G et al (2011) Ability of multicellular salt glands in *Tamarix* species to secrete Na+ and K+ selectively. Sci China Life Sci 54:282–289. https://doi.org/10.1007/s11427-011-4145-2
Ma D, Sun D, Wang C, Ding H, Qin H, Hou J, Huang X, Xie Y, Guo T (2017) Physiological responses and yield of wheat plants in zinc-mediated alleviation of drought stress. Plant Sci 8:860. https://doi.org/10.3389/fpls.2017.00860
Manousaki E, Kalogerakis N (2009) Phytoextraction of Pb and Cd by the Mediterranean saltbush (Atriplex halimus L.): metal uptake in relation to salinity. Environ Sci Pollut Res 16(7):844–854
Marius NG, Toma C (2017) Anatomical adaptation of halophytes. Springer, Berlin
Marlin D, Newete SW, Mayonde SG, Smit ER, Bayrne MJ (2017) Invasive Tamarix (Tamaricaceae) in South Africa: current research and the potential for biological control. Biol Invasions 19:2971–2992. https://doi.org/10.1007/s10530-017-1501-6
Mauseth JD (1988) Plant anatomy. Benjamin Cumming, Menlo Park, p 560

Meinhardt KA, Gehring CA (2013) Tamarix and soil ecology. Oxford University Press

Muller N (1995) River dynamics and floodplain vegetation and their alterations due to human impact. Arch Hydrobiol Suppl 101:477–512

Nawwar MAM, Buddrus J, Bauer H (1982) Dimeric phenolic constituents from the roots of Tamarix nilotica. Phytochemistry 21:1755–1758

Nedjimi B, Beladel B, Guit B (2012) Biodiversity of halophytic vegetation in Chott Zehrez Lake of Djelfa (Algeria). Am J Plant Sci 3:1527–1534. https://doi.org/10.4236/ajps.2012.311184

Newet, S.W., Allem, S.M., Venter, N., Byrne, M.J. 2018. The efficiency of Tamarix genotypes in salt excretion in South Africa. In: The 15th international phytotechnology conference. University of Novi Sad Serbia. 1–5 October

Newete SW, Allem SM, Venter N, Byrne MJ (2020) Tamarix efficiency in salt excretion and physiological tolerance to salt-induced stress in South Africa. Int J Phytoremediation 22(1):3–9. https://doi.org/10.1080/15226514.2019.1633997

Quezel P, Santa S (1963) Nouvelle flore de l'Algérie et des régions désertiques méridionales. CNRS, Paris, 1170 p

Rancic D, Pecinar I, Acic S, Stevanovic ZD (2019) Morpho-anatomic traits of halophytes species. In: Hasanuzaman et al. (ed) Hallophytes and climates changes adaptatives mechanisms and potentials uses

Rozema J, Gude H (1981) An ecophysiological study of salt secretion of four halophytes. New Phytol 89:201–2017

Saïdana D, Mahjoub MA, Boussaada O, Chriaa J, Chéraif I, Daami M, Mighri Z, Helal AN (2008) Chemical composition and antimicrobial activity of volatile compounds of *Tamarix boveana* (Tamaricaceae). Microbiol Res 163:445–455

Saoud Orfali R (2005) Phytochemical and biological study of Tamarix nilotica growing in Saudi Arabia. Master's degree report, B.Sc. Pharm, King Saud University

Sellosse MA (2017) Jamais seul. Ces microbes qui construisent les plantes, les animaux et les civilisations Ed ACTES SUD

Sghaier DB, Duarte B, Bankaji I, Caçador I, Sleimi N (2015) Growth, chlorophyll fluorescence and mineral nutrition in the halophyte Tamarix gallica cultivated in combined stress conditions: arsenic and NaCl. J Photochem Photobiol Biol 149:204–214. https://doi.org/10.1016/j.jphotobiol.2015.06.003

Sharma SK, Parmar VS (1998) Novel constitutes of Tamarix species. J Sci

Smith S, Read D (2008) Mycorrhizal symbiosis, 3rd edn. Academic Press

Smith SD et al (1998) Water relations of riparian plants from warm desert regions. Wetlands 18 (4):687–696

Sookbirsingh R, Castillo K, Gill TE, Chianelli RR (2010) Salt separation process in the saltcedar *Tamarix ramosissima* (Ledeb.). Commun Soil Sci Plant Anal 41:1271–1281

Sorensen MA, Parker DR, Trumble JT (2009) Effects of pollutant accumulation by the invasive weed saltcedar (Tamarix ramosissima) on the biological control agent Diorhabdaelongata (Coleoptera: Chrysomelidae). Environ Pollut 157:384–391

Stefani F, Bencherif K, Sabourin S, Lounès Hadj-Sahraoui A, Banchini C, Séguin S, Dalpé Y (2020) Taxonomic assignment of arbuscular mycorrhizal fungi in an 18S metagenomic dataset: a case study with saltcedar (*Tamarix aphylla*). Mycorrhiza. https://doi.org/10.1007/s00572-020-00946-y

Sultanova N, Makhmoor T, Abilov ZA, Parwee Z, Omurkamzinova VB, Atta-ur-Rahman MIC (2001) Antioxidant and antimicrobial activities of *Tamarix ramosissima*. J Ethnopharmacol 78:201–205

Surowka E, Latowski D, Libik-Konieczny, Miszalski Z (2019) ROS signaling and antioxidant defense network in halophytes. In: Hasanuzaman et al (ed) Hallophytes and climates changes adaptatives mechanisms and potentials uses

Swaminathan P, Ohrtman M, Carinder A, Deuja A, Wang C, Gaskin J, Fennell A, Clay S (2020) Water deficit transcriptomic responses dier in the invasive Tamarix chinensis and

T. ramosissima established in the Southern and Northern United States. Plants 9:86. https://doi.org/10.3390/plants9010086

Thomson WW, Liu LL (1967) Ultra structural feature of the salt gland of *Tamarix aphylla* L. Planta 73:201–220

Urbansky ET, Magnuson ML, Kelty CA, Brown SK (2000) Perchlorate uptake by salt cedar (Tamarix ramosissima) in the Las Vegas wash riparian ecosystem. Sci Total Environ 256:227–232

Waisel Y (1961) Ecological studies on Tamarix aphylla (L.) Karst., III. The salt economy. Plant Soil 13:356–364

Wang J, Xu M, Gu Y, Xu LA (2016) Differentially expressed gene analysis of Tamarix chinensis provides insights into NaCl-stress response. In: Trees. Springer, Berlin

Willdenow K (1816) Beschreibung der Gattung Tamarix. Abh Akad Berlin Phys 1812–1813:76–85

Zhang DY, Yin LK, Pan BR (2002) Biological and ecological characteristics of Tamarix L. and its effect on the ecological environment. Sci China Ser D-Earth Sci 45:18–22

Plant Synthetic Biology: A Paradigm Shift Targeting Stress Mitigation, Reduction of Ecological Footprints and Sustainable Transformation in Agriculture

15

Priyanka Singla

Abstract

At the existing growth rate of population, the necessity of agriculture-based commodities is far outstripping the yields owing to approaches employed in the Green Revolution. Moreover, current agricultural practices are excessively exploiting non-renewable energy inputs leading to global climate change and degradation of natural resources. Exploitation of modern biotechnology techniques for the developing crops that are better acclimatized to the adverse environment is significant for sustainable cultivation and harvesting a sufficient amount of food. The disparity between the use of inputs and proficient production drives the emerging influx of innovation in the field of plant biology to integrate knowledge from diverse fields spanning from molecular to ecological scales. To address these global challenges, Synthetic Biology (SynBio) has come into the forefront as an engineering discipline for the next generation of crops. SynBio potency to converge with bio- and nanotechnology, plant genetics, biochemistry, microbiology and systems biology upsurges its transformative ascendancy on the agriculture and food industry. The implementation of SynBio within agriculture promises to deliver benefits that judiciously utilize the inputs for growth thus limiting negative impacts on the environment. In this chapter, we present recent applications of SynBio in enhancing carbon fixation competence of crops, plant–microbiome interactions for efficient nutrient uptake, sustainable fertilization, bioremediation and stress mitigation, the nutritional value of crops, integrated pest management, food safety, developing green sensors, bio-manufacturing of commercially important products, and production of valuable plant metabolites in microbes. Progress and innovation of SynBio approaches in these areas promise

P. Singla (✉)
Department of Botany, Mount Carmel College (Autonomous), Bengaluru, Karnataka, India

B. Giri, M. P. Sharma (eds.), *Plant Stress Biology*,
https://doi.org/10.1007/978-981-15-9380-2_15

to unlock the masked beneficial traits in crop plants to set forth an unprecedented leap in productivity and sustainability across primary industries.

Keywords

Biofuels · Chassis · N_2 fixation · Photosynthesis · Phytosensors · Stress · Synthetic Biology

15.1 Introduction

The universal population is likely to go up to 9.8 billion by 2050 (UN DESA 2017). Although the intensification in global food production (supported by appropriate research, policies, adoption of high-yielding crop varieties, industrial revolt) has kept pace with growing population size over the past five decades, almost a billion people, most of who reside in the developing countries, suffer from inadequate diets and insecure food supplies (Varshney et al. 2011). Moreover increasing revenues and urbanization are driving a shift from traditional diets to processed foods, fortified with superior sugars, fats, oils and meats. This global dietary transition if remains unchecked necessitates an upsurge in the field of agriculture (nearly 70% augment in global food production by 2050) at a substantially faster speed than that approximated solely considering global population size (Ray et al. 2013). Unfortunately, this would be a principal sector contributing an approximated 80 per cent augment in the emission of agricultural greenhouse gases and to land clearing at a global scale (Tilman and Clark 2014). Further challenges to agriculture include desertification, salination, increased use of chemical fertilizers, exhaustion of aquifers used for irrigation, marginal soils, ecological and animal safety regulation, the rising trend of 'food as medicine'. These challenges along with competing claims for arable land, water, labour, energy and capital suggest that existing agricultural practices are also ecologically unsustainable (Varshney et al. 2011; Alexandratos and Bruinsma 2012; South et al. 2019). However, exploiting natural resources even further, for example, expanding cropland will lead to additional loss of biodiversity (Ort et al. 2015), thus directing toward technological innovations facilitating amplification of crop yield per unit area land as a principal objective (Long et al. 2006). Under existing agricultural practices planting density has by now reached its limit and yield-expansion rate of main crop plants has achieved their maxima, thus an improvement in the performance of plants at individual levels inside a crop canopy is a key focus to meet the concomitant increase in demand for food and agricultural products (Weber and Bar-Even 2019).

There is an alarming call for improvement in plant characteristics for better crop performance, chiefly to augment their yield and stress tolerance to adapt to adverse environments. Moreover, there are many de novo functions for plants to perform, such as biosensing and producing valuable compounds (Liu et al. 2013a, b). Researchers are working toward the next generation of crops that more economically use inputs for growth and alleviate environmental stress. In this biological era, the seismic shifts caused by the convergence of biotechnology, metabolic engineering, bioinformatics and synthetic biology have proved to be the key players in the front position to sustainably transform agriculture and food industries (Ort et al. 2015; Moses et al. 2017). *Synthetic Biology (SynBio),* one of the fastest-growing next-generation innovation, is defined as: 'The engineering of biology: the synthesis of complex, biologically based (or inspired) systems, which display functions that do not exist in nature' (Serrano 2007). SynBio, the growing 'green chemical' trade, is expected to be a main recipient of the growing global bio-economy where according to the BCC Research report, SynBio's economic potential is expected to reach 22% of the trillion-dollar global chemical market by 2025 (Paul and Steinbrecher 2010). SynBio propelled into eminence when the costs of DNA synthesis and sequencing significantly dropped down and key-enabling skills like versatile genome engineering tools, standardization of DNA assembly protocols, and novel sensor-reporter coordination got recognition (Flores Bueso and Tangney 2017). It entails teamwork of molecular biologists, software developers, engineers and mathematical modellers to assemble artificial building blocks for novel objectives in existing living beings or to create fully functioning de novo biological systems (Gould 2020). SynBio groundwork is deep rooted in system biology (quantitative understanding of existing biological systems) which provides defined parts (amino acids, bases, proteins, genes, circuits, cells, etc.) that can be fabricated (united and substituted) by harmonized workflows and usually in sync with modelling and computational tools to build artificial genetic code circuits that will initiate or modify existing biological functions (Di Ventura et al. 2006; Serrano 2007; Sticklen 2015; Moses et al. 2017). From an engineer's prospective, SynBio is a reductionist outlook of biology, where a cell is a circuit of inputs, processes and outputs, for example, a leaf cell can be scaled down to the light input with the output of carbohydrates. Two chief approaches used in SynBio are:

a. Top-down approach which redesign existing individual genes/molecules/enzymes/tissues/complete cells/organisms. For example, comparatively more resistant crops can be developed which can withstand changing climates and biosensors can be designed to facilitate crops with intelligence to sense and respond to their environment.
b. Bottom-up approach chases to redesign the entire programme so as to surmount tricky blockades that cannot be tackled by making changes in already active systems. Basically, a *chassis cell,* i.e. a cell with minimal complexity has to be designed for one or a number of biotechnological functions and can be customized and regulated with accuracy in a prognostic mode. It generates novel biochemical systems and organisms, e.g. a global group is synthetically designing a yeast genome to understand its fundamental properties and components (Paul and Steinbrecher 2010; Way et al. 2014).

Recognition of the indispensable elements for cellular functioning can trim down the metabolic load of secondary genetic circuits, thereby passing on the carbon to pathways bringing the best outcomes for a particular organism (Tessa Moses et al.

2017; Goold et al. 2018). In this view pioneering examples of SynBio, like a genome created to encode 387 protein-coding and 43 structural RNA genes for the growth of minimalist unicellular bacterium *Mycoplasma laboratorium* (Glass et al. 2006), play a significant role. In addition, genome transplantation of *Mycoplasma mycoides* into *Mycoplasma capricolum* cells by polyethylene glycol-mediated transformation (Lartigue et al. 2007), synthesis of an artificial bacterial (*Mycoplasma genitalium*) genome (Gibson et al. 2008) and relocating them again into bacterial cells have provided innumerable, formerly unworkable researches a ray of hope. Interestingly first eukaryotic cell equipped with fully functional 16 synthetic chromosomes is *Saccharomyces cerevisiae* are being designed now under the Synthetic Yeast Genome Project—Sc2.0 (for details read Richardson et al. 2017; Flores Bueso and Tangney 2017; Pretorius and Boeke 2018). The 'synthetic chromosome rearrangement and modification by LoxP-mediated evolution' (SCRaMbLE) system in the Yeast 2.0 genome project allows the insertion of symmetrical Cre-recombinase sites flanked by all non-essential genes. This permits for genome-wide rearrangements enabling the cell de novo ability as a chassis for genome engineering (Richardson et al. 2017; Liu et al. 2018a; Wu et al. 2018). Moreover, the grouping of a biosensor with SCRaMbLE facilitates synthetic strains to engender numerous commercial metabolites even under adverse conditions (Goold et al. 2018). SynBio technology is reshaping research and development (R&D) however development in plant SynBio is lagging behind (Xie et al. 2017). Designing a synthetic nuclear plant genome is difficult due to its intricate organization, huge size, their polyploidy poor homology directed recombination (HDR) mechanisms (Goold et al. 2018). Thus, smaller genomes within plastids like mitochondria and chloroplasts in plant cells are receiving attention from SynBio researchers (Puchta 2016).

In Planta, the organization of modular cloning tools and standardization of genetic parts were the earliest steps towards a further comprehensive execution of SynBio strategies (Engler et al. 2014). The ideal design cycles of plant SynBio include five stages: conceptualization, design, modelling, construction, probing and testing (Liu and Stewart Jr. 2015). Tools like Recombination-mediated genetic engineering (recombineering), Zinc finger nucleases (ZFNs), transcription activator-like effector nucleases (TALENs), and, more recently, the CRISPR (clustered regularly interspaced short palindromic repeats)/Cas (CRISPR-associated) system are the building blocks providing resources to achieve synthetic modulations in crops (Shan et al. 2013; Long et al. 2015; Puchta 2016; Gao et al. 2018). Most of the metabolic pathways in plants are so outsized that for the synthesis of their circuits, scientists employ 'decoupling and abstraction' techniques, i.e. breaking of the complete circuit into smaller modules, and analysing each module prior to their congregation. Mathematical modelling of all modules is vital in order to assure the rationality of the designed genetic circuits. The complication of the mathematical modelling in SynBio depends on the intricacy of the coding circuits, for example, circuits consisting of transcription factors controlling gene expression might need a more complex mathematical modelling than a circuit consisting of transcription factor independent expression of enzymes (Sticklen 2015). Moreover, to make the technology economically affordable, the modules of any circuits should be potent enough to be utilized in diverse SynBio hosts such as different crops (Bowen et al. 2008). Conversely, having a repository of modules from different organisms does not guarantee their functioning in versatile organisms; rather unexpected properties may emerge while amalgamating modules that have been characterized in a different context. Thus existing as well as novel modules must be adequately characterized for predictable utility in diverse genetic backgrounds. DNA assembly system called 'Golden Braid' containing DNA building modules (for SynBio techniques) to be used in different crop is being commercially sold (Sarrion-Perdigones et al. 2013). Engler et al. (2014) have placed 'Golden Gate Modular Cloning (MoClo) Plant Parts Kit' along with a 'Golden Gate MoClo Plant Tool Kit' in the AddGene repository. It contains all the promoters, untranslated sequences, reporters, antigenic tags, localization signals, selectable markers and terminators required to build new sequences and assembly into single and multigene binary constructs that can be used to engineer modules in many plants systems. Furthermore, SynBio involves extensive high-performance computing (HPC) for in silico predictions of the behaviour of the synthetic metabolic routes integrated within the native metabolic network and recognizing forthcoming blockages by flux balance analysis [FBA] (Küken and Nikoloski 2019). For example, assimilating sophisticated HPC to the engineering approach can foresee the consequences of photosynthesis manipulation in cross-scale models linking photosynthesis with crop field performance (Roell and Zurbriggen 2020).

Despite the fact that SynBio is a broad domain comprising numerous diverse fields, in this chapter we aim to provide an overview of diverse spectrum of SynBio applications in the field of agriculture. Potential of SynBio in optimization and re-designing of photosynthetic efficiency, generating ecologically sustainable fertilization, designing green sensors, bioremediation and stress mitigation, using photo-autotrophic bio-manufacturing, increasing crop nutritional value, producing valuable plant metabolites in microorganisms, and synthetic genomes will be discussed one after another (Fig. 15.1).

15.2 Optimizing and Re-Designing Photosynthetic Efficiency

In the present situation, in order to circumvent further ecological strain resulting from extensive agriculture, chemical fertilization and land clearance, it is crucial to fulfil global food requirements by improving crop yield without expanding the arable land. Moreover, such yield enhancement should also conjugate with efforts to reduce the global CO_2 levels (Le Quire et al. 2009). Photosynthesis, a unique feature of autotrophic organisms, falls far short of its biological limit (~11% estimated theoretically) thus providing many potential objectives for SynBio aspirants to work upon (Goold et al. 2018). Photosynthetic efficiency is characteristically influenced by three major mechanisms: a) proficiency of light (photons) captured, b) competence of photons translation into chemical energy, and c) the portion of chemical energy translocated to the harvestable organs, i.e. Harvest Index (Roell and Zurbriggen 2020). Conventional plant breeding in the past few decades

Fig. 15.1 Potentials of SynBio being exploited in eight areas of plant biology to achieve sustainability and competence in the next generation of agricultural crops

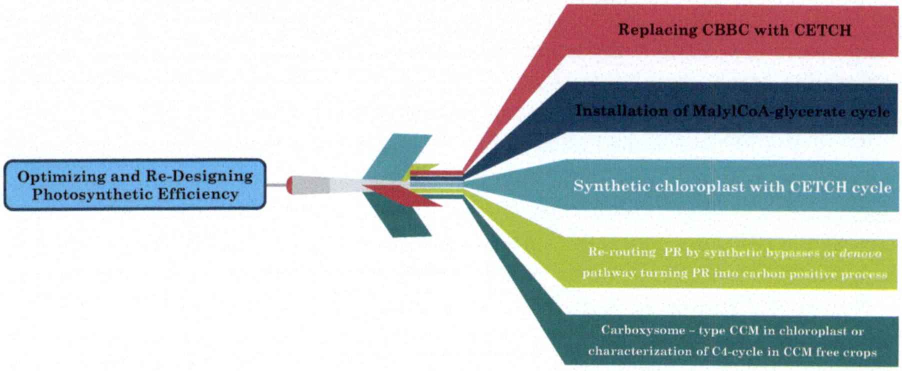

Fig. 15.2 Different approaches exploited by SynBio aspirants to enhance photosynthetic efficiency in crop plants

has significantly improved interception efficiency and harvest index but not light energy conversion competence, which depends on the effectiveness of photosynthesis and respiratory losses by the crop (Long et al. 2015). Moreover, the application of recent plant biology approaches on the road to recover carbon fixation by photosynthesis can provide solutions for the sequestration of anthropogenic CO_2 (Jez et al. 2016).

Different approaches for optimizing and re-designing photosynthetic efficiency include engineering Rubisco for higher specificity and activity, modulating Calvin–Benson–Bassham cycle (CBBC) enzymes to getaway kinetic blockage, introducing carbon-concentrating mechanisms (CCMs), and rewiring photorespiration (PR) into energetically competent routes or routes that do not discharge CO_2. Using de novo synthetic pathways including tailored-engineered enzymes offer a platform for improving competence and yield of photosynthesis (Zhu et al. 2010) (Fig. 15.2).

Schwander et al. (2016) redesigned photosynthetic carboxylation efficiency by replacing RUBISCO and CBBC with the in vitro synthetic CETCH [the crotonyl–coenzyme A (CoA)/ethylmalonyl-CoA/hydroxybutyryl-CoA] cycle. In vitro reconstitution of a synthetic enzymatic network, utilizing 17 enzymes that originate from nine organisms, yielded superior to native CO_2-fixation. Synthetic CETCH cycle can be transplanted into lithotrophic organisms, opening the route to improved CO_2 fixation (Ort et al. 2015); can be used in the development of artificial photosynthetic processes (Yadav et al. 2012), and can be the part of completely synthetic carbon metabolism module in artificial or minimal cells (Hutchison III et al. 2016). However, Bar-Even (2018) and Cotton et al. (2018) advocated that instead of replacing the too central CBBC, a parallel route to assist its activity, i.e. malylCoA-glycerate cycle can be installed, which can produce acetyl-CoA from C3 sugars without releasing CO_2 and can assimilate photorespiratory glycolate without loss of carbon (Yu et al. 2018). The most promising strategy to increase photosynthetic productivity is to entirely rewire CO_2-fixation through the amalgamation of known enzyme reactions (Erb 2011). In the course of evolution, nature has made-up five alternative microbial CO_2-fixation pathways, which have several advantages with respect to energy prerequisite and competence compared to CBBC (Berg 2011). For example, engineering *Methanococcoides burtonii* archaeal RuBisCO to tobacco chloroplasts could address the kinetic restrictions of key enzymes (Wilson et al. 2016). Recently Miller et al. (2020) designed a chloroplast imitator by encapsulating and operating photosynthetic membranes in microdroplets and by using light as an external signal. They created 'synthetic chloroplast' by incorporating the CETCH cycle which could photosynthesise from inorganic carbon, providing the basis to design an independent, entirely synthetic 'designer' carbon metabolism in the artificial organelle.

However, engineering RuBisCOs with superior CO_2-specificities and/or better catalytic rates is not easy as it is trapped in an intrinsic exchange of activity with specificity. Superior specificity for CO_2 usually results in a lower enzyme activity and vice versa is also factual (Savir et al. 2010). Hence, there is a need to look into the more appropriate alternate routes, such as cutting down wasteful reactions of CBBC, to improve photosynthetic efficiency. With the increase in temperature, the specificity of RuBisCO for CO_2 declines, and unproductive energy-intensive process of photorespiration (PR) results from RuBisCO catalysed oxygenation of Ribulose 1,5-bisphosphate (RuBP), releasing toxic by-products 2-phosphoglycolate and glycolate (Long et al. 2006). Moreover in PR, recycling of toxic products of glycine decarboxylation reaction into nontoxic products comes at the expenditure of energy and net loss of fixed carbon (Peterhansel et al. 2010; South et al. 2019). Thus alternative approach is to reduce the rate of photorespiration (PR) by increasing the concentration of CO_2 at the site of RuBisCO or rerouting PR by opening synthetic bypasses or de novo pathways that turn PR into a carbon-positive process (Ort et al. 2015; Erb and Zarzycki 2016; Eisenhut et al. 2019).

Inspired from *E. coli* glycolate oxidizing pathway, Kebeish et al. (2007) introduced bacterial glycolate dehydrogenase complex together with glyoxylate carboxylase and tartronate reductase into *Arabidopsis* chloroplast stroma. These plants could fix more carbon than the wild type, as the route adopted by them prevented the wasteful release of ammonia (NH_3) in mitochondria and hence increased the energy efficiency of the pathway (Peterhansel and Maurino 2011). Furthermore, instead of mitochondria, CO_2 is released in chloroplasts (avoided mitochondrial glycine decarboxylation to serine), thereby concentrating more CO_2 in propinquity to RuBisCO and suppressing additional oxygenation. Dalal et al. (2015) characterised the same bypass in the chloroplasts of *Camelina sativa* (biofuel crop) which proved to augment vegetative and reproductive yields by more than 50%. In an in vitro study, Trudeau et al. (2018) engineered an acetyl-CoA synthetase to convert glycolate into glycolyl-CoA and a propionyl-CoA reductase for higher glycolyl-CoA selectivity and NADPH specificity, to be reduced to D-glycerate and re-assimilated into the CBBC. These engineered enzymes were then combined with downstream condensation and assimilation enzymes (native ones): an aldolase, an isomerase and the native kinase. Together, this enzymatic sequence converted glycolate to RuBisCO's substrate RuBP, thus avoiding loss of carbon via PR. South et al. (2019) engineered mitochondrial glycolate dehydrogenase from green alga *Chlamydomonas reinhardtii* and malate synthase from peroxisomes of *Cucurbita maxima* into chloroplasts of tobacco plants. In these transgenic tobacco lines, glycolate is oxidized to glyoxylate in chloroplast by glycolate dehydrogenase and glyoxylate reacts with acetyl-CoA to form malate by the catalytic activity of malate synthase. Indigenous chloroplast enzymes malic enzyme and pyruvate dehydrogenase convert malate to acetyl-CoA and consequently to two molecules of CO_2. In addition, they maximized flux by introducing transcriptional downregulation by RNA interference of the plastidic glycerate/glycolate transporter 1 (PLGG1) that inhibited glycolate export from the chloroplast. Computational modelling of these alternative pathways highlighted the importance of optimized expression of non-native genes in preventing photoinhibition and subsequently attaining significantly improved harvestable photosynthetic quantum yield. Shen et al. (2019) designed the bypass for PR in rice by redirecting the native enzymes of peroxisomes (glycolate oxidase, oxalate oxidase and catalase) to chloroplast stroma and thus enabled superior carbon fixation in these plants over wild ones. Beyond photosynthesis, PR is associated with other essential functions such as $C1^-$, N and S metabolism, thus demanding a holistic approach while using advanced engineering strategies that completely substitute endogenous PR with synthetic ones (Eisenhut et al. 2019).

Cyanobacterial CO_2 and bicarbonate transporters add to inorganic carbon concentration in the cytoplasm, which then diffuses into carboxysomes and gets dehydrated to generate CO_2. Interestingly, low permeability of carboxysomes toward CO_2 and O_2 further guarantees high CO_2/O_2 ratio at the vicinity of RuBisCO, thereby dropping the oxygenation reaction. Thus, constructing self-assembling proteinaceous microcompartments like carboxysomes, that co-localize the RuBisCO and carbonic anhydrase (CA), can increase RuBisCO turnover by reducing PR. Using a diffusion-reaction model, McGrath and Long (2014) suggested that the execution of a carboxysome-type CCM in plant chloroplasts can potentially boost photosynthetic competence. Lin et al. (2014) used the agroinfiltration technique, to transfer multiple β-carboxysomal proteins (CcmK2, CcmM, CcmL, CcmO and CcmN) from *Synechococcus elongatus* PCC7942 to the chloroplast of *Nicotiana benthamiana* with sensors that perceive and provide fluorescent labels for visualizing the resultant structures. The shell proteins assemble in plant chloroplasts into highly organized microcompartments, thus establishing the likelihood of introducing carboxysomes into chloroplasts for probable compartmentalization of RuBisCO. Giessen and Silver (2017) proposed a synthetic minimal organelle, utilizing encapsulin nanocompartments with engineered catalytic RuBisCO and CA components to engender a CO_2-fixing arrangement. Further, encapsulin pores need to be engineered to imitate the selective permeability of natural carboxysomes. The intended route also includes implantation of bicarbonate transporters targeted to the chloroplast inner membrane, encapsulin nanocompartments into chloroplast stroma replacing endogenous lower turnover RuBisCO and minimizing stromal carbonic anhydrase (CA) activity to attain the best possible carbon flux to an engineered CO_2-fixing organelle. Long et al. (2018) characterised simplified carboxysomes, isometric with those in *Cyanobium*, within tobacco chloroplasts where they replaced the endogenous RuBisCO large subunit gene with cyanobacterial Form-1A RuBisCO large and small subunit genes, along with genes for two key α-carboxysome structural proteins. This minimal gene set organizes carboxysomes, which encapsulate the introduced RuBisCO and allow autotrophic growth at elevated CO_2.

Another CCM, i.e. C4 photosynthesis, in some photosynthetic algae, bacteria and plants have evolved mechanisms to decrease the oxygenation reaction by RuBisCO, thus signifying its incorporation into C3 crop plants (Price et al. 2013; Long et al. 2016; Schuler et al. 2016). A lot of research has been done to express the genes of the C4 pathway in rice; however, it requires many changes in both anatomy and expression of CBBC enzymes, i.e. characterisation of C4-plants into CCM-free crops (von Caemmerer et al. 2012). Wang et al. (2017) transferred the ectopic expression of *Golden-like (GLK)* transcription factor genes from *Zea mays* for establishing C4 photosynthesis in rice. In these rice plants, enlarged organelle volume of cells neighbouring the vasculature (attribute of proto-Kranz anatomy) is considered an imperative groundwork and anatomical enabler for further engineering of rice plants with C4 photosynthetic mechanism (Roell and Zurbriggen 2020). Stomatal kinetics is another factor serving a dual role in promoting photosynthetic carbon assimilation (facilitating carbon dioxide influx) and Water Use Efficiency (WUE—restricting water efflux via transpiration). Papanatsiou et al. (2019) expressed the synthetic, blue light-gated K^+ channel (BLINK1) in guard cells surrounding stomatal pores in Arabidopsis to modulate guard cell K^+ conductance. Introduction of BLINK1 accelerated both stomatal opening under light exposure and closing after irradiation and drove enhancement in stomatal kinetics to recover WUE leading to a twofold raise in biomass. Lastly, reducing redundant respiratory activity leading to unnecessary CO_2 release is the long-established scheme utilised in the engineering revolution to raise crop yield. Respiratory enzyme and transporter genes that can be engineered to (1) suppress redundant protein turnover, (2) reform respiratory metabolic activities, (3) cut futile cycles, and (4) make ion transport more efficient, can reduce respiratory costs are extensively reviewed by Amthor et al. (2019).

15.3 Ecologically Sustainable Fertilization

For sustainable food production and to prevent soil nutrient depletion, soil needs to be constantly replenished with essential nutrients like nitrogen (N), phosphorus (P) and potassium (K) [Goulding et al. 2008]. The growing demand of food production was partly dealt by the Green Revolution of the 1960s when crop yields were significantly improved through the massive implementation of chemical fertilizers, high-yielding varieties and breeding strategies to exploit plant light-harvesting capacity. However, this fails to keep pace with the increasing food demands of the growing world population, which calls for a minimum of 2.4% increase in annual yield (Ray et al. 2013). Moreover, there are many masked hazards related to chemical fertilization, for example, their toxic bioaccumulation, the release of toxic by-products causing air pollution, water eutrophication and degradation of soil health and quality (Chandini et al. 2019). Thus present scenario insists consideration of rhizospheric microbiota, which is also considered as an extension of the plant genome as it can influence plant fitness by transforming root architecture, nutrient acquisition, or biotic and abiotic stress resistance (Almario et al. 2017).

Nitrogen (N_2), although abundant in the earth's atmosphere, is a limiting nutrient and its availability often limits ecosystem productivity (Perchlik and Tegeder 2017). Unreactive, bio-unavailable N_2 in the atmosphere can be reduced to ammonium (NH_4) by only a few bacteria and Archaea, which possess the molybdenum (Mo)-dependent nitrogenase enzyme (Boyd et al. 2015). Legume plants fix a substantial proportion of the biosphere's available N_2 by developing mutualistic relation with N_2-fixing rhizobia that reside in root nodules (Rogers and Oldroyd 2014; Geddes et al. 2019). However, the rising demand for food and fodder has made chemical fertilization, an obligatory exercise adopted in today's agriculture (Perchlik and Tegeder 2017). Chemical synthesis of N fertilizers by means of the Haber–Bosch process burns 3% of the world's natural gas and adds the same percentage to global carbon emissions. The energy-intensive manufacture of nitrogen fertilizers from fossil fuels, runoff or leaching of applied chemical fertilizers, loss of fixed N_2 in the form of nitrous oxides (potent greenhouse gases) and as soluble nitrates into aquatic systems (causing eutrophication) is expected to devour approximately 2% of global energy by 2050 (Vicente and Dean 2017). Thus for a sustainable environment and economy, agricultural reliance on chemical fertilization needs to be scaled down. Moreover, the requirement of modern agriculture far outstrips both Legume-*Rhizobium* symbiotic N_2 fixation and industrially-produced nitrogenous fertilizers (Heuer et al. 2017; Perchlik and Tegeder 2017). Collectively these aspects have enlightened a renewed focus towards developing approaches to trim down the global reliance on N fertilizers. The development of plants with enhanced N use efficiency would aid in reducing fertilizer applications, lower energy costs and greenhouse gas emissions associated with its synthesis, and help mitigate the consequences of N loss into soil and water sources (Jez et al. 2016).

Since the preponderance of global calories is in cereals; hence the concept of engineering them to reconstitute N_2-fixing capability is a long-standing vision (Allen et al. 2017; Vicente and Dean 2017). This perception has been an eye-opener since the 1970s when nitrogenase was transferred from *Klebsiella pneumoniae* to *Escherichia coli* (Geddes et al. 2015). This section highlights how SynBio can use biological N_2 fixation as a blueprint for developing symbiotic N_2 fixation in cereals and agriculturally important eudicots. Three approaches have been proposed to achieve a common objective of enabling cereal crops with N_2 fixing ability (Rogers and Oldroyd 2014; Geddes et al. 2015) (Fig. 15.3).

15.3.1 Engineering Perception of N_2-Fixing Bacteria in Cereals

Four coordinated genetic programmes would provide a useful framework for structuring N_2 symbiosis in cereals: a) Nod factors recognition; b) bacterial infection; c) root nodule organogenesis and d) instituting an appropriate atmosphere for

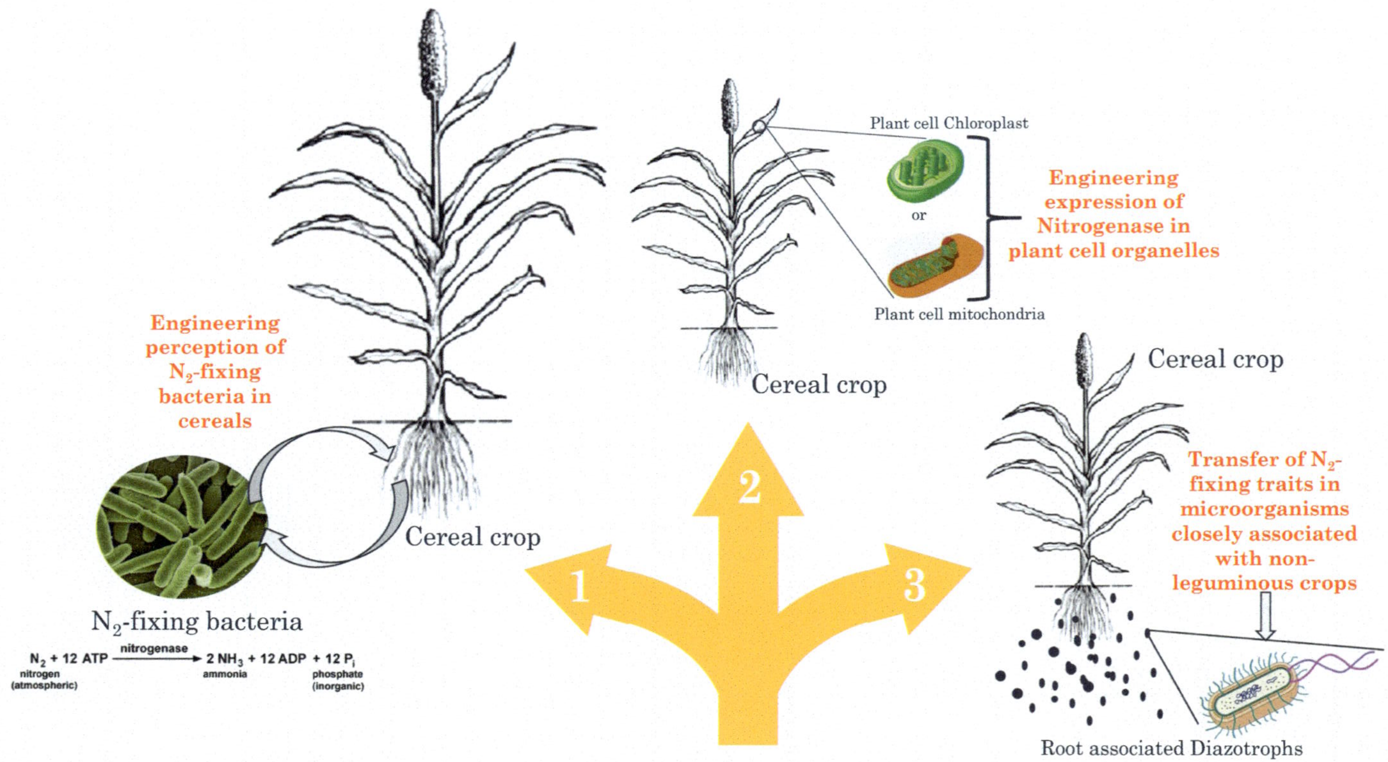

Fig. 15.3 Three approaches exploited by SynBio aspirants to enable cereal crops with N2 fixing ability

nitrogenase activity within the nodule. Engineering Nod factors perception basically means to turn on the genes of the SYM pathway, which also get activated by Myc factor perception in the arbuscular mycorrhizal symbiosis (Mus et al. 2016). Although the N_2-fixing symbiosis is restricted to legumes, several components of the legume symbiotic signalling (SYM) pathway are conserved in cereals. This indicates that cereals have an inherent potential for SYM pathway engineering to permit recognition of N_2-fixing bacteria. Zhu et al. (2006) found that rice is having genes involved in nitrogen-fixing root nodule symbioses [*NFR1*, *NFR5*, *SYMRK*, *POLLUX*, *CCaMK* (*DMI3*), *CYCLOPS* and *nodule inception* (*NIN*)] with the exception of OsSYMRK. OsSYMRK in monocots has significantly different domain structure and length as compared to that in legumes (the only longer versions allowing nodulation signalling in legumes). Thus, it is a key target of engineering nodulation signalling in cereals (Markmann and Parniske 2009). Rogers and Oldroyd (2014) suggested that the second nodulation-specific component which can be engineered into cereals is the group of transcription factors that are explicit for nodulation and are turned on downstream of the SYM pathway.

15.3.2 Engineering Expression of Nitrogenase in Plant Cells Organelles

This approach necessitates the implantation of nitrogenase-encoding atleast 16 bacterial *nif* genes into nonlegume plants (Curatti and Rubio 2014). The complexity of nitrogenase synthesis and functioning, oxygen-sensitivity and metal co-factor dependence (iron and molybdenum) present significant challenges to the employment of a functional nitrogenase in crop plants (Vicente and Dean 2017). Given these requirements for nitrogenase function, the presence of plastids and mitochondria as a potent subcellular niche to express active nitrogenase makes this engineering line effective (Mus et al. 2016). Nitrogenase can only function in niche rich in reducing power and energy, but at the same time keeping oxygen levels low. Although chloroplast permits gene expression similar to that of prokaryotes, however oxygen concentration in chloroplast can be a potent inhibitor (Burén and Rubio 2018). Thus to maintain spatiotemporal partition of photosynthetic O_2 evolution and N_2 fixation, mitochondrion (matrix possess oxygen-consuming enzymes) can be the appropriate organelle for nitrogenase reconstitution and activity in the plant cell (Curatti and Rubio 2014; Vicente and Dean 2017). Moreover, mitochondria are the chief site of plant metalloenzyme synthesis and contain biosynthetic assembly proteins, thus can offer proper functioning of equivalent Nif proteins (Balk and Pilon 2011). López-Torrejón et al. (2016) successfully isolated an ex vivo dynamic Nif Fe subunit from the mitochondrial matrix of aerobic yeast, indicating that the matrix can sustain the assembly and activity of an entire nitrogenise complex. On the other hand, Ivleva et al. (2016) study could detect only slight NifH activity in chloroplasts of plants incubated at low O_2 levels. Allen et al. (2017) re-engineered 16 Nif proteins from diazotrophic *Klebsiella pneumoniae* for targeting to the mitochondrial matrix of *Nicotiana benthamiana*. Four mitochondrial targeting peptide

(MTP)-Nif fusion proteins (B, S, H, Y) were effectively co-expressed, thus establishing the viability of reconstituting the complete complex of nitrogenase in plant cells. Yang et al. (2017) reported that few of the plant electron transport chains (ETCs) can be recruited to provide the reducing equivalents crucial for N_2 fixation. They used *E. coli* as a chassis to find out about the compatibility of Mo and Fe-only nitrogenases with ETC modules from a range of organelle of the target plant. They replaced bacterial ETC module with genes encoding ferredoxin-NADPH oxidoreductases (FNRs) and their cognate ferredoxin counterparts from plant organelles and observed compatibility of FNR-ferredoxin module from chloroplasts and root plastids with both types of nitrogenase. However, an analogous mitochondrial ETC module could not transfer electrons to nitrogenase. According to the authors, this incompatibility of mitochondrial ETC could be overcome with hybrid modules consisting of mitochondrial NADPH-dependent adrenodoxin oxidoreductase and the *Anabaena* ferredoxins FdxH or FdxB. Thus, Burén and Rubio (2018) pointed out that while engineering nitrogenase it is significant to consider the integration of *nif* gene components from different origins. Beside confirming expression and organelle targeting, it is also important to analyse whether Nif proteins accrue in soluble forms within the eukaryotic cell.

Simplification of the N_2 fixation challenge is decisively significant because it reduces the number of modules that have to be transferred into plant organelles. Additionally, it will assist in finding lost or limiting steps in the course of enabling plant cells with N_2 fixing ability. Ten proteins are mandatory to sustain N_2 fixation in *E. coli*. Plant host can provide proteins necessary for coupling cellular metabolism to N_2 fixation (NifJ and NifF) and proteins mandatory for organizing Fe and sulphur for metal co-factor assembly (NifU and NifS). NifV, already produced by a few eukaryotes, catalyzes the formation of homocitrate making the active site of nitrogenase accessible (extensively reviewed by Vicente and Dean 2017). Thus, expression of the rest of the five microbial proteins is essential to assemble a first-generation N_2-fixing non-leguminous plant, for example, nifB, an S-adenosylmethionine–dependent enzyme associated with nitrogenase catalytic site (Vicente and Dean 2017).

15.3.3 Transfer of N_2 Fixing Traits to Microorganisms Closely Associated With Non-Leguminous Crops

To achieve host control of a microbe, cereals have to manage N_2 fixation by epiphytic bacteria (growing on the root surface), endophytic bacteria (growing inside roots) or by bacteria that inhabit engineered nodules (Mueller et al. 2012). One of the approaches is to enhance the competitiveness of a particular microbe in the rhizosphere is to provide specialized carbon source that the broad-spectrum microbiota cannot catabolise (Savka et al. 2002). Oger et al. (1997) and Savka and Farrand (1997) confirmed that expression of opine synthetic genes in tobacco along with the introduction of opine catabolism genes in *Pseudomonas* provided competitive gain for colonization to opine catabolising strain over the wild-type strains. Plant growth-promoting microbe can be harmonized by engineering plants to fabricate a synthetic transkingdom signal to reroute bacteria towards roots. In analogy to natural transkingdom signalling flanked by legumes and rhizobia, rhizopine transkingdom signalling can direct synthetic symbioses to deliver N to cereals (Geddes et al. 2019). They established synthetic rhizopine scyllo-inosamine 1 (SIA)-mediated transkingdom signalling from transgenic *Medicago truncatula* and barley plants to bacteria in their rhizospheres. Another strategy is to transfer the N_2-fixing system from N_2-fixing bacteria to non-N_2-fixing bacteria in the vicinity of cereal crop. Transfer of large N_2 fixation arrangement from *Pseudomonas stutzeri* to the aerobic associative bacterium *Pseudomonas protegens* Pf-5, resulted in constitutive *nif* expression and imparted the ability to nurture micro-aerobically using N_2 as a lone N source. In N-limited surroundings, inoculation of *Arabidopsis*, alfalfa, tall fescue and wheat with transgenic *P. protegens* imparted substantial growth improvement in contrast to the near-isogenic wild type (Setten et al. 2013). Recently, Ryu et al. (2020) successfully engineered inducible nitrogenase activity in two cereal endophytes (*Azorhizobium caulinodans* ORS571 and *Rhizobium* sp. IRBG74) and an epiphyte *Pseudomonas protegens* Pf-5. As constant nitrogenase production can confer an energy burden in plants, they placed nitrogenase expression under the direction of agriculturally significant signals such as root exudates, biocontrol agents and phytohormones. Interestingly, Brown et al. (2016) described synthetic nanoparticles—cadmium sulphide nanocrystal/nitrogenase molybdenum-iron protein hybrid—as light-driven catalyst (instead of adenosine triphosphate hydrolysis) to fix nitrogen. Though it needs a thorough investigation, it is an intriguing future direction that can compete with the Haber–Bosch process (Jez et al. 2016). There are several challenges in designing symbiotic plant–microbe interactions in cereal crops. However, advanced information of the determinant and needs of biological N_2 fixation has converged with the development in SynBio technologies to make synthetic plant–microbe N_2 fixing associations a tractable endeavour (Mus et al. 2016). Thus collectively these approaches can carry the vision of self-sustainable N_2 fixing cereal crops nearer to certainty (Geddes et al. 2015).

Designing synthetic consortia comprising a microbial community fabricated by mixing selected strains using bottom-up combinations is an appropriate approach to synchronously improve the exploitation of nutrients in plants (Vorholt et al. 2017). Crop rhizosphere, when inoculated with this consortium, can inform about a range of aspects of plant–microbe exchanges that can be decoded into the agricultural appliance. Synthetic communities, with lesser intricacy as compared to native microbiota, can be comprehensively studied for its components under controlled conditions, for example, microbes can even be added, eliminated or substituted at the strain level and via in situ gene expression or silencing, particular functions can be explicitly added or removed. Further, by varying plant host genotypes for the unchanged microbial community, the potential of novel plant factors affecting the microbiota can be determined. For example, a synthetic community of seven strains was constructed to analyse the variation in phyllosphere communities as a function of *Arabidopsis thaliana* genotypes, where four out of nine *A. Thaliana* accessions tested anchoraged different community as compared to reference accession Col-0

(Bodenhausen et al. 2014). In the presence of a 35-member synthetic community, Castrillo et al. (2017) found that *A. thaliana* mutants with an altered Phosphate Stress Response (PSR) assembled different root microbiota than wild-type plants. The synthetic community activated PSR and repressed expression of plant immunity genes under phosphate-limiting conditions by enhancing the activity of the master transcriptional regulator of the PSR, i.e. PHR1 (Castrillo et al. 2017). Niu et al. (2017) studied seven-species synthetic community inoculating maize roots and by eliminating bacterial species one at a time, they concluded that *Enterobacter* species is the keystone for community assembly. While in the absence of *E. cloacae*, the unlimited abundance of *Curtobacterium pusillum* increased over other species, whereas a small synthetic community was superior at resistance against late blight disease caused by *Fusarium verticillioides*.

Interestingly, Arkin (2017) conceptualized a single synthetic *Pseudomonas* by using two strains of a diverse clade, i.e. *P. stutzeri* PDA—for perchlorate reduction and *P. stutzeri* (strain A1501)—capable of nitrogen fixation. In manned Mars missions, this synthetic strain will potentiate soil-based agriculture (by plants or microbes) on Mars. The advantages of this architecture include low initialization collection of microbial cells, on-demand cell growth with in situ resources, perchlorate wash off to cleanse soil and the exclusion of toxic wastewater.

Another very important sector of agriculture is to develop integrated pest management wherein SynBio-enhanced engineering can be utilized to develop crop resistance against many evolving pathogens (Pixley et al. 2019). Park et al. (2015) used SynBio technology to develop abscisic acid (ABA) receptors, engineered to be turned on by mandipropamid instead of ABA. When sprayed with mandipropamid, reprogrammed plants successfully survived drought stress by activating the ABA pathway, which led to the closing up of the stomata to avert water loss. Ostrov et al. (2016) removed seven codons from the *E. coli* genome by more than 62,000 mutations to totally recode the *E. coli* genome for virus resistance. Bennett (2017) suggested that designing genetically engineered agricultural pests like fruit flies and bollworm, can substantially drop the application rate of broad-spectrum insecticide. One of the approaches in the direction can be to insert a self-limiting gene to combat the pest. Offsprings of pest can inherit the gene in the wild, and die before adulthood stage, subsequently declining the wild populations. Later, the same self-limiting gene can be turned off with an antidote to facilitate further construction of the engineered pests. Since plastid is the target of multiple disease-causing viruses in plants, a futuristic and transformative approach for broad pest resistance, is the production and installation of a synthetic chloroplast genome, a synplastome (Piatek et al. 2018). If replacement of plastome with synplastome becomes successful, it could facilitate extensive metabolic engineering to recode the genome to endow broad resistance against all plastid-targeted viruses, for example, against bacteriophage infectivity while utilising bacteria in bioreactors and laboratory cultures. Recoded genomes make the codons of attacking viruses untranslatable within host, for example, if a viral stop codon is removed from or recoded to a different stop codon, the virus cannot transmit a disease to the host (Ma and Isaacs 2016).

15.4 Green Sensors

Timely detection and anticipation of abiotic and biotic stresses in plants either during pre- or post-harvesting stage can combat the challenge of sufficient food availability throughout the world. Critical research is essential for the development of sensors to timely monitor plant ecology and health (Kumar and Arora 2020). Green sensors are genetically-encoded sensors, may be a promoter or a protein that can act in response to external stimuli, and generate an output, characteristically gene expression (Shetty et al. 2003). 'Smart plant' are the ones programmed with a synthetic signal transduction pathway with a modular receptor to act in response to a wide variety of environmental factors (Pouvreau et al. 2018). The variety of exogenous stimuli accessible to green sensors (commonly called as biosensors) is far-reaching, for example, biosensors are reported for carbohydrates, coenzyme B12, amino acids, organic acids, heavy metals, light, pathogens and plant hormones (Goold et al. 2018). There are two categories of biosensors, i.e. cell-free and whole cell-based systems. In cell-free systems, purified enzymes, proteins and antibodies act as probes to detect exogenous signal molecules quickly and specifically. However this approach is expensive as the expression and refinement must be optimized, and the active agents have to be immobilized (Mehta et al. 2016). Conversely, whole cell-based biosensors are economical to cultivate and can be used in manifold assays, i.e. depending on target analytes, biosensors can be modulated through genetic manipulation (Hynninen and Virta 2010). For constructing cell-based biosensors, standardized bioparts are recognized from genome databases. Such bioparts are engineered to synthetically improve sensing modules such as transcription factors and promoters, by increasing exogenous inputs into chassis cells or by bringing in artificial genetic circuits or logic gates. On the other hand, cross-reactive bioparts should be removed from the genome of chassis cells (Kim et al. 2018). The specificity and sensitivity of constructed biosensors to exogenous analytes should be analysed by a range of experimental methods such as measurement of total bioluminescence and fluorescence of engineered chassis.

In order to assess the toxicity and mutagenicity of polluted soils, a whole plant-based sensing system act as cost-effective and real-time plant sentinels, or 'phytosensors' (Mazarei et al. 2008; Liu et al. 2013a, b). Some good examples of sentinels include synthetic circuits for phytosensing of explosives (Antunes et al. 2011) or bacterial pathogens, nematodes or elicitors in transgenic tobacco and *Arabidopsis* (Liu et al. 2013a, b, 2014). A histidine kinase (HK) based signalling system, evolutionarily conserved between plants and bacteria, can be designed and rapidly tested by SynBio tools in bacteria to be consequently implanted in plants with little modifications such as codon optimization and proper targeting. Antunes et al. (2011) assembled an absolute synthetic signal transduction pathway in Arabidopsis root cells that connects input from computationally redesigned periplasmic binding proteins (PBPs) like ribose-binding protein (RBP) to a visual response (de-greening). PBPs on the apoplast were engineered to sense and bind small extracellular ligands, such as the explosive 2,4,6-trinitrotoluene (TNT) and develop an affinity for the extracellular domain of a chemotactic protein, Trg. This complex turns on intracellular PhoR, the HK cognate of PhoB, which translocates to the nucleus and activates transcription of an output gene under the control of a synthetic plant promoter. When such plants are exposed to TNT, the de-greening circuit hinders new chlorophyll synthesis rather upregulate the expression of chlorophyll degradation genes and produces a visual response that is even distantly detectable and quantifiable (Antunes et al. 2011). Morey et al. (2011) engineered a partial synthetic pathway in which a synthetic promoter (PlantPho) is turned on using a plant-adapted PhoB (PhoB-VP64) and the endogenous HK-based cytokinin signalling pathway. Periplasmic binding protein (PBP) computationally redesigned to bind TNT, interacts with Trg which fuses with HK, PhoR allowing signal transduction resulting in transcriptional activation of an output gene in the nucleus (Morey et al. 2011). Peng et al. (2014) developed a set of 'Fukusensor' transgenic phytosensors to examine the loss-of-function scheme using a range of gamma radiations. They harboured a marker gene (*gfp* gene) encoding a fluorescent protein in agenetic background of defective DNA repair system and hypothesised that these mutant plants would have relatively higher probability of accruing mutations in the marker gene relative to engineered plants in a nonmutant background. Their study indicated that latter plants could serve in an appropriate sensor system to report the genetic effects to organisms in response to radionuclide contamination. Modular G-protein coupled receptor (GPCR) system developed using chimeric BRET-biosensors (Bioluminescence Resonance Energy Transfer) is a broad-spectrum biosensor used by Dacres et al. (2011) to sense the femtomolar quantity of diacetyl using an odorant receptor protein from *Caenorhabditis elegans*. This approach in sentinel plants would offer agricultural farming systems a better on-site judgement making aptitude, thus dropping down the application of chemical fertilizers (Goold et al. 2018; Wurtzel et al. 2019).

Auxin-dependent formation of a co-receptor complex between TIR1/AFB (F-box proteins, constituents of an SCF E3 ubiquitin-ligase complex) and Aux/IAA (family of negative regulators of the auxin response) results in ubiquitylation and proteolysis of Aux/IAA, thereby relieving the suppression of auxin-responsive genes (Tan et al. 2007). In order to investigate time and space-resolved quantitative monitoring of auxin dynamics, two types of biosensors have been synthetically designed one relying on the ubiquitylation of fluorescent proteins fused to Aux/IAA and the other on the co-receptor complex directly like chemi-luminescent ratiometric auxin sensor (Wend et al. 2013). Green fluorescent protein (GFP) synthetic sensor has been constructed in *Arabidopsis* and *Zea mays* to scrutinize known cytokinin function and to also reveal a few novel functions. Cytokinin binding commences a multistep phosphorelay signalling cascade that eventually phosphorylates nuclear transcription activators [i.e. the type-B nuclear ARABIDOPSIS RESPONSE REGULATORs (ARRs)]. Type-B ARRs recognises concatemer of 24 repeats of the consensus sequence which drives GFP expression patterns (Zürcher et al. 2013). A red light-regulated synthetic switch has been designed by utilising the operator site of *Arabidopsis* phytochrome interacting factor 6 (PIF6) upstream of a minimal promoter. Red light illumination switch phytochrome B (PhyB) into the active far-red form and consequently encourage heterodimerization with its PIF6 and turns on the reporter gene in tobacco protoplasts. Interestingly, this switch can be efficiently turned off by far-red light (Muller et al. 2014). Muller et al. (2014) constructed a light-inducible system by using an interdisciplinary SynBio approach embracing mammalian and plant cell systems to typify a split transcription factor based on the plant photoreceptor phytochrome B and one of its interacting factors (PIF6). This system in tobacco protoplasts gave toggle switch-like characteristic which could be turned ON when illuminated with red light (660 nm) and could be instantly returned to the OFF state by successive illumination with far-red light (740 nm). They also utilised this interplay for tuning auxin signalling networks in *Nicotiana tabacum* protoplasts and for chemical-inducer free manufacture of therapeutic proteins in the moss *Physcomitrella patens* (Muller et al. 2014). Similarly, Chatelle et al. (2018) engineered a green light-responsive system based on the light-sensitive bacterial transcription factor CarH and its related DNA operator sequence CarO from *Thermus thermophilus* to regulate gene expression in *A. thaliana* protoplasts.

Smart plants harmonize with decision support systems by assisting farmers to screen crop phosphorous (P) status by expressing in situ nontoxic marker gene. Marker gene expression should commence specifically in reaction to P deficit and well before P starvation restricts growth and development of the plant. Keeping this in mind, Hammond et al. (2003) constructed transgenic Arabidopsis bearing a marker gene (GUS) under the control of the promoter sequence for one of the P-sensitive genes (SQD1) encoding synthesis of sulfolipids. Leaves of these plants had GUS activity after P withdrawal, which was noticeable earlier than P starvation could limit plant growth. Thus, this technology can allow meticulous execution of P fertilization, preservation of natural resources, and improve sustainability (Hammond et al. 2003). In addition, ABA receptors can be engineered to get triggered by a synthetic molecule (that can be sprayed on a crop) which can facilitate intervention to decrease water use in reaction to such inputs that plants themselves cannot perceive like weather forecasts (Peterson et al. 2010).

Yeast *Saccharomyces cerevisiae,* growing at low pH, is favoured microbial host producer and chassis cell for the production of economically important organic acids (building blocks of a variety of products such as plastics, polymers and animal feed, nylons, flavours and fragrances) (Sauer et al. 2008; Abbott et al. 2009; Williams et al. 2017). Williams et al. (2017) recognized the *S. cerevisiae* WAR1 transcriptional regulator and PDR12 promoter as an organic acid biosensor that can be utilized to perceive varying levels of para-hydroxybenzoic acid (PHBA) production from the shikimate pathway as an input to express the output of green fluorescent protein (GFP) expression. PHBA dependent GFP expression significantly increased, by engineering positive feedback expression of the WAR1 from its target PDR12. They used this biosensor in combination with a rational strain engineering approach to precisely isolate PHBA-producing strains of *S. cerevisiae* even when they corresponded to only 0.01% of a population. Fowler et al. (2010) constructed intracellular sensors dependent on an RNA regulatory element known as a riboswitch to detect the coenzyme B12 levels in *E. coli* cells. These probes sensitively detect their target using colorimetric, fluorescent or luminescent reporters. To assess their utility in related cellular processes, the sensors were applied to scrutinize the synthesis as well as the import of coenzyme B12. These sensors also monitored the effects of genetic deletions, recombinant expression of foreign genes, and varied growth conditions on both of these processes. Verhounig et al. (2010) also employed synthetic riboswitches in tobacco chloroplast genomes as 'translational regulators' of the gene expressing in response to exogenous ligands.

Recently, Kumar and Arora (2020) have extenisvely reviewed the nano-inspired biosensors for the detection of pathogens, viruses, bacteria, fungi and abiotic stress-induced plant diseases. Uses of a varied variety of nanomaterials in the form of nanoparticles/tubes/rods/wires have facilitated quicker detection and reproducibility of many biotic and abiotic factors (Kumar and Arora 2020). The localized surface plasmon resonance (LSPR) of gold nanoparticles—AuNPs were used to construct a colorimetric nano-biosensing system to spot Tomato yellow leaf curl virus genome in infected host (Razmi et al. 2019). Wang et al. (2019a, b) designed a visual colour change and electrochemical differential pulse voltammetric biosensor for the detection of Cucumber green mottle mosaic virus and Watermelon mosaic respectively. Hong and Lee (2018) enlisted nano-material based biosensors for detecting viral infection in plants. Certain biosensors for the detection of fungal infections like surface plasmon resonance-based immunosensor to *Pseudocerocospora fijiensis* infection in banana (Luna-Moreno et al. 2019), hyperspectral analysis for detection of rusts and blights in monocots (Mahlein et al. 2013; Das et al. 2015; Zheng et al. 2018), gold nanoparticle (AuNP)-based lateral flow biosensor for detection of *Phytophthora infestans* causing late blight in potatoes and tomatoes (Zhan et al. 2018) have been developed. In order to determine ABA (hormone regulating plant abiotic stress management) in vivo in plants, bioprobes like 'Abaleons', 'ABACUS' and 'Cameleon' can be used, where they change colour of fluorescence or emit green fluorescence when ABA binds to receptor or biosensor for expression of GUS under stress can be utilised (Choi and Gilroy 2014; Pandey et al. 2018)

15.5 Bioremediation and Stress Mitigation

All anthropogenic activities have the prospective to generate ecological contaminants like organic molecules from fuel and chemical spills, cultivation, industry, and forestry and inorganic contaminants from mining, irrigation, and geochemical cycles. The use of microbes and plants for remediation proposes noninvasive, economical and eco-friendly alternatives to clean up the polluted area (Pilon-Smits 2005). de Lorenzo (2017) suggested that for the future sustainability of agriculture, bio-processes like (1) development of non-photosynthetic microbial capturing of CO_2 can help in reducing global warming and greenhouse gases, (2) characterisation of H_2O-capturing proteins in draught-resistant soil microbes can increase humidity of arid ecosystems, (3) designing of bacteria with complete plastic mineralization pathway and spreading of plastic-flocculation surface adhesions can clean up plastic waste in marine ecosystems, (4) engineering of phosphate hyperaccumulators can help in recovering diluted phosphorus from marine ecosystems, (5) construction of microbial strains engineered for absolute de-polymerization of lignin into building blocks and designing microbial biosynthetic pathway of recalcitrant lignin forms can manage lignocellulosic compounds.

The fundamental principle of bioremediation is that microorganisms are able to generate energy for their growth and productivity by degrading hazardous contaminants. Besides spontaneous bioremediation, there is a need for conscious in situ release of engineered or tailor-designed microbiota to degrade organics and convert heavy metal(oid)s into less toxic forms (Ravikumar et al. 2017). Microbial whole-cell biosensors produce a quantifiable real-time signal (growth characteristics, enzymatic activity) enabling detection and quantification of chemical effluents in soil (Lagarde and Jaffrezic-Renault 2011; Aracic et al. 2015). SynBio development has allowed a reporter gene to be placed under the regulation of a promoter that is transcriptionally active in the incidence of an explicit pollutant. CSIRO presented a very interesting example where engineered microorganisms can act as biosensors to timely detect contaminants in soil or feedstock. CSIRO constructed CYBERNOSE® sensor by implanting a protein replicating the smell receptors of the nematode into yeast or bacteria. In the presence of toxins or contaminants in food, CYBERNOSE changes shape and emits a mixture of blue and green light indicating the binding of the certain foreign molecule, thus providing food safety (Vickers 2016). SynBio strategy can even assemble the necessary genetic circuits which not only perform real-time scrutiny of the exogenous toxin but can also trim down its level. These synthetic genetic circuits can be assembled using a two-component regulatory system (TCRS) in bacteria (Casino et al. 2010). TCRSs identifying organic compounds (benzene, toluene, ethylbenzene, biphenyl, styrene, fumarate and malate) also regulate the gene expression of catabolic enzymes for these compounds and subsequently the breakdown product can be a carbon resource for microorganism (Tropel and Van Der Meer 2004). However, Ravikumar et al. (2017) advised that the engineered bacterial system should modulate bacterial genes only in the presence of the target in the environment. In today's era, numerous genetic switches (assembled using transcriptional repressors or activators) are accessible to turn on gene expression once an exogenous toxin has reached its activation threshold. Few switches developed to control the cellular responses are (1) inverter switches producing a reciprocal output (Yokobayashi et al. 2002); (2) biphasic switches using both negative and positive regulation and responding to even minimal inputs (Isaacs et al. 2003); (3) toggle switches using two repressors that cross-regulate each other's promoters (Gardner et al. 2000); and (4) riboswitches regulating gene expression by inhibiting protein synthesis (Mandal and Breaker 2004). Similarly, many logic gate types have been developed for biological circuits, including 'NAND', 'NOT IF' and 'NOR' (Guet et al. 2002). The incorporation of nanomaterials (gold particles, magnetic beads and carbon nanotubes) and electron mediators in electrochemical biosensors can improve the detection range of numerous water contaminants (Lagarde and Jaffrezic-Renault 2011). For example, the arrest of electroactive *Pseudomonas putida* cells on electrodes using carbon nanotubes resulted in an 80-fold augment in sensing accuracy and 2.8-fold rise in reaction time to trichloroethylene (Hnaien et al. 2011). Cyanobacteria strains have been engineered as in situ bioindicators of ecotoxicity, N, P and metal(oid)s concentration using luciferase in a light-on/-off response (Mateo et al. 2015). A *Shewanella oneidensis* biosensor uses outer membrane cytochrome complexes for concentration-dependent detection of arabinose as electrochemical signals (Golitsch et al. 2013). Beside bioremediation, another potential strategy to address land use problem is to engineer plants to grow even in marginal land. Crop plants can be reverse engineered (Friedel et al. 2012), using the sequences from plants like the aquatic and halophilic angiosperm *Zostea marina*, to not only cultivate in non-arable land but also aid in its restoration (Olsen et al. 2016).

Development in plant-based remediation and the designing of biosensors can offer economical tools against ecological stresses. Metabolism of 2,4,6-trinitrotoluene (TNT) in plants produces a nitro-radical which generates toxic superoxide by reacting with oxygen. Engineered plants, having a knockout of gene encoding monodehydroascorbate reductase (generates the nitro-radical), exhibit superior TNT tolerance (Johnston et al. 2015). In addition, regulating enzymatic activity like phytochelatin synthase can also modify plant adaptation to toxic metals like cadmium (Cahoon et al. 2015). Plants can also act as biosensors for scrutinizing polluted sites (see details as sentinel plants in the section of biosensors). Plants produce many dedicated metabolites as defence against attack from herbivores and pests and to acclimatize to abiotic stresses. Engineering secondary metabolites by escalating the flux in particular pathways and by understanding the regulation of biosynthetic metabolons can be of incredible significance for developing resistant crop varieties (Jirschitzka et al. 2013; Laursen et al. 2016). Important examples in this regard are metabolic engineering of a cyanogenic glycoside—dhurrin (Kristensen et al. 2005) and of raffinose family oligosaccharides in *A. thaliana* in order to enhance resistance to biotic stress like *Myzus persicae* feeding (Cao et al. 2013; Barah and Bones 2015).

Setting up of synthetic promoters in conjunction with transcriptional tools to substitute plant's original promoters can enhance the development of stress-resistant transgenic plants. Chen et al. (2010) proposed that WRKY transcription factors form an intricate regulatory network that regulates gene expression in plant defence responses against stress by acting as either transcription activators (WRKY18 and WRKY60) or repressor (WRKY40). CRISPR–Cas9 is a genome editing tool (editing—point mutation, deletions/insertions of genes or gene fragments; transcriptional regulation—CRISPRi, activation or repression, epigenomic modulations; forward genetic screens—loss of function, knock down or activation mutants), targeted to achieve improved biotic and abiotic stress tolerance in crop plants (Liu et al. 2013a, b; Jain 2015; Ricroch et al. 2017). Wang et al. (2014) used TALEN and CRISPR-Cas9 to incorporate targeted mutations in the three homoeoalleles that encode MILDEW-RESISTANCE LOCUS (MLO) proteins in hexaploid bread wheat to confer heritable broad-spectrum resistance to powdery mildew. The mutation of *OsSWEET13* (one of the five SWEET genes) with CRISPR/Cas9 technology specified the prerequisite of *OsSWEET13* expression for the state of PthXo2-dependent disease (bacterial blight) susceptibility of rice to γ-proteobacterium *Xanthomonas oryzae* pv. Oryzae (Zhou et al. 2015). Li et al. (2016) used this approach to bring in natural double amino acid substitutions (T102I + P106S

(TIPS)) to rice 5-enolpyruvylshikimate-3-phosphate synthase (EPSPS), which conferred resistance to glyphosate. They reported intron-mediated site-specific gene replacement and insertion (endogenous exon 2 with a new exon containing the TIPS substitutions in endogenous *OsEPSPS* gene) approach to engender mutations using the non-homologous end joining pathway CRISPR-Cas9 system. Butler et al. (2016) used a geminivirus replicon (GVR) to deliver sequence-specific nucleases targeting the potato *ACETOLACTATE SYNTHASE1 (ALS1*—a key target for important herbicides including chlorsulfuron and bispyribac sodium*)* gene and repair templates designed to fit in herbicide-impeding point mutations within the *ALS1* locus. These point mutations were able to support a reduced herbicide susceptibility phenotype of potato. Sun et al. (2016) generated homozygous herbicide-resistant rice plants by using two gRNAs and providing repair templates from plasmid and free double-stranded DNAs to substitute two distinct amino acid residues in the ALS1. Shimatani et al. (2017) designed a fusion of CRISPR-Cas9 and activation-induced cytidine deaminase (Target-AID) for point mutagenesis. In rice, they encouraged multiple herbicide (Imazamox)-resistance point mutations in *C287T* gene, resulted in an A96V amino acid substitution. While in tomato they produced marker-free plants with homozygous heritable DNA substitutions, signifying the possibility of base editing for improving crop's stress resistance. Nekrasov et al. (2017) generated a tomato variety resistant to the powdery mildew fungal pathogen *Oidium neolycopersici* ('Tomelo') using the CRISPR/Cas9 technology. They used whole-genome sequencing to illustrate that Tomelo is non-transgenic and only carries a 48 bp deletion that is identical to naturally existing mutations. By utilising RNAi (RNA interference) and CRISPR-Cas9 strategies, Li et al. (2017) demonstrated that two Arabidopsis glycosyltransferase genes—*UGT79B2* and *UGT79B3* are regulated by CBF1 (CRT/DRE-binding factor 1, also named DREB1B) and confer abiotic stress tolerance via modulating anthocyanin accumulation. Five pCas9/CsLOB1sgRNA constructs were designed by Peng et al. (2017) to modify the effector binding element (EBEPthA4) of the *CsLOB1* promoter, so as to enhance citrus canker resistance in Wanjincheng orange. Earlier in 2016, canker-resistant citrus varieties via Cas9/sgRNA were also generated by Jia et al. (2016). They modified the PthA4 effectors binding elements (EBEs) in the CsLOB1 Promoter of the *CsLOB1* (*Citrus sinensis* Lateral Organ Boundaries) gene in Duncan grapefruit. Hahn et al. (2017) designed a generic vector system to be exploited to clone any sgRNA sequence in a plant T-DNA vector containing a ubiquitously expressed Cas9 gene. With this vector, they explored *BIALAPHOS RESISTANCE (BAR)* and *GLABROUS1 (GL1*), where BAR, as a positive selection marker, confers resistance to glufosinate and GL1 is required for the formation of protective trichomes. Shi et al. (2017) employed a CRISPR-Cas technology to insert native maize GOS2 promoter into the 5′-untranslated region of the native *ARGOS8* gene (negative regulator of ethylene responses). The ARGOS8 variants had elevated levels of ARGOS8 transcripts relative to the native allele and compared to the WT, these variations generated drought-tolerant crops. Thus, CRISPR-Cas9 system assures its enormous involvement in exploration of the gene regulatory networks fundamental for biotic and abiotic stress adaptation and crop improvement schemes to raise stress-tolerant plants.

Addition of *cis* element is supportive for effectual expression of synthetic promoters to: i) reduce its off-target gene expression, ii) make the synthetic promoter more functional, iii) improve synthetic promoter strength. EL17—a W-box cis-acting element (in the promoter of pathogen responding gene ELI17 related to parsley) and cis-acting element F against fungal infection are important for designing synthetic promoters to regulate transcription of genes linked with pathogenic defence mechanisms in plants (Aysha et al. 2018). Shokouhifar et al. (2011) used F and EL17 in dimers to be placed upstream of the minimal *CaMV 35S* promoter. The resulting constructs of synthetic pathogen-inducible promoters in canola (*Brassica napus)* plants responded significantly to phytohormone treatments and fungal elicitors. Moradyar et al. (2016) demonstrated that the synthetic pathogen-inducible promoter SP-DDEE (having parsley D, E17 elements and a minimal promoter) significantly reduced the development of pathogenic attack by *Sclerotinia sclerotiorum*. Pathogen-inducible promoters like WUN-box, MYB, bZIP, ERF, WRKY and MYC TFs are appropriate candidates in providing defence during the primary stages of pathogen colonization or subsequent to the invasion of insects. Cold stress-inducible promoters like rd29A, *Arabidopsis COR* gene promoters like *cor47, cor6.6*, and *core 15a* are characterized to cut down unfavourable effects under diverse environmental circumstances. *Wsi18* and *OsNCED3* genes promoters have been found to be extremely inducible by ABA, drought, and high salt doses in transgenic rice (Bang et al. 2013). Hou et al. (2012) demonstrated the activity of three inducible synthetic promoters—*EKCM, EKCRM* and *ECCRM* (linked to the *GUS* reporter gene), having multiple *cis*-acting elements that drive cold and salt stress-inducible transgene expression with negligible negative impact on transgenic Arabidopsis (GUS activity under stress was significantly higher in transgenic plants). Ranjan and Dey (2012) engineered the EFCFS promoter by inserting extra copies of the stress-inducible 'AAAG' cis-motif (Dof-1) to engender hybrid–synthetic promoters—EFCFS-HS-1 containing 10 'AAAG' motif. In the presence of abiotic stress elicitors, salicylic acid and jasmonic acid, these promoters demonstrated improved activity compared to conventional constitutive promoters like EFCFS and the CaMV35S promoters. This study signifies the importance of de novo hybrid–synthetic promoter in the upcoming engineering of wide range of agronomically significant plants for sustainable agriculture in marginal soils facing abiotic and biotic stresses. Lepidopteran insects resistant transgenic tomatoes were designed from the modified insecticidal *cry1Ac* gene of *Bacillus thuringiensis*. Pcec predicts higher GUS expression in these transgenic plants as the *Pcec* expression cassette comprised of the synthetic constitutive promoter *Pmec* and transcription activation module for superior GUS expression (Sawant et al. 2001). Widespread DNA binding sites can be placed in different synthetic promoters for exploiting a regulated and selective expression pattern, (Petolino and Davies 2013; Aysha et al. 2018). Engineering C3 plants with CAM-related genes can be potent enough to improve their water use efficiency (WUE) and drought resistance (Borland et al., 2014; Yang et al., 2015). Relocating CAM carboxylation and decarboxylation genes from the facultative CAM species *Mesembryanthemum crystallinum* or engineering of CAM on-demand systems, for example, engineering of drought-responsive TFs in various gene families like AP2/ERF, MYB, WRKY, NAC, NF-Y, bZIP from the facultative and obligate CAM plant to the C3 species can enhance their drought resistance with improved WUE. In the upcoming times, genetic circuits (e.g. toggle switches, feedback loops, Boolean logic gates) linking CAM genes and drought-responsive TFs through SynBio cycles will be necessary for drought avoidance and tolerance (Yang et al. 2020). For improving salinity tolerance in plants, SynBio can modify the expression of native genes of ion transport proteins (Shan et al. 2013; Kumar and Jain 2015) and/or design novel artificial ion channels and transporters with desired properties of ion selectivity filters and voltage sensors. Lot of research is being carried out in this direction and de novo synthesis of functional Zn^{2+}-transporting four-helix transmembrane protein bundle by Joh et al. (2014) further strengthens the feasibility of constructing artificial proteins with predefined transport properties and for designing de novo regulatory networks [to design artificial plants with desired salinity tolerance] (Volkov 2015). Moreover, salinity tolerance pathways—including osmolyte synthesis, degradation of ROS, and altered signalling systems—are also being explored for engineering plants with the ability to grow on saline soils (Roy et al. 2014). In another study by Ge et al. (2018), the combination of functional genes for cytosolic ABA receptor kinase 1 and regulatory components of ABA receptor 11 with synthetic promoters (Ap, Dp and ANDp) generated drought stress tolerance, as seen by application of exogenous ABA or co-transformation with the effector dehydration-responsive element-binding protein 2A.

15.6 Sustainable Bio-Manufacturing

Access to uncontaminated, inexpensive and trustworthy natural resources has been a biology-driven keystone solution to the globally increasing affluence/carbon foot printing since the establishment of the industrial revolution. Commercially promising host organisms, i.e. prokaryotic blue-green algae (BGA) are appropriate for the fabrication of tiny secretive molecules such as alcohols, organic acids and fatty acids. In addition, eukaryotic microalgae can be engineered to produce storage commodities such as proteins, lipids, starch and alkanes. Although, a number of biofuels can be produced from BGA and eukaryotic microalgae (Wijffels et al. 2013), there are still many blockages that need to be resolved by SynBio tools prior to their commercialisation. Currently, plant-based manufacturing, i.e. *Molecular Farming* is being used for the production of commercially important products such as enzymes, antibodies, biofuel, biopharmaceuticals, biocatalysts and research-grade growth factors, cosmetic ingredients and many more. Members of planta have come out as noteworthy fabrication platform for manufacturing recombinant proteins with the target to use them directly as the product in either plant itself or in the form of purified or crude extract, rather than looking for a phenotypic or metabolic transformation (Tschofen et al. 2016). In other words, superior production scalability (As their products do not need expensive additives and ecologically damaging pre- as well as post-treatment processes) and no endotoxin production in photoautotrophic organisms make them commercially appropriate hosts for SynBio approaches (Roell and Zurbriggen 2020). Standardized methods have been recognized to improve the design of molecular farming products and furthermore, plant cell suspension cultures can be screened to opt the most prolific cells to be used for high-yielding monoclonal cell lines (Kirchhoff et al. 2012). The moss *Physcomitrella patens,* a model organism in SynBio, is an excellent green metabolic cell factory for the characterisations of novel biosynthetic pathways and sustainable production of highly valuable compounds such as terpenoids (Pan et al. 2015), biopharmaceuticals (Reski et al. 2015) and cosmetic products (Henes et al. 2018). Mosses have delivered the efficiency in the delivery of taxadiene synthase from *Taxus brevifolia* (Anterola et al. 2009), an enzyme required for the synthesis of a precursor of an anticancer drug—paclitaxel (Baird et al. 2010) and to produce the antimalarial drug artemisinin (Ikram and Simonsen 2017). *P. patens* have also been engineered for the delivery of sesquiterpenoids like sanatlol, patchoulol synthase alpha/beta-santalene synthase, a diterpenoid Sclareol (sclareol) in fragrance industry (Zhan et al. 2014; Pan et al. 2015). Rerouting the synthetic pathway of patchoulol from the cytosol to plastids enhanced the metabolic flux and consequently increased the production yield (Zhan et al. 2014). Another aspect is the difficulty to retrieve fragrance compounds because of the volatility and hydrophobicity. In this regard, Peramuna et al. (2018) showed that the introduction of patchoulol synthase on the lipid droplet membrane of *P. Patens* enhances the storage ability of the cell, thus recovering its assortment. Recently, insight into an efficient growth of bryophytes in bioreactors is being looked into so as to enable its application in the manufacture of valuable bio-products like pharmaceuticals (Tschofen et al. 2016). Several human proteins are being produced in moss *P. patens* as potential biopharmaceuticals like tumour-directed monoclonal antibodies with enhanced antibody-dependent cytotoxicity, vascular endothelial growth factor and many more. The plant made pharmaceuticals (PMPs) are not only analogous to those produced in mammalian systems, but are of superior quality, thus called as Biobetters. Clinical trials are being carried out to assess the feasibility of the moss system as chassis for next-generation biobetter pharmaceuticals (Reski et al. 2015). However, for broad-spectrum profit, mosses should be engineered for tolerance against strong light and drought. Secondly, the establishment of the first synthetic plant cell, for which *P. patens* would be perfect, with a minimal genome would probably direct towards simpler engineering (Reski et al. 2018).

Woodard et al. (2003) characterized the expression of bovine trypsin in maize seeds by inserting inactive trypsinogen under the regulation of the embryo-preferred globulin-I promoter and the optimized barley α-amylase signal sequence. Inactive trypsinogen gets activated by autocatalytic processing and/or endopeptidases in the seeds. Kim et al. (2014) also produced recombinant bovine trypsin in rice cell suspension cultures under the regulation of sucrose starvation–inducible rice α-amylase 3D promoter. Plant-based tough, elastic, fibrous animal proteins can be utilized to produce pathogen-free sustainable biopolymers, thus substituting oil-based plastics. CollPlant (Israeli biotechnology company) has developed fully functional recombinant human collagen 'CollageTM' in tobacco lines by expressing procollagen α1(I) and α2(I) along with a human proline-4-hydroxylase and a human lysyl hydroxylase. It is commercially used for tissue patch up and injury management and since it is free of cross-links, it can be customized to meet the demands of industry (Stein et al. 2009). Maize is being exploited for the manufacture of enzymes and technical reagents (Tschofen et al. 2016), for example, molecular biology research tools—Avidin (naturally found in egg whites) and β-glucuronidase—GUS (naturally found in *E. coli*). Plants are amenable to glycoengineering, in which the synthesis pathway is synthetically modified to manufacture mammalian-like glycosylated monoclonal antibodies (mAbs), for example, a plant produced therapeutic mAb is ZMappTM, a capable cure against the Ebola virus (Wec et al. 2017). Three mouse/human chimeric mAbs and ElelysoTM (taliglucerase-a—mAb for Gaucher's disease) were produced in a strain of tobacco—DXF and carrot cell culture, respectively (Mortimer 2019). Antibodies designed from molecular farming, have been analysed for extensive analytical applications like on-site revealing of allergens, pathogens and toxins and food verification by important biomarkers, selective sorbents for competent sample preparation and production of antibody-functionalized detection probes (Tschofen et al. 2016). Microbial feed additives when added to animal feeds, perk up feed digestibility or its quality by break down of indigestible fibres (by hemicellulases) and/or eliminate antinutritional factors (by phytases). Xu et al. (2013) introduced fungal beta-mannanase from *Bispora* sp. MEY-1 in maize seeds and Zhang et al. (2013) introduced an acidic endo-beta-1,3-1,4-glucanase in transgenic maize seed to be utilized in animal feed. Similarly, Gupta et al. (2015) implanted endosperm-specific *Aspergillus niger* phytase in *Zea mays* to express an enzyme activity adequate to digest phytate to release adequate phosphorus in the diet. Laccase of white-rot fungus *Trametes versicolor*, can oxidize phenolic compounds like lignin in the papermaking process (Widsten and Kandelbauer 2008), treat textile mill effluents, eradicate phenolic compounds from beverages, eradicate sulphur from fossil fuels, and develop biosensors (Singh et al. 2011), remove pesticides and xenohormones from the soil (Singh et al. 2015). However, extraction of laccase from fungi is expensive, molecular farming can implant the functional enzyme in the trees providing raw material. Singh et al. (2015) reported that laccase introduction in maize but with not much substantial output, however, plants can be manipulated to secrete laccase from their roots (Singh et al. 2015).

In order to drop-in sustainable substitutes for fuels like petrol and diesel, photosynthetic bacteria, algae and plants can utilise solar energy to produce next-generation green-/biofuels. At present, bioethanol from starch, sugar or lignocelluloses rich plants and biodiesel from oil crops are the most commonly accessible forms of biofuel (Scott et al. 2010). Engineering autotrophic organisms to facilitate with biofuel production include i) diverting energy-rich CBBC intermediates from glucose production to the formation of energy-dense biofuel precursor-like terpene, ii) developing unconventional pathways to fix CO_2, for example, reverse tricarboxylic-acid cycle, the Woods–Ljungdahl cycle, the hydroxypropionate–hydroxybutyrate cycle, iii) metabolically engineering carbon products into molecules such as acetyl-CoA, which is an originator for a lot of fuels (Chu and Majumdar 2012). Engineering pathways for novel biofuel molecules has been employed widely in yeasts, for example, a monoterpene synthase from sweet basil was introduced into *S. cerevisiae* resulted in a strain synthesizing and secreting monoterpene geraniol into the medium, thus letting go the cost of extraction (Oswald et al. 2007; Scott et al. 2010). Starch, feedstock required for bioethanol fabrication, can be produced from nutritionally less valuable cellulose and hemicellulose present in the bagasse of many crops. The implantation of polymer-degrading enzymes in plants β-glucanase for hydrolysis of cellulose to glucose and β-D-xylanase, α-glucuronidase for hydrolysis of xylan into fermentable pentoses has been effectively executed (Li et al. 2014). For example, US company Agrivida produced *Zea mays* lines individually expressing endo-β-1,4-glucanase and endo-β-1,4-xylanase, to amplify bioethanol production (Zhang et al. 2011). Shen et al. (2012) engineered temperature resistant xylanase (with a bacterial self-splicing intein) that prevents autohydrolysis during development but can be induced by heat treatment post-harvest (Shen et al. 2012). Aquatic phototrophic organism, such as cyanobacteria and green algae, is also the candidate organisms for producing biofuel at large scale, as their complete biomass can be digested into methane or converted into crude by thermochemical processes (reviewed by Wijffels et al. 2013). Since utilising lignin rich hosts for biofuel production is challenging, SynBio offers designing lignin which is more open to biotechnological applications. A ferulolyl-CoA:monolignol transferase (FMT from *Angelica sinensis*) was implanted in poplar to synthesize ferulated monolignol subunits, which can then be incorporated into the lignin polymer to construct so-called Zip-ligninTM (Wilkerson et al. 2014). Engineered poplar trees had superior cell wall digestibility under mild alkaline pre-treatment (Zhou et al. 2017). In gain-of-function approach, bacterial 3-dehydroshikimate dehydratase (QsuB), under the regulation of the plant Cinnamate-4-Hydroxylase (C4H) promoter was targeted in plastids of *Arabidopsis* to reduce the accessibility of an intermediate in the shikimate pathway (Eudes et al. 2015). This in turn reduced monolignol production for lignin synthesis in non-essential plant parts. Since hexose sugars are preferred (over pentose and lignin) as a carbon source, thus there is a need to optimize the biomass to match the microbial preference. The cell wall content of the pectin galactan (composed of the hexose galactose) was increased in *A. thaliana* by overexpressing Galactan Synthase1 (GALS1), cytosolic UDP-Glucose/UDP-Galactose-4-Epimerase2 (UGE2) synthesizing UDP-galactose and a gene for UDP-galactose transporter moving the substrate to Golgi lumen to be utilized by GALS1 (Aznar et al. 2018). This study was extended by using 3-dehydroshikimate dehydratase (QsuB) strategy via SynBio technology, where high-galactan lines were combined with cell-specific low xylan and low lignin (Aznar et al. 2018). Production of lignocellulosic-based biofuels is less preferred because sequestration of cell wall polysaccharides by lignin polymers makes the sugar extraction process considerable complex, thus demanding effective techniques for decreasing lignin content (Eudes et al. 2015; Shih et al. 2016). Techniques of RNAi genetic engineering in corn resulted in plants with less lignin in cell wall, thus shifted energy to produce more cellulose (Park et al. 2014). They also engineered crops to express microbial cellulase enzymes (endoglucanase, exoglucanase and beta-glucosidase) needed to convert the plant cellulose into fermentable sugars in stovers (Park et al. 2011). SynBio can also offer a bioenergy crop with cell walls rich in cellulose and hemicelluloses; but with more bio-degradable lignin contents (Yang et al. 2013; Sticklen 2015). However, over-accumulation of sucrose can persuade negative feedback on the photosynthetic mechanism, thus metabolic engineering also offers sink-side strategies on transporting more carbon into a different sink, for example, sucrose transfer into isomers like isomaltulose and trehalulose has been effectively engineered to augment total soluble sugar content in sugarcane (Wu and Birch 2007; Hamerli and Birch 2011). In addition, bioenergy crops accumulating huge amounts of soluble fermentable sugars should instantly process sugars post-harvest to avoid spoilage by various opportunistic microbes.

Plants can also be exploited for the fabrication of lipases and phospholipases, the enzymes used in the production of a nonpolluting and carbon-neutral fuel—biodiesel. These plant-based enzymes can be either expressed directly in oil crops or extracted in pure/crude form. Via advanced SynBio a crop, with synthetic genetic circuits for lipid biosynthesis along with regulatory systems that can generate such oil only in the non-edible or wasteful parts, can be designed (Ferry et al. 2012). With efficient editing tools such as CRISPR/CAS9 (Puchta 2016), oil crops can be engineered to suppress endogenous carbon sinks during the seed filling period. Since the majority of the plant is composed of vegetative tissue, triacylglycerol (TAG) production in vegetative tissue can be increased, for example, metabolically engineering by stacking genes to increase TAG production in leaf tissue of *Nicotiana tabacum* (Vanhercke et al. 2014) and in vegetative biomass of sugarcane (Zale et al. 2016). An increase in TAG levels in Arabidopsis leaf tissue has been achieved by mutating genes like Trigalactosyldiacylglycerol1 (TGD1) and Sugar-Dependent1 (SPD1) to decrease fatty acid beta-oxidation (Fan et al. 2014). Synthetically introduced tolerance traits in feedstock crops would push geographical/climate-dependence borders of bioenergy plants, would boost the accessibility of renewable products, and thus would increase the cost-competitiveness of biofuels (Shih et al. 2016). Algae with higher productivities than land plants and superior capability to accumulate triacylglycerides (TAGs) are another appropriate host for bioengineering related to biodiesel (Scott et al. 2010). The Commonwealth Scientific and Industrial Research Organisation (CSIRO), Australia has synthetically engineered safflower for yielding super high oleic acid-containing oils (SHOSO), which can replace some petroleum-based raw materials used in bio-lubricants, biochemical and bioplastics industries (Wood et al. 2018). In the current scenario, the concept of biorefinery in which all raw materials are transformed into useful products, for example, sugar cane processes utilising the bagasse and wood pulping processes utilizing even the bark, are getting highlighted. Molecular farming offers potent implantation of low-cost recombinant enzymes, like xylanase and oxidation/reduction enzymes such as laccase and peroxidase (the imperative prerequisite in biorefineries) in plants (Bussamra et al. 2015).

Unfortunately, till recently fossil fuels are being exploited for short chain unsaturated hydrocarbons to produce bulk chemicals, materials and polymers and are efficiently produced from fossil fuels. However, SynBio can offer utilisation of renewable sources, for example, dicarboxylic acid cis,cis-muconic acid for fabricating precursors of plastics such as nylon and polyethylene terephthalate (PET), or can be directly incorporated into polyesters (Rorrer et al. 2016). Microbial polyhydroxyalkanoates, poly-3-hydroxypropionic acid (a precursor of acrylic) and poly-3-hydroxybutarate (a precursor of propylene) were synthetically engineered in *Arabidopsis* leaf tissue (Bohmert et al. 2000). A complex engineering strategy introducing four bacterial enzymes and the regulation of their expression using spatially- and temporally-specific promoters could produce muconic acid in *Arabidopsis* (Eudes et al. 2018).

Another SynBio approach where photosynthesis can be exploited is to (1) couple the reducing power of light reactions to previously unconnected pathways to produce large amounts of high-value compounds in vivo and (2) design hybrids of PSI with (a)biotic components to generate hydrogen, simple carbon-based solar fuels or electricity in vitro (for details read the review by Leister 2019). Interestingly by the bottom-up approach, entirely synthetic systems like artificial leaves and synthetic photosynthetic cells are evolving and they are utilising solar energy as driving strength (Roell and Zurbriggen 2020). Artificial leaf is a wireless economical and highly distributed solar-to-fuels arrangement that employs low-cost systems engineering and manufacturing by earth-abundant materials. Artificial leaf comprises a triple junction—amorphous silicon photovoltaic with hydrogen- and oxygen-evolving catalysts made from a ternary alloy (NiMoZn) and a cobalt-phosphate cluster (Co-OEC), respectively. Analogous to PS II-OEC, artificial leaf performs the oxygen-evolving reaction in water and mimics a leaf in storing solar energy by water splitting, thereby producing hydrogen (H_2) energy in a clean way and leaving virtually no pollutants. Unlike green leaf, in artificial leaf, the H_2 is available for combination with CO_2 as new catalysts for this process (Nocera 2012). To develop the photosynthetic artificial cell into the energetically independent system, a biomimetic artificial organelle, i.e. giant unilamellar vesicle (GUV) producing adenosine triphosphate (ATP), by collaborating ATP two membrane-embedding component—a bacteriorhodopsin (bR) and a membrane portion of ATP synthase (Fo), is a rational energy generating system. Recently, Lee et al. (2018) performed ATP synthesis by means of photosynthetic artificial organelle (PAO), where they demonstrated in vitro carbon fixation and actin polymerization within giant unilamellar vesicle (GUV). Berhanu et al. (2019) inserted this artificial organelle into the reconstructed cell-free system (PURE system) and constructed an artificial photosynthetic cell that permits the self-constitution of its own fragments in a positive feedback loop, i.e. produces ATP for the synthesis of essential components of the proteoliposome. Promising applications of this technique can be in the study of drug delivery that can control spatiotemporal production of aptamer or single-chain Fv within a vesicle capsule or as the phosphate recycling system, where free phosphates accumulated in the system post-reaction can be recharged onto ADP to be again utilised in the system (Berhanu et al. 2019). Using a unicellular green alga (*Chlamydomonas reinhardtii*), Kanygin et al. (2020) created a new photosystem, i.e. an in vivo fusion of PSI and the [FeFe] hydrogenase from a gene transplanted into the chloroplast chromosome (by insertion of the HydA sequence into the PsaC subunit). Photosynthetic flow in the re-engineered chloroplast is diverted from CO_2 fixation (i.e. using electrons from water splitting and the photosynthetic electron transport chain) to proton reduction and resulted in biohydrogen production at high rates in a light-dependent fashion. To maximize the usefulness of such engineered systems, Kanygin et al. (2020) have positioned redox enzymes to directly capture electrons at the most reductive potential from PSI, i.e. FA/FB clusters before they enter the general cellular pool. These chimeric proteins also constrain O_2 evolution from PSII, thereby preserving hydrogenase activity, proton pumping and ATP production carried out by the PETC, thus maintaining cell viability. Thus, re-engineering the elementary route in photosynthetic microorganisms offers an inexpensive and renewable proposal for creating biofactories capable of driving novel redox chemistries, from renewable resources like sun and water.

15.7 Increasing the Nutritional Value of Crop Plants

At present agriculture system is incapable to meet global requirements for the production of reasonable quantity of nutritious foods vital for healthy diets. To ensure that the world's population has access to abundant, secure and nourishing food, future research needs to focus on breeding for the nutritional value of crop plants under the context of climate change. Nutrition-sensitive cellular agriculture has endless prospects that seek to guarantee the production of diverse nutrient-packed foods with an extended shelf life, as well as foods tailored for specific preferences to fulfil the nutritional needs of populations in a sustainable approach (FAO 2017). Plants can provide all dietary macro and micronutrients, however, micronutrients are generally unevenly distributed between plant species and even in different tissues of a particular plant. For example, a good quantity of micronutrients like iron (Fe), folate, provitamin A and vitamin E are present in non-edible rice lamina while in minimal amounts in the edible endosperm. Thus, deficiency of Vitamin A prevails in places where rice is the chief ingredient of not so diverse diet. The Golden Rice technology—known as GR2—employs two pathways genes, one coding for phytoene synthase (from maize) and second for carotene desaturase (from the bacterium *Pantoea ananatis*), leading to higher synthesis and amassing of beta-carotene (provitamin A) in the endosperm (Beyer 2010). Moreover, the industrial manufacture of vitamin B12 (primarily from the fermentation of *Pseudomonas denitrificans* and *Propionibacterium freudenreichii*) is not keeping pace with the demands of the growing population. Fang et al. (2018) engineered fast-growing, genetically tractable species *E. coli* to fabricate vitamin B12 (adenosylcobalamin) by heterologously expressing 28 genes from *Rhodobacter capsulatus, Brucella melitensis, Sinorhizobium meliloti* 32,020, *S. typhimurium* and *Rhodopseudomonas palustris* that are separated into six engineered modules.

Secondary metabolites like carotenoids and their oxygen-containing derivatives, xanthophylls have multifaceted roles in both plant and animal kingdom. In plants, they mediate photosynthetic function, responses to biotic and abiotic signals, and control plant architecture, while in animals they are allied with key concerns, like vision and cardiovascular matters, efficient immune system, cognitive function and antioxidant functioning (Wurtzel 2019). Thus, versatile carotenoids are one of the important SynBio targets while taking into account ways to modify agriculture to meet the needs of food security. Especially bioactivities of apocarotenoids, enzymatically cleaved carotenoids, are gaining attention as new toolkits that will transform SynBio applications in agriculture. They can potentially reconstruct plant architecture by designing vigorous root systems promoting growth in poor quality soils, inserting mycorrhizal colonization in presently non-mycorrhizal zones, reforming plant-based chemical factories for synthesising de novo phytochemicals that are easily harvestable from exudates or producing novel aromatic chemicals to generate a defensive barrier against pathogens (Gao et al. 2018; Stauder et al. 2018). SynBio can be utilised to design an adjustable biological circuit delivering a novel strigolactone structure (Liu et al. 2018b) that functions in plants but is no longer decipherable by unsafe rhizospheric biota and parasitic competitors. Wurtzel et al. (2019) pointed out harnessing biological circuits to design synthetic carotenoids can also generate stress-resilient plants to perk up light energy capturing efficiency and photoprotection, or diverse plants (irrespective of their genetic background or plant species) with novel carotenoids, or to create synthetic plant membranes with novel carotenoid structures for improving mineral and oxygen uptake by roots or creating a proficient blockade to plant pathogens (Wurtzel 2019). Synthetic production of plant hormones like carotenoid-derived root exudates strigolactones can offer opportunities to influence crop nutrient uptake, phytomorphology and reduce the application of chemical fertilisers. By using the 'bottom-up' approach of SynBio, isoprenoids including strigolactones were synthetically engineered and propagated in yeast (Vickers 2016). Astaxanthin in conjunction with other ketocarotenoids, the most expensive ingredient of salmon and trout feed (Moretti et al. 2006), can be biologically extracted from only bacterium *Paracoccus carotinifaciens*, the alga *Haematococus pluvialis* and the fungus *Xanthophyllomyces dendrorhous* (Ambati et al. 2014). Thus attempts have been made to metabolically engineer the synthesis of astaxanthin in crop plants like maize has been effectively engineered for high b-carotene and zeaxanthin production in seed endosperm (Zhu et al. 2008). In another maize line, lycopene e-cyclase was knocked down but phytoene synthase gene was over-expressed (Farré et al. 2016) and these plants were crossed into a high oil hybrid to yield maize seeds which can be fed to chicken as a source of astaxanthin or can be processed for commercial astaxanthin (Breitenbach et al. 2016).

In the most commonly eaten fruits and vegetables, pigments anthocyanin, guarding from wide range of diseases, are inadequate to confer optimal benefits. Tissue-specific expression of an entire synthetic bacterial biosynthetic pathway containing three *Erwinia* genes encoding phytoene synthase and desaturase along with lycopene beta-cyclase under the regulation of golden tubers of potato (*Solanum tuberosum*)-specific promoter resulted in many folds augment in carotenoids and beta-carotene (i.e. provitamin A) concentration (Diretto et al. 2007). Expression of the two *Antirrhinum majus* transgenes *Delila (Del)* and *Roseal (Rosl)* genes in tomato fruit enhanced the hydrophilic antioxidant capacity of tomato fruit and the fruits accumulated substantially higher levels of anthocyanins (approximately at par with the concentration found in blackberries and blueberries). These engineered tomatoes could substantially add to the hydrophilic antioxidant content in meals and might substitute antioxidant supplements, such as NAC or vitamins (Butelli et al. 2008). Majer et al. (2017) used *Tobacco* etch virus vector to express an entire heterologous metabolic pathway (consisting of three enzymes from the soil bacteria *Pantoea ananatis*) to synthesise health-promoting carotenoid lycopene in tobacco tissues in a cost-effective and scalable manner. In this study, phytoene synthase (crtB) not only activated the build up of endogenous carotenoids but also acted as an exceptional wide-spectrum yellow visual indicator (by impacting chloroplast metabolism) for tracking viral infection.

Very-long-chain polyunsaturated fatty acids (VLC-PUFAs) such as arachidonic acid, eicosapentaenoic acid (EPA) and docosahexaenoic acid (DHA) are beneficial for human brain development and cardiovascular functioning. Since VLC-PUFAs are abundantly present in microalgae and hardly ever in plants, the only non-sustainable source is marine fish consuming microalgae. In the earliest demonstrations of Omega-3 long-chain polyunsaturated fatty acids (LC-PUFA) engineering, Petrie et al. (2012) introduced a transgenic pathway to increase the C18 Alpha-linolenic acid (ALA) which was then converted to DHA by a microalgal D6-desaturase pathway in *Arabidopsis thaliana* seed oil. Later Petrie et al. (2014) described the fabrication of fish oil-like levels of omega-3 LC-PUFA EPA and DHA in *Camelia sativa* by the introduction of a transgenic D6-desaturase pathway by utilising yeast and microalgal genes. Later, the introduction of VLC-PUFA biosynthetic genes from moss species (known to accumulate in higher levels), has increased their concentration in model plants like *Camelina sativa* (Ruiz-Lopez et al. 2014), *Brassica napus* (Napier et al. 2019). These transgenic plants can be future sustainable, terrestrial sources of omega-3 fish oils. Metabolic engineering of customary seed crops with genes to improve the nutritional excellence of fatty acids can perk up the qualities of harvested oils. This has been achieved in *A. thaliana*, using seven enzymes from five organisms, the yeasts *Lanchancea kluyveri*, and *Pichia pastoris*, and the algae *Micromonas pusilla*, *Pyramimonas cordata* and *Pavlova salina* (Petrie et al. 2012). This strategy was reused in *Camelina sativa*, and in both the cases significant improvement in the nutritional profiles of oils was recorded (Petrie et al. 2014). Synthetic production of one of the world's most expensive spices saffron can cut manufacturing costs and ensure a steady supply. At present, a Swiss company is engineering yeast to harvest the vital flavour and colour compounds of saffron (Vickers 2016).

Another important facet connected to the nutritional value of food includes the exclusion of ill-favoured secondary metabolites like cyanogenic glycoside. For example in cassava (*Manihot esculenta* Crantz), RNA interference technique aiming two cytochrome P450 genes—*CYP79D1* and *CYP79D2*, encoding the first committed enzymes in linamarin and lotaustralin, can block the synthesis of cyanogenic glucoside (Jørgensen et al. 2005). Zero-calorie natural extract of *Stevia rebaudiana* offers steviol glycosides which are natural sweetener. Although it is very important for today's global food industry, its organoleptic qualities giving a bitter aftertaste limit their widespread recognition in the food industry (Prakash et al. 2014), wherein the field of synthetic biology can step in. SynBio can also contribute in reducing the allergenicity of allergens—*Ara* proteins which are negatively influencing the health of global peanut consumers (Palladino and Breiteneder 2018). According to Gould (2020), SynBio technology can generate novel foods like algae butter and hypoallergenic peanuts with superior taste and nutritional properties. Gluten proteins in wheat grains are accountable for the exclusive visco-elastic properties of wheat-derived foods. However a-gliadin family of gluten, triggers coeliac disease and nonceliac gluten sensitivity to humans (Sapone et al. 2011). In this regard, Sánchez-León et al. (2018) used CRISPR/Cas9 expertise to design two sgRNAs targeting a conserved sequence next to the coding region for the 33-mer of a-gliadin genes. This precisely reduced a-gliadins levels in bread and durum wheat kernels and hence reduced immune-reactivity for gluten-intolerant consumers. These low-gluten, transgene-free wheat lines could also serve as source material to introgress this trait into elite wheat varieties (Sánchez-León et al. 2018).

Combining synthetic genetic variability method with de novo domestication can exploit the genetic diversity present in the wild plant (disease resistance and stress tolerance) in order to quickly engineer superior crops. Zsögön et al. (2018) devised a CRISPR–Cas9 genome engineering approach to amend six loci [general plant growth habit (SELFPRUNING), fruit shape (OVATE), size (FASCIATED and FRUIT WEIGHT), fruit number (MULTIFLORA), and nutritional quality (LYCOPENE BETA-CYCLASE)] and combined these traits with valuable traits of wild *Solanum pimpinellifolium*. Compared with the wild parent, engineered *S. pimpinellifolium* showed improvement in all the above-mentioned traits within a single transformation experiment.

15.8 Valuable Plant Metabolites in Microorganism

Natural products and their derivatives are a collection of bioactive structurally diverse molecules, such as therapeutic drugs (Paddon et al. 2013), flavourings (Pretorius 2017), and oils (Pouvreau et al. 2018); and much more. There is a huge metabolic capability with plants to synthesize these metabolites de novo from renewable natural resources like water, CO_2 and sunlight. However, there are limitations like the dependency of plants on arable land, water and season as well as long generation times, large polyploidy genomes and other constraints, such as abiotic and biotic stresses and food safety; leading to insufficient production which subsequently raise their market values (Moses et al. 2017; Zimin et al. 2017). Promising progress to combat this drift is the synthetic transplantation of multigene pathways to fast-growing or easy-to-cultivate heterologous host, where plant-based carbon feedstocks can be transformed with high competence to explicit plant metabolites (Arendt et al. 2016; Li et al. 2018). Although algae, as an aquatic photoautotrophic alternative, has a quicker generation rate and no dependency on arable land; however, the cultivation cost of algae are excessively high (Scott et al. 2010). Consequently, there is a need to find a suitable host organism in which both indigenous genes and the ones introduced from other organisms work in the coordinated fashion for the commercial production of high-value products (Moses et al. 2017).

Microbes can be considered as classic platform to rationally design new-to-nature metabolites because they have a rapid generation rate and broad engineering toolkits, are vigorous and readily scalable, and offer a simplified product purification technology. Most importantly production process in microbes is cost-effective as inexpensive carbon sources (e.g. agricultural wastes) are converted to high-value compounds. The most broadly used heterologous host organisms are the bacterium *E. coli* and the yeast *S. cerevisiae*, due to their extremely well-understood cell biology, capacity to hold bulky genetic constructs as well as ability to tolerate industrial growth conditions (Awan et al. 2016; Goold et al. 2018). Yeast *Pichia pastoris*, able to grown on low-cost carbon sources, and the fungus *Aspergillus oryzae*, with metabolism appropriate for the manufacture of polyketides and non-ribosomal peptides can also be utilized as heterologous hosts (Keasling 2010; Awan et al. 2016). Besides host selection, genes encoding enzymes catalysing synthesis of natural product and their regulation to optimally balance the yield of high-value metabolites and growth of the host must be selected. Depending on the host organism selected, for a particular natural product, one or more biosynthetic routes can be involved (Keasling 2010) and the proportionate contribution of native and heterologous genes can vary (Awan et al. 2016). Moses et al. (2017) reviewed a road map of the most recent SynBio tools under consideration and listed the unicellular microbial hosts accessible for rational engineering of valuable plant products.

The World Health Organization (WHO) has recommended *artemisinin-based combination therapies (ACTs)* as an antimalarial treatment against the parasite *Plasmodium falciparum*. Artemisinin (a sesquiterpenoid endoperoxide) is originally sourced from Sweet wormwood (*Artemisia annua*) (Miller and Su 2011). However, the delivery of plant-derived artemisinin is uneven, ensuing in frequent shortages and cost instability. Superior production of artemisinic acid, a precursor of artemisinin, in yeast was attained using numerous synthetic pathways: (1) overexpression of a truncated 3-hydroxy-3-methyl-glutaryl-coenzyme A reductase (tHMGR) and downregulation of squalene synthase (ERG9) by a methionine repressible promoter to increase farnesyl pyrophosphate (FPP) production and decrease its use for sterols; (2) combination of a semidominant mutant allele of a Zn(II)2Cys6 binuclear cluster transcription factor (upc2-1) with a downregulating ERG9 and the addition of the amorphadiene synthase gene (ADS) from the *Artemisia annua* and an additional copy of tHMGR for the conversion of FPP to amorphadiene; and (3) expression of a cytochrome P450 (CYP71AV1) and its redox partner CPR under the control of galactose-inducible promoters for artemisinic acid construction from amorphadiene (Ro et al. 2006). Paddon et al. (2013) introduced the entire biosynthetic pathway (including plant dehydrogenase and cytochrome) to design *S. cerevisiae* strains for high-yielding biological production of artemisinic acid. They also developed an alternate conversion of artemisinic acid to artemisinin using singlet oxygen instead of dedicated photochemical kit. Since production in yeast requires enormous quantities of high-priced synthetic culture medium and huge bioreactors to work in a sterile environment, plant-based production can potentially provide an inexpensive, renewable and effortlessly measurable artemisinic acid. Fuentes et al. (2016) developed a new SynBio approach—Combinatorial Super transformation of Transplastomic Recipient Lines (COSTREL), to introduce artemisinic acid synthesis pathway into chloroplast genome of tobacco. The transplastomic plants were supertransformed with enzymes altering flux through the synthetic pathway, thereby further optimizing the metabolic productivity of engineered artemisinin (Fuentes et al. 2016).

Raspberry ketone, chief aroma compound in raspberries, is a valuable flavouring agent. Again the high production cost along with poor yields from plant tissue, makes it an excellent objective for heterologous construction in microbial strains. Lee et al. (2016) combined the best possible enzymes of ketone synthetic pathway from a wide range of plant genomes to synthetically design a yeast strain that produced more than double ketone when compared to host plants. Lee et al. (2016) represented a de novo pathway by assembling four heterologous genes, encoding phenylalanine/tyrosine ammonia lyase, cinnamate-4-hydroxlase, coumarate-CoA ligase and benzalacetone synthase for the production of raspberry ketone in an industrial strain of *S. cerevisiae* (Lee et al. 2016). Growing chocolate demands and limited Cocoa butter (CB—raw material for chocolate production extracted from cocoa beans) manufacture has resulted in a scarce supply of CB. CB is largely composed of C16 and C18 triacylglycerols (TAGs). Interestingly, Wei et al. (2017) suggested that *S. cerevisiae* have potential as CB-like lipids (CBL) production chassis, as storage lipids (mainly TAGs) of yeasts, also have relatively high-level of C16 and C18 fatty acids. In their study, they cultivated six different yeasts, *S. cerevisiae*, *Trichosporon oleaginosus*, *Rhodotorula graminis*, *Lipomyces starkeyi*, *Rhodosporidium toruloides* and *Yarrowia lipolytica* in N-limited medium. In all the six yeasts TAGs were the main lipids but comparatively *T. oleaginosus* could yield the highest amount of TAGs. However, Wei et al. (2018) reported that major TAG forms in yeast were not similar to CBL, thus necessitating the modulation of yeast TAG biosynthetic pathway for better CBL production. They cloned seven glycerol-3-phosphate acyltransferase (GPAT) genes and three lysophospholipid acyltransferase (LPAT) genes from cocoa cDNA. By expressing these cloned cocoa genes with two synthesized cocoa diacylglycerol acyltransferase (DGAT) genes in *S. cerevisiae*, they significantly increased CBL production in some of the strains, suggesting that this strategy might be accountable for CBL biosynthesis in the future.

Cannabinoids are a group of more than 100 chemical compounds (Radwan et al. 2008) mainly isolated from the strictly regulated *Cannabis sativa* L. Δ9-tetrahydrocannabinolic acid (THCA) and cannabidiol (CBD) are important drug candidates (Carvalho et al. 2017). The supply of cannabinoids is inadequate since most of the cannabinoids can only be extracted in very low amounts.

Characterization of all enzymes required for 9-THC and CBD biosynthesis, permits the fabrication of these compounds in heterologous host organisms. Thus, Carvalho et al. (2017) explored an alternative strategy where microbial production can support the design of novel cannabinoids with enhanced properties by the incorporation of tailored enzymes. They suggested that a chassis organism can also be designed for the scalable biosynthesis of less-abundant cannabinoids whose efficacy as medical drugs can be further tested. Zirpel et al. (2017) also reconstituted biosynthetic pathway for THCA production from *C. sativa* in microbial host—*Komagataella phaffii* by combining the bacterial prenyltransferase NphB with the THCA forming enzyme. The WHO classifies opioids as a significant class of medicines that consist of medication like analgesic morphine and the antitussive codeine (World Health Organization 2013). All natural opiates (morphine, codeine) and semi-synthetic opioids (oxycodone, hydrocodone, hydromorphone) are presently obtained from the opium poppy (*Papaver somniferum*). Susceptibility of poppy farming to biotic factors and increasing market demands are growing pressure on the only resource of opioids. Even, their industrial synthesis has been reported to be commercially uncompetitive (Reed and Hudlicky 2015). This challenge related to poppy supply can be addressed by microbial-based synthesis of opioids or their precursor benzylisoquinoline alkaloids (BIAs). With SynBio technique, biosynthesis of BIAs downstream of (S)-reticuline, can be reconstructed in *Escherichia coli* (Nakagawa et al. 2011) and yeast (Trenchard et al. 2015) from sustainable carbon and nitrogen resources. DeLoache et al. (2015) and Trenchard et al. (2015) engineered the biosynthetic pathway, from tyrosine to (S)-reticuline, (Thodey et al. 2014) engineered the pathway from thebaine to morphine and (Fossati et al. 2015) from (R)-reticuline to codeine. However, it is difficult to design fully functional entire biosynthetic pathway of 20 heterologous genes essential for their absolute synthesis. Galanie et al. (2015) engineered yeast with 21 (thebaine) and 23 (hydrocodone) enzyme activities from plants, mammals, bacteria and yeast itself, to produce the selected opioid compounds starting from sugar. They combined enzyme discovery, pathway and engineering along with strain optimization to comprehend full opiate biosynthesis in yeast. Synthetic scaffolds, enabling better liberty to systematize enzymes in a rational approach have been applied to attain higher production of phytochemicals. *E. coli* synthetic protein scaffolds were used to co-localize enzymes converting naringenin or eriodictyol to improved concentrations of catechin. Similarly, DNA and RNA scaffolds have been utilized to co-localize enzymes related to enhanced production of l-threonine, pentadecan and succinate in *E. coli*. Progress in synthesizing hybrid transcription factors or synthetic RNA switches can assist in the dynamic regulation of heterologous and endogenous enzymes to fine-tune phytochemical biosynthesis (Li et al. 2018).

Terpenoids play essential roles in maintaining plant cell membrane fluidity, resistance to herbivores or plant–environment interactions. They are economically significant pharmaceuticals, aromatics and budding next-generation biofuels. With the increasing demand for novel terpenoids with superior bioactivities, tailor-made terpenoid biosynthesis is another important objective for SynBio projects. For example, de novo monoglycosylated triterpene saponin is synthetically fabricated in yeast by using genes of five different plants (Moses et al. 2014). In contrast to microbes requiring fermentable sugars from plant-derived lignocellulose as an energy source, utilizing photosynthetic hosts is better. Plants synthesize the desired metabolites using renewable resources (Kim et al. 2013; Sanchez and Karhumaa 2015), allow the post-translational modifications necessary for accurate protein folding and enzyme functionality (Ikram et al. 2015) and avoid the need for metabolite purification as the desired metabolites can be directly consumed via food or feed crops. Hsu et al. (2014) suggested that targeted knockout of genes is extremely needed in SynBio programmes when challenging side branches of metabolic pathways or negative feedback loops are to be silenced or removed. Sterol side chain reductase 2 (SSR2) was knocked out using TALENs to reduce cholesterol and steroidal glycoalkaloids levels in tetraploid potatoes (Sawai et al. 2014). The ability of homologous recombination in mosses such as *P. patens* (Schaefer and Zryd 1997) offers an advantage to mosses as chassis over vascular plant hosts. Another considerable host can be algae and cyanobacteria, which are cultivated in water or nutrient-laden agricultural wastewater, thus avoiding exhaustion of arable land for their cultivation (Dismukes et al. 2008). Furthermore, this can lead to the consumption of concentrated aquatic CO_2 derived from industrial emissions (Nozzi et al. 2013; Arendt et al. 2016). Crocetin, naturally existing in *Crocus sativus* L., has potent applications in medical and food fields because of antitumor, antioxidation, antihypertension, antiatherosclerotic and antidepressant activities. To promote crocetin production in a heterologous host, key enzymes like β-carotene hydroxylase (CrtZ), carotenoid cleavage dioxygenase (CCD) and aldehyde dehydrogenase (ALD) have to be engineered combinatorially. Chai et al. (2017) proposed superior crocetin production by combinatorial manipulation of CrtZ, CCD and ALD and regulating protein expression level in *S. cerevisiae* (Chai et al. 2017).

Thus it is likely that agricultural traditions could shift from extensive cultivation of crops for a single product, to the cultivation of feedstock crops to supply carbon source for microbial growth, which promises to engender a wider variety of synthetic natural commodities, even in unbefitting season. These innovative alterations in agriculture can lead to better water use efficiencies and improvement in the resilience and productivity of agriculture (Goold et al. 2018).

15.9 Synthetic Plant Genomes

Deployment of de novo genomes into plants is still one of the far-reaching objectives of SynBio. Ground-breaking efforts in the bacterial cell construction having a synthesized *Mycoplasma* genome (Gibson et al. 2010) and the de novo creation of synthetic chromosomes in yeast (Dymond et al. 2011 have pointed out few principles for SynBio aspirants to follow. Firstly, the synthetic genomes should have approximately wild-type phenotypes and fitness; secondly, it should not include tRNA genes and transposons; and lastly, there should be genome editing sites for upcoming researches (Dymond et al. 2011). A streamlined synthetic genome: the *Plastome* (prokaryotic in nature) can be used as a vector for building synthetic circuits in plant chloroplasts. A plastome designed for the complete mevalonate pathway (All genes in the conduit were coordinately controlled by a single promoter) has been implanted in the chloroplast of tobacco leaves (Kumar et al. 2012). Liu and Stewart Jr. (2015) have anticipated that advances in synthetic plastomes would permit economical and extensive fabrication of enzymes, biopharmaceuticals, bioactive natural products of commercial interest. While the mainstream approaches in SynBio expertise focus on microbial species, the growth of new developing disruptive technologies (like CRISPR/Cas9-mediated genome editing) is taking the field closer to plants, where whole synthetic genome approach seems to hold great potential for agriculture. Moreover, The Genome Project-write (GP-write), a worldwide project run by a multidisciplinary scientists, focuses on the expansion of approaches to synthesize and analyse genomes of many diverse species. One of the overarching goals is to offer whole-genome design and synthesis by delivering well-structured projects to agriculture (Goold et al. 2018).

15.10 Conclusions and Future Perspectives

In the next green movement, SynBio is enabling mankind to apply design and engineering principles to directly convert sunlight, water, methane or waste products into economically significant, sustainable, and efficient solutions to looming global agricultural and environmental challenges. These developments will provide new markets for farmers, beyond principal food and fibre, such as designing crops to meet the aspirations of increasingly affluent consumers. Looking into the increasing demand, R&D funding and initiatives, and wide range of applications of SynBio, it is expected that its market is going to climb at approximately 23.9% CAGR (Compound Annual Growth Rate). Although a set of next-generation innovative concepts are available now, a very few (like synthetic microbiota as sustainable fertilizers) of these concepts have yet reached the stage of commercialization. Progressing from the theoretical design of a biological function to a realistic application does come with risks that must be mitigated or controlled. Technical challenges such as unexpected complexity, unavailability of most naturally existing biological genetic pathways, the ambiguity of certain genes, the erratic evolution of gene's regulatory sequences, scalability, scarcity of tools, the survival of synthetic organisms in natural environments, and ethical issues have to be addressed. The scientific community understands that societal acceptance of SynBio commodities will be based on the transparency of the research to the common man and the breakthroughs must be financially viable as well as adaptable to diverse global agroecosystems. In this viewpoint, foundations like BioBricks, are providing safe, ethical, inexpensive standardized biological parts (marketed as BioBricks™) to ensure that biological engineering is being conducted in a see-through manner to benefit the sustainability of the planet. Concepts and applications of SynBio approaches should be published in much advance to give lead time for discussion on associated ethical, environmental and biosecurity concerns, research into areas of uncertainty, and development and testing of biosafety features to avoid the pitfalls.

In the digital revolution dynamic active engagement of society is significant, thus budding do-it-yourself biology (DIYBio) movement is an encouraging paradigm to win society's confidence. However, the legislation will have to step in to ensure that new capacities that can be delivered by SynBio technologies are being accessed securely and judiciously. So, while walking the strenuous road of SynBio, patience, and a collaborative spirit of researchers, entrepreneurs, social scientists, ethicists and society as a whole can assure the beneficial adoption of this technology for sustainable agriculture.

References

Abbott DA, Zelle RM, Pronk JT, van Maris AJA (2009) Metabolic engineering of *Saccharomyces cerevisiae* for production of carboxylic acids: current status and challenges. FEMS Yeast Res 9 (8):1123–1136

Alexandratos N, Bruinsma J (2012) World agriculture towards 2030/2050. The 2012 revision. Agricultural development economics division, ESA working paper number 12-03, Food and Agriculture, Organization of the United Nations. http://www.fao.org/docrep/016/ap106e/ap106e.pdf

Allen RS, Tilbrook K, Warden AC, Campbell PC, Rolland V, Singh SP, Wood CC (2017) Expression of 16 nitrogenase proteins within the plant mitochondrial matrix. Front Plant Sci 8:287

Almario J, Jeena G, Wunder J, Langen G, Zuccaro A, Coupland G, Bucher M (2017) A. alpina root fungal microbiota. Proc Natl Acad Sci USA 114(44):E9403–E9412

Ambati RR, Moi PS, Ravi S, Aswathanarayana RG (2014) Astaxanthin: sources, extraction, stability, biological activities and its commercial applications—a review. Mar Drugs 12:128–152

Amthor JS, Bar-Even A, Hanson AD, Millar AH, Stitt M, Sweetlove LJ, Tyerman SD (2019) Engineering strategies to boost crop productivity by cutting respiratory carbon loss. Plant Cell 31(2):297–314

Anterola A, Shanle E, Perroud P-F, Quatrano R (2009) Production of taxa-4(5),11(12)-diene by transgenic *Physcomitrella patens*. Transgenic Res 18:655–660

Antunes MS, Morey KJ, Smith JJ, Albrecht KD, Bowen TA, Zdunek JK, Troupe JF, Cuneo MJ, Webb CT, Hellinga HW, Medford JI (2011) Programmable ligand detection system in plants through a synthetic signal transduction pathway. PLoS One 6(1):e16292

Aracic S, Manna S, Petrovski S, Wiltshire JL, Mann G, Franks AE (2015) Innovative biological approaches for monitoring and improving water quality. Front Microbiol 6:826

Arendt P, Pollier J, Callewaert N, Goossens A (2016) Synthetic biology for production of natural and new-to-nature terpenoids in photosynthetic organisms. Plant J 87(1):16–37

Arkin A (2017) A synthetic biology architecture to detoxify and enrich Mars soil for agriculture. University of California, Berkeley

Awan AR, Shaw WM, Ellis T (2016) Biosynthesis of therapeutic natural products using synthetic biology. Adv Drug Deliv Rev 105:96–106

Aysha J, Noman M, Wang F et al (2018) Synthetic promoters: designing the cis regulatory modules for controlled gene expression. Mol Biotechnol 60(8):608–620

Aznar A, Chalvin C, Shih PM, Maimann M, Ebert B, Birdseye DS, Loque D, Scheller HV (2018) Gene stacking of multiple traits for high yield of fermentable sugars in plant biomass. Biotechnol Biofuels 11:2

Baird RD, Tan DSP, Kaye SB (2010) Weekly paclitaxel in the treatment of recurrent ovarian cancer. Nat Rev Clin Oncol 7:575–582

Balk J, Pilon M (2011) Ancient and essential: the assembly of iron-sulfur clusters in plants. Trends Plant Sci 16:218–226
Bang SW, Park SH, Jeong JS, Kim YS, Jung H, Ha SH, Kim JK et al (2013) Characterization of the stress-inducible OsNCED3 promoter in different transgenic rice organs and over three homozygous generations. Planta 237:211–224
Barah P, Bones AM (2015) Multidimensional approaches for studying plant defence against insects: from ecology to omics and synthetic biology. J Exp Bot 66(2):479–493
Bar-Even A (2018) Daring metabolic designs for enhanced plant carbon fixation. Plant Sci 273:71–83
Bennett C (2017) Genetically engineered insects next for agriculture? Cornell University, Matt Hayes
Berg IA (2011) Ecological aspects of the distribution of different autotrophic CO_2 fixation pathways. Appl Environ Microbiol 77:1925–1936
Berhanu S, Ueda T, Kuruma Y (2019) Artificial photosynthetic cell producing energy for protein synthesis. Nat Commun 10:1325
Beyer P (2010) Golden Rice and 'Golden' crops for human nutrition. New Biotechnol 27 (5):478–481
Bodenhausen N, Bortfeld-Miller M, Ackermann M, Vorholt JA (2014) A synthetic community approach reveals plant genotypes affecting the phyllosphere microbiota. PLoS Genet 10(4): e1004283
Bohmert K, Balbo I, Kopka J, Mittendorf V, Nawrath C, Poirier Y, Tischendorf G, Trethewey RN, Willmitzer L (2000) Transgenic *Arabidopsis* plants can accumulate polyhydroxybutyrate to up to 4% of their fresh weight. Planta 211:841–845
Bowen TA, Zdunek JK, Medford JI (2008) Cultivating plant synthetic biology from system biology. New Phytol 179:583–587
Boyd ES, Garcia Costas AM, Hamilton TL, Mus F, Peters JW (2015) Evolution of molybdenum nitrogenase during the transition from anaerobic to aerobic metabolism. J Bacteriol 197 (9):1690–1699
Borland AM, Hartwell J, Weston DJ, Schlauch KA, Tschaplinski TJ, Tuskan GA, Yang X, Cushman JC (2014) Engineering crassulacean acid metabolism to improve water-use efficiency. Trends Plant Sci 19(5):327–338
Breitenbach J, Nogueira M, Farré G, Zhu C, Capell T, Christou P, Fleck G, Focken U, Fraser PD, Sandmann G (2016) Engineered maize as a source of astaxanthin: processing and application as fish feed. Transgenic Res 25(6):785–793
Brown KA, Harris DF, Wilker MB, Rasmussen A, Khadka N, Hamby H, Keable S, Dukovic G, Peters JW, Seefeldt LC, King PW (2016) Light-driven dinitrogen reduction catalyzed by a CdS: nitrogenase MoFe protein biohybrid. Science 352:448–450
Burén S, Rubio LM (2018) State of the art in eukaryotic nitrogenase engineering. FEMS Microbiol Lett 365(2):fnx274
Bussamra BC, Freitas S, Costa AC (2015) Improvement on sugar cane bagasse hydrolysis using enzymatic mixture designed cocktail. Bioresour Technol 187:173–181
Butelli E, Titta L, Giorgio M, Mock H-P, Matros A, Peterek S, Schijlen EGWM, Hall RD, Bovy AG, Luo J, Martin C (2008) Enrichment of tomato fruit with health-promoting anthocyanins by expression of select transcription factors. Nat Biotechnol 26:1301–1308
Butler NM, Baltes NJ, Voytas DF, Douches DS (2016) Geminivirus-mediated genome editing in potato (*Solanum tuberosum* L.) using sequence-specific nucleases. Front Plant Sci 7:1045
Cahoon RE, Lutke WK, Cameron JC et al (2015) Adaptive engineering of phytochelatin-based heavy metal tolerance. J Biol Chem 290(28):17321–17330
Cao T, Lahiri I, Singh V, Louis J, Shah J, Ayre BG (2013) Metabolic engineering of raffinose-family oligosaccharides in the phloem reveals alterations in carbon partitioning and enhances resistance to green peach aphid. Front Plant Sci 4:263
Carvalho Â, Hansen EH, Kayser O, Carlsen S, Stehle F (2017) Designing microorganisms for heterologous biosynthesis of cannabinoids. FEMS Yeast Res 17(4):fox037

Casino P, Rubio V, Marina A (2010) The mechanism of signal transduction by two-component systems. Curr Opin Struct Biol 20:763–771

Castrillo G, Teixeira PJ, Paredes SH, Law TF, de Lorenzo L, Feltcher ME, Finkel OM, Breakfield NW, Mieczkowski P, Jones CD, Paz-Ares J, Dangl JL (2017) Root microbiota drive direct integration of phosphate stress and immunity. Nature 543(7646):513–518

Chai F, Wang Y, Mei X, Yao M, Chen Y, Liu H, Xiao W, Yuan Y (2017) Heterologous biosynthesis and manipulation of crocetin in *Saccharomyces cerevisiae*. Microb Cell Factories 16:54

Chandini, Kumar R, Kumar R, Prakash O (2019) The impact of chemical fertilizers on our nvironment and ecosystem. In: Research trends in environmental sciences, 2nd edn

Chatelle C, Ochoa-Fernandez R, Engesser R, Schneider N, Beyer HM, Jones AR, Timmer J, Zurbriggen MD, Weber W (2018) A green-light-responsive system for the control of transgene expression in mammalian and plant cells. ACS Synth Biol 7:1349–1358

Choi WG, Gilroy S (2014) Plant biologists FRET over stress. elife 3:e02763

Chu S, Majumdar A (2012) Opportunities and challenges for a sustainable energy future. Nature 488:294–303

Chen H, Lai Z, Shi J, Xiao Y, Chen Z, Xu X (2010) Roles of arabidopsis WRKY18, WRKY40 and WRKY60 transcription factors in plant responses to abscisic acid and abiotic stress. BMC Plant Biol 10:281

Cotton CA, Edlich-Muth C, Bar-Even A (2018) Reinforcing carbon fixation: CO_2 reduction replacing and supporting carboxylation. Curr Opin Biotech 49:49–56

Curatti L, Rubio LM (2014) Challenges to develop nitrogen-fixing cereals by direct *nif*-gene transfer. Plant Sci 225:130–137

Dacres H, Wang J, Leitch V, Horne I, Anderson AR, Trowell SC (2011) Greatly enhanced detection of a volatile ligand at femtomolar levels using bioluminescence resonance energy transfer (BRET). Biosens Bioelectron 29:119–124

Dalal J, Lopez H, Vasani NB, Hu Z, Swift JE, Yalamanchili R, Dvora M, Lin X, Xie D, Qu R, Sederoff HW (2015) A photorespiratory bypass increases plant growth and seed yield in biofuel crop *Camelina sativa*. Biotechnol Biofuels 8:175

Das PK, Laxman B, Rao SVCK, Seshasai MVR, Dadwal VK (2015) Monitoring of bacterial late blight in rice using ground based hyperspectral and LISS IV satellite dated in Kurnool, Andhra Pradesh, India. Int J Pest Manage 61:259–368

de Lorenzo V (2017) Seven microbial bio-processes to help the planet. Microb Biotech 10 (5):995–998

DeLoache WC, Russ ZN, Narcross L, Gonzales AM, Martin VJJ, Dueber JE (2015) An enzyme-coupled biosensor enables (S)-reticuline production in yeast from glucose. Nat Chem Biol 11:465–471

di Ventura B, Lemerle C, Michalodimitrakis K, Serrano L (2006) From *in vivo* to in silico biology and back. Nature 443:527–533

Diretto G, Al-Babili S, Tavazza R, Papacchioli V, Beyer P, Giuliano G (2007) Metabolic engineering of potato carotenoid content through tuber-specific overexpression of a bacterial minipathway. PLoS One 2:e350

Dismukes GC, Carrieri D, Bennette N, Ananyev GM, Posewitz MC (2008) Aquatic phototrophs: efficient alternatives to land-based crops for biofuels. Curr Opin Biotechnol 19:235–240

Dymond JS, Richardson SM, Coombes CE, Babatz T, Muller H, Annaluru N, Blake WJ, Schwerzmann JW, Dai J, Lindstrom DL, Boeke AC, Gottschling DE, Chandrasegaran S, Bader JS, Boeke JD (2011) Synthetic chromosome arms function in yeast and generate phenotypic diversity by design. Nature 477:471–476

Eisenhut M, Roell MS, Weber APM (2019) Mechanism understanding of photorespiration paves the way to a new green revolution. New Phytol 15:330–338

Engler C, Youles M, Gruetzner R, Ehnert T-M, Werner S, Jones JDG, Patron NJ, Marillonnet S (2014) A golden gate modular cloning toolbox for plants. ACS Synth Biol 3(11):839–843

Erb TJ (2011) Carboxylases in natural and synthetic microbial pathways. Appl Environ Microbiol 77:8466–8477
Erb TJ, Zarzycki J (2016) Biochemical and synthetic biology approaches to improve photosynthetic CO_2-fixation. Curr Opin Chem Biol 34:72–79
Eudes A, Sathitsuksanoh N, Baidoo EEK, George A, Liang Y, Yang F, Singh S, Keasling JD, Simmons BA, Loqué D (2015) Expression of a bacterial 3-dehydroshikimate dehydratase reduces lignin content and improves biomass saccharification efficiency. Plant Biotechnol J 13:1241–1250
Eudes A, Berthomieu R, Hao Z, Zhao N, Benites VT, Baidoo EEK, Loque D (2018) Production of muconic acid in plants. Metab Eng 46:13–19
Fan J, Yan C, Roston R, Shanklin J, Xu C (2014) Arabidopsis lipins, *PDAT1* acyltransferase, and *SDP1* triacylglycerol lipase synergistically direct fatty acids toward β-oxidation, thereby maintaining membrane lipid homeostasis. Plant Cell 26(10):4119–4134
Fang H, Li D, Kang J, Jiang P, Sun J, Zhang D (2018) Metabolic engineering of *Escherichia coli* for *de novo* biosynthesis of vitamin B12. Nat Commun 9(1):4917
FAO (2017) Nutrition-sensitive agriculture and food systems in practice Options for intervention Revised edition Food and Agriculture Organization of the United Nations Rome
Farré G, Perez-Fons L, Decourcelle M, Breitenbach J, Hem S, Zhu C, Capell T, Christou C, Fraser PD, Sandmann G (2016) Metabolic engineering of astaxanthin biosynthesis in maize endosperm and characterization of a prototype high oil hybrid. Transgenic Res 25:477–489
Ferry MS, Hasty J, Cookson NA (2012) Synthetic biology approaches to biofuel production. Editorial Biofuels 3(1):9–12
Flores Bueso Y, Tangney M (2017) Synthetic biology in the driving seat of the bioeconomy. Trends Biotechnol 35(5):373–378
Fossati E, Narcross L, Ekins A, Falgueyret JP, Martin VJ (2015) Synthesis of morphinan alkaloids in *Saccharomyces cerevisiae*. PLoS One 10:e0124459
Fowler CC, Brown ED, Li Y (2010) Using a riboswitch sensor to examine coenzyme B(12) metabolism and transport in *E. coli*. Chem Biol 17(7):756–765
Friedel S, Usadel B, von Wirén N, Sreenivasulu N (2012) Reverse engineering: a key component of systems biology to unravel global abiotic stress cross-talk. Front Plant Sci 3:1–16
Fuentes P, Zhou F, Erban A, Karcher D, Kopka J, Bock R (2016) A new synthetic biology approach allows transfer of an entire metabolic pathway from a medicinal plant to a biomass crop. elife 5: e13664
Galanie S, Thodey K, Trenchard IJ, Filsinger Interrante M, Smolke CD (2015) Complete biosynthesis of opioids in yeast. Science 349(6252):1095–1100
Gao J, Zhang T, Xu B, Jia L, Xiao B, Liu H, Liu L, Yan H, Xia Q (2018) CRISPR/Cas9-mediated mutagenesis of carotenoid cleavage dioxygenase 8 (CCD8) in *Tobacco* affects shoot and root architecture. Int J Mol Sci 19(4):1062
Gardner TS, Cantor CR, Collins JJ (2000) Construction of a genetic toggle switch in *Escherichia coli*. Nature 403:339–342
Ge H, Li X, Chen S, Zhang M, Liu Z, Wang J et al (2018) The expression of CARK1 or RCAR11 driven by synthetic promoters increases drought tolerance in *Arabidopsis thaliana*. Int J Mol Sci 19(7):1945
Geddes BA, Ryu MH, Mus F, Costas AG, Peters JW, Voigt CA, Poole P (2015) Use of plant colonizing bacteria as chassis for transfer of N_2-fixation to cereals. Curr Opin Biotechnol 32:216–222
Geddes BA, Paramasivan P, Joffrin A, Thompson AL, Christensen K, Jorrin B, Brett P, Conway SJ, Oldroyd GED, Poole PS (2019) Engineering transkingdom signalling in plants to control gene expression in rhizosphere bacteria. Nat Commun 10:3430
Gibson DG, Benders GA, Andrews-Pfannkoch C, Denisova EA, Baden-Tillson H, Zaveri J, Stockwell TB, Brownley A, Thomas DW, Algire MA, Merryman C, Young L, Noskov VN, Glass JI, Venter JC, Hutchison CA III, Smith HO (2008) Complete chemical synthesis, assembly, and cloning of a *Mycoplasma genitalium* genome. Science 319:1215

Gibson DG, Glass JI, Lartigue C, Noskov VN, Chuang R-Y, Algire MA, Benders GA, Montague MG, Ma L, Moodie MM, Merryman C, Vashee S, Krishnakumar R, Assad-Garcia N, Andrews-Pfannkoch C, Denisova EA, Young L, Qi Z-Q, Segall-Shapiro TH, Calvey CH, Parmar PP, Hutchison CA III, Smith HO, Venter JC (2010) Creation of a bacterial cell controlled by a chemically synthesized genome. Science 329(5987):52–56

Giessen TW, Silver PA (2017) Engineering carbon fixation with artificial protein organelles. Curr Opin Biotech 46:42–50

Glass JI, Assad-Garcia N, Alperovich N, Yooseph S, Lewis MR, Maruf M, Hutchison CA, Smith HO, Venter JC (2006) Essential genes of a minimal bacterium. Proc Natl Acad Sci USA 103 (2):425–430

Golitsch F, Bücking C, Gescher J (2013) Proof of principle for an engineered microbial biosensor based on *Shewanella oneidensis* outer membrane protein complexes. Biosens Bioelectron 47:285–291

Goold HD, Wright P, Hailstones D (2018) Emerging opportunities for synthetic biology in agriculture. Genes (Basel) 9(7):341

Gould C (2020) 5 inspiring ways synthetic biology will revolutionize food and agriculture. pp 1–6. https://medium.com/age-of-awareness/5-inspiring-ways-synthetic-biology-will-revolutionize-food-and-agriculture-3601c25438b5

Goulding K, Jarvis S, Whitmore A (2008) Optimizing nutrient management for farm systems. Philos Trans R Soc Lond Ser B Biol Sci 363(1491):667–680

Guet CC, Elowitz MB, Hsing W, Leibler S (2002) Combinatorial synthesis of genetic networks. Science 296:1466–1470

Gupta RK, Gangoliya SS, Singh NK (2015) Reduction of phytic acid and enhancement of bioavailable micronutrients in food grains. J Food Sci Technol 52:676–684

Hahn F, Mantegazza O, Greiner A, Hegemann P, Eisenhut M, Weber AP (2017) An efficient visual screen for CRISPR/Cas9 activity in *Arabidopsis thaliana*. Front Plant Sci 8:39

Hamerli D, Birch RG (2011) Transgenic expression of trehalulose synthase results in high concentrations of the sucrose isomer trehalulose in mature stems of field-grown sugarcane. Plant Biotechnol J 9:32–37

Hammond JP, Bennett MJ, Bowen HC, Broadley MR, Eastwood DC, May ST, Rahn C, Swarup R, Woolaway KE, White PJ (2003) Changes in gene expression in *Arabidopsis* shoots during phosphate starvation and the potential for developing smart plants. Plant Physiol 132 (2):578–596

Henes B, Zülli F, Niederkrüger H, Schaaf A, Frischmuth T, Decker EL, Reski R (2018) The magic of moss. Soap Perfum Cosmet 91:64–66

Heuer S, Gaxiola R, Schilling R, Herrera-Estrella L, López-Arredondo D, Wissuwa M, Delhaize E, Rouached H (2017) Improving phosphorus use efficiency: a complex trait with emerging opportunities. Plant J 90(5):868–885

Hnaien M, Lagarde F, Bausells J, Errachid A, Jaffrezic-Renault N (2011) A new bacterial biosensor for trichloroethylene detection based on a three-dimensional carbon nanotubes bioarchitecture. Anal Bioanal Chem 400:1083–1092

Hong S, Lee C (2018) The current status and future outlook of quantum dot-based biosensors for plant virus detection. Plant Pathol J 34:85–92

Hou L, Chen L, Wang J, Xu D, Dai L, Zhang H et al (2012) Construction of stress responsive synthetic promoters and analysis of their activity in transgenic *Arabidopsis thaliana*. Plant Mol Biol Rep 30(6):1496–1506

Hsu PD, Lander ES, Zhang F (2014) Development and applications of CRISPR-Cas9 for genome engineering. Cell 157:1262–1278

Hutchison CA III, Chuang R-Y, Noskov VN, Assad-Garcia N, Deerinck TJ, Ellisman MH, Gill J, Kannan K, Karas BJ, Ma L, Pelletier JF, Qi Z-Q, Richter RA, Strychalski EA, Sun L, Suzuki Y, Tsvetanova B, Wise KS, Smith HO, Glass JI, Merryman C, Gibson DG, Venter JC (2016) Design and synthesis of a minimal bacterial genome. Science 351(6280):aad6253

Hynninen A, Virta M (2010) Whole-cell bioreporters for the detection of bioavailable metals. Adv Biochem Eng Biotechnol 118:31–63

Ikram NKBK, Simonsen HT (2017) A review of biotechnological artemisinin production in plants. Front Plant Sci 8:1966

Ikram NKBK, Zhan X, Pan X-W, King BC, Simonsen HT (2015) Stable heterologous expression of biologically active terpenoids in green plant cells. Front Plant Sci 6:129

Isaacs FJ, Hasty J, Cantor CR, Collins JJ (2003) Prediction and measurement of an autoregulatory genetic module. Proc Natl Acad Sci USA 100:7714–7719

Ivleva NB, Groat J, Staub JM, Stephens M (2016) Expression of active subunit of nitrogenase via integration into plant organelle genome. PLoS One 11:e0160951

Jain M (2015) Function genomics of abiotic stress tolerance in plants: a CRISPR approach. Front Plant Sci 6:375

Jez JM, Lee SG, Sherp AM (2016) The next green movement: plant biology for the environment and sustainability. Science 353(6305):1241–1244

Jia H, Orbovic V, Jones JB, Wang N (2016) Modification of the PthA4 effector binding elements in type I CsLOB1 promoter using Cas9/sgRNA to produce transgenic Duncan grapefruit alleviating XccΔpthA4:dCsLOB1.3 infection. Plant Biotechnol J 14(5):1291–1301

Jirschitzka J, Mattern DJ, Gershenzon J, D'Auria JC (2013) Learning from nature: new approaches to the metabolic engineering of plant defense pathways. Curr Opin Biotech 24:320–328

Joh NH, Wang T, Bhate MP, Acharya R, Wu Y, Grabe M et al (2014) *De novo* design of a transmembrane Zn^{2+}-transporting four-helix bundle. Science 6216:1520–1524

Johnston EJ, Rylott EL, Beynon E, Lorenz A, Chechik V, Bruce NC (2015) Monodehydroascorbate reductase mediates TNT toxicity in plants. Science 349(6252):1072–1075

Jørgensen K, Bak S, Busk PK, Sørensen C, Olsen CE, Puonti-Kaerlas J, Møller BL (2005) Cassava plants with a depleted cyanogenic glucoside content in leaves and tubers. Distribution of cyanogenic glucosides, their site of synthesis and transport, and blockage of the biosynthesis by RNA interference technology. Plant Physiol 139(1):363–374

Kanygin A, Milrad Y, Thummala C, Reifschneider K, Baker P, Marco P, Yacoby I, Kevin E (2020) Redding rewiring photosynthesis: a photosystem I-hydrogenase chimera that makes H_2 *in vivo*. Energy Environ Sci. https://doi.org/10.1039/C9EE03859K

Keasling JD (2010) Manufacturing molecules through metabolic engineering. Science 330:1355–1358

Kebeish R, Niessen M, Thiruveedhi K, Bari R, Hirsch HJ, Rosenkranz R, Stabler N, Schonfeld B, Kreuzaler F, Peterhansel C (2007) Chloroplastic photorespiratory bypass increases photosynthesis and biomass production in *Arabidopsis thaliana*. Nat Biotechnol 25:593–599

Kim SR, Park Y-C, Jin Y-S, Seo J-H (2013) Strain engineering of *Saccharomyces cerevisiae* for enhanced xylose metabolism. Biotechnol Adv 31:851–861

Kim N-S, Yu H-Y, Chung N-D, Kwon T-H, Yang M-S (2014) High-level production of recombinant trypsin in transgenic rice cell culture through utilization of an alternative carbon source and recycling system. Enzym Microb Technol 63:21–27

Kim HJ, Jeong H, Lee SJ (2018) Synthetic biology for microbial heavy metal biosensors. Anal Bioanal Chem 410:1191–1203

Kirchhoff J, Raven N, Boes A, Roberts JL, Russell S, Treffenfeldt W, Fischer R, Schinkel H, Schiermeyer A, Schillberg S (2012) Monoclonal tobacco cell lines with enhanced recombinant protein yields can be generated from heterogeneous cell suspension cultures by flow sorting. Plant Biotechnol J 10(8):936–944

Kristensen C, Morant M, Olsen CE, Ekstrom CT, Galbraith DW, Moller BL, Bak S (2005) Metabolic engineering of dhurrin in transgenic Arabidopsis plants with marginal inadvertent effects on the metabolome and transcriptome. Proc Natl Acad Sci U S A 102:1779–1784

Küken A, Nikoloski Z (2019) Computational approaches to design and test plant synthetic metabolic pathways. Plant Physiol 179(3):894–906

Kumar V, Arora K (2020) Trends in nano-inspired biosensors for plants. Mater Sci Energy Technol 3:255–273

Kumar V, Jain M (2015) The CRISPR-Cas system for plant genome editing: advances and opportunities. J Exp Bot 66:47–57

Kumar S, Hahn FM, Baidoo E, Kahlon TS, Wood DF, McMahan CM, Cornish K, Keasling JD, Daniell H, Whalen MC (2012) Remodeling the isoprenoid pathway in tobacco by expressing the cytoplasmic mevalonate pathway in chloroplasts. Metab Eng 14:19–28

Lagarde F, Jaffrezic-Renault N (2011) Cell-based electrochemical biosensors for water quality assessment. Anal Bioanal Chem 400:947–964

Lartigue C, Glass JI, Alperovich N, Pieper R, Parmar PP, Hutchison CA III, Smith HO, Venter JC (2007) Genome transplantation in bacteria: changing one species to another. Science 317:632–638

Laursen T, Borch J, Knudsen C et al (2016) Characterization of a dynamic metabolon producing the defense compound dhurrin in sorghum. Science 354(6314):890–893

Le Quere C, Raupach MR, Canadell JG et al (2009) Trends in the sources and sinks of carbon dioxide. Nat Geosci 2:831–836

Lee D, Lloyd NDR, Pretorius IS, Borneman AR (2016) Heterologous production of raspberry ketone in the wine yeast *Saccharomyces cerevisiae* via pathway engineering and synthetic enzyme fusion. Microb Cell Factories 15:49

Lee KY, Park S-J, Lee KA, Kim S-H, Kim H, Meroz Y, Mahadevan L, Jung K-H, Ahn TK, Parker KK, Shin K (2018) Photosynthetic artificial organelles sustain and control ATP-dependent reactions in a protocellular system. Nat Biotechnol 36:530–535

Leister D (2019) Genetic engineering, synthetic biology and the light reactions of photosynthesis. Plant Physiol 179(3):778–793

Li Q, Song J, Peng S, Wang JP, Qu GZ, Sederoff RR, Chiang VL (2014) Plant biotechnology for lignocellulosic biofuel production. Plant Biotechnol J 12:1174–1192

Li J, Meng X, Zong Y et al (2016) Gene replacements and insertions in rice by intron targeting using CRISPR–Cas9. Nat Plants 2:16139

Li P, Li YJ, Zhang FJ et al (2017) The Arabidopsis UDP-glycosyltransferases *UGT79B2* and *UGT79B3*, contribute to cold, salt and drought stress tolerance via modulating anthocyanin accumulation. Plant J 89(1):85–103

Li S, Li Y, Smolke CD (2018) Strategies for microbial synthesis of high-value phytochemicals. Nat Chem 10(4):395–404

Lin MT, Occhialini A, Andralojc PJ, Devonshire J, Hines KM, Parry MAJ, Hanson MR (2014) β-carboxysomal proteins assemble into highly organized structures in *Nicotiana* chloroplasts. Plant J 79(1):1–12

Liu W, Stewart CN Jr (2015) Plant synthetic biology. Trends Plant Sci 20(5):309–317

Liu W, Mazarei M, Rudis MR, Fethe MH, Peng Y, Millwood RJ, Schoene G, Burris JN, Stewart CN Jr (2013a) Bacterial pathogen phytosensing in transgenic tobacco and *Arabidopsis*. Plant Biotechnol J 11:43–52

Liu W, Yuan JS, Stewart CN Jr (2013b) Advanced genetic tools for plant biotechnology. Nat Rev Genet 14(11):781–793

Liu W, Mazarei M, Peng Y, Fethe MH, Rudis MR, Lin J, Millwood RJ, Arelli PR, Stewart CN Jr (2014) Computational discovery of soybean promoter cis-regulatory elements for the construction of soybean cyst nematode inducible synthetic promoters. Plant Biotechnol J 12:1015–1026

Liu W, Luo Z, Wang Y, Pham NT, Tuck L, Pérez-Pi I, Liu L, Shen Y, French C, Auer M, Marles-Wright J, Dai J, Cai Y (2018a) Rapid pathway prototyping and engineering using *in vitro* and *in vivo* synthetic genome SCRaMbLE-in methods. Nat Commun 9:1936

Liu D, Mannan AA, Han Y, Oyarzún DA, Zhang F (2018b) Dynamic metabolic control: towards precision engineering of metabolism. J Ind Microbiol Biotechnol 45:535–543

Long SP, Zhu XG, Naidu SL, Ort DR (2006) Can improvement in photosynthesis increase crop yields? Plant Cell Environ 29:315–330

Long SP, Marshall-Colon A, Zhu XG (2015) Meeting the global food demand of the future by engineering crop photosynthesis and yield potential. Cell 161(1):56–66

Long BM, Rae BD, Rolland V, Förster B, Price GD (2016) Cyanobacterial CO_2-concentrating mechanism components: function and prospects for plant metabolic engineering. Curr Opin Plant Biol 31:1–8
Long BM, Hee WY, Sharwood RE, Rae BD, Kaines S, Lim Y-L, Nguyen ND, Massey B, Bala S, von Caemmerer S, Badger MR, Price GD (2018) Carboxysome encapsulation of the CO_2-fixing enzyme Rubisco in tobacco chloroplasts. Nat Commun 9:3570
López-Torrejón G, Jiménez-Vicente E, Buesa JM, Hernandez JA, Verma HK, Rubio LM (2016) Expression of a functional oxygen-labile nitrogenase component in the mitochondrial matrix ofaerobically grown yeast. Nat Commun 7:11426
Luna-Moreno D, Sánchez-Álvarez A, Islas-Flores I, Canto-Canche B, Carrillo-Pech M, Villarreal-Chiu JF, Rodríguez-Delgado M (2019) Early detection of the fungal banana black Sigatoka pathogen *Pseudocercospora fijiensis* by an SPR immunosensor method. Sensors (Basel) 23: E465
Ma NJ, Isaacs FJ (2016) Genomic recoding broadly obstructs the propagation ofhorizontally transferred genetic elements. Cell Syst 3(2):199–207
Mahlein AK, Rumpf T, Welke P, Dehne HW, Plumer L, Steiner U, Oerke EC (2013) Development of spectral indices for detecting and identifying plant diseases. Remote Sens Environ 128:21–30
Majer E, Llorente B, Rodríguez-Concepción M, Daròs JA (2017) Rewiring carotenoid biosynthesis in plants using a viral vector. Sci Rep 7:41645
Mandal M, Breaker RR (2004) Gene regulation by riboswitches. Nat Rev Mol Cell Biol 5:451–463
Markmann K, Parniske M (2009) Evolution of root endosymbiosis with bacteria: how novel are nodules? Trends Plant Sci 14:77–86
Mateo P, Leganés F, Perona E, Loza V, Fernández-Piñas F (2015) Cyanobacteria as bioindicators and bioreporters of environmental analysis in aquatic ecosystems. Biodivers Conserv 24:909–948
Mazarei M, Teplova I, Hajimorad MR, Stewart CN Jr (2008) Pathogen phytosensing: plants to report plant pathogens. Sensors 8:2628–2641
McGrath JM, Long SP (2014) Can the cyanobacterial carbon-concentrating mechanism increase photosynthesis in crop species? A theoretical analysis. Plant Physiol 164(4):2247–2261
Mehta J, Bhardwaj SK, Bhardwaj N, Paul AK, Kumar P, Kim KH (2016) Progress in the biosensing techniques for trace-level heavy metals. Biotechnol Adv 34(1):47–60
Miller LH, Su X (2011) Artemisinin: discovery from the Chinese herbal garden. Cell 146 (6):855–858
Miller TE, Beneyton T, Schwander T, Diehl C, Girault M, Mclean R, Chotel T, Claus P, Cortina NS, Baret J-C, Erb TJ (2020) Light-powered CO_2 fixation in a chloroplast mimic with natural and synthetic parts. Science 368(6491):649–654
Moretti VM, Mentasti T, Bellagamba F, Luzzana U, Caprino F, Turchini GM, Giani I, Valfrè F (2006) Determination of astaxanthin stereoisomers and colour attributes in flesh of rainbow trout (*Oncorhynchus mykiss*) as a tool to distin- guish the dietary pigmentation source. Food Addit Contam 23:105610–105663
Morey KJ, Antunes MS, Albrecht KD, Bowen TA, Troupe JF, Havens KL, Medford JI (2011) Developing a synthetic signal transduction system in plants. Methods Enzymol 497:581–602
Mortimer JC (2019) Plant synthetic biology could drive a revolution in biofuels and medicine. Exp Biol Med (Maywood) 244(4):323–331
Moradyar M, Motallebi M, Zamani MR, Aghazadeh R (2016) Pathogen-induced expression of chimeric chitinase gene containing synthetic promoter confers antifungal resistance in transgenic canola. In Vitro Cell Dev Biol Plant 52(2):119–129
Moses T, Pollier J, Almagro L, Buyst D, Van Montagu M, Pedreno MA, Martins JC, Thevelein JM, Goossens A (2014) Combinatorial biosynthesis of sapogenins and saponins in *Saccharomyces cerevisiae* using a C-16a hydroxylase from *Bupleurum falcatum*. Proc Natl Acad Sci USA 111:1634–1639
Moses T, Mehrshahi P, Smith AG, Goossens A (2017) Synthetic biology approaches for the production of plant metabolites in unicellular organisms. J Exp Bot 68(15):4057–4074

Mueller ND, Gerber JS, Johnston M, Ray DK, Ramankutty N, Foley JA (2012) Closing yield gaps through nutrient and water management. Nature 490:254–257

Muller K, Siegel D, Jahnke FR, Gerrer K, Wend S, Decker EL, Reski R, Weber W, Zurbriggen MD (2014) A red light-controlled synthetic gene expression switch for plant system. Mol Plant 10:1679–1688

Mus F, Crook MB, Garcia K, Costas AG, Geddes BA, Kouri ED, Paramasivan P, Ryu M-H, Oldroyd GED, Poole PS, Udvardi MK, Voigt CA, Ané M-H, Peters JW (2016) Symbiotic nitrogen fixation and the challenges to its extension to nonlegumes. Appl Environ Microbiol 82 (13):3698–3710

Nakagawa A, Minami H, Kim J-S, Koyanagi T, Katayama T, Sato F, Kumagai H (2011) A bacterial platform for fermentative production of plant alkaloids. Nat Commun 2:326

Napier JA, Olsen R-E, Tocher DR (2019) Update on GM canola crops as novel sources of omega-3 fish oils. Plant Biotechnol J 17:703–705

Nekrasov V, Wang C, Win J et al (2017) Rapid generation of a transgene-free powdery mildew resistant tomato by genome deletion. Sci Rep 7:482

Niu B, Paulson JN, Zheng X, Kolter R (2017) Simplified bacterial community of maize. Proc Natl Acad Sci 114(12):E2450–E2459

Nocera DG (2012) The artificial leaf. Acc Chem Res 45(5):767–776

Nozzi NE, Oliver JWK, Atsumi S (2013) Cyanobacteria as a platform for biofuel production. Front Bioeng Biotechnol 1:7

Oger P, Petit A, Dessaux Y (1997) Genetically engineered plants producing opines alter their biological environment. Nat Biotechnol 15:369–372

Olsen JL, Rouzé P, Verhelst B, Lin YC, Bayer T, Collen J, Dattolo E, De Paoli E, Dittami S, Maumus F, Michel G, Kersting A, Lauritano C, Lohaus R, Töpel M, Tonon T, Vanneste K, Amirebrahimi M, Brakel J, Boström C, Chovatia M, Grimwood J, Jenkins JW, Jueterbock A, Mraz A, Stam WT, Tice H, Bornberg-Bauer E, Green PJ, Pearson GA, Procaccini G, Duarte CM, Schmutz J, Reusch TBH, van de Peer Y (2016) The genome of the seagrass *Zostera marina* reveals angiosperm adaptation to the sea. Nature 530:331–335

Ort DR, Merchant SS, Alric J, Barkan A, Blankenship RE, Bock R, Croce R, Hanson MR, Hibberd JM, Long SP, Moore TA, Moroney J, Niyogi KK, Parry MAJ, Peralta-Yahya PP, Prince RC, Redding KE, Spalding MH, Van Wijk KJ, Vermaas WFJ, von Caemmerera S, Weberb APM, Yeates TO, Yuand JS, Zhu XG (2015) Redesigning photosynthesis to sustainably meet global food and bioenergy demand. Proc Natl Acad Sci USA 112(28):8529–8536

Ostrov N, Landon M, Guell M, Kuznetsov G, Teramoto J, Cervantes N, Zhou M, Singh K, Napolitano MG, Moosburner M, Shrock E, Pruitt BW, Conway N, Goodman DB, Gardner CL, Tyree G, Gonzales A, Wanner BL, Norville JE, Lajoie MJ, Church GM (2016) Design, synthesis, and testing toward a 57-codon genome. Science 353(6301):819–822

Oswald M, Fischer M, Dirninger N, Karst F (2007) Monoterpenoid biosynthesis in *Saccharomyces cerevisiae*. FEMS Yeast Res 7:413–421

Paddon CJ, Westfall PJ, Pitera DJ, Benjamin K, Fisher K, McPhee D, Leavell MD, Tai A, Main A, Eng D, Polichuk DR, Teoh KH, Reed DW, Treynor T, Lenihan J, Fleck M, Bajad S, Dang G, Dengrove D, Diola D, Dorin G, Ellens KW, Fickes S, Galazzo J, Gaucher SP, Geistlinger T, Henry R, Hepp M, Horning T, Iqbal T, Jiang H, Kizer L, Lieu B, Melis D, Moss N, Regentin R, Secrest S, Tsuruta H, Vazquez R, Westblade LF, Xu L, Yu M, Zhang Y, Zhao L, Lievense J, Covello PS, Keasling JD, Reiling KK, Renninger NS, Newman JD (2013) High-level semi-synthetic production of the potent antimalarial artemisinin. Nature 496:528–532

Palladino C, Breiteneder H (2018) Peanut allergens. Mol Immunol 100:58–70

Pan X-W, Han L, Zhang Y-H, Chen D-F, Simonsen HT (2015) Sclareol production in the moss Physcomitrella patens and observations on growth and terpenoid biosynthesis. Plant Biotechnol Rep 9:149–159

Pandey R, Teig-Sussholz O, Schuster S, Avni A, Shacham-Diamand Y (2018) Integrated electrochemical chip-on-plant functional sensor for monitoring gene expression under stress. Biosens Bioelectron 117:493–500

Papanatsiou M, Petersen J, Henderson L, Wang Y, Christie JM, Blatt MR (2019) Optogenetic manipulation of stomatal kinetics improves carbon assimilation, water use, and growth. Science 363(6434):1456–1459

Park S-H, Ransom C, Mei C, Sabzikar R, Qi C, Chundawat S, Dale B, Sticklen M (2011) In the quest of alternatives to microbial cellulase mix production: corn Stover-produced heterologous multi-cellulases readily deconstruct lignocellulosic biomass to fermentable sugars. J Chem Technol Biotechnol 86(5):633–641

Park S-H, Ong RG, Mei C, Sticklen M (2014) Lignin down-regulation of *Zea mays* via dsRNAi and klason lignin analysis. J Vis Exp 89:51340

Park SY, Peterson FC, Mosquna A, Yao J, Volkman BF, Cutler SR (2015) Agrochemical control of plant water use using engineered abscisic acid receptors. Nature 520(7548):545–548

Paul H, Steinbrecher R (2010) Synthetic biology. New Scientists 207(2772):24–25

Peng Y, Allen S, Millwood RJ, Stewart CN Jr (2014) 'Fukusensor:' a genetically engineered plant for reporting DNA damage in response to gamma radiation. Plant Biotechnol J 12 (9):1329–1332

Peng A, Chen S, Lei T, Xu L, He Y, Wu L, Yao L, Zou X (2017) Engineering canker-resistant plants through CRISPR/Cas9-targeted editing of the susceptibility gene *CsLOB1* promoter in citrus. Plant Biotechnol J 15(12):1509–1519

Peramuna A, Bae H, Rasmussen EK, López CQ, Fromberg A, Simonsen HT (2018) Attaching patchoulol synthase to lipid droplet-associated proteins increase the level of patchoulol in the lipid droplets. Plant Biotechnol J. (in press)

Perchlik M, Tegeder M (2017) Improving plant nitrogen use efficiency through alteration of amino acid transport processes. Plant Physiol 175(1):235–247

Peterhansel C, Maurino VG (2011) Photorespiration redesigned. Plant Physiol 155(1):49–55

Peterhansel C, Horst I, Niessen M, Blume C, Kebeish R, Kürkcüoglu S, Kreuzaler F (2010) Photorespiration. Arabidopsis Book 8:e0130

Peterson FC, Burgie ES, Park S-Y, Jensen DR, Weiner JJ, Bingman CA, Chang C-E A, Cutler SR, Phillips GN Jr, Volkman BF (2010) Structural basis for selective activation of ABA receptors. Nat Struct Mol Biol 17:1109–1113

Petolino JF, Davies JP (2013) Designed transcriptional regulators for trait development. Plant Sci 201-202:128–136

Petrie JR, Shrestha P, Zhou XR, Mansour MP, Liu Q, Belide S, Nichols PD, Singh SP (2012) Metabolic engineering plant seeds with fish oil-like levels of DHA. PLoS One 7(11):e49165

Petrie JR, Shrestha P, Belide S, Kennedy Y, Lester G, Liu Q, Divi UK, Mulder RJ, Mansour MP, Nichols PD, Singh SP (2014) Metabolic engineering *Camelina sativa* with fish oil-like levels of DHA. PLoS One 9:e85061

Piatek AA, Lenaghan SC, Stewart CN Jr (2018) Advanced editing of the nuclear and plastid genomes in plants. Plant Sci 273:42–49

Pilon-Smits E (2005) Phytoremediation. Annu Rev Plant Biol 56:15–39

Pixley KV, Falck-Zepeda JB, Giller KE, Glenna LL, Gould F, Mallory-Smith CA, Stelly DM, Stewart CN Jr (2019) Genome editing, gene drives, and synthetic biology: will they contribute to disease-resistant crops, and who will benefit? Annu Rev Phytopathol 57:165–188

Pouvreau B, Vanhercke T, Singh S (2018) From plant metabolic engineering to plant synthetic biology: the evolution of the design/build/test/learn cycle. Plant Sci 273:1–10

Prakash I, Markosyan A, Bunders C (2014) Development of next generation *Stevia* sweetener: Rebaudioside M. Foods 3:162–175

Pretorius IS (2017) Synthetic genome engineering forging new frontiers for wine yeast. Crit Rev Biotechnol 37:112–136

Pretorius IS, Boeke JD (2018) Yeast 2.0—connecting the dots in the construction of the world's first functional synthetic eukaryotic genome. FEMS Yeast Res 18(4):foy032

Price GD, Pengelly JJL, Forster B, Du J, Whitney SM, von Caemmerer S, Badger MR, Howitt SM, Evans JR (2013) The cyanobacterial CCM as a source of genes for improving photosynthetic CO_2 fixation in crop species. J Exp Bot 64:753–768

Puchta H (2016) Using CRISPR/Cas in three dimensions: towards synthetic plant genomes, transcriptomes and epigenomes. Plant J 87:5–15

Radwan MM, Ross SA, Slade D, Ahmed SA, Zulfiqar F, Elsohly MA (2008) Isolation and characterization of new *Cannabis* constituents from a high potency variety. Planta Med 74:267–272

Ranjan R, Dey N (2012) Development of vascular tissue and stress inducible hybrid-synthetic promoters through *dof-1* motifs rearrangement. Cell Biochem Biophys 63(3):235–245

Ravikumar S, Baylon MG, Park SJ, Choi J-I (2017) Engineered microbial biosensors based on bacterial two-component systems as synthetic biotechnology platforms in bioremediation and biorefinery. Microb Cell Factories 16:62

Ray DK, Mueller ND, West PC, Foley JA (2013) Yield trends are insufficient to double global crop production by 2050. PLoS One 8:e66428

Razmi A, Golestanipour A, Nikkhah M, Bagheri A, Shamsbakhsh M, Malekzadeh-Shafaroudi S (2019) Localized surface plasmon resonance biosensing of tomato yellow leaf curl virus. J Virol Methods 267(267):1–7

Reed JW, Hudlicky T (2015) The quest for a practical synthesis of morphine alkaloids and their derivatives by chemoenzymatic methods. Acc Chem Res 48:674–687

Reski R, Parsons J, Decker EL (2015) Moss-made pharmaceuticals: from bench to bedside. Plant Biotechnol J 13(8):1191–1198

Reski R, Bae H, Simonsen HT (2018) *Physcomitrella patens*, a versatile synthetic biology chassis. Plant Cell Rep 37:1409–1417

Richardson SM, Mitchell LA, Stracquadanio G, Yang K, Dymond JS, DiCarlo JE, Lee D, Huang CLV, Chandrasegaran S, Cai Y, Boeke JD, Bader JS (2017) Design of a synthetic yeast genome. Science 355:1040–1044

Ricroch A, Clairand P, Harwood W (2017) Use of CRISPR systems in plant genome editing: toward new opportunities in agriculture. Emerg Top Life Sci 1(2):169–182

Ro D-K, Paradise EM, Ouellet M, Fisher KJ, Newman KL, Ndungu JM, Ho KA, Eachus RA, Ham TS, Kirby J, Chang MCY, Withers ST, Shiba Y, Sarpong R, Keasling JD (2006) Production of the antimalarial drug precursor artemisinic acid in engineered yeast. Nature 440:940–943

Roell M-S, Zurbriggen MD (2020) The impact of synthetic biology for future agriculture and nutrition. Curr Opin Biotech 61:102–109

Rogers C, Oldroyd GED (2014) Synthetic biology approaches to engineering the nitrogen symbiosis in cereals. J Exp Bot 65(8):1939–1946

Rorrer NA, Dorgan JR, Vardon DR, Martinez CR, Yang Y, Beckham GT (2016) Renewable unsaturated polyesters from muconic acid. ACS Sustain Chem Eng 4:6867–6876

Roy SJ, Negrão S, Tester M (2014) Salt resistant crop plants. Curr Opin Biotechnol 26:115–124

Ruiz-Lopez N, Haslam RP, Napier JA, Sayanova O (2014) Successful high-level accumulation of fish oil omega-3 long-chain polyunsaturated fatty acids in a transgenic oilseed crop. Plant J 77:198–208

Ryu MH, Zhang J, Toth T, Khokhani D, Geddes BA, Mus F, Garcia-Costas A, Peters JW, Poole PS, Ané J-M, Voigt CA (2020) Control of nitrogen fixation in bacteria that associate with cereals. Nat Microbiol 5(2):314–330

Sánchez-León S, Gil-Humanes J, Ozuna CV, Giménez MJ, Sousa C, Voytas DF, Barro F (2018) Low-gluten, nontransgenic wheat engineered with CRISPR/Cas9. Plant Biotechnol J 16 (4):902–910

Sapone A, Lammers KM, Casolaro V, Cammarota M, Giuliano MT, De Rosa M, Stefanile R, Mazzarella G, Tolone C, Russo MI, Esposito P, Ferraraccio F, Cartenì M, Riegler G, de Magistris L, Fasano A (2011) Divergence of gut permeability and mucosal immune gene expression in two gluten-associated conditions: celiac disease and gluten sensitivity. BMC Med 9:23

Sarrion-Perdigones A, Vazquez-Vilar M, Palaci J, Castelijns B, Forment J, Ziarsolo P, Blanca J, Granell A, Orzaez D (2013) GoldenBraid 2.0: a comprehensive DNA assembly framework for plant synthetic biology. Plant Physiol 162:1618–1631

Sauer M, Porro D, Mattanovich D, Branduardi P (2008) Microbial production of organic acids: expanding the markets. Trends Biotechnol 26(2):100–108
Savir Y, Noor E, Milo R, Tlusty T (2010) Cross-species analysis traces adaptation of Rubisco toward optimality in a low-dimensional landscape. Proc Natl Acad Sci USA 107:3475–3480
Savka MA, Farrand SK (1997) Modification of rhizobacterial populations by engineering bacterium utilization of a novel plant-produced resource. Nat Biotechnol 15:363–368
Savka MA, Dessaux Y, Oger P, Rossbach S (2002) Engineering bacterial competitiveness and persistence in the phytosphere. Mol Plant-Microbe Interact 15:866–874
Sawai S, Ohyama K, Yasumoto S, Seki H, Sakuma T, Yamamoto T, Takebayashi Y, Kojima M, Sakakibara H, Aoki T, Muranaka T, Saito K, Umemotog N (2014) Sterol side chain reductase 2 is a key enzyme in the biosynthesis of cholesterol, the common precursor of toxic steroidal glycoalkaloids in potato. Plant Cell 26:3763–3774
Sawant S, Singh PK, Madanala R, Tuli R (2001) Designing of an artificial expression cassette for the high-level expression of transgenes in plants. Theor Appl Genet 102:635–644
Schaefer DG, Zryd J-P (1997) Efficient gene targeting in the moss *Physcomitrella patens*. Plant J 11:1195–1206
Schuler ML, Mantegazza O, Weber AP (2016) Engineering C4 photosynthesis into C3 chassis in the synthetic biology age. Plant J 87:51–65
Schwander T, Von Borzyskowski LS, Burgener S, Cortina NS, Erb TJ (2016) A synthetic pathway for the fixation of carbon dioxide *in vitro*. Science 354(6314):900–904
Scott SA, Davey MP, Dennis JS, Horst I, Howe CJ, Lea-Smith DJ, Smith AG (2010) Biodiesel from algae: challenges and prospects. Curr Opin Biotechnol 21:277–286
Serrano L (2007) Synthetic biology: promises and challenges. Mol Syst Biol 3(158):158
Setten L, Soto G, Mozzicafreddo M, Fox AR, Lisi C, Cuccioloni M, Angeletti M, Pagano E, Díaz-Paleo A, Ayub ND (2013) Engineering *Pseudomonas protegens* Pf-5 for nitrogen fixation and its application to improve plant growth under nitrogen-deficient conditions. PLoS One 8:e63666
Shan Q, Wang Y, Li J, Zhang Y, Chen K, Liang Z, Zhang K, Liu J, Xi JJ, Qiu J-L, Gao C (2013) Targeted genome modification of crop plants using a CRISPR-Cas system. Nat Biotechnol 31:686–688
Shen B, Sun X, Zuo X, Shilling T, Apgar J, Ross M, Bougri O, Samoylov V, Parker M, Hancock E, Lucero H, Gray B, Ekborg NA, Zhang D, Johnson JCS, Lazar G, Raab RM (2012) Engineering a thermoregulated intein-modified xylanase into maize for consolidated lignocellulosic biomass processing. Nat Biotechnol 30:1131–1136
Shen BR, Wang LM, Lin XL, Yao Z, Xu HW, Zhu CH, Teng HY, Cui LL, Liu EE, Zhang JJ, He Z-H, Peng X-X (2019) Engineering a new chloroplastic photorespiratory bypass to increase photosynthetic efficiency and productivity in rice. Mol Plant 12:199–214
Shetty RS, Deo SK, Shah P, Sun Y, Rosen BP, Daunert S (2003) Luminescence-based whole-cell-sensing systems for cadmium and lead using genetically engineered bacteria. Anal Bioanal Chem 376:11–17
Shi J, Gao H, Wang H et al (2017) ARGOS8 variants generated by CRISPR-Cas9 improve maize grain yield under field drought stress conditions. Plant Biotechnol J 15(2):207–216
Shih PM, Liang Y, Loqué D (2016) Biotechnology and synthetic biology approaches for metabolic engineering of bioenergy crops. Plant J 87(1):103–117
Shimatani Z, Kashojiya S, Takayama M et al (2017) Targeted base editing in rice and tomato using a CRISPR-Cas9 cytidine deaminase fusion. Nat Biotechnol 35:441–443
Shokouhifar F, Zamani MR, Motallebi M, Mousavi A, Malboobi MA (2011) Construction and functional analysis of pathogen-inducible synthetic promoters in *Brassica napus*. Biol Plant 55 (4):689
Singh S, Dutt D, Tyagi CH, Upadhyaya JS (2011) Bio-conventional bleaching of wheat straw soda-AQ pulp with crude xylanases from SH-1 NTCC-1163 and SH-2 NTCC-1164 strains of *Coprinellus disseminatus* to mitigate AOX generation. New Biotechnol 28(1):47–57
Singh RL, Singh PK, Singh RP (2015) Enzymatic decolorization and degradation of azo dyes – a review. Int Biodeterior Biodegradation 104:21–31

South PF, Cavanagh AP, Liu HW, Ort DR (2019) Synthetic glycolate metabolism pathways stimulate crop growth and productivity in the field. Science 363(6422):10

Stauder R, Welsch R, Camagna M, Kohlen W, Balcke GU, Tissier A, Walter MH (2018) Strigolactone levels in dicot roots are determined by an ancestral symbiosis-regulated clade of the phytoene synthase gene family. Front Plant Sci 9:255

Stein H, Wilensky M, Tsafrir Y, Rosenthal M, Amir R, Avraham T, Ofir K, Dgany O, Yayon A, Shoseyov O (2009) Production of bioactive, post-translationally modified, heterotrimeric, human recombinant type-I collagen in transgenic tobacco. Biomacromolecules 10 (9):2640–2645

Sticklen M (2015) Advances in crop synthetic biology: a platform that bridges the expertise of crop molecular biologists, engineers, software developers and mathematical modelers. Adv Crop Sci Tech 3:171

Sun Y, Zhang X, Wu C et al (2016) Engineering herbicide-resistant rice plants through CRISPR/Cas9-mediated homologous recombination of acetolactate synthase. Mol Plant 9(4):628–631

Tan X, Calderon-Villalobos LIA, Sharon M, Zheng C, Robinson CV, Estelle M, Zheng N (2007) Mechanism of auxin perception by the TIR1 ubiquitin ligase. Nature 446:640–645

Thodey K, Galanie S, Smolke CD (2014) A microbial biomanufacturing platform for natural and semisynthetic opioids. Nat Chem Biol 10:837–844

Tilman D, Clark M (2014) Global diets link environmental sustainability and human health. Nature 515:518–522

Trenchard IJ, Siddiqui MS, Thodey K, Smolke CD (2015) *De novo* production of the key branch point benzylisoquinoline alkaloid reticuline in yeast. Metab Eng

Tropel D, van der Meer JR (2004) Bacterial transcriptional regulators for degradation pathways of aromatic compounds. Microbiol Mol Biol Rev 68:474–500

Trudeau DL, Edlich-Muth C, Zarzyckic J, Scheffen M, Goldsmith M, Khersonsky O, Avizemer Z, Fleishman SJ, Cotton CAR, Erb TJ, Tawfik DS, Bar-Even A (2018) Design and *in vitro* realization of carbon conserving photorespiration. PNAS 115(49):E11455–E11464

Tschofen M, Knopp D, Hood E, Stöger E (2016) Plant molecular farming: much more than medicines. Ann Rev Anal Chem 9(1):271–294

UN DESA (2017). https://www.un.org/development/desa/en/news/population/world-population-prospects-2017.html

Vanhercke T, El Tahchy A, Liu Q, Zhou X-R, Shrestha P, Divi UK, Ral J-P, Mansour MP, Nichols PD, James CN, Horn PJ, Chapman KD, Beaudoin F, Ruiz-López N, Larkin PJ, de Feyter RC, Singh SP, Petrie JR (2014) Metabolic engineering of biomass for high energy density: oilseed-like triacylglycerol yields from plant leaves. Plant Biotechnol J 12:231–239

Varshney RK, Bansal KC, Aggarwal PK, Datta SK, Craufurd PQ (2011) Agricultural biotechnology for crop improvement in a variable climate: hope or hype? Trends Plant Sci 16(7):363–371

Verhounig A, Karcher D, Bock R (2010) Inducible gene expression from the plastid genome by a synthetic riboswitch. Proc Natl Acad Sci USA 107:6204–6209

Vicente EJ, Dean DR (2017) Keeping the nitrogen-fixation dream alive. Proc Natl Acad Sci USA 114:3009–3011

Vickers C (2016) Synthetic biology. Rural Industries Research & Development Corporation, C/-Charles Sturt University, Wagga Wagga

Volkov V (2015) Salinity tolerance in plants. Quantitative approach to ion transport starting from halophytes and stepping to genetic and protein engineering for manipulating ion fluxes. Front. Plant Sci 6:873

von Caemmerer SV, Quick WP, Furbank RT (2012) The development of C4 rice: current progress and future challenges. Science 336:1671–1672

Vorholt JA, Vogel C, Carlström CI, Müller DB (2017) Establishing causality: opportunities of synthetic communities for plant microbiome research. Cell Host Microbe 22(2):142–155

Wang Y, Cheng X, Shan Q et al (2014) Simultaneous editing of three homoeoalleles in hexaploid bread wheat confers heritable resistance to powdery mildew. Nat Biotechnol 32(9):947–951

Wang P, Khoshravesh R, Karki S, Tapia R, Balahadia CP, Bandyopadhyay A, Quick WP, Furbank R, Sage TL, Langdale JA (2017) Re-creation of a key step in the evolutionary switch from C3 to C4 leaf anatomy. Curr Biol 27(21):3278–3287

Wang Y, Liu J, Zhou H (2019a) Visual detection of cucumber green mottle mosaic virus based on terminal deoxynucleotidyl transferase coupled with DNAzymes amplification. Sensors (Basel) 14:1–12.E1298

Wang Y, Li B, Liu J, Zhou H (2019b) T4 DNA polymerase-assisted upgrade of a nicking/ polymerization amplification strategy for ultrasensitive electrochemical detection of watermelon mosaic virus. Anal Bioanal Chem 411:2915–2924

Way JC, Collins JJ, Keasling JD, Silver PA (2014) Integrating biological redesign: where synthetic biology came from and where it needs to go. Cell 157(1):151–161

Weber APM, Bar-Even A (2019) Update: improving the efficiency of photosynthetic carbon reactions. Plant Physiol 179:803–812

Wec AZ, Herbert AS, Murin CD, Nyakatura EK, Abelson DM, Fels JM, He S, James RM, de La Vega MA, Zhu W, Bakken RR, Goodwin E, Turner HL, Jangra RK, Zeitlin L, Qiu X, Lai JR, Walker LM, Ward AB, Dye JM, Chandran K, Bornholdt ZA (2017) Antibodies from a human survivor define sites of vulnerability for broad protection against ebolaviruses. Cell 169:878–890

Wei Y, Siewers V, Nielsen J (2017) Cocoa butter-like lipid production ability of non-oleaginous and oleaginous yeasts under nitrogen-limited culture conditions. Appl Microbiol Biotechnol 101:3577–3585

Wei Y, Bergenholm D, Gossing M, Siewers V, Nielsen J (2018) Expression of cocoa genes in *Saccharomyces cerevisiae* improves cocoa butter production. Microb Cell Factories 17(1):11

Wend S, Dal Bosco C, Kämpf MM, Ren F, Palme K, Weber W, Dovzhenko A, Zurbriggen MD (2013) A quantitative ratiometric sensor for time-resolved analysis of auxin dynamics. Sci Rep 3:2052

Widsten P, Kandelbauer A (2008) Laccase applications in the forest products industry: a review. Enzyme Microbiol Technol 42:293–307

Wijffels RH, Kruse O, Hellingwerf KJ (2013) Potential of industrial biotechnology with cyanobacteria and eukaryotic microalgae. Curr Opin Biotech 24(3):405–413

Wilkerson CG, Mansfield SD, Lu F, Withers S, Park JY, Karlen SD, Gonzales-Vigil E, Padmakshan D, Unda F, Rencoret J, Ralph J (2014) Monolignol ferulate transferase introduces chemically labile linkages into the lignin backbone. Science 344:90–93

Williams TC, Xu X, Ostrowski M, Pretorius IS, Paulsen IT (2017) Positive-feedback, ratiometric biosensor expression improves high-throughput metabolite-producer screening efficiency in yeast. Syn Biol 2(1):ysw002

Wilson R, Alonso H, Whitney S (2016) Evolving *Methanococcoides burtonii* archaeal Rubisco for improved photosynthesis and plant growth. Sci Rep 6:22284

Wood CC, Okada S, Taylor MC, Menon A, Mathew A, Cullerne D, Stephen SJ, Allen RS, Zhou X-R, Liu Q, Oakeshott JG, Singh SP, Green AG (2018) Seed-specific RNAi in safflower generates a superhigh oleic oil with extended oxidative stability. Plant Biotechnol J 16 (10):1788–1796

Woodard SL, Mayor JM, Bailey MR, Barker DK, Love RT, Lane JR, Delaney DE, McComas-Wagner JM, Mallubhotla HD, Hood EE, Dangott LJ, Tichy SE, Howard JA (2003) Maize-derived bovine trypsin: characterization of the first large-scale, commercial protein product from transgenic plants. Biotechnol Appl Biochem 38:123–130

World Health Organization (2013) 18th WHO essential medicines list, Geneva, Switzerland

Wu L, Birch RG (2007) Doubled sugar content in sugarcane plants modified to produce a sucrose isomer. Plant Biotechnol J 5:109–117

Wu Y, Zhu R, Mitchell LA, Ma L, Liu R, Zhao M, Jia B, Xu H, Li Y-X, Yang Z-M, Ma Y, Li X, Liu H, Liu D, Xiao W-H, Zhou X, Li B-Z, Yuan Y-J, Boeke JD (2018) *In vitro* DNA SCRaMbLE. Nat Commun 9:1935

Wurtzel ET (2019) Changing form and function through carotenoids and synthetic biology. Plant Physiol 179(3):830–843

Wurtzel ET, Vickers CE, Hanson AD, Millar AH, Cooper M, Voss-Fels KP, Nikel PI, Erb TJ (2019) Revolutionizing agriculture with synthetic biology. Nat Plants 5:1207–1210

Xie Z-X, Li B-Z, Mitchell LA, Wu Y, Qi X, Jin Z, Jia B, Wang X, Zeng B-X, Liu H-M et al (2017) "Perfect" designer chromosome V and behavior of a ring derivative. Science 355(6329): eaaf4704

Xu X, Zhang Y, Meng Q, Meng K, Zhang W, Zhou X, Luo H, Chen R, Yang P, Yao B (2013) Overexpression of a fungal β-mannanase from *Bispora* sp. MEY-1 in maize seeds and enzyme characterization. PLoS One 8(2):e56146

Yadav RK, Baeg JO, Oh GH, Park N-J, Kong K-J, Kim J, Hwang DW, Biswas SK (2012) A photocatalyst-enzyme coupled artificial photosynthesis system for solar energy in production of formic acid from CO_2. J Am Chem Soc 134(28):11455–11461

Yang F, Mitra P, Zhang L, Prak L, Verhertbruggen Y, Kim J-S, Sun L, Zheng K, Tang K, Auer M, Scheller HV, Loqué D (2013) Engineering secondary cell wall deposition in plants. Plant Biotechnol J 11:325–335

Yang X, Cushman JC, Borland AM, Edwards EJ, Wullschleger SD, Tuskan GA, Owen NA, Griffiths H, Smith JA, De Paoli HC, Weston DJ, Cottingham R, Hartwell J, Davis SC, Silvera K, Ming R, Schlauch K, Abraham P, Stewart JR, Guo HB, Albion R, Ha J, Lim SD, Wone BW, Yim WC, Garcia T, Mayer JA, Petereit J, Nair SS, Casey E, Hettich RL, Ceusters J, Ranjan P, Palla KJ, Yin H, Reyes-García C, Andrade JL, Freschi L, Beltrán JD, Dever LV, Boxall SF, Waller J, Davies J, Bupphada P, Kadu N, Winter K, Sage RF, Aguilar CN, Schmutz J, Jenkins J, Holtum JA (2015) A roadmap for research on crassulacean acid metabolism (CAM) to enhance sustainable food and bioenergy production in a hotter, drier world. New Phytol 207(3):491–504

Yang J, Xie X, Yang M, Dixon R, Wang YP (2017) Modular electron-transport chains from eukaryotic organelles function to support nitrogenase activity. Proc Natl Acad Sci USA 114 (12):E2460–E2465

Yang X, Cushman JC, Borland AM, Liu Q (2020) Editorial: systems biology and synthetic biology in relation to drought tolerance or avoidance in plants. Front Plant Sci 11:394

Yokobayashi Y, Weiss R, Arnold FH (2002) Directed evolution of a genetic circuit. Proc Natl Acad Sci USA 99:16587–16591

Yu H, Li X, Duchoud F, Chuang DS, Liao JC (2018) Augmenting the Calvin–Benson–Bassham cycle by a synthetic malyl-CoA-glycerate carbon fixation pathway. Nat Commun 9:2008

Zale J, Jung JH, Kim JY, Pathak B, Karan R, Liu H, Chen X, Wu H, Candreva J, Zhai Z, Shanklin J, Altpeter F (2016) Metabolic engineering of sugarcane to accumulate energy-dense triacylglycerols in vegetative biomass. Plant Biotechnol J 14:661–669

Zhan X, Zhang Y-H, Chen D-F, Simonsen HT (2014) Metabolic engineering of the moss *Physcomitrella patens* to produce the sesquiterpenoids patchoulol and α/β-santalene. Front Plant Sci 5:636

Zhan F, Wang T, Iradukunda L, Zhan J (2018) A gold nanoparticle-based lateral flow biosensor for sensitive visual detection of the potato late blight pathogen, *Phytophthora infestans*. Anal Chim Acta 1036:153–161

Zhang DC, VanFossen AL, Pagano RM, Johnson JS, Parker MH, Pan SH, Gray BN, Hancock E, Hagen DJ, Lucero HA, Shen B, Lessard PA, Ely C, Moriarty M, Ekborg NA, Bougri O, Samoylov V, Lazar G, Raab RM (2011) Consolidated pretreatment and hydrolysis of plant biomass expressing cell wall degrading enzymes. Bioenergy Res 4:276–286

Zhang Y, Xu X, Zhou X, Chen R, Yang P, Meng Q, Meng K, Luo H, Yuan J, Yao B, Zhang W (2013) Overexpression of an acidic endo-β-1,3-1,4-glucanase in transgenic maize seed for direct utilization in animal feed. PLoS One 8(12):e81993

Zheng Q, Huang W, Cui X, Dong Y, Shi Y, Ma H, Liu L (2018) Identification of wheat yellow rust using optimal three-band spectral indices in different growth stages. Sensors (Basel) 19:E35

Zhou J, Peng Z, Long J et al (2015) Gene targeting by the TAL effector PthXo2 reveals cryptic resistance gene for bacterial blight of rice. Plant J 82(4):632–643

Zhou S, Runge T, Karlen SD, Ralph J, Gonzales-Vigil E, Mansfield SD (2017) Chemical pulping advantages of zip-lignin hybrid poplar. Chem Sus Chem 10:3565–3573

Zhu H, Riely BK, Burns NJ, Ane JM (2006) Tracing nonlegume orthologs of legume genes required for nodulation and arbuscular mycorrhizal symbioses. Genetics 172:2491–2499

Zhu C, Naqvi S, Breitenbach J, Sandmann G, Christou P, Capell T (2008) Combinatorial genetic transformation generates a library of metabolic phenotypes for the carotenoid pathway in maize. Proc Natl Acad Sci USA 105:18232–18237

Zhu XG, Long SP, Ort DR (2010) Improving photosynthetic efficiency for greater yield. Ann Rev Plant Biol 61:235–261

Zimin AV, Puiu D, Hall R, Kingan S, Clavijo BJ, Salzberg SL (2017) The first near-complete assembly of the hexaploid bread wheat genome, *Triticum aestivum*. Gigascience 6:1–7

Zirpel B, Degenhardt F, Martin C, Kayser O, Stehle F (2017) Engineering yeasts as platform organisms for cannabinoid biosynthesis. J Biotechnol 259:204–212

Zsögön A, Čermák T, Naves ER, Notini MM, Edel KH, Weinl S, Freschi L, Voytas DF, Kudla J, Peres LEP (2018) *De novo* domestication of wild tomato using genome editing. Nat Biotechnol 10:1038

Zürcher E, Tavor-Deslex D, Lituiev D, Enkerli K, Tarr PT, Müller B (2013) A robust and sensitive synthetic sensor to monitor the transcriptional output of the cytokinin signaling network in Planta. Plant Physiol 161(3):1066–1075

Role of Calcium Signalling During Plant–Herbivore Interaction

16

Amit Kumar, Ruby Panwar, Archana Singh, and
Indrakant Kumar Singh

Abstract

Plants and herbivores have co-evolved since long and plants have established sophisticated defence mechanisms for their protection. Plant defence response against herbivory involves activation of multiple signalling cascades due to wound-induced change in ion concentration and rapid deviation in the membrane potential (V_m). Amongst all the early events, Ca^{2+} signalling is one of the most crucial incidents that form a complex network with numerous components. During stress, cytosolic Ca^{2+} level upsurges due to Ca^{2+} influx and Ca^{2+} efflux by Ca^{2+}-ATPases and H^+/Ca^{2+}, respectively. This transient change in Ca^{2+} concentration is identified by Ca^{2+}-sensors [Ca^{2+}-binding proteins such as calmodulin (CaM), CaM-related proteins (CML), calcineurin B-like proteins (CBL)], which relay the signal by stimulating calcium dependent protein kinases (CDPKs). Further, these CDPKs regulate the downstream processes involved in plant defence mechanism. Herbivory not only triggers local Ca^{2+} signals, rather it also fluctuates systemic Ca^{2+} levels, which plays a significant role in systemic plant defence. The present chapter highlights the recent advancements in early signalling events during plant–herbivore interaction with special emphasis on Ca^{2+} signalling, Ca^{2+} channels, Ca^{2+}-sensors, and Ca^{2+}-responders.

A. Kumar · R. Panwar · A. Singh (✉)
Department of Botany, Hansraj College, University of Delhi, New Delhi, India
e-mail: archanasingh@hrc.du.ac.in

I. K. Singh (✉)
Molecular Biology Research Lab, Department of Zoology, Deshbandhu College, University of Delhi, New Delhi, India
e-mail: iksingh@db.du.ac.in

B. Giri, M. P. Sharma (eds.), *Plant Stress Biology*,
https://doi.org/10.1007/978-981-15-9380-2_16

Keywords

Herbivory · Calcium signalling · Ca^{2+}-binding proteins · Ca^{2+} channels CDPKs · Plant defence

16.1 Introduction

Plants encounter herbivores very often in their life span, and this interaction is continuous since millions of years ago. The myriad of interactions between them are so unique and specific that they are evolving their mutualism as well as defence mechanism simultaneously. To deal with insect attack, plants have developed highly specific defence strategies that include both constitutive as well as inducible means. Constitutive defence mechanism includes preformed barriers such as cuticle, trichomes, spines, defensive metabolites, etc., whereas induced plant defence involves cell wall reinforcement, enhanced synthesis of insecticidal/insect deterrent proteins and secondary metabolites, etc. Induced defence is cost effective and is achieved by induction of elaborate defence signalling pathways. During plant–herbivore interactions, certain consecutive events are detected such as membrane depolarization, V_m changes, elevation in cytoplasmic Ca^{2+} levels, and generation of reactive oxygen species (ROS) (Aldon et al. 2018; Erb and Reymond 2019). Further, kinases are activated and biosynthesis of phytohormones [Jasmonic Acid (JA), Salicylic acid (SA), Ethylene (ET)] is upregulated, followed by transcriptional, proteomic, and metabolic changes (Singh et al. 2008, 2018, 2020). These events are highly regulated and coordinated by chemical signals.

Among signalling molecules, calcium (Ca^{2+}) is small sized, versatile entity that participates in the transfer of signals during biotic and abiotic stresses in plants. Hence Ca^{2+} is also known as the universal second messenger or universal signalling molecule (Song et al. 2008; Batistič and Kudla 2012; Cheval et al. 2013; Moreno et al. 2014). Ca^{2+} is the most frequent and successfully studied molecule during plant–herbivore interaction. V_m polarization and elevation in $(Ca^{2+})_{cyt}$ occurs within seconds upon environmental stress. It relays the information intracellularly as well as intercellularly. In plants Ca^{2+} is stored in various organelles and subcellular compartments, which is released during stress via specific Ca^{2+} channels/pumps. Influx and efflux of Ca^{2+} is mediated by specific channels and pumps such as Ca^{2+}-ATPases and H^+/Ca^2. In normal cells, higher amount of stored Ca^{2+} is found in apoplast, mitochondria, endoplasmic reticulum, and vacuole in mM range (~1 mM), whereas its lower levels are maintained in cytosol (100–200 nM) (~0.0001 mM) (Dodd et al. 2010; Lecourieux et al. 2006). This high concentration of Ca^{2+} in the apoplast is necessary to maintain in order to notice the warning signal of environmental stress conditions. Any deviation from this Ca^{2+} concentration is recognized as stress in plant cells. In response to stress, the Ca^{2+} levels in cells change completely and it increases up to μM range in the cytosol (Messerli et al. 2006).

During herbivory, as soon as the "enemy" starts scrawling and causes damage to plants, herbivory-specific/ -induced stimulus (HAMPs, DAMPs, and SDSs) is recognized by the plant receptors (such as PRR), and the first response which is observed is the change in V_m and generation of fast electrical signal (Maffei et al. 2007a, b). Further the activity of channels and transporters is triggered, such as Ca^{2+}-channels leading to spiking of Ca^{2+} in the form of repetitive oscillation (Evans et al. 2001; Tian et al. 2020); the period of frequency, sinusoidal or square shape wave, and amplitude specify the stimulus specificity (Kosuta et al. 2008). These repetitive oscillations and spiking of Ca^{2+} form the code called as "Ca^{2+} signature" (Dodd et al. 2010; Webb et al. 1996; Mithöfer et al. 2009). Ca^{2+} signature reveals the nature and intensity of the stimulus and specific information of stress. In *Phaseolus lunatus, Spodoptera littoralis* larvae start feeding on their leaf, which results in sharp spike of Ca^{2+} (Maffei et al. 2004, 2006). This highly evoked Ca^{2+} levels during herbivore attack help in transmitting the attack message. This change in cytosolic free calcium is perceived and decoded by a myriad of Ca^{2+}-sensor proteins, which stimulates a conformational change and assists their interaction with downstream effector molecules. Ca^{2+}-sensor proteins either work as relay sensor (CaM, CML, CBL) or as relay responders (CDPKs). CaM is known to interact with target proteins and modulate their activity. CMLs are closely related to CaM family and have been deciphered only in plants. CBLs and CDPKs are also plant specific proteins and an important component of Ca^{2+} signalling. At present, specific function of most of the Ca^{2+}-sensors remains unknown, but recent studies have highlighted their contribution in development, biotic and abiotic stresses (Aldon et al. 2018). This activation of Ca^{2+} signalling as well as ROS and MAPK signalling together stimulates JA-Ile formation, which binds to SCF^{COI1} complex, which degrades JAZ repressors, and triggers the transcription factors to upregulate the production of defence-related proteins and secondary metabolites.

16.2 Generation of Calcium Signatures During Plant–Herbivore Interaction

Plants recognition of herbivores attack occurs initially when herbivores start crawling (crawling pattern) in leaf and also by chewing of the leaf (chewing behaviour) which release oral secretion (OS) (Zebelo and Maffei 2015). OS possesses elicitors such as herbivore-associated molecular patterns (HAMPs) and damage associated molecular pattern (DAMP), (Acevedo et al. 2015). Herbivory introduces these insect-derived elicitors that trigger a cascade of signals. Among them Ca^{2+} due to its smaller ionic size (231 pm) responds rapidly to the herbivore attack and travels throughout the cell to convey signal. Ca^{2+} transported to the cytosol through various Ca^{2+} specific channels and ions. The Ca^{2+} ions are not directly involved in membrane depolarization proved by *pdko3* plasmodesmatal protein in *Arabidopsis* (Bricchi et al. 2010). Ca^{2+} signalling was visualized in vivo by using Ca^{2+} sensitive like fluorescent probes or the bioluminescence based aequorin technology (Kanchiswamy et al. 2010).

In a study, it was revealed that when lepidopteran larvae fed on plants, V_m depolarization occurred in the vicinity of wounding zone, which was followed by

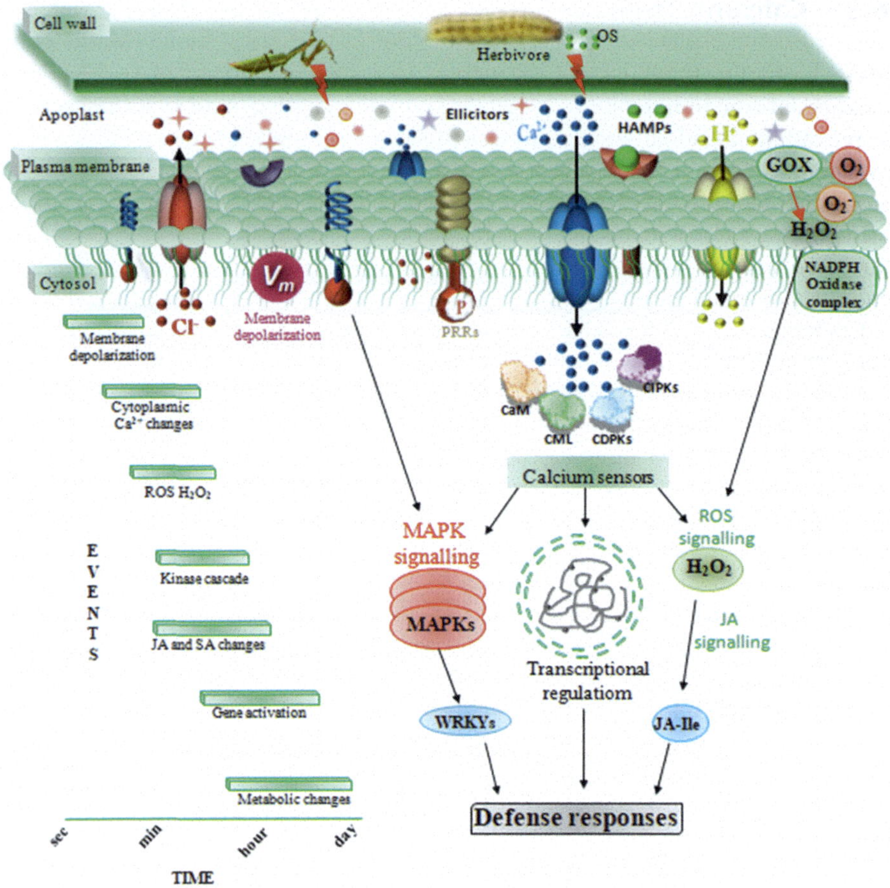

Fig. 16.1 Role of Ca^{2+} signalling in activation of plant defence against insect attack

frequent increase in $(Ca^{2+})_{cyt}$. Herbivore OSs induce $(Ca^{2+})_{cyt}$ concentration whereas no change noted in control only with mechanical damage. Herbivore derived elicitors, inceptin and volicitin, were found to be associated with Ca^{2+} influx which triggers ROS (H_2O_2) (Schmelz et al. 2006). Other responses are Ca^{2+} ion efflux and influx, activation of NADPH oxidase, production of radical molecules (ROS), synthesis of metabolites like jasmonic acid and ethylyne, change in proteomics of cell, emission of volatile organic compounds for help, and finally expression of late response gene against herbivore (Miller and Mittler 2006) (Fig. 16.1).

16.3 Calcium: Transporters During Plant–Herbivore Attack

Survival and fitness of plants during any environmental stress depends on the quickness of recognition of the "enemy" and induction of effective response against the prevailing situation. For any quick response, channels/transporters of a signal have significant contribution on transportation of ions (Erb and Reymond 2019). As soon as herbivory is perceived by the plant cell, transportation of Ca^{2+} ions takes place into the cytosol and other subcellular compartments through channels (White 2000). To restore normal condition in cells, channels efflux the ions to create concentration gradient and ready for perceiving next signal. Electrophysiological studies suggested variety of Ca^{2+} ion transporters are present in the cell that are localized in the plasma membrane and membrane of subcellular organelles (Schiøtt et al. 2004; Urquhart et al. 2007). Their mode of activation is also different like voltage based, voltage insensitive, and membrane polarization based. Different Ca^{2+} transporters have varied affinity towards ions so they work according to the situation. Binding affinity of Ca^{2+}-ATPases is high ($K_m = 1–10$ mM) but it has low capacity for transportation, maintains cellular Ca^{2+} homeostasis in cytosol. Whereas, Ca^{2+}/H^+ antiporter has low affinity of binding ($K_m = 10–15$ mM), but it quickly transports ions and are active during signal perception (Hirschi 2001).

16.3.1 Ca^{2+}-ATPases

Different Ca^{2+}-ATPases are involved in active Ca^{2+} transportation and maintenance of Ca^{2+} homeostasis in cytosol (Bose et al. 2011; Maffei et al. 2007a, b). The ATP-dependent P-type ATPases (ADPA) are one of them delivering the Ca^{2+} ions across the plasma membrane. The integrants of signalling pathway activates transient of ADPA. In *Arabidopsis*, two clades of ATP-dependent P-type ATPases, P-type ATPase II A and P-type ATPase II B were studied (Sanders et al. 2002). The P-type ATPase II A (ECA) is present in endoplasmic reticulum, whereas P-type ATPase II B (ACA) functions as an autoinhibitory function (Boursiac and Harper 2007). ECA is a broad nucleotide specific ATPase. In ACA, calmodulin-binding domain is present. In *Arabidopsis*, isoforms of ACA are localized in different subcellular membranes in plasma membrane, vacuolar membrane, and plastid membrane. This indicates that ACA has the capabilities for transducing Ca^{2+} signals during various stresses (Sarwat and Tuteja 2007). On pathogen attack transcripts of ACA12 and ACA13 are highly upregulated (Boursiac and Harper 2007). ACA of different membrane has different calmodulin-binding affinities and activity. Upon insect attack a temporary complex calmodulin-Ca^{2+} is formed, which institutes signal transfer (Huda et al. 2013). Upon insect attack, the initial Ca^{2+} burst is followed by a consistent decrease in the $(Ca^{2+})_{cyt}$, which implies the involvement of a Ca^{2+} efflux-mediated, increased Ca^{2+}-ATPase activity (Maffei et al. 2007a, b).

16.3.2 Ca^{2+}/Proton Exchangers

The Ca^{2+}/H^+ is an antiporter channel, simultaneously transports the Ca^{2+} and H^+ antagonistically to each side of plasma membrane. The driven energy for antiporter is proton motive force (PMF) (Sanders et al. 2002). Ca^{2+}/H^+ antiporters have less affinity for Ca^{2+} than Ca^{2+}-ATPases. This antiporter is also known as cation exchanger (CAX) because of their function. CAX1 and CAX3 have high specificity for Ca^{2+} binding and maintenance of Ca^{2+} homeostasis than any other CAX. External supply of Ca^{2+} increases the expression of CAX1 and CAX3. CAX1 expressed in shoot, whereas CAX3 expressed in roots strongly (Cheng et al. 2003). This indicates that CAX1 is responsible for manoeuvre of above-ground stresses and CAX3 controls underground stresses. CAX proteins are expressed during abiotic stresses such as heavy metal, drought, and salinity stress (Zhang et al. 2008). Role of CAX during herbivore stress is still unknown and yet to be discovered.

16.3.3 Ca^{2+} Channels

The transportation of Ca^{2+} into the cytosol occurs during the various stages of simulation. The Ca^{2+} ions are transported via depolarized activated Ca^{2+} channels, hyperpolarized activated Ca^{2+} channels, voltage-dependent channels, voltage-independent channels, and ligated gated channels (Sanders et al. 2002). Subcellular organelles have also Ca^{2+} channels and they also participate in signal transduction. Voltage-dependent and ligate gated channels are localized in subcellular organelles, membranes of vacuoles, and endoplasmic reticulum (Sanders et al. 2002; Demidchik and Maathuis 2007). The channels are stimulated either by phytohormone such as abscisic acid, jasmonic acid or by ROS like H_2O_2. Other types of channels include cyclic nucleotide-gated channels, glutamate receptor-like, two-pore channels, and annexins (Dodd et al. 2010).

16.3.4 Cyclic Nucleotide: Gated Channels (CNGCs)

In plants cyclic nucleotide-gated channels (CNGCs) have crucial role in Ca^{2+} translocation in plant cells (Leng et al. 1999; Urquhart et al. 2007). CNGCs include 6 trans-membrane and a pore domain. Trans-membrane domains assemble tetramerically for the formation of pore and are localized to the plasma membrane (K Jha et al. 2016). They are permeable to monovalent ions and Ca^{2+} ions when expressed heterologously. With cyclic nucleotide-binding domain, CNGCs are also able to react to calmodulin sensors (Romeis and Herde 2014; Schuurink et al. 1998). This property helps in transduction of Ca^{2+} signal cascade during herbivore attack (Meena and Vadassery 2015; Breeze 2019). The role of one of the Ca^{2+} CNGC, CNGC 19 has been implicated in *Arabidopsis* defence signalling against *Spodoptera litura* attack. CNGC19 expresses in vascular tissue of leaves and signal propagates through vascular system (Meena et al. 2019).

16.3.5 Glutamate Receptor: Like Channels (GLRCs)

Glutamate receptor-like complex (GLRC) ion channels belong to members of large gene families to form hetero-multimeric complexes (Davenport 2002). GLRC has been activated during defence (Forde and Roberts 2014). This channel is an non-selective Ca^{2+} permeable channel which regulates $(Ca^{2+})_{cyt}$ and induces the transcription factors responsible for action against herbivore attack (Toyota et al. 2018). Channels that depolarize the membrane when activated might contribute to $(Ca^{2+})_{cyt}$ responses merely by activating voltage-sensitive pathway (Hirschi 2001).

16.3.6 Two: Pore Channels

Electrophysiological studies revealed that two-pore channels (TPC) are Ca^{2+} activated channels (Pottosin et al. 2009). This protein is characterized by having 12 predictable trans-membrane domains; out of them two are incorporated domains and form a homodimer. Two putative calcium-binding EF-hands and 14-3-3 binding domain are incorporated between 6 and 7 trans-membrane domains forming cytosolic loop (Peiter et al. 2005). TPC1 vacuolar membrane has active role in defence against *S. littoralis* feeding (Bonaventure et al. 2007). The *tpc1* results in demotion of $(Ca^{2+})_{cyt}$ during plant aphid interaction (Vincent et al. 2017).

16.3.7 Annexins Channels

Annexins are considered as novel mechanosensitive plant Ca^{2+} permeable channel during stress condition. This protein is isolated and purified from *Zea mays* plant. Experimentally it has been proven that purified form of annexins incorporated into the planar lipid bilayers, elicits $(Ca^{2+})_{cyt}$ as voltage-gated channel (Laohavisit et al. 2009). It is homodimer, having two putative calcium-binding EF-hands and a 14-3-3 binding domain (Peiter et al. 2005). It has peroxidase activity during ROS metabolism. It was hypothesized that it also transports calcium into the nucleus in different ways (Alonso et al. 2006).

16.4 Calcium Sensors: Perception, Decoding, and Relaying of Ca^{2+} Signatures

During signal transduction, plethora of proteins get involved in directing Ca^{2+} ions to reach their target place in correct way. After Ca^{2+} signature, Ca^{2+} sensors contribute to further specific transmission of signal. Sensor proteins have Ca^{2+}

binding specific region, which undergoes conformational changes and passes it to downstream effector signal molecules (Clapham et al. 2007; Gifford et al. 2007). In plants variety of sensory proteins are discovered such as calmodulin (CaM), calmodulin-like (CML), calcineurin B-like (CBL), and Ca^{2+}-dependent protein kinases (CDPK). The sensor proteins possess Ca^{2+} binding region and EF-hand domain, which undergoes conformational changes and further passes the message to downstream players. These sensor proteins have varying number of EF-hands depending on the length of amino acid residues. Some other type of Ca^{2+} binding proteins are calnexin, calreticulin, annexins, and C2 domain containing proteins (Reddy and Reddy 2004; Sarwat and Tuteja 2007; Davies 2014). These other proteins lack EF-hand, hence they have different mechanism to bind with Ca^{2+}. Ca^{2+}-sensor proteins are categorized into two classes based on their mechanism: sensor responder and sensor relayer (Sanders et al. 2002). Ca^{2+} sensor relayer proteins lack enzymatic activity and only conformational changes govern their interaction with Ca^{2+} and downstream factors. CaM, CML, and CBL have this type of activating–deactivating mechanism (Luan et al. 2002). CDPKs have dual characteristics of sensing as well as relaying. Ca^{2+} binding domain induces conformational change in the protein and Ser-Thr domain has kinase activity (Sanders et al. 1999; Luan et al. 2002; Hrabak et al. 2003). So CDPKs solely activate and transduce the signal through phosphorylation cascade. The Ca^{2+}-binding sensors play a significant role of relaying signals upon herbivory attack. Ca^{2+} binds with their sensors in cytosol in a way coming from plasma membrane in the spike of Ca^{2+} ions. Sensors measure the intensity and specific signature to decipher the Ca^{2+} code. As per the signature specificity it regulates downstream to rectify the signal cascade. In Plants CaM, CBL, and CDPK are the main class of Ca^{2+} sensors during herbivory attack (Erb and Reymond 2019).

16.4.1 Calmodulin

Calmodulin (CaM) is a highly conserved protein sensor in plant species (Perochon et al. 2011). Plants encode multiple genes for these proteins, hence many similar isoforms of CaMs are present in the cytosol (Snedden and Fromm 1998). Seven different genes of 4 CaM studied in *Arabidopsis* (CaM &CaM 4; CaM 2, CaM 3 & CaM 5; CaM 6 and CaM 7) (Batistič and Kudla 2012). The isoforms of CaMs respond according to different developmental processes and stress signals. Studies suggested its other locations than cytosol such as in nucleus (Levy et al. 2004), chloroplast (Chigri et al. 2012; Yang and Poovaiah 2000), peroxisome (Yang and Poovaiah 2002), and organellar membranes (Schiøtt et al. 2004; Urquhart et al. 2007). CaM is small sized, positively charged protein of ~150 amino acid residues. They have two EF-hands motifs with globular domains connected by pliable central helix (Gifford et al. 2007). When Ca^{2+} binds to the CaM, its tertiary structure changes and its hydrophobic clefts expose, which sequentially interacts with its next downstream target. Upon herbivory, CaM binds with Ca^{2+} and regulates downstream targets to rectify signal cascade.

16.4.2 Calmodulin-Like [CaM-Like]

Calmodulin-like (CML) activity is like a CaM-Ca^{2+} sensor, but its functions are not fully characterized (hence named calmodulin-like). CML contains 1 to 6 EF-hands, but most of them are characterized having four EF-hands in the length of 83–330 residues. No functional motifs are identified in this protein (Snedden and Fromm 1998). In *Arabidopsis thaliana* (Vadassery et al. 2012a, b), *Oryza sativa* CML had been reported. This sensor present in cytosol also grabs Ca^{2+} ions. Some CMLs bind with plasma membrane and modify lipid during membrane maintenance (Batistič and Kudla 2012). CML11, CML12, CML16, CML17, and CML23 are highly induced by elicitors of insect OS, which indicates their possible involvement in defence against herbivore (Vadassery et al. 2012a, b). Among CMLs, 3 CMLs; CML37, CML 42, and CML 43 are well known characterized during plant–herbivory interaction and for strengthening plant immunity against biotic stress.

CML 37 is highly upregulated during *Spodoptera litura* damage in *Arabidopsis*. CML37 intermediates during signal cascade (Scholz et al. 2014). CML 37 and CML 42 work antagonistically to their function. CML 42 functions as a Ca^{2+} sensor during insect herbivory defence as well as during abiotic stress responses. CML 42 protein is upregulated during biotic stresses and localizes to the cytosol and nucleus and suppresses the plant defence mechanism. Jasmonic acid receptors negatively regulate its upregulation, which indicates that CML 42 connects Ca^{2+} and jasmonic acid signalling (Vadassery et al. 2012a). CML 42 is also involved in handling abiotic stress response such as ultraviolet rays and drought stress. When Arabidopsis plant is supplied with defence regulator, SA, CML 43 Ca^{2+} sensor activity and β-glucuronidase reporter activity are highly enhanced. This indicates that CML 43 actively functions as a plant immunity responser (Bender et al. 2014). CML9 is not upregulated during herbivore *S. littoralis* attack and in response of spider mite *Tetranychus urticae*. It is a specialized calcium sensor; it evokes SA-dependent defence against phytopathogenic bacteria *Pseudomonas syringae* and fungus *Alternaria brassicicola* independent of JA. It also involves in ion regulation during drought response (Heyer et al. 2018; Leba et al. 2012). *Zea mays* induced responses against *Chilo partellus*'s eggs were studied. It was found that it induces CML 312, CML442, CML159, and CML204 from maize genotype in order of attractiveness for defence against herbivore (Amanuel et al. 2020).

16.4.3 Calcineurins B-Like

Calcineurin B-like (CBL) is an important Ca^{2+} sensor for land plants (Batistič and Kudla 2009). CBL is similar to neuronal calcium sensor of animal (Luan et al. 2002; Liu and Zhu 1998; Kudla et al. 1999). Once CBL interacts with Ca^{2+} ions, they come in contact with their regulatory Ser-Thr interacting protein kinase of SnRK3 subgroup and form a CBL-interacting protein kinase (CBL-CIPK) (CIPK) complex which helps in deciphering Ca^{2+} signals (Kim 2013; Zhu et al. 2013). In plants, CBL present throughout the cell in plasma membrane, tonoplast, cytosol, and nucleus.

CBL and CaM have sequence similarity due to Ca^{2+} binding EF-hands. The size of these proteins ranges between 23 and 26 kDa, which include 4 EF-hands (Batistic et al. 2011). Structurally CBL consists of catalytic N-terminal region containing targeting sites, myristoylation and palmitoylation motifs, and C-terminal domain of regulatory nature. These motifs reported in CBL1 and CBL9 which anchor CIPK towards the plasma membrane. Experimental studies revealed that when CBL1 and CBL9 interact with CIPK26, then this complex enhances functionality of NADPH oxidase RBOHF through phosphorylation (Drerup et al. 2013; Kimura et al. 2013). It also contains region of 24 hydrophobic amino acid residues, NAF/FISL motif is highly conserved in CIPK (Meena and Vadassery 2015). Membrane associated protein substrates undergo phosphorylation due to the formation of CIPK complex. CBL targeted CIPK have cellular location based function, hence same kinases able to perform different function according to location and condition. CBL–CIPK complex becomes very specific in response to different stresses. Its signalling pathway is regulated by complex mechanisms in plant cells involving crosstalk with other signalling pathways (Yu et al. 2014).

Genes involved in the CBL–CIPK pathway were highly upregulated in *Camellia sinensis* upon mild infestation by green leafhopper (*Empoasca vitis*) by herbivory attack. The expression of a CBL-interacting protein kinase 19 gene *CsEv9* significantly increased in the Fuzao2 tea cultivar infested by the tea leafhopper (by approximately 18-fold with respect to the controls), providing the first evidence for the involvement of the CBL–CIPK pathway in response to herbivory (Yang et al. 2011).

16.4.4 Calcium Dependent Protein Kinases

CDPKs are multifunctional proteins with Ca^{2+} binding domain and kinase activity (Tena et al. 2011). CDPK has a noteworthy role of Ca^{2+} binding and herbivore-induced signalling cascades during plant–herbivory interaction. It converts the event of Ca^{2+} signals into phosphorylation and kinase activities in plants. CDPKs investigated in Protista kingdom also other than Plantae kingdom (Ozturk et al. 2002; Seki et al. 2002). It is reported in many plant species mainly in *Arabidopsis thaliana* (Cheng et al. 2002), *Oryza sativa* (Ray et al. 2007), *Triticum aestivum* (Li et al. 2008), *Zea mays* (Kong et al. 2013), *Populus trichocarpa* (Zuo et al. 2013), *Gossypium raimondii* (Liu et al. 2014), *Solanum lycopersicum* (Hu et al. 2016), *Cucumis sativus* (Xu et al. 2015), and *Ananas comosus* (Zhang et al. 2020). In plants, this sensor is localized in Ca^{2+} storing subcellular organelles plasma membrane, cytosol, mitochondria, peroxisome, endoplasmic reticulum, and nucleus for quick response (Chehab et al. 2004; Dammann et al. 2003; Sugiyama et al. 2000). CDPKs proteins size ranges from ~40 to 90 kDa. It contains 5 domains with highly conserved Ser-Thr kinase catalytic region. Next to the kinase region, junction region continued which has pretented substrate activity and autoinhibitory function. Junction region interacts with active site and impedes its kinase activity. Adjacent to it, CaM-like domain (CLD) is present, which consists of 4 EF-hands to bind with the

Ca^{2+} ion. N-terminal region ranges from 21 to 185 peptide length (Yip Delormel and Boudsocq 2019; Klimecka and Muszyńska 2007) having myristoylation and palmitoylation sites at glycine and cysteine residues, respectively. The C-terminal domain is of variable short stretches.

Due to multi-substrate and phosphorylation motifs, CDPKs participated in numerous stress signal transduction processes to tackle them. During stress in *Nicotiana tabacum*, *Nt*CDPK2 and *Nt*CDPK3 are phosphorylated at different residues at Ser40 & Thr65 and Ser54, respectively, of N-terminal region (Witte et al. 2010). Hence CDPKs are most extensively studied Ca^{2+}-sensor during plant stress. During *Spodoptera litura* attack in *Arabidopsis thaliana*, CDPK3 and CDPK13 respond for Ca^{2+} binding. It regulates the transcription of plant defensin gene (*PDF1.2*) during damage, which is independent of phytohormone signalling (jasmonic acid, abscisic acid, and ethylene signaling) (Arimura and Maffei 2010; Kanchiswamy et al. 2010). This regulation happens when phosphorylation of transcription factor HSFB2A occurs. Regulation of negative-feedback mechanism during herbivory associated Ca^{2+} induction is regulated by CDPK3. This shows that CDPKs play redundant and specific contribution in plant defence mechanism (Kanchiswamy et al. 2010).

In transgenic *Nicotiana attenuata* plant, 2 CDPKs: *Na*CDPK4 and *Na*CDPK5 have significant role for providing resistance against *Manduca sexta* in JA- and JA-signalling dependent manner. Silencing of these 2 CDPKs results in accumulation of JA during herbivory. High accumulation of JA increases resistance as well as enhanced activity of SA-induced protein kinase and wound-induced protein kinase (Yang et al. 2012). *Gm*CPK3 and *Gm*CPK31 are involved in resistance against wounding stress by herbivore *S. litura* in *Glycine max* (Liu et al. 2016). *Zm*CPK11 from the *Zea mays* gets activated during wounding. Activation of this CDPK causes elevation of Ca^{2+}, phosphorylation of CDPK, changes in subcellular localization, and binding of lipid in the membrane (Szczegielniak et al. 2012).

16.4.5 CaM-Dependent Protein Kinase (CCaMK)

CaM-dependent protein kinase (CCaMK) are known in few plants such as *Zea mays*, *Medicago truncaulata*, *Nicotiana tabacum*, and *Lotus japonicus*. These are also known in Cryptograms such as mosses, liverworts, and hornworts (Wang et al. 2015). All plants consist of single gene for this protein (Harper et al. 2004). CCaMKs are close to CDPKs having similar structure (Harmon 2007). It contains a catalytic domain, Ca^{2+}-binding domain, CaM-binding domain, and autoinhibitory domain (Swainsbury et al. 2012). Its structure consists of extended N-terminal of variable length, conserved Ser-Thr kinase domain in the centre, and C-terminus region consisting of Ca^{2+} binding domain. Adjacent to the Ser-Thr kinase domain it contains CaM-binding domain, with overlapping autoinhibitory domain (Ramachandiran et al. 1997). With 3 EF-hands, Ca^{2+}-binding domain is a visinin-like domain. When Ca^{2+} comes in close proximity and binds to visinin-like domain, auto-phosphorylation of CCaMKs occurs whereas CCaMKs causes substrate phosphorylation when binding to Ca^{2+}/CaM (Takezawa et al. 1996). Ca^{2+} binding to Ca^{2+}/CaM is ineffectual for CCaMKs auto-phosphorylation, which reveals that visinin-like domain is an important region for Ca^{2+}/CaM-dependent substrate phosphorylation (Ramachandiran et al. 1997; Takezawa et al. 1996). Its activity is completely coordinated by Ca^{2+} and CaM (Miller et al. 2013). In maize, *Zm*CCaMK activates brassinosteroid induced antioxidant defence mechanism in leaves. Brassinosteroid induction leads to increase in Ca^{2+} concentration in the protoplast from mesophyll leaves (Yan et al. 2015; Zhu et al. 2016).

16.5 Role of Ca^{2+} Signalling at Subcellular Organelles During Herbivory

Herbivory attack and damage severely affect the subcellular compartment of cells. The organelles co-ordinate functions with cytosol and other organelles to elevate response against stress. It affects the respiratory mechanism, photosynthesis efficiency and increases the catalytic activity. Mitochondria, chloroplast, and peroxisomes are majorly affected organelles during insect attack. During biotic stress, stress associated plant oxidative activity increases by Ca^{2+} signalling; respiration activity results in increase in ATP activity and ROS formation takes place within Mitochondria.

Double membranous organelle such as mitochondria, chloroplast, and nucleus has the capacity to generate their own Ca^{2+} signals during stress condition (Xiong et al. 2006). Influx of Ca^{2+} ions from chloroplast has been studied in wheat and spinach. Influx of ions related to photosynthesis electron transport system (Muto et al. 1982; Kreimer et al. 1985). Ca^{2+}-channel H^+/Ca^{2+} antiporter energized by ATP in thylakoid membrane and uniporter uptakes Ca^{2+} due to change in potential of the inner membrane (Xiong et al. 2006). In *Phaseolus lunatus* green leaves damage induced by MecWorm induces broad range of signals. Early signals of wounding result in depolarization of membrane, production of H_2O_2, and influx of Ca^{2+} ions (Maffei et al. 2006). TPC is actively involved in transferring of ions in *Arabidopsis* plastid membrane (Bonaventure et al. 2007). This results in activation of CDPK sensors for Ca^{2+} signalling (Ludwig et al. 2005).

Experimental studies in *Arabidopsis* suggested that mitochondria are more sensitive to oxidative stress. When seedlings treated with abiotic stress such as cold, oxidative burst (H_2O_2) cause the same Ca^{2+} signature in mitochondria as well as in cytosol. In cytosol, resting phase is achieved in 20 s, whereas it takes longer time in mitochondria. Ca^{2+} ions induced prolonged partial depolarization of the membrane. When Ca^{2+} inhibitors (ruthenium red and EGTA) are applied repolarization of membrane takes place (Mironova et al. 2007). Similarly, Ca^{2+} signature is also involved in modulations of nuclear-localized genes. Nucleus has two compartments: nucleoplasm and inner membrane, inner membrane stores Ca^{2+} (Brière et al. 2006). For upregulation and downregulation of genes this calcium storage is crucial. This ion transducts the kinases and phosphorylate signalling into the nucleus (Xiong et al. 2006; Carafoli 2002).

16.6 Ca^{2+}-Mediated Local and Systemic Signalling During Herbivory

Ca^{2+} ions travel a long distance from the place of wounding on plasma membrane to nucleus via many transporters and sensors. As Ca^{2+} signals convey the signal it converts into the systematic defence mechanism to counterattack the herbivory-related injury. Ca^{2+}-transporter and sensor maintains the spiking and oscillation, i.e. Ca^{2+} signature to confer accurate message. There is an integrative "crosstalk" that happens between the induced plant defence against herbivore and the signalling molecules initiated by Ca^{2+}, followed by activation of reactive oxygen species (H_2O_2), protein kinases (MAPKs), jasmonic acid, salicylic acid, ethylene, oxophytodienoic acid, and other obscure members of the octadecanoid family (Maffei et al. 2007a, b; Mithöfer et al. 2018). Ca^{2+} ions trigger the accumulation of JA (Yan et al. 2018). GLCRs channels regulate Ca^{2+} for rapid systematic induction of JA against herbivore wounding (Mousavi et al. 2013). The Ca^{2+} kinetics conveyances downstream by interacting with sensor relay proteins with targeted proteins. These proteins initiate gene expression and activate direct and indirect defence. These proteins also influence ionic changes via other membrane transporters by activating transcriptional factors. The molecular mechanisms arbitrate by Ca^{2+} responsive gene expression is still concealed. The reason behind this ambiguity is unable to distinguish between Ca^{2+}-dependent stress response and Ca^{2+}-independent stress response. A study conducted using antagonists of CaM-WP7 and SKF-7171, revealed that only ~3.3% (out of 230 genes) of *Arabidopsis* genes are Ca^{2+} regulated and most of them were induced during early time points of stress. Recent experimental studies suggested that besides secondary phytometabolites, mobile plant hormones such as auxin and cytokinin were also involved in plant defense against herbivore. When wounding is caused in Nicotiana *attenuata,* caffeoylputrescine accumulated in leaves due to mutation in the homologs of cytokinin receptor CHASE-DOMAIN CONTAINING HIS KINASE 2 (Schäfer et al. 2015). A plant produces many specialized peculiar secondary metabolites which function as defence against herbivores. Ca^{2+} induces ROS production for transmission of damaging signal to nearby cells (Gilroy et al. 2016).

16.7 Conclusion

Plants perceive and respond to insect attack by activating defence responses through activation of early signalling components. One of the most important early signalling events is Ca^{2+} signalling. During herbivory, due to membrane depolarization, an elevated level of Ca^{2+} is observed within seconds to minutes, which serves as a signal. Further, different types of Ca^{2+}-binding proteins that are part of a larger regulatory network identify this signal and decode $[Ca^{2+}]_{cyt}$ elevation after wounding/feeding. This relay of signal by Ca^{2+}relayers and Ca^{2+}responders reaches to nucleus and specific defence reprogramming proceeds via signalling networks that include phytohormones, secondary metabolites, and transcription factors, ultimately leading to defence response. The topologies of Ca^{2+}signalling networks are emerging, but require further investigation for a clear picture of the pathway and all the components.

Acknowledgement AS and IKS acknowledge Science and Engineering Research Board (SERB) [SB/FT/LS-188/2012 (AS); SB/YS/LS-59/2013 (IKS); ECR/2017/002478 (AS)], Department of Science and Technology and University Grant Commission, Govt. of India, New Delhi, India, for the financial support. AK acknowledges SERB for fellowship.

References

Acevedo FE, Rivera-Vega LJ, Chung SH, Ray S, Felton GW (2015) Cues from chewing insects—the intersection of DAMPs, HAMPs, MAMPs and effectors. Curr Opin Plant Biol 26:80–86

Aldon D, Mbengue M, Mazars C, Galaud JP (2018) Calcium signalling in plant biotic interactions. Int J Mol Sci 19(3):665

Alonso MT, Villalobos C, Chamero P, Alvarez J, García-Sancho J (2006) Calcium microdomains in mitochondria and nucleus. Cell Calcium 40(5–6):513–525

Amanuel T, Rajneesh P, Manthi SJ, Odeny DA, AO MC, Khan ZR, Pickett JA, JA BT (2020) Genome wide association analysis of a stemborer egg induced "call-for-help" defence trait in maize. Sci Rep (Nat Publ Group) 10(1)

Arimura GI, Maffei ME (2010) Calcium and secondary CPK signaling in plants in response to herbivore attack. Biochem Biophys Res Commun 400(4):455–460

Batistic O, Kim KN, Kleist T, Kudla J, Luan S (2011) The CBL–CIPK network for decoding calcium signals in plants. In: Coding and decoding of calcium signals in plants. Springer, Heidelberg, pp 235–258

Batistič O, Kudla J (2009) Plant calcineurin B-like proteins and their interacting protein kinases. Biochim Biophys Acta Mol Cell Res 1793(6):985–992

Batistič O, Kudla J (2012) Analysis of calcium signaling pathways in plants. Biochim Biophys Acta Gen Subj 1820(8):1283–1293

Bender KW, Dobney S, Ogunrinde A, Chiasson D, Mullen RT, Teresinski HJ, Singh P, Munro K, Smith SP, Snedden WA (2014) The calmodulin-like protein CML43 functions as a salicylic-acid-inducible root-specific Ca^{2+} sensor in *Arabidopsis*. Biochem J 457(1):127–136

Bonaventure G, Gfeller A, Proebsting WM, Hörtensteiner S, Chételat A, Martinoia E, Farmer EE (2007) A gain of function allele of TPC1 activates oxylipin biogenesis after leaf wounding in *Arabidopsis*. Plant J 49(5):889–898

Bose J, Pottosin I, Shabala SSS, Palmgren MG, Shabala S (2011) Calcium efflux systems in stress signaling and adaptation in plants. Front Plant Sci 2:85

Boursiac Y, Harper JF (2007) The origin and function of calmodulin regulated Ca^{2+} pumps in plants. J Bioenerg Biomembr 39(5–6):409–414

Breeze E (2019) State of (In) flux: action of a CNGC Ca2+ channel in defense against Herbivory. Am Soc Plant Biol:1423–1424

Bricchi I, Leitner M, Foti M, Mithöfer A, Boland W, Maffei ME (2010) Robotic mechanical wounding (MecWorm) versus herbivore-induced responses: early signaling and volatile emission in Lima bean (*Phaseolus lunatus* L.). Planta 232(3):719–729

Brière C, Xiong TC, Mazars C, Ranjeva R (2006) Autonomous regulation of free Ca^{2+} concentrations in isolated plant cell nuclei: a mathematical analysis. Cell Calcium 39 (4):293–303

Carafoli E (2002) Calcium signaling: a tale for all seasons. Proc Natl Acad Sci 99(3):1115–1122

Chehab EW, Patharkar OR, Hegeman AD, Taybi T, Cushman JC (2004) Autophosphorylation and subcellular localization dynamics of a salt-and water deficit-induced calcium-dependent protein kinase from ice plant. Plant Physiol 135(3):1430–1446

Cheng SH, Willmann MR, Chen HC, Sheen J (2002) Calcium signaling through protein kinases. The Arabidopsis calcium-dependent protein kinase gene family. Plant Physiol 129:469–485

Cheng NH, Pittman JK, Barkla BJ, Shigaki T, Hirschi KD (2003) The *Arabidopsis* cax1 mutant exhibits impaired ion homeostasis, development, and hormonal responses and reveals interplay among vacuolar transporters. Plant Cell 15(2):347–364

Cheval C, Aldon D, Galaud JP, Ranty B (2013) Calcium/calmodulin-mediated regulation of plant immunity. BBA Mol Cell Res 1833(7):1766–1771

Chigri F, Flosdorff S, Pilz S, Kölle E, Dolze E, Gietl C, Vothknecht UC (2012) The Arabidopsis calmodulin-like proteins AtCML30 and AtCML3 are targeted to mitochondria and peroxisomes, respectively. Plant Mol Biol 78(3):211–222

Clapham DE (2007) Calcium signaling. Cell 131(6):1047–1058

Dammann C, Ichida A, Hong B, Romanowsky SM, Hrabak EM, Harmon AC, Pickard BG, Harper JF (2003) Subcellular targeting of nine calcium-dependent protein kinase isoforms from Arabidopsis. Plant Physiol 132(4):1840–1848

Davenport R (2002) Glutamate receptors in plants. Ann Bot 90(5):549–557

Davies JM (2014) Annexin-mediated calcium signalling in plants. Plan Theory 3(1):128–140

Demidchik V, Maathuis FJ (2007) Physiological roles of nonselective cation channels in plants: from salt stress to signalling and development. New Phytol 175(3):387–404

Dodd AN, Kudla J, Sanders D (2010) The language of calcium signaling. Annu Rev Plant Biol 61:593–620

Drerup MM, Schlücking K, Hashimoto K, Manishankar P, Steinhorst L, Kuchitsu K, Kudla J (2013) The calcineurin B-like calcium sensors CBL1 and CBL9 together with their interacting protein kinase CIPK26 regulate the *Arabidopsis* NADPH oxidase RBOHF. Mol Plant 6 (2):559–569

Erb M, Reymond P (2019) Molecular interactions between plants and insect herbivores. Annu Rev Plant Biol 70:527–557

Evans NH, McAinsh MR, Hetherington AM (2001) Calcium oscillations in higher plants. Curr Opin Plant Biol 4(5):415–420

Forde B, Roberts M (2014) Glutamate receptor-like channels in plants: an amino acid sensing role in plant defence? F1000 Prime Rep, London, pp 6

Gifford JL, Walsh MP, Vogel HJ (2007) Structures and metal-ion-binding properties of the Ca^{2+}-binding helix–loop–helix EF-hand motifs. Biochem J 405(2):199–221

Gilroy S, Białasek M, Suzuki N, Górecka M, Devireddy AR, Karpiński S, Mittler R (2016) ROS, calcium, and electric signals: key mediators of rapid systemic signaling in plants. Plant Physiol 171(3):1606–1615

Harmon AC (2007) Calcium-regulated protein kinases of plants. Gravitat Space Res 16(2)

Harper JF, Breton G, Harmon A (2004) Decoding Ca^{2+} signals through plant protein kinases. Annu Rev Plant Biol 55:263–288

Heyer M, Scholz SS, Voigt D, Reichelt M, Aldon D, Oelmüller R, Boland W, Mithöfer A (2018) Herbivory-responsive calmodulin-like protein CML9 does not guide jasmonate-mediated defenses in *Arabidopsis* thaliana. PLoS One 13(5):e0197633

Hirschi K (2001) Vacuolar H^+/Ca^{2+} transport: who's directing the traffic? Trends Plant Sci 6 (3):100–104

Hrabak EM, Chan CW, Gribskov M, Harper JF, Choi JH, Halford N, Kudla J, Luan S, Nimmo HG, Sussman MR, Thomas M (2003) The *Arabidopsis* CDPK-SnRK superfamily of protein kinases. Plant Physiol 132(2):666–680

Hu Z, Lv X, Xia X, Zhou J, Shi K, Yu J, Zhou Y (2016) Genome-wide identification and expression analysis of calcium-dependent protein kinase in tomato. Front Plant Sci 7:469

Huda KM, Banu MS, Tuteja R, Tuteja N (2013) Global calcium transducer P-type Ca^{2+}-ATPases open new avenues for agriculture by regulating stress signalling. J Exp Bot 64(11):3099–3109

K Jha S, Sharma M, K Pandey G (2016) Role of cyclic nucleotide gated channels in stress management in plants. Curr Genomics 17(4):315–329

Kanchiswamy CN, Takahashi H, Quadro S, Maffei ME, Bossi S, Bertea C, Zebelo SA, Muroi A, Ishihama N, Yoshioka H, Boland W (2010) Regulation of *Arabidopsis* defense responses against Spodoptera littoralis by CPK-mediated calcium signaling. BMC Plant Biol 10(1):97
Kim KN (2013) Stress responses mediated by the CBL calcium sensors in plants. Plant Biotechnol Rep 7(1):1–8
Kimura S, Kawarazaki T, Nibori H, Michikawa M, Imai A, Kaya H, Kuchitsu K (2013) The CBL-interacting protein kinase CIPK26 is a novel interactor of *Arabidopsis* NADPH oxidase AtRbohF that negatively modulates its ROS-producing activity in a heterologous expression system. J Biochem 153(2):191–195
Klimecka M, Muszyńska G (2007) Structure and functions of plant calcium-dependent protein kinases. Acta Biochim Pol 54(2):219–233
Kong X, Lv W, Jiang S, Zhang D, Cai G, Pan J, Li D (2013) Genome-wide identification and expression analysis of calcium-dependent protein kinase in maize. BMC Genomics 14(1):1–15
Kosuta S, Hazledine S, Sun J, Miwa H, Morris RJ, Downie JA, Oldroyd GE (2008) Differential and chaotic calcium signatures in the symbiosis signaling pathway of legumes. Proc Natl Acad Sci 105(28):9823–9828
Kreimer G, Melkonian M, Holtum JA, Latzko E (1985) Characterization of calcium fluxes across the envelope of intact spinach chloroplasts. Planta 166(4):515–523
Kudla J, Xu Q, Harter K, Gruissem W, Luan S (1999) Genes for calcineurin B-like proteins in *Arabidopsis* are differentially regulated by stress signals. Proc Natl Acad Sci 96(8):4718–4723
Laohavisit A, Mortimer JC, Demidchik V, Coxon KM, Stancombe MA, Macpherson N, Brownlee C, Hofmann A, Webb AA, Miedema H, Battey NH (2009) Zea mays annexins modulate cytosolic free Ca2+ and generate a Ca2+−permeable conductance. Plant Cell 21(2):479–493
Leba LJ, Cheval C, Ortiz-Martín I, Ranty B, Beuzón CR, Galaud JP, Aldon D (2012) CML9, an *Arabidopsis* calmodulin-like protein, contributes to plant innate immunity through a flagellin-dependent signalling pathway. Plant J 71(6):976–989
Lecourieux D, Ranjeva R, Pugin A (2006) Calcium in plant defence-signalling pathways. New Phytol 171(2):249–269
Leng Q, Mercier RW, Yao W, Berkowitz GA (1999) Cloning and first functional characterization of a plant cyclic nucleotide-gated cation channel. Plant Physiol 121(3):753–761
Lévy J, Bres C, Geurts R, Chalhoub B, Kulikova O, Duc G, Journet EP, Ané JM, Lauber E, Bisseling T, Dénarié J (2004) A putative Ca2+ and calmodulin-dependent protein kinase required for bacterial and fungal symbioses. Science 303(5662):1361–1364
Li AL, Zhu YF, Tan XM, Wang X, Wei B, Guo HZ, Zhang ZL, Chen XB, Zhao GY, Kong XY, Jia JZ (2008) Evolutionary and functional study of the CDPK gene family in wheat (Triticum aestivum L.). Plant Mol Biol 66(4):429–443
Liu J, Zhu JK (1998) A calcium sensor homolog required for plant salt tolerance. Science 280 (5371):1943–1945
Liu W, Li W, He Q, Daud MK, Chen J, Zhu S (2014) Genome-wide survey and expression analysis of calcium-dependent protein kinase in *Gossypium raimondii*. PLoS One 9(6):e98189
Liu H, Che Z, Zeng X, Zhou X, Sitoe HM, Wang H, Yu D (2016) Genome-wide analysis of calcium-dependent protein kinases and their expression patterns in response to herbivore and wounding stresses in soybean. Funct Integr Genomics 16(5):481–493
Luan S, Kudla J, Rodriguez-Concepcion M, Yalovsky S, Gruissem W (2002) Calmodulins and calcineurin B–like proteins: calcium sensors for specific signal response coupling in plants. Plant Cell 14(Suppl 1):S389–S400
Ludwig AA, Saitoh H, Felix G, Freymark G, Miersch O, Wasternack C, Boller T, Jones JD, Romeis T (2005) Ethylene-mediated cross-talk between calcium-dependent protein kinase and MAPK signaling controls stress responses in plants. Proc Natl Acad Sci 102(30):10736–10741
Maffei M, Bossi S, Spiteller D, Mithöfer A, Boland W (2004) Effects of feeding *Spodoptera littoralis* on lima bean leaves. I. Membrane potentials, intracellular calcium variations, oral secretions, and regurgitate components. Plant Physiol 134(4):1752–1762

Maffei ME, Mithöfer A, Arimura GI, Uchtenhagen H, Bossi S, Bertea CM, Cucuzza LS, Novero M, Volpe V, Quadro S, Boland W (2006) Effects of feeding *Spodoptera littoralis* on lima bean leaves. III. Membrane depolarization and involvement of hydrogen peroxide. Plant Physiol 140 (3):1022–1035

Maffei ME, Mithöfer A, Boland W (2007a) Before gene expression: early events in plant–insect interaction. Trends Plant Sci 2(7):310–316

Maffei ME, Mithöfer A, Boland W (2007b) Insects feeding on plants: rapid signals and responses preceding the induction of phytochemical release. Phytochemistry 68(22–24):2946–2959

Meena MK, Vadassery J (2015) Channels hold the key: cyclic nucleotide gated channels (CNGC) in plant biotic stress signaling. J Endocyt Cell Res:25–30

Meena MK, Prajapati R, Krishna D, Divakaran K, Pandey Y, Reichelt M, Mathew MK, Boland W, Mithöfer A, Vadassery J (2019) The Ca^{2+} channel CNGC19 regulates *Arabidopsis* defense against *Spodoptera* herbivory. Plant Cell 31(7):1539–1562

Messerli MA, Robinson KR, Smith PJ (2006) Electrochemical sensor applications to the study of molecular physiology and analyte flux in plants. In: Plant electrophysiology. Springer, Berlin, pp 73–107

Miller GA, Mittler RO (2006) Could heat shock transcription factors function as hydrogen peroxide sensors in plants? Ann Bot 98(2):279–288

Miller JB, Pratap A, Miyahara A, Zhou L, Bornemann S, Morris RJ, Oldroyd GE (2013) Calcium/Calmodulin-dependent protein kinase is negatively and positively regulated by calcium, providing a mechanism for decoding calcium responses during symbiosis signaling. Plant Cell 25(12):5053–5066

Mironova GD, Belosludtsev KN, Belosludtseva NV, Gritsenko EN, Khodorov BI, Saris NE (2007) Mitochondrial Ca^{2+} cycle mediated by the palmitate-activated cyclosporin A-insensitive pore. J Bioenerg Biomembr 39(2):167–174

Mithöfer A, Mazars C, Maffei ME (2009) Probing spatio-temporal intracellular calcium variations in plants. In: Plant signal transduction. Humana Press, Totowa, pp 79–92

Mithöfer A, Boland W, Maffei ME (2018) Chemical ecology of plant–insect interactions. Annu Rev *Plant* Biol:261–291

Moreno SN, Pereira LB, Pace DA (2014) Calcium storage and homeostasis in Toxoplasma gondii. In: Toxoplasma gondii. Academic Press, pp 351–374

Mousavi SA, Chauvin A, Pascaud F, Kellenberger S, Farmer EE (2013) GLUTAMATE RECEPTOR-LIKE genes mediate leaf-to-leaf wound signalling. Nature 500(7463):422–426

Muto S, Izawa S, Miyachi S (1982) Light-induced Ca^{2+} uptake by intact chloroplasts. FEBS Lett 139(2):250–254

Ozturk ZN, Talame V, Deyholos M, Michalowski CB, Galbraith DW, Gozukirmizi N, Tuberosa R, Bohnert HJ (2002) Monitoring large-scale changes in transcript abundance in drought- and salt-stressed barley. Plant Mol Biol 48:551–573

Peiter E, Maathuis FJ, Mills LN, Knight H, Pelloux J, Hetherington AM, Sanders D (2005) The vacuolar Ca^{2+}-activated channel TPC1 regulates germination and stomatal movement. Nature 434(7031):404–408

Perochon A, Aldon D, Galaud JP, Ranty B (2011) Calmodulin and calmodulin-like proteins in plant calcium signaling. Biochimie 93(12):2048–2053

Pottosin I, Wherrett T, Shabala S (2009) SV channels dominate the vacuolar Ca^{2+} release during intracellular signaling. FEBS Lett 583(5):921–926

Ramachandiran S, Takezawa D, Wang W, Poovaiah BW (1997) Functional domains of plant chimeric calcium/calmodulin-dependent protein kinase: regulation by autoinhibitory and visinin-like domains. J Biochem 121(5):984–990

Ray S, Agarwal P, Arora R, Kapoor S, Tyagi AK (2007) Expression analysis of calcium-dependent protein kinase gene family during reproductive development and abiotic stress conditions in rice (*Oryza sativa* L. ssp. *indica*). Mol Genet Genomics 278(5):493–505

Reddy VS, Reddy AS (2004) Proteomics of calcium-signaling components in plants. Phytochemistry 65(12):1745–1776

Romeis T, Herde M (2014) From local to global: CDPKs in systemic defense signaling upon microbial and herbivore attack. Curr Opin Plant Biol 20:1–10

Sanders D, Brownlee C, Harper JF (1999) Communicating with calcium. Plant Cell 11(4):691–706

Sanders D, Pelloux J, Brownlee C, Harper JF (2002) Calcium at the crossroads of signaling. Plant Cell 14(suppl 1):S401–S417

Sarwat M, Tuteja N (2007) Calnexin: a versatile calcium binding integral membrane-bound chaperone of endoplasmic reticulum. Calcium Bind Proteins 2:36–50

Schäfer M, Meza-Canales ID, Brütting C, Baldwin IT, Meldau S (2015) Cytokinin concentrations and CHASE-DOMAIN CONTAINING HIS KINASE 2 (NaCHK2)-and N a CHK 3-mediated perception modulate herbivory-induced defense signaling and defenses in *Nicotiana attenuata*. New Phytol 207(3):645–658

Schiøtt M, Romanowsky SM, Bækgaard L, Jakobsen MK, Palmgren MG, Harper JF (2004) A plant plasma membrane Ca^{2+} pump is required for normal pollen tube growth and fertilization. Proc Natl Acad Sci 101(25):9502–9507

Schmelz EA, Carroll MJ, LeClere S, Phipps SM, Meredith J, Chourey PS, Alborn HT, Teal PE (2006) Fragments of ATP synthase mediate plant perception of insect attack. Proc Natl Acad Sci 103(23):8894–8899

Scholz SS, Vadassery J, Heyer M, Reichelt M, Bender KW, Snedden WA, Boland W, Mithöfer A (2014) Mutation of the *Arabidopsis* calmodulin-like protein CML37 deregulates the jasmonate pathway and enhances susceptibility to herbivory. Mol Plant 7(12):1712–1726

Schuurink RC, Shartzer SF, Fath A, Jones RL (1998) Characterization of a calmodulin-binding transporter from the plasma membrane of barley aleurone. Proc Natl Acad Sci 95(4):1944–1949

Seki M, Ishida J, Narusaka M et al (2002) Monitoring the expression pattern of around 7,000 Arabidopsis genes under ABA treatments using a full-length cDNA microarray. Funct Integr Genomics 2:282–291

Singh A, Singh IK, Verma PK (2008) Differential transcript accumulation in Cicer arietinum L. in response to a chewing insect Helicoverpa armigera and defence regulators correlate with reduced insect performance. J Exp Bot 59(9):2379–2392

Singh A, Jain D, Tyagi C, Singh S, Kumar S, Singh IK (2018) In silico prediction of active site and in vitro DNase and RNase activities of Helicoverpainducible pathogenesis related-4 protein from Cicer arietinum. Int J Biol Macromol 113:869–880

Singh A, Kumar A, Hartley S, Kumar Singh I (2020) Silicon: its ameliorative effect on plant defense against herbivory. J Exp Bot. https://doi.org/10.1093/jxb/eraa300

Snedden WA, Fromm H (1998) Calmodulin, calmodulin-related proteins and plant responses to the environment. Trends Plant Sci 3(8):299–304

Song WY, Zhang ZB, Shao HB, Guo XL, Cao HX, Zhao HB, Fu ZY, Hu XJ (2008) Relationship between calcium decoding elements and plant abiotic-stress resistance. Int J Biol Sci 4(2):116

Sugiyama KI, Mori IC, Takahashi K, Muto S, Shihira-Ishikawa I (2000) A calcium-dependent protein kinase functions in wound healing in ventricaria ventricosa (chlorophyta). J Phycol 36(6):1145–1152

Swainsbury DJ, Zhou L, Oldroyd GE, Bornemann S (2012) Calcium ion binding properties of *Medicago truncatula* calcium/calmodulin-dependent protein kinase. Biochemistry 51 (35):6895–6907

Szczegielniak J, Borkiewicz L, Szurmak B, Lewandowska-Gnatowska E, Statkiewicz M, Klimecka M, Cieśla J, Muszyńska G (2012) Maize calcium-dependent protein kinase (ZmCPK11): local and systemic response to wounding, regulation by touch and components of jasmonate signaling. Physiol Plant 146(1):1–14

Takezawa D, Ramachandiran S, Paranjape V, Poovaiah BW (1996) Dual regulation of a chimeric plant serine/threonine kinase by calcium and calcium/calmodulin. J Biol Chem 271 (14):8126–8132

Tena G, Boudsocq M, Sheen J (2011) Protein kinase signaling networks in plant innate immunity. Curr Opin Plant Biol 14(5):519–529

Tian W, Wang C, Gao Q, Li L, Luan S (2020) Calcium spikes, waves and oscillations in plant development and biotic interactions. Nat Plants:1–10

Toyota M, Spencer D, Sawai-Toyota S, Jiaqi W, Zhang T, Koo AJ, Howe GA, Gilroy S (2018) Glutamate triggers long-distance, calcium-based plant defense signaling. Science 361 (6407):1112–1115

Urquhart W, Gunawardena AH, Moeder W, Ali R, Berkowitz GA, Yoshioka K (2007) The chimeric cyclic nucleotide-gated ion channel ATCNGC11/12 constitutively induces programmed cell death in a Ca^{2+} dependent manner. Plant Mol Biol 65(6):747–761

Vadassery J, Reichelt M, Hause B, Gershenzon J, Boland W, Mithöfer A (2012a) CML42-mediated calcium signaling coordinates responses to Spodoptera herbivory and abiotic stresses in *Arabidopsis*. Plant Physiol 159(3):1159–1175

Vadassery J, Scholz SS, Mithöfer A (2012b) Multiple calmodulin-like proteins in *Arabidopsis* are induced by insect-derived (*Spodoptera littoralis*) oral secretion. Plant Signal Behav 7 (10):1277–1280

Vincent TR, Avramova M, Canham J, Higgins P, Bilkey N, Mugford ST, Pitino M, Toyota M, Gilroy S, Miller AJ, Hogenhout SA (2017) Interplay of plasma membrane and vacuolar ion channels, together with BAK1, elicits rapid cytosolic calcium elevations in *Arabidopsis* during aphid feeding. Plant Cell 29(6):1460–1479

Wang JP, Munyampundu JP, Xu YP, Cai XZ (2015) Phylogeny of plant calcium and calmodulin-dependent protein kinases (CCaMKs) and functional analyses of tomato CCaMK in disease resistance. Front Plant Sci 6:1075

Webb TJ, Powls R, Rees HH (1996) Characterization, fractionation and kinetic properties of the enzymes of ecdysteroid 3-epimerization and phosphorylation isolated from the midgut cytosol of the cotton leafworm, Spodoptera littoralis. Insect Biochem Mol Biol 26(8–9):809–816

White PJ (2000) Calcium channels in higher plants. BBA-Biomembranes 1465(1–2):171–189

Witte CP, Keinath N, Dubiella U, Demoulière R, Seal A, Romeis T (2010) Tobacco calcium-dependent protein kinases are differentially phosphorylated in vivo as part of a kinase cascade that regulates stress response. J Biol Chem 285(13):9740–9748

Xiong TC, Bourque S, Lecourieux D, Amelot N, Grat S, Brière C, Mazars C, Pugin A, Ranjeva R (2006) Calcium signaling in plant cell organelles delimited by a double membrane. BBA Mol Cell Res 1763(11):1209–1215

Xu X, Liu M, Lu L, He M, Qu W, Xu Q, Qi X, Chen X (2015) Genome-wide analysis and expression of the calcium-dependent protein kinase gene family in cucumber. Mol Genet Genomics 290(4):1403–1414

Yan J, Guan L, Sun Y, Zhu Y, Liu L, Lu R, Jiang M, Tan M, Zhang A (2015) Calcium and ZmCCaMK are involved in brassinosteroid-induced antioxidant defense in maize leaves. Plant Cell Physiol 56(5):883–896

Yan C, Fan M, Yang M, Zhao J, Zhang W, Su Y, Xiao L, Deng H, Xie D (2018) Injury activates Ca^{2+}/calmodulin-dependent phosphorylation of JAV1-JAZ8-WRKY51 complex for jasmonate biosynthesis. Mol Cell 70(1):136–149

Yang T, Poovaiah BW (2000) Molecular and biochemical evidence for the involvement of calcium/calmodulin in auxin action. J Biol Chem 275(5):3137–3143

Yang T, Poovaiah BW (2002) A calmodulin-binding/CGCG box DNA-binding protein family involved in multiple signaling pathways in plants. J Biol Chem 277:45049–45058

Yang H, Xie S, Wang L, Jing S, Zhu X, Li X, Zeng W, Yuan H (2011) Identification of up-regulated genes in tea leaves under mild infestation of green leafhopper. Sci Hortic 130(2):476–481

Yang DH, Hettenhausen C, Baldwin IT, Wu J (2012) Silencing *Nicotiana attenuata* calcium-dependent protein kinases, CDPK4 and CDPK5, strongly up-regulates wound-and herbivory-induced jasmonic acid accumulations. Plant Physiol 159(4):1591–1607

Yip Delormel T, Boudsocq M (2019) Properties and functions of calcium-dependent protein kinases and their relatives in *Arabidopsis* thaliana. New Phytol 224(2):585–604

Yu Q, An L, Li W (2014) The CBL–CIPK network mediates different signaling pathways in plants. Plant Cell Rep 33(2):203–214

Zebelo SA, Maffei ME (2015) Role of early signalling events in plant–insect interactions. J Exp Bot 66(2):435–448

Zhang H, Yin W, Xia X (2008) Calcineurin B-Like family in populus: comparative genome analysis and expression pattern under cold, drought and salt stress treatment. Plant Growth Regul 56(2):129–140

Zhang M, Liu Y, He Q, Chai M, Huang Y, Chen F, Wang X, Liu Y, Cai H, Qin Y (2020) Genome-wide investigation of calcium-dependent protein kinase gene family in pineapple: evolution and expression profiles during development and stress. BMC Genomics 21(1):1–16

Zhu S, Zhou X, Wu X, Jiang Z (2013) Structure and function of the CBL–CIPK Ca^{2+}-decoding system in plant calcium signaling. Plant Mol Biol Rep 31(6):1193–1202

Zhu Y, Yan J, Liu W, Liu L, Sheng Y, Sun Y, Li Y, Scheller HV, Jiang M, Hou X, Ni L (2016) Phosphorylation of a NAC transcription factor by ZmCCaMK regulates abscisic acid-induced antioxidant defense in maize. Plant Physiol

Zuo R, Hu R, Chai G, Xu M, Qi G, Kong Y, Zhou G (2013) Genome-wide identification, classification, and expression analysis of CDPK and its closely related gene families in poplar (*Populus trichocarpa*). Mol Biol Rep 40(3):2645–2662

The manufacturer's authorised representative in the EU is Springer Nature Customer Service Centre GmbH, Europaplatz 3, 69115 Heidelberg, Germany. If you have any concerns regarding our products, please contact ProductSafety@springernature.com

Printed and bound by CPI Group (UK) Ltd, Croydon, CR0 4YY
30/09/2026
02225247-0018